AF353145

Power Converters in Power Electronics

Power Converters in Power Electronics

Special Issue Editor

Minh-Khai Nguyen

MDPI • Basel • Beijing • Wuhan • Barcelona • Belgrade • Manchester • Tokyo • Cluj • Tianjin

Special Issue Editor
Minh-Khai Nguyen
Wayne State University
USA

Editorial Office
MDPI
St. Alban-Anlage 66
4052 Basel, Switzerland

This is a reprint of articles from the Special Issue published online in the open access journal *Electronics* (ISSN 2079-9292) (available at: https://www.mdpi.com/journal/electronics/special_issues/power_ponverters).

For citation purposes, cite each article independently as indicated on the article page online and as indicated below:

LastName, A.A.; LastName, B.B.; LastName, C.C. Article Title. *Journal Name* **Year**, *Article Number*, Page Range.

ISBN 978-3-03936-132-8 (Hbk)
ISBN 978-3-03936-133-5 (PDF)

© 2020 by the authors. Articles in this book are Open Access and distributed under the Creative Commons Attribution (CC BY) license, which allows users to download, copy and build upon published articles, as long as the author and publisher are properly credited, which ensures maximum dissemination and a wider impact of our publications.

The book as a whole is distributed by MDPI under the terms and conditions of the Creative Commons license CC BY-NC-ND.

Contents

About the Special Issue Editor

Minh-Khai Nguyen completed his Bachelor of Science degree in electrical engineering at the Ho Chi Minh City University of Technology, Ho Chi Minh City, Vietnam, in 2005. He then completed his Master of Science degree and Ph.D. in electrical engineering at Chonnam National University, Gwangju, South Korea, in 2007 and 2010, respectively. He was a lecturer at the Ho Chi Minh City University of Technology and Education in Ho Chi Minh City. He was also an Assistant Professor at Chosun University, Gwangju, South Korea, and at the School of Electrical Engineering and Robotics, Queensland University of Technology, Australia. He currently works in the Department of Electrical and Computer Engineering at Wayne State University, USA. His current research interests include impedance-source inverters and power converters for renewable energy systems. He is also currently an Associate Editor for the *Journal of Power Electronics* and the *International Journal of Power Electronics*.

 *electronics*

Editorial

Power Converters in Power Electronics: Current Research Trends

Minh-Khai Nguyen

Department of Electrical Engineering and Computer, Wayne State University, Detroit, MI 48202, USA; mknguyen@wayne.edu

Received: 12 April 2020; Accepted: 15 April 2020; Published: 16 April 2020

1. Introduction

In recent years, power converters have played an important role in power electronics technology for different applications, such as renewable energy systems, electric vehicles, pulsed power generation, and biomedical. Power converters in power electronics are becoming essential for generating electrical power energy in various ways. Voltage source and current source inverters are well-known as the conventional topologies to convert a DC source to an AC source. For single-phase current source inverters, DC-side low-frequency power oscillation is an inherent limitation. Various active power decoupling circuits for single-phase current source converters are reviewed in [1]. During the last decade, the multilevel inverter has become very popular in medium and high-power applications, with some advantages, such as the reduced power dissipation of switching elements, low harmonics, and low electromagnetic interferences (EMIs). An overview of the multi-level inverter topologies is presented in [2]. The development of the advanced power converters is necessary for the various applications. This special issue focuses on the development of novel power converter topologies in power electronics.

2. The Current Research Trends

Each of the 28 articles collected in this special issue proposes a solution to a specific problem related to the power converter topologies. The impedance source inverters (ISIs) overcome the limitation of the conventional inverters, because they can operate in single-stage power conversion with buck-boost voltage and shoot-through immunity. In Reference [3], a three-phase cascaded active ISI is proposed to reduce the number of passive elements. A common ground Z-source SEPIC inverter is proposed in [4] to eliminate leakage current. In order to increase the voltage gain, a multiplier cell technique is applied to the quasi-switched boost inverter with low input current ripple [5]. The applications of the ISI for grid connection [6] and photovoltaic [7] are presented.

New concepts of pulse-width modulation (PWM) control techniques and control theories for the inverters are another focus of this issue. A modified model predictive power control for grid-connected T-type inverter with reduced computational complexity is presented in [8]. A robust two-layer model predictive control for three-level inverter is shown in [9]. Current source AC-side clamped inverter for leakage current reduction in grid-connected photovoltaic (PV) system is shown in [10]. A control design of the LCL filter [11] and PWM method [12] for single-phase inverter are carried out. The space vector modulation technique is applied to the quasi-switch boost t-type inverter for common mode voltage elimination [13]. In addition, the PWM techniques for five-level inverter are presented in [14,15], with a minimal number of commutations. The inverter topology for the DC motor is presented in [16].

In addition to the research in the inverter topologies, this issue has collected important research on isolated [17,18] and non-isolated [19–22] DC-DC power converters and rectifiers [23,24]. In Reference [17], a comparative evaluation of some wide-range soft-switching full-bridge modular multilevel DC–DC converters is discussed. To realize the functions of reduced primary current loss and balanced voltage and current, a DC-DC converter with the series/parallel connection on the high-voltage/low-voltage side is presented in [18]. A bipolar bidirectional DC/DC converter and its interleaved-complementary modulation scheme are introduced in [19]. A multiple three-phase low-voltage and high-current permanent magnet synchronous generation system is proposed in [20] for 5 V/10 kA DC power supply. In Reference [21], a new technique for enlarging the stable range of peak-current-mode-controlled DC-DC converters is presented. In Reference [22], a combination of the one-comparator counter-based PWM control with pulse frequency modulation (PFM) control is presented for buck DC-DC converter. A power-based space vector modulation technique for a matrix rectifier is proposed in [23], where the modulation index and applied phase are calculated, based on the active and reactive power of the rectifier for intuitive power factor control. In Reference [24], a modeling method is introduced to establish a parametric-conducted emission model of a switching model power supply chip through a developed vector fitting algorithm.

Various power converter topologies for different applications, such as wireless power transfer [25,26], battery charging [27,28], static synchronous compensator (STATCOM) [29], and gate driver [30] are presented in this issue. In Reference [24], an inductive power transfer (IPT) system with three-bridge switching compensation topology is proposed, to achieve the load-independent constant current (CC) and load-independent constant voltage (CV) outputs. In Reference [25], a generalized fractional-order wireless power transmission is established for medium and long-range wireless power transmission. A fast-charging balancing circuit for LiFePO4 battery is proposed in [26], to address the voltage imbalanced problem of a lithium battery string. In Reference [27], an add-on type pulse charger is introduced, to shorten the charging time of a lithium ion battery. In Reference [28], an individual phase full-power testing method for high-power STATCOM is presented with reconfiguration the port connection of three-phase STATCOM. In Reference [30], an intelligent control method for suppressing electromagnetic interference (EMI) sources in the fast switching power converters is proposed, with a combination of open loop and closed loop methods.

3. Future Trends

The future research in the advanced power converter topologies will continue to find the solutions in applications of current power electronics for different disciplines. New power converters will be investigated to enhance or replace the current converter topologies. A new growing interest is the integration of impedance-source converters for renewable energy system applications.

Author Contributions: M.-K.N. worked during the whole editorial process of the special issue entitled "Power Converters in Power Electronics", published in the MDPI journal *Electronics*. M.-K.N. drafted, reviewed, edited, and finalized this editorial summary. Author has read and agreed to the published version of the manuscript.

Acknowledgments: I thank all the authors who submitted excellent research works to this Special Issue. I also thank Xiaoqiang Guo from Yanshan University, China for his contribution as a technical program committee member. I am very grateful to all reviewers for their evaluations of the merits and quality of the articles and valuable comments to improve the articles in this issue. I would also like to thank the editorial board and staff of MDPI journal *Electronics* for the opportunity to guest-edit this Special Issue.

Conflicts of Interest: The author declares no conflict of interest.

References

1. Zhang, J.; Ding, H.; Wang, B.; Guo, X.; Padmanaban, S. Active Power Decoupling for Current Source Converters: An Overview Scenario. *Electronics* **2019**, *8*, 197. [CrossRef]
2. Rana, R.A.; Patel, S.A.; Muthusamy, A.; Lee, C.W.; Kim, H.J. Review of Multilevel Voltage Source Inverter Topologies and Analysis of Harmonics Distortions in FC-MLI. *Electronics* **2019**, *8*, 1329. [CrossRef]
3. Tran, V.T.; Nguyen, M.K.; Ngo, C.C.; Choi, Y.O. Three-Phase Five-Level Cascade Quasi-Switched Boost Inverter. *Electronics* **2019**, *8*, 296. [CrossRef]
4. Wang, B.; Tang, W. A Novel Three-Switch Z-Source SEPIC Inverter. *Electronics* **2019**, *8*, 247. [CrossRef]
5. Nguyen, M.K.; Choi, Y.O. Voltage Multiplier Cell-Based Quasi-Switched Boost Inverter with Low Input Current Ripple. *Electronics* **2019**, *8*, 227. [CrossRef]

6.	Choi, W.Y.; Yang, M.K. Transformerless Quasi-Z-Source Inverter to Reduce Leakage Current for Single-Phase Grid-Tied Applications. *Electronics* **2019**, *8*, 312. [CrossRef]

7.	Sandoval, J.M.; Cardenas, V.; Barrios, M.A.; Garcia, M.G.; Janeth, A. Multiport Isolated link with Current-Fed Z-Source Converters to Manage Power Imbalance in PV Applications. *Electronics* **2020**, *9*, 280. [CrossRef]

8.	Ngo, V.Q.B.; Nguyen, M.K.; Tran, T.T.; Choi, J.H.; Lim, Y.C. A Modified Model Predictive Power Control for Grid-Connected T-Type Inverter with Reduced Computational Complexity. *Electronics* **2019**, *8*, 217. [CrossRef]

9.	Wei, X.; Wang, H.; Wang, K.; Li, K.; Li, M.; Luo, A. Robust Two-Layer Model Predictive Control for Full-Bridge NPC Inverter-Based Class-D Voltage Mode Amplifier. *Electronics* **2019**, *8*, 1346. [CrossRef]

10.	Li, X.; Wang, N.; San, G.; Guo, X. Current Source AC-Side Clamped Inverter for Leakage Current Reduction in Grid-Connected PV System. *Electronics* **2019**, *8*, 1296. [CrossRef]

11.	Yang, L.; Feng, C.; Liu, J. Control Design of LCL Type Grid-Connected Inverter Based on State Feedback Linearization. *Electronics* **2019**, *8*, 877. [CrossRef]

12.	Li, G.; Liu, C.; Fu, Z.; Wang, Y. A Novel RPWN Selective Harmonic Elimination Method for Single-Phase Inverter. *Electronics* **2020**, *9*, 489. [CrossRef]

13.	Do, D.T.; Nguyen, M.K.; Ngo, V.T.; Quach, T.H.; Tran, V.T. Common Mode Voltage Elimination for Quasi-Switch Boost T-Type Inverter Based on SVM Technique. *Electronics* **2020**, *9*, 76. [CrossRef]

14.	Becker, F.; Jamshidpour, E.; Poure, P.; Saadate, S. Modulation Strategy with a Minimal Number of Commutations for a Five-Level H-Bridge NPC Inverter. *Electronics* **2019**, *8*, 454. [CrossRef]

15.	Quach, T.H.; Do, D.T.; Nguyen, M.K. A PWM Scheme for Five-Level H-Bridge T-Type Inverter with Switching Loss Reduction. *Electronics* **2019**, *8*, 702. [CrossRef]

16.	Hernández-Márquez, E.; Avila-Rea, C.A.; García-Sánchez, J.R.; Silva-Ortigoza, R.; Marciano-Melchor, M.; Marcelino-Aranda, M.; Roldán-Caballero, A.; Márquez-Sánchez, C. New "Full-Bridge Buck Inverter–DC Motor" System: Steady-State and Dynamic Analysis and Experimental Validation. *Electronics* **2019**, *8*, 1216. [CrossRef]

17.	Chen, J.; Li, X.; Dang, H.; Shi, Y. Comparative Evaluation of Wide-Range Soft-Switching PWM Full-Bridge Modular Multilevel DC–DC Converters. *Electronics* **2020**, *9*, 231. [CrossRef]

18.	Lin, B.R. Analysis of a DC Converter with Low Primary Current Loss and Balance Voltage and Current. *Electronics* **2019**, *8*, 439. [CrossRef]

19.	Li, P.; Zhang, C.; Padmanaban, S.; Zbigniew, L. Multiple Modulation Strategy of Flying Capacitor DC/DC Converter. *Electronics* **2019**, *8*, 774. [CrossRef]

20.	Liu, J.; Tan, X.; Wang, X.; Ho-Ching IU, H. Application of the Lyapunov Algorithm to Optimize the Control Strategy of Low-Voltage and High-Current Synchronous DC Generator Systems. *Electronics* **2019**, *8*, 871. [CrossRef]

21.	Chen, Y.; Xie, F.; Zhang, B.; Qiu, D.; Chen, X.; Li, Z.; Zhang, G. Improvement of Stability in a PCM-Controlled Boost Converter with the Target Period Orbit-Tracking Method. *Electronics* **2019**, *8*, 1432. [CrossRef]

22.	Hwu, K.I.; Wang, C.W.; Yau, Y.T. Enhancement of System Stability Based on PWFM. *Electronics* **2019**, *8*, 399. [CrossRef]

23.	Kim, J.C.; Kim, D.; Kwak, S.S. Direct Power-Based Three-Phase Matrix Rectifier Control with Input Power Factor Adjustment. *Electronics* **2019**, *8*, 1427. [CrossRef]

24.	Hao, X.; Xie, S.; Chen, Z. A Parametric Conducted Emission Modeling Method of a Switching Model Power Supply (SMPS) Chip by a Developed Vector Fitting Algorithm. *Electronics* **2019**, *8*, 725. [CrossRef]

25.	Luo, B.; Shou, Y.; Lu, J.; Li, M.; Deng, X.; Zhu, G. A Three-Bridge IPT System for Different Power Levels Conversion under CC/CV Transmission Mode. *Electronics* **2019**, *8*, 884. [CrossRef]

26.	Zhang, G.; Ou, Z.; Qu, L. A Fractional-Order Element (FOE)-Based Approach to Wireless Power Transmission for Frequency Reduction and Output Power Quality Improvement. *Electronics* **2019**, *8*, 1029. [CrossRef]

27.	Wu, S.T.; Chang, Y.N.; Chang, C.Y.; Cheng, Y.T. A Fast Charging Balancing Circuit for LiFePO4 Battery. *Electronics* **2019**, *8*, 1144. [CrossRef]

28.	Kwak, B.; Kim, M.; Kim, J. Add-On Type Pulse Charger for Quick Charging Li-Ion Batteries. *Electronics* **2020**, *9*, 227. [CrossRef]

29. Huang, Q.; Li, B.; Tan, Y.; Mao, X.; Zhu, S.; Zhu, Y. Individual Phase Full-Power Testing Method for High-Power STATCOM. *Electronics* **2019**, *8*, 754. [CrossRef]
30. Xu, C.; Ma, Q.; Xu, P.; Cui, T. Shaping SiC MOSFET Voltage and Current Transitions by Intelligent Control for Reduced EMI Generation. *Electronics* **2019**, *8*, 508. [CrossRef]

 © 2020 by the author. Licensee MDPI, Basel, Switzerland. This article is an open access article distributed under the terms and conditions of the Creative Commons Attribution (CC BY) license (http://creativecommons.org/licenses/by/4.0/).

Review

Active Power Decoupling for Current Source Converters: An Overview Scenario

Jianhua Zhang [1], Hao Ding [1], Baocheng Wang [1], Xiaoqiang Guo [1,*] and Sanjeevikumar Padmanaban [2]

[1] Department of Electrical Engineering, Yanshan University, Qinhuangdao 066004, China; zjh_163com@163.com (J.Z.); dinghao@ysu.edu.cn (H.D.); bcwang@ysu.edu.cn (B.W.)

[2] Department of Energy Technology, Aalborg University, 6700 Esbjerg, Denmark; san@et.aau.dk

* Correspondence: gxq@ysu.edu.cn

Received: 29 December 2018; Accepted: 2 February 2019; Published: 8 February 2019

Abstract: For single-phase current source converters, there is an inherent limitation in DC-side low-frequency power oscillation, which is twice the grid fundamental frequency. In practice, it transfers to the DC side and results in the low-frequency DC-link ripple. One possible solution is to install excessively large DC-link inductance for attenuating the ripple. However, it is of bulky size and not cost-effective. Another method is to use the passive LC branch for bypassing the power decoupling, but this is still not cost-effective due to the low-frequency LC circuit. Recently, active power decoupling techniques for the current source converters have been sparsely reported in literature. However, there has been no attempt to classify and understand them in a systematic way so far. In order to fill this gap, an overview of the active power decoupling for single-phase current source converters is presented in this paper. Systematic classification and comparison are provided for researchers and engineers to select the appropriate solutions for their specific applications.

Keywords: current source converter; power decoupling; power ripple

1. Introduction

The current source converters have unique features, such as inherent short-circuit capability, no electrolytic capacitor, step-up voltage capability and high reliability, and they have been widely used in applications such as smart microgrids [1], industrial uninterrupted power supplies (UPS) [2], Superconductor Magnetic Energy Storage (SMES) [3], photovoltaic power systems [4], high voltage direct current (HVDC) transmission systems and flexible AC transmission (FACT) systems [5]. There are many key technique issues that need to be solved for the converter, such as the leakage current problem, the efficiency problem, and so on [6–11]. For the single-phase current source converter, there is an inherent limitation of AC-side low-frequency power oscillation, which is twice the grid fundamental frequency. It transfers to the DC side and results in the low-frequency DC-link ripple. It has many unfavorable disadvantages. For example, it may reduce the efficiency of the maximum power point (MPPT) tracking in photovoltaic applications. Therefore, it could lead the light emitting diode (LED) lamps to flicker [12]. In addition, it may cause the battery to overheat and reduce the life of fuel cells [13,14]. This is the reason why the power decoupling techniques are popular in both academic and industrial fields.

Basically, the above-mentioned power ripple can be attenuated by increasing the DC-link inductance. However, it is of bulky size and not cost-effective. Another method is to use the passive LC branch for bypassing the power ripple, but it is still not cost-effective due to the low-frequency LC circuit. Recently, active power decoupling techniques have been reported to deal with this problem. Many interesting topologies have been reported in the last decades [15–24]. For example, a buck type circuit, boost type circuit or buck-boost type circuit is applied as the auxiliary power decoupling circuit [25]. Another method reported in [26,27] uses two additional capacitors placed on the AC side to achieve power decoupling without any auxiliary switches. The decoupling topology using the center tap of the isolation transformer has been proposed [28–30]. The above-mentioned topologies are the voltage-source converters. On the other hand, the active power decoupling techniques for the current source converters have been sparsely discussed in the literature. For example, Sun et al. proposed an interesting circuit, which used a decoupling circuit connect in series with the main converter [31]. This power decoupling method has the advantage of control flexible. For the current source AC/DC/AC converter, the low-frequency ripple also exists on the DC-side. There is a kind of power decoupling circuit proposed in [32]. The capacitor is used as a power compensation component, while the H bridge circuit is applied for power decoupling. Another circuit in [33] uses the buffer capacitor to achieve the power decoupling on the DC-side. A circuit structure connected in parallel with the current source converter on the DC-side is proposed in [34]. For this circuit structure, the buffer capacitor also connects in series with the current source converter when the circuit works in power decoupling mode. Aside from that, the power decoupling circuit can be installed on the AC side [35], which needs an extra bridge arm. A similar circuit structure is proposed in [36–38]. A modified differential connected circuit structure is proposed in [39]. To decrease the number of auxiliary switches, a kind of power decoupling circuit with the lower switches multiplexing has been proposed [40]. The modified power decoupling circuits, which need no extra switch, are proposed in [41,42]. A decoupling topology using the isolation transformer for the current source converter is derived from [43]. The power decoupling circuit is independent with the main converter. However, it needs a transformer to isolate the decoupling circuit and the main current source converter.

In summary, the power decoupling method can be divided into the capacitive and inductive compensations [44]. For the active power decoupling for the current source converters, the capacitor is generally applied as the ripple power compensation device. According to the relation between the auxiliary power decoupling circuit and the main circuit, the ripple power decoupling circuit can be divided into independent and non-independent structures [45]. The following will present the details, and the rest of the paper is organized as follows. The theoretical analysis of power oscillation and ripple for the single-phase current source converter is presented in Section 2. The versatile power decoupling circuits are illustrated and discussed in Sections 3 and 4. Aside from that, a comparison of different power decoupling circuits is provided in Section 5.

2. Problem Definition

The basic single-phase current source converter is illustrated in Figure 1.

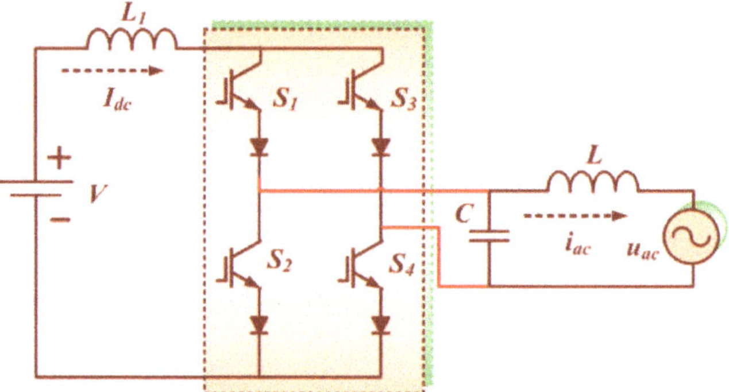

Figure 1. Basic single-phase current source converter.

In Figure 1, the grid voltage u_{ac} and grid current i_{ac} can be represented as:

$$u_{ac} = V \cdot \cos(\omega \cdot t),\tag{1}$$

$$i_{ac} = I \cdot \cos(\omega \cdot t + \varphi),\tag{2}$$

In Equation (2), φ represents the phase-angle between grid voltage and current. The output power P_{ac} can be represented as

$$P_{ac} = u_{ac} \cdot i_{ac} = \frac{VI\cos(\varphi)}{2} + \frac{VI\cos(2\omega \cdot t + \varphi)}{2},\tag{3}$$

From (3), it can be observed that P_{ac} has the ripple power at twice line frequency.
The constant power P_o can be represented as:

$$P_o = \frac{VI\cos(\varphi)}{2},\tag{4}$$

Therefore, the ripple power P_r can be represented as:

$$P_r = \frac{VI\cos(2\omega \cdot t + \varphi)}{2}\tag{5}$$

According to the power conservation law, the ripple power also exists on the DC side.

Figure 2 reflects the relation between the output power and ripple power. The AC power P_{ac} consists of the constant power P_o and ripple power P_r. According to the power conservation law, the auxiliary circuit is needed to attenuate the ripple. When the AC power P_{ac} is larger than power P_o, the ripple power is absorbed by the compensating circuit. The auxiliary circuit will emit the energy to compensate the AC power P_{ac} when the AC power P_{ac} is less than P_o. It is the basic mechanism for the power decoupling method.

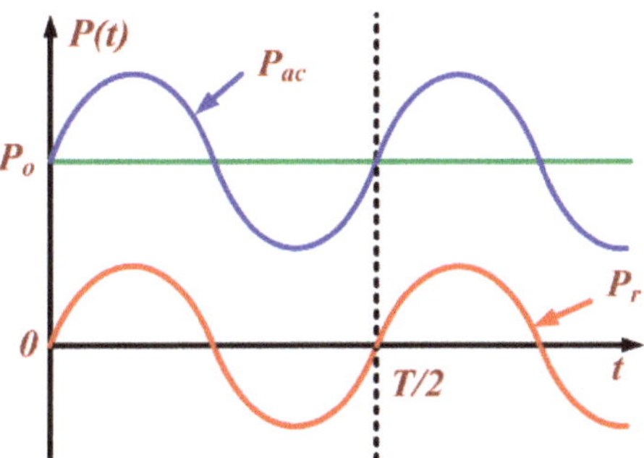

Figure 2. Output power and ripple power.

3. DC-Side Power Decoupling Solutions

The structure shown in Figure 3a connects in series with the main converter on the DC side to compensate for the ripple power. The capacitor is used as a passive component to bypass the ripple power. The circuit shown in Figure 3b is a simplified circuit in which the number of the auxiliary switch is reduced. However, the complexity of the circuit control will increase.

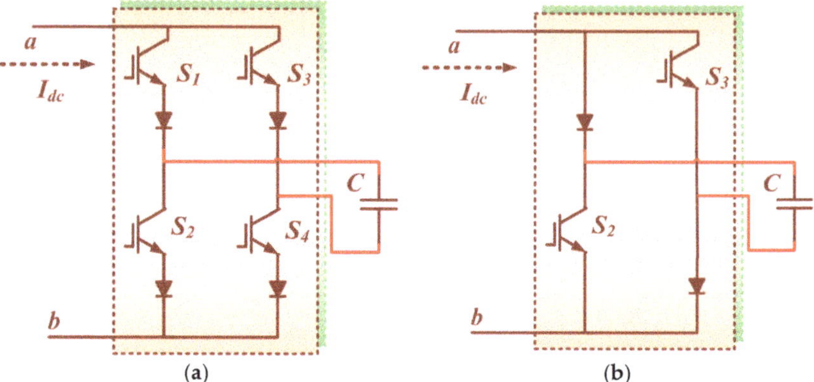

(a)   (b)

Figure 3. (a) H-bridge decoupling circuit; (b) Simplified H-bridge decoupling circuit.

The converter applies the H bridge structure shown in Figure 4. The voltage of the buffer capacitor is lower than the DC-side voltage. For this circuit, the circuit modes and power decoupling process is introduced as follows. When the auxiliary switches S_5 and S_6 are turned on, the buffer capacitor is charged. When switches S_5 and S_6 are turned off, the buffer capacitor discharges through the auxiliary diode D_1 and D_2. When only the switch S_5 or S_6 is turned on, the buffer capacitor is bypassed. The voltage of the buffer capacitor controlled to compensate the ripple power according the real time ripple power on the system. The auxiliary decoupling circuit is controlled independently with the main converter. For the function of the power decoupling circuit, the ripple power is eliminated on the DC side, and the DC-side inductance can be relatively small.

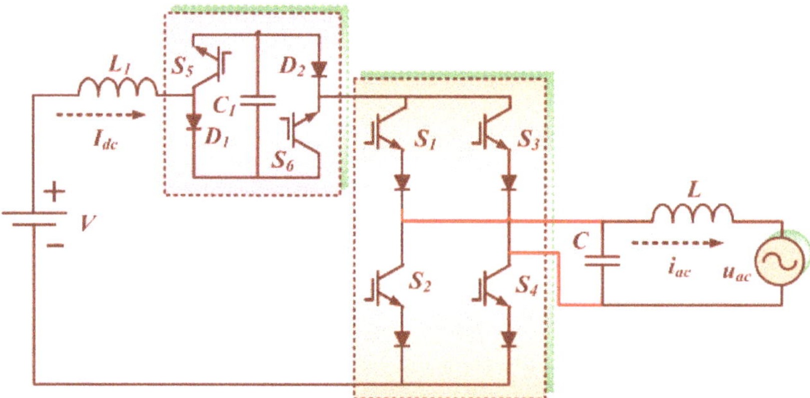

Figure 4. Power decoupling solution in [31].

Another power decoupling circuit applying the series power decoupling circuit is proposed as shown in Figure 5. The AC/DC/AC converter has three bridge arms, which play the role of rectification and inversion with sharing a bridge arm. The bridge arm '*a*' and bridge arm '*c*' work as rectification. And the bridge arm '*b*' and bridge arm '*c*' work as inversion. The decoupling circuit is in series with the DC-link in which capacitor applied to compensate the ripple power.

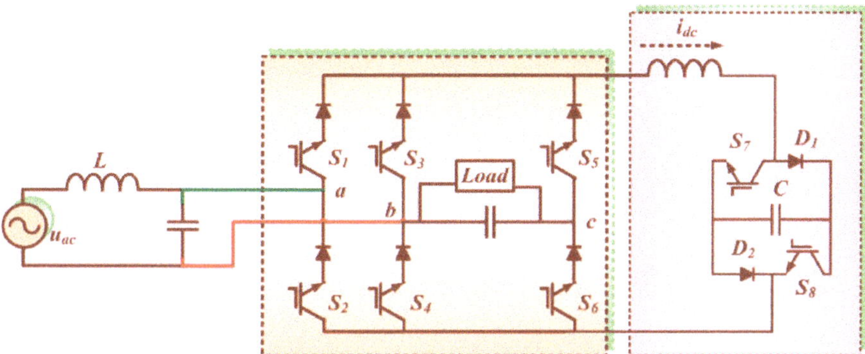

Figure 5. Power decoupling solution in [32].

Compared with a conventional AC/DC/AC converter, the circuit needs fewer switches. However, the system may be broken down when the auxiliary decoupling circuit is out of order, since the decoupling circuit is connected in series with the main circuit on the DC-side.

This type of power decoupling structure shown in Figure 3 has the advantages of a simple circuit structure, being independent with a main converter and flexible control. There are several other types of circuit structures, introduced as follow.

As shown in Figure 6, the auxiliary decoupling circuit is connected in parallel with the main converter. The buffer capacitor C is used to compensate the ripple power. Taking the positive period for example, the operation modes and power decoupling process are discussed as follows.

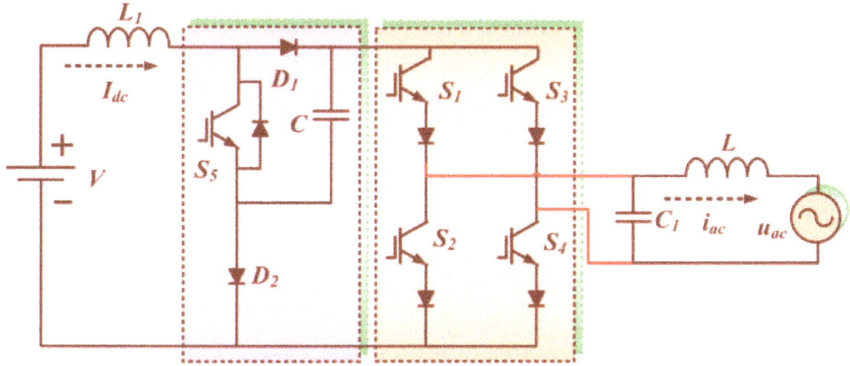

Figure 6. Power decoupling solution in [33].

In the positive period, the switches S_3 and S_2 are turned off, the switches S_1, S_4 and extra switch S_5 work to implement the energy transfer and power decoupling. When S_5 is turned on and the capacitor voltage u_c of C is lower than the grid voltage, the diode D_2 will be turned on. The DC current will flow through the circuit consisting of switch S_5 and diode D_2. When S_5 is turned on and the capacitor voltage of C is higher than the grid voltage, the diode D_2 will turn off and the capacitor C will discharge connected in series with the main converter. When S_5 is turned off and the main converter switches S_1, S_4 are turned off, the capacitor C is charged through the diodes D_1 and D_2. When S_5 is turned off and the main converter switches S_1 and S_4 are turned on, the capacitor C is bypassed. Meanwhile, the grid current and the voltage of the power compensated capacitor are controlled. In contrast to the circuit structure shown in Figure 4, the auxiliary decoupling circuit is depended on the main converter. When the grid voltage is negative, the circuit state is similar with the positive period.

As shown in Figure 7, the auxiliary decoupling circuit consists of one bridge arm and a bidirectional buck-boost circuit, which work to compensate the ripple power. The voltage of capacitor C_f, is controlled to keep constant. The capacitor C is used to compensate the ripple power.

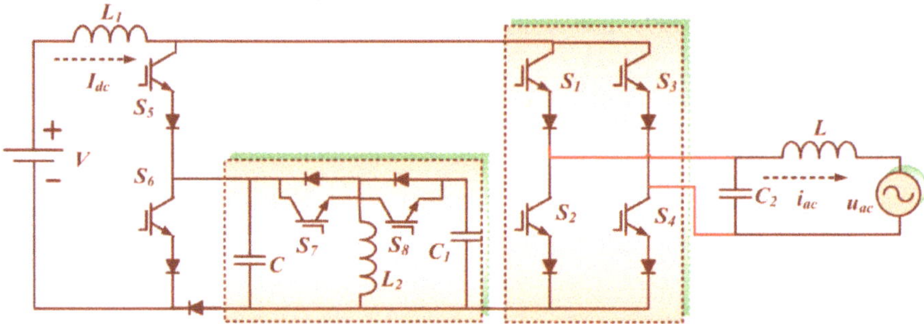

Figure 7. Power decoupling solution in [34].

For this circuit, the operation modes and power decoupling process are discussed as follows. When all the main converter switches S_1–S_4 are turned off, the extra switch S_5 remains on. In this state, the DC current is bypassed by the auxiliary bridge while the switch S_6 is on. The buffer circuit supplies zero voltage to the DC side. While switch S_6 is turned off, the buffer circuit connects in series with the DC side to compensate the ripple power. When the main converter switches S_1 and S_4 (or S_2 and S_3) are turned on, the switch S_5 keep off. In this state, when switch S_6 is turned off, the buffer circuit is isolated with the source. When switch S_6 is turned on, the buffer circuit connects in series with DC side. The switch states of S_5 and S_6 are dependent with the main circuit.

The decoupling circuit shown in Figures 6 and 7 connects in parallel with the main converter but the buffer capacitor is connected in series with the main converter.

4. The Typical Circuit Structure of AC Side Decoupling

An interesting circuit structure to decouple the ripple power on the AC side is proposed in [35].

As shown in Figure 8, there is an auxiliary bridge arm of the converter. The capacitor C_1, C_2 and C_3 work as power compensation capacitor connected in star structure. The capacitor C_2 and C_3 also work as AC filter capacitor. To compensate the ripple power, the decoupling circuit shares two arm bridges with the main converter to control the capacitor voltage effectively. The circuit shown in Figure 9 is one of the simplified circuits.

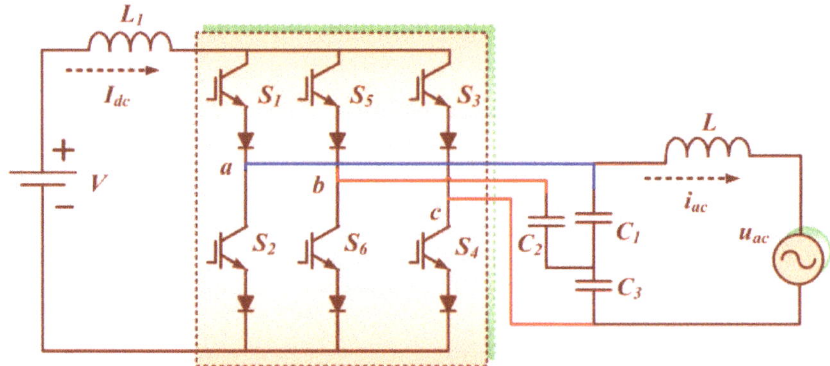

Figure 8. Power decoupling solution in [35].

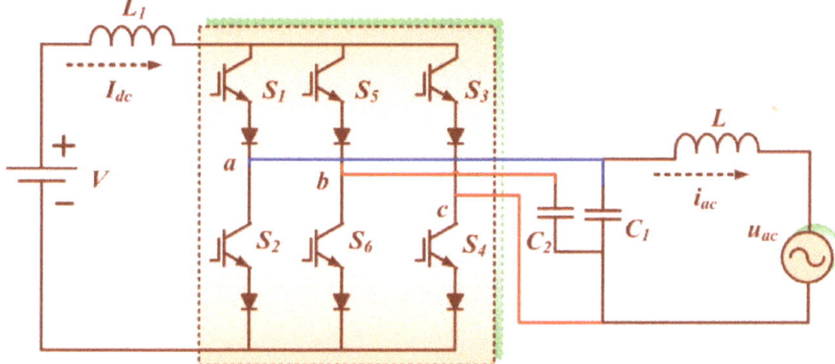

Figure 9. Power decoupling solution in [35].

The decoupling capacitor C_3 of Figure 8 can be seen as a wire in the circuit of Figure 9. The decoupling capacitor C_1 works as the AC filter capacitor.

The literature [36] introduces a circuit which also compensates the ripple power on the AC-side. It is a duality with the voltage source converter introduced in [37]. It is similar to Figure 9 in circuit structure.

As shown in Figure 10, the circuit topology consists of three bridge arms. The bridge arm '*b*' is connected with bridge arm '*c*' through buffer capacitor C. By controlling the current flowing through capacitor C, the ripple power can be compensated. Different from the circuit shown in Figure 9, the circuit only has one decoupling capacitor. Hence, it is easy to control.

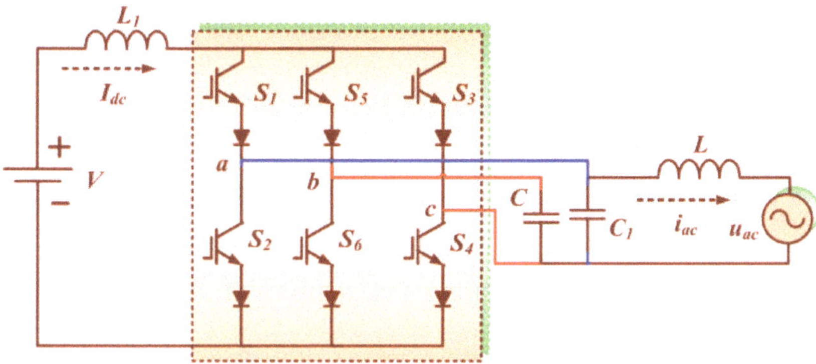

Figure 10. Power decoupling solution in [36].

For the single-phase current source AC/DC/AC converter, the ripple power exists at the source and load sides. To eliminate the low frequency DC-link energy, an interesting circuit is proposed in [38], as shown in Figure 11. The auxiliary power decoupling circuits were applied on the AC side to compensate for the ripple power, which is similar to the decoupling method shown in Figure 10. Compared with the power decoupling method shown in Figure 11, this circuit needs more extra switches. Henceforth, it has more switch losses.

A kind of differently-connected circuit was applied to achieve the ripple power decoupling as shown in Figure 12a.

As shown in Figure 12a, the current source inverter is composed of two H-type inverter which is connected in a differential way. The power decoupling circuit applies two capacitors as the power compensation device, which is placed at AC side. The number of switches is high as shown in

Figure 12a. A modified circuit structure is proposed in [39]. The middle two bridge arms of Figure 12a were synthesized into one bridge arm, as shown in Figure 12b. The power compensation capacitors C_1 and C_2 shown in Figure 12b were used also as the AC filter capacitors. The phase of the ripple voltage part of the two capacitors was reversed. So, the total voltage of the two capacitors synchronizes with the grid voltage.

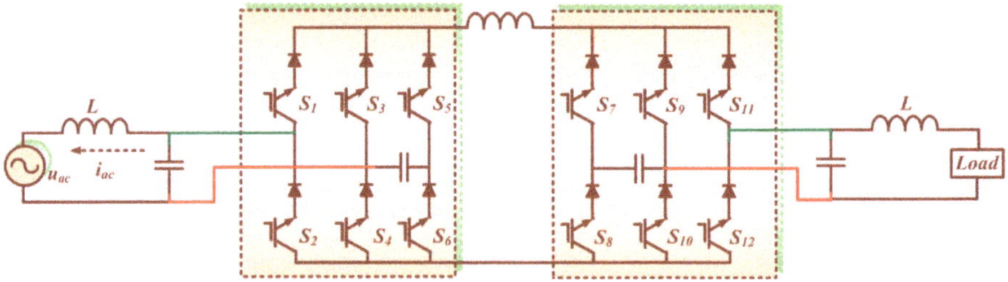

Figure 11. Current source converter with two auxiliary decoupling circuits [38].

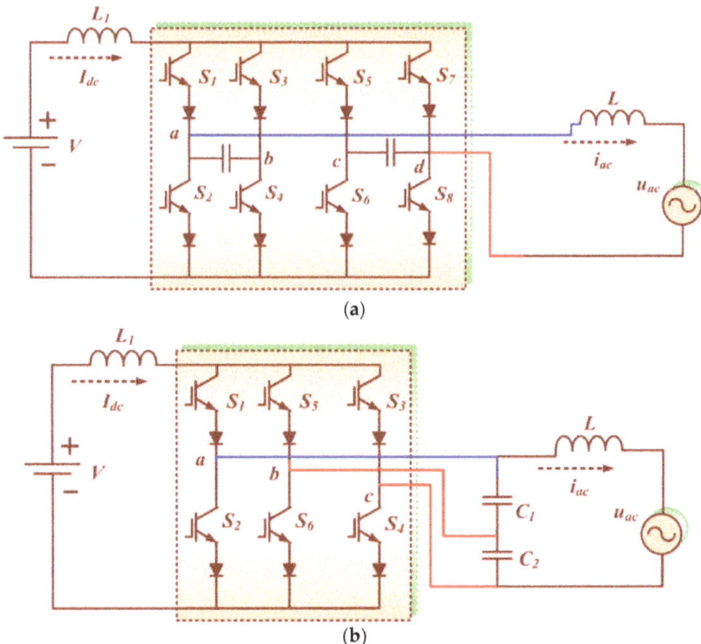

Figure 12. Topology of current source converter with differential-connected structure [39]. (**a**) four leg configuration; (**b**) three leg configuration.

As stated above, the auxiliary power decoupling circuit shares the bridge arm with the main converter. The decoupling circuit, which shares the lower switches with the main inverter is proposed in [40]. As shown in Figure 13, the decoupling circuit part is marked by the lower dotted line. The main converter is marked by the upper dotted line. It is obvious that the multiplexed switches are the lower switches of the main converter. Compared to a conventional independent circuit, the dependent circuit needs fewer switches. To decrease the number of extra switches, the modified circuit of Figure 13 is proposed in [41,42].

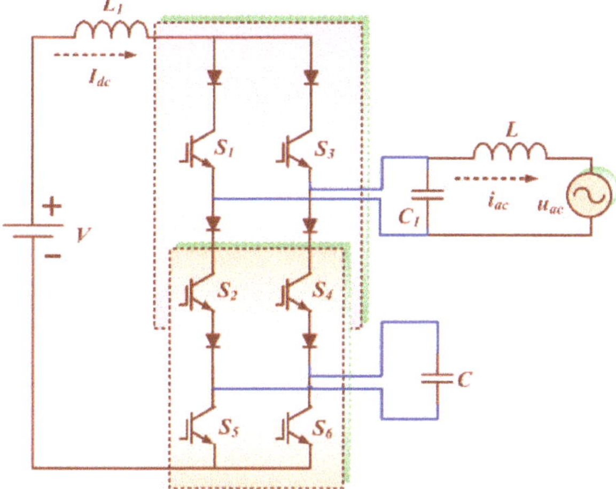

Figure 13. Power decoupling solution in [40].

As shown in Figure 14a, there is an auxiliary capacitor which connects between the bridge arms. There are two circuit states to charge the capacitor voltage. When the switches S_1, S_3 and S_4 are turned on, the capacitor C is charged. While the switch S_2 is turned on, the capacitor C discharged. The voltage of capacitor C is controlled to compensate for the ripple power. This circuit only needs one auxiliary diode to limit the discharging loop of the buffer capacitor. Compared with Figure 14a, the circuit shown in Figure 14b has two auxiliary capacitors. Thus, the circuit has more optional circuit states.

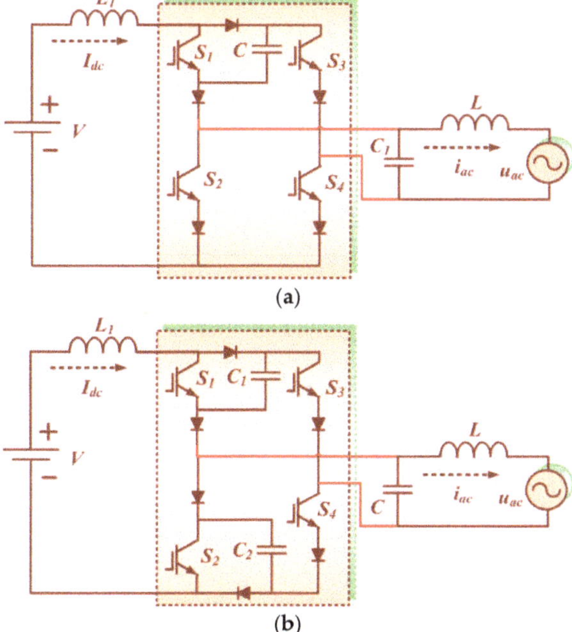

Figure 14. Power decoupling solution with no auxiliary switch [41,42]. (a) 1-capacitor configuration; (b) 2-capacitor configuration.

A decoupling topology using the isolation transformer for current source converter is derived from [43]. In Figure 15, it can be observed that the auxiliary power decoupling circuit is separated from the output circuit. The auxiliary power decoupling circuit is independent from the main converter. Compared to the dependent power decoupling circuit, control of independent circuit is simple. However, it needs more auxiliary switches and an extra transformer.

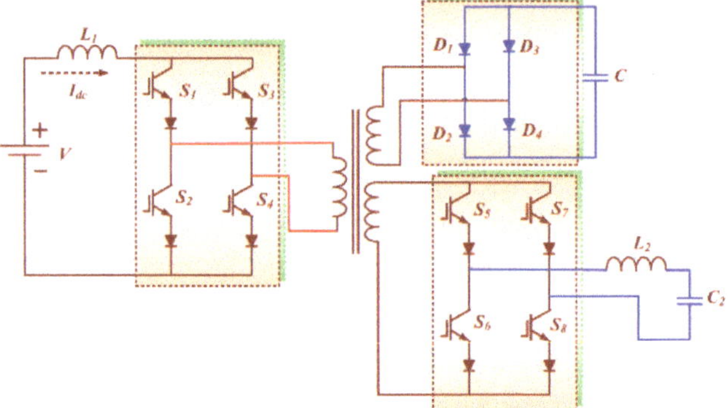

Figure 15. Power decoupling solution with three-port circuit [43].

5. Discussions and Conclusions

The ripple power in the single-phase current source converter has an adverse effect on the performance of the system. That is the reason why power decoupling is important for single-phase current source converters. This paper has presented a systematic review of the power decoupling methods for single-phase current source converters. Typically, the capacitor is used as a buffer device for power decoupling [46–49]. According the relationship between the decoupling circuit and the main converter, the topology is divided into the independent and the dependent circuits. The dependent circuit needs few auxiliary switches. However, it is complex to control. The decoupling circuit can be placed on the DC-side or AC-side of the main converter. One of the DC-side decoupling circuits is shown in Figure 4. The H-bridge decoupling circuit is connected in series with the main converter. For this circuit structure, the power decoupling circuit is controlled independently with the main converter. Other circuit structures for DC-side power decoupling are shown in Figures 6 and 7. The auxiliary power decoupling circuit is in parallel with the main converter. The AC-side decoupling circuit shares bridge arms with the main converter. The typical structure is described as follows.

The circuit structure in Figure 16a needs two or more buffer capacitors to decouple the ripple power. The typical circuit structures are shown in Figures 8, 9 and 12. The circuit structures in Figures 9 and 12 can be considered as the simplified circuit of Figure 8. For the circuit structure shown in Figure 16b, the decoupling circuit shares a bridge arm with the main converter. The typical circuit is shown as Figure 10. Only one buffer capacitor is needed for power decoupling. It can be considered similar to the circuit structure in Figure 8.

Regarding the two circuit structures shown above, the decoupling circuit is dependent on the main converter. They will interfere with each other, which results in output current distortion and increases the control complexity.

A comparison of these topologies is shown in Table 1. The structures of power decoupling circuits have different characteristics. For example, compared to the dependent decoupling circuit, the independent decoupling circuit is not influenced by the main converter. However, the independent decoupling circuit needs more extra switches. Readers might wonder which could be the best choice among these versatile solutions. The answer to this question is difficult because each of them has its own disadvantages and advantages. In this paper, an overview and comparison has carried out, and it is expected to provide a useful guide for researchers and engineers to select the appropriate power decoupling solution for their specific applications.

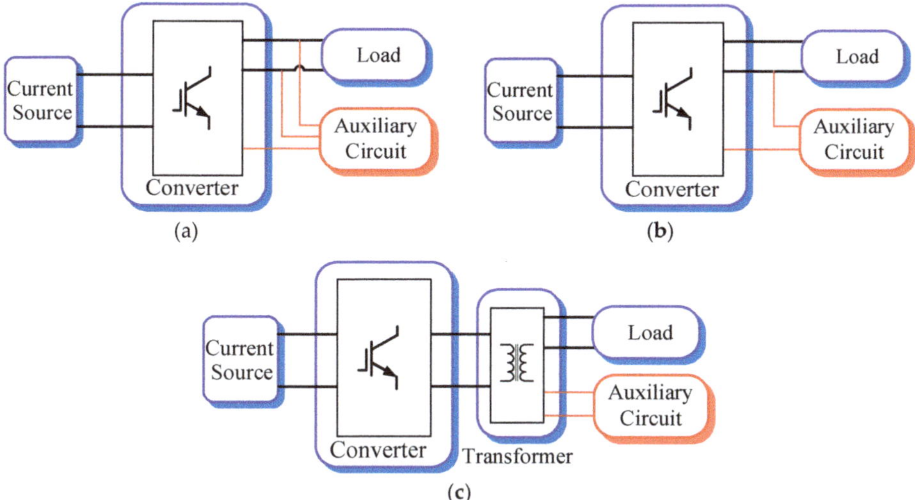

Figure 16. AC-side power decoupling circuit structure. (**a**) Share two bridge arms; (**b**) Share a bridge arm; (**c**) Three-port circuit.

Table 1. Comparison of versatile power decoupling circuits.

	Power Rating	Auxiliary Components	Connection Method	Circuit Performance	Advantages and Disadvantages
Figure 4 [31]	139.2 W	1 Capacitor (91.8 μF) + 2 Switches + 2 Diodes	In series (DC side)	**Efficiency**: 84%. **Effect**: Second-order harmonic current ripple is reduced by 87.99%. **Dynamic response**: within 5 ms.	**Advantages**: Easy control and simple circuit structure. **Disadvantages**: Endangering the system stability due to the connected power decoupling circuit.
Figure 5 [32]	300 W	1 Capacitor (100 μF) + 2 Switches + 2 Diodes	In series (DC side)	**Effect**: Second-order harmonic current ripple reduced to 1.72% of DC link current. **Dynamic response**: within 1 ms.	**Advantage**: Less auxiliary switches **Disadvantages**: Control is complex, Endangering the system stability due to connected power decoupling circuit
Figure 6 [33]	500 W	1 Capacitor (50 μF) + 1 Switch + 2 Diodes	In parallel (DC side)	**THD**: Less than 5%. The output power factor can reach to 99.9% **Efficiency**: Maximum efficiency of 94.9% **Effect**: Voltage ripple is reduced to 8.87%	**Advantages**: System stability. **Disadvantage**: Control is complex.
Figure 7 [34]	/	1 Capacitor + 4 Switches + 1 Diodes	In parallel (DC side)	**Effect**: MPPT performance can be improved for the PV converter.	**Advantages**: System stability. **Disadvantage**: Circuit structure complex, Much auxiliary switch.
Figure 8 [35]	/	3 Capacitors + 2 Switches	In parallel (AC side)	/	**Advantage**: Bridge arms multiplexing, Less auxiliary switches **Disadvantage**: Control is complex.
Figure 9 [35]	/	2 Capacitors + 2 Switches	In parallel (AC side)		

Table 1. *Cont.*

	Power Rating	Auxiliary Components	Connection Method	Circuit Performance	Advantages and Disadvantages
Figure 10 [36]	1 kW	1 Capacitor (10 μF) + 2 Switches	In parallel (AC side)	**Effect:** The input current ripple is reduced to 1.86% of DC-link current.	**Advantages:** Bridge arms multiplexing, Control is easy relatively. **Disadvantage:** Control is complex.
Figure 11 [38]	515 W	2 Capacitor (100 μF, 100 μF) + 4 Switches	In parallel (AC side)	**Efficiency:** 85.05%. **Effect:** The input voltage ripple is reduced to 3.9% of DC-link voltage.	**Advantage:** Simple control. **Disadvantages:** Extra switches for decoupling circuits exist on both AC-sides.
Figure 12b [39]	1 kW	2 Capacitors (32 μF, 32 μF) + 2 Switches	In parallel (AC side)	**Effect:** The input current ripple is reduced to 3% of DC-link current.	**Advantage:** Bridge arm multiplexing. **Disadvantage:** Control is complex.
Figure 13 [40]	1 kW	1 Capacitor (50 μF) + 2 Switches	/	**Effect:** The input current ripple is reduced to 4.275% of DC-link current. The input voltage ripple can be reduced to 8.6% of DC-link voltage.	**Advantage:** Switch multiplexing. **Disadvantages:** Endangering system stability due to a bridge arm with three switches.
Figure 14a [41]	217.5 W	1 Capacitor (90 μF)	/	**THD:** 4.63%. **Efficiency:** The power losses caused by decoupling circuit is about 1.5% **Dynamic response:** within 1 ms.	**Advantage:** No extra switches. **Disadvantage:** Control is complex.
Figure 14b [42]	217.5 W	2 Capacitors (90 μF total)	/	**THD:** 4.47%. **Effect:** Second-order harmonic current ripple is reduced to 2.36% **Dynamic response:** within 1 ms.	**Advantages:** No extra switches, More optional circuit states. **Disadvantage:** complex control
Figure 15 [43]	/	1 Capacitor + 4 switches + transformer	Isolated with main converter (AC side)	**Efficiency:** The power losses are high due to the losses of auxiliary switches and transformer.	**Advantage:** Independent structure. **Disadvantages:** Need a transformer, high power loss.

Author Contributions: This paper was a collaborative effort among all authors.

Funding: This work is supported by the National Natural Science Foundation of China (51777181), the Lite-On Research Funding, the State Key Laboratory of Reliability and Intelligence of Electrical Equipment (No. EERIKF2018002), Hebei University of Technology, Postdoctoral Research Program of Hebei Province (B2018003010), and Hundred Excellent Innovation Talents Support Program of Hebei Province (SLRC2017059).

Conflicts of Interest: The authors declare no conflicts of interest.

Abbreviations

Acronyms

AD	active damping
UPS	uninterruptible power supply
SMES	superconductor magnetic energy storage
HVDC	high voltage direct current
FACT	flexible AC transmission
MPPT	maximum power point tracking
LED	light emitting diode

Nomenclature

u_{ac}	grid voltage
i_{ac}	grid current
φ	phase-angle between grid voltage and current
P_{ac}	output power
P_o	constant power
P_r	ripple power

References

1. Senthilnathan, K.; Annapoorani, I. Multi-Port Current Source Inverter for Smart Microgrid Applications: A Cyber Physical Paradigm. *Electronics* **2019**, *8*, 1. [CrossRef]
2. Colombi, S.; Balblano, C. Clean Input UPS with Fast Rectifier Control and Improved Battery Life. U.S. Patent US7768805, 3 August 2010.
3. Wang, Z.; Jiang, L.; Zou, Z.; Cheng, M. Operation of SMES for the current source inverter fed distributed power system under islanding mode. *IEEE Trans. Appl. Superconduct.* **2012**, *23*, 5700404. [CrossRef]
4. Guo, X. A novel CH5 inverter for single-phase transformerless photovoltaic system applications. *IEEE Trans. Circ. Syst. II Express Briefs* **2017**, *64*, 1197–1201. [CrossRef]
5. Liang, J.; Nami, A.; Dijkhuizen, F.; Tenca, P.; Sastry, J. Current source modular multilevel converter for HVDC and FACTS. In Proceedings of the European Conference on Power Electronics and Applications, Lille, France, 2–6 September 2013; pp. 1–10.
6. Kerekes, T.; Teodorescu, R.; Liserre, M.; Klumpner, C.; Sumner, M. Evaluation of Three-Phase Transformerless Photovoltaic Inverter Topologies. *IEEE Trans. Power Electron.* **2009**, *24*, 2202–2211. [CrossRef]
7. Graditi, G.; Adinolfi, G.; Femia, N.; Vitelli, M. Comparative analysis of Synchronous Rectification Boost and Diode Rectification Boost converter for DMPPT applications. In Proceedings of the 2011 IEEE International Symposium on Industrial Electronics, Gdansk, Poland, 27–30 June 2011; pp. 1000–1005.
8. Yang, B.; Li, W.; Gu, Y.; Cui, W.; He, X. Improved Transformerless Inverter with Common-Mode Leakage Current Elimination for a Photovoltaic Grid-Connected Power System. *IEEE Trans. Power Electron.* **2012**, *27*, 752–762. [CrossRef]
9. Cecati, C.; Khalid, H.A.; Tinari, M.; Adinolfi, G.; Graditi, G. DC NanoGrid with Renewable Energy Sources and Modular DC/DC LLC Converter Building Block. *IET Power Electron.* **2016**, *10*, 536–544. [CrossRef]
10. Adinolfi, G.; Graditi, G.; Siano, P.; Piccolo, A. Multi-Objective Optimal Design of Photovoltaic Synchronous Boost Converters Assessing Efficiency, Reliability and Cost Savings. *IEEE Trans. Ind. Inform.* **2015**, *11*, 1038–1048. [CrossRef]
11. Cavalcanti, M.C.; De Oliveira, K.C.; De Farias, A.M.; Neves, F.A.; Azevedo, G.M.; Camboim, F.C. Modulation Techniques to Eliminate Leakage Currents in Transformerless Three-Phase Photovoltaic Systems. *IEEE Trans. Ind. Electron.* **2010**, *57*, 1360–1368. [CrossRef]
12. Chen, W.; Hui, S.R. Elimination of an electrolytic capacitor in AC/DC light-emitting diode (LED) driver with high input power factor and constant output current. *IEEE Trans. Power Electron.* **2011**, *27*, 1598–1607. [CrossRef]
13. Kim, H.; Shin, K.G. DESA: Dependable, Efficient, Scalable Architecture for management of large-scale batteries. *IEEE Trans. Ind. Inform.* **2011**, *8*, 406–417. [CrossRef]
14. Shimizu, T.; Fujita, T.; Kimura, G.; Hirose, J. A unity power factor PWM rectifier with DC ripple compensation. *IEEE Trans. Ind. Electron.* **1997**, *44*, 447–455. [CrossRef]
15. Hu, H.; Harb, S.; Kutkut, N.; Batarseh, I.; Shen, Z.J. A review of power decoupling techniques for microinverters with three different decoupling capacitor locations in PV systems. *IEEE Trans. Power Electron.* **2012**, *28*, 2711–2726. [CrossRef]
16. Guo, X.; Zhang, X.; Guan, H.; Kerekes, T.; Blaabjerg, F. Three phase ZVR topology and modulation strategy for transformerless PV system. *IEEE Trans. Power Electron.* **2019**, *34*, 1017–1021. [CrossRef]
17. Schimpf, F.; Norum, L. Effective use of film capacitors in single-phase PV-inverters by active power decoupling. In Proceedings of the 36th Annual Conference on IEEE Industrial Electronics Society, Glendale, AZ, USA, 7–10 November 2010; pp. 2784–2789.
18. Yang, Y.; Ruan, X.; Zhang, L.; He, J.; Ye, Z. Feed-forward scheme for an electrolytic capacitor-less AC/DC LED driver to reduce output current ripple. *IEEE Trans. Power Electron.* **2014**, *29*, 5508–5517. [CrossRef]
19. Cao, X.; Zhong, Q.; Ming, W. Ripple eliminator to smooth DC-Bus voltage and reduce the total capacitance required. *IEEE Trans. Power Electron.* **2015**, *62*, 2224–2235. [CrossRef]
20. Krein, P.T.; Balog, R.S.; Mirjafari, M. Minimum energy and capacitance requirements for single-phase inverters and rectifiers using a ripple port. *IEEE Trans. Power Electron.* **2012**, *27*, 4690–4698. [CrossRef]
21. Harb, S.; Mirjafari, M.; Balog, R.S. Ripple-port module-integrated inverter for grid-connected PV applications. *IEEE Trans. Ind. Appl.* **2013**, *49*, 2692–2698. [CrossRef]

22. Nussbaumer, T.; Kolar, J.W. Improving mains current quality for three-phase three-switch buck-type PWM rectifiers. *IEEE Trans. Power Electron.* **2006**, *21*, 967–973. [CrossRef]

23. Guo, X.; Yang, Y.; Zhu, T. ESI: A novel three-phase inverter with leakage current attenuation for transformerless PV systems. *IEEE Trans. Ind. Electron.* **2018**, *65*, 2967–2974. [CrossRef]

24. Nussbaumer, T.; Baumann, M.; Kolar, J.W. Comprehensive design of a three-phase three-switch buck-type PWM rectifier. *IEEE Trans. Power Electron.* **2007**, *22*, 551–562. [CrossRef]

25. Tang, Y.; Blaabjerg, F.; Loh, P.C.; Jin, C.; Wang, P. Decoupling of fluctuating power in single-phase systems through a symmetrical half-bridge circuit. *IEEE Trans. Power Electron.* **2015**, *30*, 1855–1865. [CrossRef]

26. Serban, I. A novel transistor-less power decoupling solution for single-phase inverters. In Proceedings of the 39th Annual Conference of the IEEE Industrial Electronics Society, Vienna, Austria, 10–13 November 2013; pp. 1496–1500.

27. Tang, Y.; Yao, W.; Wang, H.; Loh, P.C.; Blaabjerg, F. Transformerless photovoltaic inverters with leakage current and pulsating power elimination. In Proceedings of the 9th International Conference on Power Electronics and ECCE Asia, Seoul, Korea, 1–5 June 2015; pp. 115–122.

28. Hu, H.; Harb, S.; Kutkut, N.; Batarseh, I.; Shen, Z.J. Power decoupling techniques for micro-inverters in PV systems-a review. In Proceedings of the 2010 IEEE Energy Conversion Congress and Exposition, Atlanta, GA, USA, 12–16 September 2010; pp. 3235–3240.

29. Takahashi, H.; Takaoka, N.; Gutierrez, R.R.R.; Itoh, J.-I. Power decoupling method for isolated dc to single-phase ac converter using matrix converter. In Proceedings of the 20th European Conference on Power Electronics and Applications, Riga, Latvia, 17–21 September 2018; pp. 1–10.

30. Cai, W.; Liu, B.; Duan, S.; Jiang, L. Power flow control and optimization of a three-port converter for photovoltaic-storage hybrid system. In Proceedings of the IEEE Energy Conversion Congress and Exposition, Raleigh, NC, USA, 15–20 September 2012; pp. 4121–4128.

31. Han, H.; Liu, Y.; Sun, Y.; Su, M.; Xiong, W. Single-phase current source converter with power decoupling capability using a series-connected active buffer. *IET Power Electron.* **2015**, *8*, 700–707. [CrossRef]

32. Liu, Y.; Sun, Y.; Su, M.; Li, X.; Ning, S. A Single Phase AC/DC/AC Converter with Unified Ripple Power Decoupling. *IET Power Electron.* **2018**, *33*, 3204–3217. [CrossRef]

33. Ohnuma, Y.; Orikawa, K.; Itoh, J.I. A single-phase current source PV inverter with power decoupling capability using an active buffer. *IEEE Trans. Ind. Appl.* **2014**, *51*, 531–538. [CrossRef]

34. Román, L.T.; Silva, L.S. A single-phase current-source inverter with active power filters for grid-tied PV systems. In Proceedings of the 3rd IEEE International Symposium on Power Electronics for Distributed Generation Systems, Aalborg, Denmark, 25–28 June 2012; pp. 349–356.

35. Wei, L.; Skibinski, G.L.; Lukaszewski, R.A. Auxiliary Circuit for Use with Three-Phase Drive with Current Source Inverter Powering a Single-Phase Load. U.S. Patent 7518891B2, 14 April 2009.

36. Vitorino, M.A.; Correa, M.B.D.R. Compensation of DC link oscillation in single-phase VSI and CSI converters for photovoltaic grid connection. *IEEE Trans. Ind. Appl.* **2014**, *50*, 2021–2028. [CrossRef]

37. Su, M.; Pan, P.; Long, X.; Sun, Y.; Yang, J. An active power-decoupling method for single-phase AC-DC converters. *IEEE Trans. Ind. Inf.* **2014**, *10*, 461–468. [CrossRef]

38. Vitorino, M.A.; Wang, R.; de Rossiter Corrêa, M.B.; Boroyevich, D. Compensation of DC-Link Oscillation in Single-Phase-to-Single-Phase VSC/CSC and Power Density Comparison. *IET Power Electron.* **2014**, *50*, 2021–2028. [CrossRef]

39. Vitorino, M.A.; Correa, M.B.R.; Jacobina, C.B. Single-phase power compensation in a current source converter. In Proceedings of the 2013 IEEE Energy Conversion Congress and Exposition, Denver, CO, USA, 15–19 September 2013; pp. 5288–5293.

40. Vitorino, M.A.; Hartmann, L.V.; Fernandes, D.A.; Silva, L.E.; Correa, M.B.R. Single-phase current source converter with new modulation approach and power decoupling. *IEEE Appl. Power Electron. Conf. Expo.* **2014**, *16*, 2200–2207.

41. Liu, Y.; Sun, Y.; Su, M. Active power compensation method for single-phase current source rectifier without extra active switches. *IET Power Electron.* **2016**, *9*, 1719–1726. [CrossRef]

42. Sun, Y.; Liu, Y.; Su, M. Active power decoupling method for single-phase current-source rectifer with no additional active switches. *IEEE Trans. Power Electron.* **2016**, *31*, 5644–5653. [CrossRef]

43. Itoh, J.I.; Fayashi, F. Ripple current reduction of a fuel cell for a single-phase isolated converter using a dc active filter with a center tap. *IEEE Trans. Power Electron.* **2010**, *25*, 550–556. [CrossRef]

44. Larsson, T.; Ostlund, S. Active DC link filter for two frequency electric locomotives. In Proceedings of the International Conference on Electric Railways in a United Europe, Amsterdam, The Netherlands, 27–30 March 1995; pp. 97–100.

45. Cai, W.; Jiang, L.; Liu, B.; Zou, C. A power decoupling method based on four-switch three-port DC/DC/AC converter in dc microgrid. *IEEE Trans. Ind. Appl.* **2015**, *51*, 336–343. [CrossRef]

46. Shimizu, T.; Fujioka, Y.; Kimura, G. DC ripples current reduction method on a single phase PWM voltage source converter. In Proceedings of the IEEE Power Conversion Conference, Nagaoka, Japan, 6 August 1998; Volume 1, pp. 525–530.

47. Shimizu, T.; Jin, Y.; Kimura, G. DC ripple current reduction on a single-phase PWM voltage-source rectifier. *IEEE Trans. Ind. Appl.* **2000**, *36*, 1419–1429. [CrossRef]

48. Guo, X.; Yang, Y.; Wang, X. Optimal space vector modulation of current source converter for dc-link current ripple reduction. *IEEE Trans. Ind. Electron.* **2019**, *66*, 1671–1680. [CrossRef]

49. Tsuno, K.; Shimizu, T.; Wada, K.; Ishii, K. Optimization of the DC ripples energy compensating circuit on a single-phase voltage source PWM rectifier. In Proceedings of the IEEE 35th Annual Power Electronics Specialists Conference (IEEE Cat. No.04CH37551), Aachen, Germany, 20–25 June 2004; pp. 316–321.

© 2019 by the authors. Licensee MDPI, Basel, Switzerland. This article is an open access article distributed under the terms and conditions of the Creative Commons Attribution (CC BY) license (http://creativecommons.org/licenses/by/4.0/).

Review

Review of Multilevel Voltage Source Inverter Topologies and Analysis of Harmonics Distortions in FC-MLI

Ronak A. Rana [1], Sujal A. Patel [1], Anand Muthusamy [1], Chee woo Lee [2,*] and Hee-Je Kim [2,*]

[1] School of Electrical Engineering, Pusan National University, Busandaehak-ro 63 beongil 2, Busan 46241, Korea; ronak.eleindia@gmail.com (R.A.R.); sujalpatel15@gmail.com (S.A.P.); anand.muthusamy07@gmail.com (A.M.)

[2] Department of Electrical Engineering and Electrical Engineering Management, University of technology Sydney, 15 Broadway Ultimo, Sydney, NSW 2007, Australia

* Correspondence: cwlee1014@pusan.ac.kr (C.w.L.); heeje@pusan.ac.kr (H.-J.K.); Tel.: +82-51-510-2364 (C.w.L. & H.-J.K.)

Received: 25 September 2019; Accepted: 6 November 2019; Published: 11 November 2019

Abstract: We review the most common topology of multi-level inverters. As is known, the conventional inverters are utilized to create an alternating current (AC) source from a direct current (DC) source. The two-level inverter provides various output voltages [(Vdc/2) and (−Vdc/2)] of the load. It is a successive method, but it makes the harmonic distortion of the output side, Electromagnetic interference (EMI), and high dv/dt. We solve this problem by constructing the sinusoidal voltage waveform. This is achieved by a "multilevel inverter" (MLI). The multilevel inverter creates the output voltage with multiple DC voltages as inputs. Many voltage levels are combined to produce a smoother waveform. During the last decade, the multilevel inverter has become very popular in medium and high-power applications with some advantages, such as the reduced power dissipation of switching elements, low harmonics, and low EMIs. We introduce the information about several multilevel inverters such as the diode-clamped multilevel inverter (DC-MLI), cascaded H-bridge multilevel inverter (CHB-MLI), and flying-capacitor multilevel inverter (FC-MLI) with Power systems CAD (PSCAD) simulation. It is shown that THD is 28.88% in three level FC-MLI while THD is 18.56% in five level topology. Therefore, we can decrease the total harmonic distortion adopting the higher-level topology.

Keywords: multilevel inverter; diode clamped multilevel inverter; flying capacitor multilevel inverter; cascade H bridge multilevel inverter; total harmonic distortion; PWM control techniques; PSCAD/MULTISIM simulation

---

1. Introduction

Inverters are very useful for various industrial applications. In the last few years, the voltage-driving method has been adopted. To reduce the semiconductor transient voltage and current rating, a series and parallel connection method is needed. Moreover, the limited standard three-phase converter is also adopted up to the maximum allowable voltage of the load. Also, both the primary and the Pulse width modulation (PWM) switching frequency can be useful. The reduced switching frequency shows the low disappearance and the higher efficiency. In order to synthesize the spectrum signals of the harmonics caused by the capacity, the multi-level inverter has received more attention in recent times. Moreover, a multilevel inverter has a key role in providing improved operating voltage beyond the voltage limits of conventional semiconductors.

The following clarification defines the significance of the multilevel converter. The explanation of a multilevel inverter is: "The multilevel inverter can demonstrate the switching skill very effectively with different voltage and current levels of its input or output nodes". It is a practical reply to raise the power capacity with a comparably low elements load and decreasing the output harmonics.

The Concept of Multilevel Inverters

Two level inverters are shown in Figure 1. The input voltage of a direct current (DC) is converted to the required alternating current (AC) voltage and frequency [1]. When an inverter operates with Vdc, a two-level inverter can create two different output voltage for a load, Vdc/2 or (−Vdc)/2. To generate an AC voltage, both voltages are generally allowable in the PWM. Compared to multilevel inverters, this two-level method creates harmonic distortion, Electromagnetic interference (EMI), and huge dv/dt [2]. The idea of a multilevel inverter is not based on the two-voltage level AC. Alternatively, some voltage levels are connected to get a better waveform and less dv/dt and harmonic distortion. If the voltage level of the inverter is higher than the waveform, it is much better. However, the designs are simple, but increasing voltage levels and more components are needed. The dividing voltage and extending its control scheme become more complex in a three-phase setup [3]. Series-connected multilevel inverters are connected to the capacitor-composing energy tank of the inverter. That provides many nodes in the inverter connecting to the different phases. The term level does not correspond to the voltage level that can be provided by the output inverter. Diodes can provide the path of reactive source from the load to the power supply at the current and voltage of Resistance Inductor load (RL) load having opposite polarity.

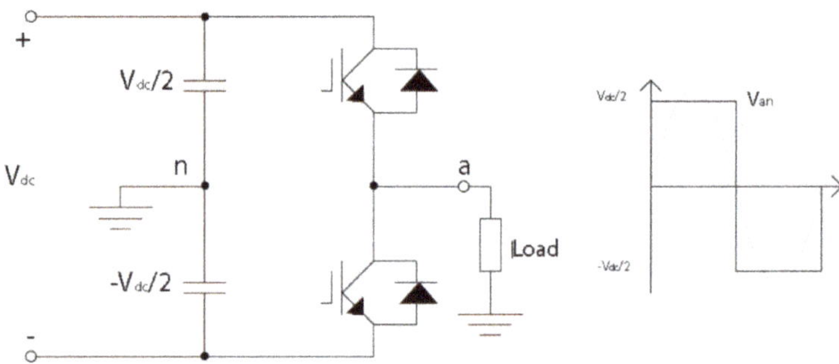

Figure 1. Two-level inverter.

A multilevel inverter has positive points compared with the two-level inverter using the PWM method. Also, a multilevel inverter produces very small distortions and dv/dt in the output side. In addition, it produces a common-mode voltage and operates at a lower switching frequency [2] than a two-stage inverter. Besides, there are several features of a multilevel inverter that can be summarized as the following: (1) the multilevel inverter does not produce low distortion output voltage, but it also decreases the dv/dt stresses. Consequently, problems with electromagnetic compatibility are cleared. (2) The multi-level inverter produces a lower voltage in a common mode; as a result, the motor attached load of the multilevel inverter may be decreased. (3) A multilevel inverter draws a small input current distortion. (4) To obtain more output voltage levels using fewer switching devices, many recently structured multilevel inverters have been presented in this publication [4–16].

In this paper, various multilevel inverter topologies are presented along with their analysis and comparative assessments. Furthermore, these topologies have been checked with the simulation results using PSCAD software. Then a comparison of different types of multilevel inverter topologies and describe PWM control techniques and we can choose one topology for further implementation and selected from the simulation results derived from the MULTISIM software.

2. Topologies of Multilevel Inverter

Normally, these inverters can be classified as voltage source or current source inverters as shown in Figure 2.

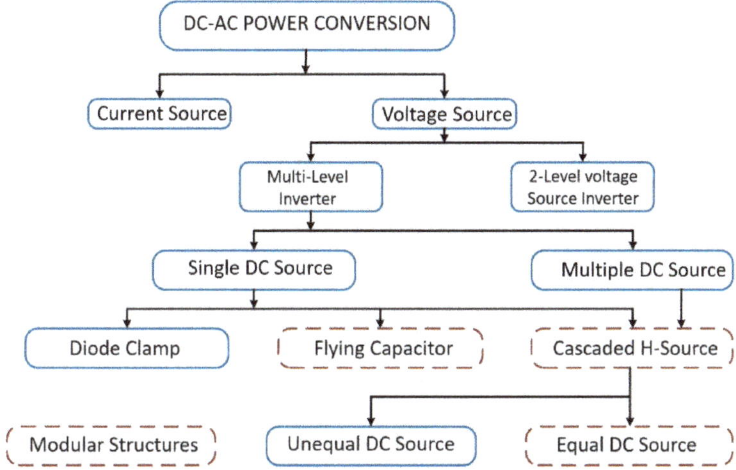

Figure 2. Voltage source or current source inverters.

Three major multilevel inverter configurations applied in industrial applications and mentioned in several literatures; cascaded H-bridges inverter with separate DC sources, Diode Clamp, and flying capacitors used virtually in low, medium and high-power applications are briefly discussed.

2.1. Diode-Clamped Multilevel Inverter (DC-MLI)

The multilevel inverter was invented in 1975, which was a cascade diode inverter. After a few years, it was called the diode-clamped multilevel inverter (DC-MLI) [2]. In DC-MLI, voltage clamping diodes are required to regulate the voltage on the DC bus to reach output voltage levels; but the multilevel inverter level should be unique because the neutral point is not available even in number levels [17]. For DC bus with a neutral point and a capacitor, there is a clamping diode attached to the number of valve pairs (m-1), where m is the level of voltage in the inverter. The explanation as to why the frequency converter is connected to the series with a diode; means that all diodes have the same rated voltage and may block a correct voltage level [18].

A 3-level DC-MLI, a voltage per capacitor and switch is Vdc/2. The equations are utilized to decide the number of devices forming a given level of a DC-MLI. If m is the number of levels, the number of capacitors on the DC side (C) can be determined through Equation (1). The number of switches (S) per phase and the clamp diodes (D) can be determined through Equations (2) and (3), respectively. Consequently, 4 switches, 2 diodes, and 2 capacitors are needed as shown in Figure 3a. The intermediate circuit voltage is divided into three phases, with two capacitors C1 and C2 which are attached in series; "n" is a neutral point which is the center point of the capacitors. There are 3 levels of output voltage Vdc/2, 0, (−Vdc)/2, for voltage level Vdc/2, S1 and S2 are activated, for voltage level (−Vdc)/2, S1' and S2' are activated, and for 0 level voltage, S2 and S1' are activated. When S2' and S1' are connected, the load is shortened, and the reactive power released.

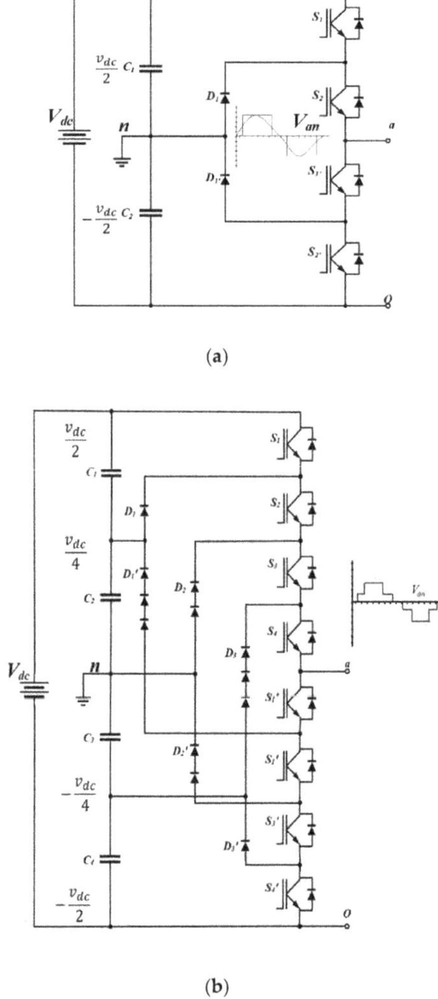

(a)

(b)

Figure 3. (a) Three-level diode-clamped multilevel inverter (DC-MLI); (b) 5-level DC-MLI.

For the 5-level diode clamped multilevel inverter, m is 5. Following Equations (1), (2) and (3), therefore, 8 switches, 4 diodes, and 4 capacitors are needed. Figure 3b shows that all diodes are designed for Vdc/4, and diodes D1′ block 3Vdc/4. Hence, 3 diodes are connected in a series. At 5 voltage levels, Vdc/2, Vdc/4, 0, (−Vdc)/4, (−Vdc)/2 are five states output. To get various output voltage levels, combinations of switching states are utilized. For Vdc/2, top all switches are activated and, for Vdc/4, S2, S3, S4, and S1′ are activated and a voltage is maintained through a clamping diodes D1 and D1′. For 0 level voltage, S3, S4, S1′ and S2′ are activated, for a voltage (−Vdc)/4, switch S4, S1′, S2′, and S3′ are activated and voltage (−Vdc)/2 the bottom all switches are activated and the diodes D2 and D2′ or D3 and D3′ are maintaining the voltage.

$$C = m - 1 \tag{1}$$

$$S = 2(m - 1) \tag{2}$$

$$D = (m-1)(m-2) \tag{3}$$

There are few advantages of DC-MLI: high efficiency, due to the main switching frequency that can be used for all devices. Capacitors are pre-charged as a group; all stages have the usual DC bus. That reduces the capacitance of the inverter. However, there are a few drawbacks of this topology: it is difficult to flow active power, and controlling is quite difficult for the capacitor voltage balance.

2.1.1. 3-Level DC-MLI PSCAD Simulations

In Figure 4a, consider the source voltage as +2 Vdc. There are total two capacitors connected to source, hence the voltage across each capacitor is Vdc. As shown in the switching Table 1 the voltage levels are available as per the switching sequence.

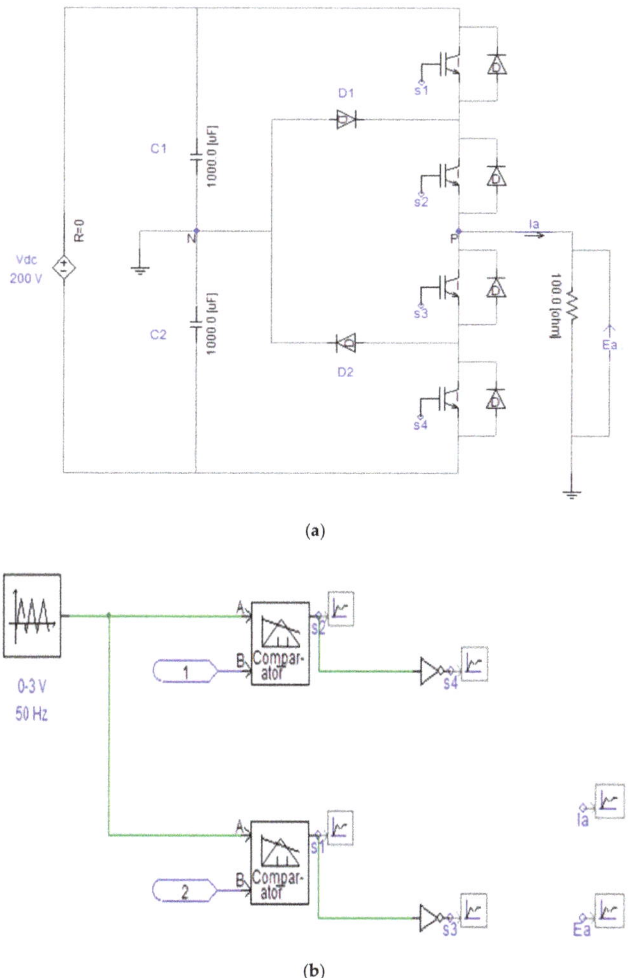

(a)

(b)

Figure 4. *Cont.*

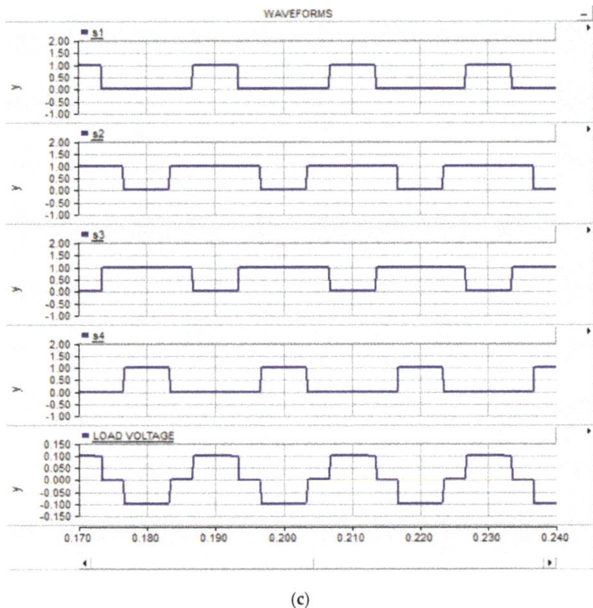

(c)

Figure 4. (**a**) Power circuit of 3- level DC-MLI; (**b**) control circuit of 3- level DC-MLI; (**c**) phase output voltage waveform of 3-level DC-MLI.

+Vdc: Switches S1, S2 are ON. In this case, current flows from source (charged capacitors C1) to load through switches S1, S2. 0 voltage: Switches S2 and S3 are ON. Hence no capacitor is discharged through the load. Here the load is directly connected to ground at both the ends (shorted). Hence the voltage across the load is 0 V. −Vdc: Switches S3 and S4 are ON. In this case, current flows from source (charged capacitors C2) to the load through switches S3 and S4.

In Figure 4b, a control circuit there is 4 switch signal generated using two comparators. 0–3 V, 50 Hz cycle, a triangular waveform is used for comparison. A different DC signal is used as a base for a comparator. This will generate four signals S1, and S2 with different widths. The other two, S3 and S4, are obtained using NOT gate. These 4 gate signals are ready to supply to IGBT which are in the 3-level DC-MLI power circuit, which is illustrated in Figure 4a.

Table 1. 3- Voltage levels of DC-MLI and switching states.

Voltage	S1	S2	S3	S4
Vdc	1	1	0	0
0	0	1	1	0
$-Vdc$	0	0	1	1

2.1.2. 5-Level DC-MLI PSCAD Simulations

In Figure 5a, consider the source voltage as +Vdc. There are total four capacitors connected to source, hence the voltage across each capacitor is +Vdc/4. As shown in the switching Table 2 the voltage levels are available as per the switching sequence.

+Vdc/2: Switches S1, S2, S3, S4 are ON. In this case, current flows from source (charged capacitors C1, C2) to load through switches S1, S2, S3, S4. +Vdc/4: Switches S2, S3, S4, and S5 are ON. Here only capacitor C2 discharges through diode D1 and the switches S2, S3, and S4 to the load. Since only half of the capacitor (only C2) discharges, hence total Vdc/4 voltage is available. 0 voltage: Switches S3, S4, S5 and S6 are ON. Hence no capacitor is discharged through the load. Here the load is directly connected to ground at both the ends (shorted). Hence the voltage across the load is 0 V. −Vdc/4: Switches S4, S5, S6, and S7 are ON. Hence the capacitor C3 discharges through the ground–load and switch S5, S6, and S7. As the discharge is in the opposite direction the voltage appeared will be negative. −Vdc/2: Switches S5, S6, S7 and S8 are ON. Capacitor C3 and C4 will discharge through the load in the opposite direction (from the ground to load to switches). Hence total voltage −Vdc/2 will appear across the load. In Figure 5b a control circuit there is 8 switch signal generated using four comparators. 0 – 10 V, 50 Hz cycle, a triangular waveform is used for comparison. And different DC signal is used as a base for a comparator. This will generate four signals S1, S2, S3 and S4 with different widths. Other four S5 to S8 are obtained using NOT gate. These 8 gate signals are ready to supply to IGBT which are in the 5-level DC-MLI power circuit, which is shown in Figure 5a.

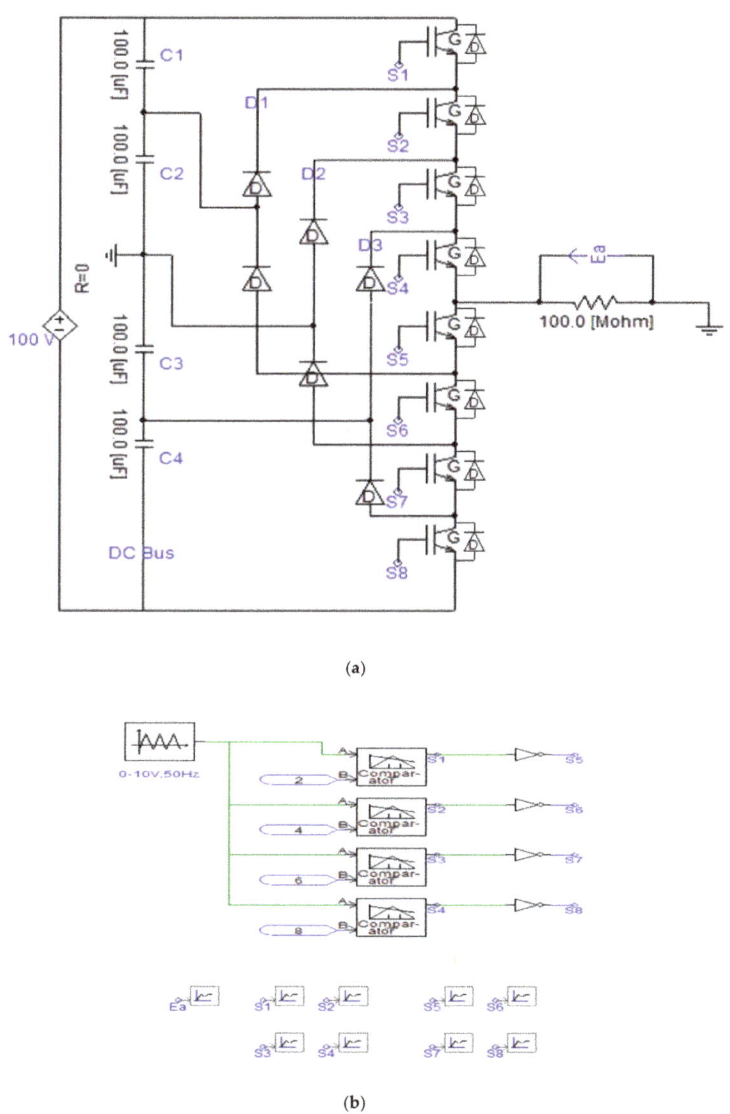

(a)

(b)

Figure 5. *Cont.*

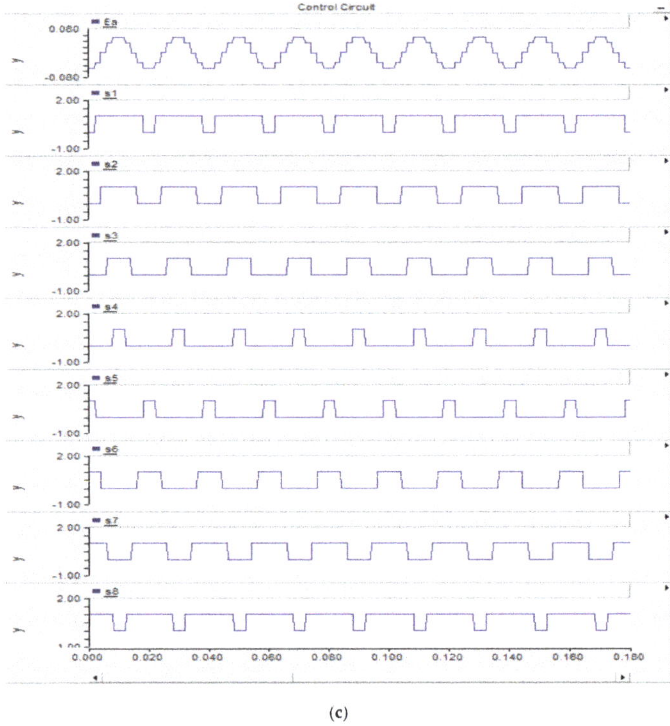

(c)

Figure 5. (a) Power circuit of 5-level DC-MLI; (b) control circuit of 5-level DC-MLI; (c) phase output voltage waveform of 5-level DC-MLI.

Table 2. 5-Voltage levels of DC-MLI and switching states.

Voltage	S1	S2	S3	S4	S5	S6	S7	S8
$\frac{V_{dc}}{2}$	1	1	1	1	0	0	0	0
$\frac{V_{dc}}{4}$	0	1	1	1	1	0	0	0
0	0	0	1	1	1	1	0	0
$\frac{-V_{dc}}{4}$	0	0	0	1	1	1	1	0
$\frac{-V_{dc}}{2}$	0	0	0	0	1	1	1	1

2.2. Cascaded H-Bridge Multilevel Inverter (CHB-MLI)

There are various inverter topologies based on a series single-phase inverter connected to the individual DC source. If an inverter is implemented in an active transfer of source applications, then the sources of each bridge must be separated. Due to this property, it is recommended to use a cascaded H-bridge multilevel inverter (CHB-MLI) in fuel cells or photovoltaic (PV) array to reach a higher level [19–21].

The resulting phase voltage has been synthesized through increasing the generated voltages through different cells. The module of a bridge is a three-level in CHB-MLI, and each module is appended to the cascade. The output voltage for an n-level cascade multilevel inverter is 2n + 1, where n is the source of DC [22]. Switching angles are obtained to reduce the overall distortion of harmonics. Maximum output voltage is decided by using Equation. (6) and minimum output voltage is decided by using Equation (7). The capacitors at the DC side (C) and several switches (S) per phase, calculated through Equations (4) and (5) respectively.

$$C = \frac{m-1}{2} \tag{4}$$

$$S = 2(m-1) \tag{5}$$

$$Maximum\ output\ voltage = \frac{m-1}{2}(Vdc) \tag{6}$$

$$Minimum\ output\ voltage = \frac{m-1}{2}(-Vdc) \tag{7}$$

Figure 6a shows 3-level CHB-MLI. Here is a full-bridge that provides 3-levels of output. It is a simple diagram of 3-level CHB-MLI. This provides 3-levels of output; Vdc, 0 and −Vdc, while battery voltage is Vdc, where E = Vdc.

As we have already discussed with Equations (4)–(7), for voltage level 3 m = 3, therefore Vdc is a maximum voltage, and (−Vdc) is minimum output voltage. One DC bus capacitor and four switches are needed. Generally, to get the output voltage, there are two activating switches. Hence, one pair of switches gives a positive result and another pair create a negative result. Therefore, to obtain the output source Vdc, T1, and T2′ are activated, and the output source −Vdc, T2 and T1′ are turned on. If the current does not pass from the full-bridge then the 0-voltage source is obtained through the switch T1 and T2 or lower two switches T1′and T2′.

Figure 6b represents 5-level CHB-MLI. One full-bridge module is a three-step cascade multilevel inverter. Two bridge modules generate five various voltage levels; 2Vdc, Vdc, 0, (−Vdc), and (−2Vdc). As a Vdc battery voltage with E = Vdc. In 5-level CHB-MLI, we take m = 5. Therefore, following Equations (4), (5), (6) and (7), 2 capacitors at the DC side, 8 switches are needed. So, the maximum output voltage is 2Vdc, and a minimum output voltage is −2Vdc.

As discussed in the three-level cascade multilevel inverter, the cross switches are activated, and upper or lower switches obtain a 0-voltage. Thus, to get the output sources 2Vdc, T11 and T12′ or T21 and T22′ to be activated and the current passing through the capacitor C1 and C2. As with a negative voltage, the other four cross switches are connected. For output voltage Vdc, the battery bypassed either top or bottom. Therefore, the current pass throughout C1, T11, a load, T22′, T21′, and T12′. Besides, for the minimum voltage (−Vdc), the current passes in the contrasting path, and its passing by C2, T22, load, T11′, T12′ and T21′. For 0 power, there are two probabilities, 1) switches T11, T12, T21, T22 are activated; 2) switches T11′, T12′, T21′, T22′ are activated. This topology is convenient especially for photovoltaic generation application, where photovoltaics modules are directly connected to an individual DC source [23].

The principle advantages of CHB-MLI are a series of H-bridges demonstrating an advanced design, this will empower the assembling procedure to be rapidly and efficiently, also voltage adjustment is easy. Although there are few disadvantages of CHB-MLI topology, each H-Bridge needs an isolated DC source, which limits its use. That has more access to the SDCS, and no ordinary DC-bus [24].

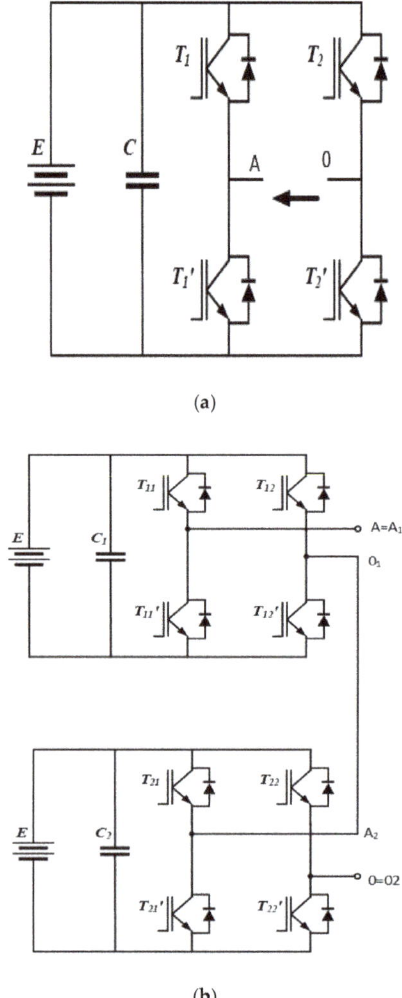

(a)

(b)

Figure 6. (**a**) Three-level cascaded H-bridge multilevel inverter (CHB-MLI); (**b**) 5-level CHB-MLI.

2.2.1. Three-Level CHB-MLI PSCAD Simulations

In Figure 7b, the control circuit has two gate signals S1 and S2. Those are available from the comparator. Another two has the NOT gate with initial two signals. These four signals are ready to supply to IGBT which are in the 3-level CHB-MLI power circuit.

In Figure 7a, all the gate pulses are applied to the different IGBT of CHB-MLI. This follows Table 3 of a sequence of a waveform for gate signals. For Vdc; switched S1 and S4 are conduct. Hence, current flow from the source, S1, load, S4, and source, for 0Vdc; S3 and S4 are conducted and for −Vdc; S2 and S3 are conducted, therefore current flow from the source, S2, load, S3, and source.

Table 3. Voltage levels of 3-level cascade H-bridge multilevel inverter and switching states.

Voltage	S1	S2	S3	S4
Vdc	1	0	0	1
0	0	0	1	1
$-Vdc$	0	1	1	0

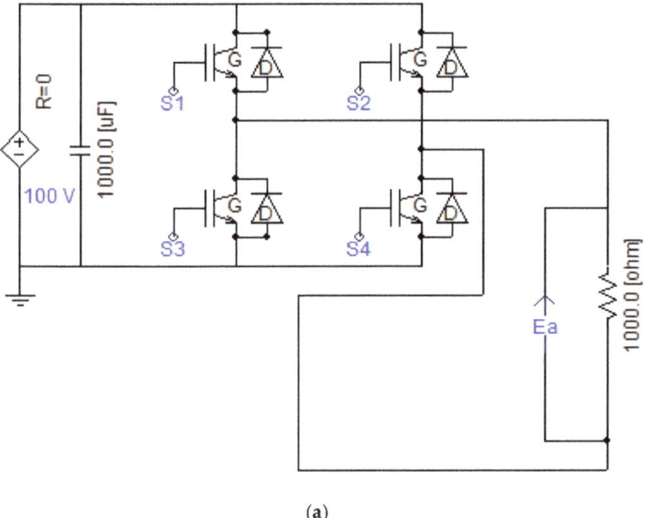

(**a**)

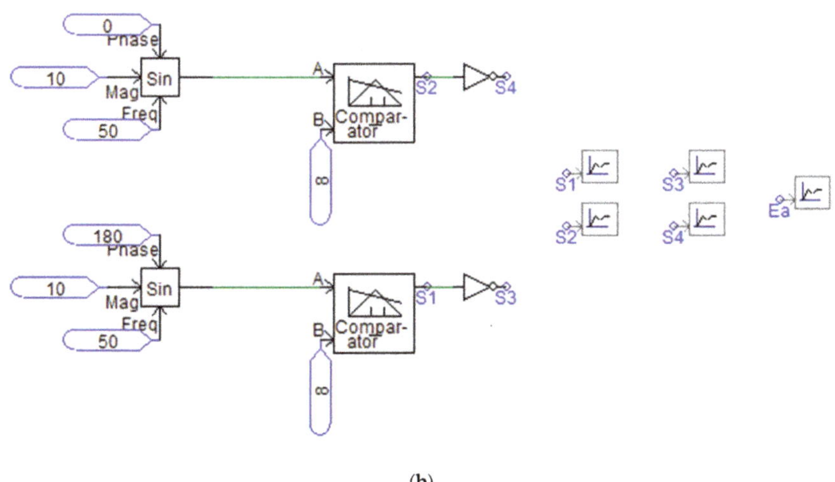

(**b**)

Figure 7. *Cont.*

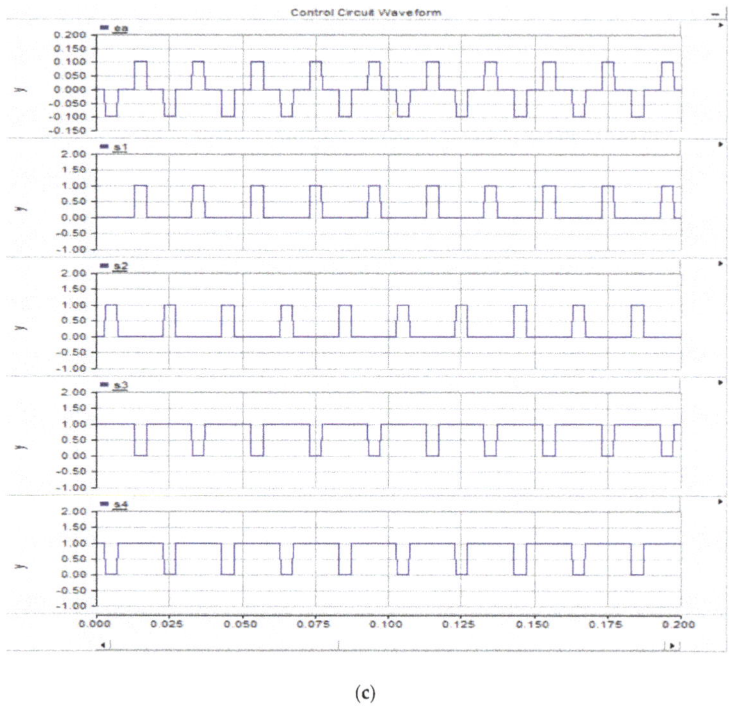

(c)

Figure 7. (**a**) Power circuit of 3- level CHB-MLI; (**b**) control circuit of 3- level CHB- MLI; (**c**) phase output voltage waveform of 3-level CHB-MLI.

2.2.2. Five-Level CHB-MLI PSCAD Simulation

In Figure 8b, the control circuit waveforms are generated by comparison of DC signal (2 V and 4V) with the triangular waveforms (−10 V to +10 V) changing the values of DC voltage will change the gate pulse widths. Four signals S1, S2, S5, and S6 are available from the comparator. The other four signals are obtained using the NOT gate with initial four signals. These 8 gate signals S1, S2, S3, S4, S5, S6, S7, and S8 are ready to supply to IGBT which are in the 5-level CHB-MLI power circuit, which is shown in Figure 8c.

In Figure 8a, all gate signals are applied to the IGBT of CHB-MLI. The following Table 4 represents the sequence of waveforms for gate signals. Considering for +2 Vdc the corresponding switches such as S1, S3, S5, and S7 are turned ON. Therefore, there is a current path takes from Source-1, S1, Load, S7, Source-2, S5, S3 and Source-1. As a result, the two sources are added to each other and so the voltage +2 Vdc is available. Similarly, for +Vdc, the switches S1, S3, S7, and S8 are turned on and the current flows through the Source-1, S1, load, S7, S8, S3 and back to Source-1. For 0 Vdc, switches S3, S4, S7, and S8 are turned on. For −Vdc, switches S2, S4, S7, S8 are turned on, Current flow through Source-1, S2, S7, S8 Load (the reverse direction), S4, back to source-1. −2Vdc: S2, S4, S6, S8 are turned ON, current flows through the Source-1, S2, S8, Source-2, S6, Load (the reverse direction), S4 and then back to Source-1.

Table 4. Voltage levels of 5-level CHB-MLI and switching states.

Voltage	S1	S2	S3	S4	S5	S6	S7	S8
$2Vdc$	1	0	1	0	1	0	1	0
Vdc	1	0	1	0	0	0	1	1
0	0	0	1	1	0	0	1	1
$-Vdc$	0	1	0	1	0	0	1	1
$-2Vdc$	0	1	0	1	0	1	0	1

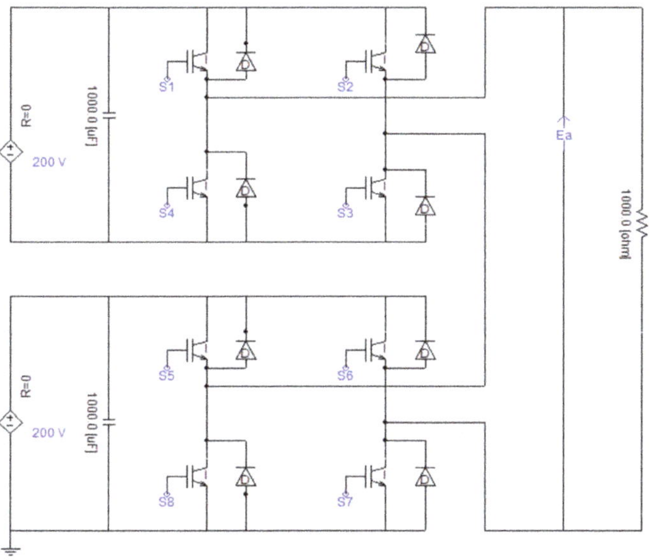

(a)

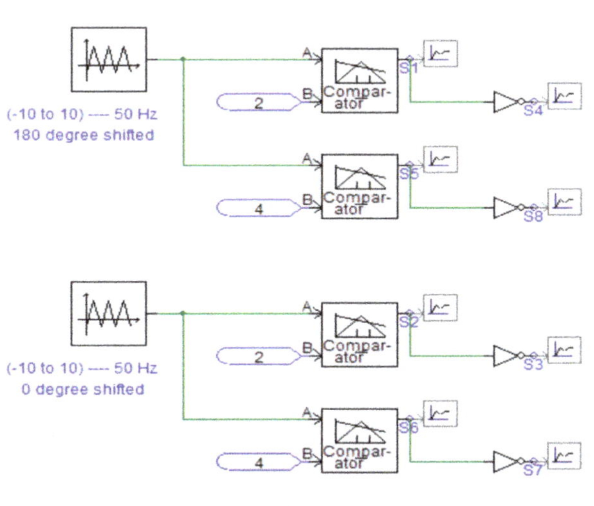

(b)

Figure 8. *Cont.*

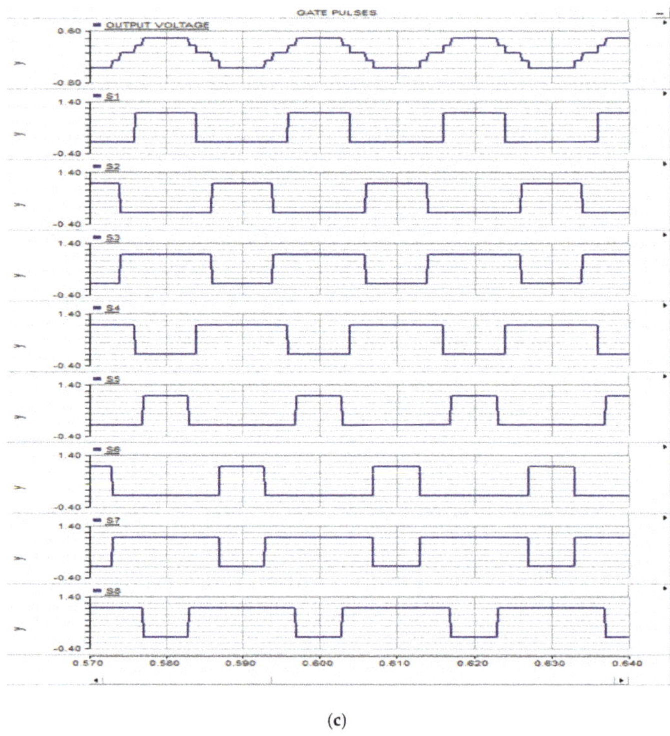

(c)

Figure 8. (**a**) Power circuit of 5-level CHB-MLI; (**b**) control circuit of 5-level CHB-MLI; (**c**) phase output voltage waveform of 5-level CHB-MLI.

2.3. Flying-Capacitor Multilevel Inverter (FC-MLI)

The same topology of the DC-MLI is a capacitor-clamping or flying-capacitor multilevel inverter (FC-MLI) topology in which clamping capacitors are utilized to clamp the voltages [25]. This topology is new compared with other topologies. The voltage between the two capacitors refers to the output voltage at the terminal [26]. The indicated voltage level in the flying-capacitor inverter is the same as like diode-clamped multilevel inverter [27]. Moreover, switching pairs may be different and asymmetrical according to the regulating policy, but the choice of the two pairs leads to the switching state redundancy, which may utilize to attain the voltage balancing of FC-MLI [2,28,29]. The number of output voltage levels is increased through appending some additional switches and a capacitor. Besides this, for FC-MLI (m-1) number of capacitors on a common DC-bus, where m is the level number of the inverter, and 2(m-1) switch-diode valve pairs are used, and [(m-1) (m-2)/2] clamping capacitors per phase are used [28,30,31].

The choice of two pairs leads to a surface moving state, which can be used to obtain the FC-MLI balancing voltage [2,28,29]. In Figure 9a, this indicates 3-level FC-MLI, which has output over a and n points such as Vdc/2, 0, (−Vdc)/2. For voltage Vdc/2, S1 and S2 are activated, while (−Vdc)/2, S1′ and S2′ are activated and for 0 voltage, one pair activated either (S1, S1′) pair or (S2, S2′) pair. If S1 and S1′ are activated, then C1 is charged, and its discharge when S2 and S2′ are deactivated. A C1 is balanced through choosing the right combination of level 0 switches.

With 5 levels of flying capacitor multilevel inverter, the output voltage is more flexible than a diode clamp inverter. Figure 9b shows a 5-level voltage through the neutral point n, Van is reached through some switching combustions. For voltage Vdc/2, Switch S1–S4 is activated, voltage level Vdc/4, there are 3 amalgamations, 1) S1, S2, S3, and S1′ are activated (Vdc/2 of top C4's and −Vdc/4 of

C1), 2) S2, S3, S4, and S4′ are activated (3Vdc/4 of C3′s and (−Vdc)/2 of lower C4), 3) S1, S3, S4, and S3′ are activated (Vdc/2 of top C4′, (−3 Vdc)/4 of C3′s or (−Vdc/)/2 of C2), and voltage 0, S3,S4,S1′ and S2′ are activated. For voltage (−Vdc)/4 S1, S1′, S2′ and S3′ are activated (Vdc/2 of top C4′s and (−3 Vdc)/4 of C3), and for (−Vdc)/2, activate all lower switches S1′ − S4′.)

A most essential superiority of FC-MLI is phase redundancy using for balancing the voltage levels or capacitors, easy to regulate an active and reactive power. In addition, using capacitors means, an inverter can move throughout small period outages. However, there are few drawbacks too. It is difficult to remark on the voltage level for capacitors, pre-charged every capacitor is at a similar voltage, starting is difficult, and switching usage and productivity are poor for actual power transmission [24].

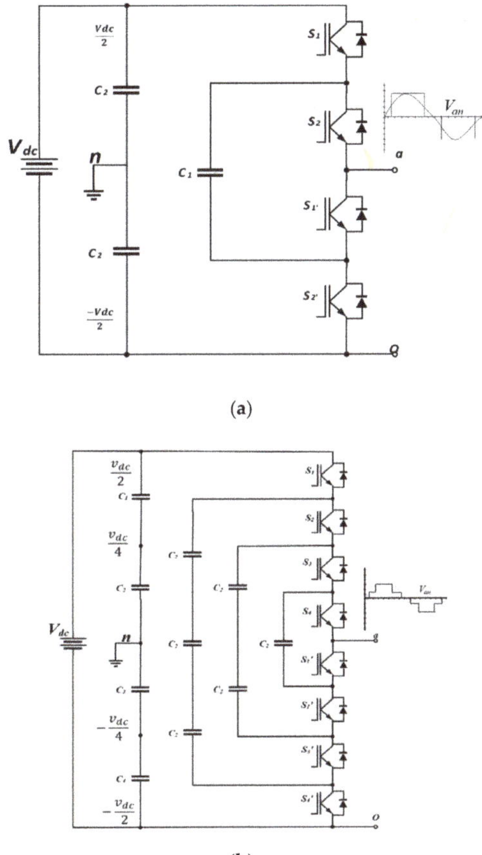

Figure 9. (**a**) Three-level flying-capacitor multilevel inverter (FC-MLI); (**b**) 5-level FC-MLI.

In the above description, positive marks of capacitors are in discharge mode and the negative capacitor marks are in charge mode. With the right choice of capacitor combinations, we can balance a capacitor charge. As with diode clamping, many bulk capacitors are needed to clamp the voltage. Specified nominal voltage of the capacitor used as the main power switch.

2.3.1. Three-Level FC-MLI PSCAD Simulations

In Figure 10b, the control circuit with two gate signals of g1 and g2 are available. Another two are obtained using the NOT gate with initial two signals. These four signals are ready to give to IGBT which are in the 3-level FC-MLI power circuit.

In Figure 10a, all the gate pulses are applied to the different IGBT of FC-MLI following Table 5 of a sequence of a waveform for gate signals. For Vdc; switched g1 and g4 are conduct. For 0Vdc; g1 and g3 are conducted and for −Vdc; g3 and g4 are conducted.

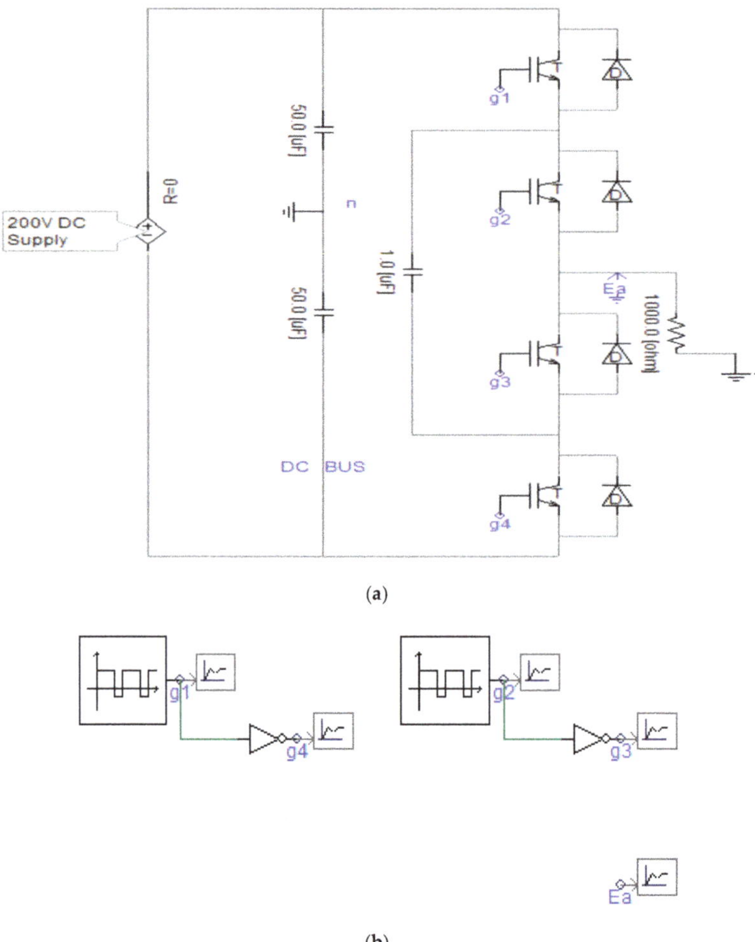

(a)

(b)

Figure 10. *Cont.*

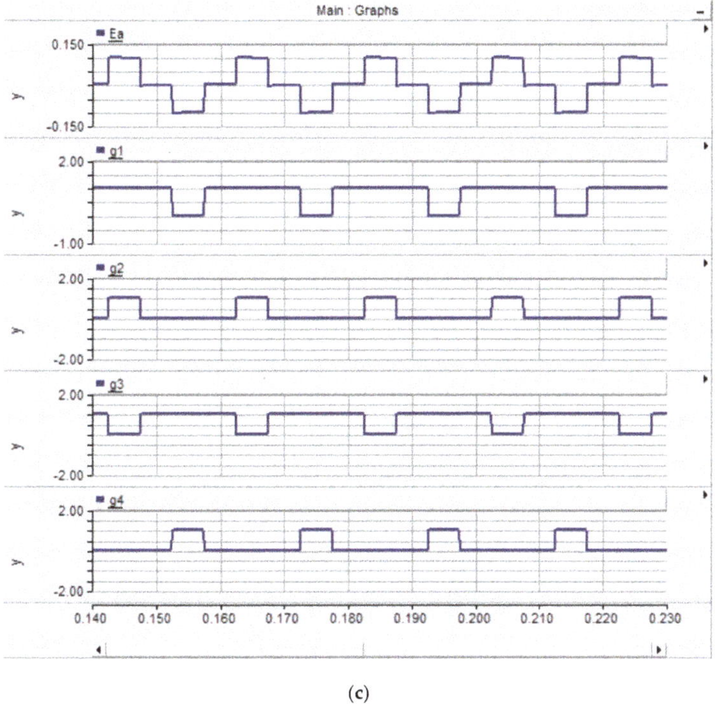

(c)

Figure 10. (**a**) Power circuit of 3- level FC-MLI; (**b**) control circuit of 3- level FC-MLI; (**c**) phase output voltage waveform of 3-level FC-MLI.

Table 5. Voltage levels of 3-level flying-capacitor multilevel inverter and switching states.

Voltage	g1	g2	g3	g4
Vdc	1	1	0	0
0	1	0	1	0
$-Vdc$	0	0	1	1

2.3.2. Five-Level FC-MLI PSCAD Simulations

Figure 11b, the control circuit waveforms are generated by comparison of the DC signal (2 V and 4 V) with the sinusoidal waveforms (5 V RMS). Changing the values of the DC voltage will change the gate pulse widths. Four signals g1, g2, g3, and g4 are available from the direct comparator. The other four signals are obtained using the NOT with initial four signals. These 8 gate signals are ready to give to IGBT which are in the FC-MLI power circuit.

In Figure 11a, all the available gate signals are given to IGBT of FC-MLI Inverter, following Table 6 of the sequence of a waveform for gate signals. Here the total voltage is divided into a capacitor bridge. Each capacitor (C1, C2, C3, and C4) of the first line gets charged with Vdc/4 voltage. Also, each intermediate capacitor (C1′, C2′, and C3′) gets charged by +Vdc/4.

+Vdc/2: Switched g1, g2, g3, and g4 are turned ON. Hence, current flow from capacitor C2, C1, g1, g2, g3, g4 to the load. Hence, by two capacitor-discharge, (+Vdc/4 +Vdc/4) = +Vdc/2 is available at load side. +Vdc/4: Similarly, g1, g2, g3, and g5 are turned on. Current flows, from capacitor C2, C1, g1, g2, g3, C3′ and g5 to the load. Hence the voltage across the load is Vdc/4. In other words, C3′ nullifies the voltage from one of the capacitors (C1 or C2) to load voltage. Hence only half voltage (of +Vdc/2) is available at the load. 0Vdc: g1, g2, g5, g6 are turned on. Similar to the previous operation, both C2′

capacitors provide the opposite voltage to C1 and C2. Hence equivalent voltage at the load is 0, −Vdc/4: g1, g5, g6, g7 are turned on. Current flows through the C2, C1, g1, C2′, C2′, C2′, g5, g6, g7, and load. Now here two capacitors charged with +Vdc/4, +Vdc/4, and three capacitors are in opposite direction and charged by +Vdc/4, +Vdc/4, +Vdc/4 hence the effective voltage at the load is (+2Vdc/4) − (+3Vdc/4) = −Vdc/4. −Vdc/2: g5, g6, g7, g8 are turned ON. Current flows from C4, C3 to load (reverse direction) Back to C4. Hence, the voltage of two capacitors is available at load side (−Vdc/4 + (−Vdc/4) = −Vdc/2.

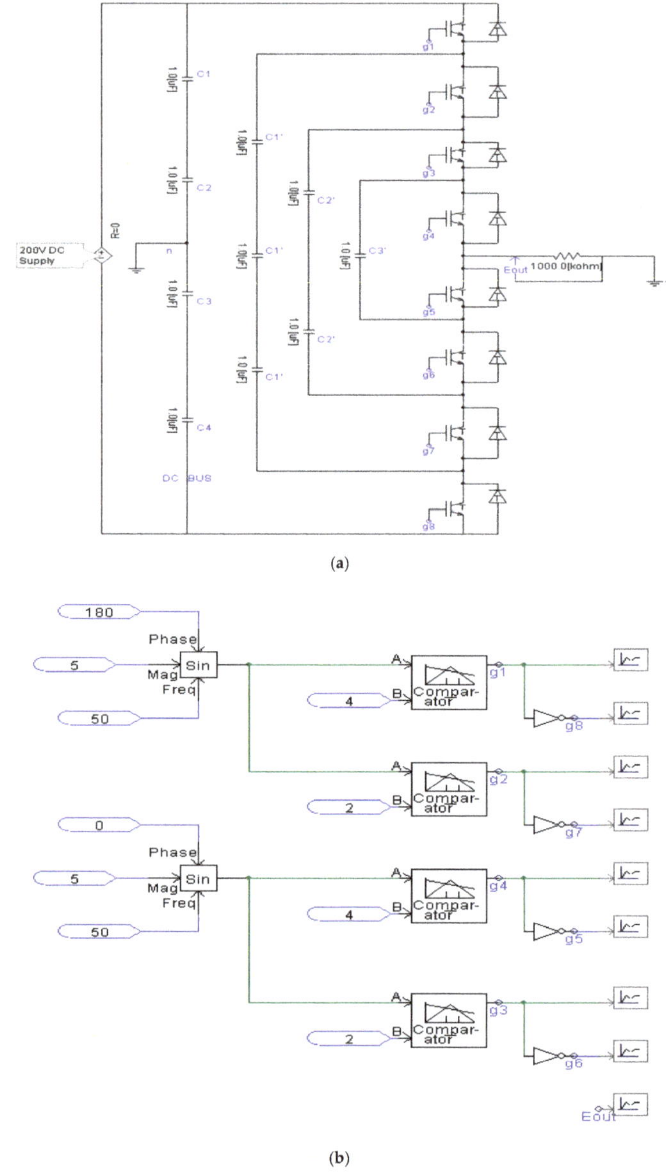

(a)

(b)

Figure 11. *Cont.*

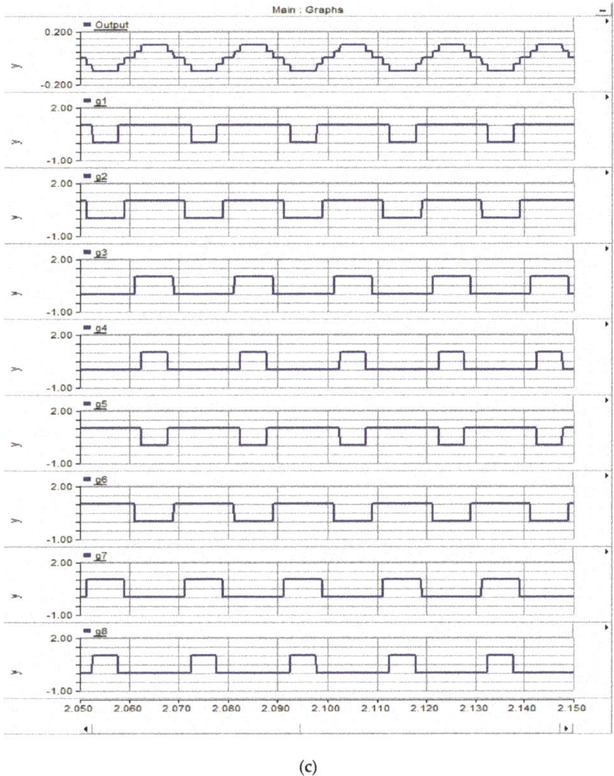

(c)

Figure 11. (**a**) Power circuit of 5-level FC-MLI; (**b**) control circuit of 5-level FC-MLI; (**c**) phase output voltage waveform of 5-level FC-MLI.

Table 6. Voltage levels of 5-level FC-MLI and switching states.

Voltage	g1	g2	g3	g4	g5	g6	g7	g8
$\frac{Vdc}{2}$	1	1	1	1	0	0	0	0
$\frac{Vdc}{4}$	1	1	1	0	1	0	0	0
0	1	1	0	0	1	1	0	0
$\frac{-Vdc}{4}$	1	0	0	0	1	1	1	0
$\frac{-Vdc}{2}$	0	0	0	0	1	1	1	1

3. Performance Evaluations of Different Multilevel Inverters and Characteristics Overview

The various topologies discussed, described, and analyzed till now are comprehensively compared in this section on the basis of different performance parameters such as number capacitors and diodes, switches, flexibility, balancing capacitor, and cost in Table 7.

For high-voltage applications, for instance high-voltage DC (HVDC), a relatively high voltage level can be chosen for a modular multilevel inverter [5,32–53] to the total required current THD, and dv/dt. For medium- and low-voltage applications, five and three-level inverters for A-NPC [54–58] and other FC-MLI inverters may be better suited to current specifications. Currently, multilevel inverters have been suggested in motor drives, flexible AC transmission systems (FACTS) and HVDC. As per the studies conducted in [2,59–67], the positive points, negative points, methods of modulation, and application of these inverters are epitomized in Table 8.

Table 7. Comparative assessment of various multilevel inverters.

IT	DC-MLI	FC-MLI	CHB-MLI	A-NPC	MMC
MS	$2(m-1)$	$2(m-1)$	$2(m-1)$	$3(m-1)$	$2(m-1)$
MD	$2(m-1)$	$2(m-1)$	$2(m-1)$	$3(m-1)$	$2(m-1)$
CD	$(m-1)(m-2)$	-	-	-	-
DBC	$(m-1)$	$(m-1)$	$\frac{(m-1)}{2}$	-	-
BC	-	$\frac{(m-1)(m-2)}{2}$	-	-	-
RD	Not Redundant	Redundant	Redundant	Redundant	Redundant
F	Not Flexible	Not Flexible	Flexible	Flexible	Not Flexible
MDV	$2\sqrt{2}$ Va	$2\sqrt{2}$ Va	$2\sqrt{2}$ Va	$2\sqrt{2}$ Va	$2\sqrt{2}$ Va
CVR	$\frac{Vdc}{m-1}$	$\frac{Vdc}{m-1}$	$\frac{Vdc}{m-1}$	$\frac{Vdc}{m-1}$	$\frac{Vdc}{m-1}$
ACC	Ia	Ia	Ia	Ia	Ia
C	M	M	M	H	H

Note: IT: inverter types, DC-MLI: diode-clamped multilevel inverter, FC-MLI: flying-capacitor multilevel inverter, CHB-MLI: cascade H-bridge multilevel inverter, A-NPC: active NPC, MMC: modular multilevel converter, MS: main switches, MD: main diodes, CD: clamping diodes, DBC: DC-bus capacitor, BC: balancing capacitor, RD: redundancy, F: flexibility, MDV: minimum DC voltage, CVR: component voltage rating, ACC: active component current, C: cost, H: high, M: medium.

Table 8. Performance summary of multilevel inverter.

R	T	A	D	MT	AP
[2,59–64]	DC-MLI	Low cost and fewer components due to a smaller number of capacitors therefore simple structure; equalized blocking voltage of power.	Unequal distribution of power losses; DC-link voltage balance limits the converter to three-level topology.	Pulse width modulation (PWM) carrier modulation (based zero-sequence injection); Space vector pulse width modulation (SVPWM) method (based space vector selection).	Renewable energy; variable speed motor drive; static var compensation; HVDC/AC transmission lines.
[64–68]	FC-MLI	Modular structure; possessing a large number of redundant states; each branch can be analyzed independently.	The poor dynamic response of dc voltage balancing; large amounts of flying capacitors reduce the system reliability; pre-charging capacitors is difficult.	Phase-shifted carrier PWM (achieves neutral balancing of flying capacitors).	Renewable energy; motor drive; induction motor control using direct torque control circuit; sinusoidal current rectifiers; static var generation.
[69–71]	CHB-MLI	Modular structure is easier to analyses; possessing fault tolerant capability; same switching frequencies for all the switches.	Separate DC sources are required; isolated transformers increase the system volume.	Phase-shifted carrier PWM (achieves equalization of power losses).	Renewable energy; DC power source utilization; Power factor compensators; flexible alternating current transmission system (FACTS); electric vehicle drives.
[32–51,72–76]	MMC	Modular structure; easy to achieve fault tolerant operation.	Low-frequency voltage oscillation of floating capacitors; complex data acquisition and communication for each power cell.	Nearest-level modulation (achieves equalization of power losses and dc voltage balance).	High-voltage DC (HVDC) transmission, A static synchronous compensator (STATCOM)
[54–58]	A-NPC	Simple structure; easily extendable to a higher level by stacking flying.	Unequal usage of switches and flying capacitors; requires series switches to handle high voltage.	Hybrid phase-shifted PWM (low frequency for high voltage switches and high frequency for flying capacitor cells).	Renewable energy

Note: R: Reference, T: Topology, A: Advantage, D: Disadvantage, MT: Modulation Technique, AP: Application.

3.1. Choosing Topology

Discussed all multilevel inverters are utilized to generate a smooth waveform, mostly for medium voltage drives. Out of three FC-MLI [19,77–82] is relatively new compared to DC-MLI and CHB-MLI. Moreover, FC-MLI has some unique benefits over DC-MLI, including the truancy of clamp diodes and the ability to adjust the voltage of the flying capacitor by selecting an excess state. Even the voltage level is higher than 3 [81]. Unlike CHB-MLI, no separate voltage sources are required. Also, there are two types having some deficiencies, such as most DC in CHB-MLI. Losses are higher in the DC-MLI. but FC-MLI has introduced more potential outcomes to regulate the DC-link capacitor's voltage contrasting to another multilevel topology that uses redundant switching arrangements. An FC-MLI topology is designed for higher voltage and lower voltage-distortion performance. Moreover, in an FC-MLI, capacitors are using for clamping the power over the devices. Also, FC-MLI is appropriate to heavy load systems. Meanwhile, induction motors are utilizing as a load in many industrial applications [78].

3.2. MULTISIM Simulations of FC-MLI

3.2.1. Three-Level FC-MLI MULTISIM Simulations

In the above Figure 12a, the control circuit has a hierarchical block which will be explained in the next section. Here, the same power circuit is used as the above explained. There are three capacitors. Among three capacitors, two capacitors make the DC bus and remaining capacitor acts as a flying capacitor which is used for the clamping voltage at different levels. Here we have used 20 V DC as source power supply. There are four MOSFET switches which are triggered by the gate pulses that generate from the control circuit. Four optocouplers MCT 2E are used for the isolation purpose.

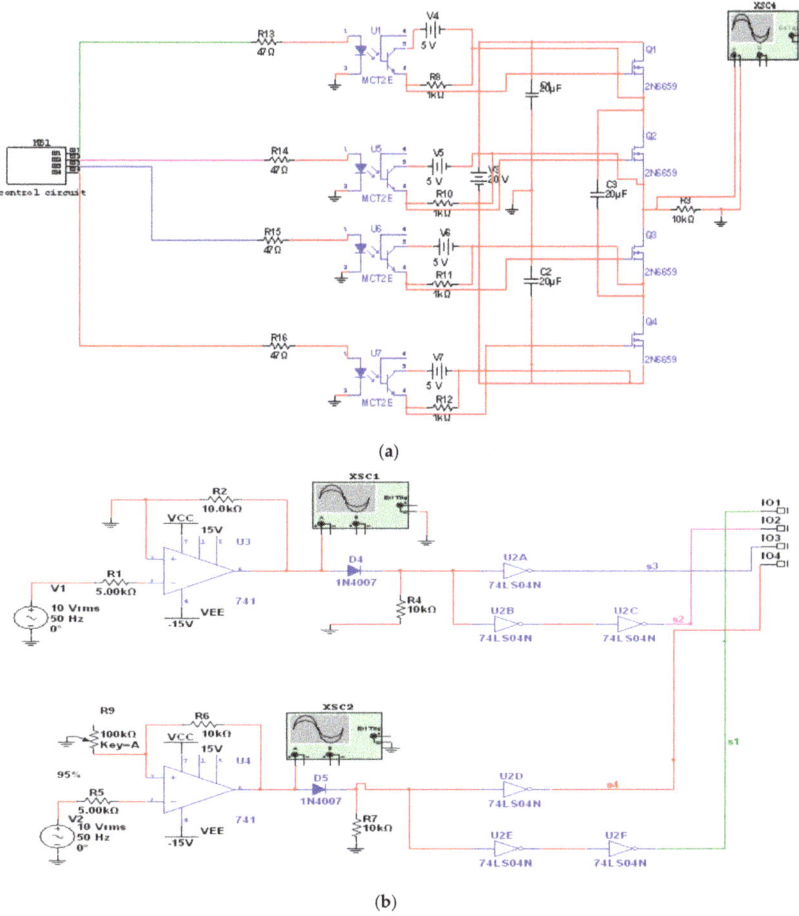

(a)

(b)

Figure 12. (a) Power circuit of 3-Level FC-MLI MULTISIM simulations; (b) control circuit of 3-level FC-MLI MULTISIM simulations.

We have generated four gate pulses using the OP-AMPs. In Figure 12b, we have used one OP-AMP as a ZERO CROSSING DETECTOR and another as a SCHMITT TRIGGER. The output of the SCHMITT TRIGGER is a square wave but shifted from the output of the ZERO CROSSING DETECTOR (ZCD). The shifting can vary by varying the POT of 50Kohm. If the shifting is higher than the output, that has the higher width of zero pulses.

As in Figure 13. we can see the all 3 levels of output. Here we used 20 V DC as a power source, therefore we have 2 different levels of (Vdc/2) =10 V, and (−Vdc/2) = −10 V. By changing the DC voltage in the OP-AMP comparator, we can change the width of output zero voltage. This width of zero voltage decides the amount of reactive power flow in the system.

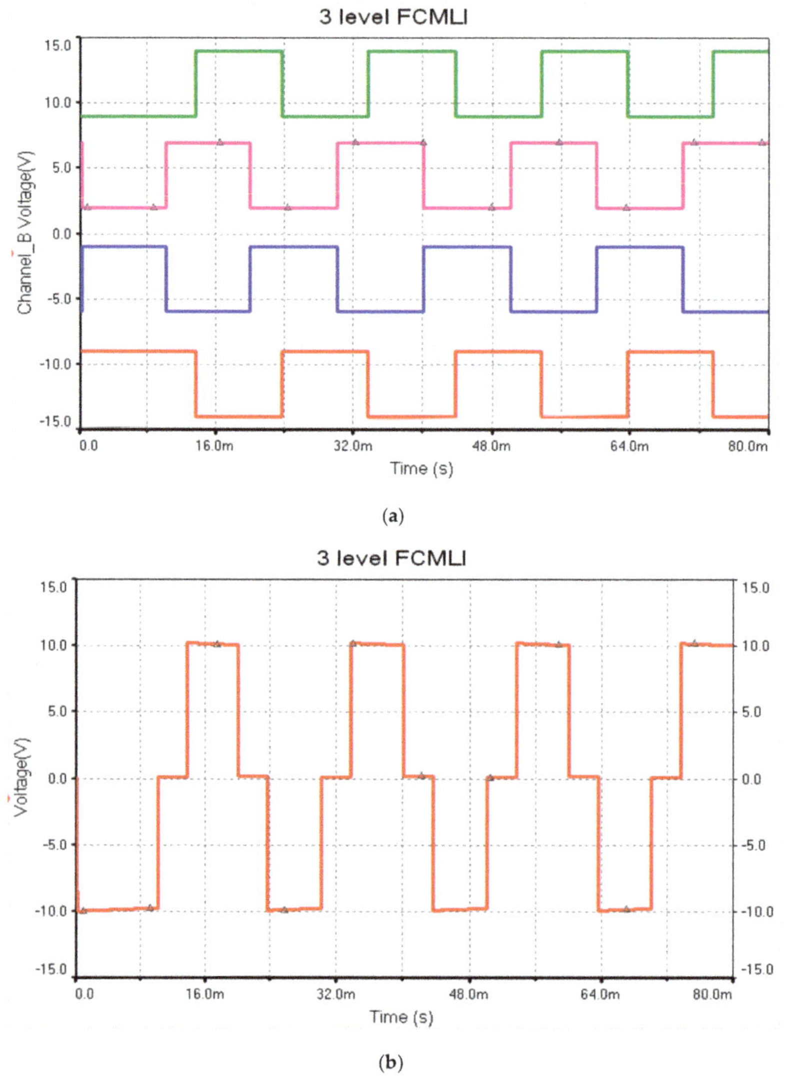

Figure 13. (**a**) Gate pules of 3-level FC-MLI; (**b**) phase output voltage waveforms of a 3-level FC-MLI using MULTISIM.

3.2.2. 5-Level FC-MLI MULTISIM Simulations

The full set-up is shown in Figure 14a. That is divided into the following two parts. That diagram is the same as explained above. Here, we have used MOSFET 2N6659 as a switching device. And the similar three-level optocoupler MCT 2E is used for isolation. We have demonstrated the purely resistive load for the simplicity. We can change the load as per the relevant applications.

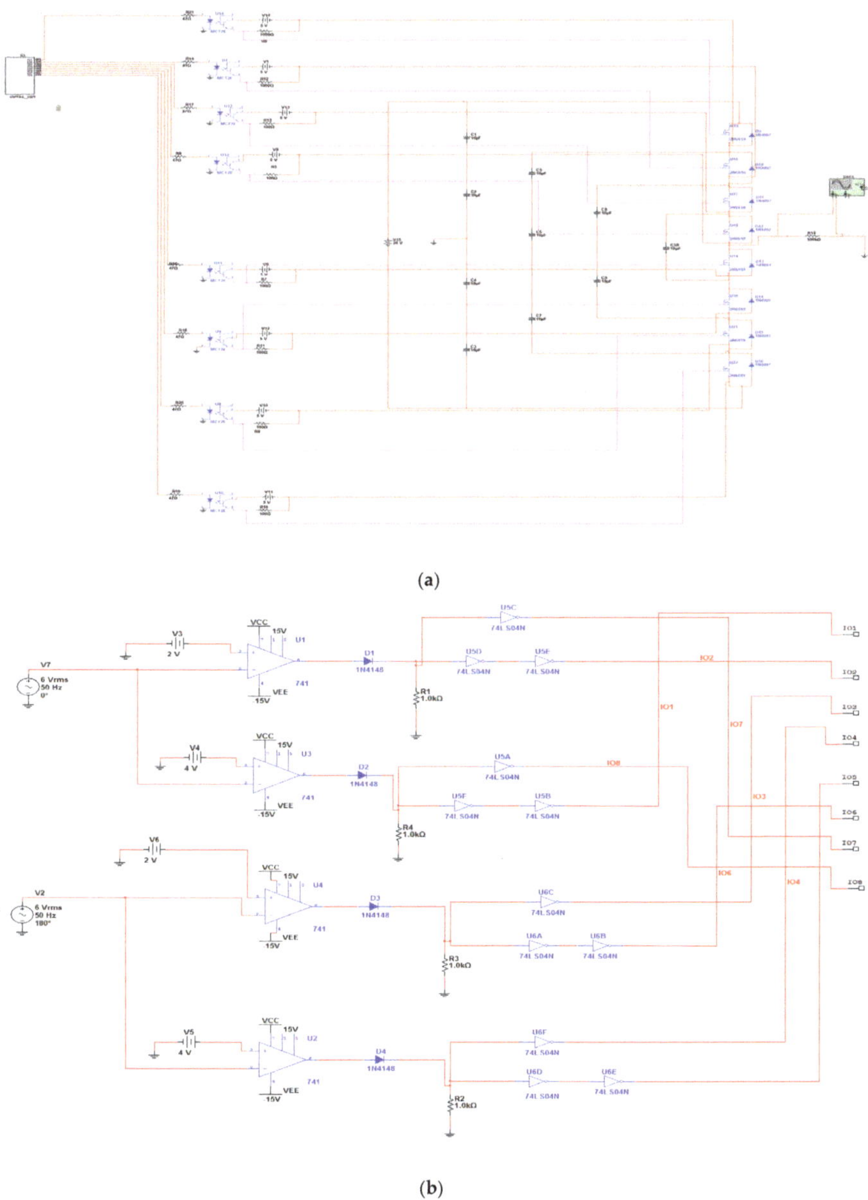

(a)

(b)

Figure 14. (**a**). Power circuit of 5-Level FC-MLI MULTISIM simulations; (**b**) control circuit of 5-Level FC-MLI MULTISIM simulations.

Now as the above explained theory, 4 switches get turning on at any time and remaining four switches get turning off at any time what you want. Therefore, each switch gets 1/4th voltage of power supply as a reverse burden.

Figure 14b for the control circuit, we have used the two sinusoidal power supplies as the reference by comparing it with DC power supply. We have four different outputs having four gates from the required eight pulses. We can obtain the NOT gate combination from the remaining four pulses. Thus, we can get eight pulses and these pulses are applied to the power circuit through an optocoupler.

In Figure 15a–c we can see all the 5-levels output and gate pulses. Here we have used 20 V DC. As a source, we have 4 different levels of (Vdc/2) = 10 V, (Vdc/4) = 5 V, 0 V, (−Vdc/4) = −5 V, (Vdc/2) = −10 V. By changing the DC. voltage in the OP-AMP comparator; we can change the width of zero for output voltage. This width of zero voltage decides the amount of reactive power flow in the system.

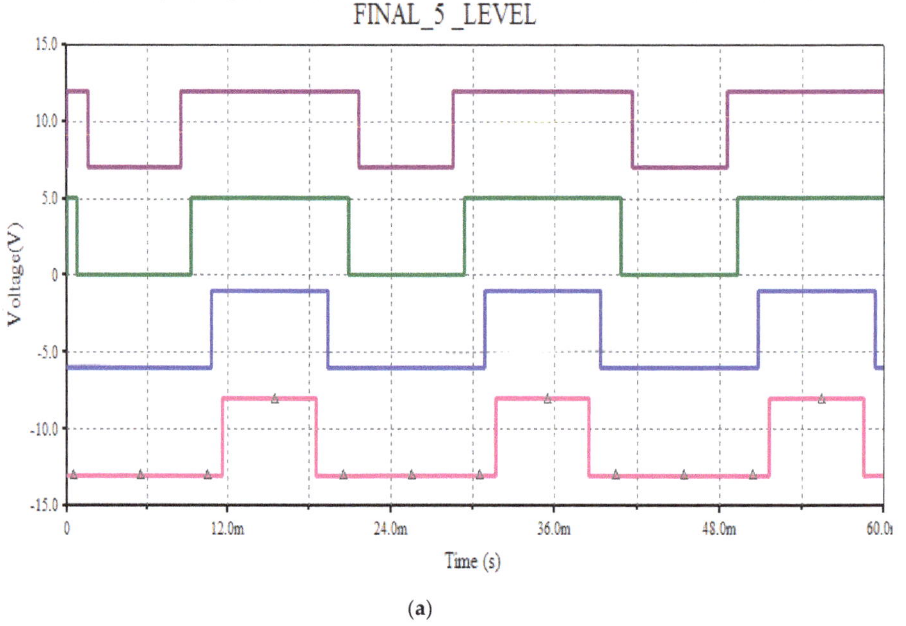

(a)

Figure 15. *Cont.*

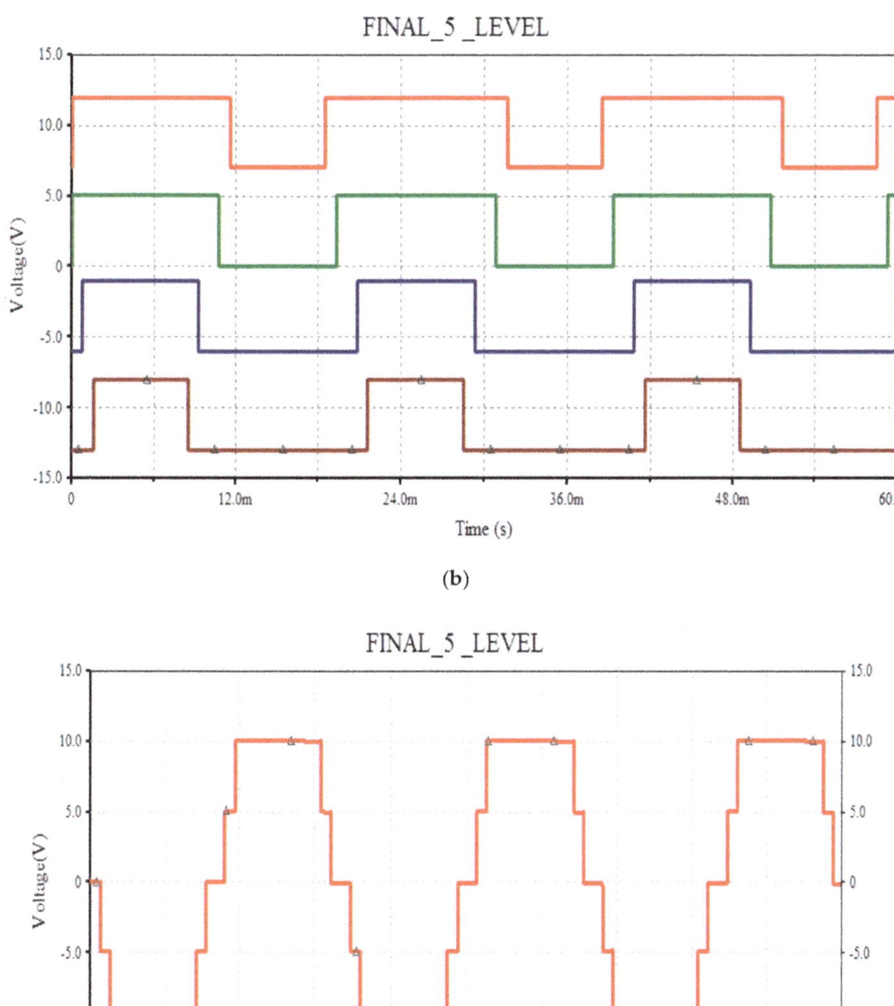

(b)

(c)

Figure 15. (**a**) S1–S4 gate pules of 5-level FC-MLI; (**b**) S5-S8 gate pules of 5-level FC-MLI; (**c**) Phase output voltage waveforms of a 5-level FC-MLI using MULTISIM.

4. Harmonics Analysis of FC-MLI

4.1. Total Harmonic Distortion (THD)

The THD is a presenting measurement of harmonic distortion and defined the proportion between harmonic components and the fundamental frequency of the power. THD is utilized to characterize the linearity of audio systems and the power quality of electric systems.

If the input signal is sinusoidal, the computation often corresponds to the proportion of the powers of all the harmonic frequencies beyond the power of the 1st harmonic and the main frequency.

$$THD = \frac{P_2 + P_3 + P_4 + \cdots + P_\infty}{P_1} = \frac{\sum_{i=2}^{\infty} P^i}{P_1} \tag{8}$$

Which can equivalently be written as:

$$THD = \frac{P_{total} - P_1}{P_1} \tag{9}$$

It is also defined as the amplitude ratio, not the power ratio which results in the determination of THD given the square root to the above equation.

$$THD = \frac{\sqrt{V_2^2 + V_3^2 + V_4^2 + \cdots + V_\infty^2}}{V_1} \tag{10}$$

4.1.1. Harmonics Analysis of 3-Level FC-MLI

As shown in Figure 16, the analysis can be seen that for lower order mainly 5th and 7th harmonics are predominant. Here the third harmonic has been reduced very much.

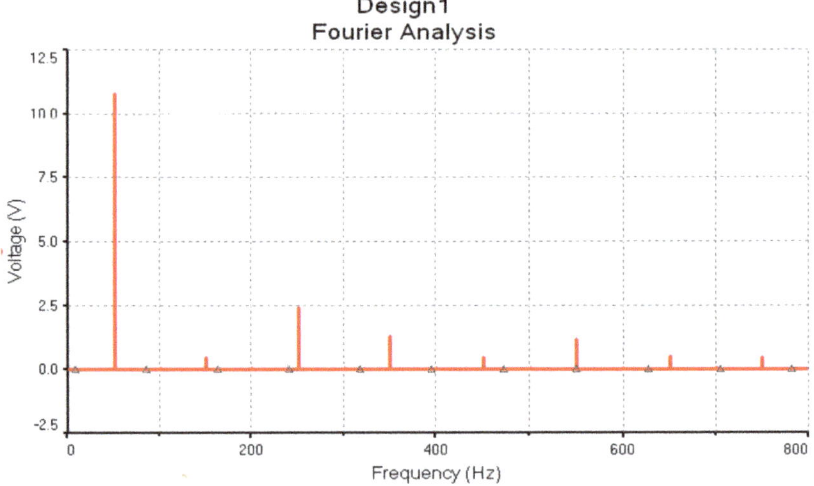

Figure 16. Fourier analysis of 3-level FC- MLI.

If we change the pulse width of our gate pulses then the harmonic reduction can be changed; more specifically, THD can be changed. Because of changing our pulse width, the starting angle of output wave changes and thus the harmonics also changes. However, if we compare this with conventional two-level inverter then we cannot find any harmonics in this topology compared with the two-level inverter.

In Table 9 shows the analysis of THD. The total harmonic distortion shows that how much percentage harmonics are present in the waveform or how much the waveform is disturbed from an actual sine wave.

Table 9. Total harmonic distortion (THD).

Fourier Analysis for V (25)					
DC Components	0.142395				
No. Harmonics	15				
THD	28.8803%				
Grid Size	512				
Interpolation Degree	1				
Harmonics	Frequency	Magnitude	Phase	Norm. Mag	Norm. Phase
1	50	10.7723	147.086	1	0
2	100	0.0359127	−87.093	0.00333381	−234.18
3	150	0.482066	−91.605	0.0447506	−238.69
4	200	0.0370429	−90.48	0.00343872	−237.57
5	250	2.40309	−164.03	0.223081	−311.11
6	300	0.0368095	−93.112	0.00341705	−240.2
7	350	1.2825	129.104	0.119056	−17.982
8	400	0.0367623	−94.406	0.00341267	−241.49
9	450	0.470903	−113.53	0.0437143	−260.61
10	500	0.0369758	−96.153	0.0034325	−243.24
11	550	1.14951	178.466	0.106709	31.3803
12	600	0.0368479	−97.806	0.00342062	−244.89
13	650	0.517375	111.212	0.0480284	−35.874
14	700	0.0369105	−99.11	0.00342644	−246.21
15	750	0.454395	−131.73	0.042818	−278.82

The following analysis is carried out by the MULTISIM software; here, we have found 15 numbers of harmonics. While analyzing; the sampling frequency is taken as 16 kHz. This analysis is also carried out with Lagrange's 1st degree of interpolation.

The above tabulated analysis gives us that mainly 5th and 7th harmonics are the predominating and has a major part in THD normal magnitude. A phase is measured by taking the fundamental 50 Hz as a reference.

4.1.2. Harmonics Analysis of 5-Level FC-MLI

In Figure 17, the analysis shows that the major 3rd harmonics is predominant but all other has very low impact. Hence, here THD is also less as given in Table 10.

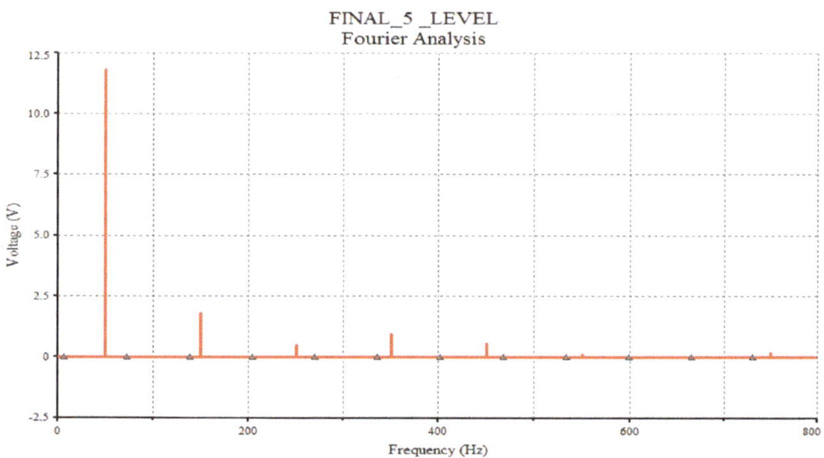

Figure 17. Fourier analysis of 5-level FC- MLI.

Table 10. Total harmonic distortion.

Fourier Analysis for V (47)					
DC components	−0.037107				
No. Harmonics	15				
THD	18.5624%				
Grid Size	512				
Interpolation Degree	1				
Harmonics	Frequency	Magnitude	Phase	Norm. Mag	Norm. Phase
1	50	11.8044	−0.86678	1	0
2	100	0.00272927	−87.038	0.0002231207	−86.172
3	150	1.82486	−3.0824	0.154591	−2.2156
4	200	0.000617738	−38.435	0.000052331	−37.568
5	250	0.489755	174.772	0.0414891	175.638
6	300	0.00209954	69.2359	0.00017786	70.1027
7	350	0.930224	172.689	0.0788029	173.556
8	400	0.00315766	75.7679	0.000267498	76.6347
9	450	0.561122	170.638	0.0475349	171.505
10	500	0.00256501	79.3973	0.000217292	80.264
11	550	0.114071	168.906	0.00966337	169.773
12	600	0.000955366	95.2264	8.09328×10^{-05}	96.0932
13	650	0.00525038	−19.574	0.000444781	−18.707
14	700	0.000845858	−137.32	7.16559×10^{-05}	−136.45
15	750	0.195151	164.189	0.016532	165.055

Table 10 shows the analysis of THD. Here, only 3rd harmonic is predominating, and all others are much lower. Now if we change the width of gate pulses the THD can be further reduced. One thing to note is that in 3-level the THD was 28.88% while in the 5-level topology the THD is 18.56%. Hence by increasing the levels of the output, we can reduce the total harmonics distortion. Therefore, if the number of levels are increased, then THD is decreased.

5. Pulse width modulation (PWM) control techniques

Depending on the control strategy, the most important two types of the PWM modulation techniques are categorized as open and closed loop. SPMW, PWM space vector, sigma-delta modulation (SDM) are the part of open-loop PWM methods, and hysteresis, optimized current control; linear methods are identified by closed-loop current control methods.

5.1. Open-Loop PWM Control Techniques

5.1.1. Sigma Delta PWM (SDM)

The SDM method was first classified by Jager in 1952 as a 1-bit PCM (pulse code modification) coding method. A 1-bit coding application is created utilizing the integrated feedback block on a pulse modulator, which is part of the encoding process, as shown in Figure 18a [83]. Sigma delta modulation originally suggested as a way to encode 1-bit audio and video signals in digital control and modulation problems. SDM has been classified to avoid a decrease in the power density of the altered signals with an increase in the sampling frequency. An SDM method was acquired by adding a test and restraint unit to the main SDM modulator, as shown in Figure 18b.

SDM modified the energy transformation operation utilizing analog to digital converters, which are widely used in some applications because of the expansion of semiconductor technologies. Using an inverter with resonant DC-link permits using soft switching applications rather than hard PWM switching. This type of inverter is mainly known as zero switching losses acquired through switching to the present time, and this technique can be adjusted according to the SDM. In Figure 18b, the modulator output signal varies between +Vo, 0 and Vo depending on the sampling period of fs, and the output signal is compared to the amplitudes of input. The result of the comparison (e) is integrated (E) and the quantizer decide the output signal across from E. The multilevel SDM creates a multi-bit data stream and the output decode creates various output states that might be utilized to regulate the activate/deactivate states of multilevel inverter switches. Figure 19 shows a schematic diagram of a multilevel inverter regulated by SDM. The interaction between the SDM modulator and the inverter is controlled by a multilevel decoding logic block, which receives the SDM signals and decodes the inverter switching signals [84–86].

Studies show that it designed to regulate DC link inverter, SDM might be expanded to regulate multilevel inverters via logical interface. For SDM-regulated multilevel inverters, output faults, for instance, unequal voltage distribution and system imbalances, might be decreased to a high switching frequency up to 200 kHz.

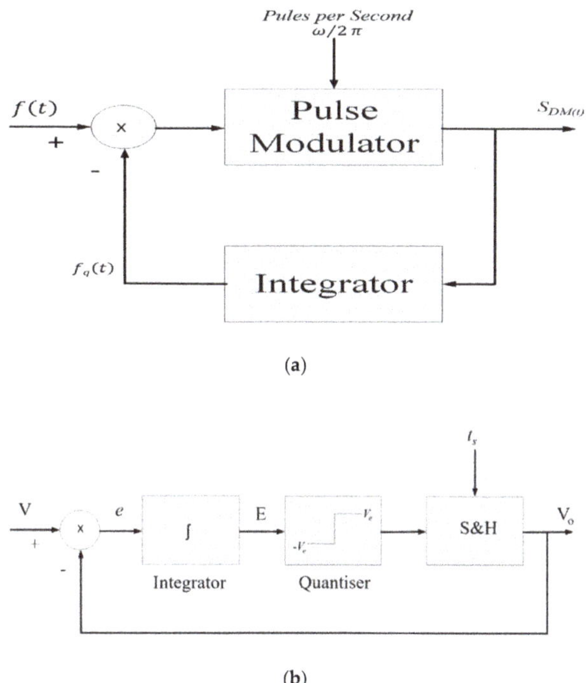

Figure 18. Schematic diagram of delta modulation (**a**) simple delta modulator; diagram of delta modulation (**b**) sigma-delta modulation (SDM).

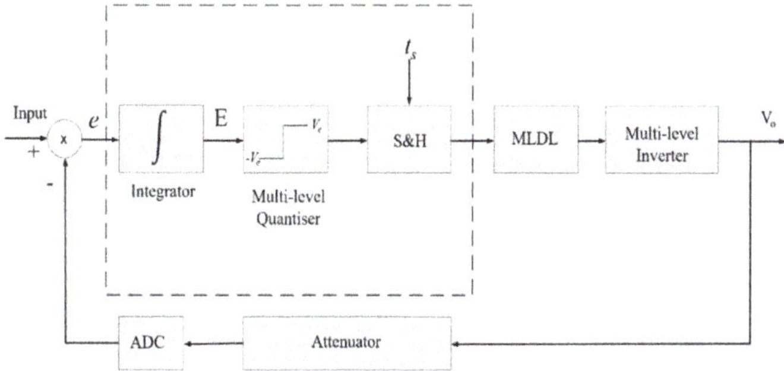

Figure 19. SDM control of a multilevel inverter.

5.1.2. Sinusoidal PWM (SPWM)

The most widely used method among many PWM techniques is the SPWM out of other power switching inverters. In the SPWM, the sine wave of the source voltage when compared to a carrier wave of triangular type to produce gate signals for the inverter switch. Energy debauchery can be considered a major problem for high-power applications. A basic SPWM frequency control method has been suggested to reduce the switching losses. SPWM multi-carrier regulate method has also been executed to enhance the performance of operated multilevel inverters and is categorized as per perpendicular or horizontal carrier signal adjustment. Perpendicular carrier distribution methods are called phase dissipation (PD), phase opposition (POD) and alternative phase dissipation (APOD); however, the horizontal arrangement is called the phase shift control method (PS). Even though PS-PWM is only useful for CHB-MLI and FC-MLI, PD-PWM is more useful for DC-MLI [87].

From the above discussion, multicarrier SPWM regulates methods is shown in Figure 20a PD, (b) POD, (c) APOD, (d) PS. SPWM is the mainly utilized PWM control technique because of its many advantages, such as the simple implementation of lower harmonic output signals per other methods and lower transmission dissipations. In the SPWM, a high-frequency triangular carrier signal is analyzed to a low-frequency sine wave signal in an analog or digital device. The frequency of the sine wave modulation signal determines the required frequency of the line voltage at the output of the inverter [88].

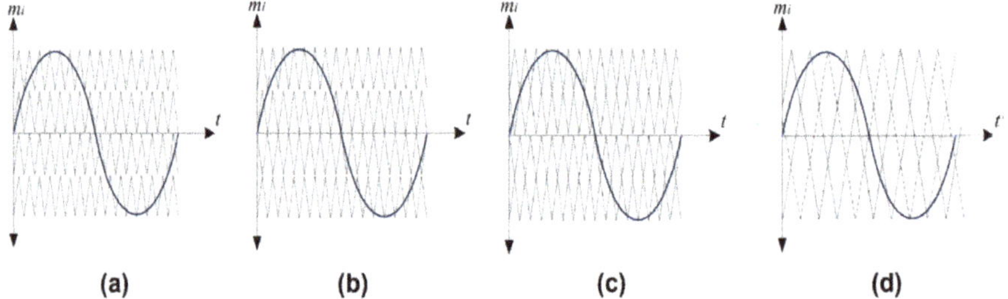

Figure 20. Multi-carrier sinusoidal PWM (SPWM) control methods: (**a**) PD, (**b**) POD, (**c**) APOD, (**d**) PS.

5.2. Closed-Loop PWM Control Techniques

Most widely used applications of PWM converters, like drives of a motor, active filters and many more, need a control form consisting of an inner current feedback loop. A multilevel inverter system uses photovoltaic (PV) or wind power sources to combine renewable power sources into the grid. Several methods have been conducted to reduce harmonics by utilizing current regulate for active power filter or by connecting a photoelectric and wind generator integrated to grid applications of a multilevel inverter. Current control methods are primarily discussed in terms of hysteresis current control and line current control [89–92].

Hysteresis modulation is a current regulate technique in which the load current follows the reference current in the hysteresis band in non-linear load application of multilevel inverter. Figure 21a is a schematic layout of the hysteresis regulation of the H-bridge, and Figure 21b illustrates the fundamental law of the hysteresis modulation. The employed controller creates a sinusoidal source current that of desired magnitude and also frequency, which is analyzed according to the actual current. If the current surpasses the higher level of the hysteresis band, it must choose the next higher voltage level to try to force the current error to zero. However, the voltage level of the inverter might be not enough to reset the current error to zero, and the inverter must move to the further higher voltage level to obtain the correct voltage level. Consequently, the current returns to the hysteresis band and the current follow the reference current in the hysteresis band. The three hysteresis controllers correctly select the voltage level are explained as offset band three-level, double band three-level, and time-based three-level hysteresis controllers [86,91–94].

Considering this to be a different method of multilevel inverter usage, the grid connection to the inverter must need current regulating. Linear current regulators are categorized as ramp comparators, stationary vector, and synchronous vector regulator. The ramp regulator uses the output ripple and feedback to regulate the switching instantly. In a 3-phase isolated neutral load method, the 3-phase current must have a zero-sum. Thus, linear compensation and the required two reference voltages of a three-phase inverter can be determined by utilizing 2 to 3 phase AB/ABC conversion blocks as shown in Figure 22.

The main linear current regulator comprises a tracking controller with an equivalent integrator for photovoltaic converters. Numerous harmonic compensation strategies can be found in the following articles on re-control and a linear resonant harmonic compensator [86,95–97].

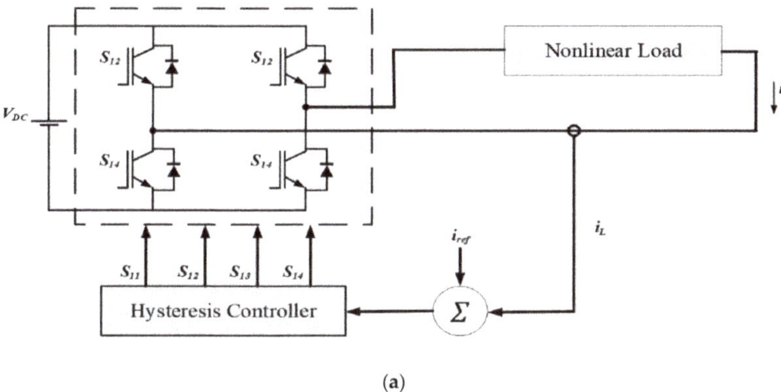

(a)

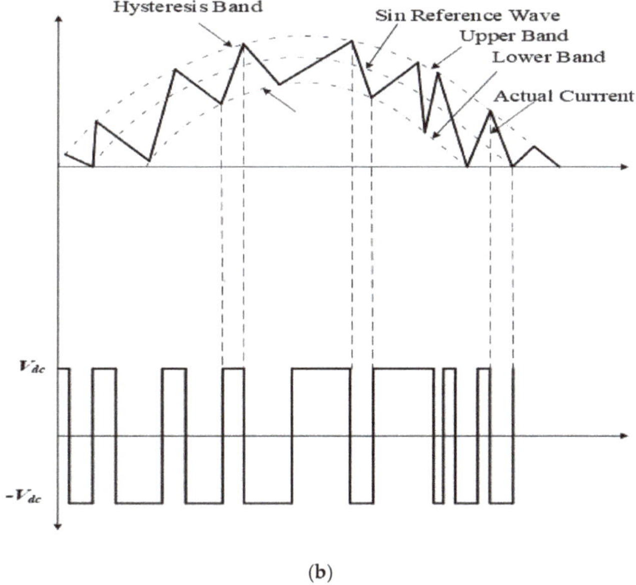

(b)

Figure 21. Hysteresis current control: (**a**) schematic diagram of a H-bridge cell with hysteresis regulator; hysteresis current control (**b**) hysteresis current band and voltage curves of load feedback.

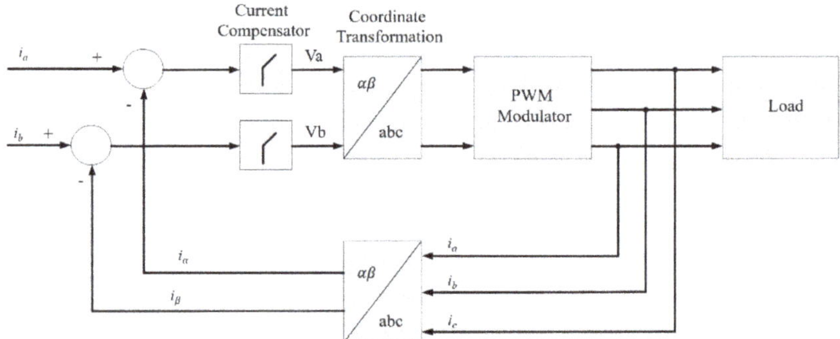

Figure 22. The schematic plot of stationary linear current regulator.

6. Conclusions

Many researchers and industries have acquired a multilevel inverter because of the capacities to regulate high- and medium-power applications. Also, the sustainable power source of DC needs to be changed over to AC. We demonstrate three different types of multilevel inverter, such as a cascade H multilevel inverter, a diode-clamp multilevel inverter, and a flying-capacitor multilevel inverter. Although all multilevel inverters are utilized to produce smoothing signals, the two types mentioned above have some drawbacks. For example, the individual DC sources are required for each inverter in CHB-MLI inverter and DC-MLI inverter switching losses are higher. These topologies are modeled in PSCADS SIMULINK. However, a flying-capacitor multilevel inverter has several features to regulate DC link capacitors compared to another multilevel topology using redundant switch configurations. From the MULTISIM simulation analysis and modeling of a flying-capacitor multilevel inverter in three levels, THD was 28.88% while in the five-level topology the THD was 18.56%. Therefore, we can decrease the total harmonic distortion adopting the higher-level topology.

Author Contributions: This paper was a collaboration effort among all authors. All authors conceived the methodology, conducted the experiment tests, and wrote the paper. R.A.R. and S.A.P. analyzed, designed and development of the experiment methodologies, A.M. and C.w.L. supported the work to perform in real time and, H.-J.K. verified the overall experiment.

Funding: This research was supported by Basic Research Laboratory through the National Research Foundations of Korea funded by the Ministry of Science, ICT and Future Planning (NRF-2015R1A4A1041584).

Conflicts of Interest: The authors declare no conflict of interest.

References

1. Rashid, M.H. *Power Electronics Circuits, Devices and Application Handbook*, 3rd ed.; Pearson Education: Upper Saddle River, NJ, USA, 2003.
2. Rodriguez, J.; Jih-Sheng, L.; Fang, P. Multilevel inverters: A survey of topologies, controls, and applications. *IEEE Trans. Ind. Electron.* **2002**, *49*, 724–738. [CrossRef]
3. Raghavendra, R.K.; Vijay, B.B.; Hiralal, M.S. A three-phase hybrid cascaded modular multilevel inverter for renewable energy environment. *IEEE Trans. Power Electron.* **2016**, *32*, 1070–1078.
4. Youssef, O.; Kamal, A.H.; Luc-Andre, G. Packed u cells multilevel converter topology: Theoretical study and experimental validation. *IEEE Trans. Ind. Electron.* **2011**, *58*, 1294–1306.
5. Sumit, K.C.; Chandan, C. A new asymmetric multilevel inverter topology suitable for solar PV applications with varying irradiance. *IEEE Trans. Sustain. Energy* **2017**, *8*, 1496–1506.
6. Vahid, D.; Arash, K.S.; Mostafa, A.; Soheila, E.; Keith, A.C. A new family of modular multilevel converter based on modified flying-capacitor multicell converters. *IEEE Trans. Power Electron.* **2015**, *30*, 138–147.
7. Vahid, D.; Saeedeh, D. Analytical modelling of single-phase stacked multicell multilevel converters exploiting kapteyn (fourier–bessel) series. *IET Power Electron.* **2013**, *6*, 1220–1238.

8.   Arash, K.; Vahid, D.; Mostafa, A.; Saeedeh, D. Reduced dc voltage source flying capacitor multicell multilevel inverter: Analysis and implementation. *IET Power Electron.* **2014**, *7*, 439–450.

9.   Arash, K.; Mostafa, A.; Keith, A.C.; Vahid, D. A new breed of optimized symmetrical and asymmetrical cascaded multilevel power converters. *IEEE J. Emerg Sel. Top. Power Electron.* **2015**, *3*, 1160–1170.

10.   Saeedeh, D.; Ebrahim, B.; Soheila, E.; Vahid, D.; Mehran, S. Flying-capacitor stacked multicell multilevel voltage source inverters: Analysis and modelling. *IET Power Electron.* **2014**, *7*, 2969–2987.

11.   Arash, K.; Vahid, D.; Keith, A.C. New flying-capacitor-based multilevel converter with optimized number of switches and capacitors for renewable energy integration. *IEEE Trans. Energy Convers* **2016**, *31*, 846–859.

12.   Margarita, N.; Samir, K.; Sibylle, D.; Jose, R. Reduced multilevel converter: A novel multilevel converter with a reduced number of active switches. *IEEE Trans. Ind. Electron.* **2018**, *65*, 3636–3645.

13.   Amir, T.; Jafar, A.; Mohammad, R. A multilevel inverter structure based on a combination of switched capacitors and dc sources. *IEEE Trans. Ind. Inform.* **2017**, *13*, 2162–2171.

14.   Meysam, S.; Jafar, A.; Seyyed, M.H. Cascaded multilevel inverter based on symmetric asymmetric dc sources with reduced number of components. *IET Power Electron.* **2017**, *10*, 1468–1478.

15.   Reza, B.; Majid, M.; Elyas, Z.; Hossein, M.K.; Frede, B. A new boost switched-capacitor multilevel converter with reduced circuit devices. *IEEE Trans. Power Electron.* **2017**, *33*, 6738–6754.

16.   Krishna, K.G.; Alekh, R.; Pallavee, B.; Lalit, K.S.; Shailendra, J. Multilevel inverter topologies with reduced device count: A review. *IEEE Trans. Power Electron.* **2016**, *31*, 135–151.

17.   Kavitha, M.; Arunkumar, A.; Gokulnath, N.; Arun, S. New Cascaded H-Bridge Multilevel Inverter Topology with Reduced Number of Switches and Sources. *IOSR-JEEE* **2012**, *2*, 26–36. [CrossRef]

18.   Ke, Ma.; Frede, B. Modulation Methods for Neutral-Point-Clamped Wind Power Converter Achieving Loss and Thermal Redistribution under Low-Voltage-Ride-Through. *IEEE Trans. on Ind. Electron.* **2013**, *61*, 835–845.

19.   Hochgra, C.; Lasseter, R.; Divan, D.; Lipo, T.A. Comparison of multilevel inverters for static VAR compensation. In Proceedings of the IEEE Industry Applications Society Annual Meeting, Denver, CO, USA, 2–6 October 1994.

20.   Kuhn, H.; Rüger, N.E.; Mertens, A. Control strategy for multilevel inverter with non-ideal dc sources. In Proceedings of the IEEE Power Electronics Specialists Conference, Orlando, FL, USA, 17–21 June 2007.

21.   Panagis, P.; Stergiopoulos, F.; Marabeas, P.; Manias, S. Comparison of state-of-the-art multilevel inverters. In Proceedings of the IEEE Annual Power Electronics Specialist Conference PESC, Rhodes Greece, 15–19 June 2008.

22.   Sumithira, T.R.; Nirmal, A. Elimination of harmonics in multilevel inverters connected to solar photovoltaic systems using ANFIS: An experimental case study. *Appl. Res. Technol.* **2013**, *11*, 124–132. [CrossRef]

23.   Jih-Sheng, L.; Fang, P. Multilevel converters-a new breed of power converters. *IEEE Trans. Ind. Appl.* **1996**, *32*, 509–517. [CrossRef]

24.   Mohammadreza, D. *Analysis of Different Topologies of Multilevel Inverter Thesis*; Chalmers University of Technology: Gothenburg, Sweden, 2010.

25.   Khairnar, D.D.; Deshmukh, V.M. Performance Analysis of Diode Clamped 3 Level MOSFET Based Inverter. *Int. Electr. Eng. J.* **2014**, *5*, 1484–1489.

26.   Mehdi, N.; Bin, W.; Navid, R. A Novel Five level voltage source inverter with sinusoidal pulse width modulator for medium voltage applications. *IEEE Trans. Power Electron.* **2016**, *31*, 1959–1967.

27.   Kai, T.; Bin, W.; Mehdi, N.; Xu, D.D.; Cheng, Z.; Navid, R. A Capacitor voltage balancing method for Nested neutral point clamped (NNPC) inverter. *IEEE Trans. Power Electron.* **2015**, *31*, 2575–2583.

28.   Song, B.M.; Kim, J.; Lai, J.S.; Seong, K.C.; Kim, H.J.; Park, S.S. A multilevel soft-switching inverter with inductor coupling. *IEEE Trans. Ind. Appl.* **2001**, *37*, 628–636. [CrossRef]

29.   Meynard, T.A.; Foch, H. Multi-level conversion: High voltage choppers and voltage-source inverters. In Proceedings of the IEEE Power Electronics Specialists Conference, Toledo, Spain, 29 June–3 July 1992.

30.   Xu, L.; Agelidis, V.G. Active capacitor voltage control of flying capacitor multilevel converters. *IEEE Electr. Power Appl.* **2004**, *151*, 313–320. [CrossRef]

31.   Feng, C.; Liang, J.; Agelidis, V.G.; Green, T.C. A multi-modular system based on parallel-connected multilevel flying capacitor converters controlled with fundamental frequency SPWM. In Proceedings of the 32nd IEEE Conference on Industrial Electronics, Paris, France, 7–10 November 2006.

32. Nami, A.; Liang, J.; Dijkhuizen, F.; Demetriades, G.D. Modular multilevel converters for HVDC applications: Review on converter cells and functionalities. *IEEE Trans. Power Electron.* **2015**, *30*, 18–36. [CrossRef]

33. Hiller, M. Method for Controlling a Polyphase Converter with Distributed Energy Stores. U.S. Patent 12/067, 555, 20 March 2008.

34. Ilves, K.; Taffner, F.; Norrga, S.; Antonopoulos, A.; Harnefors, L.; Nee, H.-P. A submodule implementation for parallel connection of capacitors in modular multilevel converters. *IEEE Trans. Power Electron.* **2015**, *30*, 3518–3527. [CrossRef]

35. Farhadi, M.; Babaei, E. Cross-switched multilevel inverter: An innovative topology. *IET Trans. Power Electron.* **2013**, *6*, 642–651. [CrossRef]

36. Mathew, E.C.; Ghat, M.B.; Shukla, A. A generalized cross-connected submodule structure for hybrid multilevel converters. *IEEE Trans. Ind. Appl.* **2016**, *52*, 3159–3170. [CrossRef]

37. Wang, Y.; Marquardt, R. Future HVDC-grids employing modular multilevel converters and hybrid DC-breakers. In Proceedings of the 15th European Conference on Power Electronics and Applications (EPE), Lille, France, 3–5 September 2013.

38. Marquardt, R. Modular multilevel converter topologies with DC-short circuit current limitation. In Proceedings of the 8th International Conference on Power Electronics—ECCE Asia, Jeju, Korea, 30 May–3 June 2011.

39. Li, X.; Song, Q.; Liu, W.; Zhu, Z.; Xu, S. Experiment on DC-fault ride through of MMC using a half-voltage clamp submodule. *IEEE J. Emerg. Sel. Top. Power Electron.* **2018**, *6*, 1273–1279. [CrossRef]

40. Elserougi, A.A.; Massoud, A.M.; Ahmed, S. A switched-capacitor submodule for modular multilevel HVDC converters with DC-fault blocking capability and a reduced number of sensors. *IEEE Trans. Power Del.* **2016**, *31*, 313–322. [CrossRef]

41. Qin, J.; Saeedifard, M.; Rockhill, A.; Zhou, R. Hybrid design of modular multilevel converters for HVDC systems based on various submodule circuits. *IEEE Trans. Power Del.* **2015**, *30*, 385–394. [CrossRef]

42. Nami, A.; Wang, L.; Dijkhuizen, F.; Shukla, A. Five level cross connected cell for cascaded converters. In Proceedings of the 2013 15th European Conference on Power Electronics and Applications (EPE), Lille, France, 3–5 September 2013.

43. Li, R.; Fletcher, J.E.; Xu, L.; Holliday, D.; Williams, B.W. A hybrid modular multilevel converter with novel three-level cells for DC fault blocking capability. *IEEE Trans. Power Del.* **2015**, *30*, 1853–1862. [CrossRef]

44. Hu, X.; Zhang, J.; Xu, S.; Jiang, Y. Investigation of a new modular multilevel converter with DC fault blocking capability. *IEEE Trans. Ind. Appl.* **2018**, *55*, 552–562. [CrossRef]

45. Chattopadhyay, S.K.; Chakraborty, C. Multilevel inverters with level doubling network: A new topological variation. In Proceedings of the Industrial Electronics Society Conference, Vienna, Austria, 10–14 November 2013.

46. Chattopadhyay, S.K.; Chakraborty, C. A new multilevel inverter topology with self-balancing level doubling network. *IEEE Trans. Ind. Electron.* **2014**, *61*, 4622–4631. [CrossRef]

47. Ji, Y.; Chen, G. A novel three-phase seven-level active clamped converter using H-bridge as a level doubling network. In Proceedings of the IEEE 8th International Power Electronics and Motion Control Conference, IPEMC-ECCE Asia, Hefei, China, 22–26 May 2016.

48. Rajesh, V.; Chattopadhyay, S.K.; Chakraborty, C. Full bridge level doubling network assisted multilevel DC link inverter. In Proceedings of the IEEE 7th Power India International Conference, PIICON 2016, Bikaner, India, 25–27 November 2017.

49. Agelidis, V.G.; Demetriades, G.D.; Flourentzou, N. Recent advances in high-voltage direct-current power transmission systems. In Proceedings of the 2006 IEEE International Conference on Industrial Technology, Mumbai, India, 15–17 December 2006.

50. Akagi, H. Classification, terminology, and application of the modular multilevel cascade converter (MMCC). *IEEE Trans. Power Electron.* **2011**, *26*, 3119–3130. [CrossRef]

51. Hu, X.; Zhang, J.; Xu, S.; Jiang, Y. Extended state observer-based fault detection and location method for modular multilevel converters. In Proceedings of the IECON 42nd Annual Conference of the IEEE Industrial Electronics Society, Florence, Italy, 23–26 October 2016.

52. Pérez, M.; Rodríguez, J.; Pontt, J.; Kouro, S. Power distribution in hybrid multi-cell converter with nearest level modulation. In Proceedings of the IEEE International Symposium on Industrial Electronics, 4–7 June 2007.

53. Ebrahimi, J.; Babaei, E.; Gharehpetian, G.B. A new topology of cascaded multilevel converters with reduced number of components for high-voltage applications. *IEEE Trans. Power Electron.* **2011**, *26*, 3109–3118. [CrossRef]

54. Barbosa, P.; Steimer, P.; Steinke, J.; Winkelnkemper, M.; Celanovic, N. Active-neutral-point-clamped (ANPC) multilevel converter technology. In Proceedings of the European Conference on Power Electronics and Applications, Dresden, Germany, 11–14 September 2005.

55. Bruckner, T.; Bernet, S.; Guldner, H. The active NPC converter and its loss-balancing control. *IEEE Trans. Ind. Electron.* **2005**, *54*, 855–868. [CrossRef]

56. Pulikanti, S.R.; Agelidis, V.G. Hybrid flying-capacitor-based active-neutral-point-clamped five-level converter operated with SHE-PWM. *IEEE Trans. Ind. Electron.* **2011**, *58*, 4643–4653. [CrossRef]

57. Wang, H.; Kou, L.; Liu, Y.-F.; Sen, P.C. A new six-switch five-level active neutral point clamped inverter for PV applications. *IEEE Trans. Power Electron.* **2017**, *32*, 6700–6715. [CrossRef]

58. Naderi, R.; Smedley, K. A new hybrid active neutral point clamped flying capacitor multilevel inverter. In Proceedings of the IEEE Applied Power Electronics Conference and Exposition (APEC), Charlotte, NC, USA, 15–19 March 2015.

59. Kouro, S.; Malinowski, M.; Gopakumar, K.; Pou, J.; Franquelo, L.G.; Wu, B.; Rodriguez, J.; Perez, M.A.; Leon, J.I. Recent advances and industrial applications of multilevel converters. *IEEE Trans. Ind. Electron.* **2010**, *57*, 2553–2580. [CrossRef]

60. Teichmann, R.; Bernet, S. A comparison of three-level converters versus two-level converters for low-voltage drives, traction, and utility applications. *IEEE Trans. Ind. Appl.* **2005**, *41*, 855–865. [CrossRef]

61. Yuan, X. Derivation of voltage source multilevel converter topologies. *IEEE Trans. Ind. Electron.* **2017**, *64*, 966–976. [CrossRef]

62. Nabae, A.; Takahashi, I.; Akagi, H. A new neutral-point-clamped PWM inverter. *IEEE Trans. Ind. Appl.* **1981**, *IA-17*, 518–523. [CrossRef]

63. Rodriguez, J.; Bernet, S.; Steimer, P.K.; Lizama, I.E. A survey on neutral-point-clamped inverters. *IEEE Trans. Ind Electron.* **2010**, *57*, 2219–2230. [CrossRef]

64. Rodriguez, J.; Bernet, S.; Wu, B.; Point, J.O.; Kouro, S. Multilevel voltage-source-converter topologies for industrial medium-voltage drives. *IEEE Trans. Ind. Electron.* **2007**, *54*, 2930–2945. [CrossRef]

65. Meynard, T.A.; Foch, H. Multi-level choppers for high voltage applications. *EPE J.* **1992**, *2*, 45–50. [CrossRef]

66. Leon, J.I.; Vazquez, S.; Franquelo, L.G. Multilevel converters: Control and modulation techniques for their operation and industrial applications. *Proc. IEEE* **2017**, *105*, 2066–2081. [CrossRef]

67. Abu-Rub, H.; Holtz, J.; Rodriguez, J.; Baoming, G. Medium-voltage multilevel converters-State of the art, challenges, and requirements in industrial applications. *IEEE Trans. Ind. Electron.* **2010**, *57*, 2581–2596. [CrossRef]

68. Meynard, T.A.; Foch, H.; Forest, F.; Turpin, C. Multicell converters: Derived topologies. *IEEE Trans. Ind. Electron.* **2002**, *49*, 978–987. [CrossRef]

69. Hammond, P.W. A new approach to enhance power quality for medium voltage ac drives. *IEEE Trans. Ind. Appl.* **1997**, *33*, 202–208. [CrossRef]

70. Malinowski, M.; Gopakumar, K.; Rodriguez, J.; Perez, M.A. A survey on cascaded multilevel inverters. *IEEE Trans. Ind. Electron.* **2010**, *57*, 2197–2206. [CrossRef]

71. Peng, F.Z.; Wang, J. A universal STATCOM with delta-connected cascade multilevel inverter. In Proceedings of the PESC Record-IEEE Annual Power Electronics Specialists Conference, Aachen, Germany, 20–25 June 2004.

72. Hammami, M.; Grandi, G. Input current and voltage ripple analysis in level doubling network (LDN) cells for H-bridge multilevel inverters. *IEEE Trans. Ind. Electron.* **2019**, *66*, 8414–8423. [CrossRef]

73. Dewangan, N.K.; Gurjar, V.S.; Ullah, U.; Zafar, S. A level-doubling network (LDN) for cross-connected sources based multilevel inverter (CCS-MLI). In Proceedings of the IEEE International Conference on Power Electronics, Drives and Energy Systems (PEDES), Mumbai, India, 16–19 December 2014.

74. Srndovic, M.; Viatkin, A.; Grandi, G. Grid-connected three-phase H-bridge inverter with level doubling network controlled by staircase modulation techniques. In Proceedings of the 10th International Conference on Integrated Power Electronics Systems, Stuttgart, Germany, 20–22 March 2018.

75. Allebrod, S.; Hamerski, R.; Marquardt, R. New transformerless, scalable modular multilevel converters for HVDC-transmission. In Proceedings of the PESC Record-IEEE Annual Power Electronics Specialists Conference PESC, Rhodes Greece, 15–19 June 2008.

76. Mohammadi, H.P.; Bina, M.T. A transformerless medium-voltage STATCOM topology based on extended modular multilevel converters. *IEEE Trans. Power Electron.* **2011**, *26*, 1534–1545.

77. Leon, J.I.; Kouro, S.; Franquelo, L.G.; Rodriguez, J.; Wu, B. The essential role and the continuous evolution of modulation techniques for voltage-source inverters in the past, present, and future power electronics. *IEEE Trans. Ind. Electron.* **2016**, *63*, 2688–2701. [CrossRef]

78. Meynard, T.A.; Foch, H.; Thomas, P.; Courault, J.; Jakob, R.; Nahrstaedt, M. Multicell converters: Basic concepts and industry applications. *IEEE Trans. Ind. Electron.* **2002**, *49*, 955–964. [CrossRef]

79. Xiaomin, K.; Corzine, K.A.; Familiant, Y.L. Full binary combination schema for floating voltage source multi-level inverters. *IEEE Trans. Power Electron.* **2002**, *17*, 891–897. [CrossRef]

80. Meynard, T.A.; Fadel, M.; Aouda, N. Modeling of multilevel converters. *IEEE Trans. Ind. Electron.* **1997**, *44*, 356–364. [CrossRef]

81. Xiaoming, Y.; Stemmler, H.; Barbi, I. Self-balancing of the clamping capacitor-voltages in the multilevel capacitor-clamping-inverter under sub-harmonic PWM modulation. *IEEE Trans. Power Electron.* **2001**, *16*, 256–263. [CrossRef]

82. Richardeau, F.; Baudesson, P.; Meynard, T. Failures-tolerance and remedial strategies of a PWM multicell inverter. In Proceedings of the IEEE Power Electronics Specialists Conference, Galway, Ireland, 18–23 June 2000.

83. Peebles, P.Z. *Digital Communication Systems*; Prentice Hall: Upper Saddle River, NJ, USA, 1987.

84. Zierhofer, C.M. Adaptive sigma-delta modulation with one-bit quantization. *IEEE Trans. Circuits Syst. II Analog Digit. Signal Process* **2000**, *47*, 408–415. [CrossRef]

85. Brukner, T.; Bernet, S. Investigation of a high-power three-level quasi-resonant dc-link voltage source inverter. *IEEE Trans. Ind. Appl.* **2001**, *37*, 619–627. [CrossRef]

86. Diaz, C.S.; Escobar, F.I.; Diotaiuti, G.C.; Arguis, V.M.G. Adaptive sigma delta modulator applied to control five level multilevel inverter. In Proceedings of the IEEE International Caracas Conference on Devices, Circuits and Systems, ICCDCS, Punta Cana, Dominican Republic, 3–5 November 2004.

87. Zhang, L.; Watkins, S.J.; Shepherd, W. Analysis and control of a multi-level flying capacitor inverter. In Proceedings of the IEEE International Power Electronics Congress-CIEP, Guadalajara, Mexico, 24–24 October 2002.

88. Ismail, B.; Taib, S.; Saad, A.R.M.; Isa, M.; Hadzer, C.M. Development of a single phase SPWM microcontroller-based inverter. In Proceedings of the 1st International Power and Energy Conference, Putra Jaya, Malaysia, 28–29 November 2006.

89. Bojoi, R.; Griva, G.; Guerriero, M.; Farina, F.; Profumo, F.; Bostan, V. Improved current control strategy for power conditioners using sinusoidal signal integrators in synchronous reference frame. In Proceedings of the PESC Record-IEEE Annual Power Electronics Specialists Conference, Aachen, Germany, 20–25 June 2004.

90. Kang, B.J.; Liaw, C.M. Random hysteresis PWM inverter with robust spectrum shaping. *IEEE Trans. Aerosp. Electron. Syst.* **2001**, *37*, 619–628. [CrossRef]

91. Escobar, G.; Martínez, P.R.; Leyva-Ramos, J. Analog circuits to implement repetitive controllers with feedforward for harmonic compensation. *IEEE Trans. Ind. Electron.* **2007**, *54*, 567–573. [CrossRef]

92. Loh, P.C.; Bode, G.H.; Holmes, D.G.; Lipo, T.A. A time-based double-band hysteresis current regulation strategy for single-phase multilevel inverters. *IEEE Trans. Ind. Appl.* **2003**, *39*, 883–892.

93. Loh, P.C.; Bode, G.H.; Tan, P.C. Modular hysteresis current control of hybrid multilevel inverters. *IEE Proc. Electr. Power Appl.* **2005**, *152*, 1–8. [CrossRef]

94. Blanco, E.; Bueno, E.; Espinosa, F.; Cobreces, S.; Rodríguez, F.J.; Ruiz, M.A. Fast harmonics compensation in VSCs connected to the grid by synchronous frame generalized integrators. In Proceedings of the IEEE International Symposium on Industrial Electronics, Dubrovnik, Croatia, 20–23 June 2005.

95. Castilla, M.; Miret, J.; Matas, J.; García de Vicuña, L.; Guerrero, J.M. Linear current control scheme with series resonant harmonic compensator for single-phase grid-connected photovoltaic inverters. *IEEE Trans. Ind. Electron.* **2008**, *55*, 2727–2733. [CrossRef]

96. Zare, F.; Nami, A. A new random current control technique for a single-phase inverter with bipolar and unipolar modulations. In Proceedings of the 4th Power Conversion Conference-NAGOYA, PCC-NAGOYA, Nagoya, Japan, 2–5 April 2007.

97. Ahmad, Z.; Singh, S.N. Comparative analysis of single-phase transformer less inverter topologies for grid connected PV system. *Sol. Energy* **2017**, *149*, 245–271. [CrossRef]

© 2019 by the authors. Licensee MDPI, Basel, Switzerland. This article is an open access article distributed under the terms and conditions of the Creative Commons Attribution (CC BY) license (http://creativecommons.org/licenses/by/4.0/).

Article

Three-Phase Five-Level Cascade Quasi-Switched Boost Inverter

Van-Thuan Tran [1], Minh-Khai Nguyen [2,*], Cao-Cuong Ngo [3] and Youn-Ok Choi [4,*]

[1] Department of Telecommunication Operation, Telecommunications University, Nha Trang 650000, Viet Nam; thuantsttq@gmail.com
[2] Faculty of Electrical and Electronics Engineering, Ho Chi Minh City University of Technology and Education, 1 Vo Van Ngan Street, Thu Duc District, Ho Chi Minh City 700000, Vietnam
[3] Faculty of Electrical and Electronics Engineering, Ho Chi Minh University of Technology (HUTECH), 475A Dien Bien Phu Street, Binh Thanh District, Ho Chi Minh City 700000, Vietnam; cuongnc@uef.edu.vn
[4] Department of Electrical Engineering, Chosun University, 309 Pilmun-daero, Dong-gu, Gwangju 61452, Korea
* Correspondence: nmkhai00@gmail.com (M.-K.N.); yochoi@chosun.ac.kr (Y.-O.C.); Tel.: +82-62-230-7175 (M.-K.N.)

Received: 30 January 2019; Accepted: 1 March 2019; Published: 6 March 2019

Abstract: This paper presents a three-phase cascaded five-level H-bridge quasi-switched boost inverter (CHB-qSBI). The merits of the CHB-qSBI are as follows: single-stage conversion, shoot-through immunity, buck-boost voltage, and reduced passive components. Furthermore, a PWM control method is applied to the CHB-qSBI topology to improve the modulation index. The voltage stress across power semiconductor devices and the capacitor are significantly lower using improved pulse-width modulation (PWM) control. Additionally, by controlling individual shoot-through duty cycle, the DC-link voltage of each module can achieve the same values. As a result, the imbalance problem of the DC-link voltage can be solved. A detailed analysis and operating principle with the modulation scheme and comprehensive comparison for the CHB-qSBI are illustrated. The experimental and simulation results are presented to validate the operating principle of the three-phase CHB-qSBI.

Keywords: cascaded H-bridge inverter; three-phase inverter; Z-source network; quasi-switched-boost network; shoot-through

---

1. Introduction

Nowadays, multilevel inverters are attractive for high power high voltage applications due to their well-known properties. The benefits of the multilevel power inverter topologies are as follows: improved output waveforms, low electromagnetic interference (EMI), and small filter size [1–3]. Neutral point clamped (NPC), flying capacitors, and cascaded H-bridge (CHB) inverters [4–14] are three basic multilevel inverter structures. Among these structures, the CHB structures [9–14] have unique merits: higher output voltage, flexibility, and power levels. Furthermore, the CHB inverter can achieve high reliability with modular configuration. The output voltage of each phase in the CHB inverter is achieved by the sum of each output voltage. The CHB structure has some merits in using cascading more H-bridge modules and independent sources. Besides that, the output voltage of the CHB inverter has a high number of levels and reaches medium voltage which results in removing the boost transformer and dropping the size of the output filter. In [14], the low-frequency transformers were used to cascade the H-bridge circuit with a single DC source, but the size of the cascaded system is increased because of using low-frequency transformers.

However, the traditional CHB inverter [9–14] is a buck DC–AC power conversion. In addition, both power switches in the same branch cannot be turned on at the same time. To overcome the limitation of the traditional CHB inverters, the CHB quasi-Z-source inverters (CHB-qZSI) were discussed in [15–17]. A battery-energy-stored CHB-qZSI-based photovoltaic power generation system was presented in [18]. To enhance the performance of the three-phase five-level CHB-qZSI for grid-connected applications, an innovative modulation technique was introduced in [19] for the voltage stress reduction. A fault-tolerant strategy for the three-phase CHB-qZSI was presented in [20]. In the CHB-qZSI, each module of the three-phase CHB-qZSI topology uses two inductors and two capacitors. As a result, the size, weight, and cost of the cascaded system are increased significantly when the number of output voltage levels is increased. The quasi-switched boost inverter (qSBI) topology had been introduced in [21] to replace the qZSI because it has the same feature as buck-boost voltage and high reliability with shoot-through (ST) immunity. A detailed comparison between qZSI and qSBI topology was discussed in [22]. The comparison results in [22] show that the qSBI topology uses one less capacitor and one less inductor; higher boost factor with the same parasitic effect; lower current rating on switches and diodes; and higher efficiency when compared to qZSI topology. However, with a simple boost control method, the modulation index is low when a voltage gain is required. To improve the modulation index, a novel pulse-width modulation (PWM) scheme for qSBI was proposed in [23]. The single-phase grid-connected CHB-qSBI was discussed in [24]. In CHB-qSBI [24], each module of the CHB-qSBI topology only uses one inductor and one capacitor. A three-phase CHB-qSBI has been proposed in [25] with only some simulation results. Besides that, with using a simple boost method, the CHB-qSBI in [24,25] must use a small modulation index to be able to achieve a high voltage gain. Consequently, the voltage stress across power semiconductor devices and capacitor is high.

To explore more features of the three-phase CHB-qSBI, this paper presents the operating theories, circuit analysis, and experimental verification of the three-phase CHB-qSBI in detail. Furthermore, a PWM control method in [23] is applied to the CHB-qSBI topology to improve the modulation index. The stress voltage across power semiconductor devices and the capacitor is significantly dropped in comparison to CHB-qSBI with using conventional PWM control. To solve the imbalance problem of the DC-link voltage, ST duty cycle of each module is controlled individually. As a result, the DC-link voltage of each module can obtain the same values. The simulation and experimental results are provided to validate the operating principle of the three-phase CHB-qSBI under improved PWM method.

2. Conventional Three-Phase CHB Inverter Topologies

Figure 1 illustrates the traditional three-phase five-level CHB inverter. Two H-bridge circuits were used in each phase to generate a five-level output voltage. Each H-bridge module was connected to the isolated DC voltage source. In the conventional CHB inverter, both upper and lower switches of the H-bridge leg cannot be switched on at the same time. The dead-time between upper and lower switches should be employed to avoid the ST phenomenon in the H-bridge circuit.

Figure 2 presents the three-phase CHB-qZSI topology, where each quasi-Z-source network module used two pair of inductor and capacitor. By adding the ST time interval to the H-bridge switches, the DC-link voltage of each H-bridge module in the CHB-qZSI was boosted to a higher value than the DC source voltage. The CHB-qZSI has the buck-boost voltage function, single-stage conversion, and ST immunity.

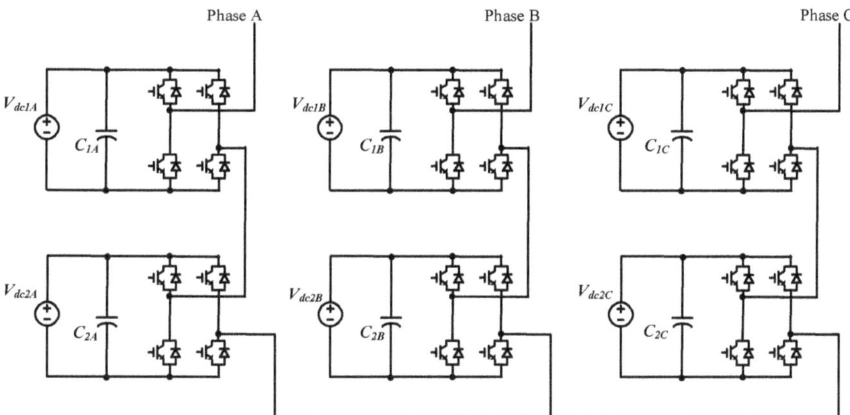

Figure 1. Construction of traditional three-phase cascaded H-bridge (CHB) inverter topology.

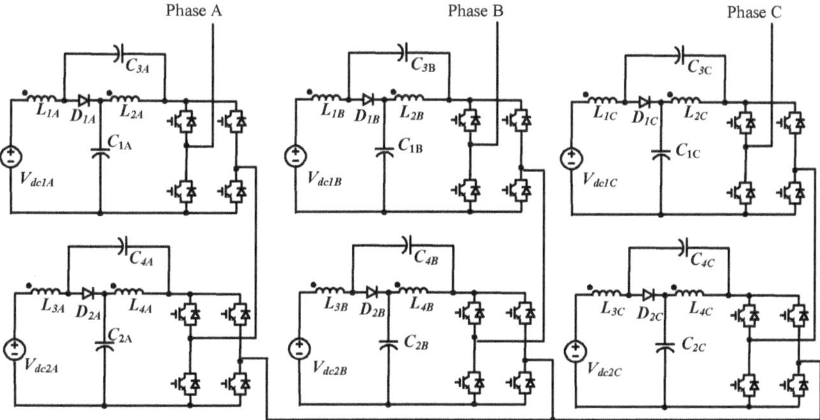

Figure 2. Construction of three-phase CHB quasi-Z-source inverters (CHB-qZSI) topology.

3. Three-Phase CHB-qSBI

The three-phase CHB-qSBI topology is indicated in Figure 3. Similar to conventional CHB inverter topology, each phase of the CHB-qSBI consists of two qSBI modules. The proposed cascaded system consists of six qSBI modules, three filter inductors, six separate DC sources, and a three-phase load. Compared to the conventional CHB module, a qSBI module includes one capacitor, one inductor, two diodes, and one active switch is added.

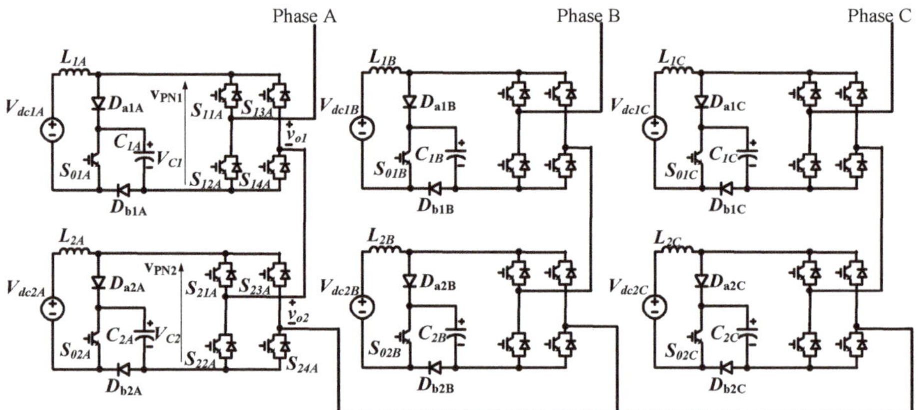

Figure 3. Proposed three-phase CHB quasi-switched boost inverter (CHB-qSBI) topology.

3.1. Operating Principle

As an example, the module 1 in the introduced system is used to analyze the circuit. As shown in Figure 4, there were three operating states: the non-shoot-through (NST) 1 state, the NST 2 state, and the shoot-through (ST) state.

In NST 1 state as shown in Figure 4a, switch S_0 is switched on. During this state, diode D_{a1} conducts while diode D_{b1} is blocked. As a result, the capacitor is discharged while the inductor stores energy. The time interval in this state is $(0.5 - D_1/2).T$, where D_1 is the duty cycle of each cycle of module 1; T is a switching period. We obtain:

$$\begin{cases} L_1 \frac{di_{L1}}{dt} = V_{dc1} \\ C_1 \frac{dv_{C1}}{dt} = -I_{PN1}, \end{cases} \tag{1}$$

where I_{PN1} is the equivalent DC-link current at load side.

In ST state as shown in Figure 4b, both 4 switches on the H-bridge circuit and switch S_0 are switched on at the same time. During this state, two diodes D_{a1} and D_{b1} are blocked. As a result, the capacitor is discharged, while the inductor stores energy. The time interval in this state is $D_1.T$. We obtain:

$$\begin{cases} L_1 \frac{di_{L1}}{dt} = V_{dc1} + V_{c1} \\ C_1 \frac{dv_{C1}}{dt} = -I_{in}, \end{cases} \tag{2}$$

where I_{in} is the average value of the input current (inductor current).

In NST 2 state as shown in Figure 4c, switch S_0 is switched off and the inverter has two active states and two zero states of the inverter main circuit for single-phase topology. During the NST state, D_{a1} and D_{b1} are turned on. As a result, the capacitor is charged from V_{dc1}, while the inductor transfers energy from the DC voltage source to the main circuit. The time interval in this state is $(0.5 - D_1/2).T$. We obtain:

$$\begin{cases} L_1 \frac{di_{L1}}{dt} = V_{dc1} - V_{c1} \\ C_1 \frac{dv_{C1}}{dt} = I_{in} - I_{PN1}. \end{cases} \tag{3}$$

Applying the volt-second balance and ampere-second balance principles to L and C in steady state, Equations (1), (2), and (3) yield:

$$\begin{cases} V_{c1} = \frac{2}{1-3D_1} V_{dc1} \\ I_{in} = \frac{2(1-D_1)}{1-3D_1} I_{PN1}. \end{cases} \tag{4}$$

The peak value of DC-link voltage that crosses the inverter of module 1 can be expressed in the NST states as:

$$V_{PN1} = V_{c1} = \frac{2}{1 - 3D_1} V_{dc1} \tag{5}$$

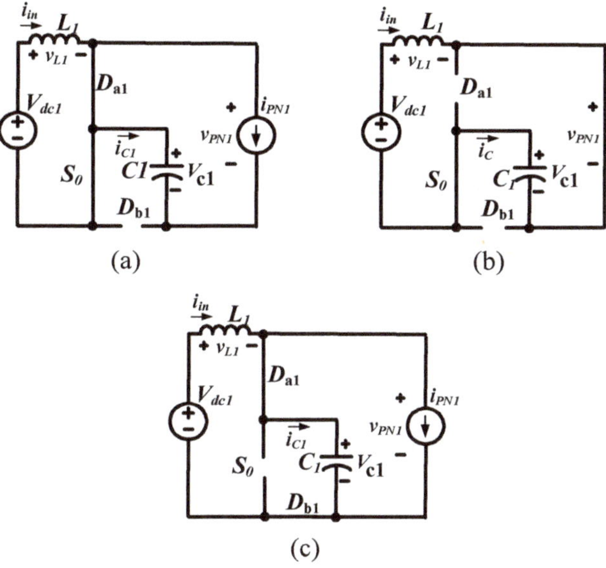

Figure 4. Operating states of qSBI module 1: (**a**) non-shoot through (NST) 1, (**b**) shoot-through (ST), and (**c**) NST 2 states.

3.2. Improved PWM Control for Three-Phase CHB-qSBI

A PWM technique for the single-phase qSBI to improve modulation index was presented in [23]. In this section, the PWM technique in [23] will be extended to the three-phase CHB-qSBI. The PWM strategy for the cascaded system for phase A with the modulation in Figure 4 is illustrated in Figure 5. For module 1, a fixed voltage V_{SH1} was compared to the triangle waveform (dashed line) with double frequency and half of the amplitude of that of V_{tri} to generate the ST state in the inverter bridge. Besides that, a square pulse signal with the same frequency as the triangle waveform (dashed line) and 50% duty cycle was used to control the S_0 switch. Two control waveforms, $V_{control}$ and $-V_{control}$, were compared to a triangle waveform, V_{tri}, to generate control signals for H-bridge switches. The additional switch S_0 signal was produced by comparing between the saw-tooth waveform, v_{saw} and V_{SH1}. Table 1 describes a truth table of the switching states of the switches in module 1. The switching status 0 and 1 in Table 1 represents the turning off and turning on of the switches, respectively.

The output voltage v_{o1} of H-bridge module 1 has three levels: $-V_{PN1}$, 0, and V_{PN1}. For module 2, the high-frequency triangle waveform, V_{tri}, was shifted in 90° and the high-frequency triangle waveform $*v_{tri}$ was shifted in 180° to generate the output voltage v_{o2} of H-bridge module 2. The output voltage of the cascaded system is total v_{o1} and v_{o2}. As a result, the 5-level output voltage of the proposed cascaded system was produced.

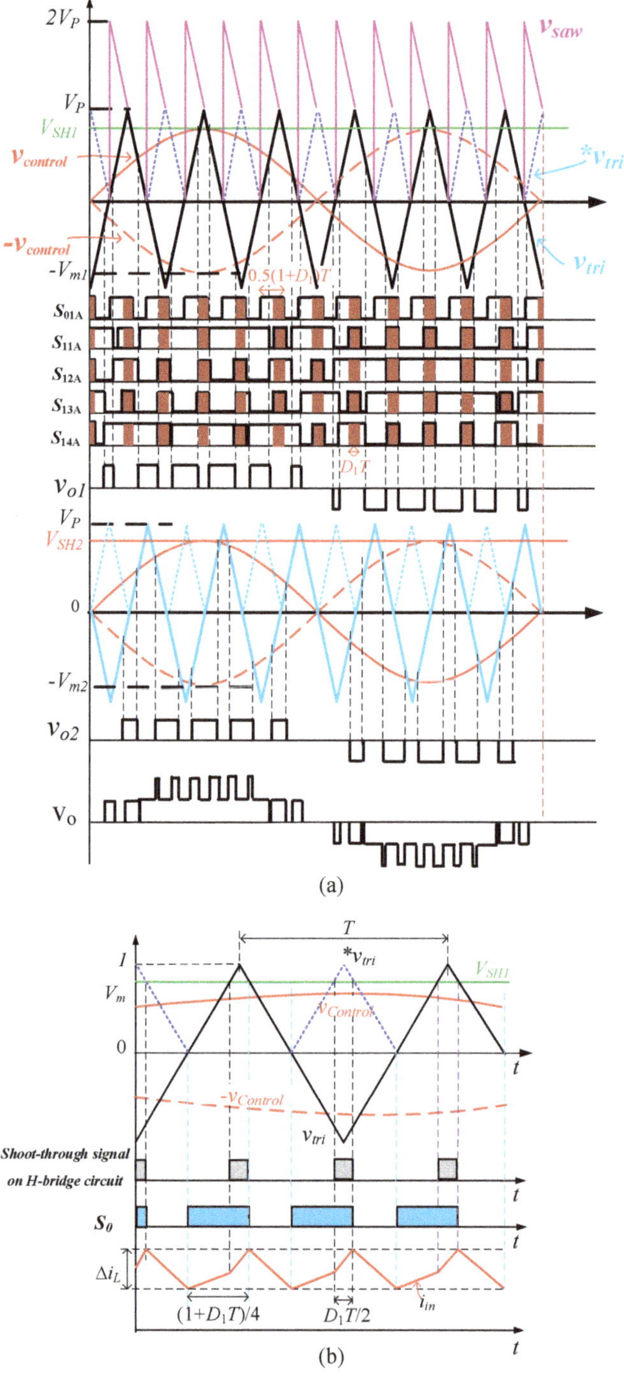

Figure 5. Improved PWM strategy for three-phase CHB-qSBI. (**a**) Improved PWM strategy for phase-A of the CHB-qSBI and (**b**) ST on H-bridge and S_0 signal generation.

Table 1. A truth table of switching signal generation in module 1.

Compared Condition	Status of Switches				
	S_{01A}	S_{11A}	S_{12A}	S_{13A}	S_{14A}
$V_{SH1} \leq {}^*v_{tri}$	1	1	1	1	1
$V_{SH1} \leq v_{saw}$	1	-	-	-	-
$V_{SH1} > v_{saw}$	0	-	-	-	-
$v_{control} > v_{tri}$	-	1	0	-	-
$v_{control} \leq v_{tri}$	-	0	1	-	-
$-v_{control} > v_{tri}$	-	-	-	1	0
$-v_{control} \leq v_{tri}$	-	-	-	0	1

"-": Undefinable state.

4. Comparison between Three-Phase CHB-qSBI under Improved PWM Method and Three-Phase CHB-qZSI

Table 2 compares the passive components, semiconductor devices, and the governing equations of the three-phase CHB-qSBI and the three-phase CHB-qZSI. From Table 2, the three-phase five-level CHB-qSBI uses six fewer capacitors, six fewer inductors, six more switches, and six more diodes. Although the three-phase CHB-qSBI reduces the number of passive components, it increases the number of active components. The three-phase CHB-qSBI under the improved PWM method uses a higher modulation index to produce the same voltage gain. Figure 6 compares the voltage stresses on diodes and switches of the CHB-qZSI and CHB-qSBI for the same voltage gain. As shown in Figure 6, the voltage stresses of the CHB-qSBI is lower than those of the CHB-qZSI. Consequently, the stress voltage across power semiconductor devices and the capacitor of the three-phase CHB-qSBI was significantly lower. Therefore, the weight, cost, and size of the three-phase CHB-qSBI under the improved PWM method were reduced in comparison with the three-phase CHB-qZSI topology.

Table 2. Comparison between CHB-qSBI and CHB-qZSI.

Parameter		Three-Phase CHB-qZSI	Three-Phase CHB-qSBI
Number of inductors		12	6
Number of capacitors		12	6
Number of diodes		30	36
Number of switches		24	30
Capacitor voltage	$V_{C1} = V_{C2}$	$\frac{1-D}{1-2D}V_{dc}$	$\frac{2}{1-3D}V_{dc}$
	$V_{C3} = V_{C4}$	$\frac{D}{1-2D}V_{dc}$	NA
DC-Link voltage each module, V_{PN}		$\frac{1}{1-2D}V_{dc}$	$\frac{2}{1-3D}V_{dc}$
Diodes voltage stresses, V_D		$\frac{1}{1-2D}V_{dc}$	$\frac{2}{1-3D}V_{dc}$
Switches voltage stresses, V_S		$\frac{1}{1-2D}V_{dc}$	$\frac{2}{1-3D}V_{dc}$
Voltage gain, G at each module		$\frac{M}{2M-1}$	$\frac{2M}{3M-2}$
ST immunity		Yes	Yes
Input current		Continuous	Continuous

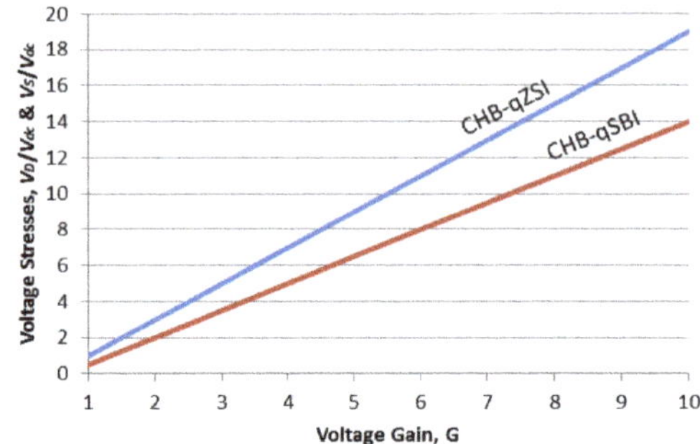

Figure 6. Voltage stress comparison for the same voltage gain.

5. Simulation and Experimental Verifications

5.1. Simulation Verification

To confirm the theoretical discussion of the three-phase CHB-qSBI in this paper, the software for power electronic named PSIM 9.1 was used to simulate the suggested inverter. The major parameters of the simulation are as follow: the output phase voltage is 220 Vrms with five-level; the AC side filter and load are 10 mH and 30 Ω, respectively; the inductor-capacitor and switching frequency for each module are 1 mH, 2200 µF, and 10 kHz. The list of parameters in the simulation is provided in Table 3. It is worth noting that the high valued DC electrolytic capacitors of 2200 µF were used in the simulation and experiment because the double-line-frequency ripple is not suppressed in the CHB-qSBI system. As analyzed in [16,22,26], the single-phase qSBI had a double-line-frequency ripple on the passive element at the DC side. The double-line-frequency ripple caused several problems related to efficiency, lifetime, cost, size, and reliability of the inverter system. To suppress the double-line-frequency ripple on quasi-switched boost network, a feedback controller for qSBI was introduced in [27].

Table 3. Parameters for the three-phase CHB-qSBI inverter.

	Parameter	Value
	Output voltage	220 Vrms
	Output frequency	50 Hz
	Inductors	1 mH
	Capacitors	2200 uF
Load	Inductor (L_f)	10 mH
	Resistor (R)	30 Ω
	Switching frequency	10 KHz

The simulation results for the three-phase CHB-qSBI are noted in Figures 7 and 8. In Figure 7, the input voltages of V_{dc1} and V_{dc2} were set the same value of 50 V. It is clear that both the capacitor voltages in all quasi-switched-boost networks were boosted to 195 V. The DC-link voltages were the square waveform and the peak DC-link voltages were equal to the capacitor voltages, as shown in Figure 7c. The DC-link voltage of two quasi-switched-boost modules was the same with the same ST duty cycle. The root mean square (RMS) value output voltage for each phase was 220 V. The peak output current in each phase was 10.3 A. The total harmonic distortion (THD) of the output phase current was 1.6%. Figure 7e shows the simulation results at the start-up process. The inrush currents have appeared in the inductors at the start-up owing to the existing passive elements in the quasi-switched-boost network. It can be seen in Figure 7a that the proposed three-phase CHB-qSBI operated in a stable state after 0.15 s. Figure 7e shows the simulation results when the load was changed from 30 Ω to 60 Ω at 0.4 s.

Next, the suggested inverter was tested with the unbalanced condition of the input voltage. The input voltage V_{dc1} was 40 V, while V_{dc2} was 50 V. In Figure 8, the simulation results for unbalanced input voltage have been shown. The capacitor voltage and the DC-link voltage are slightly different between the two quasi-switched-boost modules; this is a small insignificant difference in the steady state. Clearly, both of the capacitor voltages in all quasi-switched-boost parts were also boosted to 195 V. The DC-link voltages were the square waveform and the peak DC-link voltages were equal to the capacitor voltages, as shown in Figure 8c. The DC-link voltage of two quasi-switched-boost modules was the same with the same ST duty cycle. The RMS value of the output voltage for each phase was 220 V. The peak output current in each phase was 10.3 A. The THD of the output phase current was 1.7%. The suggested inverter can improve the unbalanced problems in comparison with the conventional CHB inverter.

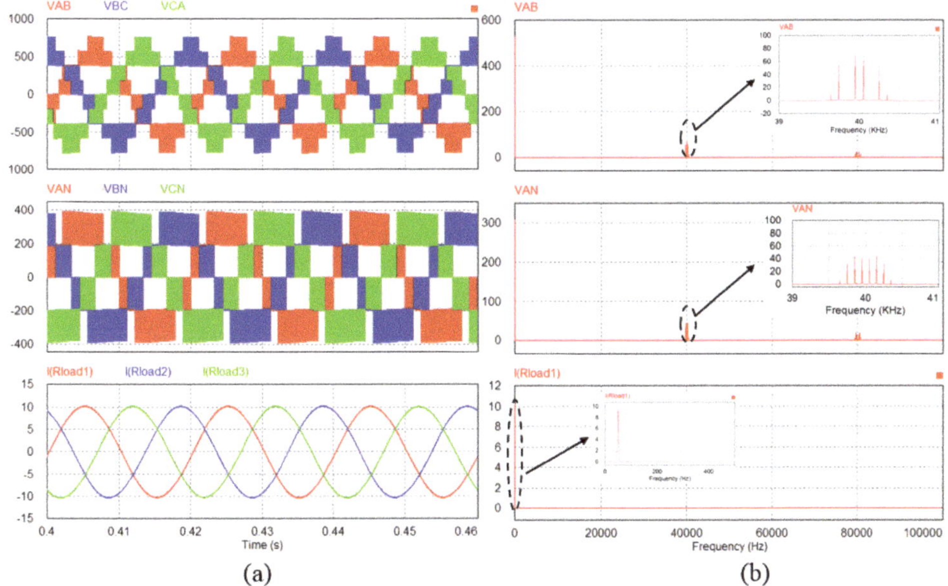

(a)

(b)

Figure 7. *Cont.*

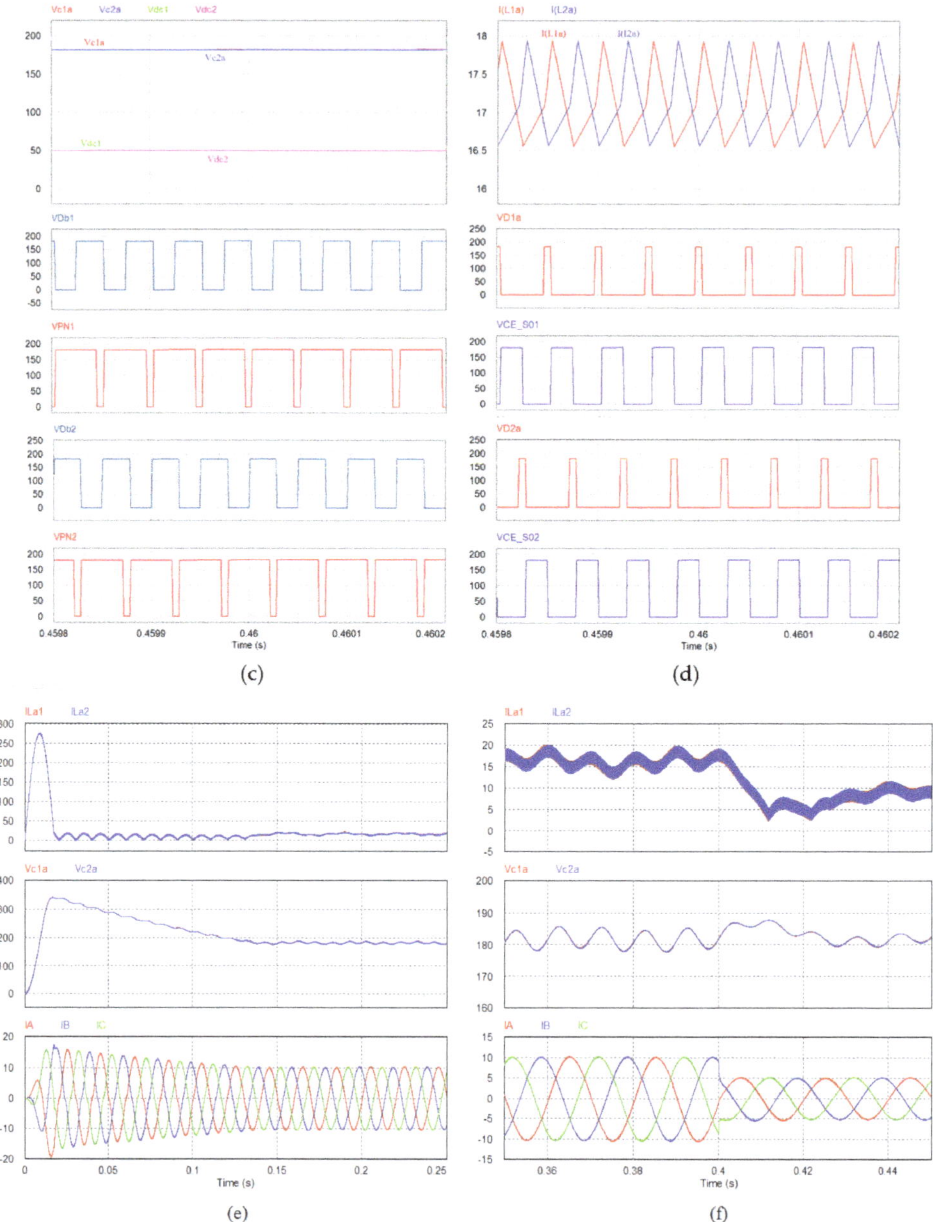

Figure 7. Simulation results for the CHB-qSBI when the input voltage $V_{dc1} = V_{dc2} = 50$ V. From top to bottom: (**a**) output line-to-line voltages, output phase voltages, and phase currents; (**b**) harmonic spectrum of output line-to-line voltages, harmonic spectrum of output phase voltages, and harmonic spectrum of phase currents; (**c**) input voltages, capacitors C_1 and C_2 voltages, diodes D_{b1} and D_{b2} voltages, and DC-link voltages; (**d**) inductor currents, diode D_{a1} voltage, switch S_{01} voltage, diode D_{a2} voltage, and switch S_{02} voltage of phase A; (**e**,**f**) inductor currents, capacitor voltages, and output phase current.

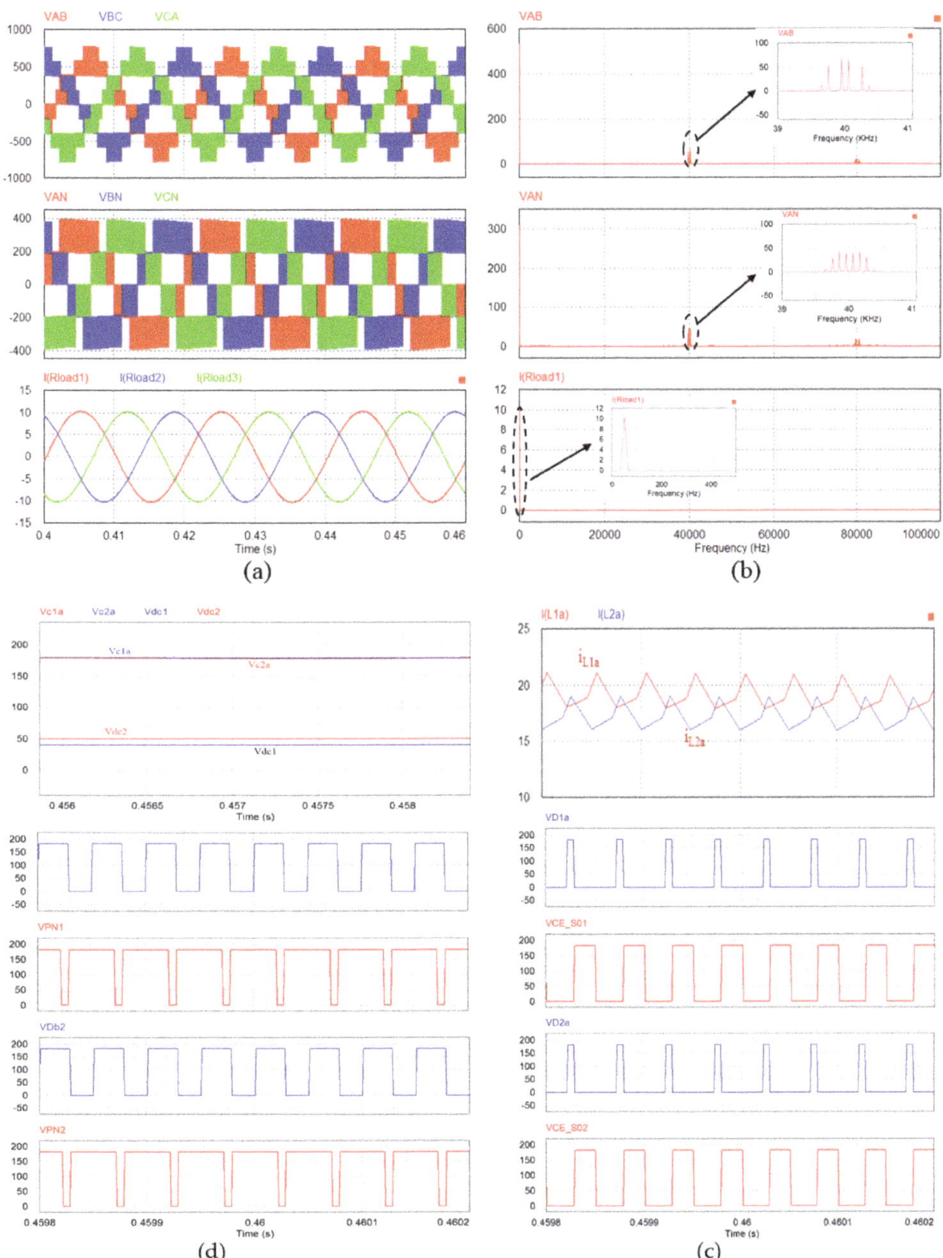

Figure 8. Simulation results for the CHB-qSBI when the input voltage $V_{dc1} = 40$ V and $V_{dc2} = 50$ V. From top to bottom: (**a**) output line-to-line voltages, output phase voltages, and phase currents; (**b**) harmonic spectrum of output line-to-line voltages, harmonic spectrum of output phase voltages, and harmonic spectrum of phase currents; (**c**) input voltages, capacitors C_1 and C_2 voltages, diodes D_{b1} and D_{b2} voltages, and DC-link voltages; and (**d**) inductor currents, diode D_{a1} voltage, switch S_{01} voltage, diode D_{a2} voltage, and switch S_{02} voltage of phase A.

5.2. Experimental Verifications

A scaled-down laboratory prototype of the three-phase CHB-qSBI inverter was built as shown in Figure 9. The microcontroller was DSP TMS320F28335. All switches were G40N120 IGBTs. The diodes were DSEI60-06A. The inductors, capacitors, and load parameters in the experiment were the same as the simulation. The switching frequency was 10 kHz. The RMS value of the output voltage for each phase was 110 V.

Figure 9. The prototype model with three-phase CHB-qSBI.

The experimental results were tested with the unbalanced input voltage condition. The input voltages V_{dc1} and V_{dc2} were 20 V and 25 V, respectively. The capacitor voltage and DC-link voltage were boosted to the same value of 92 V for all modules, shown in Figure 10a. The DC-link voltage of the two modules was kept on the same value by controlling the ST duty cycle even though the input voltage of the two modules was different. The output phase voltage, V_{AN}, had five levels: −180 V, −90 V, 0, 90 V, and 180 V, as shown in Figure 10b. The maximum output voltage had a lower value than the total voltage of DC-link voltages because of the voltage drop in the H-bridge circuit. The measured output phase voltage after the filter was 110.5 Vrms. The measured output current was 3.68 Arms, as shown in Figure 10e. The three-phase output voltage and harmonic spectrum of phase-A voltage are seen in Figure 10c,d. The THD of output voltage after the filter was 2.67%. The experimental results were consistent with the simulation.

The efficiency of the proposed inverter was measured at the unbalanced condition input voltage, V_{dc1} = 20 V and V_{dc2} = 25 V, as shown in Figure 11. The maximum efficiency of the inverter was 86.8% at the output power of 630 W. The efficiency of the inverter was not high because the selection of semiconductor devices in the experiment was not optimal.

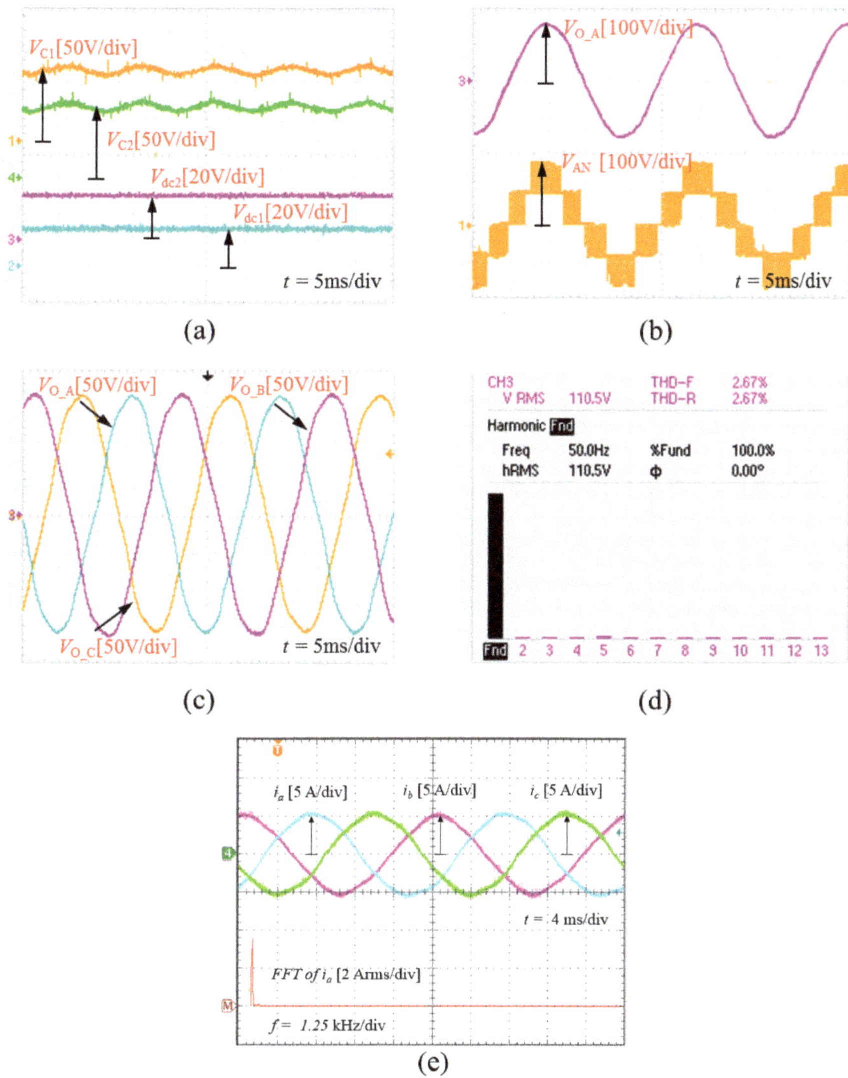

Figure 10. Experimental results for the suggested inverter with the unbalanced condition input voltage, $V_{dc1} = 20$ V and $V_{dc2} = 25$ V. (**a**) Input voltage and capacitor voltage in qSBI stages; (**b**) five-level output voltage before and after the filter of phase A (V_{AN}); (**c**) three-phase output voltage; (**d**) harmonic spectrum of output voltage, and (**e**) output currents and harmonic spectrum of output phase-A current.

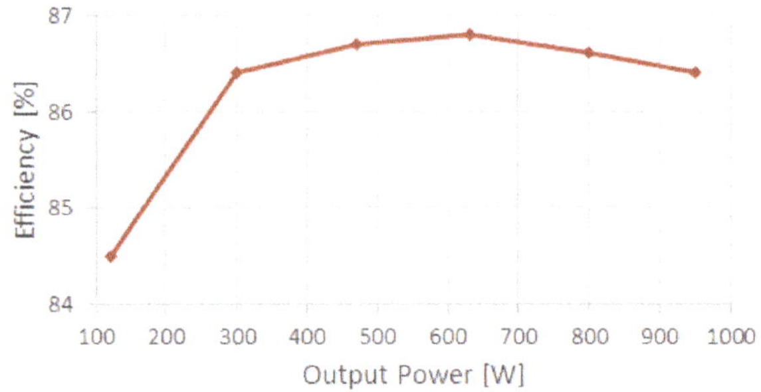

Figure 11. The efficiency of the proposed inverter at the unbalanced condition input voltage, V_{dc1} = 20 V and V_{dc2} = 25 V.

6. Conclusions

The configuration three-phase CHB-qSBI is presented in this paper. The three-phase CHB-qSBI can buck-boost voltage with single-stage power conversion. Furthermore, the three-phase CHB-qSBI immunized the ST phenomenon. In comparison to the three-phase CHB-qZSI, the CHB-qSBI dropped a large number of inductors and capacitors. The DC-link voltage of each module can achieve the same values by controlling the ST duty cycle. The paper describes the circuit analysis, operating theories, and PWM strategy of the introduced topology. Simulation and experimental results prove the validity of the improved PWM strategy for controlling the three-phase CHB-qSBI. The three-phase CHB-qSBI can be applicable.

Author Contributions: All authors contributed equally to this work and all authors have read and approved the final manuscript.

Funding: This research was funded by the Korea Institute of Energy Technology Evaluation and Planning (KETEP), and the Ministry of Trade, Industry and Energy (MOTIE) of the Republic of Korea grant number 20184010201650 and the APC was funded by KETEP and MOTIE.

Acknowledgments: This work was supported by the Korea Institute of Energy Technology Evaluation and Planning (KETEP) and the Ministry of Trade, Industry and Energy (MOTIE) of the Republic of Korea (NO. 20184010201650).

Conflicts of Interest: The authors declare no conflict of interest.

Abbreviations

CHB	Cascaded H-bridge
D_1	Shoot-through duty cycle
DSP	Digital signal processing
EMI	Electromagnetic interference
FC	Flying capacitor
NPC	Neutral point clamped
NST	Non-shoot-through
PWM	Pulse-width modulation
qSBI	quasi-switched boost inverter
ST	Shoot-through
T	Period time
qZSI	Quasi-Z-source inverter

References

1. Malinowski, M.; Gopakumar, K.; Rodriguez, J.; Pérez, M.A. A survey on cascaded multilevel inverters. *IEEE Trans. Ind. Electron.* **2010**, *57*, 2197–2206. [CrossRef]
2. Su, G.J. Multilevel dc-link inverter. *IEEE Trans. Ind. Appl.* **2005**, *41*, 848–854. [CrossRef]

3.   Calais, M.; Borle, L.J.; Agelidis, V.G. Analysis of multicarrier PWM methods for a single-phase five-level inverter. In Proceedings of the 2001 IEEE 32nd Annual Power Electronics Specialists Conference, Vancouver, BC, Canada, 17–21 June 2001; Volume 3, pp. 1173–1178.

4.   Pou, J.; Pindado, R.; Boroyevich, D. Voltage-balance limits in four-level diode-clamped converters with passive front ends. *IEEE Trans. Ind. Electron.* **2005**, *52*, 190–196. [CrossRef]

5.   Hammami, M.; Rizzoli, G.; Mandrioli, R.; Grandi, G. Capacitors voltage switching ripple in three-phase three-level neutral point clamped inverters with self-balancing carrier-based modulation. *Energies* **2018**, *11*, 3244. [CrossRef]

6.   Son, Y.; Kim, J. A novel phase current reconstruction method for a three-level neutral point clamped inverter (NPCI) with a neutral shunt resistor. *Energies* **2018**, *11*, 2616. [CrossRef]

7.   Meynard, T.A.; Foch, H.; Thomas, P.; Courault, J.; Jakob, R.; Nahrstaedt, M. Multilevel converters: Basic concepts and industry applications. *IEEE Trans. Ind. Electron.* **2002**, *49*, 955–964. [CrossRef]

8.   Kang, K.P.; Cho, Y.; Kim, H.S.; Baek, J.W. DC-link capacitor voltage imbalance compensation method based injecting harmonic voltage for cascaded multi-module neutral point clamped inverter. *Electronics* **2019**, *8*, 155. [CrossRef]

9.   Kang, J.W.; Hyun, S.W.; Ha, J.O.; Won, C.Y. Improved neutral-point voltage-shifting strategy for power balancing in cascaded NPC/H-bridge inverter. *Electronics* **2018**, *7*, 167. [CrossRef]

10.   Noman, A.M.; Al-Shamma'a, A.A.; Addoweesh, K.E.; Alabduljabbar, A.A.; Alolah, A.I. cascaded multilevel inverter topology based on cascaded H-bridge multilevel inverter. *Energies* **2018**, *11*, 895. [CrossRef]

11.   Villanueva, E.; Correa, P.; Rodriguez, J.; Pacas, M. Control of a single-phase cascader H-bridge multilevel converter for grid-connected photovoltaic systems. *IEEE Trans. Ind. Electron.* **2009**, *56*, 4399–4406. [CrossRef]

12.   Kouro, S.; Moya, A.; Villanueva, E.; Correa, P.; Wu, B.; Rodriguez, J. Control of a cascaded H-bridge converter for grid-connected photovoltaic systems. In Proceedings of the IEEE 35th Annual Conference of the Industrial Electronics Society, Porto, Portugal, 3–5 November 2009; Volume 9, pp. 1–7.

13.   Viola, F. Experimental evaluation of the performance of a three-phase five-level cascaded H-bridge inverter by means FPGA-based control board for grid connected applications. *Energies* **2018**, *11*, 3298. [CrossRef]

14.   Suresh, Y.; Pand, A.K. Research on a cascaded multilevel inverter by employing three-phase transformers. *IET Power Electron.* **2012**, *5*, 561–570. [CrossRef]

15.   Zhou, Y.; Liu, L.; Li, H. A high-performance photovoltaic module-integrated converter (MIC) based on cascaded quasi-Z-source inverters (qZSI) using eGaN FETs. *IEEE Trans. Power Electron.* **2013**, *28*, 2727–2738. [CrossRef]

16.   Sun, D.; Ge, B.; Yan, X.; Bi, D.; Zhang, L.H.Y.; Abu, H.; Ben, L.; Feng, F.Z. Modeling, impedance-design, and efficiency analysis of quasi-Z source module in cascaded multilevel photovoltaic power system. *IEEE Trans. Ind. Electron.* **2014**, *61*, 6108–6117. [CrossRef]

17.   Sun, D.; Ge, B.; Peng, F.Z.; Haitham, A.R.; Bi, D.; Liu, Y. A new grid-connected PV system based on cascaded H-bridge quasi-Z source inverter. In Proceedings of the IEEE International Symposium on Industrial Electronics (ISIE), Hangzhou, China, 28–31 May 2012; pp. 951–956.

18.   Ge, B.; Liu, Y.; Abu-Rub, H.; Peng, F.Z. State-of-charge balancing control for a battery-energy-stored quasi-Z-source cascaded-multilevel-inverter-based photovoltaic power system. *IEEE Trans. Ind. Electron.* **2018**, *65*, 2268–2279. [CrossRef]

19.   Miceli, R.; Schettino, G.; Viola, F. Performance evaluation of a three-phase five-level quasi-Z-source cascaded h-bridge for grid-connected applications. In Proceedings of the IEEE International Power Electronics and Application Conference and Exposition (PEAC), Shenzhen, China, 4–7 November 2018; pp. 1–6.

20.   Aleenejad, M.; Mahmoudi, H.; Moamaei, P.; Ahmadi, R. A fault-tolerant strategy based on fundamental phase shift compensation for three phase multilevel converters with quasi-Z-Source networks. In Proceedings of the IEEE Power and Energy Conference at Illinois (PECI), Urbana, IL, USA, 19–20 February 2016; pp. 1–6.

21.   Nguyen, M.K.; Le, T.V.; Park, S.J.; Lim, Y.C. A class of quasi-switched boost inverters. *IEEE Trans. Ind. Electron.* **2015**, *62*, 1526–1536. [CrossRef]

22.   Nguyen, M.K.; Lim, Y.C.; Park, S.J. A comparison between single-phase quasi-Z-source and quasi-switched boost inverters. *IEEE Trans. Ind. Electron.* **2015**, *62*, 6336–6344. [CrossRef]

23.   Nguyen, M.K.; Choi, Y.O. PWM control scheme for quasi-switched-boost inverter to improve modulation index. *IEEE Trans. Power Election.* **2018**, *33*, 4037–4044. [CrossRef]

24. Tran, T.T.; Nguyen, M.K. Cascaded five-level quasi-switched-boost inverter for single-phase grid-connected system. *IET Power Electron.* **2017**, *10*, 1896–1903. [CrossRef]
25. Tran, V.T.; Nguyen, M.K.; Yoo, M.H.; Choi, Y.O.; Cho, G.B. A three-phase cascaded H-bridge quasi switched boost inverter for renewable energy. In Proceedings of the IEEE International Conference on Electrical Machines and Systems (ICEMS), Sydney, Australia, 11–14 August 2017; pp. 1–5.
26. Liu, Y.; Abu-Rub, H.; Ge, B.; Peng, F.Z. Impedance design of 21-kW quasi-Z-source H-bridge module for MW-scale medium-voltage cascaded multilevel photovoltaic inverter. In Proceedings of the IEEE 23rd International Symposium on Industrial Electronics (ISIE), Istanbul, Turkey, 1–4 June 2014; pp. 2490–2495.
27. Gautam, A.R.; Rathore, N.; Fulwani, D. Second-order harmonic ripple mitigation: A solution for the micro-inverter applications. In Proceedings of the IEEE Industry Applications Society Annual Meeting (IAS), Portland, OR, USA, 23–27 September 2018; pp. 1–6.

 © 2019 by the authors. Licensee MDPI, Basel, Switzerland. This article is an open access article distributed under the terms and conditions of the Creative Commons Attribution (CC BY) license (http://creativecommons.org/licenses/by/4.0/).

Article

A Novel Three-Switch Z-Source SEPIC Inverter

Baocheng Wang * and Wei Tang

Department of Electrical Engineering, Yanshan University, Qinhuangdao 066004, China;
weitang@stumail.ysu.edu.cn
* Correspondence: bcwang@ysu.edu.cn

Received: 14 January 2019; Accepted: 19 February 2019; Published: 21 February 2019

Abstract: In this paper, a novel single-phase transformerless Z-source inverter (ZSI) derived from the basic SEPIC topology, which is named SEPIC-based ZSI, is proposed. The negative end of the input DC voltage of this topology is directly connected to the load and grounded, which can completely eliminate leakage current. Furthermore, this topology has some attractive characteristics such as buck–boost capability, impressive voltage gain, linear voltage gain is realized by a simple control method, and so on. The theoretical design and simulation results are demonstrated by corresponding experiments carried out on a 500 W laboratory prototype controlled by using a DSP TMS320F28335 controller combined with a FPGA SPARTAN-6.

Keywords: transformerless; SEPIC converter; single phase; Z-source inverter

---

1. Introduction

At present, with the consumption of fossil fuels and environmental pollution, people are increasingly persistent in the development of renewable distributed energy generation systems such as photovoltaic (PV), fuel cell (FC), and so on [1–5]. Among them, photovoltaic power generation, which is used to convert solar energy into electricity, is the most widely applied [6–8]. However, the output electricity is DC, and therefore an inverter is necessary for the photovoltaic power generation system [9].

Based on the availability of transformers, inverters are divided into isolated inverters and non-isolated inverters [10]. Isolated inverters have the advantage of isolating, which can eliminate leakage current and ensure the safety of staff. However, the isolated inverters will have a large volume and loss due to the existing transformers [11], which do not meet the concept of energy saving. As such, more people have turned their attention to non-isolated inverters [12].

Many interesting non-isolated topologies, such as H5 [13], H6 [14], and ZVR [15], have been presented in the literature. Although constantly improving topologies or control methods can reduce their leakage currents to a certain extent, the existing leakage currents in their topologies cannot be eliminated from the source. Furthermore, the voltage gain of these topologies is often unsatisfied.

A great deal of attention has been paid to the Z-source inverter (ZSI) since it was first proposed [16]. The characteristic of the Z-source inverter is that it has very high boost capacity. Therefore, many quasi and semi Z-source inverters have been developed [17–19], but they still cannot deal thoroughly with the problem of leakage current. As a result, some inverter topologies that are combined with Z-source topologies and have the feature of dual-grounding are proposed [20]. Based on the analysis of leakage current [21], dual-grounding can completely solve the problem of leakage current. Although the proposed inverters in [20] have their advantages, they are not suitable for some field applications.; the buck–boost-based type of three-switch three-state (TSTS) ZSIs cannot produce reactive power and the boost-based type of TSTS ZSIs cannot filter well, due to the lack of a filter inductor, which results in higher total harmonic distortion (THD). A new Z-source inverter-based on a CUK converter is proposed in [22], which has a better filter compared to the inverters in [20], but there is a current spark in the S1 switch when the topology starts. The SEPIC-based Z-source inverter is proposed in [23], but its performance is not verified by experiments.

According to the above analysis, this paper puts forward a novel Z-source inverter based on a SEPIC converter. Compared with traditional semi and quasi Z-source inverters, the proposed inverter has the feature of dual-grounding, which is valid for thoroughly eliminating leakage current. Furthermore, the voltage gain of the proposed inverter is more than 1, which can be applied well for a flexible output situation. Furthermore, the proposed inverter is based on a SEPIC converter. The SEPIC converter has the superiority of having the same polarity of input and output, the isolation of input and output, and a complete turn-off. At the same time, the proposed topology has solved the current spark problem in the S1 switch. Therefore, a new topology named "SEPIC-based ZSI" is presented. This topology can be used in situations where flexible control voltage is required. At the same time, the sine output is negative first and then positive, which can meet some application requirements.

This paper is organized as follows: Firstly, the operation modes of the proposed inverter are analyzed and the design of the components is presented in Section 2. Section 3 displays the control diagram of the proposed inverter and some key waveforms under the driven signal. Secondly, the theoretical analysis is proved through corresponding simulations and experiments, which are shown in Section 4. Finally, a complete summary of the proposed topology is given in Section 5.

2. Operation Mode and Analysis of the Novel Inverter

The operation mode, including the mode analysis and the design of the proposed SEPIC-based ZSI, including the parameters calculation, is discussed in this section.

2.1. Structure and Operation Mode

The structure of the proposed SEPIC-based ZSI, which is derived from a combination of a Z-source inverter and a SEPIC converter, is shown in Figure 1. Z-source inverters have an outstanding buck–boost capacity and the SEPIC converter has advantages including the same polarity of input and output, the isolation of input and output, and a complete turn-off. Furthermore, the negative terminal of the PV array is connected to the load side, which is necessary for eliminating leakage current, as shown in Figure 1. Three operation modes are shown in Figure 2.

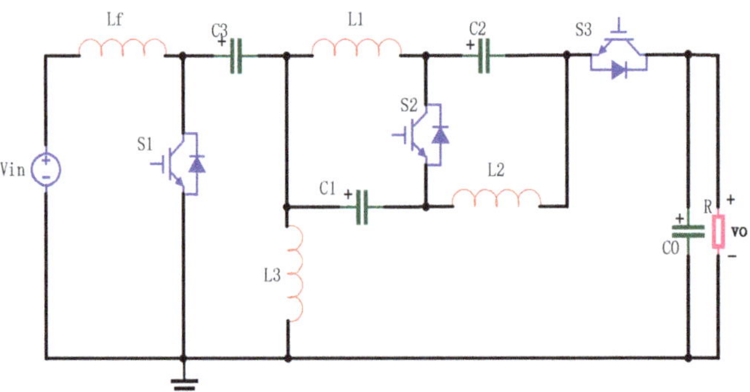

Figure 1. Proposed SEPIC-based Z-source inverter (ZSI).

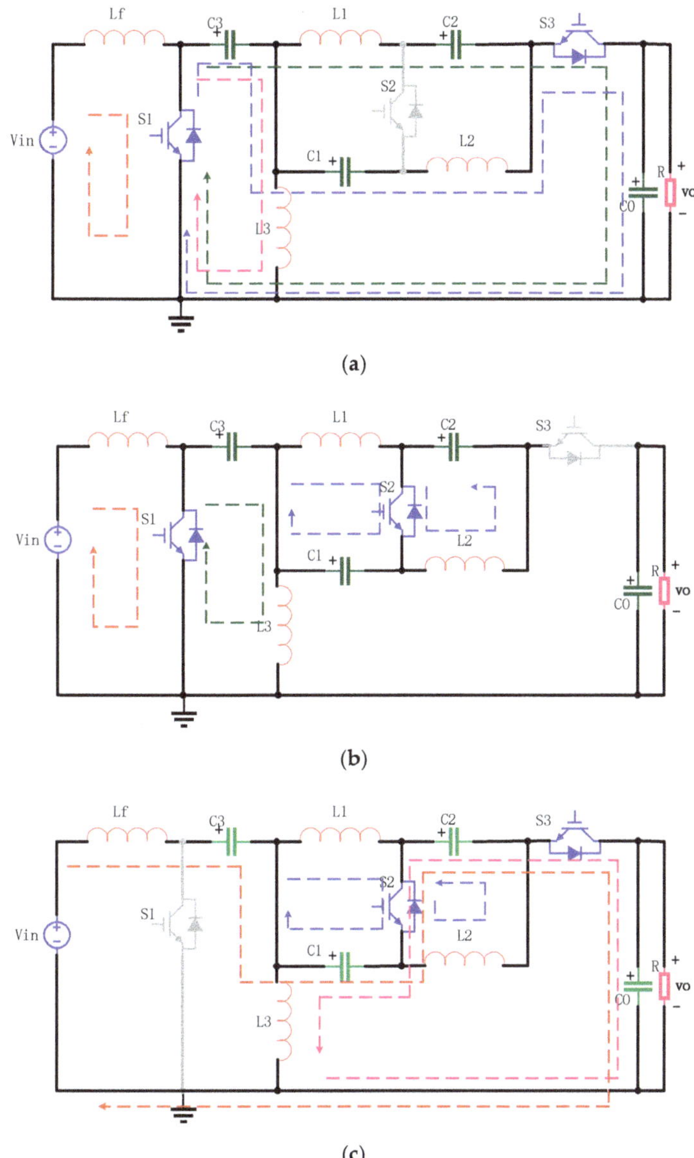

Figure 2. Equivalent circuits of the SEPIC-based ZSI in one switching period (**a**) S1 and S3 are ON, S2 is OFF; (**b**) S1 and S2 are ON, S3 is OFF; and (**c**) S2 and S3 are ON, S1 is OFF.

The operation modes are discussed as follows: In mode I, as shown in Figure 2a, switch S1 and switch S3 are on, whereas switch S2 is off. The inductor, Lf, is magnetized by input voltage V_{in}, and capacitors $C1$, $C2$, and $C3$ are charged. According to Kirchhoff's law of voltage and current, the expression of this mode is shown in Equations (1) and (2). In mode II, switch S3 is off and switches S1 and S2 are on, as shown in Figure 2b. The inductor, Lf, is magnetized by input voltage V_{in}, and capacitors $C1$, $C2$, and $C3$ are discharged. Equations (3) and (4) show the expression of mode II. In mode III, switches S2 and S3 are turned on while switch S1 is turned off, as depicted in Figure 2c. Capacitors $C1$, $C2$, and

C3 are charged. The equations of this mode are expressed as shown in Equations (5) and (6). In Figure 2, the dashed line direction represents the current direction of the inductors.

$$\begin{cases} V_{Lf} = V_{in} \\ V_{L1} = -(V_{C0} + V_{C2} + V_{C3}) \\ V_{L2} = -(V_{C0} + V_{C1} + V_{C3}) \\ V_{L3} = -V_{C3} \end{cases} \tag{1}$$

$$\begin{cases} i_{C1} = i_{L2} \\ i_{C2} = i_{L1} \\ i_{C3} = i_{L1} + i_{L2} + i_{L3} \\ i_{C0} = i_{L1} + i_{L2} - i_o \end{cases} \tag{2}$$

where V_{in} denotes the input voltage; V_{C0}, V_{C1}, V_{C2}, and V_{C3} are the voltage of capacitors C0, C1, C2, and C3; and i_{C1}, i_{C2}, and i_{C3} are the current of capacitors C1, C2, and C3. Similarly, V_{L1}, V_{L2}, V_{L3}, and V_{Lf} are the voltage of inductors L1, L2, L3, and input inductor Lf, and i_{L1}, i_{L2}, and i_{L3} are the current of inductors L1, L2, and L3. i_o is the output current.

$$\begin{cases} V_{Lf} = V_{in} \\ V_{L1} = V_{C1} \\ V_{L2} = V_{C2} \\ V_{L3} = -V_{C3} \end{cases} \tag{3}$$

$$\begin{cases} i_{C1} = -i_{L1} \\ i_{C2} = -i_{L2} \\ i_{C3} = -i_{L3} \\ i_{C0} = -i_o \end{cases} \tag{4}$$

$$\begin{cases} V_{Lf} = V_{in} - (V_{C0} + V_{C1} + V_{C2} + V_{C3}) \\ V_{L1} = V_{C1} \\ V_{L2} = V_{C2} \\ V_{L3} = V_{C0} + V_{C1} + V_{C2} \end{cases} \tag{5}$$

$$\begin{cases} i_{C1} = i_{Lf} - i_{L1} - i_{L3} \\ i_{C2} = i_{Lf} - i_{L2} - i_{L3} \\ i_{C3} = i_{Lf} \\ i_{C0} = i_{Lf} - i_{L3} - i_o \end{cases} \tag{6}$$

where i_{Lf} is the current of inductor Lf.

2.2. Voltage in Capacitors and Current in Inductors

In order to simplify calculation, we suppose that the value of inductor L1 is equal to the value of inductor L2, similarly, we suppose the value of capacitor C1 is equal to capacitor C2. At the same time, all passive components are ideal. Then, based on the volt-second balance principle, the voltage of capacitors and the current of inductors can be expressed easily as follows:

$$\begin{cases} \frac{V_{C1}}{V_{in}} = \frac{V_{C2}}{V_{in}} = \frac{1-D_2}{1-D_1} \\ \frac{V_{C3}}{V_{in}} = 1 \\ \frac{V_o}{V_{in}} = \frac{D_1 + 2D_2 - 2}{1-D_1} \end{cases} \tag{7}$$

$$\begin{cases} i_{L1} = i_{L2} = i_o \\ i_{L3} = -i_o \\ i_{Lf} = \frac{2 - D_1 - 2D_2}{D_1 - 1} i_o \end{cases} \tag{8}$$

where D_1 is the duty cycle of switch S1, D_2 is the duty cycle of switch S2, and I_o is the output current.

2.3. Design of Inductors

According to mode I, the input inductor Lf is magnetized by input voltage, so the current ripple of inductor Lf can be calculated by combining the equations in mode I with the expression of inductor voltage, $V_L = L di_L / dt$. The equation of the current ripple of inductor Lf is shown as follows:

$$\Delta i_{Lf} = \frac{V_{in} D_1 T_s}{Lf}, \tag{9}$$

where Δi_{Lf} is the current ripple of inductor Lf and T_s denotes the switching period. Δi_{Lf} is related to the input voltage V_{in}, duty cycle D_1, switching frequency, and the value of inductor Lf.

Then, according to Equation (9), inductor Lf can be calculated as follows:

$$Lf = \frac{V_{in} D_1 T_s}{\Delta i_{Lf}}. \tag{10}$$

Similarly, inductors $L1$ and $L2$ are related to D_2. Then, combining Equation (7), inductors $L1$ and $L2$ can be calculated as follows:

$$L1 = L2 = \frac{V_{C1} D_2 T_s}{\Delta i_{L1}} = \frac{V_{in}(1 - D_2) D_2 T_s}{\Delta i_{L1}(1 - D_1)}, \tag{11}$$

where Δi_{L1} denotes the current ripple of inductor $L1$.

At the same time, by combining Equation (8), inductor $L3$ can be expressed as follows:

$$L3 = \frac{V_{C3} D_1 T_s}{\Delta i_{L3}} = \frac{V_{in} D_1 T_s}{\Delta i_{L3}}, \tag{12}$$

where Δi_{L3} denotes the current ripple of inductor $L3$.

2.4. Design of Capacitors

According to the above calculations of the inductors, the voltage ripple of the capacitor $C1$ is affected by the current of inductor $L1$, the duty cycle of switch S2, switching frequency, and the value of $C1$. Therefore, the same principle is applied to capacitors and, based on $i_c = C dV_c / dt$, the value of capacitors can be calculated as follows:

$$C1 = C2 = \frac{i_{L1}(1 - D_2) T_s}{\Delta V_{C2}}, \tag{13}$$

where ΔV_{C2} is the voltage ripple of capacitor $C2$.

As for the value of capacitor $C3$, it is associated with the current of inductor $L3$. The equation is shown as follows:

$$C3 = \frac{i_{L3}(1 - D_2) T_s}{\Delta V_{C3}}, \tag{14}$$

where ΔV_{C3} is the voltage ripple of capacitor $C3$.

Finally, the value of output capacitor $C0$ can be calculated as follows:

$$C0 = \frac{2 I_o D_3 T_s}{\Delta V_{C0}}, \tag{15}$$

where ΔV_{C0} is the voltage ripple of capacitor C0.

2.5. Peak Voltage and Current in Switches and Analysis

According to Figure 2 and Equations (1), (2), (7) and (8), the peak voltage and current of switches can be concluded as follows:

$$V_{S-\max} = (1+k)V_{in},\tag{16}$$

$$I_{S-\max} = (A+1)I_o,\tag{17}$$

where $V_{S-\max}$ and $I_{S-\max}$ are the maximum voltage stresses and current stresses, respectively. k and A represent maximum boost ratio and voltage gain, respectively. I_o is the output current, which can be expressed as follows:

$$I_o = I_m \sin \omega t,\tag{18}$$

where I_m is the maximum output current.

3. Control Method

The expression of the duty cycle, key waveforms, and control diagram are displayed in this section. At the same time, the implementation of control in MATLAB is also shown.

3.1. Expression of Duty Cycle

The output voltage of the SEPIC-based TSTS Z-source inverter is defined as follows:

$$v_o = V_o \sin \omega t = AV_{in} \sin \omega t,\tag{19}$$

where A, or the peak voltage gain, is defined as $A = V_o/V_{in}$ and the maximum output voltage is V_o.

Regarding boost, D_1 is set as a constant value and k as the maximum boost ratio, defined as follows [20]:

$$k = \frac{D_1}{1-D_1} \Rightarrow D_1 = \frac{k}{1+k}.\tag{20}$$

Regarding the inversion part, the sinusoidal output voltage, v_o, is generated by D_2 as a varied sinusoidal value. D_3 can be written from D_1 and D_2, and they are defined as follows [20]:

$$D_2 = \frac{k+2}{2(k+1)} - \frac{A}{2(k+1)} \sin \omega t,\tag{21}$$

$$D_3 = 2 - D_1 - D_2.\tag{22}$$

3.2. Key Waveforms in the Switching Cycle and Analysis

According to the above analysis, it is easy to get the key theoretical waveforms of the proposed topology, as shown in Figure 3. The values v_{gs1}, v_{gs2}, and v_{gs3} are the switching statuses of switches S1, S2, and S3, respectively. The values V_{s1}, V_{s2}, and V_{s3} are the voltages of switches S1, S2, and S3. When the switch is turned on, the voltage of the switch will be equal to zero. In contrast, there will be voltage in the switch when it is turned off and the value of voltage is equal to $V_{in}(1+k)$, based on Equation (16). Furthermore, V_{Lf}, V_{L1}, and V_{L3} are the voltages of inductors Lf, L1, and L3, respectively. According to the operation modes in Figure 2, when D_1 is at a high level, Lf is magnetized by V_{in}, so $V_{Lf} = V_{in}$. When D_1 is at a low level, according to the loop analysis, $V_{Lf} = V_{in} - V_{s1}$. As for V_{L1} and V_{L3}, the same analysis method is applied and the value is shown in Figure 3. Similarly, i_{Lf}, i_{L1}, and i_{L3} represent the inductors' currents and they are changed following their voltage.

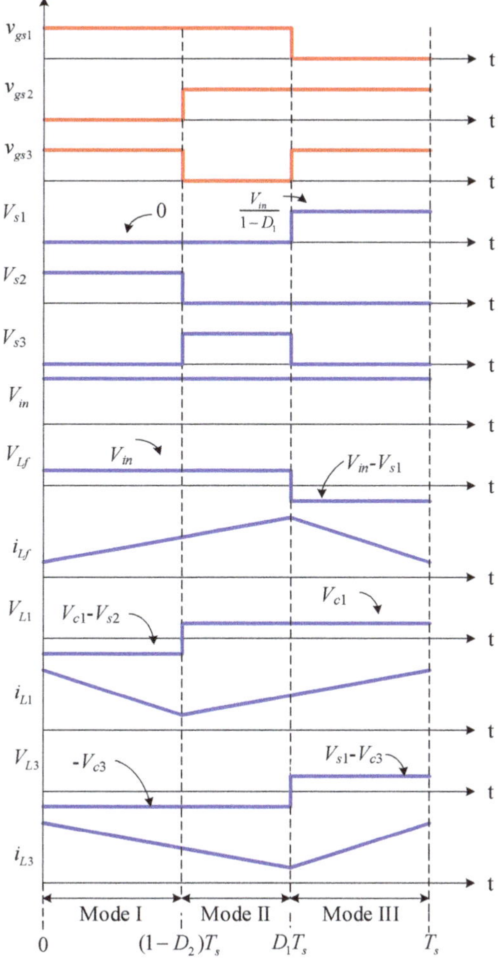

Figure 3. Key waveforms in one switching period.

3.3. Control Diagram

Figure 4 shows the control diagram of the proposed inverter. The result of Equation (20) is a constant, it is then compared with the carrier signal to produce the switch S1 pulse, which is used to boost the input voltage. However, the result of Equation (21) is a sinusoidal variable. The driven signal of switch S2 is generated by comparing the resulting varying duty cycle with the carrier signal, which generates a sinusoidal output. According to the operation modes, only two switches are turned on at the same time. Therefore, the pulse of switch S3 can be determined by the driven signal of switches S1 and S2. The pulse of switch S3 is obtained by the XOR gate. Through the XOR gate, it is guaranteed that only two switches are turned on at a time.

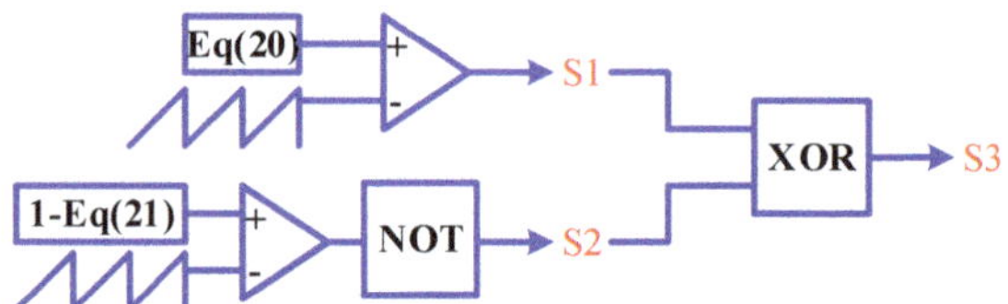

Figure 4. Control block diagram of the proposed inverter.

4. Simulation and Experimental Results

The simulation and experimental results of the proposed inverter are displayed in this section.

4.1. Simulation Results

The proposed inverter, under resistive load, was simulated in MATLAB/Simulink, assuming the current ripple of all inductors was calculated by $\Delta i_L = 20\% I_L$. Similarly, $\Delta V_C = 7\% V_C$ was used to calculate the voltage ripple of all capacitors. Therefore, based on Equations (7)–(15), the value of inductors and capacitors could be calculated as shown in Table 1. Considering the laboratory conditions, in order to experiment conveniently, the switches were IGBT and the switching frequency, f_s, was 20 kHz. The output of the simulation was 124 V, 50 Hz under the condition of 100 V input voltage.

Table 1. Parameters selected for the inverter simulation.

Parameters	V_{in} (V)	V_o (rms, V)	f_s (kHz)	A (Voltage Gain)	k	$L3$ (mH)	Lf (mH)	$L1, L2$ (mH)	$C0$ (μF)	$C1, C2$ (μF)	$C3$ (μF)
value	100	124	20	1.75	2	1.5	0.5	1.65	37.6	15	9

Figure 5 shows the key waveforms of the proposed inverter in MATLAB/Simulink. Figure 6 shows the output voltage, output current, and Fast Fourier transform (FFT) analysis of the proposed inverter. The voltage of the capacitors is shown in Figure 7. Additionally, to better test the proposed inverter the output results of the proposed inverter are displayed in Figure 8, under the condition that the load changed suddenly.

Based on the above results, the simulation results verified the theoretical analysis. According to Figure 5, the key waveforms were consistent with the theoretical waveforms shown in Figure 3 and, at the same time, the values of the voltage satisfied the theoretical calculation. The output voltage satisfied the voltage gain and the voltage of the switches was about 300 V, which is consistent with Equation (16). Similarly, Figure 7 demonstrates Equation (7). Furthermore, the THD = 1.48%, which is well below 5%. According to Figure 8 it still worked normally, although there will be fluctuations and harmonics in the process of change. In summary, the proposed inverter can operate with satisfying performance under complex conditions.

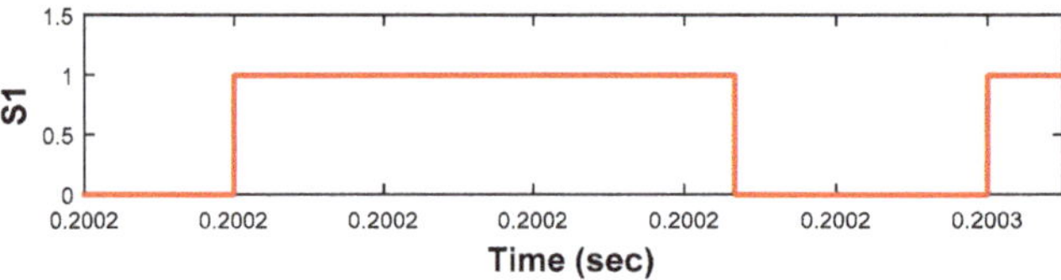

Figure 5. *Cont.*

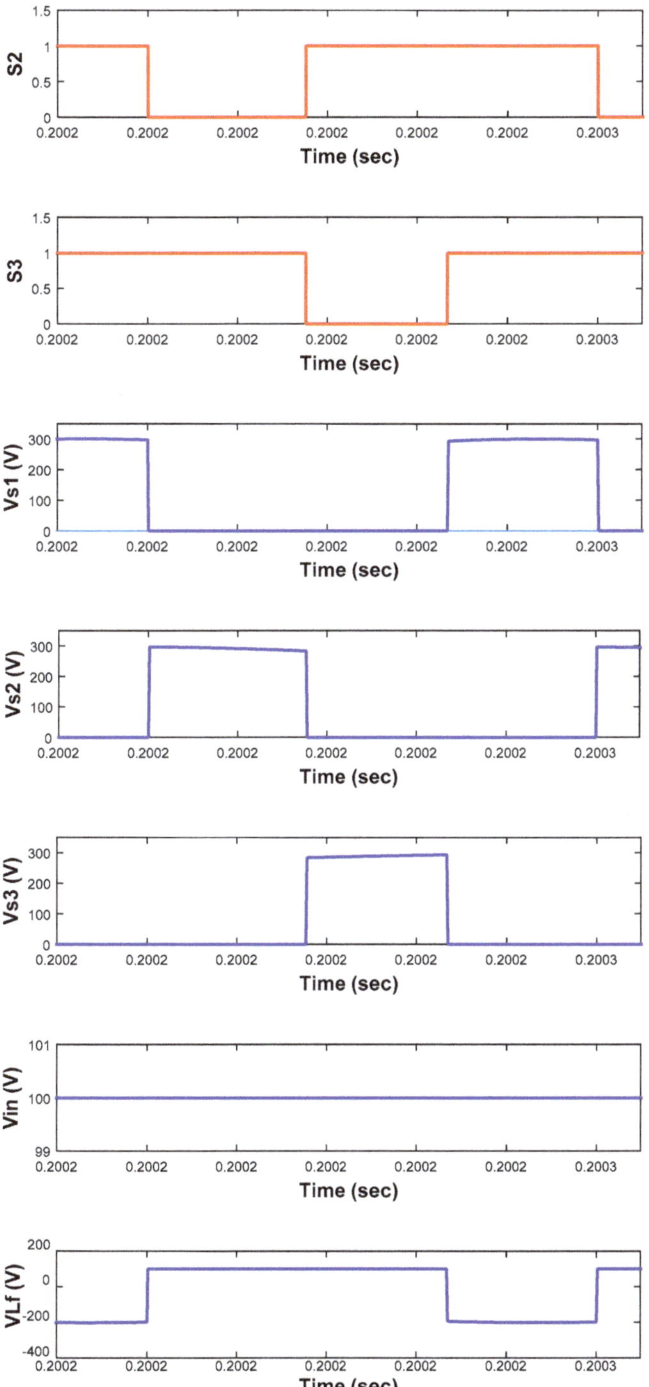

Figure 5. *Cont.*

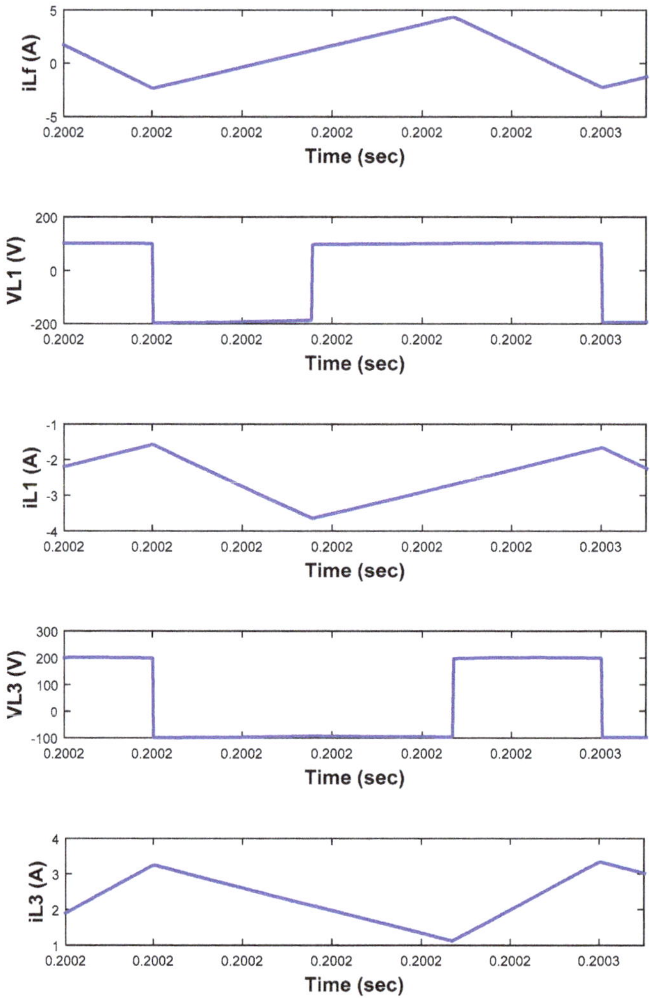

Figure 5. Key waveforms of the proposed inverter in MATLAB.

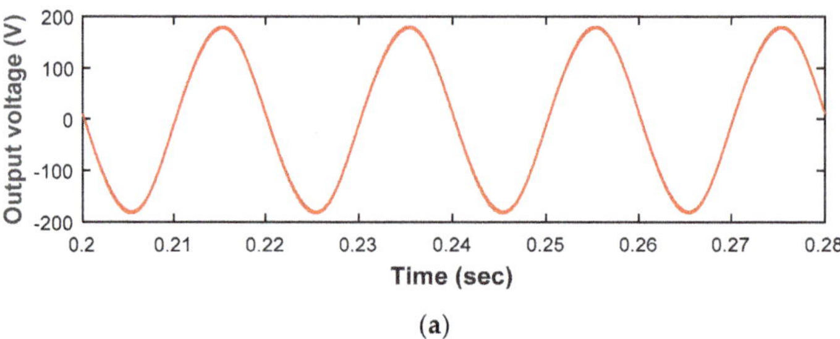

(**a**)

Figure 6. *Cont.*

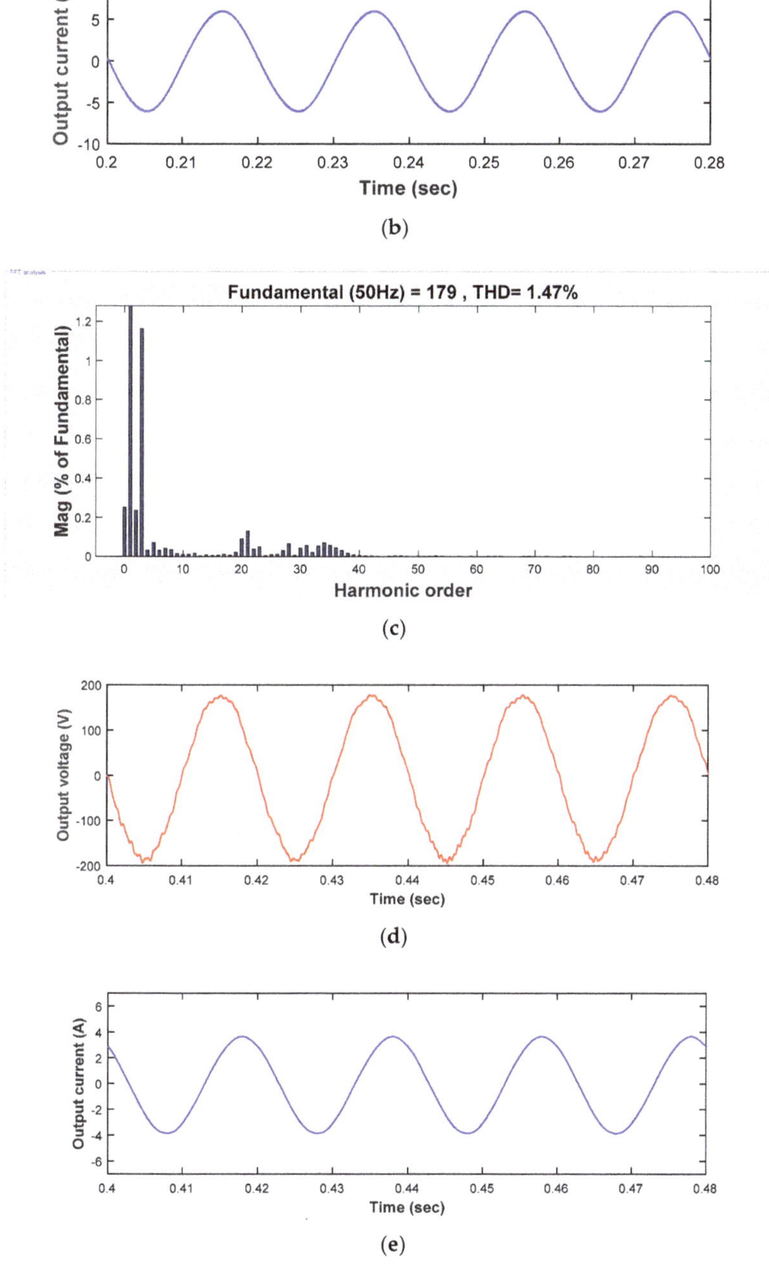

(b)

(c)

(d)

(e)

Figure 6. Input voltage at 100 V and load at 30 Ω for SEPIC-based ZSI: (**a**) load voltage, (**b**) load current, (**c**) Fast Fourier transform (FFT) analysis, (**d**) output voltage with inductive load, and (**e**) output current with inductive load.

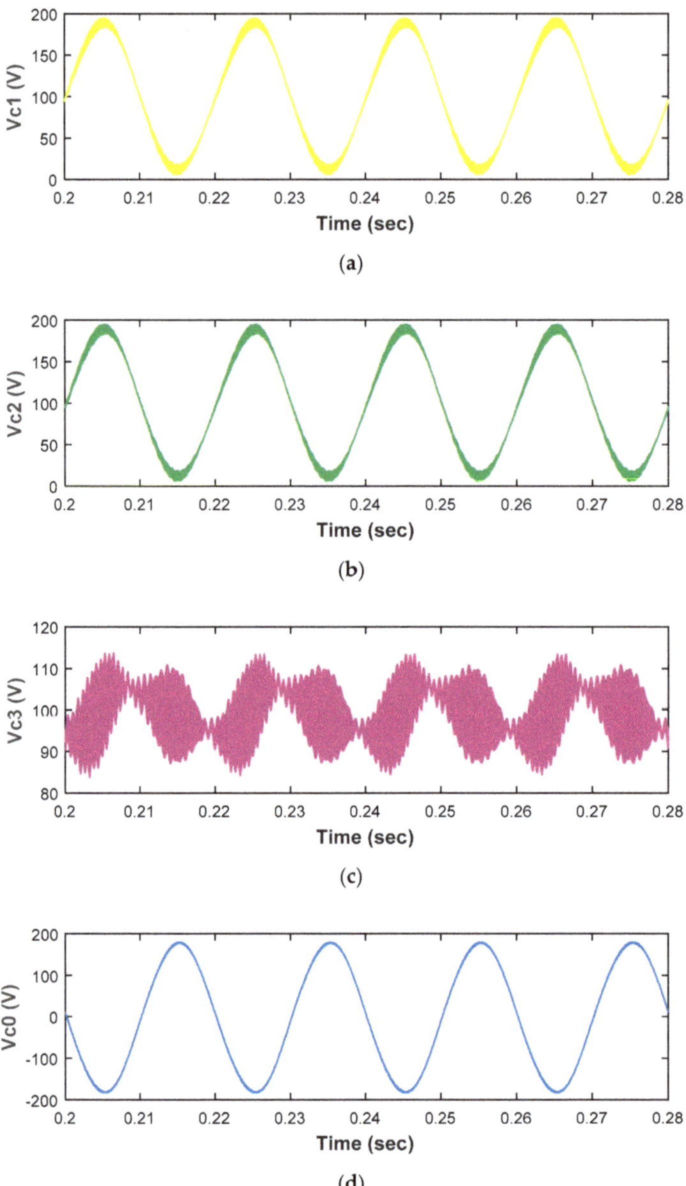

Figure 7. Input voltage at 100 V and load at 30 Ω for SEPIC-based ZSI. Voltage wave of (**a**) capacitor C1, (**b**) capacitor C2, (**c**) capacitor C3, and (**d**) capacitor C0.

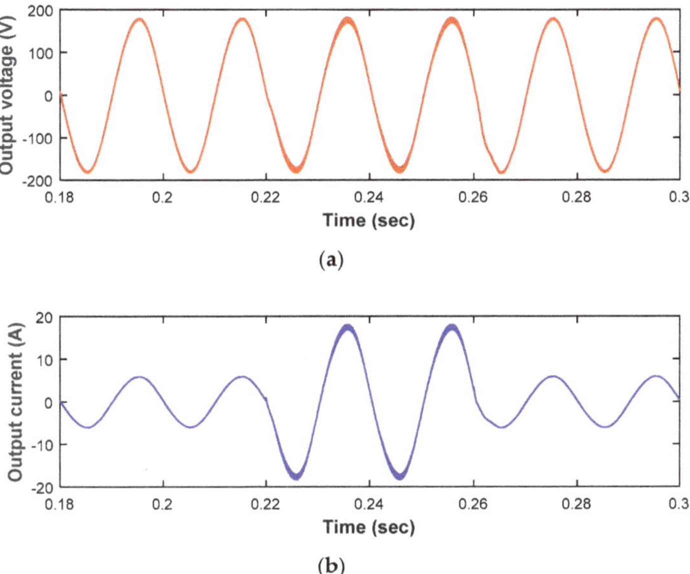

Figure 8. Input voltage at 100 V and load changed (30–10–30 Ω) for SEPIC-based ZSI: (**a**) load voltage and (**b**) load current.

4.2. Experimental Results

The corresponding experiments were also done. The IGBT, named K40T1202 IGBT, was used for each switch in the experiment. The experimental key waveforms are shown in Figure 9. Figure 10 shows the experimental results of output voltage and output current, which correspond with Figure 6. Similarly, Figure 11 shows the experimental results of the voltage of capacitors. The experimental output waveforms of half load changed to full load and full load changed to half load are shown in Figure 12. Furthermore, in order to verify that the proposed inverter can suppress the leakage current well, a capacitor with the value of 0.15 µF was used for C_{PV}. Figure 13 shows the experimental waveform of the leakage current.

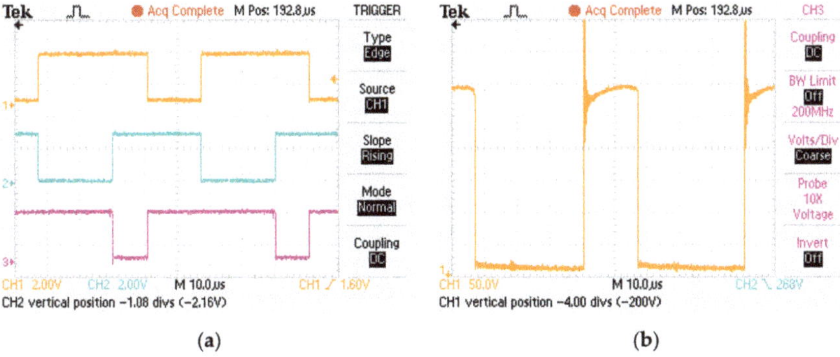

(**a**)   (**b**)

Figure 9. *Cont.*

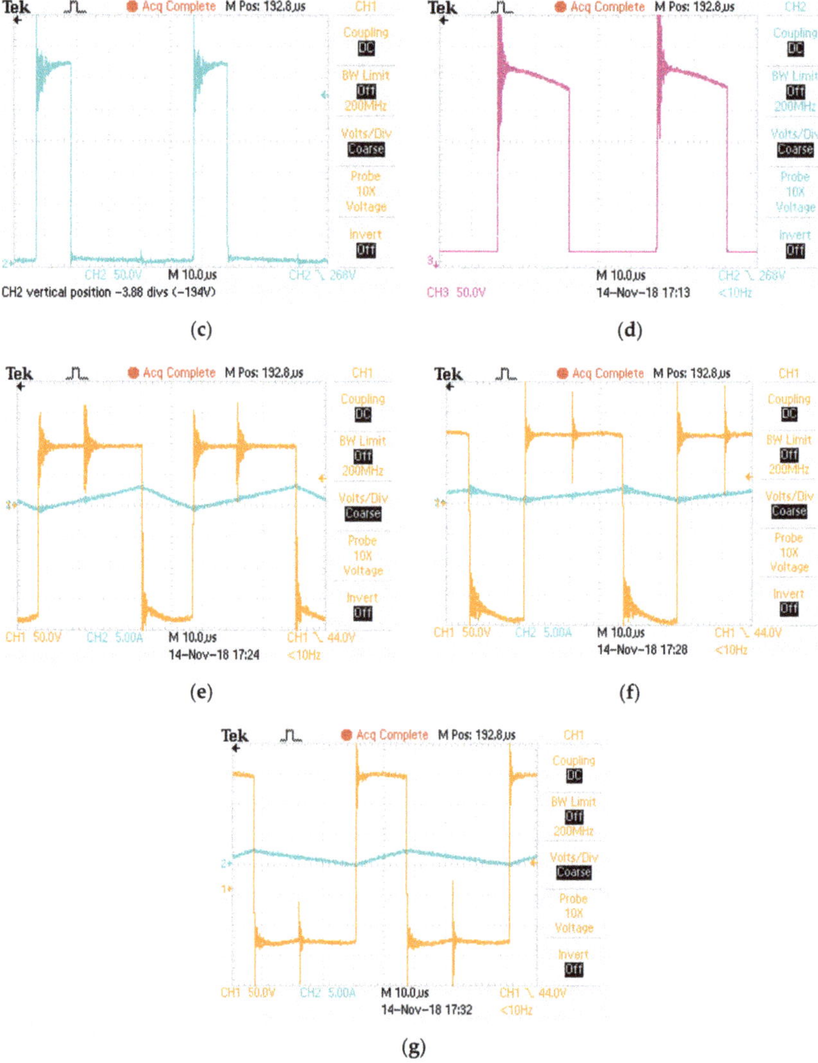

Figure 9. Experimental waveforms: (**a**) driven signal of the switches; (**b**) the voltage of switch S1 (CH1: Time (10 μs/div), V_{S1} (50 V/div)); (**c**) the voltage of switch S2 (CH2: Time (10 μs/div), V_{S2} (50 V/div)); (**d**) the voltage of switch S3 (CH3: Time (10 μs/div), V_{S3} (50 V/div)); (**e**) the voltage and current of inductor Lf (CH1: Time (10 μs/div), V_{Lf} (50 V/div), CH2: Time (10 μs/div), i_{Lf} (5 A/div)); (**f**) the voltage and current of inductor L1 (CH1: Time (10 μs/div), V_{L1} (50 V/div)), CH2: Time (10 μs/div), i_{L1} (5 A/div)); and (**g**) the voltage and current of inductor L3 (CH1: Time (10 μs/div), V_{L3} (50 V/div), CH2: Time (10 μs/div), i_{L3} (5 A/div)).

The experimental waveforms in Figure 9 prove the theoretical key waveforms in Figure 3 and the simulation waveforms in Figure 5. Since the switches in the simulation were ideal and the experimental switches were not ideal, there are voltage spikes in Figure 9. The output voltage and current satisfied the analysis and simulation results. Furthermore, it can be seen from Figure 12 that the proposed inverter still worked well when the load changed suddenly. According to Table II in [24], compared to the traditional single-phase H4 inverter with UPWM and BPWM, the HERIC and H5 topology has lower leakage current. According to the experimental results of the HERIC topology, the max value of leakage current was more than 100 mA under the output power of 160 W. However, the leakage current of the proposed inverter was only about 48 mA under the output power of 500 W which had a good performance for eliminating leakage current and benefited from the dual-grounding.

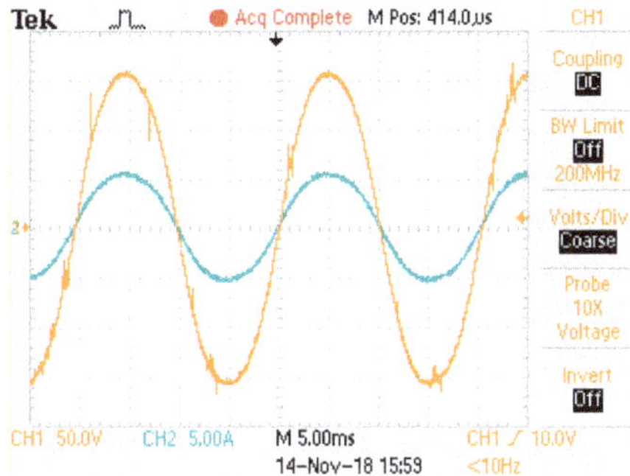

Figure 10. Input voltage at 100 V and load at 30 Ω. Experimental waveforms of v_o (CH1: Time (5 ms/div), v_o (50 V/div)) and I_o (CH2: Time (5 ms/div), I_o (5 A/div)).

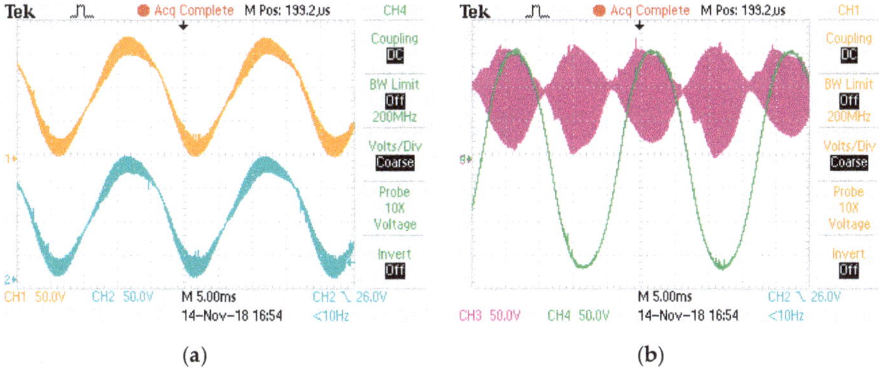

(**a**)　　　　　　　　　　　　(**b**)

Figure 11. Input voltage at 100 V and load at 30 Ω. Experimental waveforms of (**a**) V_{c1} (CH1: Time (5 ms/div), V_{c1} (50 V/div)) and V_{c2} (CH2: Time (5 ms/div), V_{c2} (50 V/div)); (**b**) V_{c3} (CH3: Time (5 ms/div), V_{c3} (50 V/div)) and V_{c0} (CH4: Time (5 ms/div), V_{c0} (50 V/div)).

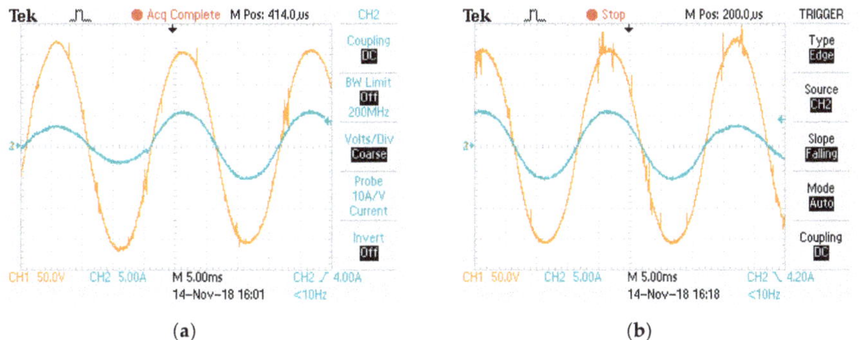

(**a**)　　　　　　　　　　　　(**b**)

Figure 12. Input voltage at 100 V and sudden load changes: (**a**) from 60 to 30 Ω, (**b**) from 30 to 60 Ω. Experimental waveforms of v_o (CH1: Time (5 ms/div), v_o (50 V/div)) and I_o (CH2: Time (5 ms/div), I_o (5 A/div)).

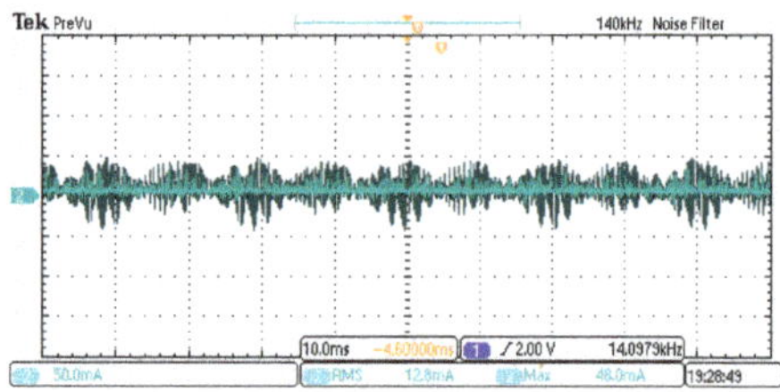

Figure 13. The waveform of leakage current (CH2: I (50 mA/div)).

Table 2 shows the efficiency of the proposed inverter. However, the efficiency of this topology was relatively low due to the use of IGBT as the switches, which are not appropriate for low–medium power converters, and the limitation of the laboratory conditions. A good device can be used to enhance the efficiency of the proposed topology.

Table 2. The measured efficiency of the proposed inverter.

Output power (W)	120	204.5	311	358	431
Efficiency (%)	89.9	90.08	91.17	91.70	89.58

In conclusion, from the above experimental results, it can be observed that the experimental results were in good agreement with the theoretical analysis and the simulation results, again verifying the effectiveness of the proposed inverter.

4.3. Comparison

In conclusion, the existing Z-source-based inverter topologies have some of the following disadvantages: complex control techniques [25–27], high semiconductor device counts [28,29], high numbers of capacitors and inductors [30], ripples in the capacitor voltage and inductor current [31–33], unsatisfactory voltage gain [34], and leakage current [35]. Table 3 shows the characteristics of different Z-source inverters. In addition, a transformer [36,37] is used to boost input voltage and isolate input and output, which will increase the cost, weight, and volume of the inverter. So, the inverter with a common ground can deal well with these problems.

Table 3. Comparison between different Z-source inverters. EMI: electromagnetic interference.

	L	C	D	S	Control Method	Voltage Gain	Common Ground	EMI	Soft Switch
Semi-ZSI [25]	2	2	0	2	nonlinear	<1	Yes	high	complex
Semi-ZS-based [34]	3	3	0	2	linear	<1	Yes	high	complex
Basic SBI [35]	1	1	2	2	linear	>1	No	high	complex
Embedded-type qSBI [35]	1	1	2	2	linear	>1	No	high	complex
DC-linked-type qSBI [35]	1	1	2	2	linear	>1	No	high	complex
CUK-based ZSI [22]	4	4	0	3	linear	>1	Yes	low	complex
Boost-based ZSI [20]	3	3	0	3	linear	>1	Yes	high	complex
Buck–boost-based [20]	3	3	0	3	linear	>1	Yes	high	complex
Proposed in [23]	4	4	0	3	linear	>1	Yes	low	simple

In Table 3, it is easy to see that the CUK converter has low input current ripple and output current ripple, and the SEPIC converter has low input current ripple, so both CUK-based ZSI and SEPIC-based ZSI have low electromagnetic interference (EMI) [38]. Furthermore, SEPIC has the possibility of a series of resonant operations between the balancing capacitor and the parallel inductor, which can be beneficial for a soft-switching operation [39]. Therefore, SEPIC-based ZSI makes it easy to achieve a soft-switching operation.

5. Conclusions

A novel SEPIC-based Z-source inverter that eliminates leakage current is proposed in this paper. The feature of the proposed topology is a combination of a Z-source inverter and a SEPIC converter, so the current inverter has the advantages of both a SEPIC converter and a Z-source inverter, which has a high voltage gain. Furthermore, the proposed inverter is controlled by a simple linear control method, which is easy to achieve, and the proposed inverter has a low voltage stress in the switch, which is beneficial for choosing switches. Furthermore, compared with traditional single-phase H4, H5, and HERIC topologies, the proposed topology has features including fewer switches and very low leakage current, which is helpful for the application of transformerless inverters. However, the proposed topology also has the merits of the SEPIC converter. It can be seen in Figures 6 and 10 that the proposed inverter has large output harmonics, due to the lack of a filter inductor. The simulation results verified the theoretical analysis. Furthermore, experiments with a laboratory prototype, using a DSP controller combined with FPGA, showed good results that were in agreement with the simulation results. The future research direction will be to study new control methods to reduce the output harmonics.

Author Contributions: This paper was a collaborative effort among all of the authors. All authors conceived the methodology, conducted the performance tests, and wrote the paper.

Funding: This work is supported by the National Natural Science Foundation of China under Grant 51777181.

Conflicts of Interest: The authors declare no conflict of interest.

Abbreviations

The following abbreviations are used in this manuscript:

Acronyms	
ZSI	Z-source inverter
VSI	Voltage source inverter
TSTS	Three-switch three-state
FFT	Fast Fourier transform
THD	Total harmonic distortion
SBI	Switched Boost Inverters
QSBI	Quasi-Switched Boost Inverters
UPWM	Unipolar Pulse Width Modulation
BPWM	Bipolar Pulse Width Modulation
HERIC	Highly Efficient Reliable Inverter Concept
IGBT	Insulated Gate Bipolar Transistor
SEPIC	Single Ended Primary Inductor Converter
Nomenclature	
A	Peak voltage gain
k	Maximum boost radio
D_1, D_2, D_3	Duty cycle functions
ω	Output voltage angular frequency
S1, S2, S3	Semiconductor switches
V_{in}	DC input voltage
V_{Li}	Voltage of inductors
V_{Ci}	Voltage of capacitors
V_{si}	Voltage of switches
i_{Li}	Current of inductors
i_{Ci}	Current of capacitors
I_o	Output peak current
Δi_{Li}	Current ripple of inductors
ΔV_{Ci}	Voltage ripple of capacitors
T_s	Switching period
f_s	Switching frequency
EMI	Electromagnetic Interference

References

1. Sulake, N.R.; Devarasetty Venkata, A.K.; Choppavarapu, S.B. FPGA Implementation of a Three-Level Boost Converter-fed Seven-Level DC-Link Cascade H-Bridge inverter for Photovoltaic Applications. *Electronics* **2018**, *7*, 282. [CrossRef]
2. Ji, Y.; Yang, Y.; Zhou, J.; Ding, H.; Guo, X.; Padmanaban, S. Control Strategies of Mitigating Dead-time Effect on Power Converters: An Overview. *Electronics* **2019**, *8*, 196. [CrossRef]
3. Nengroo, S.H.; Kamran, M.A.; Ali, M.U.; Kim, D.-H.; Kim, M.-S.; Hussain, A.; Kim, H.J. Dual Battery Storage System: An Optimized Strategy for the Utilization of Renewable Photovoltaic Energy in the United Kingdom. *Electronics* **2018**, *7*, 177. [CrossRef]
4. Yu, B. An Improved Frequency Measurement Method from the Digital PLL Structure for Single-Phase Grid-Connected PV Applications. *Electronics* **2018**, *7*, 150. [CrossRef]
5. Filippini, M.; Molinas, M.; Oregi, E.O. A Flexible Power Electronics Configuration for Coupling Renewable Energy Sources. *Electronics* **2015**, *4*, 283–302. [CrossRef]
6. Renaudineau, H.; Donatantonio, F.; Fontchastagner, J.; Petrone, G.; Spagnuolo, G.; Martin, J.P.; Pierfederici, S. A PSO-Based Global MPPT Technique for Distributed PV Power Generation. *IEEE Trans. Ind. Electron.* **2015**, *62*, 1047–1058. [CrossRef]
7. Xiao, G. A novel CH5 inverter for single-phase transformerless photovoltaic system applications. *IEEE Trans. Circuits Syst. II Express Briefs* **2017**, *64*, 1197–1201.
8. Sangwongwanich, A.; Yang, Y.; Blaabjerg, F. High-Performance Constant Power Generation in Grid-Connected PV Systems. *IEEE Trans. Power Electron.* **2016**, *31*, 1822–1825. [CrossRef]
9. Kim, K.; Cha, H.; Kim, H.G. A New Single-Phase Switched-Coupled-Inductor DC–AC Inverter for Photovoltaic Systems. *IEEE Trans. Power Electron.* **2017**, *32*, 5016–5022. [CrossRef]
10. Meraj, M.; Iqbal, A.; Brahim, L.; Alammari, R.; Abu-Rub, H. A high efficiency and high reliability single-phase modified quasi Z-Source inverter for non-isolated grid-connected applications. In Proceedings of the IECON 2015—41st Annual Conference of the IEEE Industrial Electronics Society, Yokohama, Japan, 9–12 November 2015.
11. Siwakoti, Y.P.; Blaabjerg, F. A novel flying capacitor transformerless inverter for single-phase grid connected solar photovoltaic system. In Proceedings of the 2016 IEEE 7th International Symposium on Power Electronics for Distributed Generation Systems (PEDG), Vancouver, BC, Canada, 27–30 June 2016.
12. Sree, K.R.; Rathore, A.K.; Breaz, E.; Gao, F. Soft-Switching Non-Isolated Current-Fed Inverter for PV/Fuel Cell Applications. *IEEE Trans. Ind. Appl.* **2016**, *52*, 351–359. [CrossRef]
13. Guo, X.; Jia, X. Hardware-based cascaded topology and modulation strategy with leakage current reduction for transformerless PV systems. *IEEE Trans. Ind. Electron.* **2016**, *63*, 7823–7832. [CrossRef]
14. Zhang, L.; Sun, K.; Xing, Y.; Xing, M. H6 transformerless full-bridge PV grid-tied inverters. *IEEE Trans. Power Electron.* **2014**, *29*, 1229–1238. [CrossRef]
15. Guo, X.; Zhang, X.; Guan, H.; Kerekes, T.; Blaabjerg, F. Three phase ZVR topology and modulation strategy for transformerless PV system. *IEEE Trans. Power Electron.* **2019**, *34*, 1017–1021. [CrossRef]
16. Peng, F.Z. Z-source inverter. In Proceedings of the 2002 IEEE Industry Applications Conference, 37th IAS Annual Meeting, Pittsburgh, PA, USA, 13–18 October 2002; pp. 775–781.
17. Chen, M.T.; Lin, S.H.; Cai, J.B.; Chou, D.Y. Implementing a single-phase quasi-Z-source inverter with the indirect current control algorithm for a reconfigurable PV system. In Proceedings of the IEEE International Conference on Industrial Technology, Taipei, Taiwan, 14–17 March 2016; pp. 323–328.
18. Li, H.B.; Kayiranga, T.; Lin, X.; Shi, Y.; Li, H. A resonance suppression method for GaN-based single-phase quasi-Z-source PV inverter with high switching frequency. In Proceedings of the IEEE Energy Conversion Congress and Exposition, Montreal, QC, Canada, 20–24 September 2015; pp. 2522–2526.
19. Ahmed, T.; Mekhilef, S.; Nakaoka, M. Single phase transformerless semi-Z-source inverter with reduced total harmonic distortion (THD) and DC current injection. In Proceedings of the IEEE ECCE Asia Downunder, Melbourne, Australia, 3–6 June 2013; pp. 1322–1327.
20. Huang, L.; Zhang, M.; Hang, L.; Yao, W.; Lu, Z. A Family of Three-Switch Three-State Single-Phase Z-Source Inverters. *IEEE Trans. Power Electron.* **2013**, *28*, 2317–2329. [CrossRef]
21. Li, W.; Gu, Y.; Luo, H.; Cui, W.; He, X.; Xia, C. Topology Review and Derivation Methodology of Single-Phase Transformerless Photovoltaic Inverters for Leakage Current Suppression. *IEEE Trans. Ind. Electron.* **2015**, *62*, 4537–4551. [CrossRef]

22. Wang, B.; Tang, W. A New CUK-Based Z-Source Inverter. *Electronics* **2018**, *7*, 313. [CrossRef]

23. Wang, B.; Tang, W. A Novel SEPIC-Based Z-Source Inverter. In Proceedings of the IECON 2018—44th Annual Conference of the IEEE Industrial Electronics Society, Washington, DC, USA, 21–23 October 2018; pp. 4347–4352.

24. Kafle, Y.R.; Town, G.E.; Guochun, X.; Gautam, S. Performance comparison of single-phase transformerless PV inverter systems. In Proceedings of the 2017 IEEE Applied Power Electronics Conference and Exposition (APEC), Tampa, FL, USA, 26–30 March 2017.

25. Cao, D.; Jiang, S.; Yu, X.; Peng, F.Z. Low-Cost Semi-Z-source Inverter for Single-Phase Photovoltaic Systems. *IEEE Trans. Power Electron.* **2011**, *26*, 3514–3523. [CrossRef]

26. Nguyen, M.K.; Tran, T.T. A single-phase single-stage switched-boost inverter with four switches. *IEEE Trans. Power Electron.* **2017**, *33*, 6769–6781. [CrossRef]

27. Ribeiro, H.; Borges, B.; Pinto, A. Single-stage DC–AC converter for photovoltaic systems. In Proceedings of the IEEE Energy Conversion Congress and Exposition, Atlanta, GA, USA, 12–16 September 2010; pp. 604–610.

28. Nguyen, M.K.; Le, T.V.; Park, S.J.; Lim, Y.C.; Yoo, J.Y. Class of high boost inverters based on switched-inductor structure. *IET Power Electron.* **2015**, *8*, 750–759. [CrossRef]

29. Oleksandr, H.; Carlos, R.C.; Enrique, R.C.; Dmitri, V.; Serhii, S. Single phase three-level neutral-point-clamped quasi-Z-source inverter. *IET Power Electron.* **2015**, *8*, 1–10.

30. Caceres, R.O.; Barbi, I. A boost dc–ac converter: Analysis, design, and experimentation. *IEEE Trans. Power Electron.* **1999**, *14*, 134–141. [CrossRef]

31. Zhou, Y.; Liu, L.; Li, H. A high performance photovoltaic module-integrated converter (MIC) based on cascaded quasi-Z-source inverters (qZSI) using eGaN FETs. *IEEE Trans. Power Electron.* **2013**, *28*, 2727–2738. [CrossRef]

32. Sun, D.; Ge, B.; Yan, X.; Bi, D.; Zhang, H.; Liu, Y.; Abu-Rub, H.; Ben-Brahim, L.; Peng, F.Z. Modeling, impedance design, and efficiency analysis of quasi-Z source module in cascade multilevel photovoltaic power system. *IEEE Trans. Ind. Electron.* **2014**, *61*, 6108–6117. [CrossRef]

33. Tang, Y.; Xie, S.; Ding, J. Pulsewidth modulation of Z-source inverters with minimum inductor current ripple. *IEEE Trans. Ind. Electron.* **2014**, *61*, 98–106. [CrossRef]

34. Vazquez, N.; Rosas, M.; Hernandez, C.; Vazquez, E.; Perez-Pinal, F.J. A New Common-Mode Transformerless Photovoltaic Inverter. *IEEE Trans. Ind. Electron.* **2015**, *62*, 6381–6391. [CrossRef]

35. Nguyen, M.-K.; Le, T.-V.; Park, S.-J.; Lim, Y.-C. A Class of Quasi-Switched Boost Inverters. *IEEE Trans. Ind. Electron.* **2015**, *62*, 1526–1536. [CrossRef]

36. Nguyen, M.K.; Lim, Y.C.; Park, S.J. Improved trans-Z-source inverter with continuous input current and boost inversion capability. *IEEE Trans. Power Electron.* **2013**, *28*, 4500–4510. [CrossRef]

37. Nguyen, M.K.; Lim, Y.C.; Park, S.J.; Shin, D.S. Family of high-boost Z-source inverters with combined switched-inductor and transformer cells. *IET Power Electron.* **2013**, *6*, 1175–1187. [CrossRef]

38. Veerachary, M. Power tracking for nonlinear PV sources with coupled inductor SEPIC converter. *IEEE Trans. Aerosp. Electron. Syst.* **2005**, *41*, 1019–1029. [CrossRef]

39. Hasanpour, S.; Baghramian, A.; Mojallali, H. A Modified SEPIC-Based High Step-Up DC-DC Converter with Quasi-Resonant Operation for Renewable Energy Applications. *IEEE Trans. Ind. Electron.* **2019**, *66*, 3539–3549. [CrossRef]

© 2019 by the authors. Licensee MDPI, Basel, Switzerland. This article is an open access article distributed under the terms and conditions of the Creative Commons Attribution (CC BY) license (http://creativecommons.org/licenses/by/4.0/).

Article

Voltage Multiplier Cell-Based Quasi-Switched Boost Inverter with Low Input Current Ripple

Minh-Khai Nguyen [1,2] and Youn-Ok Choi [2,*]

[1]   Faculty of Electrical and Electronics Engineering, Ho Chi Minh City University of Technology and Education,
       1 Vo Van Ngan Stress, Thu Duc District, Ho Chi Minh City 700000, Vietnam; nmkhai00@gmail.com
[2]   Department. of Electrical Engineering, Chosun University, 309 Pilmun-daero, Dong-gu,
       Gwangju 61542, Korea
*   Correspondence: yochoi@chosun.ac.kr; Tel.: +82-62-230-7175

Received: 21 January 2019; Accepted: 15 February 2019; Published: 18 February 2019

Abstract: A novel single-phase single-stage voltage multiplier cell-based quasi-switched boost inverter (VMC-qSBI) is proposed in this paper. By adding the voltage multiplier cell to the qSBI, the proposed VMC-qSBI has the following merits; a decreased voltage stress on an additional switch, a high voltage gain, a continuous input current, shoot through immunity, and a high modulation index. A new pulse-width modulation (PWM) control strategy is presented for the proposed inverter to reduce the input current ripple. To improve the voltage gain of the proposed inverter, an extension is addressed by adding the VMCs. The operating principle, steady-state analysis, and impedance parameter design guideline of the proposed inverter are presented. A comparison between the proposed inverter and other impedance source-based high-voltage gain inverters is shown. Simulation and experimental results are provided to confirm the theoretical analysis.

Keywords: input current ripple; voltage multiplier; shoot through state; quasi-switched boost inverter; Z-source inverter

---

1. Introduction

In recent years, many dc–ac power conversion topologies have attracted the interest of researchers for various applications, such as ac motor drives, uninterruptible power supplies, hybrid electric vehicles, and renewable energy systems [1]. Voltage source inverters (VSIs) and current source inverters (CSIs) are two basic dc–ac power conversion devices. In [2], the design of a single-phase photovoltaic VSI model and the simulation of its performance were presented. Two problems associated with VSIs are that the output voltage cannot exceed the input dc voltage, and both power switches in the same leg cannot be simultaneously turned on. Similarly, the output voltage of CSIs cannot be lower than the source voltage, and both power switches in the same leg cannot be turned off at the same time.

To solve these problems associated with VSIs and CSIs, impedance source inverters [3–10] have been proposed. The classical Z-source inverter (ZSI) uses two inductors and two capacitors in the impedance source network to step-up/down the input voltage. In order to improve the discontinuous input current disadvantage of the ZSI, a class of quasi-ZSIs (qZSIs) was proposed in [3]. Nevertheless, with ZS/qZSIs, it is very difficult to achieve a high voltage gain owing to the limitation of the modulation index and the shoot through (ST) duty cycle. The voltage gain in the ZS/qZSIs can be improved by adding an inductor, capacitor, and diode to the impedance source network, as presented in the continuous/ripple input current switched-inductor (SL) qZSIs [5,6], the enhanced-boost qZSI [7], and the modified switched-capacitor ZSI [8]. However, passive element-based qZSIs [4–8] increase the volume and loss of the power inverter because of the use of a large number of passive components. Coupled-inductor-based qZSIs [9] can reduce the size of the inverter, but the coupled inductor must be well designed to avoid voltage spikes on the dc-bus. A current ripple damping control scheme for single-phase quasi-Z-source system was introduced in [10] to minimize passive elements in the impedance source network.

To decrease the volume and loss of the power inverter, active impedance source inverters have been recently proposed in [11–20]. The switched boost inverter (SBI) [11] uses fewer passive elements and more semiconductors to produce a smaller voltage gain than ZSIs. A family of quasi-switched boost inverters (qSBIs) was presented in [12] to overcome the disadvantages of the SBI. Compared with the qZSI, the qSBI in [12] produces the same voltage gain. In [13], the qSBI was compared with the qZSI in terms of size, loss, and voltage/current ripple on passive elements. Similar to the ZSIs, passive elements, including the capacitor, inductor, coupled inductor, and diode, were also added to the qSBIs to increase the voltage gain. For instance, the topologies in [14–16] were introduced by applying the SL structure to the switched-boost network. In [17], a high-voltage gain switched-ZSI (SZSI) was introduced by using one more inductor, one more capacitor, and two more diodes when compared with the qSBI. To reduce the voltage stress on the capacitor, diode, and switch of the impedance-source network, a high voltage gain qSBI was proposed in [18]. An active switched-capacitor qZSI (ASC-qZSI) was proposed in [19], where only one inductor and one capacitor were added to the qSBI to obtain the same voltage gain as the SZSI. Coupled-inductor-based qSBIs were also presented in [20], with a spike generated on the dc-bus voltage owing to the effect of the leakage inductance in the coupled inductor. To enhance the performance of the qSBIs, a family of PWM control strategies for a single-phase qSBIs has been introduced in [21]. Because the additional states are inserted in the extra switch, the switching loss of the qSBIs under these PWM control methods is increased. A control-based approach to suppress the low-frequency harmonic component in the input inductor current of the single-phase qSBI has been presented in [22]. Even though the active impedance source inverters in [11–22] have good performance, where there is a reduced volume and loss with a high voltage gain, the voltage stress on the additional switch is very high because it equals the dc-bus voltage. To decrease the voltage stress on devices, switched-capacitor qSBIs were investigated in [23].

A voltage multiplier cell (VMC) is used in the dc–dc power conversion process [24–26] to provide a high voltage gain and reduce the voltage stress on semiconductor devices. In [27], the pulsating dc voltage is rectified by the voltage multiplier to boost the output voltage without using magnetic elements. In this study, the VMC is applied to the qSBI to achieve a high dc–ac voltage gain with a high modulation index. Compared with other active impedance source inverters, the voltage stress on the extra switch is reduced significantly. A new PWM control method is presented for the proposed inverter to reduce the input current ripple. By adding the VMC, the proposed inverter can extend to n-cells in order to achieve a high voltage gain requirement. Because of the advantages of the proposed solution, the proposed inverter can replace the qZSI for the photovoltaic generator applications where a low dc voltage needs to invert into a high grid-connected ac voltage. Section 2 proposes the inverter with the operating principle, PWM control technique, and steady-state analysis. The design of the proposed inverter is presented in Section 3. A comparison with other high voltage gain qZS/qSBIs is made in Section 4, while simulation and experimental results are shown in Section 5.

2. Proposed Inverter Topologies

The single-phase qSBI with a continuous input current [12] is presented in Figure 1a. It includes one boost inductor (L_B), one capacitor (C_0), six diodes (D_a, D_0, and four body diodes in the H-bridge circuit), five active switches (S_1–S_5), a passive filter (L_f and C_f), and a resistive load (R). The qSBI uses the ST mode to step-up the input voltage. The following formulas of the qSBI are obtained as given in [12].

$$\begin{cases} V_{PN} = V_{S5} = \dfrac{1}{1-2D_{ST}}V_{dc} = B \cdot V_{dc} \\ \hat{v}_o = M \cdot B \cdot V_{dc} = \dfrac{M}{1-2D_{ST}}V_{dc}, \end{cases} \tag{1}$$

where V_{PN}, V_{S5}, $\hat{v}_o$, B, M, and D_{ST} are the dc-bus voltage, the voltage stress on the additional switch S_5, the amplitude of the output voltage, the boost factor, the modulation index, and the ST duty cycle, respectively.

From Equation (1), it can be seen that the voltage gain of the qSBI is low because $D_{ST} \leq (1-M)$. Moreover, the voltage stress on the additional switch S_5 is high because it equals the dc-link voltage. Therefore, the qSBI is not suitable for high voltage gain applications.

Figure 1b shows the proposed single-phase VMC-qSBI with a single VMC, where two capacitors (C_{11} and C_{12}) and two diodes (D_{11} and D_{12}) are added to the switched-boost network. The proposed single-VMC-qSBI uses one boost inductor (L_B), three capacitors (C_0, C_{11}, and C_{12}), eight diodes (D_a, D_0, D_{11}, D_{12}, and four body diodes in the H-bridge circuit), five active switches (S_1–S_5), a passive filter (L_f and C_f), and a resistive load (R). A combination of C_{11}–C_{12}–D_{11}–D_{12}–S_5 plays a role as a VMC. It should be noted that the proposed single-VMC-qSBI will become the 2-cell SC-qSBI in [23] if the negative node of C_0 is connected to the positive node of C_{11} in Figure 1b. Like other VMC-based converters, the proposed inverter has high current transients of the capacitor across S_5, D_{11}, D_{12}, and D_0. To reduce the current transients of capacitors and achieve soft-switching operation, a resonant inductor can be used, as reported in [24] and [25]. Figure 1c shows an n-cell extension of the proposed VMC-qSBI. Each VMC, including two capacitors and two diodes, is connected in cascade to obtain an n-cell topology. When $n > 1$, a multiple of two capacitors and two diodes is added to the proposed inverter. Consequently, the size, weight, and loss of the proposed inverter are increased. Note that, a three-phase H-bridge circuit can be used in the proposed converter for the three-phase inverter system.

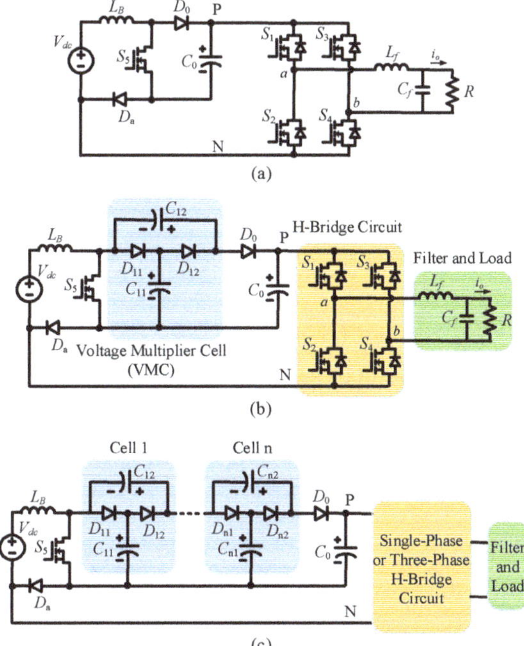

Figure 1. Conventional and proposed inverter topologies. (**a**) Conventional quasi-switched boost inverter (qSBI), (**b**) proposed single voltage multiplier cell (VMC-qSBI), and (**c**) proposed n-VMC-qSBI extension topologies.

2.1. Operating States

As an example, the operating principle and steady-state analysis of the proposed single-phase single-VMC-qSBI are presented in this paper. Figure 2 shows the three operating states of the proposed single-phase single-VMC-qSBI. In the non-ST states, as shown in Figure 2a,b, the H-bridge circuit and the load side are equivalent to a current source. When the switch S_5 is turned on, diodes D_{11} and

D_0 are reversed-biased, while the diodes D_a and D_{12} are forward-biased. The inductor L_B and the capacitor C_{12} are charged, while the capacitors C_{11} and C_0 are discharged. The inverter operates in the non-ST state 1, as shown in Figure 2a. The time interval in the non-ST state 1 is $D_5 \cdot T$, where D_5 is the duty cycle of the switch S_5 during one switching period, T. The following equations can be written as

$$\begin{cases} L_B \frac{di_{LB}}{dt} = V_{dc} \\ V_{C11} = V_{C12} \\ V_{PN} = V_{C0} \end{cases} \text{ and } \begin{cases} C_{11} \frac{dv_{C11}}{dt} = -i_{C12\_Non1} \\ C_{12} \frac{dv_{C12}}{dt} = i_{C12\_Non1} \\ C_0 \frac{dv_{C0}}{dt} = -I_{PN}, \end{cases} \tag{2}$$

where I_{PN} and i_{C12_Non1} are the average dc-link current and the instantaneous current through capacitor C_{12} in the non-ST state 1.

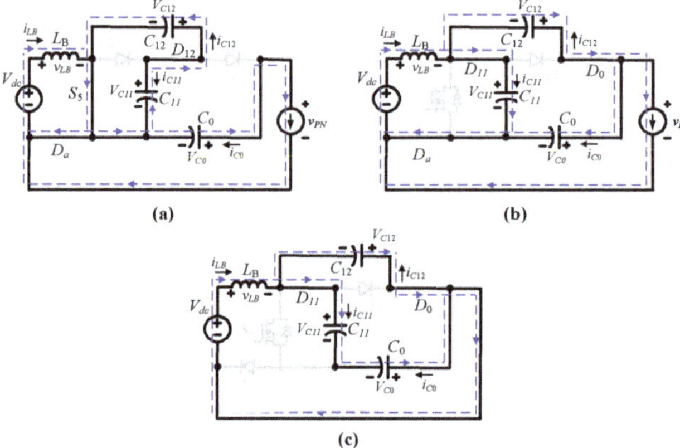

Figure 2. Operating states of proposed single-VMC-qSBI. (**a**) Non-ST state 1, (**b**) non-ST state 2, and (**c**) ST state.

When the switch S_5 is turned off and the H-bridge circuit generates the active or zero vectors and the inverter operates in the non-ST state 2, as shown in Figure 2b. During this state, diode D_{12} is reversed-biased, while the diodes D_a, D_{11}, and D_0 are forward-biased. Inductor L_B and capacitor C_{12} are discharged, while capacitors C_{11} and C_0 are charged. The following formulas can be obtained as

$$\begin{cases} L_B \frac{di_{LB}}{dt} = V_{dc} - V_{C11} \\ V_{C0} = V_{PN} = V_{C11} + V_{C12} \end{cases} \text{ and } \begin{cases} C_{11} \frac{dv_{C11}}{dt} = I_{LB} + i_{C12\_Non2} \\ C_{12} \frac{dv_{C12}}{dt} = i_{C12\_Non2} \\ C_0 \frac{dv_{C0}}{dt} = -I_{PN} - i_{C12\_Non2}, \end{cases} \tag{3}$$

where i_{C12_Non2} is the instantaneous current through capacitor C_{12} in the non-ST state 2.

In the ST state, as shown in Figure 2c, the H-bridge circuit is shorted by the switches on the same H-bridge leg. Meanwhile, switch S_5 is turned off during the ST state. Diodes D_{11} and D_0 are forward-biased, while diodes D_a and D_{12} are reversed-biased. The time interval is $D_{ST} \cdot T$. The inductor L_B and capacitor C_{11} are charged, while the capacitors C_{12} and C_0 are discharged. The following equations can be obtained as

$$\begin{cases} L_B \frac{di_{LB}}{dt} = V_{dc} + V_{C12} \\ V_{C0} = V_{C11} + V_{C12} \\ V_{PN} = 0 \end{cases} \text{ and } \begin{cases} C_{11} \frac{dv_{C11}}{dt} = I_{LB} + i_{C12\_ST} \\ C_{12} \frac{dv_{C12}}{dt} = i_{C12\_ST} \\ C_0 \frac{dv_{C0}}{dt} = -I_{LB} - i_{C12\_ST}, \end{cases} \tag{4}$$

where i_{C12_ST} is the instantaneous current through capacitor C_{12} in the ST state.

2.2. PWM Technique with Low Input Current Ripple

Figure 3 shows a PWM control technique for the proposed single-phase VMC-qSBI. As shown in Figure 3a, two sinusoidal voltages—v_{sin} and $-v_{sin}$—are compared to a high-frequency triangle waveform, v_{tri1}, to produce PWM signals for the H-bridge switches (S_1–S_4). A fixed voltage, V_5, is compared to the triangle waveform (v_{tri2}) with the double frequency of v_{tri1}, as shown in Figure 3, to produce the control signal for the switch S_5. To produce an ST control signal, another fixed voltage, V_{ST}, is also compared to v_{tri2}. This ST control signal (ST in Figure 3b) is then inserted into the H-bridge switches through OR logic gates.

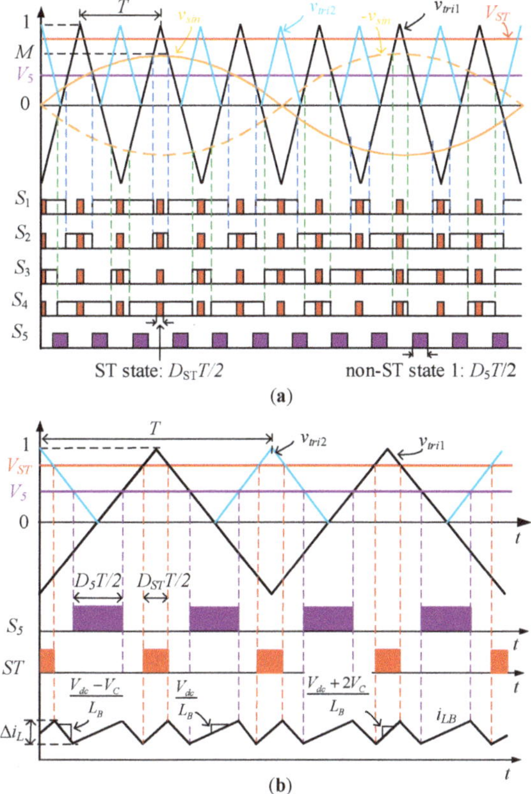

Figure 3. Pulse-width modulation (PWM) control technique for the proposed single-phase inverter. (**a**) Signal generation and (**b**) enlargement waveforms.

2.3. Steady-State Analysis

By using the volt-second balance to the inductor L_B, in steady-state, from Equations (2) to (4), the voltage on the capacitors is calculated as

$$\begin{cases} V_C = V_{C11} = V_{C12} = \frac{1}{1-2D_{ST}-D_5} V_{dc} \\ V_{C0} = 2V_C = \frac{2}{1-2D_{ST}-D_5} V_{dc}. \end{cases} \tag{5}$$

Applying the charge-second balance principle to the capacitors C_{11}, C_{12}, and C_0, from Equations (2) to (4), the following equations are obtained as Equation (6).

$$\begin{cases} -D_5 i_{C12\_Non1} + (1 - D_{ST} - D_5)i_{C12\_Non2} + D_{ST}i_{C12\_ST} + (1 - D_5)I_{LB} = 0 \\ D_5 i_{C12\_Non1} + (1 - D_{ST} - D_5)i_{C12\_Non2} + D_{ST}i_{C12\_ST} = 0 \\ -(1 - D_{ST} - D_5)i_{C12\_Non2} - D_{ST}i_{C12\_ST} - D_{ST}I_{LB} - (1 - D_{ST})I_{PN} = 0. \end{cases} \tag{6}$$

Solving Equation (6), the instantaneous current through capacitor C_{12} in the non-ST state 1 and the average inductor L_1 current are given as

$$\begin{cases} i_{C12\_Non1} = \frac{1 - D_5}{2D_5} I_{LB} \\ I_{LB} = \frac{2(1 - D_{ST})}{1 - 2D_{ST} - D_5} I_{PN}. \end{cases} \tag{7}$$

The inductor is charged during both non-ST state 1 and ST state. From Equations (2) and (4), the inductor current ripple can be rewritten as follows.

$$\begin{cases} \Delta I_{L\_NST1} = \frac{V_{dc}}{L_B} \frac{D_5 T}{2} \\ \Delta I_{L\_ST} = \frac{V_{dc} + V_C}{L_B} \frac{D_{ST} T}{2}. \end{cases} \tag{8}$$

From Equation (8), it can be seen that the slope of the inductor current in the ST state is higher than that in the non-ST state 1. The inductor current ripple depends on the ST duty cycle. Further, the inductance selection is based on the maximum ST duty cycle. If we select $D_5 = D_{ST}$ or $2D_{ST}$, the high-frequency (HF) inductor current ripple is too high, and the voltage gain of the inverter is not high at a high modulation index. In this paper, we select $D_5 = 3D_{ST}$ to obtain a low inductor current ripple with a high voltage gain. Substituting $D_5 = 3D_{ST}$ in to Equation (5), the peak dc-link voltage in the non-ST states is

$$V_{PN} = V_{C0} = 2V_C = \frac{2}{1 - 5D_{ST}} V_{dc}. \tag{9}$$

The boost factor of the proposed VMC-qSBI is calculated as

$$B = \frac{V_{PN}}{V_{dc}} = \frac{2}{1 - 5D_{ST}}. \tag{10}$$

The peak ac voltage is defined as

$$\hat{v}_o = M \cdot B \cdot V_{dc} = \frac{2M}{1 - 5D_{ST}} V_{dc}. \tag{11}$$

For the proposed n-cell VMC-qSBI, as shown in Figure 1c, the capacitor voltage is determined in the steady-state as

$$\begin{cases} V_C = V_{C11} = V_{Cn2} = \frac{V_{dc}}{1 - (n+1)D_{ST} - D_5} \\ V_{Cn1} = nV_C \\ V_{C0} = (n + 1)V_C = \frac{(n+1)V_{dc}}{1 - (n+1)D_{ST} - D_5}. \end{cases} \tag{12}$$

The peak ac voltage of the n-cell VMC-qSBI is expressed as

$$\hat{v}_o = \frac{(n+1)M}{1 - (n+1)D_{ST} - D_5} V_{dc}. \tag{13}$$

3. Parameter Design Guideline

For the single-phase inverter, there will be a pulsating power at double the fundamental frequency, which will pose a significant challenge to the front-end boost converter. Similar to other single-phase inverter topologies, the proposed single-phase VMC-qSBI also generates the double fundamental frequency ripple at the dc side. The double fundamental frequency ripple on the inductors and capacitors at the dc side can be mitigated by using a feedback control method, as reported in [10] for qZSI and [22] for qSBI. Therefore, in this study, the effect of the pulsating power can be ignored in the design stage. Then, the inductance and capacitance are only selected according to the HF ripple. The ac side circuit of the proposed VMC-qSBI is described by its equivalent dc load [12]. The average dc-link current depends on the equivalent dc load (R_l), and is calculated as [12]

$$I_{PN} = \frac{(1-D_{ST})V_{PN}}{R_l}. \tag{14}$$

3.1. Parameter Design of Inductor

The inductor current waveform of the proposed VMC-qSBI is shown in Figure 3b. The peak-to-peak current ripple of the inductor L_B is calculated as Equation (8). To limit the peak-to-peak inductor current ripple by $r_{LB}\% \cdot I_{LB}$, the inductance of L_1 should be

$$L_1 > \frac{3D_{ST}(1-5D_{ST})^2 R_l}{8r_{LB}\%(1-D_{ST})^2 f}, \tag{15}$$

where $f = 1/T$ is the switching frequency of the inverter.

3.2. Parameter Design of Capacitor

To select the C_{11}, C_{12}, and C_0 capacitances of the proposed VMC-qSBI, the peak-to-peak capacitor voltage ripples can be rewritten from Equation (2) as

$$\begin{cases} \Delta V_{C11} = \frac{-i_{C12\_non1}}{C_{11}} \cdot \frac{D_5 T}{2} \\ \Delta V_{C12} = \frac{i_{C12\_non1}}{C_{12}} \cdot \frac{D_5 T}{2} \\ \Delta V_{C0} = \frac{I_{PN}}{C_0} \cdot \frac{D_5 T}{2}, \end{cases} \tag{16}$$

Substituting Equations (6) and (14) into Equation (16), the C_{11}, C_{12}, and C_0 capacitances of the proposed VMC-qSBI are calculated as

$$\begin{cases} C_{11} = C_{12} = \frac{2(1-3D_{ST})(1-D_{ST})^2}{r_{C1}\%(1-5D_{ST})R_l f} \\ C_0 = \frac{3D_{ST}(1-D_{ST})}{2r_{C0}\%R_l f}, \end{cases} \tag{17}$$

where $r_{C1}\%$ and $r_{C0}\%$ are the voltage ripple percentages of capacitors C_{11} and C_{12} and capacitor C_0, respectively.

3.3. Parameter Design of Switches

From Figure 3, the voltage stress of the diodes is calculated as

$$\begin{cases} V_{S1-S4} = V_{C0} = \frac{2V_{dc}}{1-5D_{ST}} \\ V_{S5} = V_{C11} = \frac{V_{dc}}{1-5D_{ST}}. \end{cases} \tag{18}$$

Because the ST state turns on all of the H-bridge switches S_1–S_4, as shown in Figure 2c, the peak current of switches S_1–S_4 equals a half of the ST current, which is the inductor current. Consequently, the current stress of the switches S_1–S_4 is

$$I_{S1-S4} = \frac{I_{LB}}{2} = \frac{2(1-D_{ST})^2}{(1-5D_{ST})^2} \frac{V_{dc}}{R_l}. \tag{19}$$

The peak current of switch S_5 is determined based on the non-ST state 1 in Figure 2a as

$$I_{S5} = I_{LB} + i_{C12\_Non1} = \frac{2(1 + 3D_{ST})(1 - D_{ST})^2}{3D_{ST}(1 - 5D_{ST})^2} \frac{V_{dc}}{R_l}. \tag{20}$$

3.4. Parameter Design of Diodes

From Figure 2, the voltage stress of the switches is calculated as

$$\begin{cases} V_{D0} = V_{D11} = V_{D12} = V_{C11} = \frac{V_{dc}}{1 - 5D_{ST}} \\ V_{Da} = V_{C0} = \frac{2V_{dc}}{1 - 5D_{ST}}. \end{cases} \tag{21}$$

The peak current of the diodes D_0, D_{11}, and D_a should be selected such that it equals the inductor L_B current, and the peak current of the diode D_{12} equals the instantaneous current through capacitor C_{12} in the non-ST state 1. Therefore, the peak current of the diodes is calculated as

$$\begin{cases} I_{Da} = I_{D0} = I_{D11} = I_{LB} = \frac{4(1 - D_{ST})^2}{(1 - 5D_{ST})^2} \frac{V_{dc}}{R_l} \\ I_{D12} = i_{C12\_Non1} = \frac{2(1 - 3D)(1 - D_{ST})^2}{3D(1 - 5D_{ST})^2} \frac{V_{dc}}{R_l}. \end{cases} \tag{22}$$

4. Comparison with Other Active Impedance Source Inverters

In this section, the proposed VMC-qSBI is compared with other active impedance source inverters. The selected topologies for comparison are the qSBI [12], the SL-qSBI [14], the SZSI [17], ASC-qZSI [19], and 2-cell SC-qSBI [23]. Table 1 shows the overall comparison between the proposed VMC-qSBI and other active impedance source inverters.

4.1. Input Current Ripple

As shown in Table 1, the input current ripple of the SL-qSBI is very high because the input current is either the inductor current in the non-ST state or the twofold inductor current in the ST state. Note that as shown in Figure 3, the PWM control method cannot be applied to the SL-qSBI [13], ASC-qZSI [19], and 2-cell SC-qSBI [23] because of the increasing harmonic distortion of the output voltage. Therefore, the input current ripple of these inverters is high. Under the same PWM control method, as shown in Figure 3, the qSBI [11], SZSI [17], and the proposed inverter have a low input current ripple.

Table 1. Overall comparison of the proposed VMC-qSBI and other active impedance source inverters using the same PWM method.

	qSBI [12]	SL-qSBI* [14]	SZSI [17]	ASC-qZSI* [19]	2-cell SC-qSBI* [23]	Single-VMC-qSBI
Boost factor (B)	$\frac{1}{1-4D_{ST}}$	$\frac{1+D_{ST}}{1-3D_{ST}}$	$\frac{1}{1-5D_{ST}+D_{ST}^2}$	$\frac{1}{1-3D_{ST}+D_{ST}^2}$	$\frac{2}{1-3D_{ST}}$	$\frac{2}{1-5D_{ST}}$
Gain voltage (G)	$\frac{M}{4M-3}$	$\frac{2M-M^2}{3M-2}$	$\frac{M}{M^2+3M-3}$	$\frac{M}{M^2+M-1}$	$\frac{2M}{3M-2}$	$\frac{2M}{5M-4}$
Input current ripple	Low	Very high	Low	High	High	Low
I_{LB}/I_{PN}	$(1-D_{ST})B$	$(1-D_{ST})B/(1+D_{ST})$	$(1-D_{ST})B$	$(1-D_{ST})B$	$(1-D_{ST})B$	$(1-D_{ST})B$
I_{L2}/I_{PN}	NA	$(1-D_{ST})B/(1+D_{ST})$	$(1-D_{ST})^2B$	$(1-D_{ST})^2B$	NA	NA
I_{S5}/I_{PN}	$(1-D_{ST})B$	$\frac{2(1-D_{ST})B}{1+D_{ST}}$	$(1-D_{ST})B$	$(1-D_{ST})B$	$\frac{(1-D_{ST})}{D_{ST}(1-3D_{ST})}$	$\frac{(1-D_{ST})(1+3D_{ST})}{3D_{ST}(1-5D_{ST})}$
ST current (I_{sh}/I_{PN})	$(1-D_{ST})B$	$\frac{2(1-D_{ST})B}{1+D_{ST}}$	$(1-D_{ST})(2-D_{ST})B$	$(1-D_{ST})(2-D_{ST})B$	$(1-D_{ST})B$	$(1-D_{ST})B$
V_{C0}/V_{dc} and V_{Da}/V_{dc}	B	B	B	B	$B/2$	B
V_{C11}/V_{dc}	B	B	B	B	$B/2$	$B/2$
V_{C12}/V_{dc}	NA	NA	$D_{ST}B$	$(1-D_{ST})B$	$B/2$	$B/2$
V_{DS5}/V_{dc} and V_{D0}/V_{dc}	B	B	B	B	$B/2$	$B/2$
Switch	5	5	5	5	5	5
Diode	6	9	8	6	8	8
Inductor	1	2	2	2	1	1
Capacitor	1	1	2	2	3	3

where V_{DS5} is the drain-source voltage stress on S_5. NA: Not applicable; *Note that the PWM control method in Figure 3 cannot be applied to these inverters.

4.2. Voltage and Current Stresses

Figure 4 shows the respective ratios of the voltage stress to the equivalent dc voltage under the same PWM control method. As shown in Figure 4a, the voltage stress on capacitor C_0 and diode D_a of the proposed VMC-qSBI is higher than that of the 2-cell SC-qSBI. However, the voltage stress on switch S_5, capacitors C_{11} and C_0, and diodes D_0 and D_a of the proposed VMC-qSBI are smaller than that of the other active impedance source inverters, as shown in Figure 4a,b. The capacitor C_{12} voltage stress of the proposed VMC-qSBI is smaller than that of the ASC-qZSI and 2-cell SC-qSBI, but it is higher than that of SZSI.

The inductor current stress comparison is shown in Figure 5a. Because the qSBI, ASC-qSBI, 2-cell SC-qSBI, and the proposed VMC-qSBI have a single inductor, their inductor current stress is highest and is equal to the source current. Figure 5b shows the ST current stress comparison. The proposed VMC-qSBI, qSBI, and 2-cell SC-qSBI have the lowest ST current stress.

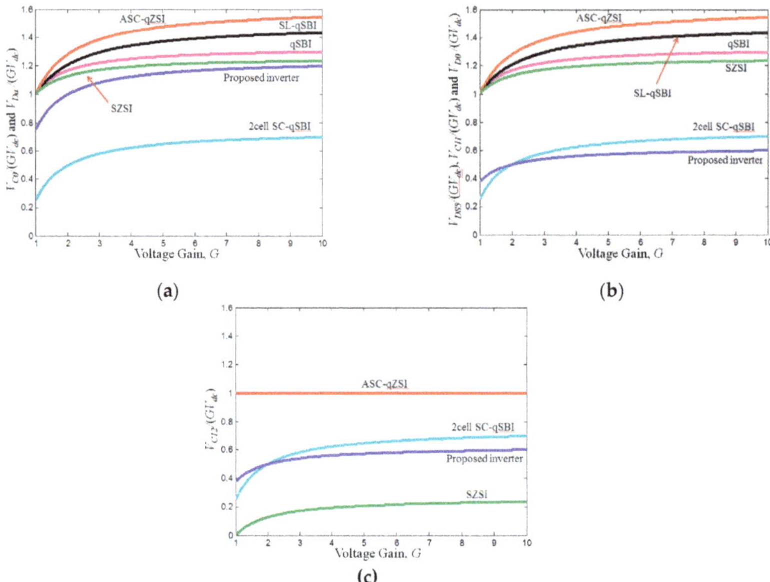

Figure 4. Voltage stress comparison. (**a**) Voltage stress on capacitor C_0 and diode D_a, (**b**) voltage stress on switch S_5, capacitor C_{11}, and diode D_0, and (**c**) capacitor C_{12} voltage stress.

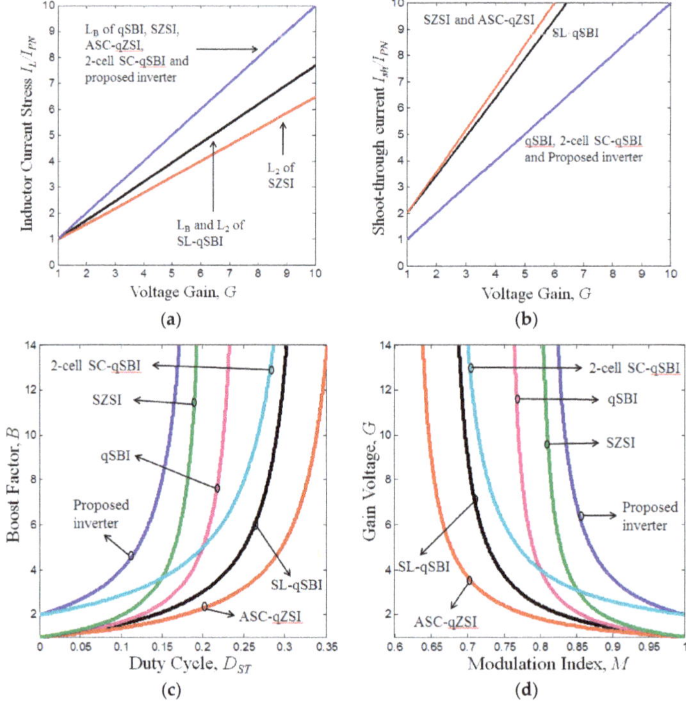

Figure 5. Voltage and current stress comparison. (**a**) Inductor current stress, (**b**) ST current stress, (**c**) boost factor, and (**d**) voltage gain.

4.3. Voltage Gain and Boost Factor

Figure 5c,d compares the boost factor and the voltage gain of the proposed VMC-qSBI with those of the active impendence source inverters, respectively. The voltage gain of the proposed VMC-qSBI is greatest for the same modulation index. The high modulation index is very important to achieve a high output waveform quality with a low total harmonic distortion (THD).

4.4. Element Count

Table 1 also compares the number of elements required for the inverters. Although the proposed VMC-qSBI adds one active switch, it results in savings of a large number of passive devices compared with the high voltage gain qZSIs in [17,19,20]. The proposed VMC-qSBI has the same number of semiconductors as the SZSI, but it uses one more capacitor and one less inductor. Compared with the SL-qSBI, the proposed VMC-qSBI uses two more capacitors, one less diode, and one less inductor. The proposed inverter uses the same number of components as the 2-cell SC-qSBI. Note that the number of diodes in Table 1 includes four body diodes of the H-bridge switches.

5. Simulation and Experiment Results

5.1. Simulation Results

To validate the performance of the proposed inverter, PSIM simulation software was used. The simulation parameters for the single-phase VMC-qSBI are given in Table 2. The drain-to-source on-resistance of the switches S_1–S_4 and the switch S_5 was set to 0.2 Ω and 8 mΩ, respectively. The body diode threshold voltage of the switches S_1–S_4 and the forward voltage of the diodes D_a, D_0, D_{11}, and D_{12} are 1.5 V and 0.73 V, respectively. The fundamental frequency of the ac output voltage and the switching frequency are 50 Hz and 20 kHz, respectively. An inductor-capacitor (LC) filter of 1 mH and 20 μF was connected to the inverter output. A purely resistive load of 40 Ω was used in the simulation.

Table 2. Parameters used for simulation and test.

Parameters		Values
Input voltage range (V_{dc})		50–72 V
Maximum input current		8 A
Output power (P_o)		350 W
Output voltage (V_o)		110 Vrms/50 Hz
Input inductor		0.37 mH
Capacitors	C_{11}, C_{12}	1000 μF/100 V
	C_0	2 $\times$ 680 μF/200 V
Output filter	L_f	1 mH
	C_f	20 μF
Switching frequency		20 kHz
Modulation index (M)		0.9
MOSFETs	S_5	IRFP4668 (200 V, 140 A)
	S_1~S_4	IRFP460 (500 V, 20 A)
Diodes	D_{11}, D_{12}, D_0	STPS60SM200C (200 V, 30 A)
	D_a	IXYS30-60A (600 V, 37 A)

Figure 6 shows the simulation results for the single-phase VMC-qSBI when V_{dc} = 50 V, D_{ST} = 0.1, D_5 = 0.3, and M = 0.9. Because of the existing passive components in the impedance-source network, an inrush current has appeared at the start-up process, as can be seen in Figure 6a. Figure 6b,c shows the simulation results in the steady-state. The voltage of capacitors C_{11}, C_{12}, and C_0 in the steady-state are boosted to 97.1 V, 96.2 V, and 193 V, respectively. The ac output voltage is 121 V_{rms}. The input current is continuous and has a small peak-to-peak ripple of 0.92 A.

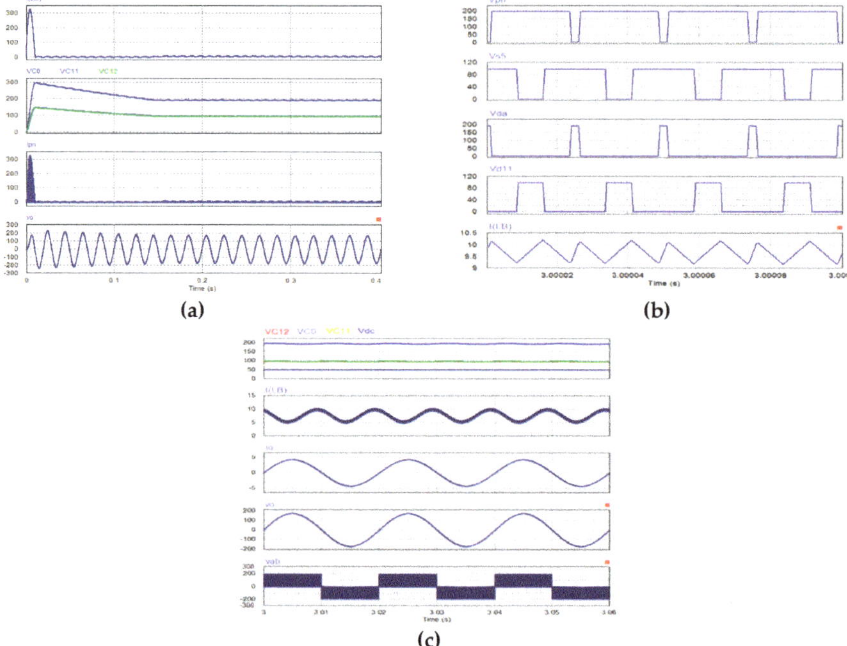

Figure 6. Simulation results for the proposed single-phase VMC-qSBI when V_{dc} = 50 V. From top to bottom: (**a**) input current, capacitor voltages, dc-link current, and output voltage; (**b**) dc-link voltage, drain-source voltage of S_5, diodes D_a–D_{11} voltage, and input current; and (**c**) input voltage and capacitors voltages, input current, output current, and output voltages before and after the L-C filter.

The three-phase VMC-qSBI as shown in Figure 7 is used to test the PWM technique with low input current ripple as presented in Section 2.2. All parameters of the three-phase VMC-qSBI including capacitance, inductance, diodes, MOSFETs, and load are the same as those of the single-phase VMC-qSBI. The maximum constant boost PWM control method in [28] is used to control three-phase H-bridge switches, while the additional switch S_5 is controlled by the constant voltage V_5 as shown in Figure 3b. Figure 8 shows the simulation results for the three-phase VMC-qSBI when V_{dc} = 50 V, D_{ST} = 0.1, D_5 = 0.3, and M = 0.9 × 1.15. As shown in Figure 8a, the capacitors C_{11}, C_{12}, and C_0 voltage are boosted to 98 V, 97 V, and 195 V in the steady-state, respectively. The three phase currents are 1.8 A in RMS. Under resistive load of 40 Ω, the simulated output phase voltage is 72 V in RMS, whereas the calculated value of the output phase voltage is 73 V in RMS. The input current is continuous. As shown in Figures 6b and 8b, the D_a diode reverse voltage is zero in non-ST states, and is equal to capacitor C_0 voltage in the ST state. The voltage stress on the additional switch S_5 is a half of dc-link voltage. All of the simulation results are in agreement with the theoretical analysis.

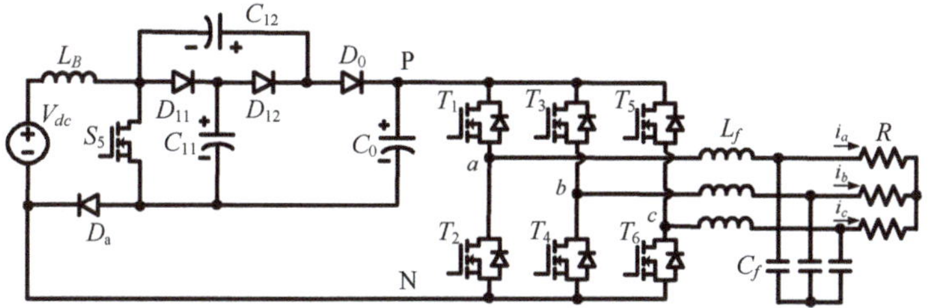

Figure 7. Proposed three-phase VMC-qSBI in simulation.

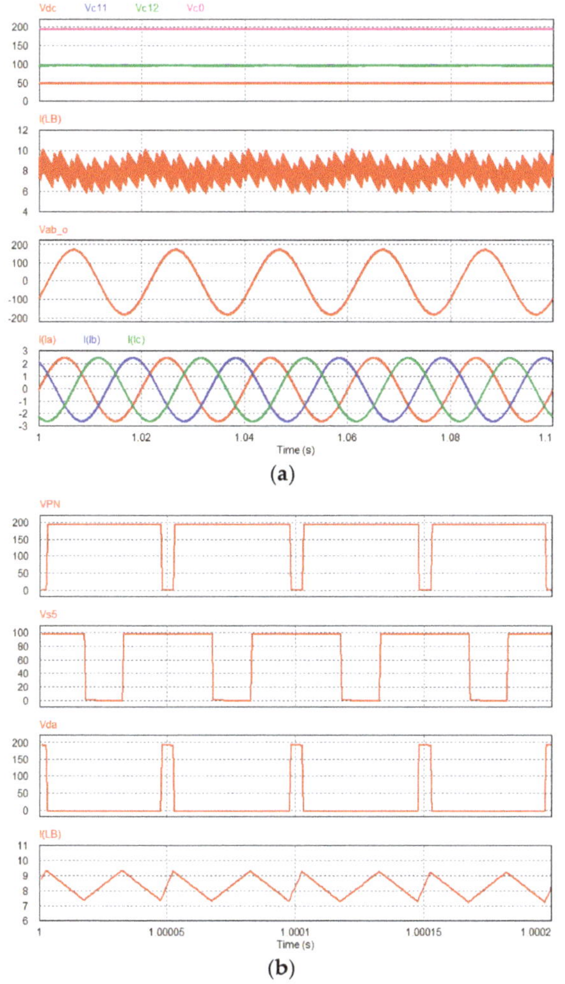

Figure 8. Simulation results for the proposed three-phase VMC-qSBI when $V_{dc} = 50$ V, $D_{ST} = 0.1$, $D_5 = 0.3$, and $M = 0.9 \times 1.15$. From top to bottom: (**a**) input voltage and capacitors voltages, input current, line-to-line voltage, and output currents and (**b**) dc-link voltage, drain-source voltage of S_5, diode D_a voltage, and input current.

5.2. Experimental Results

The experimental prototype was built to test the proposed VMC-qSBI. A 0.37 mH boost inductor was used, and the dc input voltage is supplied by 61604 Chroma Programmable ac Sources with a current limit of 8 A. The voltage stress on switch S_5 is equal to the capacitor C_{11} and C_{12} voltages; thus, for switch S_5, the IRFP4668 MOSFET was chosen with a voltage limit of 200 V. The switches on the H-bridge circuit are IRFP460 MOSFETs. Three diodes D_{11}, D_{12}, and D_0 are Schottky STPS60SM200C diodes, and one IXYS30-60A is used as the diode D_a. Capacitors C_{11} and C_{12} are 1000 µF/100 V. The capacitor C_0 was obtained by connecting in parallel two 680 µF/200 V capacitors. The parameters of the experiment are listed in Table 2. The switching frequency of the H-bridge switches is 20 kHz, and the switching frequency of switch S_5 is 40 kHz. The PWM control signals of the switches are generated by TMS320F28335 DSP, and are driven by isolated TLP250 amplifiers.

Figure 9 shows the experimental results of the proposed VMC-qSBI when V_{dc} = 50 V, M = 0.9, V_o = 110 V_{rms}, and P_o = 300 W. The dc-link voltage is boosted to 181 V from the input voltage of 50 V. The voltages of capacitors C_{11}, C_{12}, and C_0 are boosted to 91.6 V, 89.7 V, and 181 V, respectively, in the steady-state. The boost factor of the experiment is 3.62, while the calculated value for D_{ST} = 0.1 from Equation (10) in the ideal case is 4. The ac output voltage is 110 V_{rms}/50 Hz. The input current is continuous. The current THD at the output is 0.9%. The HF peak-to-peak inductor current of 1.1 A is presented in Figure 9c.

Then, the input voltage is increased to 72 V. To obtain the same 110-V_{rms} output voltage while the modulation index is kept at 0.9, the ST duty cycle is decreased to 0.05. Figure 10 shows the experimental results of the VMC-qSBI when V_{dc} = 72 V, M = 0.9, V_o = 110 V_{rms}, and P_o = 300 W. The voltages of capacitors C_{11}, C_{12}, and C_0 are boosted to 91.2 V, 90.2 V, and 181 V, respectively, from a 72 V input voltage. The HF peak-to-peak inductor current is 0.8 A. The measured THD of the output current is 1%.

Figure 11a,b shows the gate-pulse waveforms for switches S_1–S_3 when V_{dc} is 50 V and 72 V, respectively. As shown in Figure 11a,b, the overlap duty cycle (D_{ST}) of the gate pulse for switches S_1–S_3 is one-third of the duty cycle of switch S_5. The waveforms from top to bottom in Figures 9a and 10a are the output voltage after the L-C filter, the load current, the input current, and the input voltage. The waveforms from top to bottom in Figures 9b and 10b are the voltages across capacitors C_{11}, C_{12}, and C_0, and the output voltage before the L-C filter. The waveforms from top to bottom in Figures 9c and 10c are the drain-source S_5 voltage, the diode D_a voltage, the input current, and the dc-link voltage. The waveforms from top to bottom in Figures 9d and 10d are the voltages across diodes D_{11}, D_{12}, and D_0, and the drain-source S_1 voltage. The waveforms in Figure 11a,b are the control gate signals of S_5 and S_1–S_3.

Figure 12a shows the boost factor comparison between the calculation and the experiment for the VMC-qSBI. In this experiment, the duty cycle was varied from 0.05 to 0.12, while the output voltage and the output power were kept at 110 V_{rms} and 300 W, respectively. Because of the parasitic elements in the experimental setup, the experimental values are lower than the calculated values. Figure 12b shows the efficiency of the VMC-qSBI with various power loads. The maximum efficiency value of the VMC-qSBI at a load power of 165 W and an input voltage of 72 V is 91.3%.

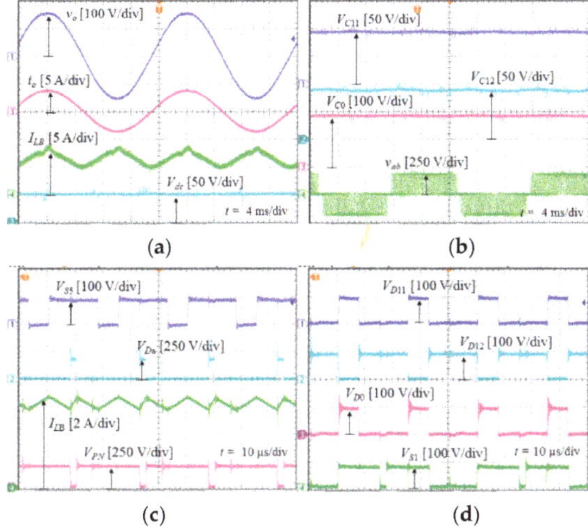

Figure 9. Experimental waveforms for the proposed inverter when V_{dc} = 50 V. From top to bottom: (**a**) output voltage after the L-C filter, load current, input current, and input voltage; (**b**) voltages across capacitors C_{11}, C_{12}, and C_0, and output voltage before the L-C filter, (**c**) drain-source S_5 voltage, diode D_a voltage, input current, and dc-link voltage; and (**d**) voltages across diodes D_{11}, D_{12}, and D_0, and the drain-source S_1 voltage.

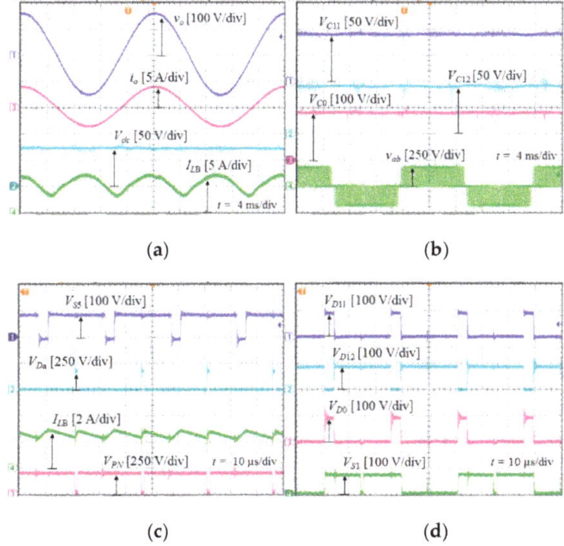

Figure 10. Experimental waveforms for the proposed inverter when V_{dc} = 72 V. From top to bottom: (**a**) output voltage after the L-C filter, load current, input current, and input voltage; (**b**) voltages across capacitors C_{11}, C_{12}, and C_0, and output voltage before the L-C filter, (**c**) drain-source S_5 voltage, diode D_a voltage, input current, and dc-link voltage; and (**d**) voltages across diodes D_{11}, D_{12}, and D_0, and the drain-source S_1 voltage.

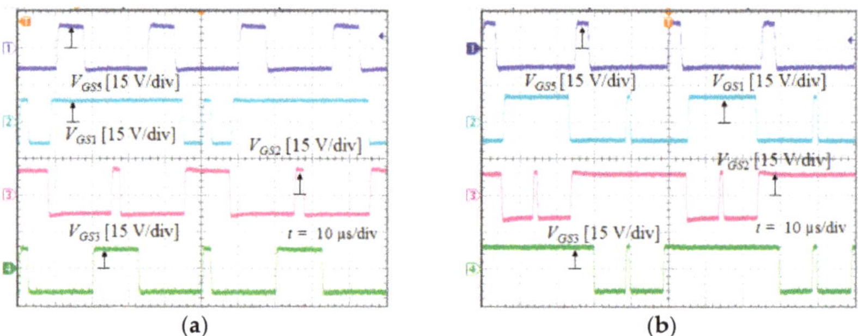

(a)   (b)

Figure 11. Control signal waveforms when (**a**) V_{dc} = 50 V and (**b**) V_{dc} = 72 V.

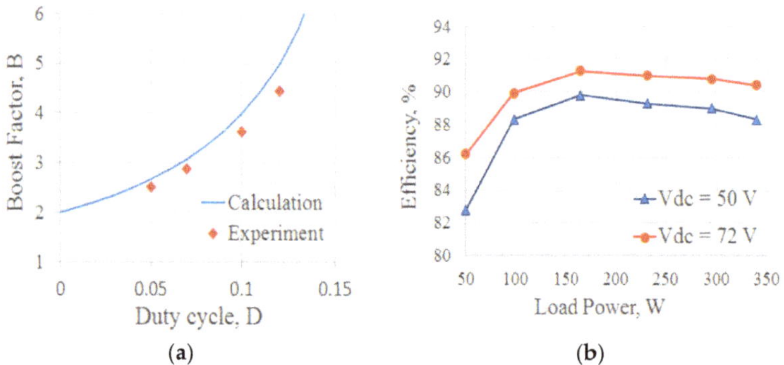

(a)   (b)

Figure 12. Measured results of (**a**) boost factor and (**b**) efficiency of the proposed inverter.

6. Discussions and Conclusion

Although ZS/qZSIs and qSBIs have a good performance with buck-boost voltage function, single-stage conversion, and ST immunity, their disadvantage is high voltage stress on capacitors, diodes, and switch. Using a small voltage rating of the devices leads to reduce the loss and cost of the power inverter system. Moreover, a small ST duty ratio or a high modulation index helps to enhance the output waveform quality of the inverter. In this paper, a new single-phase single-stage boost inverter based on the VMC structure was proposed. A new PWM control strategy was used for the proposed inverter to achieve a high voltage gain with a low input current ripple. Compared to the other active impedance source inverters, the proposed VMC-qSBI has a high voltage gain, low input current ripple, low voltage stress on the switch and diode, low ST current, and high modulation index. The extension was presented to improve the voltage gain of the proposed inverter by adding the VMCs. The operating modes, steady-state analysis, and design guideline were presented. A laboratory prototype was tested to verify the accuracy of the proposed VMC-qSBI. Simulation and experimental results were shown.

Because the proposed VMC-qSBI has high reliability with ST immunity, low voltage stress on devices, and high voltage gain inversion with multi-VMCs, it is suitable for renewable energy system applications such as photovoltaic and wind power.

Author Contributions: M.-K.N. conducted topology, experimental work, data analysis, and writing the original draft. Y.-O.C. reviewed and revised the whole work.

Funding: This research was funded by the Korea Institute of Energy Technology Evaluation and Planning (KETEP) and the Ministry of Trade, Industry& Energy (MOTIE) of the Republic of Korea grant number 20184010201650 and the APC was funded by KETEP and MOTIE.

Acknowledgments: This work was supported by the Korea Institute of Energy Technology Evaluation and Planning (KETEP) and the Ministry of Trade, Industry& Energy (MOTIE) of the Republic of Korea (NO. 20184010201650).

Conflicts of Interest: The authors declare no conflict of interest.

References

1. Meneses, D.; Blaabjerg, F.; Garcia, O.; Cobos, J.A. Review and comparison of step-up transformerless topologies for photovoltaic ac-module application. *IEEE Trans. Power Electron.* **2013**, *28*, 2649–2663. [CrossRef]
2. Maris, T.I.; Kourtesi, S.; Ekonomou, L.; Fotis, G.P. Modeling of a single phase photovoltaic inverter. *Solar Energy Mater. Sol. Cells.* **2007**, *91*, 1713–1725. [CrossRef]
3. Anderson, J.; Peng, F. A class of quasi-Z-source inverters. In Proceedings of the 2008 IEEE Industry Applications Society Annual Meeting, Edmonton, AB, Canada, 5–9 October 2008; pp. 1–7.
4. Kadwane, S.G.; Shinde, U.K.; Gawande, S.P.; Keshri, R.K. Symmetrical shoot-through based decoupled control of Z-source inverter. *IEEE Access.* **2017**, *5*, 11298–11306. [CrossRef]
5. Nguyen, M.K.; Lim, Y.C.; Choi, J.H. Two switched-inductor quasi-Z-source inverters. *IET Power Electron.* **2012**, *5*, 1017–1025. [CrossRef]
6. Nguyen, M.K.; Lim, Y.C.; Cho, G.B. Switched-inductor quasi-Z-source inverter. *IEEE Trans. Power Electron.* **2011**, *26*, 3183–3191. [CrossRef]
7. Jagan, V.; Kotturu, J.; Das, S. Enhanced-boost quasi-Z-source inverters with two-switched impedance networks. *IEEE Trans. Ind. Electron.* **2017**, *64*, 6885–6897. [CrossRef]
8. Ho, A.V.; Chun, T.W. Modified capacitor-assisted Z-source inverter topology with enhanced boost ability. *J. Power Electron.* **2017**, *17*, 1195–1202.
9. Qian, W.; Peng, F.Z.; Cha, H. Trans-Z-source inverters. *IEEE Trans. Power Electron.* **2011**, *26*, 3453–3463. [CrossRef]
10. Ge, B.; Liu, Y.; Abu-Rub, H.; Balog, R.S.; Peng, F.Z.; McConnell, S.; Li, X. Current ripple damping control to minimize impedance network for single phase quasi-Z source system. *IEEE Trans. Ind. Informat.* **2016**, *12*, 1054–4848. [CrossRef]
11. Mishra, S.; Adda, R.; Joshi, A. Inverse Watkins–Johnson topology-based inverter. *IEEE Trans. Power Electron.* **2012**, *27*, 1066–1070. [CrossRef]
12. Nguyen, M.K.; Le, T.V.; Park, S.J.; Lim, Y.C. A class of quasi-switched boost inverters. *IEEE Trans. Ind. Electron.* **2015**, *62*, 1526–1536. [CrossRef]
13. Nguyen, M.K.; Lim, Y.C.; Park, S.J. A comparison between single-phase quasi-Z-source and quasi-switched boost inverters. *IEEE Trans. Ind. Electron.* **2015**, *62*, 6336–6344. [CrossRef]
14. Nguyen, M.K.; Le, T.V.; Park, S.J.; Lim, Y.C.; Yoo, J.Y. A class of high boost inverters based on switched-inductor structure. *IET Power Electron.* **2015**, *8*, 750–759. [CrossRef]
15. Ho, A.V.; Chun, T.W.; Kim, H.G. Development of multi-cell active switched-capacitor and switched-inductor Z-source inverter topologies. *J. Power Electron.* **2014**, *14*, 834–841. [CrossRef]
16. Chub, A.; Liivik, L.; Zakis, J.; Vinnikon, D. Improved switched-inductor quasi-switched-boost inverter with low input current ripple. In Proceedings of the 56th International Scientific Conference on Power and Electrical Engineering of Riga Technical University (RTUCON), Riga, Latvia, 14 October 2015; pp. 1–6.
17. Nozadian, M.H.B.; Babaei, E.; Hosseini, S.H.; Asl, E.S. Steady-state analysis and design considerations of high voltage gain switched Z-source inverter with continuous input current. *IEEE Trans. Ind. Electron.* **2017**, *64*, 5342–5350. [CrossRef]
18. Nguyen, M.K.; Duong, T.D.; Lim, Y.C.; Choi, J.H. High voltage gain quasi-switched boost inverters with low input current ripple. *IEEE Trans. Ind. Informat. (Early Access)* **2018**, 1. [CrossRef]
19. Ho, A.V.; Hyun, J.S.; Chun, T.W.; Lee, H.H. Embedded quasi-Z-source inverters based on active switched-capacitor structure. In Proceedings of the IECON 2016—42nd Annual Conference of the IEEE Industrial Electronics Society, Florence, Italy, 23–26 October 2016; pp. 3384–3389.
20. Nguyen, M.K.; Lim, Y.C.; Choi, J.H.; Choi, Y.O. Trans-switched boost inverters. *IET Power Electron.* **2016**, *9*, 1065–1073. [CrossRef]

21. Nguyen, M.K.; Tran, T.T.; Lim, Y.C. A family of PWM control strategies for single-phase quasi-switched-boost inverter. *IEEE Trans. Power Electron.* **2019**, *34*, 1458–1469. [CrossRef]
22. Gambhir, A.; Mishra, S.K.; Joshi, A. Power frequency harmonic reduction and its redistribution for improved filter design in current-fed switched inverter. *IEEE Trans. Ind. Electron.* **2019**, *66*, 4319–4333. [CrossRef]
23. Nguyen, M.K.; Duong, T.D.; Lim, Y.C.; Kim, Y.G. Switched-capacitor quasi-switched boost inverters. *IEEE Trans. Ind. Electron.* **2018**, *65*, 5105–5113. [CrossRef]
24. Prudente, M.; Pfitscher, L.L.; Emmendoerfer, G.; Romaneli, E.F.; Gules, R. Voltage multiplier cells applied to non-isolated dc–dc converters. *IEEE Trans. Power Electron.* **2008**, *23*, 871–887. [CrossRef]
25. Deng, Y.; Rong, Q.; Li, W.; Zhao, Y.; Shi, J.; He, X. Single-switch high step-up converters with built-in transformer voltage multiplier cell. *IEEE Trans. Power Electron.* **2012**, *27*, 3557–3567. [CrossRef]
26. Alcazar, Y.J.A.; Oliveira, D.D.S.; Tofoli, F.L.; Torrico-Bascope, R.P. DC-DC nonisolated boost converter based on the three-state switching cell and voltage multiplier cells. *IEEE Trans. Ind. Electron.* **2013**, *60*, 4438–4449. [CrossRef]
27. Berkovich, Y.; Shenkman, A.; Axelrod, B.; Golan, G. Structures of transformerless step-up and step-down controlled rectifiers. *IET Power Electron.* **2008**, *1*, 245–254. [CrossRef]
28. Shen, M.; Wang, J.; Joseph, A.; Peng, F.Z.; Tolbert, L.M.; Adams, D.J. Constant boost control of the Z-source inverter to minimize current ripple and voltage stress. *IEEE Trans. Ind. Appl.* **2006**, *42*, 770–778. [CrossRef]

© 2019 by the authors. Licensee MDPI, Basel, Switzerland. This article is an open access article distributed under the terms and conditions of the Creative Commons Attribution (CC BY) license (http://creativecommons.org/licenses/by/4.0/).

Article

Transformerless Quasi-Z-Source Inverter to Reduce Leakage Current for Single-Phase Grid-Tied Applications

Woo-Young Choi * and Min-Kwon Yang

Division of Electronic Engineering, Chonbuk National University, Jeonju 561-756, Korea; mkyang@jbnu.ac.kr
* Correspondence: wychoi@jbnu.ac.kr; Tel.: +82-063-270-4218

Received: 13 February 2019; Accepted: 4 March 2019; Published: 12 March 2019

Abstract: The conventional single-phase quasi-Z-source (QZS) inverter has a high leakage current as it is connected to the grid. To address this problem, this paper proposes a transformerless QZS inverter, which can reduce the leakage current for single-phase grid-tied applications. The proposed inverter effectively alleviates the leakage current problem by removing high-frequency components for the common-mode voltage. The operation principle of the proposed inverter is described together with its control strategy. A control scheme is presented for regulating the DC-link voltage and the grid current. A 1.0 kW prototype inverter was designed and tested to verify the performance of the proposed inverter. Silicon carbide (SiC) power devices were applied to the proposed inverter to increase the power efficiency. The experimental results showed that the proposed inverter achieved high performance for leakage current reduction and power efficiency improvement.

Keywords: quasi-z-source inverter; grid-tied; leakage current; power efficiency

---

1. Introduction

Quasi-Z-source (QZS) inverters have been widely used for grid-tied applications, due to their advantages over the traditional voltage-source inverters (VSIs) using a DC-DC converter [1–7]. Figure 1 shows the circuit diagram of the QZS inverter for single-phase grid-tied applications [2–4]. It has a QZS circuit (L_1, L_2, C_1, C_2, D_1) and a full-bridge inverter (S_1, S_2, S_3, S_4, L_3, L_4). As it is connected to the grid v_g without an isolation transformer, a leakage current i_p flows through the parasitic capacitance C_p between the inverter and the grid [5]. This leakage current originated from the common-mode voltage v_p, which changes rapidly as the inverter operates with high switching frequency [6]. Due to shoot-through states, the common-mode voltage in the QZS inverter can be higher than that in the conventional VSIs [7]. This leads to a higher leakage current, which decreases the power efficiency of the inverter.

A simple way to reduce the leakage current is to use bipolar pulse-width modulation (PWM) [8]. Power switches are diagonally operated when bipolar PWM is adopted. The common-mode voltage requires only a grid frequency component, yielding a low leakage current. However, as the voltage v_{AB} has two levels V_{PN} and $-V_{PN}$, the output filter inductors L_3 and L_4 should have high current ripples and high core losses. Despite its low leakage current characteristic, a QZS inverter with bipolar PWM is not suitable for grid-tied applications because of its reduced power efficiency.

Another method is to use a decoupling circuit to disconnect the inverter from the grid [9–15]. The decoupling circuit technique has been described for single-phase transformerless inverter applications in [9]. Many advanced transformerless inverters have been developed for voltage source inverters, such as the H5 inverter [10], the highly efficient and reliable inverter concept (HERIC) inverter [11], and the H6 inverter [12]. Many efforts have also been made to develop transformerless inverters by applying decoupling circuits to the current source inverters [16,17].

To date, however, only a few studies [13–15] have been reported that apply decoupling circuits to ZS inverters. In [13], a symmetric ZS HERIC inverter was proposed. It uses an additional two power switches for maintaining the common-mode voltage constant. In [14,15], a QZS HERIC inverter was suggested. It uses an additional two power switches and two power diodes for clamping the common-mode voltage. The inverters in [13–15] commonly have three distinct switching states, namely powering, freewheeling, and shoot-through states. During powering states, the inverter delivers the DC power into the grid. During shoot-through states, the inverter is decoupled from the grid by turning on all power switches simultaneously in full-bridge inverter legs. After every powering state and shoot-through state, freewheeling states are necessary. During freewheeling states, the grid current circulates through extra power switches, which operate with the grid frequency. However, the previous inverters [13–15] have drawbacks such as (1) they still use the bipolar PWM for power switches in full-bridge inverter legs; and (2) they always need freewheeling states for every powering state and shoot-through state. These drawbacks increase the number of switching times for the power switches, causing high switching power losses.

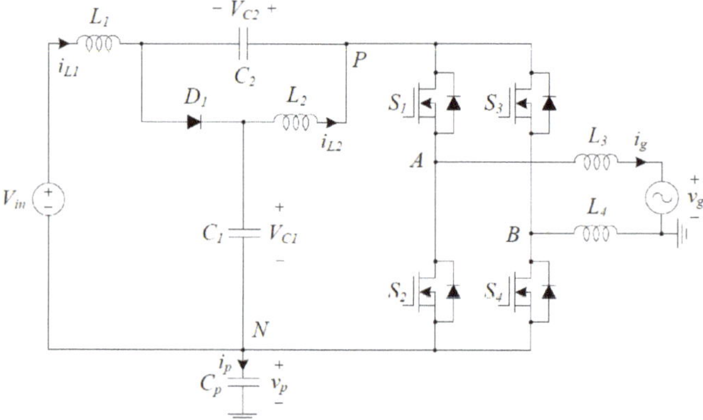

Figure 1. Circuit diagram of the QZS inverter for single-phase grid-tied applications.

To address the above-mentioned drawbacks, this paper proposes a transformerless QZS inverter, which can effectively reduce the leakage current for single-phase grid-tied applications. Figure 2 shows the circuit diagram of the proposed inverter. The idea behind the proposed inverter was to relate two high-frequency switching legs (S_1, S_2 and S_3, S_4) with the grid using a bidirectional switch (S_5, S_6) and two inductors (L_3, L_4). The bidirectional switch operates with the grid frequency, providing a current path for clamping the common-mode voltage v_p to the neutral and the grid voltage v_g, for positive and negative grid cycles, respectively. The leakage current i_p can be reduced due to the absence of high-frequency components for v_p. Shoot-through states are implemented by turning on two power switches simultaneously in only one switching leg. This leads to low switching power losses compared to the previous inverters in [13–15]. A control scheme is suggested for regulating the DC-link voltage V_{PN} and the grid current i_g. Since the peak DC-link voltage is controlled by a voltage controller, the grid current control can be directly implemented to regulate i_g. In this paper, Section 2 describes the operation principle and the control strategy of the proposed inverter. Section 3 presents the experimental results for a 1.0 kW prototype inverter. Silicon carbide (SiC) power devices were applied to the proposed inverter to increase the power efficiency. Section 4 ends the paper by presenting the conclusion.

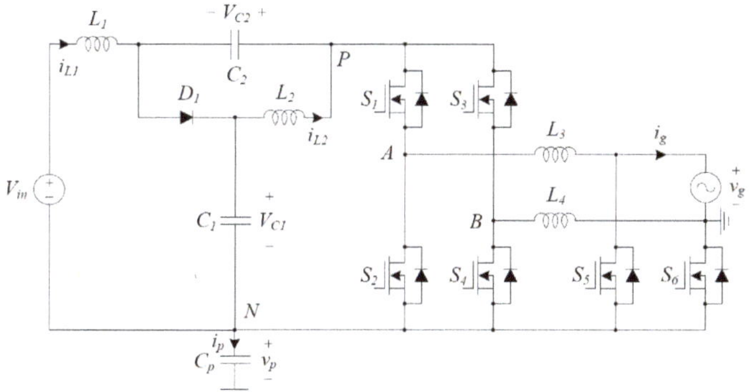

Figure 2. Circuit diagram of the proposed inverter.

2. Proposed Inverter

2.1. Operation Principle

Figure 2 shows a circuit diagram of the proposed inverter. It has a QZS circuit (L_1, L_2, C_1, C_2, D_1), a full-bridge inverter (S_1, S_2, S_3, S_4, L_3, L_4), and a bidirectional switch (S_5, S_6). C_p is modeled as the parasitic capacitance between the inverter and the grid voltage v_g. V_{in} is the DC voltage. V_{PN} is the DC-link voltage. V_{C1} and V_{C2} are the capacitor voltages for C_1 and C_2, respectively. i_{L1} and i_{L2} are the inductor currents for L_1 and L_2, respectively. i_{L3} and i_{L4} are the inductor currents for L_3 and L_4, respectively. $i_{L1} \sim i_{L4}$ are assumed to be continuous. C_1 and C_2 are assumed to have large capacitance so that their ripple components are negligible.

The proposed inverter has three switching states, namely the powering, freewheeling, and shoot-through states. For a positive grid cycle, S_6 is always turned on. S_1 and S_2 operate complementarily during the non-shoot-through states. Figure 3 shows the switching circuit diagrams of the proposed inverter for a positive grid cycle. During the powering state in Figure 3a, S_1 is turned on. The DC voltage source supplies electric power to the grid as D_1 is turned on. i_g flows through v_g, S_6, V_{PN}, S_1, and L_3. The following voltage equation is obtained as in:

$$- V_{PN} + V_{L3} + v_g = 0 \tag{1}$$

i_{L1} flows through L_1, D_{Sb}, C_1, and V_{in}. i_{L2} flows through L_2, C_2, and D_1. The following voltage equations are obtained as in:

$$V_{L1} = V_{in} - V_{C1} \tag{2}$$

$$V_{L2} = -V_{C2} \tag{3}$$

During the freewheeling state in Figure 3b, S_2 is turned on. The energy stored in L_3 is transferred to v_g. i_g freewheels through v_g, S_6, S_2, and L_3. The following voltage equation is obtained as in:

$$V_{L3} + v_g = 0 \tag{4}$$

As L_1 and L_2 are discharged, C_1 and C_2 are charged. During the shoot-through state in Figure 3c, S_1 and S_2 are turned on simultaneously. As D_1 is turned off, L_1 and L_2 are charged. The following voltage equations are obtained as in:

$$V_{L1} = V_{in} + V_{C2} \tag{5}$$

$$V_{L2} = V_{C1} \tag{6}$$

i_g flows through v_g, S_6, S_2, and L_3, as in the freewheeling state.

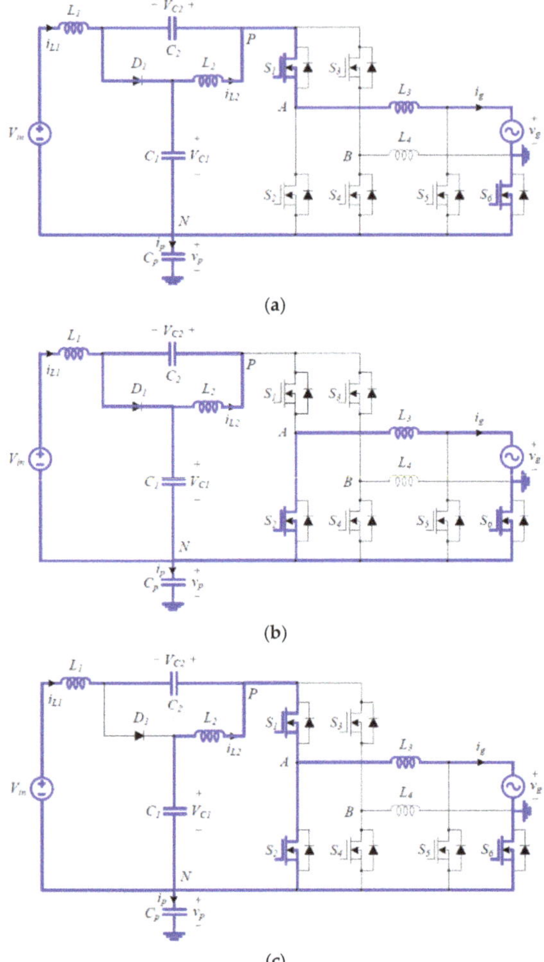

Figure 3. Switching circuit diagrams of the proposed inverter for a positive grid cycle: (**a**) powering state; (**b**) freewheeling state; (**c**) shoot-through state.

Figure 4 shows the switching circuit diagrams of the proposed inverter for a negative grid cycle. S_5 is always turned on for a negative grid cycle. S_3 and S_4 operate complementarily during the non-shoot-through states. The operation principle for a negative grid cycle is not described here because it can be analogously explained as the operation principle for a positive grid cycle.

As shown in Figure 3, S_6 provides a closed path for C_p to be clamped to the zero voltage for a positive grid cycle. On the other hand, as shown in Figure 4, S_5 provides a closed path for C_p to be clamped to the grid voltage for a negative grid cycle for a negative grid cycle. Then, the common-mode voltage v_p can be represented as in:

$$v_p = \begin{cases} 0 & \text{when } v_g > 0 \\ v_g & \text{when } v_g \leq 0. \end{cases} \tag{7}$$

The parasitic capacitance C_p can be free from the high-frequency components for both positive and negative grid cycles. This leads to the low leakage current i_p, regardless of the high-frequency switching operation of the inverter.

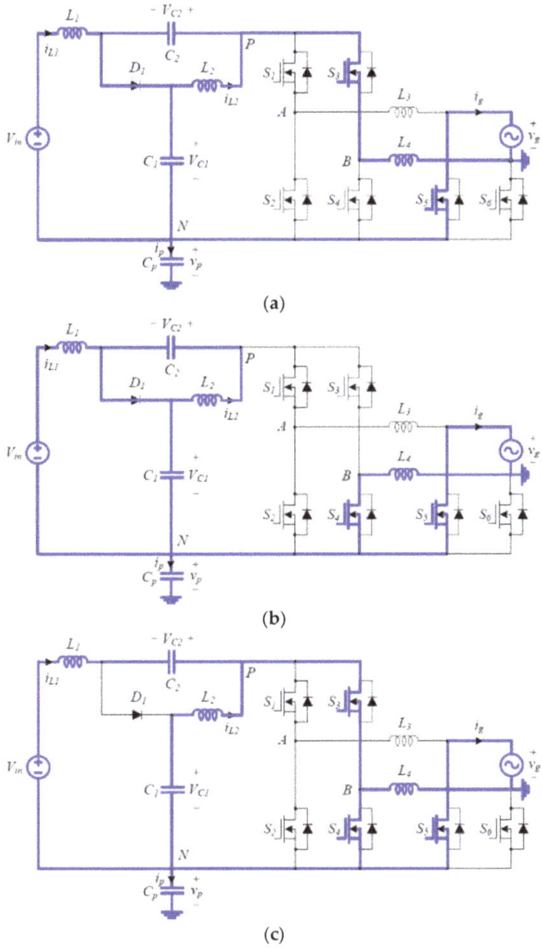

Figure 4. Switching circuit diagrams of the proposed inverter for a negative grid cycle: (**a**) powering state; (**b**) freewheeling state; (**c**) shoot-through state.

Figure 5 shows the signal diagrams for $S_1 \sim S_4$ when they operate with high switching frequency. S_1 and S_3 are the main control switches for positive and negative grid cycles, respectively. S_1 (S_3) and S_2 (S_4) operate complementarily during the non-shoot-through states, respectively. A simple boost modulation scheme is adopted for generating the shoot-through duty cycles [18]. The shoot-through state is equally distributed into two parts adjacent with the on-time of the main control switch. Shoot-through states are implemented by turning on two power switches simultaneously in only one switching leg, which leads to low switching power losses, compared to the previous inverters in [13–15]. Given that the on-time interval for the main control switch is T_{ON} for one switching period T_S, from (1) and (4), the average voltage $V_{L3,avg}$ for L_3 over T_S should be zero, respectively, as in:

$$V_{L3,avg} = \frac{(V_{PN} - v_g)\,T_{ON} - v_g(T_S - T_{ON})}{T_S} = 0. \tag{8}$$

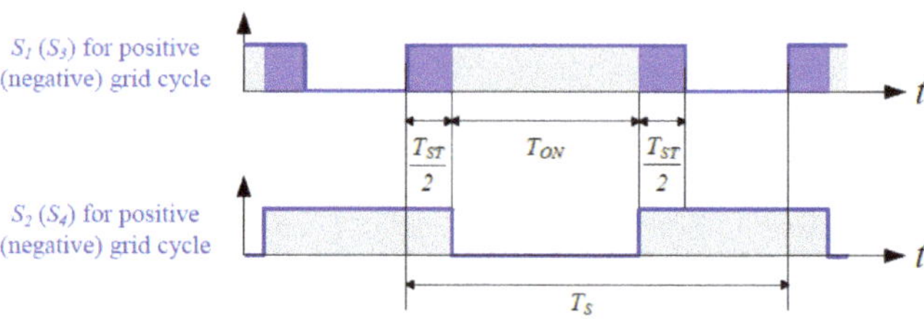

Figure 5. Signal diagrams for $S_1 \sim S_4$ when they operate with high switching frequency.

The ratio between T_{ON} and T_S is obtained as in:

$$\frac{T_{ON}}{T_S} = \frac{v_g}{V_{PN}} = \frac{V_g|\sin \omega t|}{V_{PN}} \tag{9}$$

where V_g is the positive peak value of v_g, and ω is the angular frequency of v_g. Given that the time interval during the shoot-through state is T_{ST} for T_S, the average voltages $V_{L1,avg}$ and $V_{L2,avg}$ for L_1 and L_2 over T_S should be zero, respectively, as in:

$$V_{L1,avg} = \frac{(V_{in} + V_{C2})\,T_{ST} + (V_{in} - V_{C1})\,(T_S - T_{ST})}{T_S} = 0, \tag{10}$$

$$V_{L2,avg} = \frac{V_{C1}T_{ST} - V_{C2}(T_S - T_{ST})}{T_S} = 0. \tag{11}$$

From Equations (10) and (11), we have:

$$V_{C1} - V_{C2} = V_{in}. \tag{12}$$

From Equation (11), the shoot-through duty cycle D_{ST} is represented as in:

$$D_{ST} = \frac{T_{ST}}{T_S} = \frac{V_{C2}}{V_{C1} + V_{C2}}. \tag{13}$$

From Equations (12) and (13), V_{C1} and V_{C2} are represented as in:

$$V_{C1} = \frac{1 - D_{ST}}{1 - 2D_{ST}}V_{in}, \tag{14}$$

$$V_{C2} = \frac{D_{ST}}{1 - 2D_{ST}}V_{in}. \tag{15}$$

Since $V_{PN} = V_{C1} + V_{C2}$, from Equations (14) and (15), we have:

$$\frac{V_{PN}}{V_{in}} = \frac{1}{1 - 2D_{ST}}. \tag{16}$$

2.2. Control Strategy

As the proposed inverter steps down the DC-link voltage V_{PN} to the level of v_g, it regulates the DC-link voltage and controls the grid current i_g. From Equations (3) and (6), the average voltage for L_2 over T_S is obtained with the inductor current deviation Δi_{L2} as in:

$$L_2 \frac{\Delta i_{L2}}{T_S} = V_{C1} D_{ST} - V_{C2}(1 - D_{ST}). \tag{17}$$

Since $V_{PN} = V_{C1} + V_{C2}$, D_{ST} is derived as in:

$$D_{ST} = \frac{V_{C2}}{V_{PN}} + L_2 \frac{\Delta i_{L2}}{V_{PN} T_S}. \tag{18}$$

Suppose that $L_1 = L_2 = L_i$ and $i_{L1} = i_{L2} = i_i$, D_{ST} can be represented as in:

$$D_{ST} = D_{ST,N} + D_{ST,C} \tag{19}$$

where $D_{ST,N}$ is the nominal shoot-through duty cycle and $D_{ST,C}$ is the controlled shoot-through duty cycle as in:

$$D_{ST,N} = \frac{V_{C2}}{V_{PN}^*}, \tag{20}$$

$$D_{ST,C} = L_i \frac{|\Delta i_i|}{V_{PN}^* T_S}. \tag{21}$$

V^*_{PN} is the reference value for the peak DC-link voltage. To make the peak DC-link voltage to track its reference V^*_{PN}, the following proportional-integral (PI) voltage control is used as in:

$$D_{ST,C} = k_p(V_{C2}^* - V_{C2}) + k_i \int (V_{C2}^* - V_{C2})\, dt \tag{22}$$

where k_p and k_i are the PI control gains, respectively. Here, V^*_{C2} is the reference value for the capacitor voltage V_{C2}, which is utilized for the DC-link voltage control as V_{PN} is a pulsating voltage [19]. V^*_{C2} is given as in:

$$V_{C2}^* = \frac{V_{PN}^* - V_{in}}{2}, \tag{23}$$

which is obtained from the following relations as $V_{PN} = V_{C1} + V_{C2}$ and $V_{in} = V_{C1} - V_{C2}$.

For the positive grid cycle, from Equations (1) and (4), the average voltage for L_3 over T_S is obtained with the grid current deviation Δi_g as in:

$$L_3 \frac{\Delta i_g}{T_S} = (V_{PN} - v_g)\, D_1 - v_g(1 - D_1) = 0 \tag{24}$$

where D_1 is the duty cycle of S_1 without considering D_{ST}. From Equation (24), D_1 is represented as in:

$$D_1 = \frac{v_g}{V_{PN}} + L_3 \frac{\Delta i_g}{V_{PN} T_S}, \text{when } v_g > 0. \tag{25}$$

Similarly, the duty cycle D_3 of S_3 without considering D_{ST} for a negative grid cycle can be represented as in:

$$D_3 = -\frac{v_g}{V_{PN}} - L_4 \frac{\Delta i_g}{V_{PN} T_S}, \text{when } v_g < 0. \tag{26}$$

Supposed that $L_3 = L_4 = L_g$, D_1 and D_3 without considering D_{ST} can be represented as the sinusoidal PWM duty cycle D_{SPWM} as in:

$$D_{SPWM} = D_{SPWM,N} + D_{SPWM,C} \tag{27}$$

where $D_{SPWM,N}$ is the nominal sinusoidal PWM duty cycle and $D_{SPWM,C}$ is the controlled sinusoidal PWM duty cycle as in:

$$D_{SPWM,N} = \frac{V_g |\sin \omega t|}{V_{PN}^*}, \tag{28}$$

$$D_{SPWM,C} = L_g \frac{|\Delta i_g|}{V^*_{PN} T_S}.$$

(29)

To make the grid current to track its reference i^*_g, the following proportional (P) current control is used as in:

$$D_{SPWM,C} = k_g (i^*_g - |i_g|)$$

(30)

where k_g is the P control gain. i^*_g is given as in:

$$i^*_g = I^*_g |\sin \omega t|$$

(31)

where I^*_g is the peak magnitude of the current reference. Figure 6 shows the control block diagrams of the proposed inverter. Figure 6a shows the control block diagram for the DC-link voltage control. Figure 6b shows the control block diagram for the grid current control. The phase-locked loop (PLL) control is used for the grid synchronization [20]. The duty cycle for the main control switch can be finally obtained by summing D_{SPWM} and D_{ST}.

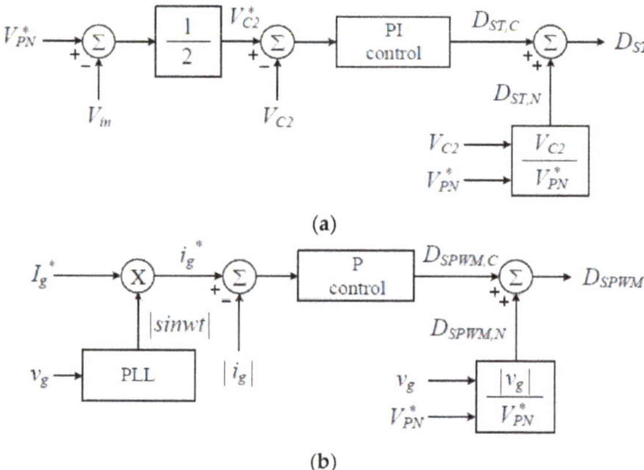

Figure 6. Control block diagrams of the proposed inverter: (**a**) DC-link voltage control; (**b**) grid current control.

3. Experimental Results

A 1.0 kW prototype inverter was designed for the proposed inverter for v_g = 60 Hz/220 V_{rms}. The valve regulated lead-acid batteries were used for the DC voltage V_{in}, the nominal voltage of which was 250 V. The V^*_{PN} was set to 500 V for D_{ST} = 0.25. SiC metal-oxide field-effect transistors (MOSFETs) (C2M0080120D, CREE) were used for S_b. A SiC Schottky diode (C4D20120D, CREE) was used for D_1. A digital signal controller (dsPIC30F6015, Microchip) was used for implementing the DC-link voltage and grid current controllers and for generating the duty cycle signals.

The inverter in Figure 1 and the inverter in [15] were designed using insulated gate bipolar transistors (IGBTs). In the conventional inverter in Figure 1, an IGBT (IKW25T120, Infineon) was used for $S_1 \sim S_4$ with a switching frequency of 10 kHz. A unipolar PWM was adopted for the conventional inverter, due to its better switching performance than a bipolar PWM. In the inverter in [15], IKW25T120 was used for $S_1 \sim S_4$ with a switching frequency of 10 kHz. IKW25T120 and C4D20120D were used for the extra power switches and extra power diodes, respectively. The proposed inverter used IKW25T120 for $S_1 \sim S_4$ with a switching frequency of 10 kHz and for S_5 and S_6 with a switching frequency of 60 Hz, respectively. Figure 7 shows a picture of the prototype inverter. The inductance for $L_1 \sim L_4$ was selected as 1.0 mH. L_3 and L_4 were implemented as a coupled inductor, which simplified the circuit layout. The capacitance for C_1 and C_2 was selected as 560 μF.

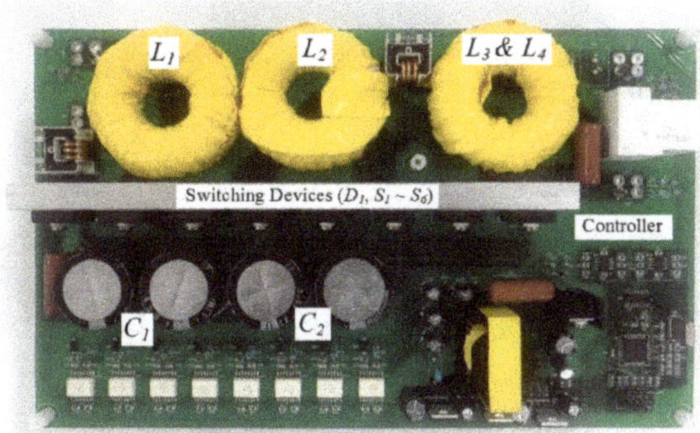

Figure 7. Picture of the prototype inverter.

Figure 8 shows the experimental waveforms of the conventional inverter in Figure 1 and the proposed inverter, respectively. Figure 8a shows v_g and v_p in the conventional inverter. Figure 8b shows v_g and i_p in the conventional inverter. Figure 8c shows v_g and v_p in the proposed inverter. Figure 8d shows v_g and i_p in the proposed inverter. The leakage current in the proposed inverter was greatly reduced compared to that in the conventional inverter.

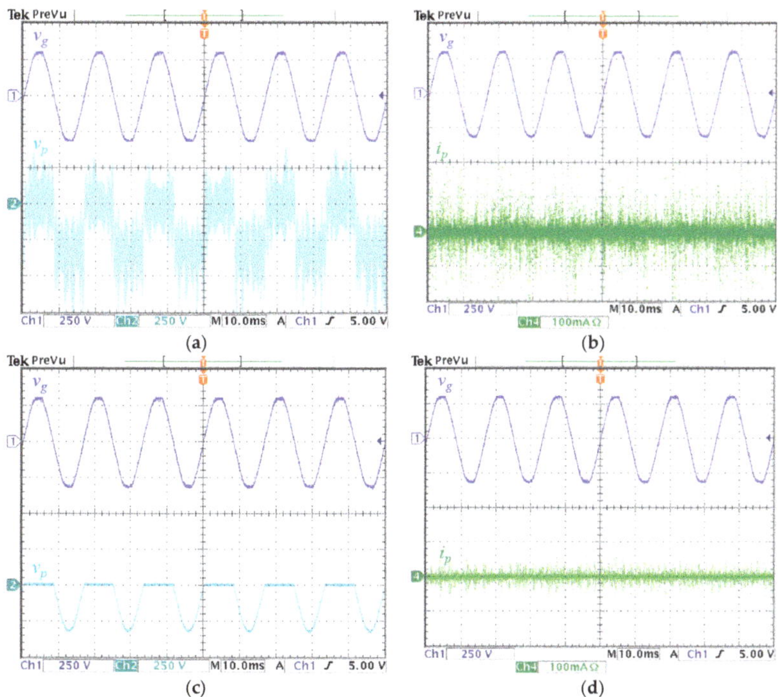

Figure 8. Experimental waveforms: (**a**) v_g and v_p in the conventional inverter; (**b**) v_g and i_p in the conventional inverter; (**c**) v_g and v_p in the proposed inverter; (**d**) v_g and i_p in the proposed inverter.

Figure 9 shows the experimental waveforms of the proposed inverter. Figure 9a shows the V_{in}, V_{PN}, v_g, and i_g. The proposed inverter steps up V_{in} to V_{PN}, and controls i_g with high power factor.

Figure 9b shows the V_{in}, V_{C2}, and V_{PN}. The peak DC-link voltage was regulated and the grid current was controlled in the proposed inverter. Figure 9c shows the V_{in}, V_{C12}, V_{C2}, and i_{L1}. Since the proposed inverter operates in the grid-tied mode, a twice grid frequency DC ripple current was observed on the DC input side. If active power decoupling schemes [17] are adopted, the ripple components as well as the reactive component sizes can be reduced.

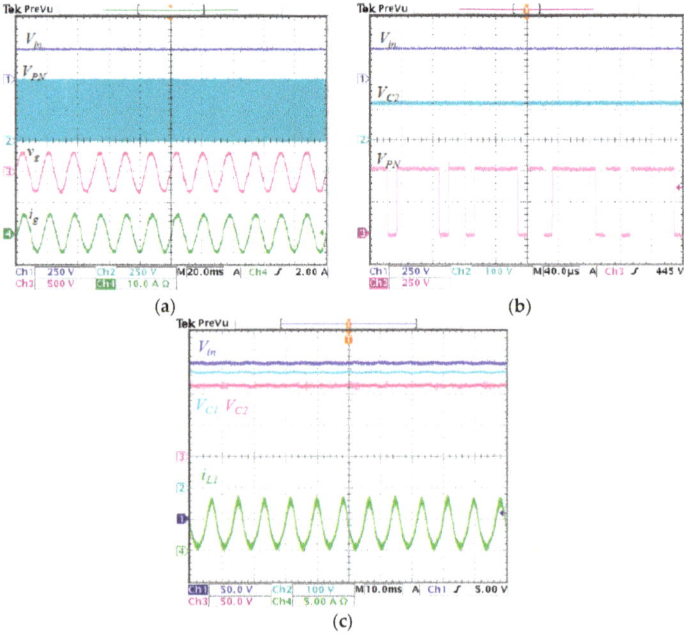

Figure 9. Experimental waveforms of the proposed inverter: (**a**) V_{in}, V_{PN}, v_g, and i_g; (**b**) V_{in}, V_{C2}, and V_{PN}; (**c**) V_{in}, V_{C1}, V_{C2}, and i_{L1}.

Figure 10 shows the power efficiency curves of the inverters. Figure 10a shows the power efficiency curves when the IGBTs have been adopted for the power switches. The inverter in Figure 1 achieved an efficiency of 91.7% at the rated power. The inverter in [15] achieved an efficiency of 92.1% at the rated power. The proposed inverter achieved an efficiency of 92.6% at the rated power, obtaining the highest efficiency of 92.9% at 0.6 kW. The proposed inverter obtained higher power efficiency than the previous inverters. The proposed inverter not only improved the power efficiency by reducing the leakage current, but also alleviated switching power losses by reducing the number of switching times for power switches. Figure 10b shows the power efficiency curves when the SiC MOSFETs were adopted for the power switches. It is observed that the power efficiency was improved when the SiC MOSFETS were used for the power switches. The inverter in Figure 1 achieved an efficiency of 93.6% at the rated power. The inverter in [15] achieved an efficiency of 94.0% at the rated power. The proposed inverter achieved an efficiency of 94.7% at the rated power, obtaining the highest efficiency of 95.1% at 0.6 kW. Regardless of the types of power switches, the proposed inverter achieved higher power efficiency than previous inverters.

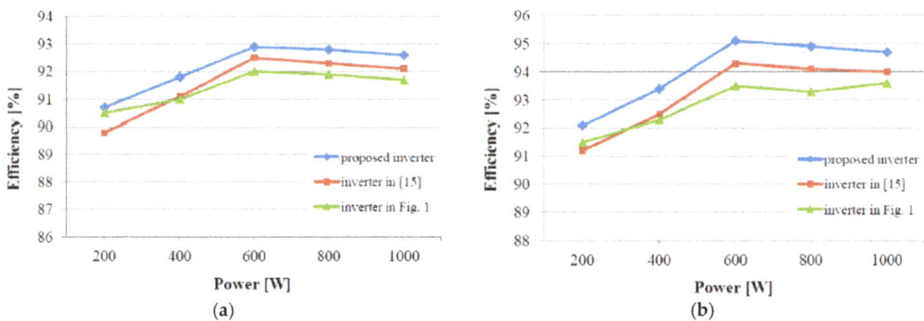

Figure 10. Power efficiency curves: (**a**) when IGBTs were adopted; (**b**) when SiC MOSFETs were adopted.

4. Discussion

4.1. Power Loss Comparison

A unipolar PWM was used for both the conventional inverter in Figure 1 and the proposed inverter. In both inverters, S_1 and S_2 (S_3 and S_4) had one turn-on switching loss and one turn-off switching loss during one switching period, respectively, for the positive (negative) grid cycle. Both inverters had two turn-on switching losses and two turn-off switching losses during one switching period. The number of switching times was identical during one switching period, as one of the switching legs operated at grid frequency. The number of conducted switching devices was also identical during one switching period. Then, both inverters had identical switching and conduction losses. Even though they had identical power losses for the switching devices, the conventional inverter in Figure 1 had a high leakage current, while the proposed inverter had a low leakage current. This is the reason why the proposed inverter achieved higher power efficiency than the conventional inverter in Figure 1.

Figure 11a shows the circuit diagram of the previous inverter in [15]. Figure 11b shows its switching signal diagrams for $S_1 \sim S_4$ during one switching period. In the previous inverter in [15], S_5 (S_6) was always turned on for the positive (negative) grid cycle. S_1 and S_4 (S_2 and S_3) operated together during one switching period. In order to generate the shoot-through switching states, S_1, S_2, S_3, and S_4 had to be simultaneously turned on during one switching period, as shown in Figure 11b. S_1 and S_4 (S_2 and S_3) had two turn-on switching losses and two turn-off switching losses during one switching period, respectively, for the positive (negative) grid cycle. Meanwhile, S_2 and S_3 (S_1 and S_4) had one turn-on switching loss and one turn-off switching loss during one switching period, respectively, for the positive (negative) grid cycle. Then, the previous inverter in [15] had six turn-on switching losses and six turn-off switching losses during one switching period. Thus, the previous inverter in [15] had higher switching losses than the proposed inverter, even though the additional switches (S_5, S_6) in both inverters operated at the grid frequency. In addition, the conduction losses of the previous inverter in [15] were higher than the proposed inverter because of the use of two additional diodes. This is the reason why the proposed inverter achieved higher power efficiency than the previous inverter in [15]. This is also the reason why the previous inverter in [15] achieved lower power efficiency than the conventional inverter in Figure 1. As the power level decreased, the switching power losses became significant, which reduced the light load efficiency.

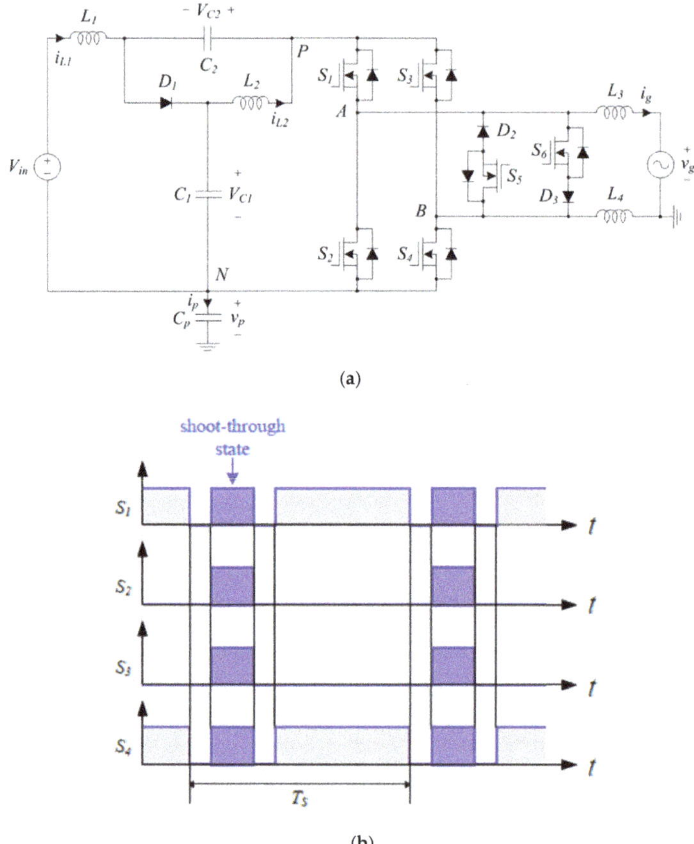

(a)

(b)

Figure 11. Circuit and signal diagrams of the inverter in [15]: (**a**) circuit diagram; (**b**) signal diagram for $S_1 \sim S_4$.

4.2. Topological Investigation

The decoupling circuits for reducing the leakage current can be divided into two concepts, namely AC-based and DC-based decoupling techniques [9]. The proposed inverter can be derived from the AC-based decoupling technique as the bidirectional switch (S_5, S_6) and two inductors (L_3, L_4) provide the current path for clamping the common-mode voltage to the neutral and the grid voltage for the positive and negative grid cycles, respectively. The DC-link voltage between the QZS circuit and the full-bridge inverter pulsates as the shoot-through state is generated. Thus, deriving a DC-based decoupling technique will be quite achievable for the ZS-based inverters. Along with the present state of the art and trend of decoupling circuit techniques [9], the embedded-switch inverter (ESI) [21] and zero-voltage state rectifier (ZVR) [22] concepts can be candidate solutions for extending leakage current reduction schemes to three-phase inverter applications. Furthermore, multilevel inverters using cascaded topologies [23–25] can be advanced approaches for QZS inverters to improve the output power quality.

5. Conclusions

A high-efficiency transformerless QZS inverter was proposed for single-phase grid-tied applications. The proposed inverter effectively reduced the leakage current, providing high power efficiency. A bidirectional switch operating with the grid frequency was used, which provided a current path to remove high-frequency components for the common-mode voltage. Switching power losses were also alleviated by reducing the number of switching times for power switches. The operation principle of the proposed inverter was described. A control scheme was suggested for regulating the DC-link voltage and the grid current. A 1.0 kW prototype inverter was designed and tested to evaluate the performance of the proposed inverter. SiC MOSFETs were applied to the proposed inverter to increase the power efficiency. The experimental results showed that the proposed inverter achieved high performance in terms of leakage current reduction and power efficiency improvement.

Author Contributions: W.-Y.C. managed the project, and wrote the manuscript. M.-K.Y. performed the experiments, analyzed the data, and edited the manuscript.

Funding: This research was supported by the National Research Foundation of Korea (NRF-2016R1D1A3B03932350). It was also supported by research funds from Chonbuk National University in 2018.

Conflicts of Interest: The authors declare no potential conflict of interest.

References

1. Anderson, J.; Peng, F.Z. A class of quasi-Z-source inverters. In Proceedings of the IEEE Industry Applications Society Annual Meeting, Edmonton, AB, Canada, 5–9 October 2008; pp. 1–7.
2. Zhou, Y.; Li, H.; Li, H. A single-phase PV quasi-Z-source inverter with reduced capacitance using modified modulation and double-frequency ripple suppression control. *IEEE Trans. Power Electron.* **2016**, *31*, 2166–2173. [CrossRef]
3. Gu, Y.; Chen, Y.; Zhang, B. Enhanced-boost quasi-Z-source inverter with an active switched Z-network. *IEEE Trans. Ind. Electron.* **2018**, *65*, 8372–8381. [CrossRef]
4. Komurcugil, H.; Bayhan, S.; Bagheri, F.; Kukrer, S.; Abu-Rub, H. Model-based current control for single-phase grid-tied quasi-Z-source inverters with virtual time constant. *IEEE Trans. Ind. Electron.* **2018**, *65*, 8277–8286. [CrossRef]
5. Xiao, H.; Xie, S. Leakage current analytical model and application in single-phase transformerless photovoltaic grid-connected inverter. *IEEE Trans. Electromagn. Compat.* **2010**, *52*, 902–913. [CrossRef]
6. Bozorgi, A.M.; Hakemi, A.; Farasat, M.; Monfared, M. Modulation techniques for common-mode voltage reduction in the Z-source ultra-sparse matrix converters. *IEEE Trans. Power Electron.* **2019**, *34*, 958–970. [CrossRef]
7. Noroozi, N.; Zolghadri, M.R. Three-phase quasi-Z-source inverter with constant common-mode voltage for photovoltaic application. *IEEE Trans. Ind. Electron.* **2018**, *65*, 4790–4798. [CrossRef]
8. Mohan, N. *Power Electronics: A First Course*; John Wiley & Sons: Hoboken, NJ, USA, 2012; pp. 1–270.
9. Li, W.; Ge, Y.; Luo, H.; Cui, W.; He, X.; Xia, C. Topology review and derivation methodology of single-phase transformerless photovoltaic inverters for leakage current suppression. *IEEE Trans. Ind. Electron.* **2015**, *62*, 4537–4551. [CrossRef]
10. Saridakis, S.; Koutroulis, E.; Blaabjerg, F. Optimization of SiC-based H5 and conergy-NPC transformerless PV inverters. *IEEE J. Emerg. Sel. Top. Power Electron.* **2015**, *3*, 555–567. [CrossRef]
11. Kerekes, T.; Teodorescu, R.; Rodriguez, P.; Vazquez, G.; Aldabas, E. A new high-efficiency single-phase transformerless PV inverter topology. *IEEE Trans. Ind. Electron.* **2011**, *58*, 184–191. [CrossRef]
12. Gonzalez, R.; Lopez, J.; Sanchis, P.; Marroyo, L. Transformerless inverter for single-phase photovoltaic systems. *IEEE Trans. Power Electron.* **2007**, *22*, 693–697. [CrossRef]
13. Li, K.; Shen, Y.; Yang, Y.; Qin, Z.; Blaabjerg, F. A transformerless single-phase symmetrical Z-source HERIC inverter with reduced leakage currents for PV systems. In Proceedings of the IEEE Applied Power Electronics Conference and Exposition (APEC), San Antonio, TX, USA, 4–8 March 2018; pp. 356–361.
14. Meraj, M.; Iqbal, A.; Brahim, L.; Alammari, R.; Abu-Rub, H. A high efficiency and high reliability single-phase modified quasi Z-source inverter for non-isolated grid connected applications. In Proceedings of the IEEE IECON, Yokohama, Japan, 9–12 November 2015; pp. 4305–4310.
15. Meraj, M.; Rahman, S.; Iqbal, A.; Ben-Brahim, L. Common mode voltage reduction in a single-phase quasi Z-source inverter for transformerless grid-connected solar PV applications. *IEEE J. Emerg. Sel. Top. Power Electron.* **2018**. [CrossRef]

16. Guo, X. A novel CH5 inverter for single-phase transformerless photovoltaic system applications. *IEEE Trans. Circ. Syst. Ii Express Briefs* **2017**, *64*, 1197–1201. [CrossRef]
17. Zhang, J.; Ding, H.; Wang, B.; Guo, X.; Padmanaban, S. Active power decoupling for current source converters: An overview scenario. *Electronics* **2019**, *8*, 197. [CrossRef]
18. Peng, F.Z. Z-source inverter. *IEEE Trans. Ind. Appl.* **2003**, *39*, 504–510. [CrossRef]
19. Ellabban, O.; Mierlo, J.V.; Lataire, P. A DSP-based dual-loop peak dc-link voltage control strategy of the Z-source inverter. *IEEE Trans. Power Electron.* **2012**, *27*, 4088–4097. [CrossRef]
20. Golestan, S.; Monfared, M.; Freijedo, F.D.; Guerrero, J.M. Design and tuning of a modified power-based PLL for single-phase grid-connected power conditioning systems. *IEEE Trans. Power Electron.* **2012**, *27*, 3639–3650. [CrossRef]
21. Guo, X.; Yang, Y.; Zhu, T. ESI: A novel three-phase inverter with leakage current attenuation for transformerless PV systems. *IEEE Trans. Ind. Electron.* **2018**, *65*, 2967–2974. [CrossRef]
22. Guo, X.; Zhang, X.; Guan, H.; Kerekes, T.; Blaabjerg, F. Three-phase ZVR topology and modulation strategy for transformerless PV system. *IEEE Trans. Power Electron.* **2019**, *34*, 1017–1021. [CrossRef]
23. Guo, X.; Jia, X. Hardware-based cascaded topology and modulation strategy with leakage current reduction for transformerless PV systems. *IEEE Trans. Ind. Electron.* **2016**, *63*, 7823–7832. [CrossRef]
24. Ge, B.; Liu, Y.; Abu-Rub, H.; Peng, F.Z. State-of-charge balancing control for a battery-energy-stored quasi-Z-source cascaded-multilevel-inverter-base photovoltaic power system. *IEEE Trans. Ind. Electron.* **2018**, *65*, 2268–2279. [CrossRef]
25. Sun, D.; Ge, B.; Liang, W.; Abu-Rub, H.; Peng, F.Z. An energy stored quasi-Z-source cascade multilevel inverter-based photovoltaic power generation system. *IEEE Trans. Ind. Electron.* **2015**, *62*, 5458–5467. [CrossRef]

© 2019 by the authors. Licensee MDPI, Basel, Switzerland. This article is an open access article distributed under the terms and conditions of the Creative Commons Attribution (CC BY) license (http://creativecommons.org/licenses/by/4.0/).

Article

Multiport Isolated link with Current-Fed Z-Source Converters to Manage Power Imbalance in PV Applications

Jose M. Sandoval [1], Victor Cardenas [1,*], Manuel A. Barrios [1], Mario Gonzalez Garcia [2] and Janeth Alcala [3]

[1] Engineering Department, Autonomous University of San Luis Potosi, Dr. Manuel Nava No. 8, Col. Zona Universitaria Poniente, San Luis Potosí 78290, Mexico
[2] Engineering Department, Conacyt-Autonomous University of San Luis Potosi, San Luis Potosí 78290, Mexico
[3] Faculty of Electromechanical Engineering, University of Colima, Manzanillo, Colima 28864, Mexico
*   Correspondence: vcardena@uaslp.mx

Received: 29 December 2019; Accepted: 30 January 2020; Published: 6 February 2020

Abstract: In order to address power imbalance in large-scale PV systems, this paper presents a multiport isolated medium-frequency (MF) link to process different power levels from PV arrays, using current-fed Z-source inverters (CZSI) modules to drive the energy from the PV array to the MF link in a single power stage. The MF link provides the galvanic isolation required by many codes and standards to integrate a system to the grid. The multiport configuration of the MF link allows the system to process different power levels on each primary port and, being all ports magnetically coupled, provide balanced power levels on secondary ports which can be used for multilevel converters applications. Additionally, the current-fed version of the Z-source topology offers natural protection against short-circuit damage enhancing reliability of the system. The analysis of CZSI cells and the study of asymmetrical power levels transference, as well as simulation and experimental results are presented.

Keywords: photovoltaics; Z-source; current-fed; medium-frequency; power-imbalance

---

1. Introduction

Since the past two decades, Distributed Energy Sources based on Renewable Energy have grabbed much attention in power transformation and distribution sectors. In particular, solar-photovoltaic (PV) systems have the highest growing rate (along with wind energy) with the addition of 109 GW in 2018–2019 [1]. In terms of installed capacity, PV energy increased from 22 to 489 GW in this decade [2]. The integration of solar energy to the utility grid requires power processing stages in order to successfully inject the gathered energy. The implementation of utility-scale solar systems requires galvanic isolation according to standards in many countries, and it is conventionally implemented by a grid frequency transformer. Even though traditional transformers are reliable and efficient, they are also heavy and bulky [3].

Some approaches for PV systems propose the implementation of Medium-Frequency (MF) isolation in order to reduce volume and weight [3–5]. The topologies with MF links usually features: a DC-AC stage consisting on a DC-DC converter and a DC-AC stage in order to drive the transformer, a Magnetic link operating in medium-high frequency, and an AC-AC stage consisting on a rectifier and a DC-AC stage. The key features of these systems are: individual maximum power point tracking (MPPT), galvanic isolation, enhanced voltage elevation ratio, reduced total harmonic distortion (THD), voltage stress reduction in medium-voltage (MV) stage, and reduced overall system volume and weight. Some drawbacks are: voltage imbalance in multilevel cells due to different power levels in each string, more power stages due to the presence of an additional low voltage inverter and MV rectifier, and increased stress in low voltage stage switches since the all the input power must be handled by a single switch. A solution for multilevel cell imbalance is presented in [3] featuring a MF link that consists on a transformer with several secondary windings which being magnetically coupled *naturally* balances the cascaded cells of the multilevel converter. Some downsides are that the system has four power stages, and reports low global efficiency levels. However, some adjustments in the topology can help improving reliability and efficiency. The PV system low voltage side in [5] consists on a PV string and a DC-AC power stage. The DC-AC power stage requires buck-boost and MPPT capability. DC-AC stage is commonly integrated by a boost converter and an inverter in order to drive the MF transformer [5]. Another approach is stated in [6], proposing a flyback converter plus a H-bridge topology for DC-AC stage features, and enhanced efficiency with high frequency isolation; however, the number of semiconductors and current stress limits the topology to low power applications (<300 W). In [7] a Cuk-SEPIC combination featuring a single switch and coupled inductors is presented in order to obtain a bipolar output enhancing efficiency, however current stressed switch can present reliability problems.

Since PV system behave as current sources, current-fed converters are considered for the DC-AC stage. The isolated current-fed push-pull converter features a low component count and simply circuitry which makes it rather reliable and simple to implement; it also benefits from a high-voltage conversion ratio [8–10]. Some identified drawbacks of the topology are: high switching losses, high-voltage spikes in switches, high start-up current, and central tapped transformer coils requires a bigger core and can be a serious limitation for a magnetically coupled multiport configuration [9,11–13].

The L-type converter is an interesting topology that features reduced current stress in switches and reduced voltage stress on output capacitor and diodes [8,11]. There are a better use of the magnetic core of the MF link and thus making the topology a good option for the multiport implementation. Some limitations of the topology are: higher voltage efforts in switches, voltage spikes in switches due to a resonance with the leakage inductance of the transformer. Current-fed topologies with boost operation also require additional start-up circuitry due to the current in inductors during starting process. The L-type converter is suitable to low voltage, high current applications [8,11].

The Z-source inverter (ZSI) has been considered for DC-AC converters [14,15]; this topology has a buck-boost DC-AC capability in a single-stage which suggests an efficiency and reliability enhancement. The operation principle of the ZSI and current-fed topologies mentioned consist in apply a shoot-through state to overlap the H-bridge branches (0V) which provides natural protection against short-circuits, making it suitable for large-scale applications. After its presentation, variations on the topology were proposed for many applications such a grid-tied converters without galvanic isolation [16–18], with galvanic isolation [19–21], and finally for grid-tied PV systems [22–24].

Z-sources topologies are interesting for PV applications because they can operate as voltage and current source. Additionally, Z-source converters operating as current source can be better applied in medium-high power applications due to the restriction of <1 kV in PV arrays terminals. A current fed Z-source Inverter (CZSI) version is proposed in [25] which features a constant source current, protecting the PV array from returning current.

The CZSI shares many features of the above mentioned topologies such as DC-AC power processing in a single stage and voltage spikes in switches due to resonance with the leakage inductance [12]. Some particular feature of the CZSI in deference to the mentioned topologies is a very low current stress in diode a capacitors, which are the most vulnerable elements on the Z network and thus enhancing the reliability of the circuit. Some drawbacks of the CZSI against the other topologies are the number of passive components and limited voltage boost. Since the PV system voltage does not presents a wide variation, voltage boost drawback is not considered of high relevance. The CZSI has similar to L-type topology power requirements for the magnetic core of the MF link, which is an important feature for multiport applications.

The system considered in this work is shown in Figure 1 and consists in three power stages: Low voltage DC-AC stage consists in a PV array, followed by an elevation stage based on a current-fed Z-source inverter (CZSI) that performs MPPT and sets the voltage for the inverter who will drive the MF link; MF Link is an elevation AC-AC stage which provides multiport configuration integrated by multiple isolated and magnetically coupled inputs and outputs. The MF Link can manage x primary ports and y secondary ports, only restricted by the physical space and power the magnetic core of the transformer can handle. In this study, three input ports where selected since is the minimum number of modules in order to validate that different power levels can be processed in x modules and not only flowing between two modules, and in order to show that output ports are balanced, two ports where selected.

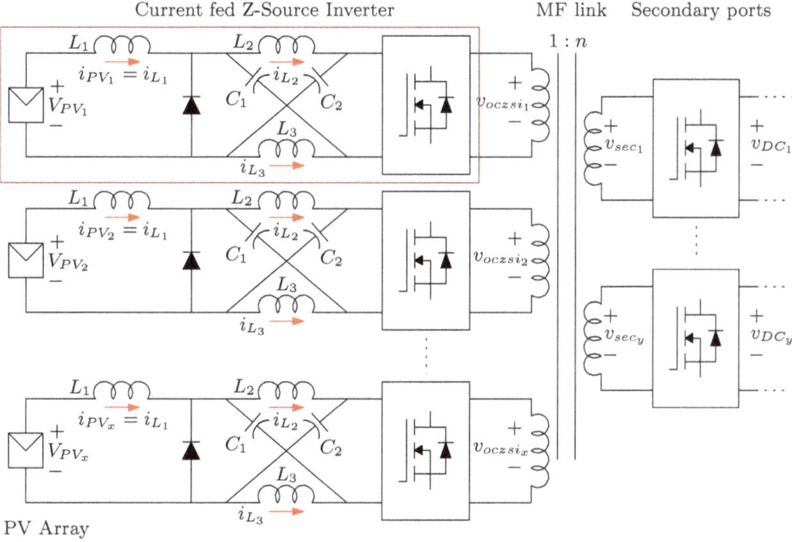

Figure 1. Multiport isolated link system with CZSI input cells to manage asymmetrical power supplies.

The focus of this paper is derived from [26] and presents the analysis of a multiport isolated link in asymmetrical input power transference conditions using CZSI modules to manage the energy. This work shows CZSI analysis, simulation and experimental results to validate the analysis performed. This proposal offers reduction of power stage by using CZSI modules energized by unbalanced PV systems, reliable magnetic isolation with a multiport link to meet standard requirements and, naturally balanced output ports and isolated sources for grid-connected oriented applications.

2. Current-Fed Z-Source Inverter Operation and Design

In Figure 2, the CZSI circuit is presented and Figure 3 shows the equivalent circuits for operation states. In steady state, average voltage in L_1, L_2 and L_3 remains zero as well as current in C_1 and C_2 over a switching period T. The CZSI presents two main operation states: conduction and shoot-through; in conduction state mode (Sw_1 and Sw_4 ON and Sw_2 and Sw_3 OFF, Figure 3a), current across inductors L_1, L_2 and L_3 is given by PV array i_{PV}, and the voltage across the capacitors is set by the PV source V_{PV}. During conduction state, L_3 is charging with V_{PV}, to supply current for the impedance network later.

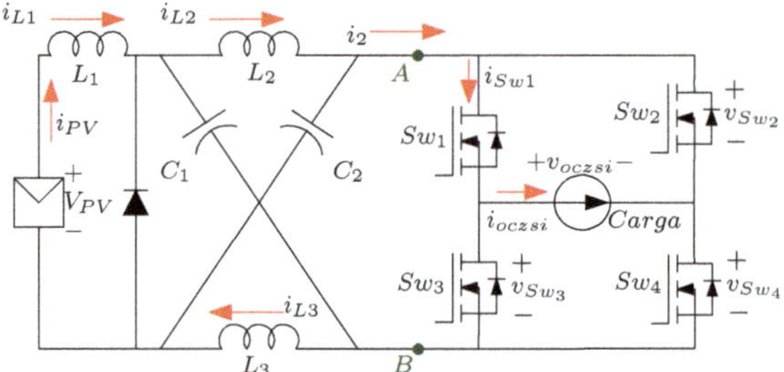

Figure 2. Current-fed Z-source inverter.

In shoot-through state (All switches are ON, Figure 3b), i_{L_1} and i_{L_2} rises as i_{L_3} falls since L_3 supplies the current, i_{PV} remains constant in Δ_{i_L}, limited by inductance L. In this time, voltages in inductors are imposed by v_C. The capacitor voltage decrease in Δ_{v_C} and is limited by a capacitance C. As it can be seen, that the Z LC circuit is always in conduction and the switches are the ones that control operation modes, an hence, the output voltage. The required shoot-through duty cycle D_z for a desired V_{ozp} can be found by the following expression:

$$D_z = 1 - \frac{V_{PV}}{V_{ozp}}, \tag{1}$$

where V_{ozp} is the peak value of v_{oczsi}. The voltage boost for the CZSI is directly modified by D_z; therefore, by manipulating this duty cycle the MPPT can be performed and the DC-AC conversion remains in a single stage.

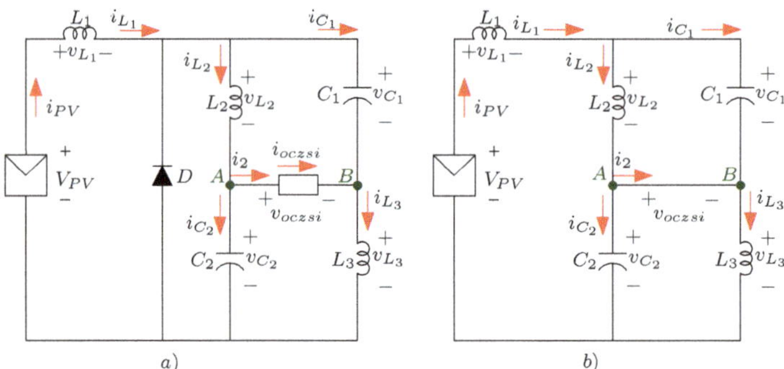

Figure 3. CZSI equivalent circuits: (**a**) CZSI conduction state, (**b**) CZSI shoot-through state.

Output voltage can be obtained through Equation (1), as it is a function of D_z; higher values of D_z produces higher gain limited to $2V_{PV}$. v_{oczsi} RMS voltage is obtained by:

$$V_{oRMS} = V_{ozp}\sqrt{1 - D_z}. \tag{2}$$

The shoot through time TD_z is shared by all of the modules due to the parallel connection, thus if one of the windings is in OV, the rest will be in OV as well. The last results in all of the windings sharing the duty cycle D_z at the higher value.

The effect of D_z in switches is that during $t = [0 \quad T - D_z T]$, half of the current i_2 given by:

$$i_2 = 2i_{L_2} - i_{L_1}, \tag{3}$$

is conducted by all switches. During $t = [T - D_z T \quad T]$, current in switches is i_2. In conduction mode, i_{oczsi} has a ripple Δi_o with a slope given by:

$$m = \frac{\Delta i_o}{\Delta_t} = \frac{I_A - I_B}{\Delta_t},$$ (4)

where I_A and I_B are peak values of i_2 (Figure 4). However, i_{oczsi} is affected by winding impedance Z composed by the leakage inductance L_p and the winding resistance, producing a non-zero current $i_{oczsi} = i_{L_p}$ (Figure 4). The RMS current for i_{oczsi} is given by:

$$I_{oRMS} = \sqrt{\frac{1}{T} \int_0^{T-D_z T} (mT + I_A)^2 dt + \int_{T-D_z T}^{T} i_{Lp}^2 dt}.$$ (5)

By solving Equation (5), I_{oRMS} is given by:

$$I_{oRMS} = \sqrt{(i_{PV}^2 + \frac{3}{4}\Delta i_L^2)(1 - D_z) + i_{Lp}^2 D_z},$$ (6)

where I_A is the upper i_2 value, I_B the lower value, and L_p the leakage inductance.

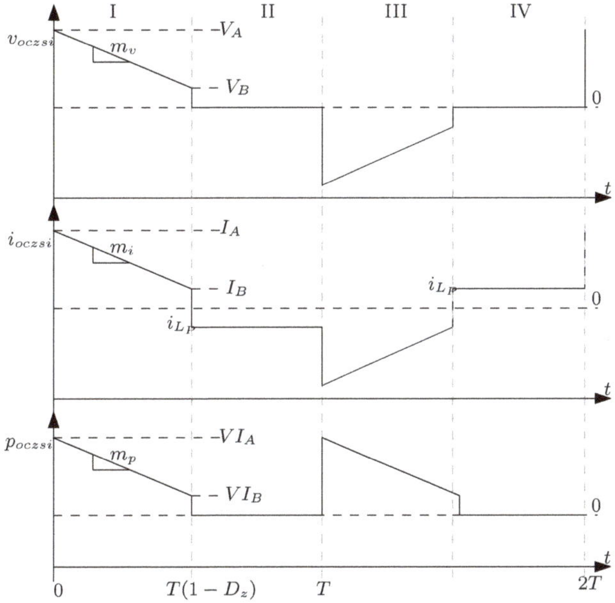

Figure 4. CZSI Output: V_{oczsi}, i_{oczsi}, and power p_{oczsi}.

In Figure 5a, a simplified representation of a multiport configuration consisting on x primary windings transformer and a single output is shown. Voltages V_1, V_2 through V_x are the RMS value of CZSI modules output voltage V_{oczsi}, and V_m is the secondary voltage v_{sec}. In the MF link, the sum of average power supplied to each primary port must match the average power on the secondary ports, thus for a single output:

$$P_1 + P_2 + ... + P_x = P_{sec},$$ (7)

expanding the concept to y secondary ports, the sum of the average power on each secondary output must also match the sum of the average input power:

$$P_1 + P_2 + ... + P_x = P_{sec_1} + P_{sec_2} + ... + P_{sec_y}. \tag{8}$$

For multilevel converters applications, it can be considered that power in each secondary port is balanced since they all have the same number of turns and the same output voltage. Secondary ports also share the same current since the all supply the same load, thereby Equation (8) can be expressed as:

$$P_1 + P_2 + ... + P_x = yP_{sec}. \tag{9}$$

The parallel configuration shows that when a module is in shoot-through state, the remaining modules will share this state, thus voltage v_{oczsi} will be zero for all of the windings, resulting in an identical equivalent D_z value for all input CZSI modules. The last implies that all the cells will have very similar values of v_{oczsi}.

The imposed duty cycle D_z will be the higher value of all CZSI modules for the MPP, thus having different D_z references is not necessary. Current levels will be in function of the power supplied by the PV array, thus different in all primary ports.

Despite the different power levels on the modules, the non-zero current associated to L_p, remains similar for all windings since it is product of parasitic elements, and it is assumed that all windings are homogeneous.

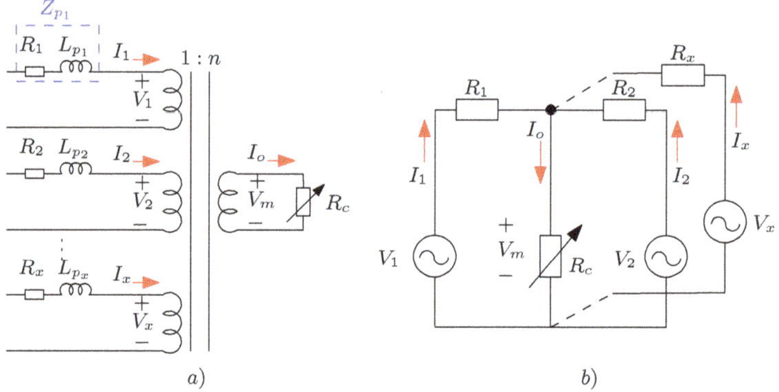

Figure 5. (a) Multiport transformer, (b) Equivalent circuit.

Semiconductor Sizing for CZSI Modules

The voltage and current stress on semiconductors are defined by the operation mode of the CZSI module. The voltage stress for the semiconductors on CZSI conduction mode, is the voltage v_{ozp} and can be found with Equation (1) and zero for shoot-through mode:

$$v_{Sw} = \begin{cases} v_{ozp} & 0 \le t < T(1 - D_z) \\ 0 & T(1 - D_z) \le t \le T. \end{cases} \tag{10}$$

For selection purposes, a security factor of 50% can be added to v_{ozp} and is given by:

$$V_{Sw_{selec}} = V_{ozp}1.5 = \left(V_{PV}\frac{1}{1 - D_z} \right)1.5. \tag{11}$$

The current across the semiconductors, also depends on operation mode and can be expressed by:

$$i_{Sw} = \begin{cases} i_2 & 0 \leq t < T(1 - D_z) \\ i_2/2 & T(1 - D_z) \leq t \leq T. \end{cases} \tag{12}$$

The maximum current that semiconductors must conduct is the peak value of i_2 plus a security factor of 25%. The current to be considered when choosing a device is given by:

$$I_{Sw_{selec}} = i_{2_{max}} = \left(I_{PV} + \frac{3\Delta i_L}{2} \right) 1.25. \tag{13}$$

3. Power Transfer under Unbalanced Conditions

An issue of having the whole installed capacity divided in strings is that the power in each array of PV panels can be unbalanced. Let's consider a two input modules scenario, where P_1 and P_2 are CZSI modules output power and are different from each other.

A simple scheme is given in Figure 5b for analysis purposes. Resistance R is considered equal for all windings and thus $R_1 = R_2 = R$. From Figure 5b, the voltage on the secondary side can be computed by:

$$V_1 n - I_1 R = V_m = V_2 n - I_2 R, \tag{14}$$

where n is the turn ratio and, V_1 and V_2 are the RMS values of v_{oczsi} in modules 1 and 2 respectively. Such voltages cannot be arbitrary set to extract two different power levels on each loop, since (14) has to be fulfilled. If the voltage V_1 is imposed, then V_2 will be in function of the desired power, thus V_2 is a variable voltage source. To find a combination of V_1 and V_2, that will allow different power transfer on modules, first V_m must be found in terms of V_1 with (14). Then P_2 can be computed from:

$$P_2 = \frac{R}{n} I_2^2 + \frac{V_m}{n} I_2. \tag{15}$$

By solving (15) for I_2, a voltage value for V_2 to transfer P_2 can be found. For more than two modules or windings, a similar analysis can be performed. One module has a fixed voltage, and all the subsequent modules adapt their voltage to meet the power transference needed.

$$V_1 n - I_1 R = V_m = V_2 n I_2 R = \ldots = V_x n I_x R. \tag{16}$$

Therefore the current for x given modules, (17) can be solved for I_x:

$$P_x = \frac{R}{n} I_x^2 + \frac{V_m}{n} I_x. \tag{17}$$

Once expressions for output voltage and current are given, power on each winding can be calculated. Based on Figure 4, the average power is given by:

$$P = \frac{1}{T} \int_0^{T-TD_z} V_o I_B + \frac{V_o(I_A - I_B)}{2} t(dt), \tag{18}$$

$$P = \frac{V_o(I_A + I_B)}{2}(1 - D), \tag{19}$$

where I_A and I_B are the upper and lower I_2 values respectively. Apparent power S is expressed by:

$$S = V_{oRMS} \sqrt{(i_{PV}^2 + \frac{3}{4}\Delta i_L^2)(1 - D_z) + i_{Lp}^2 D_z}. \tag{20}$$

From (20) it can be observed that the $i_{Lp}^2 D_z$ component is associated to reactive power, and since all cells have similar values of L_p (because all windings are identical) the component in $i_{Lp}^2 D_z$ will be the same for all modules, an thus, have similar reactive power levels on all cells despite power level.

4. Results

Once the operation principle has been presented and the design equations given, simulation studies in PSIM, and experimental tests were performed with the parameters shown in Table 1. Simulations were performed considering the schematic in Figure 1; a PV system with a MF link including 3 primary ports and 2 secondary ports with a turn ratio of 1:7. A variable DC load is applied to each secondary port in order to extract 300 W from each PV array to get a total output power of 900 W, also, each CZSI module has a switching frequency ok 20 kHz with an input voltage of 50 V.

After the validation through simulations, a prototype has been built. The experimental prototype consists in two input CZSI modules energized by a PV array of two 250W PV panel in order to get a total power of 500 W and 50 V input voltage at MPP. Turn ration in the multiport transformer and D_z reference parameters were retained from simulation studies.

4.1. Simulation Results

In first instance, the system's performance was evaluated under balanced conditions using a PV array of 2 panels in series obtaining 50 V, 6.6 A, 300 W with an irradiance of $1\ kW/m^2$ applied to each CZSI module, and a variable load is applied to each secondary port in order to extract the maximum power around 900 W. In Figure 6 the voltage v_{oczsi} on modules 1, 2 and 3 are presented, and it can be see that they have a v_{ozp} value of 80 V. In the secondary side, the voltages v_{sec_1} and v_{sec_2} shows the same waveform with the voltage gain due to MF link's turn ratio, an thus the voltage in all of the ports remains balanced. From Equation (1), the voltage in conduction mode can be estimated in 80 V for the V_{PV} in the simulation, and from Equation (2), its RMS value of 62.3 V can be found.

Table 1. Simulation and experimental parameters.

Parameters	Simulation	Experimental
Module output power	300 W	300 W
Total power	900 W	600 W
Irradiance	$1000\ W/m^2$	unknow
Input voltage (V_{PV})	50 V	50 V
CZSI module output voltage (V_{ozp})	80 V	80 V
Secondary ports voltage (v_{sec})	560 V	560V
Duty cycle (D_z)	0.37	0.37
L_1, L_2 and L_3	0.38 mH	0.38 mH
C_1 and C_2	$2\ \mu F$	$2\ \mu F$
Switching frequency	20 kHz	20 kHz
Primary ports	3	2
Secondary ports	2	2
MF link turn ratio	1:7	1:7
Primary leakage inductance (L_p)	$1.18\ \mu H$	$1.18\ \mu H$
Secondary leakage inductance (L_s)	$37.7\ \mu H$	$37.7\ \mu H$
Primary parasitic resistance (R_p)	$45\ m\Omega$	$45\ m\Omega$
Secondary parasitic resistance (R_s)	$1.59\ \Omega$	$1.59\ \Omega$
Magnetizing inductance (L_m)	2 mH	2 mH

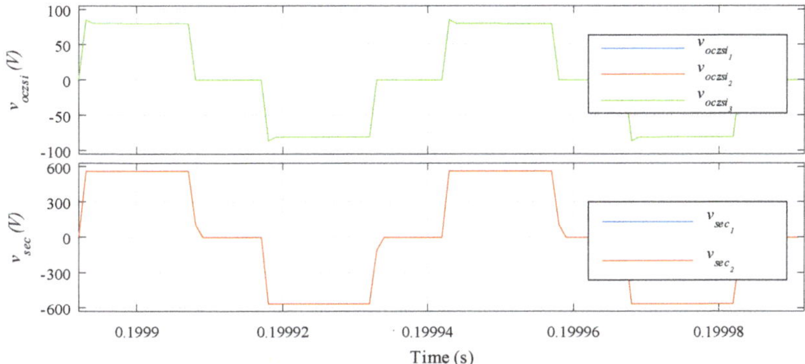

Figure 6. Simulation results: CZSI modules output voltages v_{oczsi_1}, v_{oczsi_2}, v_{oczsi_3} and secondary ports voltages v_{sec_1} and v_{sec_2} under balanced conditions.

Since the study is in balanced power conditions, the same current i_{oczsi} is expected on the three CZSI modules and RMS value can be estimated from Equations (3) and (6) to be around 5.6 A and 4.8 A respectively. Figure 7 validates that under balanced conditions the current on primary modules are equal. It can be observed that during shoot-through times, there is a non-zero current on CZSI modules output and in conduction mode. On the other hand, the current on the secondary ports, does not present non-zero current in shoot-through times and thus the effect of parasitic elements on MF link are decoupled from primary side.

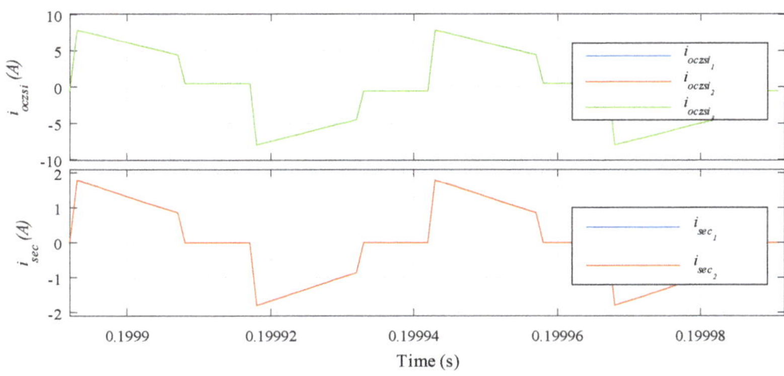

Figure 7. Simulation results: CZSI modules output current i_{oczsi_1}, i_{oczsi_2}, i_{oczsi_3} and secondary ports currents i_{sec_1} and i_{sec_2} under balanced conditions.

Figure 8 presents the average power on primary and secondary ports. First, the average output power on CZSI modules are shown in the first graphic and all of them are overlapped as expected for balanced conditions. In the same fashion, the average power in secondary side is balanced in both secondary outputs, and finally it can be observed that the sum of the average power of the three CZSI modules, matches with the sum of the average power on the secondary ports, and the difference between both is attributed to parasitic resistance in the MF link.

Once the operation under balanced conditions are validated, power imbalance studies were carried out. In order to appreciate the power imbalance, different irradiance were applied to each CZSI module: 900 W/m^2 for module 1, 500 W/m^2 in module 2, and 300 W/m^2 to module 3. Variable loads were kept on each secondary output in order to obtain the maximum power extraction. Also a fixed 0.37 D_z in all modules was set to put the system in MPP; since the higher D_z is imposed to all coupled modules, a master-slave configuration is used, selecting the module with higher irradiance as the master.

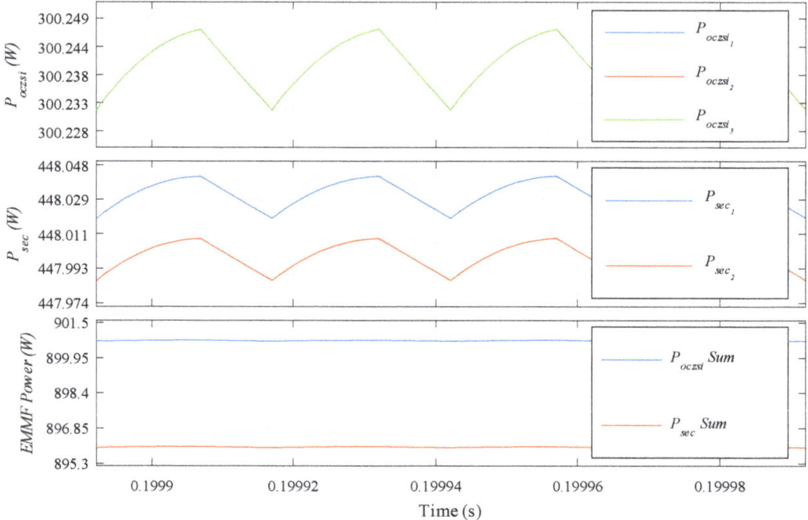

Figure 8. CZSI simulation results: Average power in primary ports and average power in secondary ports under balanced power operation.

Voltages in primary and secondary ports in steady state are shown in Figure 9. Voltage in primary ports are the output of CZSI modules and all three of them presents the same v_{oczsi} value of 80 V during conduction state, and zero voltage in shoot-through states despite the imbalance applied. Similarly, the voltage on the secondary side has the same waveform with the gain of the MF link in the magnitude, passing from 80 V peak on primary side to 560 V peak on the secondary side. This validates that voltage balance is kept in primary and secondary ports despite power imbalance.

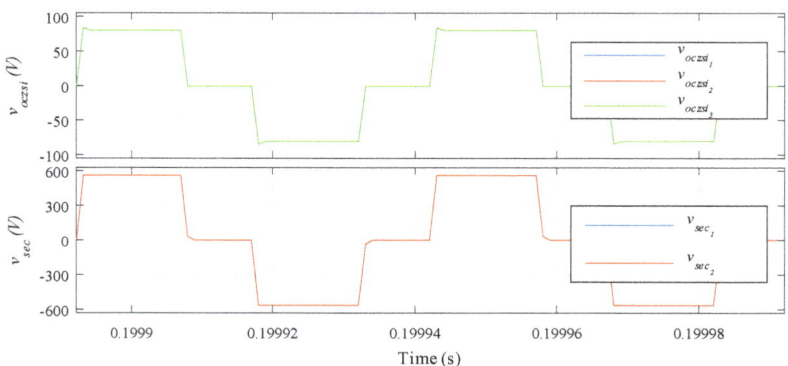

Figure 9. Simulation results: CZSI modules output voltages v_{oczsi_1}, v_{oczsi_2}, v_{oczsi_3} and secondary ports voltages v_{sec_1} and v_{sec_2} under unbalanced conditions.

Figure 10 shows the steady state current on primary and secondary ports and it can be observed that currents i_{oczsi_1}, i_{oczsi_2} and i_{oczsi_3} are different in function of the power supplied by the corresponding PV array. The non-zero current can be see during shoot-through state produced by parasitic elements in MF link. Considering, all primary ports have the same voltage, and each one has different current values, it validates that each port is managing different power levels. Despite the imbalance in CZSI modules, the second graphic shows that the current on secondary ports are equal without non-zero current during shoot-through states, which validates a balanced and decoupled power on secondary ports.

Figure 11 shows the power in primary ports as well as the average power in secondary ports. It can be see that each module has a different average power in primary side, in contrast to the secondary side that has exactly the same average power in both secondary ports.

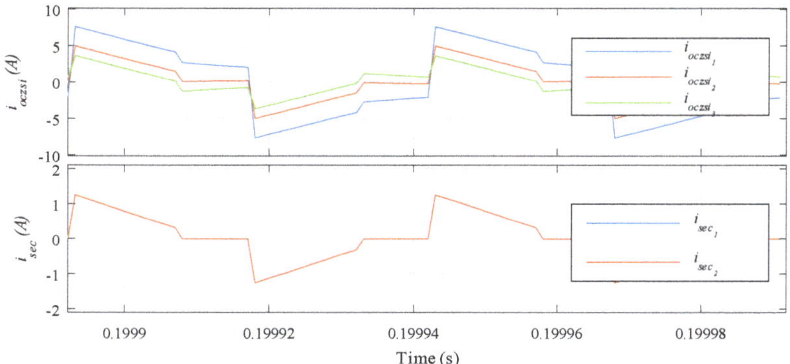

Figure 10. Simulation results: CZSI modules output current i_{oczsi_1}, i_{oczsi_2}, i_{oczsi_3} and secondary ports currents i_{sec_1} and i_{sec_2} under unbalanced conditions.

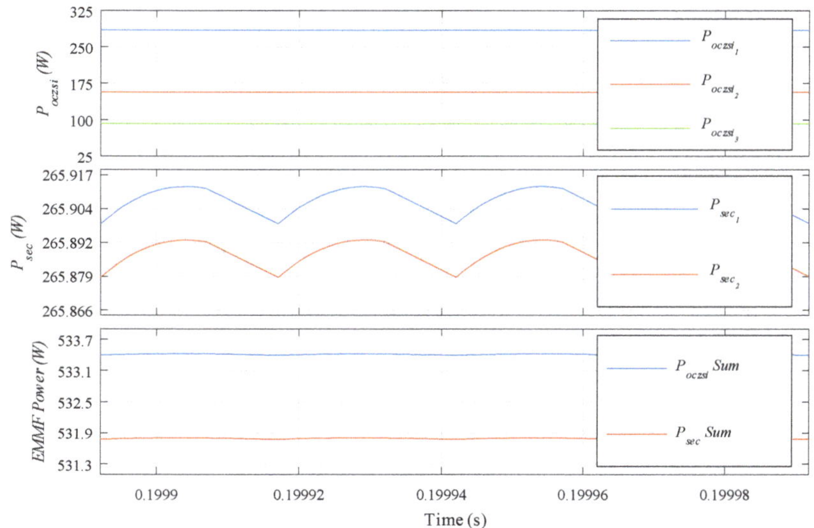

Figure 11. CZSI simulation results: Average power in primary ports and average power in secondary ports under unbalanced power operation.

When comparing the sum of the average power on primary and secondary side it is basically the same, but similar to balanced conditions, a small difference can be observed due to leakage inductance which produces the non-zero current present in primary side but not in secondary side. Finally a summary of the results obtained in the simulation studies is presented in Table 2.

Table 2. Simulation results.

Parameters	Primary Port 1	Primary Port 2	Primary Port 3	Secondary Port 1	Secondary Port 2
Balanced					
RMS Voltage	62.3 V	62.3 V	62.3 V	436 V	436 V
RMS Current	4.8 A	4.8 A	4.8 A	1.03 A	1.03 A
Output Power	300 W	300 W	300 W	448 W	448.3 W
Unbalanced					
RMS Voltage	62.3 V	62.3 V	62.3 V	436 V	436 V
RMS Current	4.56 A	2.51 A	1.48 A	0.646 A	0.646 A
Output Power	284.4 W	156.7 W	92.3 W	265.9 W	265.8 W

4.2. Experimental Results

Once the CZSI modules and MF link were validated through simulations, an experimental prototype has been tested in multiport configuration. The prototype is shown in Figure 12 and consists in 2 input CZSI modules, a magnetic link and two secondary cells with a diode full-wave rectifier on the output. The MF link consist on a multiport transformer with 15 turns on each primary port and 106 turns on each secondary port, built in a N27 E-80 ferrite core. Each CZSI module is energized by a PV panel array consisting on two 30 V, 8,33 A, 250 W panels connected in series achieving 60V at its MPP. Tests were carried out applying a 940 Ω resistive load to the output in order to demand 470 W at MPP. The CZSI modules were operated with a fixed D_z as marked in Table 1, but due to ambient conditions, only 60% of the array power was produced. Despite the last, tests validates the behavior of the system with a PV system supply.

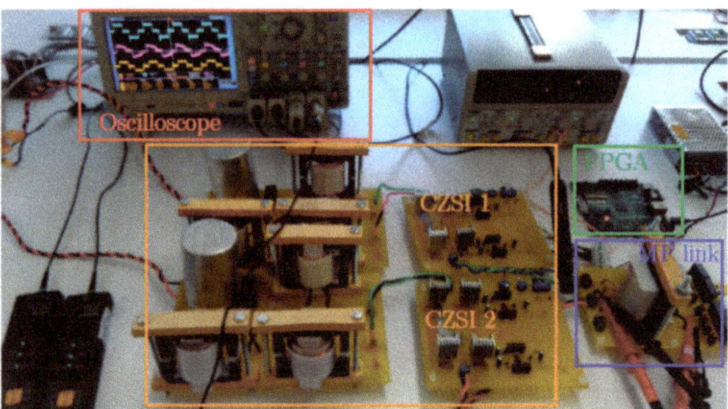

Figure 12. Experimental prototype.

In Figure 13 the performance of the prototype under balanced conditions is shown. The currents i_{oczsi_1} and i_{oczsi_2} are on channels 1 and 3 respectively, it can be see that they are overlapped with very small variations with a peak value of 7.5 A and 7.3 A on conduction mode, giving an approximate RMS value of 5 A; it can be noticed that the values are consistent with the simulation presented in Figure 7 with a peak value around 7.5 A and descending to 4.5 A. A high-frequency component oscillation due to a resonance between the leakage inductance of the MF link and the parasitic capacitance of the MOSFET can be appreciated during shoot-through times. In channels 2 and 4, voltages v_{oczsi_2} and v_{oczsi_2} are shown and are also equal with a v_{ozp} near to 75 V for a more-less 60 V RMS on both modules. The v_{oczsi} values obtained are similar to the ones presented in Figure 6 and the behavior is consistent showing a small spike during conduction state. The similar values of both CZSI values confirm that both PV systems are in a balanced state. In balanced state, both modules extracted 300 W from each PV array for a 600 W output power.

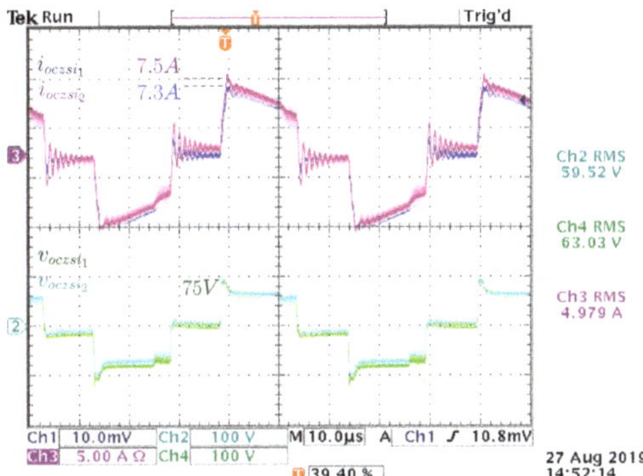

Figure 13. System test under balanced conditions. Ch1 (10 mV/div=5 A/div): i_{oczsi_1}, Ch3 (5 A/div): i_{oczsi_2}, Ch2 (100 V/div): v_{oczsi_1} and Ch4 (100 V/div): v_{oczsi_2}.

Once the performance of the prototype was validated in balanced conditions, a test under power imbalance was performed using the same two PV panels arrays of the balanced test. The resistive load and D_z duty cycle was also kept the same as before. Additionally, partial shade was applied to the PV panel array in module 2 in order to introduce the power imbalance.

In Figure 14, current i_{oczsi} in module 1 and 2 can be observed on channel 1 and 3 respectively. It can be noted that the same high frequency oscillation due to resonance between MOSFET's parasitic capacitance and the leakage inductance of the MF link in Figure 13 is also present under unbalanced conditions. Different current levels can be clearly appreciated, with 7.2 A peak on i_{oczsi_1} and 6 A peak on i_{oczsi_2}. Despite both cells have different current levels, v_{oczsi} stays on the same level as in balanced condition with 75 V v_{ozp}. A non-zero current in i_{oczsi} during shoot-through times can be observed and is related to the parasitic elements of the transformer.

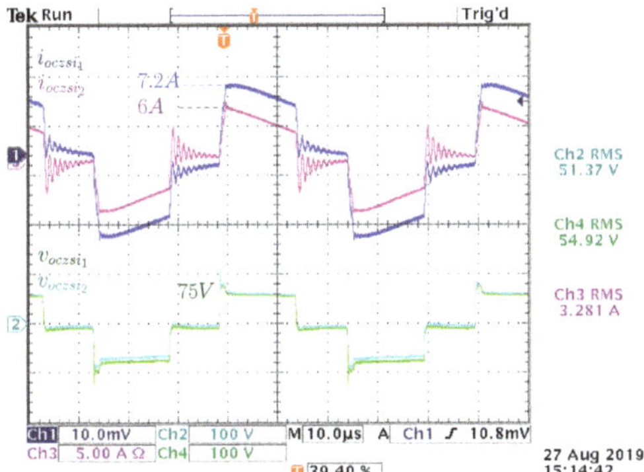

Figure 14. Multiport power imbalance test. Ch1 (10 mV/div=5 A/div): i_{oczsi_1}, Ch3 (5 A/div): i_{oczsi_2}, Ch2 (100 V/div): v_{oczsi_1} and Ch4 (100 V/div): v_{oczsi_2}.

In Figure 15 power in primary and secondary ports are presented. The first and second waveforms were taken after the diode full-wave rectifier and correspond to power in secondary ports 1 and 2 respectively; it can be see that they are equal in magnitude. The third and fourth waveform are the power in CZSI module 1 and 2 respectively, observe that they have different power levels and the sum matches the sum of power in the secondary side. The last validates that a balanced output can be generated despite power imbalances in the MF link inputs.

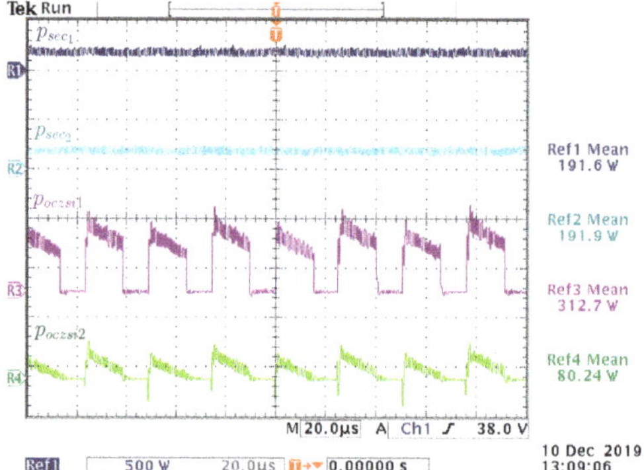

Figure 15. Multiport power imbalance test. Ch1 (500 W/div): output power in secondary port 1 p_{sec_1}, Ch2 (500 W/div): output power in secondary port 2 p_{sec_2}, Ch3 (500 W/div): output power in CZSI module 1 p_{oczsi_1}, and output power in CZSI module 2 Ch4 (500 W/div): p_{oczsi_1}.

5. Discussion

From both simulation an experimental results, it was shown that current and voltage stress was kept in reasonable levels and only switches were affected by a voltage transitory. The last, is also present in boost-HB topologies, however over-current peaks are not present in CZSI as can be seen in Figures 13 and 14, which suggest a lower current stress in semiconductor devices. In addition the shoot through operation gives more time to react to faults related to gate signals in switches: if the gate signals are lost due to external factor like irradiated noise from MF link or processor damage, the switches will be protected by the Z-source by limiting current until a protection flag is enabled. On the other hand, boost-HB topologies will just be damaged in case of losing gate signal, and thus suggesting that the CZSI enhance the reliability of the system.

For unbalanced power conditions, it can be seen that voltage between all PV arrays terminals remains similar for all the ports. However current supplied to each CZSI modules is different and is a function of the power supplied by the corresponding PV array. The voltage remains similar for all CZSI modules, as shown in simulation and experimental results. The current is the parameter that changes for the primary modules and thus the one that validates the transfer of different power levels in primary ports as shown in Figures 10 and 13.

Despite the unbalanced power levels of the primary modules, the voltage is the same in secondary ports, but also the current is the same and thus the power on secondary side is balanced without additional actions, or modulation taken. The last is particularly useful for multilevel inverter applications where multiple isolated and balanced sources are required for proper operation. Topologies with galvanic isolation previously reported, usually implement multiple single-input single-output transformers in order to manipulate strings individually, this allows to follow MPP on each PV array and drive the transformer. However this produce different output on secondary side for each string and thus requires to be balanced by some added control strategy.

6. Conclusions

A multiport scheme with CZSI modules to manage unbalanced power conditions was introduced in this paper. The operation principle of the CZSI was presented as well as the design equations to size CZSI components. An analysis of the multiport operation under unbalanced conditions was given. In this scheme, the CZSI was able to drive power from a PV array to the MF link in a single stage. Buck-boost operation of the CZSI modules is achieved by controlling the shoot-through time. CZSI has the limitation that it only has the duty cycle as a freedom degree and thus MPPT can only be achieved by manipulating this parameter. However, by operating as a current-fed converter, the CZSI modules can operate at similar voltage levels and managing a wider difference between the current levels which make this converter a feasible alternative for PV applications. Simulation and experimental studies were performed in balanced and unbalanced conditions, confirming that this multiport scheme can manage different power levels form PV arrays on its primary side, at the time it generates isolated and balanced outputs on the secondary ports, making it suitable for multilevel converter applications.

Author Contributions: Formal analysis, J.M.S. and V.C.; Investigation, J.M.S. and M.G.G.; Methodology, J.M.S., M.A.B. and J.A.; Project administration, V.C.; Supervision, V.C.; Writing – original draft, J.M.S.; Writing – review & editing, V.C. and J.A. All authors have read and agreed to the published version of the manuscript.

Funding: This research work was supported by the project UASLP-TRAFOS SLP 252557. The authors would like to thank The National Council of Science and Technology of Mexico (CONACyT) for the academic scholarship under the grant number 415079.

Acknowledgments: The authors acknowledge to the UASLP Electrical Power Quality and Motor Control Laboratory (LABCEECM) staff for technical and material support during experimental stage.

Conflicts of Interest: The authors declare no conflicts of interest.

References

1. International Renewable Energy Agency. Roadmap to 2050 (2019 edition). In *Global Energy Transformation*; IRENA; 2019; p. 16. Available online: https://www.irena.org/-/media/Files/IRENA/Agency/Publication/2019/Apr/IRENA_Global_Energy_Transformation_2019.pdf (accessed on 30 January 2020)
2. Int. Renew. Energy Agency. Solar-Photovoltaic Statistics. In *Renewable Capacity Statistics 2019*; IRENA; 2019; p. 21. Available online: https://www.irena.org/-/media/Files/IRENA/Agency/Publication/2019/Mar/IRENA_RE_Capacity_Statistics_2019.pdf (accessed on 30 January 2020)
3. Islam, R.M.; Youguang G.; Jianguo Z. A Multilevel Medium-Voltage Inverter for Step-Up-Transformer-Less Grid Connection of Photovoltaic Power Plants. *Trans. Ind. Electron.* **2014**, *4*, 881–889. [CrossRef]
4. Elsayad N.; Osama M. A cascaded high frequency AC link system for large-scale PV-assisted EV fast charging stations. In Proceedings of the 2017 IEEE Transportation Electrification Conference and Expo (ITEC), Chicago, IL, USA, 22–24 June 2017; pp. 90–94. Available online: https://ieeexplore.ieee.org/document/7993252 (accessed on 30 January 2020)
5. Islam, R.M.; Mahfuz-Ur-Rahman A.M. Modular Medium-Voltage Grid-Connected Converter With Improved Switching Techniques for Solar Photovoltaic Systems. *Trans. Ind. Electronics.* **2017**, *64*, 8887–8896. [CrossRef]
6. Zaim, R.; Jamaludin J.; Rahim N.A. Photovoltaic Flyback Microinverter With Tertiary Winding Current Sensing. *Trans. Power Electron.* **2019**, *34*, 7588–7602. [CrossRef]
7. Nathan, K.; Ghosh, S. A New DC-DC Converter for Photovoltaic Systems: Coupled-Inductors Combined Cuk-SEPIC Converter. *Trans. Energy Conversion.* **2019**, *34*, 191–201. [CrossRef]
8. Kim, H.; Yoon, C.; Choi, S. An Improved Current-Fed ZVS Isolated Boost Converter for Fuel Cell Applications. *IEEE Trans. Power Electron.* **2010**, *25*, 2357–2364. [CrossRef]
9. Wu,Q.; Wang, Q.; Xu, J.; Xu, Z. Active-clamped ZVS current-fed push–pull isolated dc/dc converter for renewable energy conversion applications. *IET Power Electron.* **2018**, *11*, 373–381. [CrossRef]
10. Ohtsu, S.; Yamashita, T.; Yamamoto, K. Stability in high-output-voltage push–pull current-fed converters. *IEEE Trans. Power Electron.* **1993**, *8*, 135–139. [CrossRef]
11. Wolfs, P.J. A current-sourced DC-DC converter derived via the duality principle from the half-bridge converter. *IEEE Trans. Ind. Electron.* **1993**, *40*, 139–144. [CrossRef]
12. Nguyen, M.-K.; Duong, T.-D; Lim, Y.-C.; and Choi, J.-H. Active CDS-clamped L-type current-Fed isolated DC-DC converter. *J. Power Electron.* **2018**, *4*, 955–964.
13. De Aragao Filho, W.C.P.; Barbi, I. A comparison between two current-fed push-pull DC-DC converters-analysis, design and experimentation. In Proceedings of the International Telecommunications Energy Conference 1996, Boston, MA, USA, 6–10 October 1996; pp. 313–320. Available online: https://ieeexplore.ieee.org/document/573330 (accessed on 30 January 2020)
14. Guisso, R.A.; Andrade, A.M.S.S. Grid-tied single source quasi-Z-source cascaded multilevel inverter for PV applications. *Electron. Lett.* **2019**, *55*, 42–343. [CrossRef]

15. Andrade, A.M.S.S.; Guisso, R.A. Quasi-Z-source network DC–DC converter with different techniques to achieve a high voltage gain. *Electron. Lett.* **2018**, *54*, 710–712. [CrossRef]

16. Liu, Y.; Ge, B.; Abu-Rub, H.; Sun, H. Hybrid Pulsewidth Modulated Single-Phase Quasi-Z-Source Grid-Tie Photovoltaic Power System. *Trans. Ind. Informatics* **2016**, *12*, 621–632. [CrossRef]

17. Choi, W.-Y.; Yang, M.-K. Transformerless Quasi-Z-Source Inverter to Reduce Leakage Current for Single-Phase Grid-Tied Applications. *Electronics* **2019**, *8*, 312. [CrossRef]

18. Wang, B.; Tang, W. A Novel Three-Switch Z-Source SEPIC Inverter. *Electronics* **2019** **8**, 247. [CrossRef]

19. Rahul, J.R.; Kirubakaran, A. Single-Phase Quasi-Z-source based Isolated DC/AC converter. In Proceedings of the IEEE International Conference on Power Systems (ICPS), New Delhi, India, 4–6 March 2016.

20. Liu, Y.; Abu-Rub, H.; Ge, B. Front-End Isolated Quasi-Z-Source DC–DC Converter Modules in Series for High-Power Photovoltaic Systems—Part I: Configuration, Operation, and Evaluation. *Trans. Ind. Electron.* **2017**, *64*, 347–358. [CrossRef]

21. Ahmed, H.F.; Cha, H. A New Class of Single-Phase High-Frequency Isolated Z-Source AC–AC Converters With Reduced Passive Components. *Trans. Power Electron.* **2018**, *33*, 1410–1419. [CrossRef]

22. Sajadian, S.; Ahmadi, R. ZSI for PV systems with LVRT capability. *IET Renew. Power Gen.* **2019**, *12*, 1286–1294. [CrossRef]

23. Uno, M.; Shinohara, T. Module-Integrated Converter Based on Cascaded Quasi-Z-Source Inverter With Differential Power Processing Capability for Photovoltaic Panels Under Partial Shading. *Trans. Power Electron.* **2019**, *34*, 11553–11565. [CrossRef]

24. Liu, J.; Wu, J.; Qiu, J.; Zeng, J. Switched Z-Source/Quasi-Z-Source DC-DC Converters With Reduced Passive Components for Photovoltaic Systems. *IEEE Access* **2019**, *7*, 40893–40903. [CrossRef]

25. Anderson, J.; Peng, F.Z. Four quasi-Z-Source inverters. In Proceedings of the 2008 IEEE Power Electronics Specialists Conference, Rhodes, Greece, 15–19 June 2008.

26. Sandoval, J.M.; Cardenas, V. Current-Fed Z-Source Converter for Medium-Voltage Medium-Frequency Isolated Solar-Photovoltaic Systems. In Proceedings of the 2019 IEEE Applied Power Electronics Conference and Expo, Anaheim, CA, USA, 17–21 March 2019.

© 2020 by the authors. Licensee MDPI, Basel, Switzerland. This article is an open access article distributed under the terms and conditions of the Creative Commons Attribution (CC BY) license (http://creativecommons.org/licenses/by/4.0/).

Article

A Modified Model Predictive Power Control for Grid-Connected T-Type Inverter with Reduced Computational Complexity

Van-Quang-Binh Ngo [1,2], Minh-Khai Nguyen [3], Tan-Tai Tran [1], Joon-Ho Choi [1,*] and Young-Cheol Lim [1,*]

[1] Department of Electrical Engineering, Chonnam National University, Gwangju 500-757, Korea; qbinhtam@gmail.com (V.-Q.-B.N.); trantantaikdd@gmail.com (T.-T.T.)
[2] Department of Physics, College of Education, Hue University, Thua Thien Hue 530000, Vietnam
[3] Faculty of Electrical and Electronics Engineering, Ho Chi Minh City University of Technology and Education, 1 Vo Van Ngan Street, Thu Duc district, Ho Chi Minh City 700000, Vietnam; nmkhai00@gmail.com
* Correspondence: joono@chonnam.ac.kr (J.-H.C.); yclim@chonnam.ac.kr (Y.-C.L.)

Received: 16 January 2019; Accepted: 13 February 2019; Published: 15 February 2019

Abstract: This study proposed a modified power strategy based on model predictive control for a grid-connected three-level T-type inverter. The controller utilizes the mathematical model to forecast the performance of the grid current, the balance of DC-bus capacitor voltages and switching frequency. The proposed method outlines a new technique to formulate a control objective. The control objective includes the absolute error of the inverter voltage reference and its possible values instead of the grid current error. By using the modified equivalent transformations in the cost function, the execution time was reduced 22% compared to the traditional model predictive control while maintaining the high dynamic performances of the power and low total harmonic distortion of the current. A comparative investigation showed that the proposed method obtains a high-performance control compared with the classical power control scheme with linear PI controllers and space vector modulation. The feasibility of the proposed method was verified by the simulation and experimental results.

Keywords: computational complexity; direct power control; finite control set model predictive control; PI controllers; space vector modulation; three-level T-type inverter

---

1. Introduction

Over the last few years, multilevel converters have been recognized as an alternative approach for high-power electronics owing to the improvement of capacity and performance compared with the two-level converter [1–3]. In particular, the T-type inverter topology is preferred due to the benefits in term of low conduction losses and high sinusoidal waveforms of the output voltage [4,5]. The unbalance of capacitor voltages is the big problem for this configuration. However, this issue has been addressed by several methods [6–10].

Grid-connected power converters have proven to be important in many industrial applications such as active power filters, distributed systems, and renewable power generation systems [11–16]. The conventional control strategy is the voltage-oriented control [11,13] which can ensure the steady-state and performance via the traditional PI current controllers with pulse width modulation (PWM) or space vector modulation (SVM). However, the completeness of the current decoupling of the internal current controller and accurate tuning parameters are the drawbacks of this method. This technique has a low dynamic response and is not suitable for a nonlinear system. Recently, to improve the performance, direct power control (DPC) [17] has been presented by employing a look-up table to decide the proper switching state of the inverter. Nonetheless, a big ripple of the active and reactive powers is the problem of this approach. Moreover, it is necessary to require a small sampling time to obtain a reasonable steady-state and good transient performances. To overcome these issues, various approaches have been developed such as using DPC established on extended-state observation with PWM [18], port-controlled Hamiltonian system [12], sliding mode control [19], and predictive control [20–23].

Over the last decade, many researchers have focused on a finite control set model predictive control (FCS-MPC) for power electronics applications due to its simplicity of configuration, easy realization, and rapid transient response [24–27]. Besides, the benefits of FCS-MPC are that the non-linearities, delay compensation and additional constraints are easily imposed on the controller. However, in the FCS-MPC, every state needs to be enumerated to achieve the optimal value, leading to the computational burden becoming a major issue, especially with a high number of switching states such as multilevel inverters and long prediction horizons. To deal with this problem, a simplified MPC is put forward in [28] for converters. This method employs switching state group for optimization loop to reduce the computation time. An alternative technique mentioned for multilevel converters [29,30] is to reduce the computational cost by developing a modified sphere decoding algorithm. In [31], the equivalent transformation in the cost function is employed in the optimization loop to reduce the time-consuming computation for converters with inductive load load. The modified algorithm in [32] incorporates the conventional FCS-MPC and steady-state assessment to decrease the computational cost. Nonetheless, the computational burden of this approach is large, similar to the conventional FCS-MPC in the unfavorable case. The proposed approach is presented in [22] to reduce the execution time by using the preselected voltage reference and DC-bus capacitor voltage balancing. Nonetheless, this method is only applied to a 3L-T-type inverter with LC filter in standalone mode.

We here refine the method in [31] by taking into account the computational delay in the control design for grid-connected to the 3L-T-type inverter. This approach not only keeps the DC-link capacitor voltage balancing and decreases the switching frequency but also guarantees the high-performance control. The cost function based on the model predictive control is utilized to achieve these purposes. A control horizon of modified one-step prediction is used to compensate the computational delay, diminish the power ripples and enhance the control performance. Compared with linear controllers, the proposed method has a fast dynamic response and no overshoot of transient response. Furthermore, to address the aforementioned challenge, this research aimed to develop the cost function for reducing the computational effort in the optimization problem. The formula of the cost function consists of the current tracking error with the conventional model predictive current control method. In this research, a predictive model of grid current was employed to obtain the required inverter voltage. Then, the revised cost function, which includes the inverter output voltage errors, neutral-point voltage and switching frequency reduction, was defined. Consequently, the computational burden of the proposed method was reduced by 22% compared to the traditional FCS-MPC, providing the practicality of the real-time implementation. Simulation and experiments were used to evaluate and verify the performance of the proposed method.

This paper is structured as follows: Section 2 introduces the dynamic model of a predictive control strategy based on voltage orientation for a grid-connected 3L-T-type inverter. Section 3 outlines the proposed modified model predictive control strategy to reduce the computational cost. In Section 4, a performance comparison between the proposed method and traditional FCS-MPC and the classical DPC with linear PI controllers and SVM (DPC-SVM) is presented. Section 5 summarizes some conclusions.

2. Dynamic Modeling for Grid-Connected 3L-T-Type Inverter

2.1. Configuration of the Grid-Connected 3L-T-Type Inverter

A simplified circuit of the grid connected 3L-T-type inverter is illustrated in Figure 1. The operating principle of each inverter branch can be described by three switching statuses [1], [0], and [−1]. During switching status [1] or [P], both switches S_{1x} and S_{2x} with $x \in \{a, b, c\}$ are turned "ON" leading to $u_{xZ} = U_{dc}/2$, while [−1] or [N] signifies two switches S_{1x} and S_{2x} are turned "OFF", resulting $u_{xZ} = -U_{dc}/2$. The switching state [0] or [O] means that the two internal switches S_{2x} and S_{3x} are "ON" and u_{xZ} is clamped to zero. Thus, taking into account the three phases of the inverter, 27 possible configurations are produced for the 3L-T-type. The summary of switching states generated by this inverter is presented in Table 1.

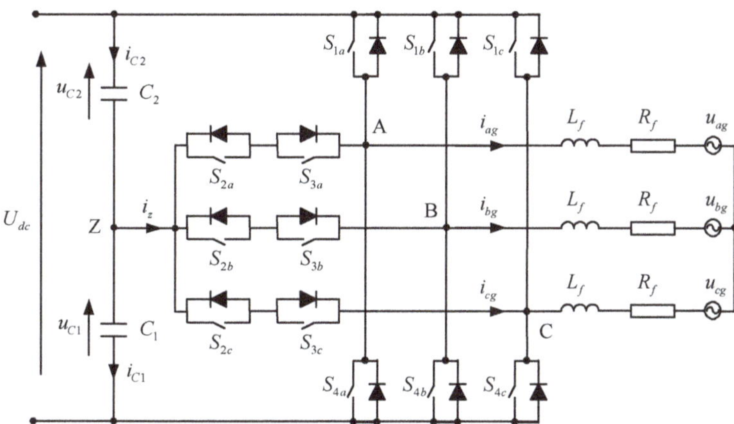

Figure 1. Simplified configuration of the grid-connected 3L-T-type inverter.

Table 1. Operating principle of inverter branch x.

Switching State	Switches				One Phase Inverter Voltage
S_x	S_{1x}	S_{2x}	S_{3x}	S_{4x}	u_{xZ}
N	0	0	1	1	$-U_{dc}/2$
O	0	1	1	0	0
P	1	1	0	0	$U_{dc}/2$

2.2. Mathematical Model

The inverter output voltage generated by the 3L-T-type can be defined:

$$u_{inv} = \frac{2}{3}\left(u_{AZ} + m u_{BZ} + m^2 u_{CZ}\right), \tag{1}$$

where u_{AZ}, u_{BZ}, and u_{CZ} are the three-phase inverter voltage, relating to the neutral-point Z of the DC-bus; and $m = e^{j2\pi/3} = -\frac{1}{2} + j\frac{\sqrt{3}}{2}$.

The phase to neutral voltages u_{xZ} of the 3L-T-type inverter are calculated in terms of the switching status S_x and DC-bus voltage U_{dc} [7,27]:

$$u_{xZ} = S_x \frac{U_{dc}}{2}, \tag{2}$$

where S_x stands for the status of an inverter branch with three possible combinations $\{-1, 0, 1\}$.

Under the assumption of a constant DC-bus voltage and $C_1 = C_2 = C_{dc}$, the dynamic of neutral-point voltage (u_z) can be phrased in terms of the switching states and the grid currents [22]:

$$\frac{du_z}{dt} = \frac{d\left(u_{c1} - u_{c2}\right)}{dt} = -\frac{1}{C_{dc}}i_z = -\frac{1}{C_{dc}}\left((1 - |S_a|)i_{ag} + (1 - |S_b|)i_{bg} + (1 - |S_c|)i_{cg}\right). \tag{3}$$

The mathematical model of grid-connected inverter is given by:

$$u_{inv} = u_g + R_f i_g + L_f \frac{di_g}{dt}, \tag{4}$$

where $u_g = \left[u_{ag}, u_{bg}, u_{cg}\right]^T$, $i_g = \left[i_{ag}, i_{bg}, i_{cg}\right]^T$ are the grid voltage and current vectors and R_f, L_f are the filter resistance and inductance, respectively.

The dynamics model in the dq synchronous reference frame is derived from the rotational transformation of Equation (4) as follows:

$$\begin{aligned} u_{inv\_d} &= R_f i_d + L_f \frac{di_d}{dt} + u_d - \omega L_f i_q, \\ u_{inv\_q} &= R_f i_q + L_f \frac{di_q}{dt} + u_q + \omega L_f i_d, \end{aligned} \tag{5}$$

where ω is the grid frequency.

Considering the grid-voltage oriented condition (Figure 2), the components of grid voltage can be achieved by:

$$u_d = \hat{U}_g, \; u_q = 0, \tag{6}$$

where $\hat{U}_g$ denotes the grid voltage magnitude that is achieved by using the phase-locked loop (PLL).

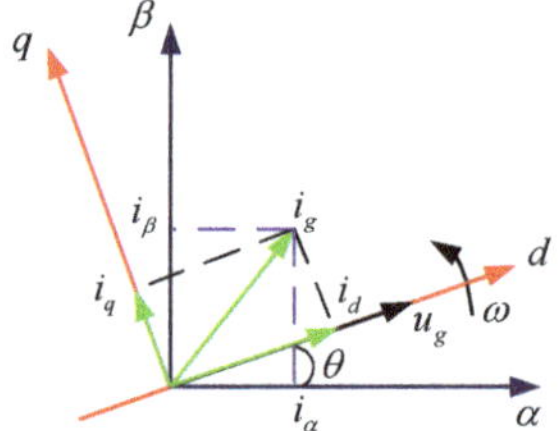

Figure 2. Coordinate reference system.

As a result, the dynamics model in continuous time is rewritten from Equations (3), (5) and (6) as:

$$\begin{aligned} \frac{di_d}{dt} &= -\frac{R_f}{L_f}i_d + \frac{1}{L_f}(u_{inv\_d} - \hat{U}_g) + \omega i_q, \\ \frac{di_q}{dt} &= -\frac{R_f}{L_f}i_q + \frac{1}{L_f}u_{inv\_q} - \omega i_d, \\ \frac{du_z}{dt} &= \frac{1}{2C_{dc}}(2|S_a| - |S_b| - |S_c|)i_\alpha + \frac{\sqrt{3}}{2C_{dc}}(|S_b|) - |S_c|)i_\beta, \end{aligned} \tag{7}$$

where the grid current components in the dq coordinate frame are achieved by their values in the $\alpha\beta$ stationary reference frame by using Park transformation:

$$\begin{aligned} i_d &= i_\alpha \cos\theta + i_\beta \sin\theta, \\ i_q &= i_\beta \cos\theta - i_\alpha \sin\theta. \end{aligned} \tag{8}$$

Similarly, the component of inverter voltage is expressed based on Equations (1) and (2) as:

$$
\begin{aligned}
u_{inv\_d} &= u_{inv\alpha} \cos\theta + u_{inv\beta} \sin\theta, \\
u_{inv\_q} &= u_{inv\beta} \cos\theta - u_{inv\alpha} \sin\theta,
\end{aligned}
\tag{9}
$$

where $u_{inv\alpha}$ and $u_{inv\beta}$ are the elements of inverter voltage which are achieved from the Clarke transformation:

$$
\begin{aligned}
u_{inv\alpha} &= \frac{U_{dc}}{6}\left(2S_a - S_b - S_c\right), \\
u_{inv\beta} &= \frac{\sqrt{3}U_{dc}}{6}\left(S_b - S_c\right).
\end{aligned}
\tag{10}
$$

We can calculate the grid active and reactive powers as [11,17]:

$$
\begin{aligned}
P_g &= \frac{3}{2}\left(u_d i_d + u_q i_q\right) = \frac{3}{2}\hat{U}_g i_d, \\
Q_g &= \frac{3}{2}\left(u_q i_g - u_d i_q\right) = -\frac{3}{2}\hat{U}_g i_q.
\end{aligned}
\tag{11}
$$

Equation (11) means that we can employ the grid current components to control the active and reactive powers.

3. Proposed Control Strategy Based on Model Predictive Control

The purposes of the proposed method are to:

- Follow the active and reactive power references.
- Ensure the capacitor voltage balancing.
- Decrease the frequency of the switches.

With the aim to accomplish the control goal, the cost function of the conventional FCS-MPC for the grid system, which takes into account the one-step prediction and delay compensation, is presented as [25,26]:

$$
g_{cnv} = \left| i_d^*(k+2) - i_d^p(k+2) \right| + \left| i_q^*(k+2) - i_q^p(k+2) \right| + \lambda_{dc}\left| u_z^p(k+2) \right| + \lambda_n sw_c,
\tag{12}
$$

where $i_d^*(k+2)$, $i_q^*(k+2)$ and $i_d^p(k+2)$, $i_q^p(k+2)$ represent the reference and predicted currents at instant $k+2$, respectively. λ_{dc} and λ_n denote the weighting elements of the balance of capacitor voltages and switching frequency reduction.

The weighting factors λ_{dc} and λ_n are considered as tuning parameters to achieve an acceptable control performance. Adjusting these values is not an easy mission for the classical PI controllers. Section 4 illustrates the details of the selection of these parameters.

sw_c in Equation (12) is the additional constraint imposed on the cost function to reduce the switching frequency of the inverter. Thus, its expression can be given by:

$$
sw_c = \sum_{x=a,b,c} \left| S_x(k+1) - S_x(k) \right|.
\tag{13}
$$

In a real-time control system, it is necessary to consider a time delay produced by computation time. To overcome this problem, in the traditional approach [25,27], the predicted variables at instant $k+1$ are estimated by using the dynamic model, measurement feedback and previous switching status at instant k. Next, the associated predicted variables at time $k+2$ are obtained from the estimated variables at time $k+1$ and all switching states of the inverter. Thus, the optimal control input is achieved from cost function optimization at time $k+2$ and implemented to the inverter at $k+1$, as shown in Figure 3.

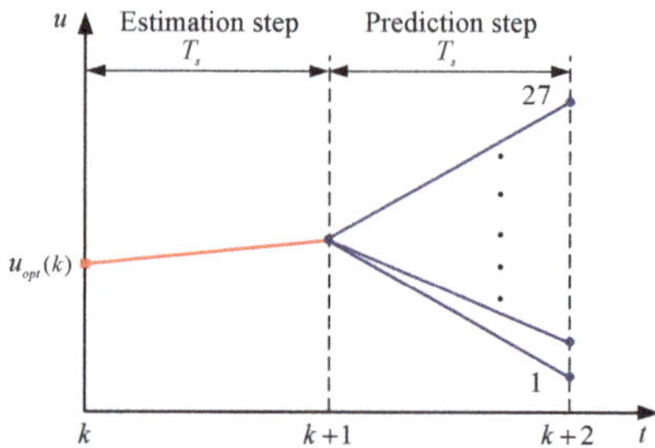

Figure 3. Control variable of one-step horizon with delay compensation.

As mentioned in Section 2.2, a control action $u = [S_a, S_b, S_c]^T$ is a finite set of feasible control associated with the switching states $u \in \mathbb{U} = \{u_1, u_2, ..., u_{27}\}$. Consequently, the optimal control action u_{opt} is implemented in the converter at time $k + 1$ by minimizing Equation (14):

$$
\begin{aligned}
\hat{i}_d(k+1) &= f_1(i_d(k), i_q(k), u_k), \ \hat{i}_q(k+1) = f_2(i_q(k), i_d(k), u_k), \\
i_d^p(k+2) &= f_1(\hat{i}_d(k+1), \hat{i}_q(k+1), u_{k+1}), \ i_q^p(k+2) = f_2(\hat{i}_d(k+1), \hat{i}_q(k+1), u_{k+1}), \\
\hat{u}_z(k+1) &= f_3(i_g(k), u_z(k), u_k), \ u_z^p(k+2) = f_3(\hat{i}_g(k+1), \hat{u}_z(k+1), u_{k+1}), \\
g_{cnv}(u_{k+1}) &= \left| i_d^*(k+2) - i_d^p(k+2) \right| + \left| i_q^*(k+2) - i_q^p(k+2) \right| + \lambda_{dc} \left| u_z^p(k+2) \right| + \lambda_n sw_c, \\
u_{opt} &= arg \left\{ \min_{u_{k+1} \in \{-1,0,1\}^3} g_{cnv}(u_{k+1}) \right\}.
\end{aligned}
\tag{14}
$$

Considering a sampling interval T_s, the discrete-time of the grid current is obtained by employing the first-order forward Euler approximation for Equation (7):

$$
\begin{aligned}
\hat{i}_d(k+1) &= i_d(k)\left(1 - \frac{T_s R_f}{L_f}\right) + \frac{T_s}{L_f}\left(u_{inv\_d}(k) - \hat{U}_g\right) + T_s \omega i_q(k), \\
\hat{i}_q(k+1) &= i_q(k)\left(1 - \frac{T_s R_f}{L_f}\right) + \frac{T_s}{L_f}u_{inv\_q}(k) - T_s \omega i_d(k),
\end{aligned}
\tag{15}
$$

where the inverter voltages $u_{inv\_d}(k)$ and $u_{inv\_q}(k)$ are achieved from Equation (9) by using the previous state $S_{opt}(k)$.

Substituting the grid current references $i_d^*(k+2)$ and $i_q^*(k+2)$ by their predicted currents $i_d^p(k+2)$, $i_q^p(k+2)$ into Equation (7), we therefore obtain the components of inverter reference voltage:

$$
\begin{aligned}
u_{inv\_d}^*(k+1) &= \hat{i}_d(k+1)\left(R_f - \frac{L_f}{T_s}\right) + \frac{L_f}{T_s}i_d^*(k+2) + \hat{U}_g - \omega L_f \hat{i}_q(k+1), \\
u_{inv\_q}^*(k+1) &= \hat{i}_q(k+1)\left(R_f - \frac{L_f}{T_s}\right) + \frac{L_f}{T_s}i_q^*(k+2) + \omega L_f \hat{i}_d(k+1),
\end{aligned}
\tag{16}
$$

where the grid current references $i_d^*(k)$ and $i_q^*(k)$ are determined from the active and reactive powers in accordance with Equation (11).

The predictive reference current can be achieved by using the second-order Lagrange extrapolation as [27]:

$$
\begin{aligned}
i_d^*(k+2) &= 6i_d^*(k) - 8i_d^*(k-1) + 3i_d^*(k-2), \\
i_q^*(k+2) &= 6i_q^*(k) - 8i_q^*(k-1) + 3i_q^*(k-2).
\end{aligned}
\tag{17}
$$

Approaching the neutral-point voltage in the same manner, we have its discrete-time expression as:

$$
\begin{aligned}
u_z^p(k+1) &= u_z(k) + \frac{T_s}{2C_{dc}}\left(2|S_a(k+1)| - |S_b(k+1)| - |S_c(k+1)|\right)i_\alpha(k) \\
&\quad + \frac{\sqrt{3}T_s}{2C_{dc}}\left(|S_b(k+1)| - |S_c(k+1)|\right)i_\beta(k).
\end{aligned}
\tag{18}
$$

A modified MPC is presented in [31] to decrease the computational cost. However, this approach is only applied to converters in standalone mode and does not take into account the delay compensation. To improve the dynamic performance due to computational delay and reduce the high computational burden, an extension of this method is employed for the grid-connected 3L-T-type inverter to regulate the grid power exchange. The aim of the proposed control strategy is to determine the best inverter voltages (u_{inv_dn} and u_{inv_qn}) in 27 voltage vectors that are closest to the inverter voltage references $u^*_{inv\_d}(k+1)$ and $u^*_{inv\_q}(k+1)$. Then, the optimal control of the system is achieved through a simple optimization technique. The proposed predictive control algorithm is divided into three main steps:

(1) **Estimate the grid current**: The grid current components $\hat{i}_d(k+1)$ and $\hat{i}_q(k+1)$ at time $k+1$ are evaluated by utilizing Equation (15) with the previous optimal switching state (u_k) at time k.

(2) **Calculate the inverter voltage reference**: According to Equation (16), the reference components of inverter voltage $u^*_{inv\_d}(k+1)$ and $u^*_{inv\_q}(k+1)$ at time $k+1$ are calculated in regard to the grid current references $i^*_d(k+2)$, $i^*_q(k+2)$ at time $k+2$ and estimated grid currents $\hat{i}_d(k+1)$ and $\hat{i}_q(k+1)$ at time $k+1$. Then, the neutral-point voltage at time $k+1$ is computed from grid current $i_g(k)$ and all admissible switching states switching states u_{k+1}.

(3) **Evaluate and select the best control input**: The optimal control input that has the lowest of the cost function g_{mdf} is implemented at time $k+1$ to the 3L-T-type inverter:

$$
\begin{aligned}
\hat{i}_d(k+1) &= f_1(i_d(k), i_q(k), u_k),\ \hat{i}_q(k+1) = f_2(i_q(k), i_d(k), u_k),\\
u^p_z(k+1) &= f_3(i_g(k), u_z(k), u_{k+1}),\ u_{inv\_dn} = f_6(u_{inv\_n});\ u_{inv\_qn} = f_7(u_{inv\_n}),\\
u^*_{inv\_d}(k+1) &= f_4(i^*_d(k+2), \hat{i}_d(k+1), \hat{i}_q(k+1)),\ u^*_{inv\_q}(k+1) = f_5(i^*_q(k+2), \hat{i}_d(k+1), \hat{i}_q(k+1)),\\
g_{mdf}(u_{k+1}) &= \left| u^*_{inv\_d}(k+1) - u_{inv\_dn} \right| + \left| u^*_{inv\_q}(k+1) - u_{inv\_qn} \right| + \lambda_{dc}\left| u^p_z(k+1) \right| + \lambda_n sw_c,\\
u_{opt} &= arg\left\{ \min_{u_{k+1}\in\{-1,0,1\}^3} g_{mdf}(u_{k+1}) \right\}.
\end{aligned}
\tag{19}
$$

where $u^*_{inv\_d}(k+1)$ and $u^*_{inv\_q}(k+1)$ represent the desired inverter voltages at instant $k+1$; and u_{inv_dn} and u_{inv_qn} stand for all possible inverter voltage components in dq synchronous reference frame.

It is worth mentioning that there are different terms of the cost function between the traditional MPC and the proposed method as shown in Equations (14) and (19). With the traditional FCS-MPC [25–27], all predicted control variables are considered in the optimization loop. In this case, 27 predictive currents $i^p_d(k+2)$ and $i^p_q(k+2)$ are needed to enumerate in the cost function ($g_{cnv}(u_{k+1})$) to obtain the best switching state, which is applied to the inverter at time $k+1$. On the other hand, only one inverter reference voltage $u^*_{inv}(k+1)$, which is achieved through the estimated grid current $\hat{i}_g(k+1)$ and grid current reference $i^*_g(k+2)$, is taken into account for the loop optimization. Then, by only one evaluating cost function ($g_{mdf}(u_{k+1})$) with simple code optimization, the optimal control input is achieved. Therefore, the proposed technique reduces the computational time of the performance criterion optimization compare with the traditional FCS-MPC. The comparison of computational cost between two controllers is summarized in Table 2, which highlights the benefit of the proposed method. To validate the efficiency of the proposed algorithm, the function tic-toc in Matlab is used to measure the computation time. Significantly, the minimum, average and maximum values of evaluation time of the conventional FCS-MPC are 17, 29, and 41 µs, respectively, and 10, 22, and 34 µs for the proposed algorithm, respectively. Figure 4 shows the impact of the computation time in two controllers. This can lead to a 24% reduction of computational complexity, resulting in an increase of sampling frequency for improving the control performance. Consequently, this method has many practical applications for power converters in term of a time-consuming optimization algorithm, thereby supporting the feasibility of real-time applications with low-cost processors. The overall strategy of the proposed control algorithm is illustrated in Figure 5.

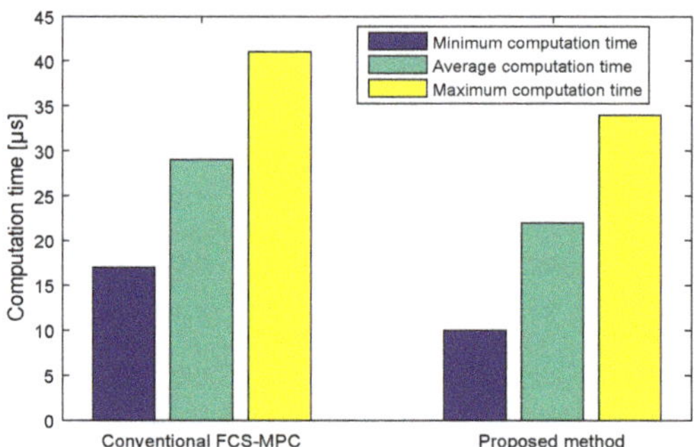

Figure 4. Computation time of two controllers.

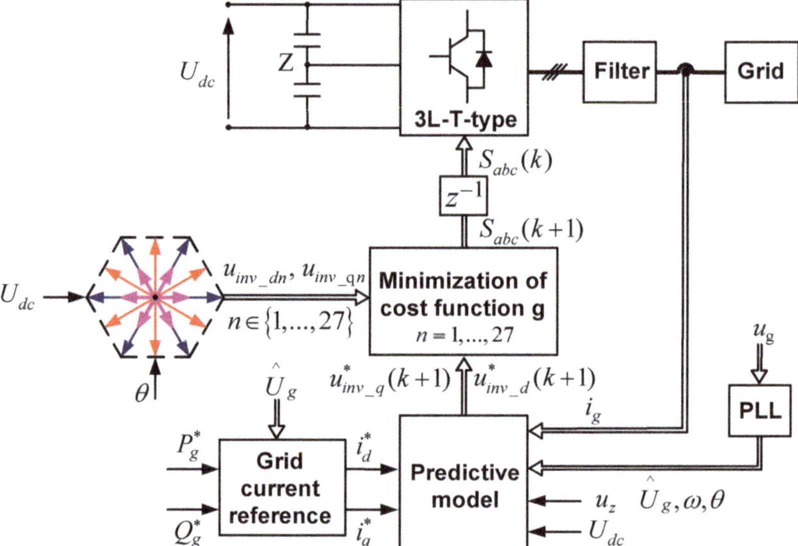

Figure 5. Block diagram of the proposed control scheme.

Table 2. Comparison of computational cost between two methods.

Variables	Conventional FCS-MPC	Proposed Method
$\hat{i}_d(k+1)$, $\hat{i}_q(k+1)$	1	1
$u_{inv\_d}(k+1)$, $u_{inv\_q}(k+1)$	27	27
$i_d^p(k+2)$, $i_q^p(k+2)$	27	0
$u_{inv\_d}^*(k+1)$, $u_{inv\_q}^*(k+1)$	0	1
$u_z^p(k+1)$	27	27
sw_c	27	27
Total	109	83

Finally, the control objectives are obtained by enumerated the proposed cost function, as demonstrated in Algorithm 1.

Algorithm 1 Algorithm of modified model predictive power control with reduced computational complexity.

Input: $i_g(k), u_z(k), u_g(k), U_{dc}(k), P_g^*(k), Q_g^*(k)$

Output: S_a, S_b, S_c.

1. Estimate the grid current $\hat{i}_{gd}(k+1)$ and $\hat{i}_{gq}(k+1)$ based on (15).

2. Compute the inverter voltage components in the dq synchronous reference frame u_{inv_dn} and u_{inv_qn}, with $n = 1$–27 according to (9).

3. Calculate the grid current references $i_{gd}^*(k), i_{gq}^*(k)$ and inverter voltage $u_{inv\_d}^*(k+1), u_{inv\_q}^*(k+1)$ based on (11) and (16).

4. Estimate sw_c for all switching states based on (13).

5. Predict the neutral-point voltage $u_z^p(k+1)$ by using (18).

6. Evaluate the cost function g_{mdf} from (19).

7. Select the optimal switching state x_{opt}: $[\sim, x_{opt}] = min(g_{mdf})$; Return to step 1.

---

4. Simulation and Experimental Results

4.1. Simulation Results

To verify the control performance of the proposed control scheme for the grid tie 3L-T-type inverter, examinations were conducted in Matlab/Simulink simulation environment with SimPowerSystems toolbox. The system parameters are as indicated in Table 3. The control algorithm was implemented by utilizing the Matlab function block.

Table 3. Simulation parameters.

System Parameters	Value	Representation
U_{dc}	600 [V]	DC-bus voltage
C_{dc}	1000 [μF]	DC-bus capacitance
R_f	80 [mΩ]	Filter resistance
L_f	10 [mH]	Filter inductance
f_s	20 [kHz]	Sampling frequency of the proposed controller
f_g	50 [Hz]	Grid voltage frequency
U_{dm}	380 [V]	Grid line-to-line voltage
λ_{dc}	20	Weighting factor of the balance of capacitor voltages
λ_n	60	Weighting factor of switching frequency

To evaluate the steady-state control performance, the mean absolute percentage error (MAPE) was used to estimate the power ripple and neutral-point voltage deviation as:

$$MAPE = \frac{1}{n} \sum_{i=1}^{n} \left| \frac{y_i^* - y_i}{y_i^*} \right|, \tag{20}$$

where y_i^* and y_i are the reference and measurement vectors, respectively.

To calculate the average switching frequency of the inverter achieved by the FCS-MPC, the following expression suggested in [27] can be used:

$$f_{sw} = \sum_{x=a,b,c} \frac{n_{sw_{1x}} + n_{sw_{2x}} + n_{sw_{3x}} + n_{sw_{4x}}}{12T_{sim}}, \tag{21}$$

where $n_{sw_{1x}}$, $n_{sw_{2x}}$, $n_{sw_{3x}}$ and $n_{sw_{4x}}$ denote the number of commutations of each leg in the time interval (T_{sim}).

In the first scenario, the proposed method was performed to demonstrate the ability of the bidirectional transmitted power with a unity power factor ($Q_{g\_ref} = 0$ Var). The reference of power factor (PF) changed from -1 to 1 at $t = 0.15$ s, corresponding to a step change of active power P_{g_ref} from -20 kW to 20 kW. In the inverting mode operation ($P_g < 0$), the active power was supplied to the grid, whereas the power was absorbed from the grid with rectifying mode operation ($P_g > 0$). As shown in Figure 6, the proposed method enhanced a fast steady-state and high control performance. The phase difference between the grid voltage and current is depicted in Figure 7a. Figure 7b shows the variable switching frequency of inverter output voltage. However, this does not impact too much the control performance.

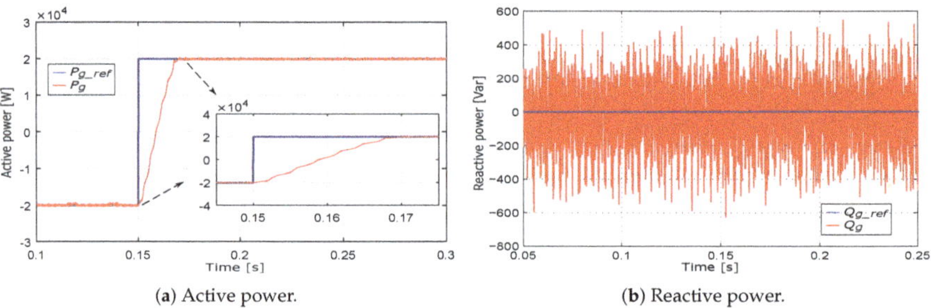

(**a**) Active power.

(**b**) Reactive power.

Figure 6. The transient power responses for a step in active power reference with a unity power factor.

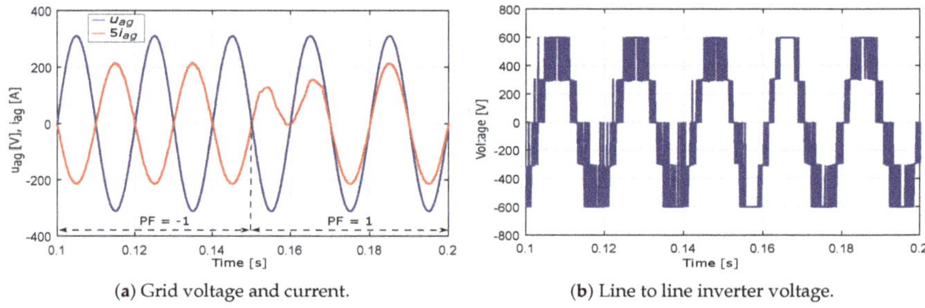

(**a**) Grid voltage and current.

(**b**) Line to line inverter voltage.

Figure 7. Inverter output voltage and grid voltage and current waveforms.

In the second scenario, the control system was conducted with different values of active and reactive powers. To confirm the effectiveness of the control scheme, a comparative analysis was carried out among the proposed method, conventional FCS-MPC [27] and direct power control with PI controllers [11]. This linear controller employs the redundant states to keep the DC-bus capacitor voltage balancing [6]. The sampling frequency of the MPC and DPC-SVM is considered $f_s = 20$ kHz and 6 kHz, respectively, to produce the equivalent average switching frequency $f_{sw} = 3$ kHz. The reference of active and reactive powers were set at 4 kW and -2 kVar, according to the $PF = -0.89$. The active power reference (P_{g-ref}) was stepped from 4 kW to 7.5 kW at $t = 0.15$ s, and then back to 4 kW at instant $t = 0.25$ s. The active power reference and its measured value are demonstrated in Figure 8a. A step change in required reactive power Q_{g-ref}) from -2 kVar to 2 kVar at $t = 0.2$ corresponded to leading power factor and lagging power factor, respectively, as illustrated in Figure 8b. The results shown in Figure 8 prove that the proposed method enhanced the dynamic performance with no overshoot compared to the DPC-SVM, as confirmed in Table 4. In fact, the active power of the conventional MPC and proposed method reached steady-state of the transient response from 4 kW to 7.5 kW in about 0.8 ms and 1.1 ms with DPC-SVM, respectively. The MAPE of the active power of the proposed method reduced from 5.22% to 3.75%, and from 11.03% to 7.98% with reactive power, compared with DPC-SVM.

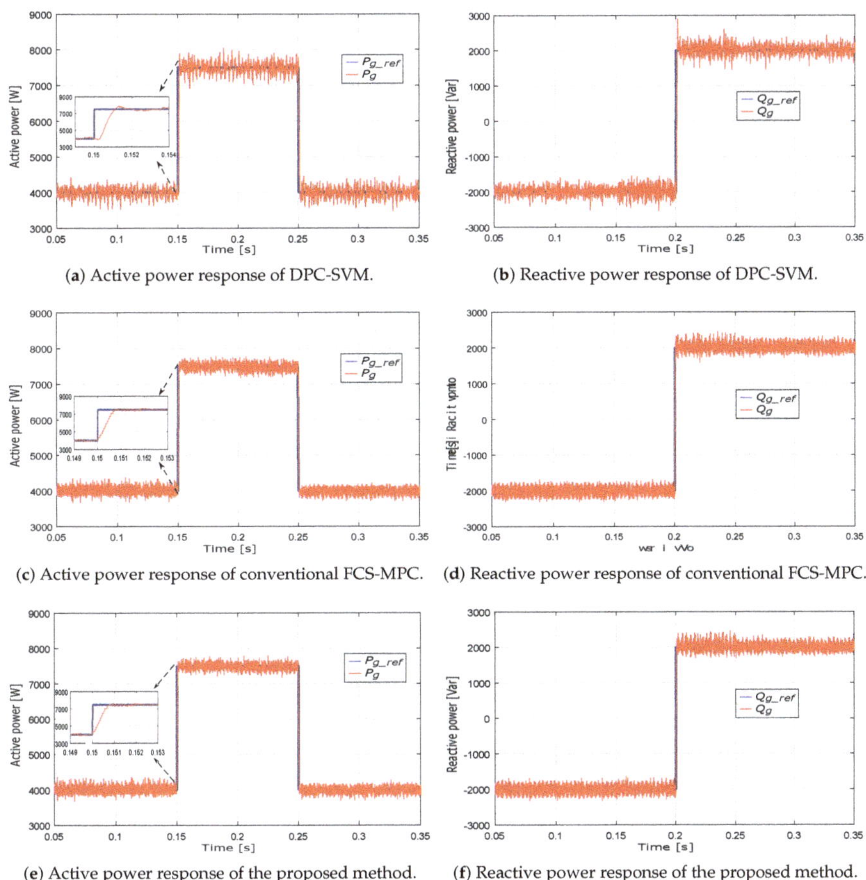

(**a**) Active power response of DPC-SVM.

(**b**) Reactive power response of DPC-SVM.

(**c**) Active power response of conventional FCS-MPC.

(**d**) Reactive power response of conventional FCS-MPC.

(**e**) Active power response of the proposed method.

(**f**) Reactive power response of the proposed method.

Figure 8. The transient responses of active and reactive powers.

Table 4. Comparative studies of transient performance for three controllers.

Active Power Step	$P_g^* = 4 \rightarrow 7.5$ (kW)		
	DPC-SVM	Conventional FCS-MPC	Proposed Method
Rise time (ms)	1.1	0.8	0.8
Settling time (ms)	4.8	0.8	0.8
Overshoot (%)	5.3	0	0
THD of grid current (%)	3.2	2.51	2.5

Figure 9 demonstrates the dynamic response and grid current harmonic spectra using the Powergui fast Fourier transform (FFT) toolbox. The results in Figure 9 indicate the advantages of the proposed control method. It was observed that the total harmonic distortion (THD) of the grid current at $t = 0.18$ s for the proposed method (2.5%) was better than DPC-SVM (3.2%). Particularly, the proposed method accomplished a similar current performance compared with the traditional FCS-MPC, as shown in Figure 9c,e. Moreover, the THD of the grid current was not only decreased but was also less time-consuming. It means that the control algorithm proved the advantage of the proposed method in terms of computation load. Consequently, it is fundamental to underline that the proposed method represents a useful alternative to implementing MPC algorithm in a real-time system.

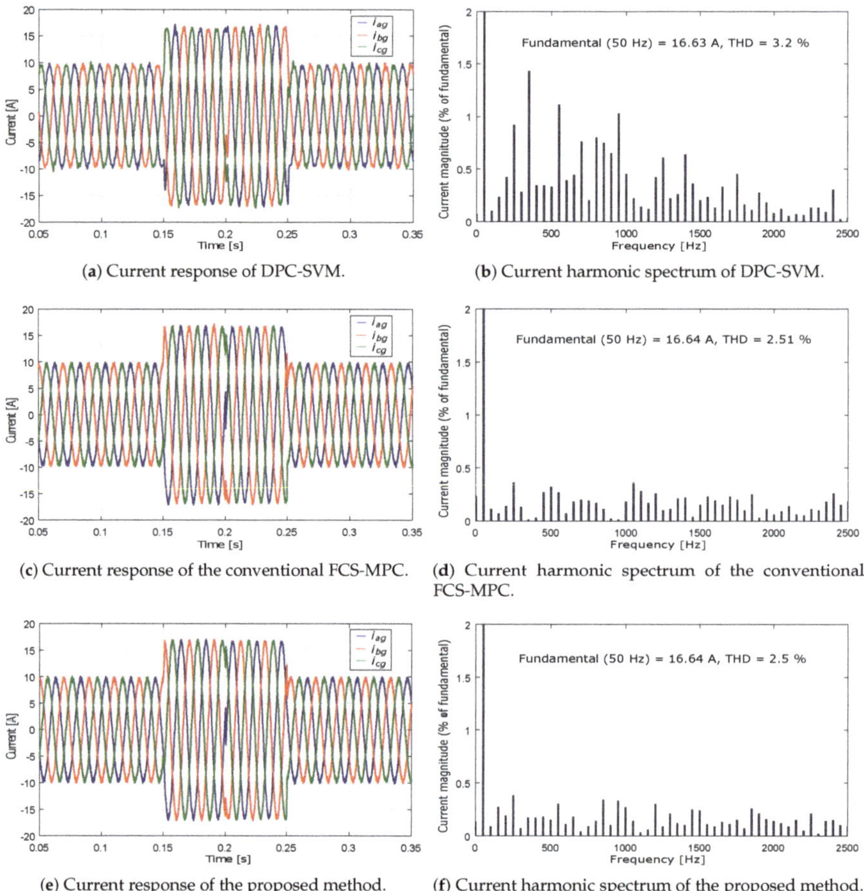

(**a**) Current response of DPC-SVM.

(**b**) Current harmonic spectrum of DPC-SVM.

(**c**) Current response of the conventional FCS-MPC.

(**d**) Current harmonic spectrum of the conventional FCS-MPC.

(**e**) Current response of the proposed method.

(**f**) Current harmonic spectrum of the proposed method.

Figure 9. The dynamic response and harmonic spectrum of the grid current for DPC-SVM, conventional FCS-MPC, and the proposed method.

In addition, the proposed method can solve the drawback of unbalancing DC-link capacitor voltage, as shown in Figure 10. It is interesting to note that the DC-link capacitor voltages remained balanced with MAPE of neutral-point voltage deviation for the proposed method of 0.48% and 0.93% with DPC-SVM.

To analyze the influence of the non-linear load, a study with diode rectifier and resistive load ($R = 60\,\Omega$) was investigated, as illustrated in Figure 11c. The active and reactive powers were set at 4 kW and 0 Var, respectively. Figure 11 demonstrates the results with stand-alone and grid-connected modes at $t = 0.1$ s. As shown in Figure 11a,b, the ripples of the grid current and powers were increased in the standard-alone mode. However, the grid current remained sinusoidal after connecting to the grid. Moreover, the active and reactive powers continued tracking their reference values with a fast transient time. Therefore, i the proposed method could adopt a non-linear load.

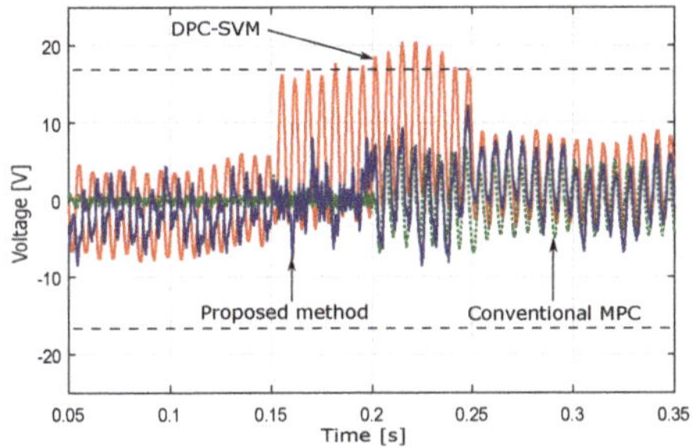

Figure 10. Performance of the neutral-point voltage u_z.

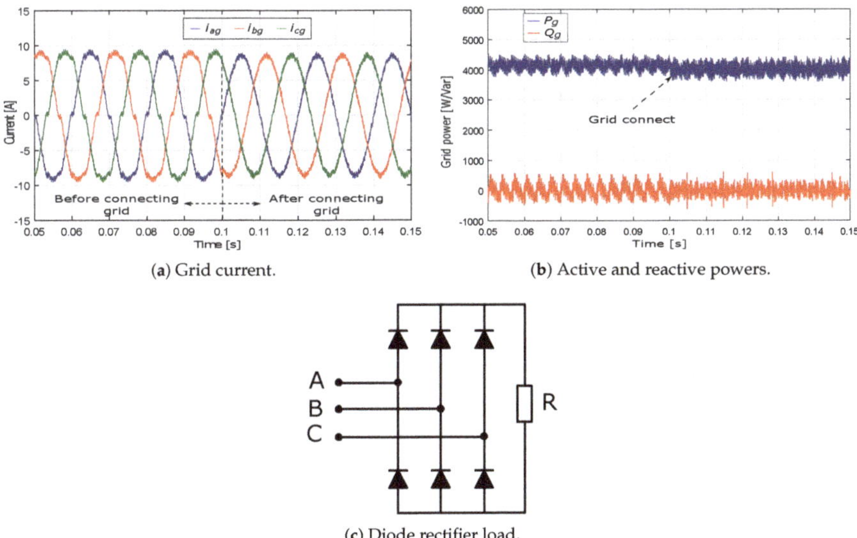

(**a**) Grid current.

(**b**) Active and reactive powers.

(**c**) Diode rectifier load.

Figure 11. Grid current and power responses with the non-linear load.

The selection of the weighting factor is not a transparent task for FCS-MPC, but this issue can be addressed by using the multi-objective optimization techniques suggested by Cortes et al. [33]. The unbalance of DC-bus capacitor voltage is guaranteed by a high value of λ_{dc}, but it augments the power ripples and the THD of the grid current. The criteria for selecting λ_{dc} is the neutral-point voltage deviation, which is considered about 3% of DC-bus voltage, as depicted in Figure 10. On the contrary, the switching frequency was reduced by increasing the weighting factor λ_n, resulting in switching loss reduction. However, the augmentation of the THD of grid current and power ripple also occurred as a result of the weighting factor λ_n. Recognizing the THD as a key to estimate the performance, a satisfactory weighting factor can be achieved by increasing its value in a simulation environment. When the weighting factor λ_n changed from 0 to 120, the switching frequency f_{sw} decreased from 4197 to 2049 Hz, corresponding to the increment in active power error (Figure 12b) and THD of the grid current from 4.02% to 15.55% and from 2.28% to 9.17%, respectively (Figure 12a). As a result, the weighting factor λ_{dc} and λ_n can be selected at 20 and 60 to obtain an adequate control performance with the low THD, the DC-bus capacitor voltage balancing and the switching frequency device at 3 kHz.

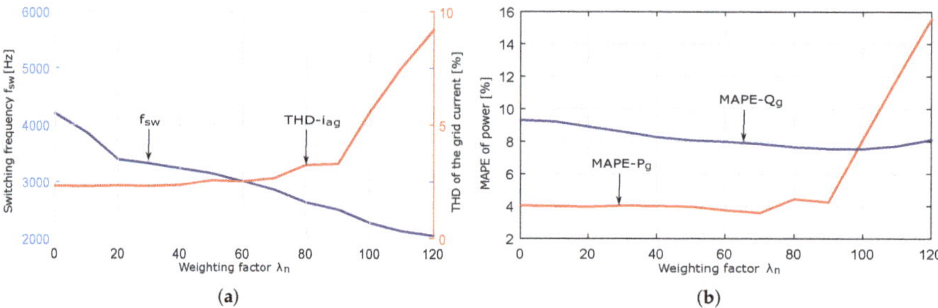

Figure 12. Weighting factor impact on: (**a**) the THD of the grid current and switching frequency; and (**b**) the power ripple.

4.2. Experimental Results

With the purpose of verifying the benefit of the proposed control strategy such as rapid dynamic transient, reasonable THD of the grid current and reduced execution time, a scale down prototype was built in the laboratory, as illustrated in Figure 13. The control algorithm was implemented in a digital signal processing (DSP) TMS320F28335 controller by using the S-function builder block in Matlab/Simulink environment. Twelve FGH40T120SMD IGBT modules were employed for the inverter. Moreover, DC-link used two capacitors LGG2G102MELC50 1000 µF-400 V for the test bench. The parameters of the filter remained the same as in the simulation. The capacitor voltage, grid voltage, and current were measured by LV 25-P and LA 25-P transducers. The switch signals were generated by general-purpose input/output (GPIO) outputs of the DSP.

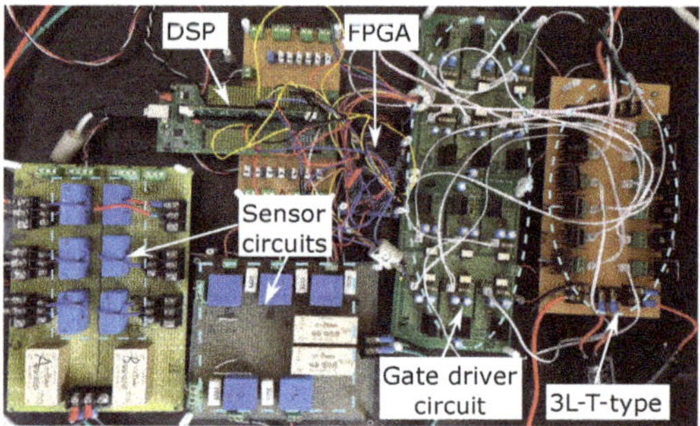

Figure 13. Experimental prototype in the laboratory.

The DC-link voltage was fixed at 600 V while the root mean squared and frequency of the grid voltage remained at 220 V and 60 Hz, respectively. The sampling time of the proposed control method was 100 µs, which is suitable for low-speed processing of DSP. The stable and dynamic transient states were examined with the change in the active and reactive powers to confirm the feasibility of the proposed control. In the first scenario, Figure 14 illustrates the steady-state performance of the proposed control strategy with active power $P_g = 1300$ W and unity power factor ($Q_g = 0$ Var). The experimental results indicate that the ability to track power and achieved sinusoidal grid current waveform. In this test, the three-phase grid voltage was not a good sinusoidal waveform, as shown in Figure 14b, leading to a decrease in the quality of the grid current.

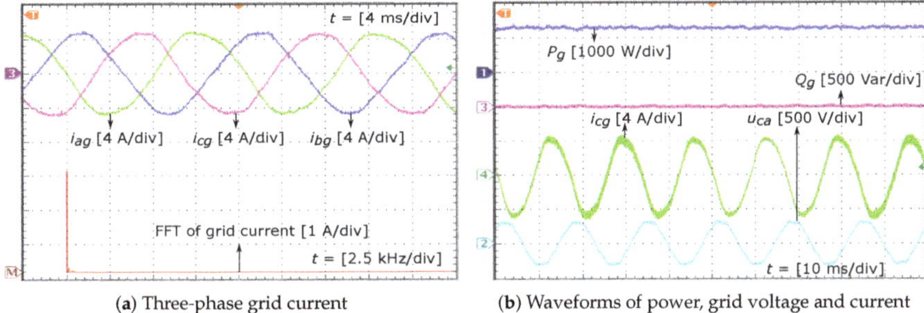

(**a**) Three-phase grid current   (**b**) Waveforms of power, grid voltage and current

Figure 14. Experimental results of the proposed method in steady-state with $P_g = 1300$ W and $Q_g = 0$ Var.

In the second scenario, two cases of the step change in the active and reactive powers were investigated to validate the ability tracking behavior of the proposed control scheme. In the first case, the active power reference was changed from 0 to 1300 W according to the change in grid current from 0 to 4.2 A. As shown in Figure 15, the active power and grid current reached the steady-state after a short transient time. Furthermore, the voltage-balancing of DC-bus capacitors and the reactive power were guaranteed even under the transient response. In the second case, the active power reference was kept at 1000 W while the reference of reactive power was stepped from 0 to 436 Var. It is apparent from the results in Figure 16 that the proposed approach obtained an accurate power tracking capacity under the change of the power factor.

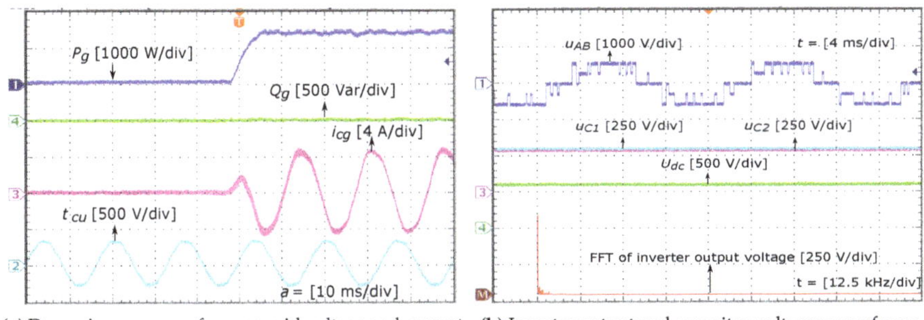

(**a**) Dynamic responses of power, grid voltage and current   (**b**) Inverter output and capacitor voltages waveforms

Figure 15. Experimental results of the step change in active power from 0 to 1300 W with $Q_g = 0$ Var.

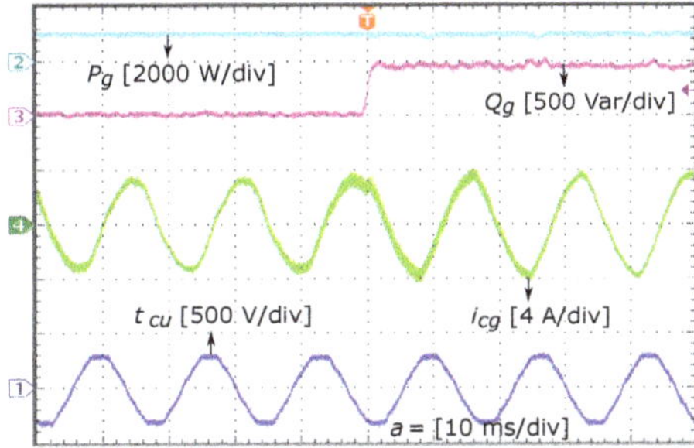

Figure 16. Transient response of the step change in reactive power from 0 to 436 Var with $P_g = 1000$ W.

Two controllers had the same performance of the THD for the grid current, as depicted in Figure 17. The THD of the grid current for the proposed method was 3.4% and 3.5% with the conventional

FCS-MPC. Moreover, the execution time of the control strategy for the proposed method decreased from 88 to 69 μs in comparison to the traditional FCS-MPC, as illustrated in Figure 18. This underlined the 22% reduction of execution time with the proposed method. Although, with the development of the microcontrollers, the execution time can be decreased by using the fast DSP or FPGA, this means that the cost of the system is increased, which becomes a challenge in industrial applications. Consequently, our study provides additional support to implement the MPC algorithm with the low-cost processor.

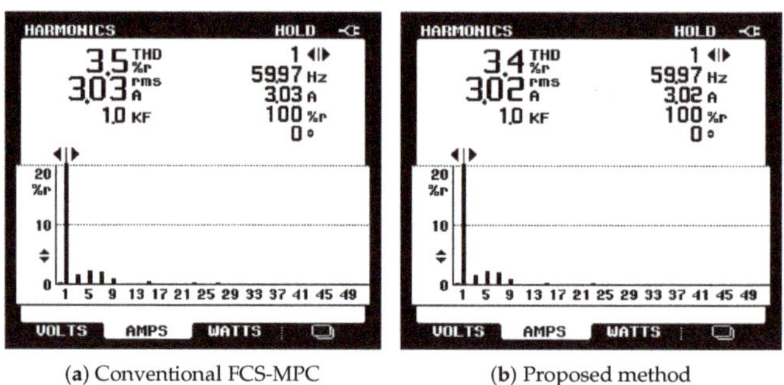

(a) Conventional FCS-MPC   (b) Proposed method

Figure 17. THD of grid current.

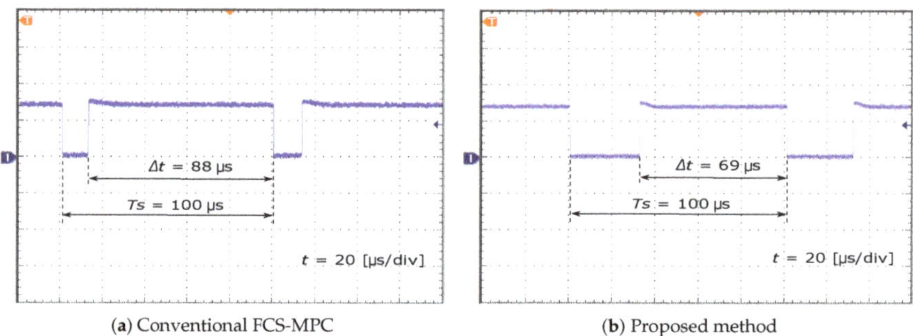

(a) Conventional FCS-MPC   (b) Proposed method

Figure 18. Execution time of the control strategy for the conventional FCS-MPC and proposed methods.

5. Conclusions

This paper presents a development power control strategy based on model predictive control for grid-tie 3L-T-type inverter while ensuring DC-bus capacitor voltage balancing and reduction of switching frequency. Moreover, the inverter voltage reference is applied to compute the equivalent cost function for determining the best control input, leading to reduce the computational effort compared with the conventional FCS-MPC method. A comparison study indicated that the proposed method achieves high-performance control of power and low THD of the current compared with the classical DPC-SVM. Simulations and experiments verified and highlighted the effectiveness of the proposed method.

Author Contributions: V.-Q.-B.N. established the major part of this paper, which includes modeling, simulation investigation, and original draft preparation. M.-K.N. contributed to review and editing. T.-T.T. contributed to validating in the real system. Y.-C.L. provided resources and supervision. J.-H.C. provided resources and funding acquisition.

Funding: This research received no external funding.

Acknowledgments: This research was supported by Korea Electric Power Corporation (Grant number: R18XA04).

Conflicts of Interest: The authors declare no conflict of interest.

Nomenclature

3L-T-type	Three-level T-type
C_1, C_2, C_{dc}	DC-bus capacitance
DPC	Direct power control
DPC-SVM	Classical DPC with linear PI controllers and SVM
FCS-MPC	Finite control set model predictive control
f_{sw}	Average switching frequency
g	Cost function
i_g	Grid current
L_f	Filter inductance
MAPE	Mean absolute percentage error
$[P], [N], [O]$	Positive, negative and zero states
P_g	Grid active power
PLL	Phase-locked loop
PWM	Pulse width modulation
Q_g	Grid reactive power
R_f	Filter resistance
SVM	Space vector modulation
S_x	Switching state
T_s	Sampling time
U_{dc}	DC-bus voltage
$\hat{U}_g$	Grid voltage magnitude
u_g	Grid voltage
u_{inv}	Inverter output voltage
u_z	Neutral-point voltage
λ_{dc}	Weighting factor of the balance of capacitor voltages
λ_n	Weighting factor of switching frequency

References

1. Blaabjerg, F.; Liserre, M.; Ma, K. Power Electronics Converters for Wind Turbine Systems. *IEEE Trans. Ind. Appl.* **2012**, *48*, 708–719. [CrossRef]
2. Portillo, R.C.; Prats, M.M.; Leon, J.I.; Sanchez, J.A.; Carrasco, J.M.; Galvan, E.; Franquelo, L.G. Modeling Strategy for Back-to-Back Three-Level Converters Applied to High-Power Wind Turbines. *IEEE Trans. Ind. Electron.* **2006**, *53*, 1483–1491. [CrossRef]
3. Kouro, S.; Malinowski, M.; Gopakumar, K.; Pou, J.; Franquelo, L.G.; Wu, B.; Rodriguez, J.; Perez, M.A.; Leon, J.I. Recent Advances and Industrial Applications of Multilevel Converters. *IEEE Trans. Ind. Electron.* **2010**, *57*, 2553–2580. [CrossRef]
4. Schweizer, M.; Kolar, J.W. Design and Implementation of a Highly Efficient Three-Level T-Type Converter for Low-Voltage Applications. *IEEE Trans. Power Electron.* **2013**, *28*, 899–907. [CrossRef]
5. Qin, C.; Zhang, C.; Chen, A.; Xing, X.; Zhang, G. A Space Vector Modulation Scheme of the Quasi-Z-Source Three-Level T-Type Inverter for Common-Mode Voltage Reduction. *IEEE Trans. Ind. Electron.* **2018**, *65*, 8340–8350. [CrossRef]
6. Wu, B.; Narimani, M. *High-Power Converters and AC Drives*; Wiley-IEEE Press: Piscataway, NJ, USA, 2017.
7. Ngo, B.Q.V.; Ayerbe, P.R.; Olaru, S. Model Predictive Control with Two-step horizon for Three-level Neutral-Point Clamped Inverter. In Proccedings of the 20th International Conference on Process Control, Strbske Pleso, Slovakia, 9–12 June 2015; pp. 215–220.
8. Kang, J.W.; Hyun, S.W.; Ha, J.O.; Won, C.Y. Improved Neutral-Point Voltage-Shifting Strategy for Power Balancing in Cascaded NPC/H-Bridge Inverter. *Electronics* **2018**, *7*, 167. [CrossRef]
9. Vahedi, H.; Labbe, P.A.; Al-Haddad, K. Balancing three-level NPC inverter DC bus using closed-loop SVM Real time implementation and investigation. *IET Power Electron.* **2016**, *9*, 2076–2084. [CrossRef]
10. In, H.C.; Kim, S.M.; Lee, K.B. Design and Control of Small DC-Link Capacitor-Based Three-Level Inverter with Neutral-Point Voltage Balancing. *Energies* **2018**, *11*, 1435. [CrossRef]

11. Wilamowski, B.M.; Irwin, J.D. *Power Electronics and Motor Drives*; Taylor and Francis Group: Oxfordshire, UK, 2011.

12. Gui, Y.; Lee, G.H.; Kim, C.; Chung, C.C. Direct power control of grid connected voltage source inverters using port-controlled Hamiltonian system. *Int. J. Control Autom. Syst.* **2017**, *15*, 2053–2062. [CrossRef]

13. Zorig, A.; Belkheiri, M.; Barkat, S. Control of three-level T-type inverter based grid connected PV system. In Proceedings of the 13th International Multi-Conference on Systems, Signals & Devices, Leipzig, Germany, 21–24 March 2016; pp. 66–71.

14. Hernández, J.C.; Sanchez-Sutil, F.; Vidal, P.G.; Rus-Casas, C. Primary frequency control and dynamic grid support for vehicle-to-grid in transmission systems. *Int. J. Electr. Power Energy Syst.* **2018**, *100*, 152–166. [CrossRef]

15. Hernandez, J.C.; Bueno, P.G.; Sanchez-Sutil, F. Enhanced utility-scale photovoltaic units with frequency support functions and dynamic grid support for transmission systems. *IET Renew. Power Gener.* **2017**, *11*, 361–372. [CrossRef]

16. Bueno, P.G.; Hernández, J.C.; Ruiz-Rodriguez, F.J. Stability assessment for transmission systems with large utility-scale photovoltaic units. *IET Renew. Power Gener.* **2016**, *10*, 584–597. [CrossRef]

17. Malinowski, M.; Kazmierkowski, M.P.; Hansen, S.; Blaabjerg, F.; Marques, G.D. Virtual-flux-based direct power control of three-phase PWM rectifiers. *IEEE Trans. Ind. Appl.* **2001**, *37*, 1019–1027. [CrossRef]

18. Song, Z.; Tian, Y.; Yan, Z.; Chen, Z. Direct Power Control for Three-Phase Two-Level Voltage-Source Rectifiers Based on Extended-State Observation. *IEEE Trans. Ind. Electron.* **2016**, *63*, 4593–4603. [CrossRef]

19. Sebaaly, F.; Vahedi, H.; Kanaan, H.Y.; Moubayed, N.; Al-Haddad, K. Design and Implementation of Space Vector Modulation-Based Sliding Mode Control for Grid-Connected 3L-NPC Inverter. *IEEE Trans. Ind. Electron.* **2016**, *63*, 7854–7863. [CrossRef]

20. Scoltock, J.; Geyer, T.; Madawala, U.K. Model Predictive Direct Power Control for Grid-Connected NPC Converters. *IEEE Trans. Ind. Electron.* **2015**, *62*, 5319–5328. [CrossRef]

21. Vazquez, S.; Marquez, A.; Aguilera, R.; Quevedo, D.; Leon, J.I.; Franquelo, L.G. Predictive Optimal Switching Sequence Direct Power Control for Grid-Connected Power Converters. *IEEE Trans. Ind. Electron.* **2015**, *62*, 2010–2020. [CrossRef]

22. Ngo, V.Q.B.; Nguyen, M.K.; Tran, T.T.; Lim, Y.C.; Choi, J.H. A Simplified Model Predictive Control for T-Type Inverter with Output LC Filter. *Energies* **2019**, *12*, 31. [CrossRef]

23. Donoso, F.; Mora, A.; Cárdenas, R.; Angulo, A.; Sáez, D.; Rivera, M. Finite-Set Model-Predictive Control Strategies for a 3L-NPC Inverter Operating With Fixed Switching Frequency. *IEEE Trans. Ind. Electron.* **2018**, *65*, 3954–3965. [CrossRef]

24. Ngo, B.Q.V.; Rodriguez-Ayerbe, P.; Olaru, S.; Niculescu, S.I. Model predictive power control based on virtual flux for grid connected three-level neutral-point clamped inverter. In Proceedings of the 18th European Conference on Power Electronics and Applications, Karlsruhe, Germany, 5–9 September 2016; pp. 1–10.

25. Kouro, S.; Cortes, P.; Vargas, R.; Ammann, U.; Rodriguez, J. Model Predictive Control: A Simple and Powerful Method to Control Power Converters. *IEEE Trans. Ind. Electron.* **2009**, *56*, 1826–1838. [CrossRef]

26. Rodriguez, J.; Kazmierkowski, M.P.; Espinoza, J.R.; Zanchetta, P.; Abu-Rub, H.; Young, H.A.; Rojas, C.A. State of the Art of Finite Control Set Model Predictive Control in Power Electronics. *IEEE Trans. Ind. Inform.* **2013**, *9*, 1003–1016. [CrossRef]

27. Rodriguez, J.; Cortes, P. *Predictive Control of Power Converters and Electrical Drives*; John Wiley & Sons, Inc.: New York, NY, USA, 2012.

28. Shi, T.; Zhang, C.; Geng, Q.; Xia, C. Improved model predictive control of three-level voltage source converter. *Electr. Power Compon. Syst.* **2014**, *42*, 1029–1138. [CrossRef]

29. Geyer, T.; Quevedo, D.E. Performance of Multistep Finite Control Set Model Predictive Control for Power Electronics. *IEEE Trans. Power Electron.* **2015**, *30*, 1633–1644. [CrossRef]

30. Aguilera, R.P.; Baidya, R.; Acuna, P.; Vazquez, S.; Mouton, T.; Agelidis, V.G. Model predictive control of cascaded H-bridge inverters based on a fast-optimization algorithm. In Proceedings of the 41st Annual Conference of the IEEE Industrial Electronics Society, Yokohama, Japan, 9–12 November 2015; pp. 4003–4008.

31. Xia, C.; Liu, T.; Shi, T.; Song, Z. A Simplified Finite-Control-Set Model-Predictive Control for Power Converters. *IEEE Trans. Ind. Inform.* **2014**, *10*, 991–1002.

32. Taheri, A.; Zhalebaghi, M.H. A new model predictive control algorithm by reducing the computing time of cost function minimization for NPC inverter in three-phase power grids. *ISA Trans.* **2017**, *71*, 391–402. [CrossRef] [PubMed]
33. Cortes, P.; Kouro, S.; Rocca, B.L.; Vargas, R.; Rodriguez, J.; Leon, J.I.; Vazquez, S.; Franquelo, L.G. Guidelines for weighting factors design in Model Predictive Control of power converters and drives. In Proceedings of the 2009 IEEE International Conference on Industrial Technology, Gippsland, VIC, Australia, 10–13 February 2009; pp. 1–7.

© 2019 by the authors. Licensee MDPI, Basel, Switzerland. This article is an open access article distributed under the terms and conditions of the Creative Commons Attribution (CC BY) license (http://creativecommons.org/licenses/by/4.0/).

Article

Robust Two-Layer Model Predictive Control for Full-Bridge NPC Inverter-Based Class-D Voltage Mode Amplifier

Xinwei Wei [1,2], Hongliang Wang [1,2], Kangliang Wang [1,2], Kui Li [1,2], Minying Li [1] and An Luo [1,2,*]

[1] Guangdong Zhicheng Champion Group Co., Ltd., Dongguan 523718, China; weixinwei@hnu.edu.cn (X.W.); wanghl123@hnu.edu.cn (H.W.); kangliangwang@hnu.edu.cn (K.W.); kuili@hnu.edu.cn (K.L.); lmy@zhicheng-champion.com (M.L.)

[2] The College of Electrical and Information Engineering, Hunan University, Changsha 410082, China

[*] Correspondence: an_luo@hnu.edu.cn

Received: 8 October 2019; Accepted: 6 November 2019; Published: 14 November 2019

Abstract: Finite control set model predictive control (FCS-MPC) is able to handle multiple control objectives and constraints simultaneously with good dynamic performance. However, its industrial application is limited by its high dependence on system model and the huge computational effort. In this paper, a novel robust two-layer MPC (RM-MPC) with strong robustness is proposed for the full-bridge neutral-point clamped (NPC) voltage mode Class-D amplifier (CDA) aiming at this problem. The errors caused by the parameter mismatches or uncertainties of the LC filter and the load current are regarded as lumped disturbance and estimated by the designed Luenberger observer. The robust control can be achieved by compensating the estimated disturbance to the used predictive model. In order to reduce computation of the controller, a two-layer MPC is proposed for the full-bridge NPC inverter with an LC filter. The first layer is used to calculate the optimal output level which minimizes the tracking error of the output voltage. The second layer is used to determine the switching state for the purpose of capacitor voltage balancing. The experimental results show that the lumped model error is observed centrally through only one observer with low complexity. The two-layer MPC further reduced the computation without affecting the dynamic performance.

Keywords: model predictive control (MPC); neutral-point clamped (NPC) inverter; disturbance observer; parameter uncertainty; stability analysis

---

1. Introduction

In the area of industrial measurement, testing, and process technology, there exist many applications of power amplifiers in order to generate current and voltage signals of special shape at high power levels. [1,2]. Voltage mode Class-D amplifiers (CDAs), composed of voltage source inverters with LC filters, are used to power voltage-driven loads, such as the piezoelectric ceramic transducer [3] and the electrostrictive transducer [4]. Commonly used inverter topologies in CDAs can be divided into three categories: the half H-bridge inverter [5], the full H-bridge inverter [6–8], and the cascaded H-bridge inverter [1]. However, the used inverter topology in this paper is the full-bridge neutral-point clamped (NPC) inverter, which features lower voltage stress on power semiconductors, lower voltage harmonics, smaller electromagnetic interference compared with the half H-bridge inverter and the full H-bridge inverter [9]. In addition, this topology costs less switch devices compared with the cascaded H-bridge. However, the closed-loop control of the output voltage is still a complex but meaningful issue when this topology and an LC filter are used together as a voltage mode Class-D amplifier. The reason can be stated as follow. First, arbitrary waveforms in a wide band may be required in the CDA [10]. This demand requires the dynamic response of the voltage controller to be fast enough.

Second, load parameters of CDAs may be complex and variable [4–11]. Then the voltage controller is also required to be robust.

In order to achieve output closed-loop control of cascaded H-bridge CDAs, a single PI voltage controller was used in [1]. However, PI gains are required to be turned repeatedly in this method, and the steady state performance and the transient response compromise each other [12]. In [13], a double closed-loop PI controller, whose bandwidth was increased compared with the single PI controller in [1], was used for cascaded H-bridge voltage mode CDAs. Limited by the dynamic performance of the existing linear controllers, nonlinear controllers, such as the sliding mode controller, was proposed in [14,15] for voltage mode CDAs. The sliding mode controller has better dynamic performance, but it suffers from finding out the sliding surface and the existing chattering phenomena.

As an another nonlinear controller, model predictive control (MPC) has the advantages of ability to handle multiple control objectives and constraints, simplicity and fast dynamic response, and has been widely concerned and studied in recent years. Moreover, it has been successfully applied to several multilevel inverters. In [16], the finite control set model predictive control (FCS-MPC) was applied for the grid-tied three-phase three-level NPC inverter. In [17], the FCS-MPC was also used to a full-bridge NPC inverter. But in [16,17], the output current or voltage tracking and the capacitor voltage balancing were achieved simultaneously by repeatedly predicting and evaluating the sum of the quadratic terms with weight factors in the cost function, which reflected the two control objectives, respectively. However, the repeated predicting and evaluating the complex cost function costed many computations. In [18], the MPC based on optimal switching sequences was proposed for the full-bridge NPC inverter, which could achieve fixed switching frequency for the switch devices. However, this method failed to balance the capacitor voltage [19]. In [19], a low-complexity MPC was also proposed for the full-bridge NPC rectifier. Although the capacitor voltage balancing could be achieved with unbalanced loads, the fixed switching frequency still limited the dynamic performance of the MPC. In [20,21], the complexity of the MPC algorithm was reduced by employing the multistep MPC for modular multilevel converter (MMC) and cascaded H-bridge inverter. But the dynamic performance would also be affected. In [22,23], only the adjacent voltage vectors or output levels were considered for the FCS-MPC algorithm, and the required computation was reduced greatly. However, both of the dynamic response and the control accuracy would be affected under the condition of load step or reference step for these methods. In [24–27], the process of evaluating the quadratic cost function was regarded as a least square problem, and was proposed to be solved by sphere decoding algorithm or its improved algorithms. But large amount of calculation was still unavoidable for the sphere decoding algorithm.

In addition, because of the high dependence on system model, the effectiveness of the MPC faces enormous challenges when there are errors between the actual system model and the established model. This issue can also be expressed as the robustness of the MPC. The robust MPC has been studied for various power electronic converters, such as three-phase three-level NPC converters [28], three-phase PWM rectifiers [29], flying capacitors inverters [30], and three-phase inverters with LC filters [31,32], etc. In [28], the robustness was achieved by a weighted average process of the measured system variables and the predicted variables. Then the control error caused by the model error could be reduced. However, it failed to deal with the dynamic changes of parameters. In [29], the robust MPC was achieved based on an online disturbance observer. But the influence of the parasitic resistances of the grid-tied inductors were not investigated in the simulation and experiment. In [30], the system robustness was improved based on an adaptive observer. But the variation of the filter inductor was also not included. In [31], the robustness was also achieved based on a disturbance observer. But the load current was obtained by an additional observer, which made the control system more complex. In [32], the output of the three phase inverter prediction model at current control instant was compensated by the modeling error of the last control instant. However, only simulation results were provided, and additional current sensors were required.

In this paper, a robust two-layer MPC is proposed for the full-bridge NPC inverter based CDAs. Based on the designed Luenberger observer, the disturbances caused by both the parameter uncertainties or mismatches of the LC filter and the load current can be centrally estimated and compensated to the prediction model in each control period, which can save computation and avoid the use of load current sensor. Moreover, layered structure is used in the proposed robust MPC, so that the output voltage tracking and the capacitor voltage balancing can be achieved simultaneously and decoupled without affecting the dynamic performance, and the required computation can be further reduced.

The rest of this paper is organized as follows. In Section 2, the discrete mathematical model of the CDA is established. In Section 3, a Luenberger disturbance observer based on Kalman filter is designed to estimate the disturbance caused by the parameter mismatch and the load current. In Section 4, a two-layer MPC for the voltage mode CDA is proposed. Section 5 reports the experiment results. In Section 6, the performance of the proposed robust two-layer MPC is focused on discussion and comparison. And the conclusions are presented in Section 7.

2. Modeling of the Voltage Mode Amplifier Using Full-Bridge NPC Inverter

The structure of the full-bridge NPC inverter-based voltage mode Class-D amplifier is shown in Figure 1. The filter inductor is denoted by L_f, and the filter capacitor is denoted by C_f. The voltage of C_f is denoted by V_o, which is also the final output voltage of the digital amplifier. The current of L_f is denoted by i_f, and the final output current of the digital amplifier is denoted by i_o. Both the defined positive directions of i_f and i_o are shown in Figure 1. The output voltage of the full-bridge NPC inverter is denoted by V_{ab}. The full-bridge NPC inverter consists of two bridges. Each bridge consists of four transistors with four antiparallel freewheeling diodes and two clamping diodes. The dc input is denoted by V_{dc}, and two identical capacitors C_1 and C_2 are connected in series to obtain two levels of $V_{dc}/2$ and $-V_{dc}/2$. Driving signals of the transistors can be denoted by S_{xi}. $x{\in}\{a, b\}$ denotes legs of the inverter, where a denotes the left one, b denotes the right one. $i{\in}\{1, 2, 3, 4\}$ denotes the number of transistor in the same bridge. In normal operation, S_{a1} and S_{a3} complement each other, and S_{a2} and S_{a4} complement each other, too. S_{b1}, S_{b2}, S_{b3}, and S_{b4} also meet this constraint. U_{C1} and U_{C2} are used to represent the voltages of capacitors C_1 and C_2, respectively. S is defined by $[S_{a1}\ S_{a2}\ S_{a3}\ S_{a4}\ S_{b1}\ S_{b2}\ S_{b3}\ S_{b4}]$ and used to denote the switching state of the inverter. M denotes the output level of the full-bridge NPC inverter. And $M{\in}\{-2, -1, 0, 1, 2\}$ is easy to be obtained.

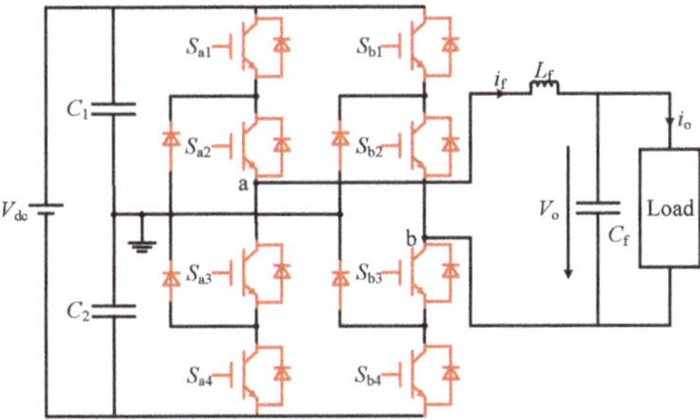

Figure 1. The structure of the full-bridge neutral-point clamped (NPC) inverter based voltage mode Class-D amplifier.

Because of the limitation of the complementary driving signals mentioned above, there are only nine effective switching states, which can be denoted by S1–S9. Table 1 shows the relationship between the output level M, the inductor current i_f, the change of U_{C1}, and the nine effective switching states.

Table 1. Relationship between M, i_o, U_C1, and S.

M	S	U_C1	
		$i_\mathrm{o} > 0$	$i_\mathrm{o} < 0$
2	S1 = [1 1 0 0 0 0 1 1]	invariant	invariant
1	S2 = [1 1 0 0 0 1 1 0]	decrease	increase
	S3 = [0 1 1 0 0 0 1 1]	increase	decrease
0	S4 = [1 1 0 0 1 1 0 0]	invariant	invariant
	S5 = [0 1 1 0 0 1 1 0]	invariant	invariant
	S6 = [0 0 1 1 0 0 1 1]	invariant	invariant
-1	S7 = [0 1 1 0 1 1 0 0]	increase	decrease
	S8 = [0 0 1 1 0 1 1 0]	decrease	increase
-2	S9 = [0 0 1 1 1 1 0 0]	invariant	invariant

Assuming that U_C1 and U_C2 are well balanced, the differential equation of the full-bridge NPC inverter-based voltage mode amplifier can be obtained as Equation (1) from Figure 1, based on the Kirchhoff's laws of voltage and current.

$$\begin{cases} L_\mathrm{f}\frac{di_\mathrm{f}}{dt} + V_\mathrm{o} = V_\mathrm{ab} \\ C_\mathrm{f}\frac{dV_\mathrm{o}}{dt} + i_\mathrm{o} = i_\mathrm{f} \end{cases} \tag{1}$$

In Equation (1), V_ab can also be represented by the output level of the full-bridge NPC inverter, M, as shown in Equation (2).

$$V_\mathrm{ab} = \frac{V_\mathrm{dc}}{2}M \tag{2}$$

By substituting Equation (2) into Equation (1), Equation (3) can be obtained.

$$\begin{cases} \frac{di_\mathrm{f}}{dt} = -\frac{1}{L_\mathrm{f}}V_\mathrm{o} + \frac{V_\mathrm{dc}}{2L_\mathrm{f}}M \\ \frac{dV_\mathrm{o}}{dt} = \frac{1}{C_\mathrm{f}}i_\mathrm{f} - \frac{1}{C_\mathrm{f}}i_\mathrm{o} \end{cases} \tag{3}$$

The filter inductor current i_f and the filter capacitor voltage V_o can be selected as the state variables of the system, and can be denoted by $x = [i_\mathrm{f}\ V_\mathrm{o}]^\mathrm{T}$. Therefore, and the model of the full-bridge NPC inverter based voltage mode amplifier can be transformed into Equation (4),

$$\dot{x} = Ax + B_1 M + B_2 i_\mathrm{o} \tag{4}$$

where $A = \begin{bmatrix} 0 & -\frac{1}{L_\mathrm{f}} \\ \frac{1}{C_\mathrm{f}} & 0 \end{bmatrix}$, $B_1 = \begin{bmatrix} \frac{V_\mathrm{dc}}{2L_\mathrm{f}} \\ 0 \end{bmatrix}$, $B_2 = \begin{bmatrix} 0 \\ -\frac{1}{C_\mathrm{f}} \end{bmatrix}$.

For the purpose of digital control, Equation (4) should be discretized. If the sampling period and the control period are denoted as T_S, the discrete model can be expressed as Equation (5),

$$x(k+1) = A_\mathrm{d}x(k) + B_\mathrm{1d}M(k) + B_\mathrm{2d}i_\mathrm{o}(k) \tag{5}$$

where $A_\mathrm{d} = e^{AT_\mathrm{S}}$, $B_\mathrm{1d} = \int_0^{T_\mathrm{S}} e^{A\tau}B_1 d\tau$, $B_\mathrm{2d} = \int_0^{T_\mathrm{S}} e^{A\tau}B_2 d\tau$, k and $k+1$ represent the kT_S and $(k+1)T_\mathrm{S}$ instant, respectively.

3. Design of the Luenberger Observer for Disturbance Estimation

Luenberger observer adopts a predictor-corrector structure. In the predictor, a predictive model is used to predict the system operation. In the corrector, a feedback signal is compensating to the predictive model to correct the error between the actual system and the used predictive model [29].

In this paper, it is used to estimate and compensate the lumped disturbance caused by the mismatch or uncertainty of the filter parameters and the load current.

3.1. Design of the Disturbance Observer

In order to achieve robust MPC, there are two great challenges. The first one is the unknown load current i_o, because the load current sensor is intentionally avoided. And it will generate large disturbance for the precise control of the output voltage V_o. The second one is the uncertain parameters of the LC filter. Based on Equations (4) and (5), it can be seen that the parameter matrices A_d, B_{1d}, and B_{2d} are calculated from L_f and C_f, which are the actual parameters of the LC filter. However, the actual parameters L_f and C_f may not equal to the nominal parameters L_{fn} and C_{fn}, which are used in the controller. The actual parameter of the filter inductor, L_f, may not be equal to the nominal parameter L_{fn}, because of several phenomena, such as the magnetic saturation. And the actual parameter of the filter capacitor, C_f, may also not be equal to the nominal parameter C_{fn}, because of the unmodeled parasitic resistance and the manufacturing tolerance.

Both of L_f and C_f are difficult to obtain accurately in practical engineering. In order to distinguish the discrete model used in the controller from the actual system model, the parameter matrices used in the controller are denoted by A_{dn}, B_{1dn}, and B_{2dn}, and can be calculated by Equations (6)–(8) based on the nominal parameters L_{fn} and C_{fn}, respectively.

$$A_{dn} = e^{A_n T_S} \tag{6}$$

$$B_{1dn} = \int_0^{T_S} e^{A_n \tau} B_{1n} d\tau \tag{7}$$

$$B_{2dn} = \int_0^{T_S} e^{A_n \tau} B_{2n} d\tau \tag{8}$$

In (6)–(8), $A_n = \begin{bmatrix} 0 & -\frac{1}{L_{fn}} \\ \frac{1}{C_{fn}} & 0 \end{bmatrix}$, $B_{1n} = \begin{bmatrix} \frac{V_{dc}}{2L_{fn}} \\ 0 \end{bmatrix}$, $B_{2n} = \begin{bmatrix} 0 \\ -\frac{1}{C_{fn}} \end{bmatrix}$.

The relationship between the actual parameter matrices A_d, B_{1d}, and B_{2d} and the nominal parameter matrices A_{dn}, B_{1dn}, and B_{2dn} can be shown in Equations (9)–(11),

$$A_d = A_{dn} + \Delta A_d \tag{9}$$

$$B_{1d} = B_{1dn} + \Delta B_{1d} \tag{10}$$

$$B_{2d} = B_{2dn} + \Delta B_{2d} \tag{11}$$

where ΔA_d, ΔB_{1d}, ΔB_{2d} denote the model errors caused by the mismatch or uncertainty of the LC filter parameters. Then the discrete system model in Equation (5) can be transformed into Equation (12) by substituting Equations (9)–(11) into (5).

$$x(k+1) = A_{dn}x(k) + B_{1dn}M(k) + B_{2dn}i_o(k) + \Delta A_d x(k) + \Delta B_{1d}M(k) + \Delta B_{2d}i_o(k) \tag{12}$$

In the right side of Equation (12), there are four uncertain terms. The first one is the third term $B_{2dn}i_o(k)$, which is uncertain because of the unknown $i_o(k)$ in the absence of the load current sensor. The second one and the third one are the fourth term $\Delta A_d x(k)$ and the fifth term $\Delta B_{1d}M(k)$, which are uncertain because of the uncertain matrices ΔA_d, and ΔB_{1d}. The fourth one is the last term $\Delta B_{2d}i_o(k)$, which is uncertain because of both the uncertain matrix ΔB_{2d} and the unknown load current $i_o(k)$. For the purpose of achieving robust control against the unknown $i_o(k)$ and the uncertain ΔA_d, ΔB_{1d},

ΔB_{2d}, the sum of the four terms is regarded as the lumped disturbance variable $N(k)$, as shown in Equation (13).

$$N(k) = B_{2dn}i_o(k) + \Delta A_d x(k) + \Delta B_{1d}M(k) + \Delta B_{2d}i_o(k) \tag{13}$$

The lumped disturbance $N(k)$ is a two-dimensional variable, and can be expressed as $N(k) = [N_1(k)\ N_2(k)]^T$. If both of $N_1(k)$ and $N_2(k)$ can be successfully estimated and compensated to the system predictive model based on the nominal parameters in real time, the disturbance-rejection approach can be implemented to achieve robustness against the uncertainty of the filter parameters and the load current i_o. And this will be still effective even if the system parameters vary during operation, which is regarded as the dynamic parameter variations.

The disturbance variables $N_1(k)$ and $N_2(k)$ can be assumed to be constant during each sampling interval [29,31], and they can be extended as the system variables. Then Equation (14) is obtained as,

$$X(k) = \begin{bmatrix} i_f(k) \\ V_o(k) \\ N_1(k) \\ N_2(k) \end{bmatrix} = \begin{bmatrix} A_{dn11} & A_{dn12} & 1 & 0 \\ A_{dn21} & A_{dn22} & 0 & 1 \\ 0 & 0 & 1 & 0 \\ 0 & 0 & 0 & 1 \end{bmatrix} \begin{bmatrix} i_f(k-1) \\ V_o(k-1) \\ N_1(k-1) \\ N_2(k-1) \end{bmatrix} + \begin{bmatrix} B_{1dn11} \\ B_{1dn21} \\ 0 \\ 0 \end{bmatrix} M(k-1) = \Phi X(k-1) + GM(k-1) \tag{14}$$

where $\Phi = \begin{bmatrix} A_{dn11} & A_{dn12} & 1 & 0 \\ A_{dn21} & A_{dn22} & 0 & 1 \\ 0 & 0 & 1 & 0 \\ 0 & 0 & 0 & 1 \end{bmatrix} = \begin{bmatrix} A_{dn} & 1 \\ 0 & 1 \end{bmatrix}$, $G = \begin{bmatrix} B_{1dn11} \\ B_{1dn21} \\ 0 \\ 0 \end{bmatrix} = \begin{bmatrix} B_{1dn} \\ 0 \end{bmatrix}$.

And the output equation of the system can be expressed as Equation (15),

$$Y(k) = \begin{bmatrix} i_f(k) \\ V_o(k) \end{bmatrix} = CX(k) \tag{15}$$

where $C = \begin{bmatrix} 1 & 0 & 0 & 0 \\ 0 & 1 & 0 & 0 \end{bmatrix}$.

Based on Equations (14) and (15), a discrete observer can be constructed as shown in Equation (16),

$$\hat{X}(k) = \Phi\hat{X}(k-1) + GM(k-1) + L\big(Y(k-1) - \hat{Y}(k-1)\big) = \Phi\hat{X}(k-1) + GM(k-1) + LC\big(X(k-1) - \hat{X}(k-1)\big) \tag{16}$$

where $\hat{X}(k)$ and $\hat{X}(k-1)$ respectively denote the estimated value of $X(k)$ and $X(k-1)$, and L denotes the gain matrix. $M(k-1)$ denotes the output level of the full-bridge NPC inverter in the $(k-1)$th control period, which is obtained by the proposed two-layer MPC in the $(k-1)$th control period.

$\hat{N}(k)$ can be used to denote the estimated value of $N(k)$, and it can be calculated by Equation (17).

$$\hat{N}(k) = \begin{bmatrix} \hat{N}_1(k) \\ \hat{N}_2(k) \end{bmatrix} = \begin{bmatrix} 0 & 0 & 1 & 1 \end{bmatrix} \begin{bmatrix} \hat{i}_f(k) \\ \hat{V}_o(k) \\ \hat{N}_1(k) \\ \hat{N}_2(k) \end{bmatrix} \tag{17}$$

3.2. Parameter Design

The Kalman filter is a commonly used method to optimally estimate the state of a dynamic system from a series of imperfect noisy measurements, especially in presence of uncertainties [31]. In this paper, the designed observer can be regarded as a discrete Kalman filter to calculate the observer gain matrix L.

As an optimal recursive data processing algorithm, the discrete Kalman filter can perform cyclic calculation according to the following five steps in each control period.

The first step is to calculate the prior state estimate value $\hat{X}_k^-$ based on Equation (18).

$$\hat{X}_k^- = \Phi\hat{X}(k-1) + GM(k-1) \tag{18}$$

The second step is to calculate the priori estimate error covariance matrix P_k^- based on Equation (19),

$$P_k^- = \Phi P_{k-1}\Phi^{\mathrm{T}} + Q \tag{19}$$

where Q is the given process noise covariance matrix, and P_{k-1} is the posteriori estimate error covariance matrix in the last control period.

The third step is to calculate the observer gain matrix L based on Equation (20),

$$L = P_k^- C^{\mathrm{T}}\left(CP_k^- C^{\mathrm{T}} + R\right)^{-1} \tag{20}$$

where R is the given measurement noise covariance matrix.

The fourth step is to update the estimated state variable $\hat{X}(k)$ based on Equation (21).

$$\hat{X}(k) = \hat{X}_k^- + L\left(Y(k) - C\hat{X}_k^-\right) \tag{21}$$

The fifth step is to calculate the posteriori estimate error covariance matrix P_k based on Equation (22).

$$P_k = (I - LC)P_k^- \tag{22}$$

where I denotes the identity matrix.

The first step and the second step can be collectively referred to as the prediction link of the discrete Kalman filter. And the last three steps can be collectively referred to as the correction link of the discrete Kalman filter.

The stability of the designed discrete Kalman filter has been proved by [31,33], and will not be discussed here.

Remark: The performance of the designed state observer is determined by the given matrix Q and R. The larger Q is, the faster the observed values converge to their actual value, but too fast convergence speed will lead to noise interference. The smaller R is, the less noise interference, but the slower convergence rate. Therefore, Q and R should be adjusted synthetically to achieve the tradeoff between convergence speed and noise suppression.

4. Two-Layer Model Predictive Control

The basic control objectives of the voltage CDA using full-bridge NPC inverter include two terms: (1) output voltage tracking; (2) capacitor voltage balancing. Traditional FCS-MPC (TFCS-MPC) requires repeated predictions and evaluations for each effective switching state, and the one which minimizes the cost function is selected as the optimal control option. Thus when it is applied to the full-bridge NPC inverter, there will be nine candidate switching states. And this places large computational burden on the digital controllers when a small control period is required.

Therefore, a two-layer MPC for the cascaded full-bridge NPC voltage mode amplifier is proposed in this paper, which is much simpler than TFCS-MPC without affecting the dynamic performance. And the proposed two-layer MPC decouples the two control objectives, which also allows the two control objectives to be achieved simultaneously without weight factors. The structure of the proposed two-layer MPC is shown in Figure 2, where the first layer is used to calculate the optimal output level for the purpose of achieving the first control objective, the second layer is used to determine the switching state for the purpose of achieving the second control objective.

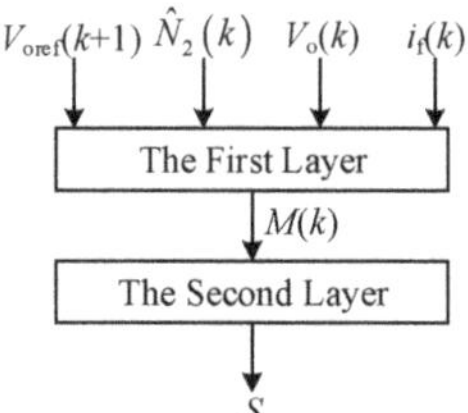

Figure 2. The structure of the proposed two-layer model predictive control (MPC).

4.1. The First Layer

The reference of the output voltage can be denoted by V_{oref}, which is also the voltage signal to be amplified. In this system, the cost function corresponding to level h can be defined as $J(h)$ in Equation (23),

$$J(h) = \left| V_{oref}(k+1) - V_{oh}(k+1) \right|,$$
$$h \in H = \{-2n, -2n+1, \cdots, 0, \cdots, 2n-1, 2n\} \tag{23}$$

where $V_{oh}(k+1)$ denotes the output current at $(k+1)T_S$ instant when h is selected in the kth control period. Based on Equations (12) and (13), $V_{oh}(k+1)$ can be predicted as given in Equation (24).

$$V_{oh}(k+1) = A_{dn21}i_f(k) + A_{dn22}V_o(k) + B_{1dn21}M(k) + N_2(k) \tag{24}$$

In Equation (24), $N_2(k)$ cannot be obtained because it is determined by the uncertain model errors and the unknown load current i_o without configured load current sensor. However, the designed Luenburger observer can successfully estimate $N_2(k)$ to $\hat{N}_2(k)$, then we are allowed to replace $N_2(k)$ with $\hat{N}_2(k)$. Thus Equation (24) can be improved to Equation (25).

$$V_{oh}(k+1) = A_{dn21}i_f(k) + A_{dn22}V_o(k) + B_{1dn21}M(k) + \hat{N}_2(k) \tag{25}$$

Another function, $J_1(h)$, with the output level h as its independent variable can be defined by Equation (26).

$$J_1(h) = V_{oh}(k+1) - V_{oref}(k+1) = A_{dn21}i_f(k) + A_{dn22}V_o(k) + B_{1dn21}M(k) + N_2(k) - V_{oref}(k+1) \tag{26}$$

The relationship between $J_1(h)$ and h will be linear if h is supposed to be continuous. For the convenience of expression, another variable, h_{sol} is defined as the solution of $J_1(h) = 0$, and can be calculated as Equation (27).

$$h_{sol} = \frac{V_{oref}(k+1) - A_{dn21}i_f(k) - A_{dn22}V_o(k) - N_2(k)}{B_{1dn21}} \tag{27}$$

Because of $J(h) = \left| J_1(h) \right| \geq 0$, the optimal output level $M(k)$, which minimizes $J(h)$, must be equal to the integer nearest to h_{sol}. Thus the optimal output level $M(k)$ is allowed to be directly obtained by Equation (28),

$$M(k) = \mathrm{argmin}_{h \in H} \left| J_1(h) \right| = round\left(h \big|_{J_1(h)=0} \right) \tag{28}$$

where $round(x)$ denotes the rounding function, which is equal to the integer nearest to x.

In order to avoid the case that the result of Equation (28) does not belong to H, Equation (29) is also required after Equation (28).

$$M(k) = \begin{cases} 2, & M(k) > 2 \\ M(k), & -2 \leq M(k) \leq 2 \\ -2, & M(k) < -2 \end{cases} \tag{29}$$

Thus the cost function is evaluated only once, which greatly reduces the computational burden.

4.2. The Second Layer

The second layer is used to determine the switching state to achieve capacitor voltage balancing. At the same time, the switching action times should also be considered when the switching state is determined, because there are multiple switching states to be selected if level −1, 0, or 1 is required.

In steady-state operation, the full-bridge NPC inverter is generally switched between adjacent levels. Thus the output level will be switched between 2, 1, and 0 when $M(k) > 0$. Table 2 shows the number of switching actions when the three levels are switched between each other. According to Table 2, if the submodule output level is 2, the switching state can only select S1. If the submodule output level is 1, in order to achieve the purpose of capacitor voltage balance, S2 should be selected when the signs of i_o and ΔU_C are the same, while S3 should be selected when they are opposite, considering the result of Table 1. If the submodule output level is 0, the switching state can only select S5, in order to minimize the switching actions when level 0 and level 1 are switched between each other.

Table 2. Switching actions when the output level switching between 2, 1, and 0.

Switching Level	Switching State	Action Times
2 and 1	S1 and S2	2
	S1 and S3	2
1 and 0	S2 and S4	2
	S2 and S5	2
	S2 and S6	6
	S3 and S4	6
	S3 and S5	2
	S3 and S6	2
2 and 0	S1 and S4	4
	S1 and S5	4
	S1 and S6	4

Similar analysis can be done when $M(k) < 0$ and the following conclusions can be drawn. If the submodule output level is −2, S9 is selected. If the submodule output level is −1, S8 is selected when the signs of i_o and ΔU_C are the same, while S7 is selected when the signs of i_o and ΔU_C are opposite. If the submodule output level is 0, S5 is selected.

Figure 3 shows the flow chart of the switching state selection process. The parallel structure of the middle and lower layer control shows that they are more suitable for implementation by a FPGA.

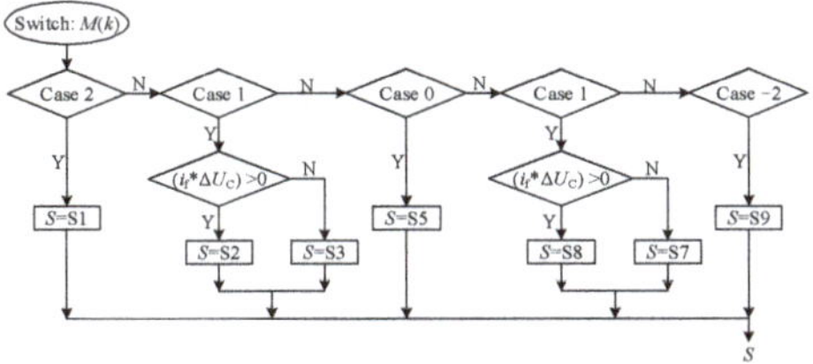

Figure 3. The flowchart of the switching state selection process.

5. Experimental Verification

In order to verify the feasibility and validity of the proposed robust multilayer MPC applied to the full-bridge NPC voltage-mode digital power amplifier, a 2 kW experimental prototype with a 50 Hz–800 Hz output band is built in the laboratory as shown in Figure 4. The actual value L_f of the filter inductor used is 2 mH, and the actual value C_f of the filter capacitor used is 10 uF. The voltage of the dc input, V_{dc}, is 300 V, so that two voltage levels of 150 V and -150 V can be obtained. The capacitance of those two capacitors C_1 and C_2 is 1070 uf. The control frequency is set to be 100 kHz, which means that the sample period is set to be 10 us. The high control frequency is used because wide range output frequency and high precision out voltage are required. Thus the maximum switching frequency of the used switch devices is 50 kHz. In fact, the designed digital power amplifier uses SGH80N60UFD type fast IGBT and DSE130–60; a type fast recovery diode, whose maximum switching frequency can be up to 100 kHz.

Figure 4. The designed 2 kW experiment prototype.

5.1. Steady State Performance

In order to study the steady state performance of the proposed RM-MPC, the output voltage reference V_{oref} is set to a sine wave with an 800 Hz frequency and a 200 V root-mean-square (RMS) value, which may be widely used in underwater electroacoustic transduction system. The load is set to a 20 Ω resistor. The experiment results are shown in Figure 5, where (a) shows the waveforms of V_o and its reference V_{oref}, (b) shows the waveform of the estimated value of N_1, (c) shows the waveform of the estimated value of N_2, (d) shows the waveforms of the two capacitor voltages in the dc side.

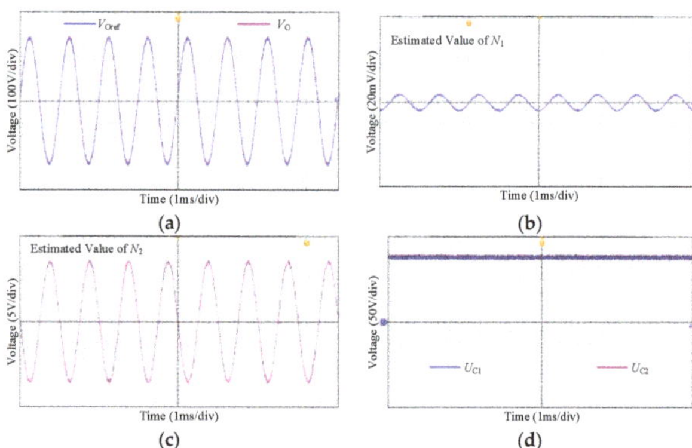

Figure 5. Steady state experiment results. (**a**) Waveforms of V_o and V_{oref}; (**b**) Waveform of the estimated value of N_1; (**c**) Waveform of the estimated value of N_2; (**d**) Waveforms of capacitor voltages.

It can be seen that the output voltage is accurately tracked with a 0.52% total harmonics distortion (THD). At the same time, the two capacitor voltages in the dc side are well balanced. In addition, the disturbance variables N_1 and N_2 are successfully estimated with little noises. In this way, the proposed RM-MPC shows good steady state performance.

5.2. Dynamic Performance

In order to study the dynamic performance of the proposed RM-MPC, the output voltage reference V_{oref} is set to a sine wave with an 50 Hz frequency and a 100V RMS value for initialization. However, the RMS value of the desired sine wave steps to 200 V at $t = 0.05$ s. And the load is still set to a 20 Ω

resistor. The experiment results are shown in Figure 6, where (a) shows the waveforms of V_o and its reference V_{oref}, (b) shows the waveform of the estimated value of N_1, (c) shows the waveform of the estimated value of N_2, (d) shows the waveforms of the two capacitor voltages in the dc side.

It can be seen that the output voltage tracks the step variation of its reference quickly, and the tracking error is reduced to 1 V within 0.54 ms. Besides, the two capacitor voltages in the dc side are also well balanced during transient variation. Thus, the fast dynamic performance of the proposed multilayer MPC is verified. Moreover, Figure 6b,c shows that the estimated values of N_1 and N_2 change quickly with the step variation of V_{oref}, so that the fast dynamic performance of the designed Luenberger observer is also verified.

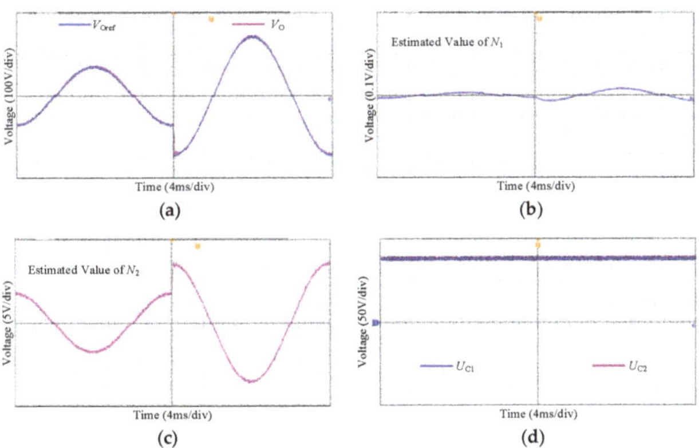

Figure 6. Dynamic experiment results. (**a**) Waveforms of V_o and V_{oref}; (**b**) waveform of the estimated value of N_1; (**c**) waveform of the estimated value of N_2; (**d**) waveforms of capacitor voltages.

5.3. Robust Performance

In order to study the robust performance of the proposed RM-MPC, two groups of experiments are carried out, where the output voltage reference V_{oref} is set to a sine wave with an 50 Hz frequency and a 200 V RMS value, and the load is also set to a 20 Ω resistor. However, in the first group, the inductance value used in the controller is set to 1 mH, and the capacitance value used in the controller is set to 5 uF, which means that there are −50% parameter mismatch. In the second group, the inductance value used in the controller is set to 3 mH, and the capacitance value used in the controller is set to 15 uF, which means that there are +50% parameter mismatch.

The results of the first group experiment are shown in Figure 7, where (a) shows the waveforms of V_o and its reference V_{oref}, (b) shows the waveform of the estimated value of N_1, (c) shows the waveform of the estimated value of N_2, (d) shows the waveforms of the two capacitor voltages in the dc side. And the results of the second group experiment are shown in Figure 8.

It can be seen that the good tracking effect of the output voltage are maintained with the help of the designed Luenberger observer, even if there are ±50% parameter mismatches. Furthermore, the well capacitor voltage balancing is also achieved against the parameter mismatches. Thus the strong robustness is also verified by the experiment results.

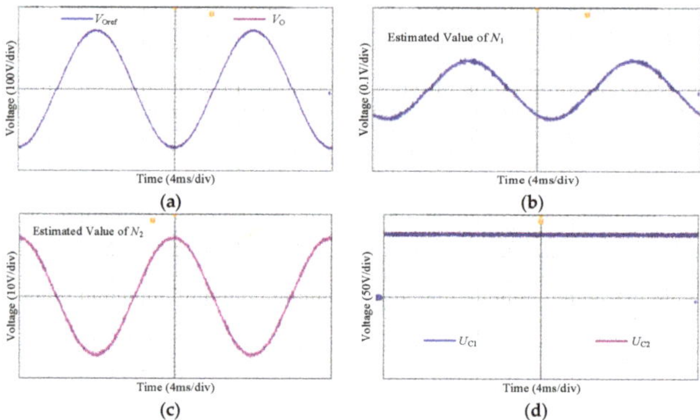

Figure 7. Robustness experiment results with −50% parameter mismatch. (**a**) Waveforms of V_o and V_{oref}; (**b**) waveform of the estimated value of N_1; (**c**) waveform of the estimated value of N_2; (**d**) waveforms of capacitor voltages.

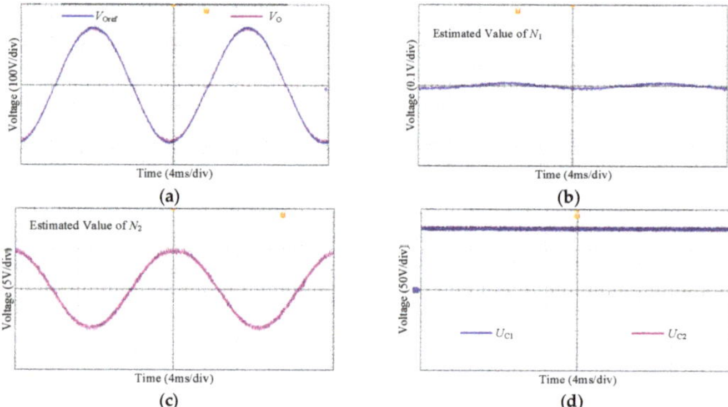

Figure 8. Robustness experiment results with +50% parameter mismatch. (**a**) Waveforms of V_o and V_{oref}; (**b**) waveform of the estimated value of N_1; (**c**) waveform of the estimated value of N_2; (**d**) waveforms of capacitor voltages.

6. Discussion

In Section 3.1, it can be seen that all the uncertain terms in Equation (12) can be divided into three categories, which are the ones related to the load current, the ones related to the uncertain parameters, and the ones related to both the load current and the uncertain parameters. In Sections 5.1 and 5.2, there are no parameter uncertainties or mismatches, then the designed observer only estimates the uncertain term related to the load current. In this way, the observer operates as a load current observer, and the two subsections of 5.1 and 5.2 focus on the steady-state and dynamic performance of the designed observer and the proposed two-layer MPC. In Section 5.3, parameter uncertainties exist, and the designed observer estimates the sum of all the uncertain terms. Thus Section 5.3 focuses on the robustness performance against the parameter mismatches of the LC filter.

Compared with the dual observers used method in [31] and the additional current sensors based method in [32], only one observer is used to observer all the uncertain terms caused by the parameter uncertainties and the load current in a centralized way. This not only improves the system robustness against both the parameter uncertainties of the filter inductor and the filter capacitor, but also avoids the use of additional load current sensors and observers. And this also leads to a reduction in both the amount of calculation and the economic cost for the hardware configuration. In addition, the layered structure of the proposed MPC further reduces the computation. It can be inferred that the proposed robust MPC can also be extended to general inverters with LC filters, which are widely used in distributed generation systems, energy storage systems, and uninterruptible power supplies.

7. Conclusions

In this paper, a robust multilayer MPC, which can achieve decoupling control of the output voltage and capacitor voltage balancing simultaneously in different layers against the parameter uncertainties or mismatches of the LC filter, is proposed for the full-bridge NPC inverter-based CDAs. The errors caused by the parameter mismatches or uncertainties of the LC filter and the load current are regarded as lumped disturbance and estimated by the designed Luenberger observer. Based on the estimated disturbance, a two-layer MPC is proposed, where the output voltage tracking and the capacitor voltage balancing are achieved in the first layer and the second layer, respectively. Finally, the steady state performance, the dynamic performance and the robust performance are verified on the designed 2 kW experiment prototype.

Compared with existing methods, the proposed robust two-layer MPC uses only one observer to observe the lumped disturbance caused by the parameter mismatch and the load current, which simplifies the control system and reduces the calculation of the system. The layered structure further reduces the computation without affecting the dynamic performance of the MPC. However, the control delay is not considered, which may affect the control effect and calls for further research.

Author Contributions: Conceptualization, X.W.; formal analysis, A.L.; funding acquisition, M.L.; investigation, X.W.; methodology, A.L.; project administration, M.L.; software, K.L.; validation, K.W.; writing—original draft, X.W.; writing—review and editing, H.W. and A.L.

Funding: The Program for Guangdong Introducing Innovative and Entrepreneurial Teams: 2017ZT07G23; Research on High-power and High-efficiency Electro-acoustic Transduction Mechanism and Control Method: 51837005; Research on Topology, Passive Current Sharing Mechanism and Control for Multiphase Resonant Converter with Coupled Resonant Tank: 51977069.

Conflicts of Interest: The authors declare no conflict of interest.

References

1. Ertl, H.; Kolar, J.W.; Zach, F.C. Analysis of a multilevel multicell switch-mode power amplifier employing the "flying-battery" concept. *IEEE Trans. Ind. Electron.* **2002**, *49*, 816–823. [CrossRef]
2. Zhou, J.; Deng, Z.; Liu, C.; Li, K.; He, J. Current ripple analysis of five-phase six-leg switching power amplifiers for magnetic bearing with one-cycle control. In Proceedings of the 2016 19th International Conference on Electrical Machines and Systems (ICEMS), Chiba, Japan, 13–16 November 2016; pp. 1–6.
3. Agbossou, K.; Dion, J.L.; Carignan, S.; Abdelkrim, M.; Cheriti, A. Class D amplifier for a power piezoelectric load. *IEEE Trans. Ultrason. Ferroelectr. Freq. Control.* **2000**, *47*, 1036–1041. [CrossRef] [PubMed]
4. Kagawa, Y.; Gladwell, G.M.L. Finite Element Analysis of Flexure-Type Vibrators with Electrostrictive Transducers. *IEEE Trans. Sonics Ultrason.* **1970**, *17*, 41–48. [CrossRef]
5. Colli-Menchi, A.I.; Torres, J.; Sánchez-Sinencio, E.A. Feed-Forward Power-Supply Noise Cancellation Technique for Single-Ended Class-D Audio Amplifiers. *IEEE J. Solid-State Circuits* **2014**, *49*, 718–728. [CrossRef]
6. Fang, J.; Ren, Y. Self-adaptive phase-lead compensation based on unsymmetrical current sampling resistance network for magnetic bearing switching power amplifiers. *IEEE Trans. Ind. Electron.* **2012**, *59*, 1218–1227. [CrossRef]
7. Wang, X.; Zhao, Z.; Li, K.; Chen, K.; Liu, F. Analysis of the Steady-State Current Ripple in Multileg Class-D Power Amplifiers Under Inductance Mismatches. *IEEE Trans. Power Electron.* **2019**, *34*, 3646–3657. [CrossRef]
8. Hung, T.; Rode, J.; Larson, L.; Asbeck, P. H-Bridge Class-D Power Amplifiers for Digital Pulse Modulation Transmitters. In Proceedings of the 2007 IEEE/MTT-S International Microwave Symposium, Honolulu, HI, USA, 3–8 June 2007; pp. 1091–1094.

9. Song, W.; Ma, J.; Feng, L.Z.X. Deadbeat Predictive Power Control of Single-Phase Three-Level Neutral-Point-Clamped Converters Using Space-Vector Modulation for Electric Railway Traction. *IEEE Trans. Power Electron.* **2016**, *31*, 721–732. [CrossRef]

10. Morozov, K.A.; Scussel, K.F.; Coryer, M.W.M.J. The new development of the autonomous sources for ocean acoustic navigation, tomography and communications. In Proceedings of the IEEE OCEANS, Anchorage, AK, USA, 19–22 September 2017; pp. 1–8.

11. Sherman, C.; Butler, J. *Transducers and Arrays for Underwater Sound*; Springer: New York, NY, USA, 2007.

12. Kakosimos, P.; Abu-Rub, H. Predictive control of a Grid-Tied cascaded full-bridge NPC inverter for reducing high-frequency common-mode voltage components. *IEEE Trans. Ind. Inf.* **2018**, *14*, 2385–2394. [CrossRef]

13. Dong, M.; Zhang, Y.; Li, M.X.J. Multilevel dual closed-loop controlling Class-D power amplifier with FPGA. *Int. Conf. Control. Eng. Comm. Tech.* **2012**, 421–425.

14. Yu, S.; Tseng, M. Optimal control of a nine-level Class-D audio amplifier using sliding-mode quantization. *IEEE Trans. Ind. Electron.* **2011**, *58*, 3069–3076. [CrossRef]

15. Zhang, Y.; Meng, Q.; Li, Y.; Cai, L. A high-efficiency cascaded multilevel class-D amplifier with FPGA based on sliding mode control. In Proceedings of the 2010 3rd International Symposium on Systems and Control in Aeronautics and Astronautics (ISSCAA 2010), Harbin, China, 8–10 June 2010; pp. 86–90.

16. Hamdi, M.; Hamouda, M.; Al-Haddad, L.S.K. FCS-MPC for grid-tied three-phase three-level NPC inverter with experimental validation. In Proceedings of the International Conference on Green Energy Conversion Systems (GECS), Hammamet, Tunisia, 23–25 March 2017; pp. 1–6.

17. Stolze, P.; Kramkowski, M.; Mouton, T.; Kennel, M.T.R. Increasing the performance of finite-set model predictive control by oversampling. In Proceedings of the IEEE Conference on Innovative Engineering Technologies, Cape Town, South Africa, 25–28 February 2013; pp. 551–556.

18. Vazquez, S.; Aguilera, R.P.; Acuna, P.; Pou, J.; Leon, J.I.; Franquelo, L.G.; Agelidis, V.G. Model Predictive Control for Single-Phase NPC Converters Based on Optimal Switching Sequences. *IEEE Trans. Ind. Electron.* **2016**, *63*, 7533–7541. [CrossRef]

19. Ma, J.; Song, W.; Wang, X.; Feng, F.B.X. Low-Complexity Model Predictive Control of Single-Phase Three-Level Rectifiers with Unbalanced Load. *IEEE Trans. Power Electron.* **2018**, *33*, 8936–8947. [CrossRef]

20. Baidya, R.; Acuna, R.A.P. Multistep model predictive control for cascaded H-bridge inverters: Formulation and analysis. *IEEE Trans. Power Electron.* **2018**, *33*, 876–886. [CrossRef]

21. Guo, P.; He, Z.; Yue, Y.; Xu, Q.; Huang, X.; Chen, Y.; Luo, A. A novel two-stage model predictive control for modular multilevel converter with reduced computation. *IEEE Trans. Ind. Electron.* **2019**, *66*, 2410–2422. [CrossRef]

22. Cortes, P.; Wilson, A.; Kouro, S.; Rodriguez, J.; Abu-Rub, H. Model predictive control of multilevel cascaded H-bridge inverters. *IEEE Trans. Ind. Electron.* **2010**, *57*, 2691–2699. [CrossRef]

23. Gong, Z.; Dai, P.; Yuan, X.; Guo, X.W.G. Design and experimental evaluation of fast model predictive control for modular multilevel converters. *IEEE Trans. Ind. Electron.* **2016**, *63*, 3845–3856. [CrossRef]

24. Karamanakos, P.; Kennel, T.G.R. Reformulation of the long horizon direct model predictive control problem to reduce the computational effort. In Proceedings of the IEEE Energy Conversion Congress and Exposition, Pittsburgh, PA, USA, 14–18 September 2014; pp. 3512–3519.

25. Karamanakos, P.; Kennel, T.G.R. A computationally efficient model predictive control strategy for linear systems with integer inputs. *IEEE Trans. Control. Syst. Technol.* **2016**, *24*, 1463–1471. [CrossRef]

26. Karamanakos, P.; Kennel, T.G.R. Constrained long-horizon direct model predictive control for power electronics. In Proceedings of the Energy Conversion Congress & Exposition, Milwaukee, WI, USA, 18–22 September 2016; pp. 1–8.

27. Karamanakos, P.; Mouton, T.G.T. Computationally efficient sphere decoding for long-horizon direct model predictive control. In Proceedings of the Energy Conversion Congress & Exposition, Milwaukee, WI, USA, 18–22 September 2016; pp. 18–22.

28. Zhang, Z.; Li, Z.; Kazmierkowski, M.; Kennel, J.R.R. Robust predictive control of three-level NPC back-to-back power converter PMSG wind turbine systems with revised predictions. *IEEE Trans. Power Electron.* **2018**, *33*, 9588–9598. [CrossRef]

29. Xia, C.; Wang, M.; Liu, Z.S.T. Robust model predictive current control of three-phase voltage source PWM rectifier with online disturbance observation. *IEEE Trans. Ind. Inf.* **2012**, *8*, 459–471. [CrossRef]

30. Ghanes, M.; Trabelsi, M.; Ben-Brahim, H.A.L. Robust adaptive observer-based model predictive control for multilevel flying capacitors inverter. *IEEE Trans. Ind. Electron.* **2016**, *63*, 7876–7886. [CrossRef]
31. Nguyen, H.; Kim, E.; Kim, I.; Jung, H.C.J. Model predictive control with modulated optimal vector for a three-phase inverter with an LC filter. *IEEE Trans. Power Electron.* **2018**, *33*, 2690–2703. [CrossRef]
32. Shen, K.; Zhang, J. Modeling error compensation in FCS-MPC of a three-phase inverter. In Proceedings of the 2012 IEEE International Conference on Power Electronics, Drives & Energy Systems (PEDES), Bengaluru, India, 16–19 December 2012; pp. 1–6.
33. Jazwinski, A. *Stochastic Processes and Filtering Theory*; Academic Press: Pittsburgh, PA, USA, 1970.

© 2019 by the authors. Licensee MDPI, Basel, Switzerland. This article is an open access article distributed under the terms and conditions of the Creative Commons Attribution (CC BY) license (http://creativecommons.org/licenses/by/4.0/).

 *electronics*

Article

Current Source AC-Side Clamped Inverter for Leakage Current Reduction in Grid-Connected PV System

Xiangli Li, Na Wang, Guocheng San and Xiaoqiang Guo *

Department of Electrical Engineering, Yanshan University, Qinhuangdao 066004, China; lxl@ysu.edu.cn (X.L.); 18712728609@163.com (N.W.); hbsgc@163.com (G.S.)
* Correspondence: gxq@ysu.edu.cn; Tel.: +86-13785070194

Received: 16 October 2019; Accepted: 29 October 2019; Published: 6 November 2019

Abstract: For the grid-connected photovoltaic inverters, the switching-frequency common-mode voltage brings the leakage current, which should be eliminated. So far, many kinds of single-phase inverters have been published for this purpose, but most of them are the conventional voltage-type ones, which have the disadvantages of poor reliability due to the DC-link electrolytic capacitor and the risk of short-through of the bridge switches. To solve this technical issue, a novel current source inverter with AC-side clamping is proposed to mitigate the switching-frequency common-mode voltage. Meanwhile, a novel modulation method is proposed for the new single-phase inverter to achieve low-frequency operation of the main switches, which reduces the switching losses. Finally, the proposed method is implemented on the TMS320F28335DSP + XC6SLX9FPGA digital hardware platform. Also, the performance comparisons are done with the traditional solution. The results prove the proposed solution.

Keywords: current source inverter; common-mode voltage; leakage current

---

1. Introduction

Photovoltaic (PV) power generation is one of the ways to effectively use energy. Through photovoltaic panels to obtain energy, photovoltaic systems can provide green sustainable solutions [1,2]. In general, energy can be obtained from photovoltaic panels by grid connection of photovoltaic inverters or by connecting transformers. However, the transformer is relatively heavy, sizable, and costly, with undesirable power loss problems. Therefore, the transformerless PV inverters are promising and attractive in industrial and academic fields [3–7]. However, when the transformerless PV inverter is connected to the grid, there are still many technical challenges to be solved, such as the leakage current or ground current. Without the transformer isolation, the electrical connection exists between the photovoltaic panel and the grid, and a leakage current will be generated on the parasitic capacitance between the photovoltaic panel and the ground. Leakage currents can adversely affect grid current, personal safety and electro-magnetic interference issues. So, the German standard VDE 0126-1-1 defines that the PV system should be off from the grid when the leakage current exceeds 300 mA.

To solve the above problems, many kinds of topologies have been published, such as Heric, H6, oH5, H5 [8–11]. In theory, a constant common-mode voltage would be achieved with the above topologies. In fact, due to the influence of the switch junction capacitance, the switching common-mode voltage changes with high frequency. Consequently, the leakage current cannot be completely eliminated. Nevertheless, most topologies are proposed from the perspective of voltage-type inverters. In a voltage source inverter, the DC-side electrolytic capacitor reduces the reliability and life of the inverter system. Moreover, the voltage source inverter has a risk of short-through, which leads to the reliability issue. In addition, the current source inverter has a unique short-circuit operation capability, which improves the reliability of the system. Aside from that, the DC-link of the current source inverter uses an inductor instead of an electrolytic capacitor, which can be designed to work at high temperatures [12–15]. The current-type inverters, as a matter of fact, have been applied in photovoltaic systems over the past few decades [16–22].

However, due to the switching-frequency common-mode voltage, the leakage current of the conventional current-source four-switch inverter is large. This characteristic limits the application of conventional current source inverters in transformerless photovoltaic systems. Inspired by the bypass-type voltage source inverter topology [23], this paper proposes a novel AC-side clamped current-source inverter topology, which effectively suppresses the switching-frequency common-mode voltage, so as to reduce the leakage current. At the same time, for the new inverter, a new modulation for reducing the switching loss is proposed. In addition, the proposal is proven by the experiment results.

2. Traditional Current Source Inverter

Figure 1 is a schematic of the conventional current-source inverter circuit. Where, L_{dc1} and L_{dc2} are DC-side inductors, S_1–S_4 are IGBTs (Insulated Gate Bipolar Transistor), C_f is the AC-side filter capacitor, v_g is the AC-side voltage, and C_{PV} is the parasitic capacitance between the PV array and the ground. The voltage change across the parasitic capacitance C_{PV} will cause a leakage current, which will affect the grid current.

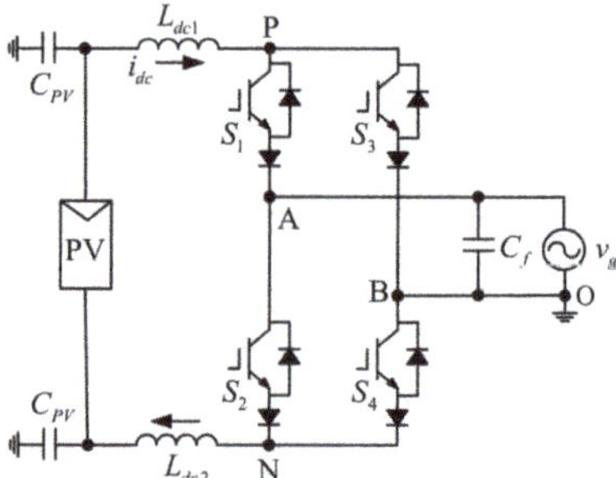

Figure 1. Conventional current-source inverter.

In order to understand the factors affecting the leakage current, a common-mode loop model, as shown in Figure 2, is established. Where, V_{CM} represents the common-mode voltage (CMV), and Z is the equivalent impedance of the common-mode loop.

$$V_{CM} = \frac{V_{PO} + V_{NO}}{2} \tag{1}$$

$$Z = \frac{sL_{dc1}L_{dc2}}{L_{dc1} + L_{dc2}} \tag{2}$$

$$i_{leakage} = 2C_{PV}\frac{dV_{CPV}}{dt} \tag{3}$$

$$V_{CPV} = \frac{1}{2sC_{PV}Z + 1}V_{CM} \tag{4}$$

where, V_{CPV} represents the voltage across the parasitic capacitance $2C_{PV}$. According to Equations (3) and (4),

$$I_{leakage}(s) = \frac{2sC_{PV}}{2sC_{PV}Z + 1}V_{CM} \tag{5}$$

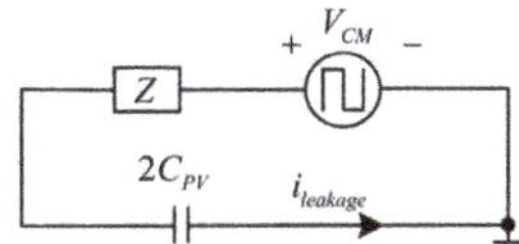

Figure 2. Common-mode model.

According to Equation (2) and Equations (3)–(5), the leakage current $I_{leakage}$ is dependent on the equivalent common-mode impedance and the rate of the common-mode voltage change. Therefore, reducing the leakage current can be considered from two points, one is to increase the common-mode loop impedance, and the other is to reduce the rate of the common-mode voltage change or to maintain the common-mode voltage constant.

Conventional current-source inverters have four switching states, as shown in Table 1. As shown in Figure 3, the driving signal waveforms of S_1 and S_3 are changed by low frequency, and the driving signals of S_2 and S_4 are changed at a high frequency.

Table 1. Current space vectors, switching states and common-mode voltage (CMV).

Current Vectors	Switching States				CMV
	S_1	S_2	S_3	S_4	
I_1	1	0	0	1	$0.5v_g$
I_2	1	1	0	0	v_g
I_3	0	1	1	0	$0.5v_g$
I_4	0	0	1	1	0

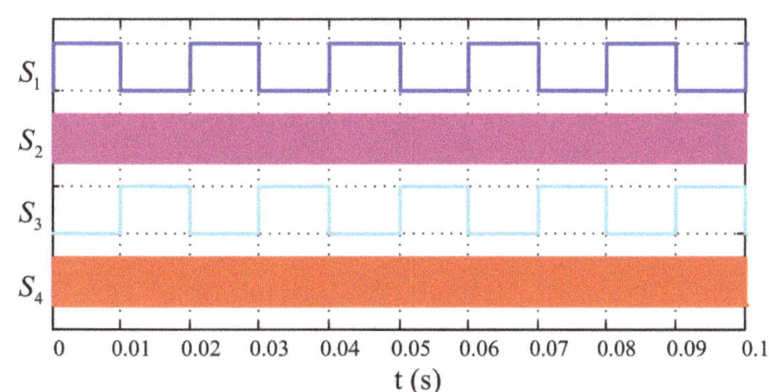

Figure 3. Gating signals of current-source inverter.

As switch S_1 and the switch S_4 are turned on, the common-mode voltage can be derived according to Equation (1):

$$V_{CM} = \frac{V_{PO} + V_{NO}}{2} = \frac{V_{AO} + V_{BO}}{2} = \frac{v_g}{2} \tag{6}$$

where, V_{PO} represents the potential of P with respect to O, and V_{NO} represents the potential of N with respect to O.

When the switch S_1 and the switch S_2 are turned on, the common-mode voltage is:

$$V_{CM} = \frac{V_{PO} + V_{NO}}{2} = V_{AO} = v_g \tag{7}$$

As the switch S_2 and the switch S_3 are turned on, the common-mode voltage is:

$$V_{CM} = \frac{V_{PO} + V_{NO}}{2} = \frac{V_{BO} + V_{AO}}{2} = \frac{v_g}{2} \tag{8}$$

When the switch S_3 and the switch S_4 are turned on, the common mode voltage is:

$$V_{CM} = \frac{V_{PO} + V_{NO}}{2} = V_{BO} = 0 \tag{9}$$

In summary, the switching-frequency common-mode voltage in the system leads to serious leakage current problems of the current-type inverter. In order to effectively suppress the leakage current, this paper will propose a new topology in the next section, which can eliminate the switching common-mode voltage variation for the leakage current attenuation.

3. New Current-Source Inverter

This section introduces a new current-source inverter topology that can eliminate high-frequency common-mode voltage variation from a topological perspective. As shown in Figure 4, a voltage source DC-bypass inverter is proposed in Reference [23]. It uses the diode to clamp the unchancommon-mode voltage unchanged at $V_{dc}/2$, and thus, the leakage current can be suppressed. The corresponding current-type topology is obtained, as illustrated in Figure 5.

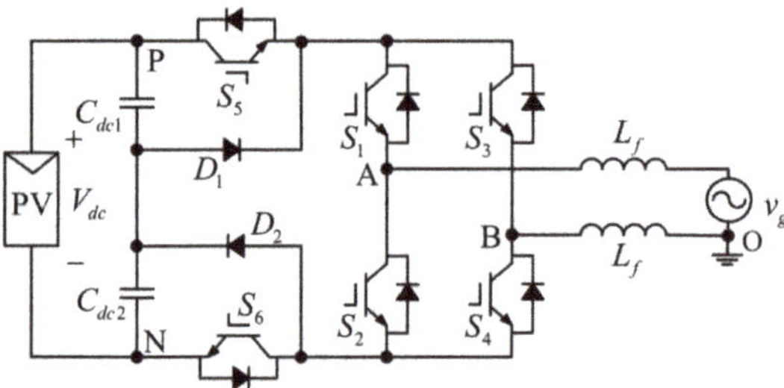

Figure 4. Voltage source inverter with DC-side clamping.

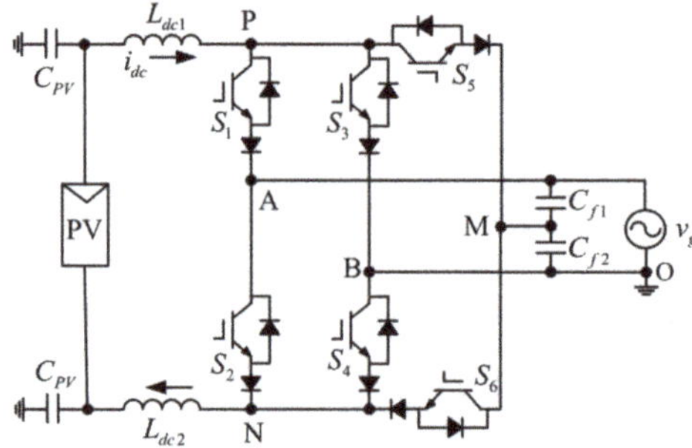

Figure 5. New current-source inverter with AC-side clamping.

According to Table 1, the common-mode voltage depends on the grid voltage, and the common-mode voltage variation is from 0 to v_g. For eliminating the high-frequency common-mode voltage, the DC-side positive bus P and the negative bus N are clamped at the midpoint of the AC-side filter capacitors during the freewheeling period. According to Equation (1), the common-mode voltage is $v_g/2$ in the freewheeling cycle.

$$V_{CM} = \frac{V_{PO} + V_{NO}}{2} = \frac{v_g}{2} \tag{10}$$

During the freewheeling period, the switch S_5 and the switch S_6 are on to establish a freewheeling path for the DC-side current. The switching states in the active state are the same as that of the conventional current-source inverter. During the half-positive cycle, the switch S_1 and the switch S_4 are turned on. In the half-negative cycle, the switch S_2 and the switch S_3 are on. The new current-source inverter switching states and corresponding common-mode voltages are given in Table 2.

Table 2. Current space vectors, switch states and CMV.

Current Vectors	Switching States						CMV
	S_1	S_2	S_3	S_4	S_5	S_6	
I_1	1	0	0	1	0	0	$0.5v_g$
I_3	0	1	1	0	0	0	$0.5v_g$
I_0	0	0	0	0	1	1	$0.5v_g$

It can be seen from Table 2 that the new current-source inverter has three switching states, and the common-mode voltages corresponding to each switching state are the same, which are $0.5v_g$, and the common-mode voltage frequency is consistent with the fundamental frequency. Since the frequency of the grid voltage v_g is smaller than the high switching frequency, the influence of the fundamental-frequency voltage on the switching frequency common-mode characteristics can be ignored. According to Equations (3)–(5), the proposed method eliminates the switching frequency common-mode voltage variation, so that the leakage current is effectively suppressed.

The driving signal waveform of the proposed inverter is shown in Figure 6. The switches S_1–S_6 are high-frequency changes and the switching loss is increased. To solve this problem, a novel modulation method for reducing the switching loss based on the characteristics of the new topology is proposed. Taking the half-positive cycle operating state as an example, the active state I_1 and the zero state I_0 alternately operate. Figures 7 and 8 are circuit diagrams of the system operating in the active state I_1 and the zero state I_0, respectively.

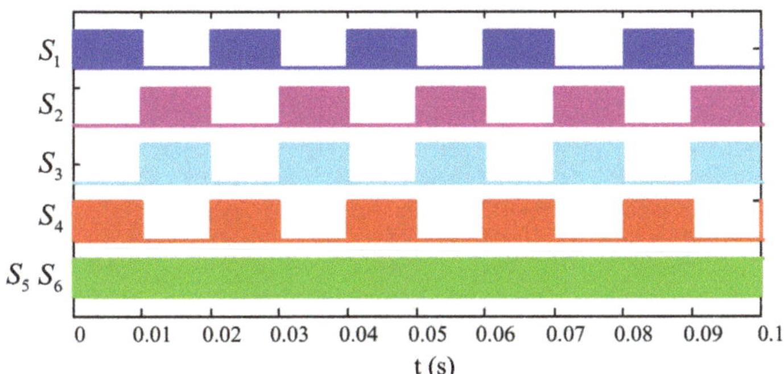

Figure 6. Gating signal of new current-source inverter.

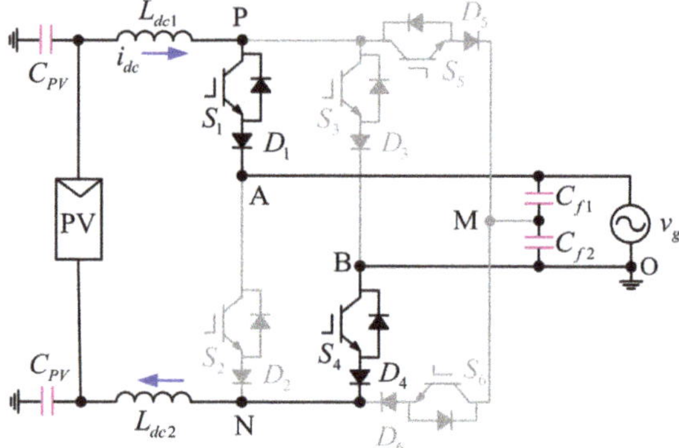

Figure 7. Schematic of the active state I_1.

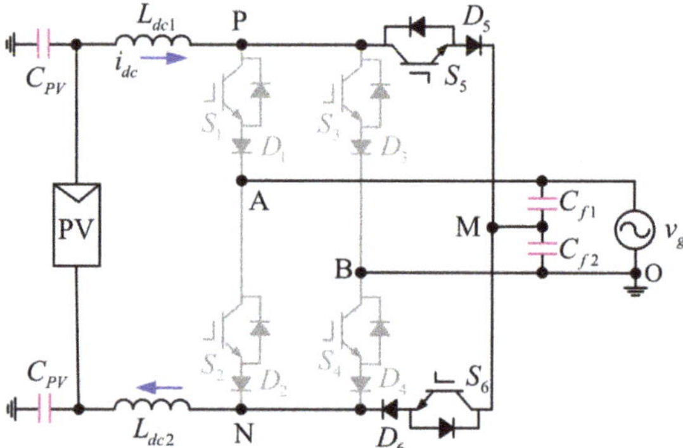

Figure 8. Schematic with the active state I_0.

According to Figure 8, when the system operates in the zero state, the potential at point A is v_g, the potential at point B is 0, the point P and the point N are clamped at point M, and the potential at point M is $v_g/2$. Therefore, in the zero state, if the switch S_1 and the switch S_4 are in the on state, the diodes D_1 and D_4 are reversely turned off, and the upper arm and the lower arm are still in the off state, and the circuit is as shown in Figure 9. The switch S_1 and the switch S_4 can always be in the on state during the half-positive cycle. In a similar manner, for the half-negative cycle, the switch S_2 and the switch S_3 can always be in the on state. The switching states of the new modulation method can be obtained, as shown in Table 3. Figure 10 shows the gating signals' waveform when using the new modulation method. It can be seen that the switches S_1–S_4 vary by low frequency, and the switching loss is reduced.

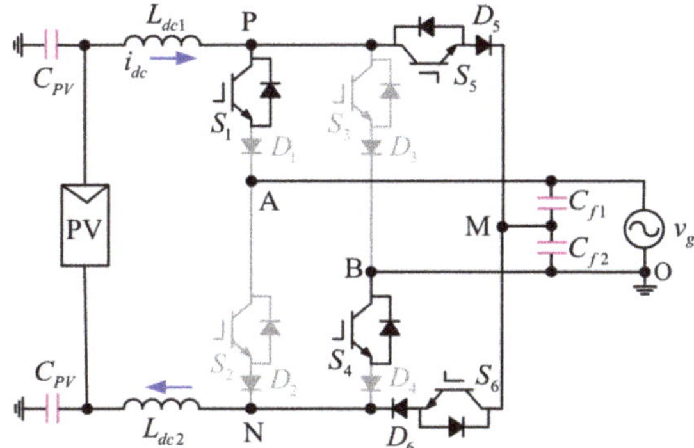

Figure 9. Schematic under new modulation method.

Table 3. Current space vectors, switching states and CMV.

Current Vectors	Switching States						CMV
	S_1	S_2	S_3	S_4	S_5	S_6	
I_1	1	0	0	1	0	0	$0.5v_g$
I_{01}	1	0	0	1	1	1	$0.5v_g$
I_3	0	1	1	0	0	0	$0.5v_g$
I_{03}	0	1	1	0	1	1	$0.5v_g$

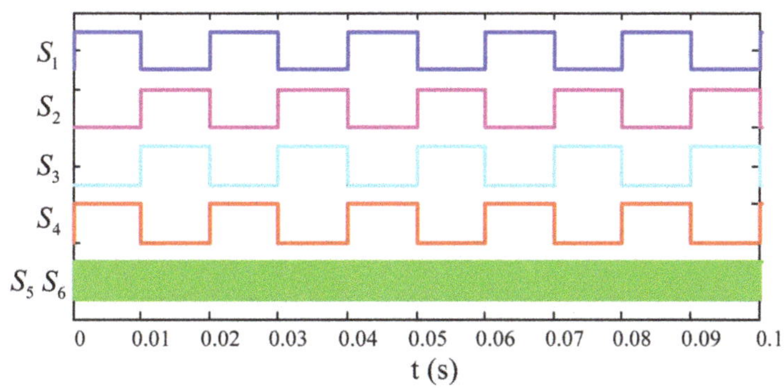

Figure 10. Gating signals with the new modulation.

4. Results

In this section, a system experiment setup was built, and the proposal was compared against the traditional method. The algorithm is realized by the DSP (TMS320F28335) and FPGA (Xilinx XC6SLX9). The experimental parameters are shown in Table 4. Figure 11 shows the control diagram of the proposal. The phase angle θ is detected to generate the reference current. The output m of the PR (Proportional Resonant) regulator is used as the input of the new modulation method. The new modulation method judges the positive or negative of m and selects the corresponding switching sequence. The specific switching sequence selection is shown in Figure 12, where T_s represents the switching period.

Table 4. Experimental parameters.

Parameters	Values
Switching frequency	10 kHz
AC-side filter capacitors (C_{f1} C_{f2})	18.8 μF
DC-side inductances (L_{dc1} L_{dc2})	5 mH
Parasitic capacitance (C_{pv})	75 nF
DC-side current	8 A
DC-side voltage	200 V
AC-side current	6 A
AC-side voltage	110 V (RMS)

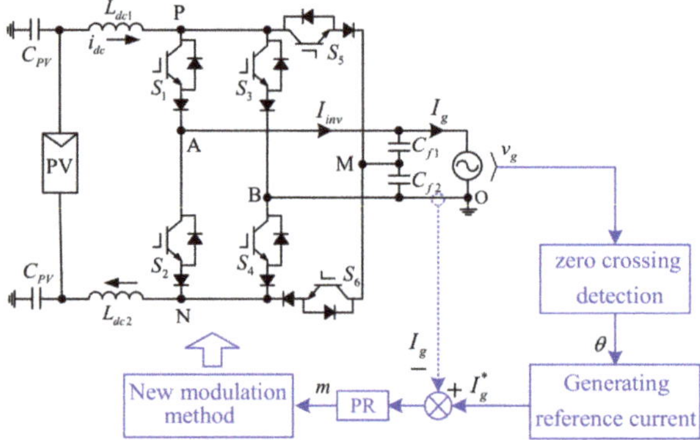

Figure 11. System control block diagram.

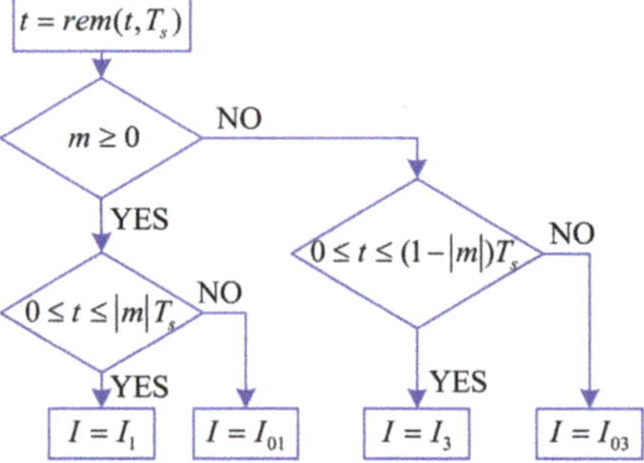

Figure 12. Switching sequence selection for new modulation method.

The experimental results for the conventional current-type inverter are shown in Figure 13. The output current I_{inv} is unipolar, and the grid current I_g is sinusoidal. Figure 13b shows the voltages V_{PO}, V_{NO} and the common-mode voltage V_{CM}. The leakage current $I_{leakage}$ waveform is shown in Figure 13c, and the maximum leakage current is 2 A. The switching-frequency variation of the common-mode voltage causes the leakage current to far exceed the 300 mA. As shown in Figure 13d, for the gating signal waveforms of the switches S_1–S_4, the switch S_1 and the switch S_3 operate at the power frequency, and the switch S_2 and the switch S_4 operate at a high frequency.

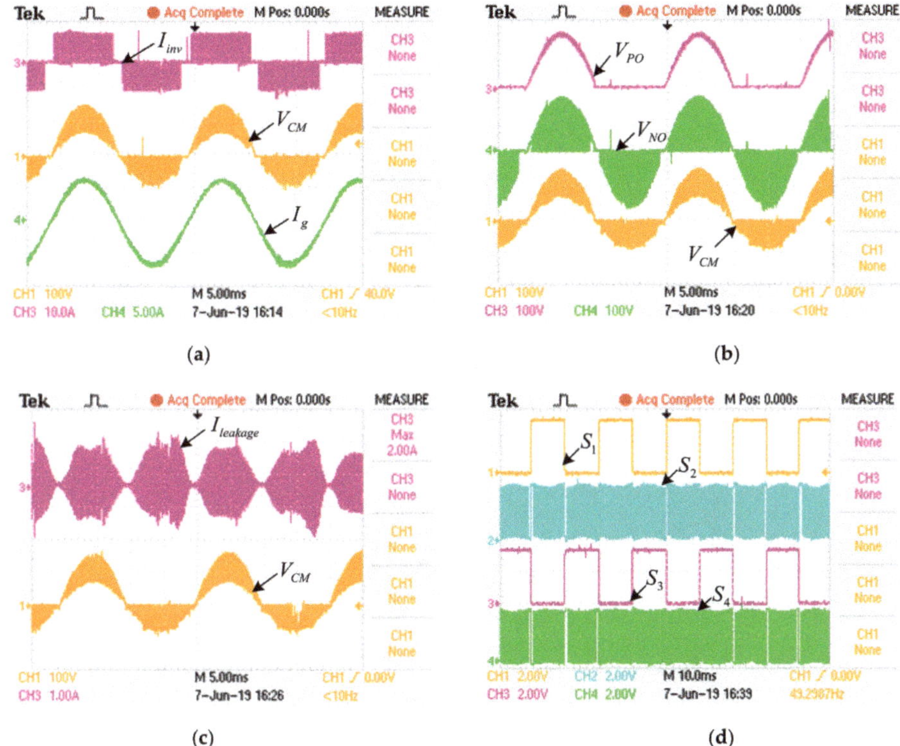

Figure 13. Results with conventional current-type inverter. (**a**) Output current I_{inv}, common-mode voltage V_{CM}, grid current I_g; (**b**) V_{PO}, V_{NO}, common-mode voltage V_{CM}; (**c**) Leakage current $I_{leakage}$, common-mode voltage V_{CM}; (**d**) Switches S_1–S_4 gating signal waveforms.

The output current I_{inv} and the grid current I_g are shown in Figure 14a. From Figure 14d, the switches S_1-S_4 operate in power frequency, which reduces the switching loss. Figure 14b is the experimental result of the voltages V_{PO}, V_{NO} and the common-mode voltage V_{CM}. The common-mode voltage is the half-grid voltage, ignoring the effect of the fundamental-frequency component of the common-mode voltage on the high-frequency common-mode behavior. According to Equation (5), the leakage current is able to be attenuated. The experimental waveform of the leakage current $I_{leakage}$ is shown in Figure 14c. The maximum value of the leakage current is 144 mA, which meets the standard and is less than 300 mA, which is satisfactory in terms of the VDE-0126-1-1.

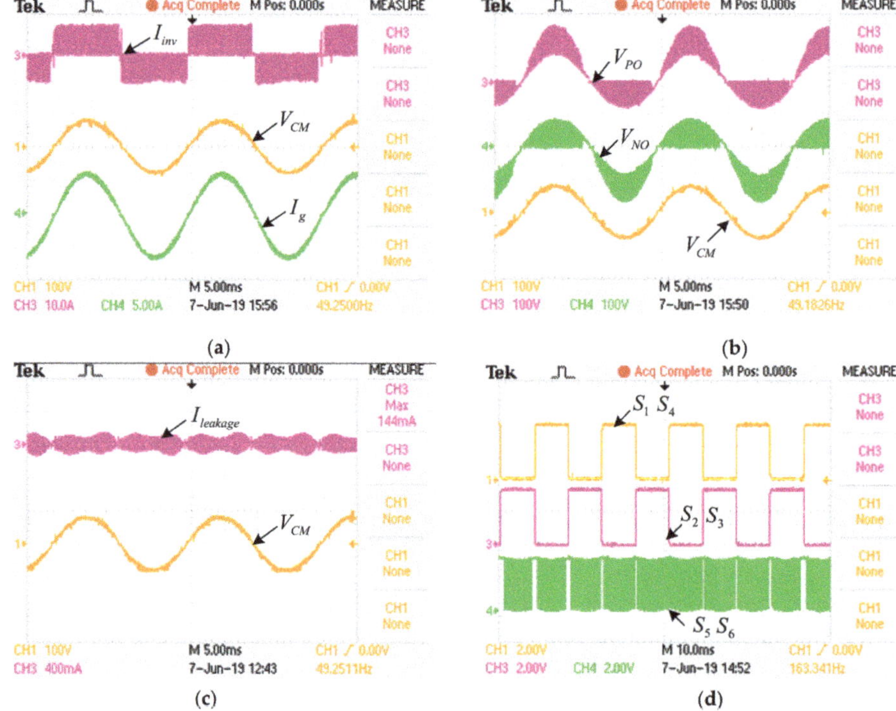

(a)

(b)

(c)

(d)

Figure 14. Results with new current-source inverter. (**a**) Output current I_{inv}, common-mode voltage V_{CM}, grid current I_g; (**b**) V_{PO}, V_{NO}, common-mode voltage V_{CM}; (**c**) Leakage current $I_{leakage}$, common-mode voltage V_{CM}; (**d**) Switches S_1–S_6 gating signal waveforms.

5. Conclusions

In this paper, a novel current-source inverter was proposed. Through the principle analysis and experimental research, the following conclusions were drawn. (1) The high-frequency variation of the common-mode voltage of the conventional current-source inverter cannot effectively suppress the leakage current. (2) The topology proposed in this paper can effectively eliminate the high-frequency common-mode voltage of the system, thus effectively suppressing the system leakage current and satisfying the VDE-0126-1-1 standard.

Author Contributions: This paper was a collaborative effort among each of authors.

Funding: This work was supported in part by the Natural Science Foundation of Hebei Province (E2019203563), and the State Key Laboratory of Reliability and Intelligence of Electrical Equipment (EERIKF2018002), Hebei University of Technology.

Conflicts of Interest: The authors declare no conflict of interest.

References

1. Xiao, G.; Yang, Y.; He, R.; Wang, B.; Blaabjerg, F. Transformerless Z-source four-leg PV inverter with leakage current reduction. *IEEE Trans. Power Electron.* **2019**, *34*, 4343–4352.
2. Ji, Y.; Yang, Y.; Zhou, J.; Ding, H.; Guo, X.; Padmanaban, S. Control strategies of mitigating dead-time effect on power converters: An overview. *Electronics* **2019**, *8*, 196. [CrossRef]
3. Siwakoti, Y.P.; Blaabjerg, F. Common-ground-type transformerless inverters for single-phase solar photovoltaic systems. *IEEE Trans. Ind. Electron.* **2018**, *65*, 2100–2111. [CrossRef]

4. Rojas, C.A.; Aguirre, M.; Kouro, S.; Geyer, T.; Gutierrez, E. Leakage current mitigation in photovoltaic string inverter using predictive control with fixed average switching frequency. *IEEE Trans. Ind. Electron.* **2017**, *64*, 9344–9354. [CrossRef]

5. Freddy, T.K.S.; Rahim, N.A.; Hew, W.; Che, H.S. Comparison and analysis of single-phase transformerless grid-connected PV inverters. *IEEE Trans. Power Electron.* **2014**, *29*, 5358–5369. [CrossRef]

6. Zeng, H.; Li, Y.; Zhang, B.; Zheng, T.Q.; Hao, R.; Yang, Z. An improved H5 topology with low common-mode current for transformerless PV grid-connected inverter. *IEEE Trans. Power Electron.* **2019**, *34*, 1254–1265.

7. Kafle, Y.R.; Town, G.E.; Guochun, X.; Gautam, S. Performance comparison of single-phase transformerless PV inverter systems. In Proceedings of the 2017 IEEE Applied Power Electronics Conference and Exposition (APEC), Tampa, FL, USA, 26–30 March 2017.

8. Xiao, G.; Jia, X. Hardware-based cascaded topology and modulation strategy with leakage current reduction for transformerless PV systems. *IEEE Trans. Ind. Electron.* **2016**, *62*, 7823–7832.

9. Yang, B.; Li, W.; Gu, Y.; Cui, W.; He, X. Improved transformerless inverter with common-mode leakage current elimination for a photovoltaic grid-connected power system. *IEEE Trans. Power Electron.* **2012**, *27*, 752–762. [CrossRef]

10. Guo, X.; Yang, Y.; Zhu, T. ESI: A novel three-phase inverter with leakage current attenuation for transformerless PV systems. *IEEE Trans. Ind. Electron.* **2018**, *65*, 2967–2974. [CrossRef]

11. Zhang, L.; Sun, K.; Xing, Y.; Xing, M. H6 transformerless full-bridge PV grid-tied inverters. *IEEE Trans. Power Electron.* **2014**, *29*, 1229–1238. [CrossRef]

12. Dai, H.; Jahns, T.; Torres, R.; Han, D.; Sarlioglu, B. Comparative evaluation of conducted common-mode EMI in voltage-source and current-source inverters using wide-bandgap switches. In Proceedings of the IEEE Transportation Electrification Conference and Expo, Long Beach, CA, USA, 13–15 June 2018.

13. Dai, H.; Jahns, T.M. Comparative investigation of PWM current source inverters for future machine drives using high-frequency wide-bandgap power switches. In Proceedings of the IEEE Applied Power Electronics Conference, San Antonio, TX, USA, 4–8 March 2018.

14. Dai, H.; Jahns, T.; Torres, R.; Liu, M.; Sarlioglu, B.; Chang, S. Development of high-frequency WBG power modules with reverse-voltage-blocking capability for an integrated motor drive using a current-source inverter. In Proceedings of the IEEE Energy Conversion Congress and Exposition, Portland, OR, USA, 23–27 September 2018.

15. Torres, R.A.; Dai, H.; Lee, W.; Jahns, T.M.; Sarlioglu, B. Current-source inverters for integrated motor drives using wide-bandgap power switches. In Proceedings of the 2018 IEEE Transportation Electrification Conference and Expo, Long Beach, CA, USA, 13–15 June 2018.

16. Chen, Y.; Smedley, K. Three-phase boost-type grid-connected inverter. *IEEE Trans. Power Electron.* **2008**, *23*, 2301–2309. [CrossRef]

17. Dash, P.P.; Kazerani, M. Dynamic modeling and performance analysis of a grid-connected current-source inverter-based photovoltaic system. *IEEE Trans. Sustain. Energy* **2011**, *2*, 443–450. [CrossRef]

18. Wang, Z.; Zou, Z.; Zheng, Y. Design and control of a photovoltaic energy and SMES hybrid system with current-source grid inverter. *IEEE Trans. Appl. Supercond.* **2013**, *23*, 5701505. [CrossRef]

19. Anand, S.; Gundlapalli, S.K.; Fernandes, B.G. Transformer-less grid feeding current source inverter for solar photovoltaic system. *IEEE Trans. Ind. Electron.* **2014**, *61*, 5334–5344. [CrossRef]

20. Barbosa, G.; Braga, C.; Rodrigues, B.; Teixeira, E. Boost current multilevel inverter and its application on single-phase grid-connected photovoltaic systems. *IEEE Trans. Power Electron.* **2006**, *21*, 1116–1124. [CrossRef]

21. Noguchi, T. A new three-level current source PWM inverter and its application for grid connected power conditioner. *Energy Convers. Manag.* **2010**, *51*, 1491–1499.

22. Garcia, L.S.; Buiatt, G.M.; de Freitas, L.C.; Coelho, E.; Farias, V.J.; de Freitas, L.C.G. Dual transformerless single-stage current source inverter with energy management control strategy. *IEEE Trans. Power Electron.* **2013**, *28*, 4644–4656. [CrossRef]

23. Gonzalez, R.; Lopez, J.; Sanchis, P.; Marroyo, L. Transformerless inverter for single-phase photovoltaic systems. *IEEE Trans. Power Electron.* **2007**, *22*, 693–697. [CrossRef]

© 2019 by the authors. Licensee MDPI, Basel, Switzerland. This article is an open access article distributed under the terms and conditions of the Creative Commons Attribution (CC BY) license (http://creativecommons.org/licenses/by/4.0/).

Article

Control Design of LCL Type Grid-Connected Inverter Based on State Feedback Linearization

Longyue Yang [1,2,*], **Chunchun Feng** [2] **and Jianhua Liu** [1,2]

[1] Jiangsu Province Laboratory of Mining Electric and Automation, China University of Mining and Technology, Xuzhou 221116, China

[2] School of Electrical and Power Engineering, China University of Mining and Technology, Xuzhou 221116, China

* Correspondence: yanglongyue@cumt.edu.cn; Tel.: +86-0516-83885605

Received: 11 July 2019; Accepted: 3 August 2019; Published: 7 August 2019

Abstract: Control strategy is the key technology of power electronic converter equipment. In order to solve the problem of controller design, a general design method is presented in this paper, which is more convenient to use computer machine learning and provides design rules for high-order power electronic system. With the higher order system Lie derivative, the nonlinear system is mapped to a controllable standard type, and then classical linear system control method is adopted to design the controller. The simulation and experimental results show that the two controllers have good steady-state control performance and dynamic response performance.

Keywords: state feedback linearization; grid-connected inverter; LCL filter

---

1. Introduction

As the main equipment of an AC power system, grid-connected voltage source inverters (VSI) have been widely used in recent years, such as photovoltaic inverters [1], Pulse Width Modulation Pulse Width Modulation (PWM) rectifiers [2], static var generators [3], active power filters [4], etc.

The filter is an important part of the inverter, the structure of which directly determines the mathematical model and control mode of the inverter. Nowadays, a Inductance-Capacitance-Inductance (LCL) filter, with the advantages of small size, low cost and high harmonic attenuation for high frequency current, is widely used in voltage source type grid-connected converters. Considering that the LCL filter is the 3-order system, the damping of the system is small, resulting in a resonant peak of the grid-side inductance current. This phenomenon will adversely affect the safe and stable operation of grid-connected system. In order to improve the damping characteristics of the system, additional system damping is required. Active damping control method is commonly used at present stage [5]. With the state variables feedback, the traditional proportional integral controller can be used to achieve the grid-side inductance current control.

However, for nonlinear power electronic devices, this design method of the controller is based on small signal modeling [6] and harmonic linearization. As the coupling and high order terms in the system are ignored in the Taylor series calculation, the obtained controller is suitable for working at steady working point with poor performance in other control domains [7].

In order to solve the global control problem, many solutions have been put forward [8,9]. Among them, the state feedback linearization method that develops from differential geometry [10,11] has become an effective way to solve the nonlinear power electronics system control problems. Based on the differential homeomorphism, Lie derivative is used to analyze the numerical relation between the state variables, the input variables, and the output variables. Then, the necessary and sufficient conditions for controllability and observability of nonlinear control systems can be established. Choi et al. proposed a feedback linearization direct torque control for the permanent magnet synchronous motor [12]. In addition, the drive flux and torque ripple were suppressed. Yang et al. combined feedback linearization and sliding mode variable structure control to complete the control of three-phase four-leg inverter [13]. This method was used to decouple the torque and stator flux of the inductive motor by Lascu et al. [14]. Yang et al. applied the state feedback control to the Modular Multilevel Converter (MMC) system and analyzed the performance characteristics of the system [15].

Different from the local approximate linearization method, the feedback decoupling is achieved by adopting this nonlinear algorithm without ignoring the higher-order terms. However, the commonly used feedback linearization algorithm at the present stage is mostly used in the 2-order or 1-order system. The research on the controller design of high-order or other complex systems is rare. In addition, when the order of the controlled object is high, whether the state feedback control can be transformed into a simple form has not been analyzed by literatures.

In order to achieve the control of high-order power electronic systems, the design of controller based on LCL filter type grid-connected inverters is studied in this paper. For the 3-order control system, two controller design methods based on state feedback linearization are proposed in this paper. Lie derivative vector field is used to solve the system relationship and design the decoupling matrix. Through the nonlinear mapping, the nonlinear system can be mapped to a controllable standard form. Then, the classical linear system control method, which can be designed easily, is applicable for the controller design. Finally, the performance of the two controllers is compared and analyzed by simulations and experiments.

2. Single Closed-Loop Controller Design Based on State Feedback Linearization

2.1. Model of the Three-Phase Three-Leg Grid-Connected Inverter

Figure 1 shows the structure of the three-phase three-leg LCL grid-connected inverter. L_1 and L_2 denote the inverter side inductance and the grid-side inductance, respectively.

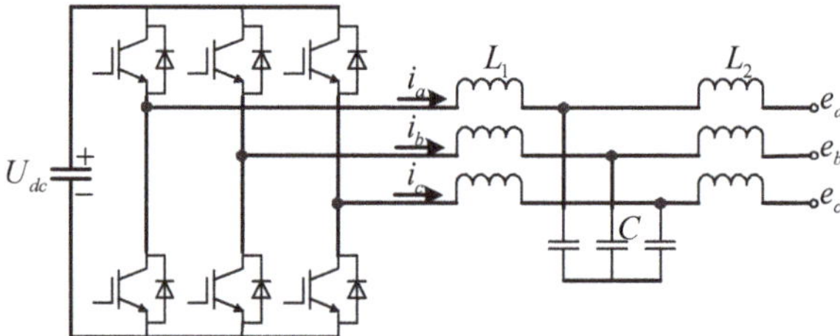

Figure 1. Schematic diagram of the three-phase three-leg grid-connected inverter.

As the model has been analyzed by many literatures, this paper does not repeat the description. The mathematical model of inverter in dq coordinate axis is given directly as

$$\frac{d}{dt}\begin{bmatrix} i_{1d} \\ i_{1q} \\ u_{cd} \\ u_{cq} \\ i_{2d} \\ i_{2q} \\ u_{dc} \end{bmatrix} = \begin{bmatrix} 0 & \omega & -1/L_1 & 0 & 0 & 0 & m_d/L_1 \\ -\omega & 0 & 0 & -1/L_1 & 0 & 0 & m_q/L_1 \\ 1/C & 0 & 0 & \omega & -1/C & 0 & 0 \\ 0 & 1/C & -\omega & 0 & 0 & -1/C & 0 \\ 0 & 0 & 1/L_2 & 0 & 0 & \omega & 0 \\ 0 & 0 & 0 & 1/L_2 & -\omega & 0 & 0 \\ -\frac{m_d}{C_{dc}} & -\frac{m_q}{C_{dc}} & 0 & 0 & 0 & 0 & 0 \end{bmatrix}\begin{bmatrix} i_{1d} \\ i_{1q} \\ u_{cd} \\ u_{cq} \\ i_{2d} \\ i_{2q} \\ u_{dc} \end{bmatrix}$$
$$+ \begin{bmatrix} 0 & 0 & 0 & 0 & 0 & -1/L_2 & 0 \\ 0 & 0 & 0 & 0 & -1/L_2 & 0 & 0 \end{bmatrix}^T \begin{bmatrix} e_d \\ e_q \end{bmatrix} \tag{1}$$

where the state variables $X = \begin{bmatrix} i_{1d} & i_{1q} & u_{cd} & u_{cq} & i_{2d} & i_{2q} & u_{dc} \end{bmatrix}^T$ represent the inverter side inductance current, the filter capacitor voltage, the grid-side inductance current and the DC voltage in the synchronous reference coordinate system, respectively. e_d and e_q represent grid voltage in the synchronous reference coordinate system, respectively. m_d and m_q represent the nonlinear pulse width modulation variables, respectively. ω represents the fundamental angular frequency of the system.

According to (1), due to the mutual inductance and mutual capacitance, the inductance current and the capacitance voltage is coupled in the synchronous reference coordinate system.

Usually, the voltage source grid-connected inverter system is a double closed loop structure. The outer loop controls DC voltage, and the inner loop controls AC current. However, for the different structure like the PV system, the DC voltage may not need to be controlled. Considering the generality, only the current inner loop is analyzed. Due to the LCL filter, the system can be regarded as a 2-input 2-output system. At the same time, the existence of LCL filter capacitor, which introduces a pair of conjugate pure imaginary roots to cause resonance, increases the order of the system. Therefore, it is necessary to consider the suppression of resonance.

2.2. Single Closed-Loop Control Strategy Based on State Feedback Linearization

Define the LCL filter inverter side inductance current, filter capacitor voltage, and grid-side inductance current as state variables, written as $X = \begin{bmatrix} x_1 & x_2 & x_3 & x_4 & x_5 & x_6 \end{bmatrix}^T = \begin{bmatrix} i_{1d} & i_{1q} & u_{cd} & u_{cq} & i_{2d} & i_{2q} \end{bmatrix}^T$. Define the 2 dimensional modulation vector in the synchronous coordinate system as the input variable, written as $U = \begin{bmatrix} u_1 & u_2 \end{bmatrix}^T = \begin{bmatrix} m_d & m_q \end{bmatrix}^T$. The deviation of grid-side inductance current (controlled object) is taken as the output variable $Y = R(X) = \begin{bmatrix} r_1(X) & r_2(X) \end{bmatrix}^T = \begin{bmatrix} i_{dref} - i_{2d} & i_{qref} - i_{2q} \end{bmatrix}^T$, in which i_{dref} and i_{qref} are the reference current.

Therefore, Equation (1) can be expressed as a nonlinear differential equation composed of polynomials of state variables and input variables:

$$\begin{cases} \dot{X} = f[X(t)] + G[X(t)]U \\ Y = R(X) \end{cases} \tag{2}$$

where
$$f(X) = \begin{bmatrix} \omega x_2 - x_3/L_1 \\ -\omega x_1 - x_4/L_1 \\ x_1/C + \omega x_4 - x_5/C \\ x_2/C - \omega x_3 - x_6/C \\ x_3/L_2 + \omega x_6 - u_{gd}/L_2 \\ x_4/L_2 - \omega x_5 - u_{gq}/L_2 \end{bmatrix}, \quad G(X) = [g_1(X) \, g_2(X)] = \begin{bmatrix} 0 & u_{dc}/L_1 & 0 & 0 & 0 & 0 \\ u_{dc}/L_1 & 0 & 0 & 0 & 0 & 0 \end{bmatrix}^T.$$

Equation (2) shows that the LCL-type three-phase three-leg converter is a 2-input 2-output affine nonlinear system with the state variables number of $n = 6$. It is nonlinear for the state vector X, but linear for the input U. Since the state variables are all dq-axis symmetric variables, the number of vector field pairs is $l = 3$. The corresponding $l - 1$ order Lie derivative can be described as

$$\begin{cases} ad_f g_1(X) = L_f g_1 - L_{g_1} f \\ = \begin{bmatrix} 0 & \frac{\omega u_{dc}}{L_1} & -\frac{u_{dc}}{L_1 C} & 0 & 0 & 0 \end{bmatrix}^T \\ ad_f^2 g_1(X) = L_f[ad_f g_1(X)] - L_{ad_f g_1(X)} f \\ = \begin{bmatrix} \frac{\omega^2 u_{dc}}{L_1} + \frac{u_{dc}}{L_1^2 C} & 0 & 0 & \frac{2\omega u_{dc}}{L_1 C} & -\frac{u_{dc}}{L_1 L_2 C} & 0 \end{bmatrix}^T \\ ad_f g_2(X) = L_f g_2 - L_{g_2} f \\ = \begin{bmatrix} -\frac{\omega u_{dc}}{L_1} & 0 & 0 & -\frac{u_{dc}}{L_1 C} & 0 & 0 \end{bmatrix}^T \\ ad_f^2 g_2(X) = L_f[ad_f g_2(X)] - L_{ad_f g_2(X)} f \\ = \begin{bmatrix} 0 & \frac{\omega^2 u_{dc}}{L_1} + \frac{u_{dc}}{L_1^2 C} & -\frac{2\omega u_{dc}}{L_1 C} & 0 & 0 & -\frac{u_{dc}}{L_1 L_2 C} \end{bmatrix}^T \end{cases} \tag{3}$$

where $ad_f g$ denotes Lie bracket operation of vector fields f and g, written as $ad_f g = L_f g - L_g f = (\partial g / \partial X) f - (\partial f / \partial X) g$. Then, the corresponding 6-dimensional vector field matrix can be expressed as

$$D_6 = \begin{bmatrix} g_1 & g_2 & ad_f g_1 & ad_f g_2 & ad_f^2 g_1 & ad_f^2 g_2 \end{bmatrix}$$

$$= \begin{bmatrix} \frac{u_{dc}}{L_1} & 0 & 0 & -\frac{\omega u_{dc}}{L_1} & \frac{\omega^2 u_{dc}}{L_1} + \frac{u_{dc}}{L_1^2 C} & 0 \\ 0 & \frac{u_{dc}}{L_1} & \frac{\omega u_{dc}}{L_1} & 0 & 0 & \frac{\omega^2 u_{dc}}{L_1} + \frac{u_{dc}}{L_1^2 C} \\ 0 & 0 & -\frac{u_{dc}}{L_1 C} & 0 & 0 & -\frac{2\omega u_{dc}}{L_1 C} \\ 0 & 0 & 0 & -\frac{u_{dc}}{L_1 C} & \frac{2\omega u_{dc}}{L_1 C} & 0 \\ 0 & 0 & 0 & 0 & -\frac{u_{dc}}{L_1 L_2 C} & 0 \\ 0 & 0 & 0 & 0 & 0 & -\frac{u_{dc}}{L_1 L_2 C} \end{bmatrix} \quad (4)$$

Obviously, it can be obtained that $Rank D_6 = 6 = n$. The rank of the matrix formed by the vector fields is n in the neighborhood of X_0.

According to Equations (3) and (4), the elements in $\begin{bmatrix} g_1 & g_2 & ad_f g_1 & ad_f g_2 & ad_f^2 g_1 & ad_f^2 g_2 \end{bmatrix}$ do not contain state variables. In other words, they are the constant vector fields, meaning that the result of Lie brackets operation between two elements is a zero vector. Define D_i as the matrix of the former i column elements in Equation (4). It can be calculated that the 6 vector fields satisfy involution relation. Therefore, according to the nonlinear control theory, state feedback can be applied to linearize the inverter system.

A MIMO nonlinear system has a relative degree $\{r_1, \ldots, r_m\}$ at a point x_0 if

1. $L_{g_j} L_f^k r_i(X) \neq 0$ for all $0 \leq j \leq m$, for all $k \leq r_i - 1$, for all $0 \leq i \leq m$ and for all x in a neighborhood of x_0.
2. the $m \times m$ matrix $E(X)$ is nonsingular at $x = x_0$.

In Equation (2), the output variable of the inverter is current error $r_1(X) = i_{dref} - i_{2d}$. In order to analyze the numerical relation of the output, the Lie derivatives need to be calculated as follows:

$$\begin{cases} L_{g_1} r_1(X) = \frac{\partial r_1(X)}{\partial X} g_1(X) = 0 \\ L_{g_2} r_1(X) = \frac{\partial r_1(X)}{\partial X} g_2(X) = 0 \\ L_f r_1(X) = \frac{\partial r_1(X)}{\partial X} f(X) = -\frac{1}{L_2} x_3 - \omega x_6 + \frac{1}{L_2} e_d \\ L_{g_1} L_f r_1(X) = \frac{\partial(L_f r_1(X))}{\partial X} g_1(X) = 0 \\ L_{g_2} L_f r_1(X) = \frac{\partial(L_f r_1(X))}{\partial X} g_2(X) = 0 \\ L_f^2 r_1(X) = \frac{\partial(L_f r_1(X))}{\partial X} f(X) = \frac{x_5 - x_1}{L_2 C} - \frac{2\omega}{L_2} x_4 + \omega^2 x_5 + \frac{\omega}{L_2} e_q \\ L_{g_1} L_f^2 r_1(X) = \frac{\partial(L_f^2 r_1(X))}{\partial X} g_1(X) = -\frac{\bar{u}_{dc}}{L_1 L_2 C} \neq 0 \\ L_{g_2} L_f^2 r_1(X) = \frac{\partial(L_f^2 r_1(X))}{\partial X} g_2(X) = 0 \end{cases} \quad (5)$$

where $L_{g_1} L_f^2 r_1(X) \neq 0$. The relation degree of the output is $d_1 = 3$.

Similarly, the output $r_2(X) = i_{qref} - i_{2q}$ is symmetrical. The relation degree of the output $r_2(X)$ is $d_2 = 3$.

The Lie derivatives can be obtained as follows:

$$
\begin{cases}
L_{g_1}r_2(X) = \frac{\partial r_2(X)}{\partial X}g_1(X) = 0 \\[4pt]
L_{g_2}r_2(X) = \frac{\partial r_2(X)}{\partial X}g_2(X) = 0 \\[4pt]
L_f r_2(X) = \frac{\partial r_2(X)}{\partial X}f(X) = -\frac{1}{L_2}x_4 + \omega x_5 + \frac{1}{L_2}e_q \\[4pt]
L_{g_1}L_f r_2(X) = \frac{\partial(L_f r_2(X))}{\partial X}g_1(X) = 0 \\[4pt]
L_{g_2}L_f r_2(X) = \frac{\partial(L_f r_2(X))}{\partial X}g_2(X) = 0 \\[4pt]
L_f^2 r_2(X) = \frac{\partial(L_f r_2(X))}{\partial X}f(X) = \frac{x_6-x_2}{L_2 C} + \frac{2\omega}{L_2}x_3 + \omega^2 x_6 - \frac{\omega}{L_2}e_d \\[4pt]
L_{g_1}L_f^2 r_2(X) = \frac{\partial(L_f^2 r_2(X))}{\partial X}g_1(X) = 0 \\[4pt]
L_{g_2}L_f^2 r_2(X) = \frac{\partial(L_f^2 r_2(X))}{\partial X}g_2(X) = -\frac{\overline{u}_{dc}}{L_1 L_2 C} \neq 0
\end{cases}
\tag{6}
$$

Based on the above analysis, the decoupling matrix can be established as follows.

$$
\begin{aligned}
E(X) &= \begin{bmatrix} L_{g_1}L_f^2 r_1(X) & L_{g_2}L_f^2 r_1(X) \\ L_{g_1}L_f^2 r_2(X) & L_{g_2}L_f^2 r_2(X) \end{bmatrix} \\[6pt]
&= \begin{bmatrix} -u_{dc}/(L_1 L_2 C) & 0 \\ 0 & -u_{dc}/(L_1 L_2 C) \end{bmatrix}
\end{aligned}
\tag{7}
$$

It can be calculated that the matrix is a non-singular matrix. The total relation degree of the system output variables is $d = d_1 + d_2 = 6 = n$. Therefore, in view of coordinate transformation, the original system can be directly transformed into a controllable linear system of Brunovsky standard type.

Define the state variable of linear standard system after feedback linearization as $Z = [z_1\; z_2\; z_3\; z_4\; z_5\; z_6]^T$. According to the relative order theory, the relationship between the new state variables and the original state variables can be expressed as

$$
\begin{cases}
z_1 = r_1(X) = i_{dref} - x_5 \\[4pt]
z_2 = L_f r_1(X) = -\frac{1}{L_2}x_3 - \omega x_6 + \frac{1}{L_2}e_d \\[4pt]
z_3 = L_f^2 r_1(X) = \frac{x_5-x_1}{L_2 C} - \frac{2\omega}{L_2}x_4 + \omega^2 x_5 + \frac{\omega}{L_2}e_q \\[4pt]
z_4 = r_2(X) = i_{qref} - x_6 \\[4pt]
z_5 = L_f r_2(X) = -\frac{1}{L_2}x_4 + \omega x_5 + \frac{1}{L_2}e_q \\[4pt]
z_6 = L_f^2 r_2(X) = \frac{x_6-x_2}{L_2 C} + \frac{2\omega}{L_2}x_3 + \omega^2 x_6 - \frac{\omega}{L_2}e_d
\end{cases}
\tag{8}
$$

Equation (8) shows that the original system state variable X is mapped into new state variable Z after the coordinate transformation of the Lie derivative. The coupling and high order terms in the system are not ignored in the coordinate transformation. The numerical relationship between new state variables can be calculated as

$$
\begin{cases}
\dot{z}_1 = \frac{\partial r_1(X)}{\partial X}\dot{X} = -\frac{1}{L_2}x_3 - \omega x_6 + \frac{1}{L_2}e_d = z_2 \\[2mm]
\dot{z}_2 = \frac{\partial(L_f r_1(X))}{\partial X}\dot{X} = \frac{x_5-x_1}{L_2 C} - \frac{2\omega}{L_2}x_4 + \omega^2 x_5 + \frac{\omega}{L_2}e_q = z_3 \\[2mm]
\dot{z}_3 = \frac{\partial(L_f^2 r_1(X))}{\partial X}\dot{X} = -\frac{3\omega}{L_2 C}x_2 + \left(\frac{L_1+L_2}{L_1 L_2^2 C} + \frac{3\omega^2}{L_2}\right)x_3 \\[2mm]
\qquad +\left(\frac{3\omega}{L_2 C} + \omega^3\right)x_6 - \left(\frac{1}{L_2^2 C} + \frac{\omega^2}{L_2}\right)e_d - \frac{u_{dc}}{L_1 L_2 C}u_1 \\[2mm]
\dot{z}_4 = \frac{\partial r_2(X)}{\partial X}\dot{X} = -\frac{1}{L_2}x_4 + \omega x_5 + \frac{1}{L_2}e_q = z_5 \\[2mm]
\dot{z}_5 = \frac{\partial(L_f r_2(X))}{\partial X}\dot{X} = \frac{x_6-x_2}{L_2 C} + \frac{2\omega}{L_2}x_3 + \omega^2 x_6 - \frac{\omega}{L_2}e_d = z_6 \\[2mm]
\dot{z}_6 = \frac{\partial(L_f^2 r_2(X))}{\partial X}\dot{X} = \frac{3\omega}{L_2 C}x_1 + \left(\frac{L_1+L_2}{L_1 L_2^2 C} + \frac{3\omega^2}{L_2}\right)x_4 \\[2mm]
\qquad -\left(\frac{3\omega}{L_2 C} + \omega^3\right)x_5 - \left(\frac{1}{L_2^2 C} + \frac{\omega^2}{L_2}\right)e_q - \frac{u_{dc}}{L_1 L_2 C}u_2
\end{cases}
\tag{9}
$$

It can be seen that information of z_1 is not directly contained in state variable z_3 and information of z_4 is not directly contained in state variable z_6. Defining the control variable after the coordinate transformation as $V = [v_1\ v_2]^T$, the relationship between the original nonlinear system control variable U and V of can be obtained as

$$
U = E^{-1}(X)[V - \Gamma(X)]
\tag{10}
$$

where $\Gamma(X) = \begin{bmatrix} L_f^{r_1} r_1(X) \\ L_f^{r_2} r_2(X) \end{bmatrix} = \begin{bmatrix} \left(-\frac{3\omega}{L_2 C}x_2 + \left(\frac{L_1+L_2}{L_1 L_2^2 C} + \frac{3\omega^2}{L_2}\right)x_3 \right. \\ \left. +\left(\frac{3\omega}{L_2 C} + \omega^3\right)x_6 - \left(\frac{1}{L_2^2 C} + \frac{\omega^2}{L_2}\right)e_d \right) \\ \left(\frac{3\omega}{L_2 C}x_1 + \left(\frac{L_1+L_2}{L_1 L_2^2 C} + \frac{3\omega^2}{L_2}\right)x_4 \right. \\ \left. -\left(\frac{3\omega}{L_2 C} + \omega^3\right)x_5 - \left(\frac{1}{L_2^2 C} + \frac{\omega^2}{L_2}\right)e_q \right) \end{bmatrix}.$

Substituting Equation (10) into Equation (2), the controller can be expressed as

$$
\begin{cases}
u_1 = \frac{1}{\bar{u}_{dc}}\left[-3L_1\omega i_{1q} + \left(3L_1 C\omega^2 + \frac{L_1+L_2}{L_2}\right)u_{cd}\right. \\[2mm]
\qquad \left. +(3L_1\omega + L_1 L_2 C\omega^3)i_{2q} - \left(L_1 C\omega^2 + \frac{L_1}{L_2}\right)e_d\right] - \frac{L_1 L_2 C}{\bar{u}_{dc}}v_1 \\[2mm]
u_2 = \frac{1}{\bar{u}_{dc}}\left[3L_1\omega i_{1d} + \left(3L_1 C\omega^2 + \frac{L_1+L_2}{L_2}\right)u_{cq}\right. \\[2mm]
\qquad \left. -(3L_1\omega + L_1 L_2 C\omega^3)i_{2d} - \left(L_1 C\omega^2 + \frac{L_1}{L_2}\right)e_q\right] - \frac{L_1 L_2 C}{\bar{u}_{dc}}v_2
\end{cases}
\tag{11}
$$

The original nonlinear system is transformed into the linear system as follows:

$$
\begin{cases}
\dot{Z} = AZ + BV \\
Y = CZ
\end{cases}
\tag{12}
$$

where $A = \begin{bmatrix} 0 & 1 & 0 & 0 & 0 & 0 \\ 0 & 0 & 1 & 0 & 0 & 0 \\ 0 & 0 & 0 & 0 & 0 & 0 \\ 0 & 0 & 0 & 0 & 1 & 0 \\ 0 & 0 & 0 & 0 & 0 & 1 \\ 0 & 0 & 0 & 0 & 0 & 0 \end{bmatrix}$, $B = \begin{bmatrix} 0 & 0 \\ 0 & 0 \\ 1 & 0 \\ 0 & 0 \\ 0 & 0 \\ 0 & 1 \end{bmatrix}$, $C = \begin{bmatrix} 1 & 0 \\ 0 & 0 \\ 0 & 0 \\ 0 & 1 \\ 0 & 0 \\ 0 & 0 \end{bmatrix}$.

The matrix A shows that the dq components of the original system are completely decoupled by nonlinear mapping. The transfer function of the system is shown in Figure 2. The block diagrams in the dotted box are the equivalent controller and the original system.

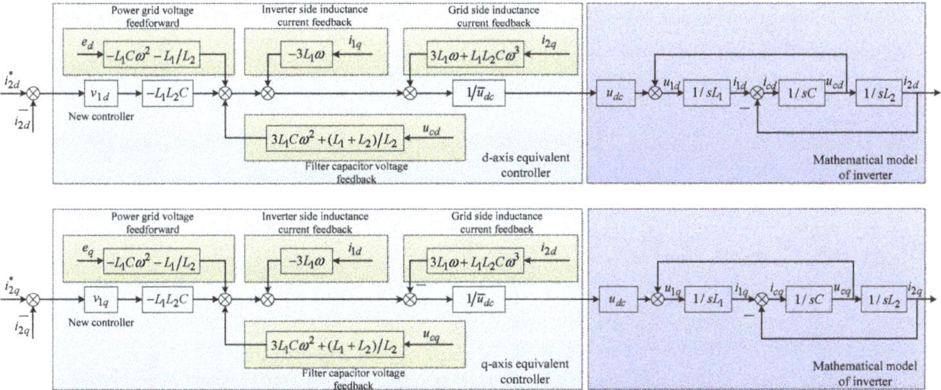

Figure 2. Current control system based on state feedback linearization.

For the system shown in Equation (12), the controller V can be designed to make it stable at the point $[z_1\ z_4]^T = [0\ 0]^T$. As the feedback decoupling of the original system has been completed, the controller can be designed based on the linear theory to achieve the control of original system.

2.3. Parameters Design of Current Single Closed Loop Controller

Through the state feedback linearization, the system is transformed into a 3-order controllable standard type. Therefore, at least two open-loop zeros should be introduced into the linear controller, that is, two first-order differentiation elements. Considering that the differentiation elements may cause system noise, the controller should contain a filter part (i.e., the first-order inertial element). To make the controller simple, the transfer function of the system controller after linearization can be written as:

$$v_1 = v_2 = -(k_2 s^2 + k_1 s + k_0)/(s + k_3) \tag{13}$$

Then the open-loop transfer function and the closed-loop transfer function of the control system can be represented as follows:

$$G_{OL}(s) = \frac{k_2 s^2 + k_1 s + k_0}{s^3(s + k_3)} \tag{14}$$

$$G_{CL}(s) = \frac{k_2 s^2 + k_1 s + k_0}{s^4 + k_3 s^3 + k_2 s^2 + k_1 s + k_0} \tag{15}$$

Compared with the original open-loop system, the controller which adds two zeros and one pole as a series correction link turns the original 3-order system into a 4-order one. Therefore, the order reduction processing of the higher order system is considered.

Assume two poles and two zeros constitute a pair of dipoles. Then, Equation (15) can be rewritten as

$$G_{CL}(s) = \frac{k_2(s^2 + 2\zeta_1 \omega_{n1} s + \omega_{n1}^2)}{(s^2 + 2\zeta_2 \omega_{n2} s + \omega_{n2}^2)(s^2 + 2\zeta_3 \omega_{n3} s + \omega_{n3}^2)} \tag{16}$$

With dipoles elimination, the system can be transformed to a typical 2-order system. For the closed-loop transfer function of the 2-order linear system, the parameter design method of the linear system can be used to construct the restrictive conditions, so as to obtain the parameters range.

1. Stability conditions of the system

According to the stability criterion, the closed loop characteristic equation of the original system should satisfy the Hurtwitz criterion as follows:

$$\begin{cases} \Delta_1 = k_2 k_3 > 0 \\ \Delta_2 = \Delta_1 - k_1 > 0 \\ \Delta_3 = k_1 \Delta_2 - k_0 k_3^2 > 0 \\ \Delta_4 = k_0 k_1 k_2 k_3 > 0 \end{cases} \tag{17}$$

2. Restrictive conditions of dipoles parameters

The closed loop system (16) can be reduced to a 2-order system with sufficient condition of dipoles existence. It is necessary to keep the dipole far from the imaginary axis (i.e., $\zeta_2 \omega_{n2} > \zeta_3 \omega_{n3}$).

$$\left| (\zeta_1 \omega_{n1} - \zeta_2 \omega_{n2}) + j(\omega_{n1} \sqrt{1 - \zeta_1^2} - \omega_{n2} \sqrt{1 - \zeta_2^2}) \right| < \frac{\zeta_2 \omega_{n2}}{10} \tag{18}$$

3. Cut-off frequency of the closed loop system

The cut-off frequency (i.e., ω_b) of current control system is generally less than 1/5 of switching frequency.

$$Gain(j\omega)\big|_{\omega = \omega_b} = \frac{k_2}{\sqrt{(\omega_{n3}^2 - \omega_b^2)^2 + (2\zeta_3 \omega_{n3} \omega_b)^2}} = 0.707 \tag{19}$$

4. Closed loop amplitude frequency characteristic at zero frequency

The closed loop system should maintain good tracking performance at low frequency band. The amplitude frequency characteristics satisfy the requirement of $Gain(j\omega)\big|_{\omega=0} \approx 1$.

$$Gain(j\omega)\big|_{\omega=0} = \frac{k_2 \omega_{n1}^2}{\omega_{n2}^2 \omega_{n3}^2} \approx 1 \tag{20}$$

Combined with the above four conditions, the optimal control shown in Equation (12) can be realized according to the classical control theory.

3. Design of Double Closed Loop Controller Based on Reduced Order State Feedback Linearization

The single closed loop feedback linearization method is used to decouple and simplify the inverter. The controller designed through this method is simple in structure, but it is too dependent on system precise model. Furthermore, the design of controller coefficient is complex, which limits the application of the algorithm.

Considering the objective of control, it is not necessary to configure all open loop poles of LCL type inverter model to the coordinate origin. Therefore, the system can be divided into two parts, which are analyzed separately. Then, a control strategy based on reduced order state feedback linearization can be designed. Consequently, the capacitance voltage in the LCL filter can be adopted as the intermediate variable of the reduced order feedback line linearization system. At the same time with system decoupling control, the single state variable active damping strategy is added to the system to achieve resonance suppression.

3.1. State Feedback of Inverter Side Inductance and Filter Capacitor Subsystem

Define the inverter side inductance current and the filter capacitor voltage as state variables $X = [x_1 \, x_2 \, x_3 \, x_4]^T = \left[i_{1d} \, i_{1q} \, u_{cd} \, u_{cq} \right]^T$, modulation variables as input variables $U = \left[u_{1d} \, u_{1q} \right]^T$, filter capacitor voltage as output variables $Y = R(X) = \left[u_{cd} \, u_{cq} \right]^T$.

The model of the inverter inductance and the filter capacitance subsystem can be written as Equation (2), where

$$f(X) = \begin{bmatrix} \omega x_2 - x_3/L_1 \\ -\omega x_1 - x_4/L_1 \\ x_1/C + \omega x_4 - i_{2d}/C \\ x_2/C - \omega x_3 - i_{2q}/C \end{bmatrix}, \; G(X) = \begin{bmatrix} g_1(X) & g_2(X) \end{bmatrix} = \begin{bmatrix} u_{dc}/L_1 & 0 \\ 0 & u_{dc}/L_1 \\ 0 & 0 \\ 0 & 0 \end{bmatrix}.$$

The grid-side inductance current can be considered as a measurable feedback variable in the system.

The model is a 2-input 2-output affine nonlinear system with 4 state variables (i.e., $n = 4$). According to the aforementioned analysis method, the 4-dimensional vector field matrix can be calculated as follows:

$$\begin{aligned} D_4 &= \begin{bmatrix} g_1 & g_2 & ad_f g_1 & ad_f g_2 \end{bmatrix} \\ &= \begin{bmatrix} u_{dc}/L_1 & 0 & 0 & -\omega u_{dc}/L_1 \\ 0 & u_{dc}/L_1 & \omega u_{dc}/L_1 & 0 \\ 0 & 0 & -u_{dc}/L_1 C & 0 \\ 0 & 0 & 0 & -u_{dc}/L_1 C \end{bmatrix} \end{aligned} \tag{21}$$

From (21), it can be obtained that $RankD_4 = 4 = n$. The calculation result shows that the vector fields are involution, which satisfies the state feedback linearization condition.

The Lie derivative of the system output (i.e., $r_1(X) = u_{cd}$) can be calculated as

$$\begin{cases} L_{g_1} r_1(X) = \frac{\partial r_1(X)}{\partial X} g_1(X) = 0 \\ L_{g_2} r_1(X) = \frac{\partial r_1(X)}{\partial X} g_2(X) = 0 \\ L_f r_1(X) = \frac{\partial r_1(X)}{\partial X} f(X) = \frac{1}{C} x_1 + \omega x_4 - \frac{1}{C} i_{2d} \\ L_{g_1} L_f r_1(X) = \frac{\partial (L_f r_1(X))}{\partial X} g_1(X) = \frac{u_{dc}}{L_1 C} \neq 0 \\ L_{g_2} L_f r_1(X) = \frac{\partial (L_f r_1(X))}{\partial X} g_2(X) = 0 \end{cases} \tag{22}$$

In Equation (22), the total relationship of the output variables of the system is $d = d_1 + d_2 = 4$, so the subsystem can be directly transformed to a linear controllable system through coordinate transformation. The corresponding decoupling matrix can be written as

$$E(X) = \begin{bmatrix} L_{g_1} L_f r_1(X) & L_{g_2} L_f r_1(X) \\ L_{g_1} L_f r_2(X) & L_{g_2} L_f r_2(X) \end{bmatrix} = \begin{bmatrix} \frac{u_{dc}}{L_1 C} & 0 \\ 0 & \frac{u_{dc}}{L_1 C} \end{bmatrix} \tag{23}$$

Then, the new state variables after the feedback linearization can be expressed as

$$\begin{cases} \dot{z}_1 = \frac{\partial r_1(X)}{\partial X} \dot{X} = \frac{1}{C} x_1 + \omega x_4 - \frac{1}{C} i_{2d} \\ \dot{z}_2 = \frac{\partial (L_f r_1(X))}{\partial X} \dot{X} = \frac{2\omega}{C} x_2 - \frac{1+L_1 C \omega^2}{L_1 C} x_3 - \frac{\omega}{C} i_{2q} + \frac{u_{dc}}{L_1 C} u_{1d} \\ \dot{z}_3 = \frac{\partial r_2(X)}{\partial X} \dot{X} = \frac{1}{C} x_2 - \omega x_3 - \frac{1}{C} i_{2q} \\ \dot{z}_4 = \frac{\partial (L_f r_2(X))}{\partial X} \dot{X} = -\frac{2\omega}{C} x_1 - \frac{1+L_1 C \omega^2}{L_1 C} x_4 + \frac{\omega}{C} i_{2d} + \frac{u_{dc}}{L_1 C} u_{1q} \end{cases} \tag{24}$$

The relationship between the new control variable V and the original one U is as (10), where

$$\Gamma(X) = \begin{bmatrix} L_f^{r_1} r_1(X) \\ L_f^{r_2} r_2(X) \end{bmatrix} = \begin{bmatrix} \frac{2\omega}{C} x_2 - \frac{1+L_1 C \omega^2}{L_1 C} x_3 - \frac{\omega}{C} i_{2q} \\ -\frac{2\omega}{C} x_1 - \frac{1+L_1 C \omega^2}{L_1 C} x_4 + \frac{\omega}{C} i_{2d} \end{bmatrix}.$$

Combining Equation (23) and Equation (10), the original control variable U can be expressed as

$$\begin{cases} u_{1d} = \frac{L_1 C}{u_{dc}} v_{1d} - \frac{L_1 C}{u_{dc}} [\frac{2\omega}{C} i_{1q} - \frac{1+L_1 C \omega^2}{L_1 C} u_{cd} - \frac{\omega}{C} i_{2q}] \\ u_{1q} = \frac{L_1 C}{u_{dc}} v_{1q} - \frac{L_1 C}{u_{dc}} [-\frac{2\omega}{C} i_{1d} - \frac{1+L_1 C \omega^2}{L_1 C} u_{cq} + \frac{\omega}{C} i_{2d}] \end{cases} \tag{25}$$

Through coordinate transformation, the original nonlinear system is transformed to two linear systems as follows:

$$\begin{cases} \dot{\mathbf{Z}}_d = \begin{bmatrix} \dot{z}_1 \\ \dot{z}_2 \end{bmatrix} = \begin{bmatrix} 0 & 1 \\ 0 & 0 \end{bmatrix}\begin{bmatrix} z_1 \\ z_2 \end{bmatrix} + \begin{bmatrix} 0 \\ v_{1d} \end{bmatrix} \\ y_d = z_1 \\ \dot{\mathbf{Z}}_q = \begin{bmatrix} \dot{z}_3 \\ \dot{z}_4 \end{bmatrix} = \begin{bmatrix} 0 & 1 \\ 0 & 0 \end{bmatrix}\begin{bmatrix} z_3 \\ z_4 \end{bmatrix} + \begin{bmatrix} 0 \\ v_{1q} \end{bmatrix} \\ y_q = z_3 \end{cases} \tag{26}$$

3.2. State Feedback of Grid-Side Inductance Subsystem

Similarly, define the grid-side inductance current of the LCL filter as the state variables $\mathbf{X} = [x_1\ x_2]^T = \begin{bmatrix} i_{2d}\ i_{2q} \end{bmatrix}^T$, the filter capacitor voltage as the input variables $\mathbf{U} = \begin{bmatrix} u_{2d}\ u_{2q} \end{bmatrix}^T = \begin{bmatrix} u_{cd}\ u_{cq} \end{bmatrix}^T$, and the grid side current as the output variable $Y = \begin{bmatrix} i_{2d}\ i_{2q} \end{bmatrix}^T$.

The model of subsystem can be written as Equation (2), where $(\mathbf{X}) = \begin{bmatrix} \omega x_2 - e_d/L_2 \\ -\omega x_1 - e_q/L_2 \end{bmatrix}$,

$\mathbf{G}(\mathbf{X}) = [g_1(\mathbf{X})\ g_2(\mathbf{X})] = \begin{bmatrix} 1/L_2 & 0 \\ 0 & 1/L_2 \end{bmatrix}$.

Then, the original control variable $\mathbf{U}$ can be expressed as

$$\begin{cases} u_{2d} = L_2 v_{2d} - L_2 \omega i_{2q} + e_d \\ u_{2q} = L_2 v_{2q} + L_2 \omega i_{2d} + e_q \end{cases} \tag{27}$$

3.3. Parameters Design of Double Closed Loop Controller

In the reduced order feedback linearization method, the filter capacitor voltage is the input and output of the two reduced order systems, respectively. Then the double loop control system is constructed that the outer control loop is the grid side current control and the inner control loop is the filter capacitor voltage control.

The control inner loop is the filter capacitor voltage control and can be designed as a unit negative feedback closed loop system. Considering the resonance suppression, differential feedback is needed. Therefore, a double control loop strategy can be designed, in which the *d*-axis system is shown in Figure 3.

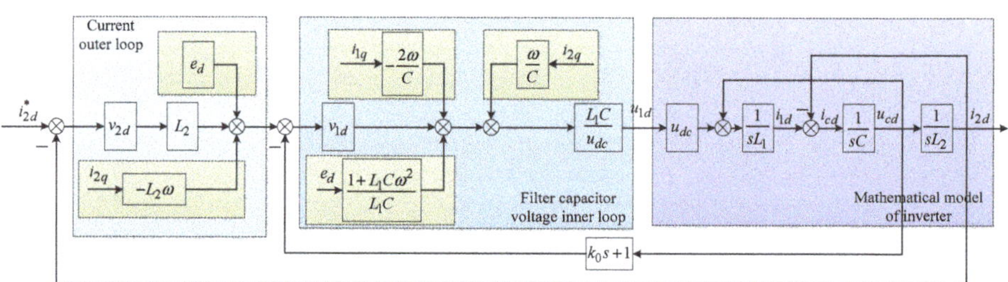

Figure 3. Block diagram of the reduced order feedback decoupling control system.

In Figure 3, the VSI system is transformed to three series integration links. The outer control loop is the grid-side inductance current control, and the conventional proportional integral controller is compatible. The inner control loop is the filter capacitor voltage control. In order to track high frequency ripple, a proportional controller can be adopted to improve the response speed of the system. Due to the influence of the inner control loop on the original system structure, the outer current control loop is no longer a precise feedback. A feedback term deviation appears. However, compared with the forward channel gain, the deviation, which can be equivalent to a feed forward interference, is smaller. Therefore, it can be compensated by a PI controller.

Define the outer loop controller parameters as k_2 and k_3, the inner loop controller parameter as k_1. Then, the control system can be represented as Figure 4.

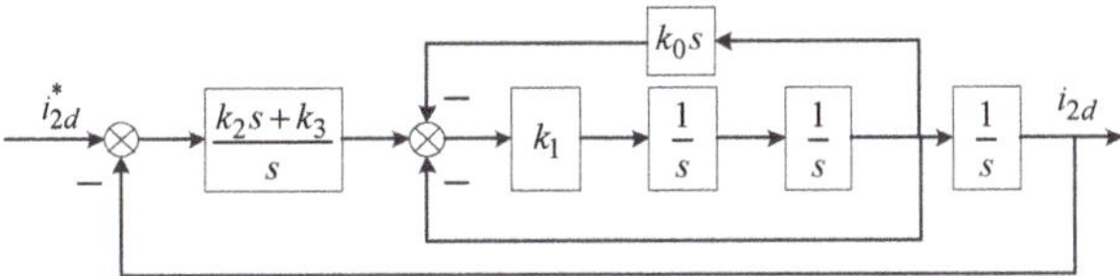

Figure 4. Equivalent block diagram of control system.

Open loop and closed loop transfer function of the system can be written as

$$G_{OL}(s) = \frac{k_1(k_2 s + k_3)}{s^2(s^2 + k_0 k_1 s + k_1)} \tag{28}$$

$$G_{CL}(s) = \frac{k_1(k_2 s + k_3)}{s^4 + k_0 k_1 s^3 + k_1 s^2 + k_1 k_2 s + k_1 k_3} \tag{29}$$

In the double closed loop current control system shown in (29), the numerator order of the transfer function is quite different from the denominator order. The parameters coupling degree is low. Therefore, the parameters can be designed directly according to the system characteristics. As the parameters design method was analyzed before, this section does not repeat the description.

4. Simulations and Experiments

4.1. Simulation and Experimental Environment

In order to verify the effectiveness of the proposed algorithm, the three-phase static var generator is used as the controlled object.

Matlab/Simulink simulation software (2010b, MathWorks, Inc., Natick, Massachusetts 01760 USA) is used to carry out numerical simulation analysis of three-phase three-leg grid-connected inverter based on feedback linearization. IGBT model in MATLAB/Simulink simulation software is selected as switch, with internal resistance $R_{on} = 1$ mΩ, snubber resistance $R_s = 500$ kΩ, snubber capacitance Cs = inf.

Furthermore, in order to verify the feasibility of the algorithm, a 50 kW prototype of three-phase three-leg inverter is built and the experiment is carried out. The Controller chip of the prototype is TMS320F28335 (Texas Instruments, Inc., Dallas, Texas 75243 USA). The IGBT module is SKM150GB12V (Semikron, Ltd., Nuremberg, Germany), with the switching frequency of *10 kHz*.

A photograph of the test rig is shown in Figure 5.

Based on the system parameters shown in Table 1, the control parameters of the two feedback linearized systems are selected, respectively.

1. Based on Equations (17)–(20), selecting the cut-off frequency of the open-loop transfer function as 750 Hz, the single closed-loop control system parameters can be designed as: $k_3 = 5000$, $k_2 = 10^7\pi$, $k_1 = 200k_2$, $k_0 = 10^4 k_2$. At this time, the corresponding phase margin is $P_m = 45°$, and the closed-loop bandwidth is 1 kHz.
2. Similarly, the parameters of the double closed loop control system can be designed as: $k_0 = 2 \times 10^{-4}$, $k_1 = 10^8$, $k_2 = 5 \times 10^3$, $k_3 = 10^2 k_3$. Then, the corresponding open loop cut-off frequency is 680 Hz, phase margin is $P_m = 43°$, and closed-loop bandwidth is 1 kHz.

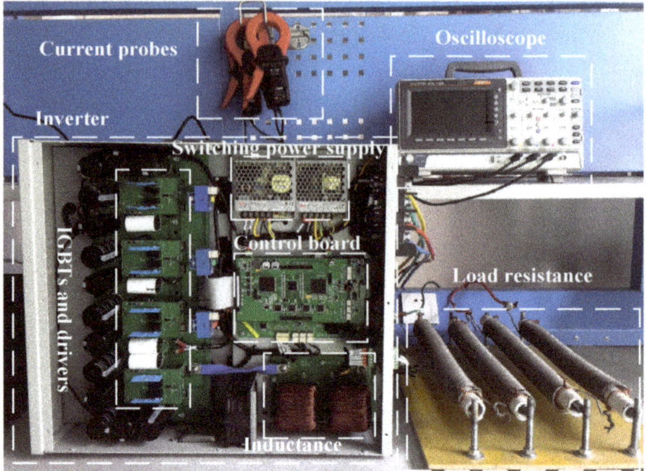

Figure 5. Photograph of the experimental test rig.

Table 1. Parameters for the system.

Variables	Symbols	Value
DC Bus Voltage	u_{dc}	650 V
Grid Voltage	u_g	380 V
Grid-side inductor	L_2	0.2 mH
Inverter side inductor	L_1	0.3 mH
Filter capacitor	C	20 uF
Switching frequency	f	10 kHz

The corresponding frequency domain characteristics of two control systems are shown in Figure 6.

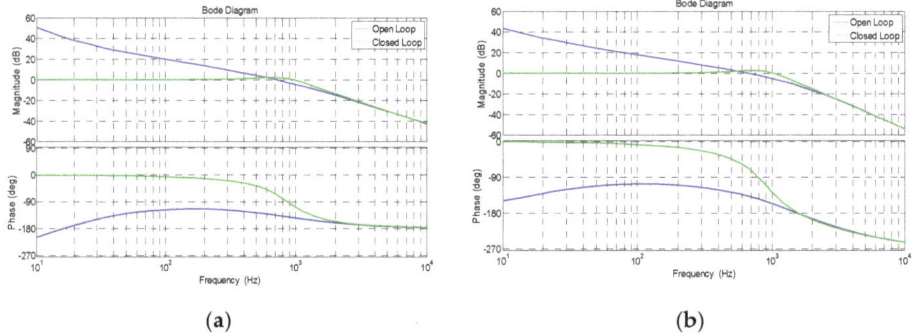

(a)   (b)

Figure 6. Bode diagrams of the closed-loop control systems: (**a**) Single closed-loop control system, (**b**) double closed-loop control system.

4.2. Steady State Control Performance

Setting VSI output current as $i_2 = 50$ A, the three-phase output current on the basis of two control methods is shown in Figure 7.

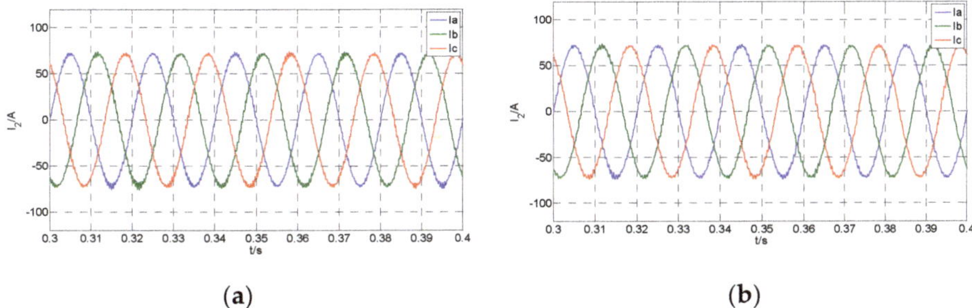

(a)   (b)

Figure 7. Steady-state current waveforms of voltage source inverters (VSI) based on two control schemes: (**a**) Output current of the single closed-loop control system, (**b**) output current of the double closed-loop control system.

According to Figure 7, both the two inverter systems can track the reference current accurately with small current distortion. Compared to single closed loop control, the output ripple of double closed loop control system is smaller.

The prototype is adopted to carry out the experiment. The output current of the system based on two kinds of control methods is shown in Figure 8. The output current Total Harmonic Distortion (THD) with the two control methods are 4.36% and 1.57%, respectively. The current ripple of single closed loop control system based on feedback linearization is larger. The main reason is that the control system is highly dependent on the accuracy of the system model. However, the sampling and control delays, which exist in real systems, cause the inaccuracy of the feedback signal. In addition, the nonlinear magnetization curve of the inductance and the equivalent series resistance of the capacitance also affect the model accuracy. It reduces the damping effect of the resonance point and increases the current ripple. The reduced order double closed loop controller takes the filter capacitor voltage as the intermediate variable. To some extent, the measure reduces the coupling relationship among the control variables and improves the robustness of the system. The steady state control performance is better.

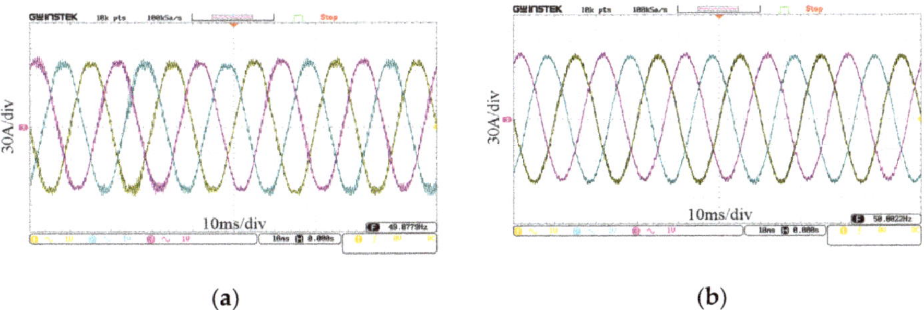

(a)   (b)

Figure 8. Experimental steady-state current waveforms: (**a**) Output current of the single closed-loop control experiment, (**b**) output current of the double closed-loop control experiment.

4.3. Dynamic Control Performance

In this paper, a dynamic test is carried out in the form of virtual load. The grid-connected current is set as 25 A at the initial time, and then doubles at 0.3 s. The output current of the system is shown in Figure 9.

When the load current changes, the output current of the two systems fluctuates with little overshoot. After a brief transient process, the system reaches a new steady state within one fundamental period. Compared to the direct feedback linearization control, the reduced order feedback linearization control lacks one closed loop zero, which reduces the response speed. However, the transition process is relatively smooth. Therefore, it can be seen that the two control schemes both have fast response and small overshoot.

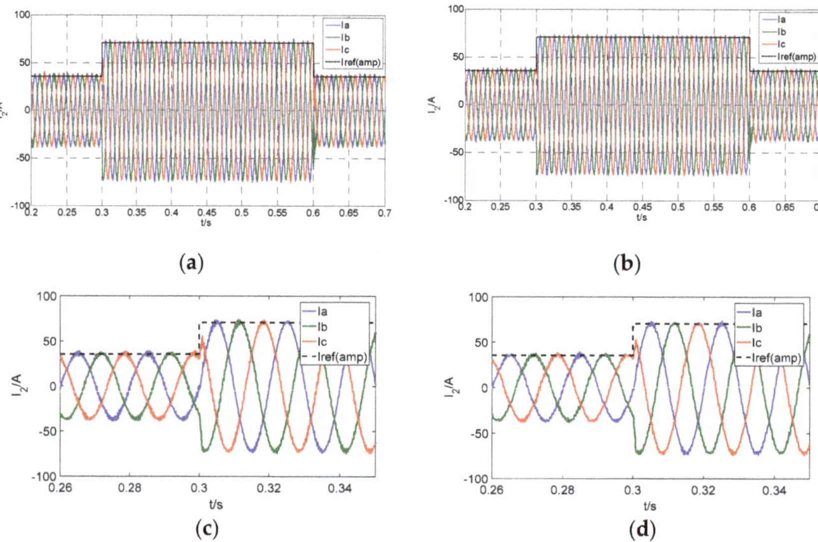

(a)

(b)

(c)

(d)

Figure 9. Dynamic current waveforms of VSI based on two control schemes: (**a**) Output current of the single closed-loop control system, (**b**) output current of the double closed-loop control system, (**c**) partial enlarged detail of (**a**), (**d**) partial enlarged detail of (**b**).

The experimental conditions are the same as those of the simulation. The experimental results are shown in Figure 10. Similar to the simulation results, it is shown that both of the two control schemes have good dynamic performance.

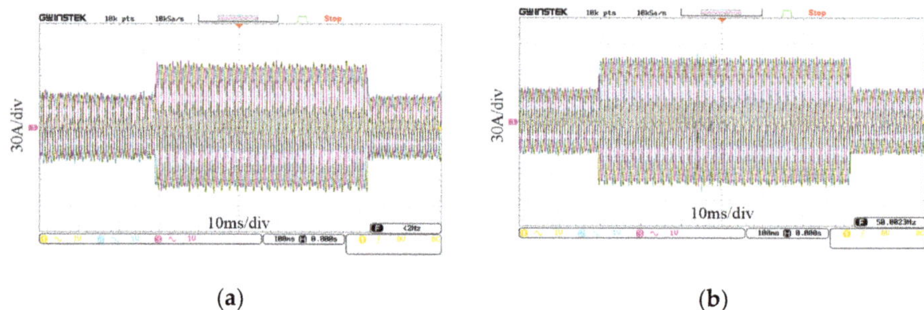

(a)

(b)

Figure 10. Experimental dynamic current waveforms: (**a**) Output current of the single closed-loop control experiment, (**b**) output current of the double closed-loop control experiment.

5. Conclusions

In order to solve the problem of poor global control performance of the controller designed by small signal modeling and harmonic linearization, two universal controller design methods based on state feedback linearization are proposed for the LCL type VSI. Relation degree of the VSI system and that of the reduced order model are analyzed, respectively. Subsequently, the coupling matrixes are designed by adopting the Lie derivative. Through the nonlinear transformation, the original system is mapped to the Brunovsky standard type, and the new system linear controller can be used to complete the control of the original complex system. Then, the parameters design ideas of two kinds of controllers are given. The proposed single closed-loop current controller has a simple structure, but it is highly dependent on the exact model of the system. The double closed loop controller sets the filter capacitor voltage as the inner loop control variable, by which a reduced order feedback linearization control can be achieved step by step, to improve the adaptability and robustness of the system. The simulation and experimental results show that the two controllers have good steady-state control performance and dynamic response performance. In practical application, the robustness of linear control system based on reduced-order feedback is better.

Author Contributions: L.Y. wrote the paper and designed the control method; C.F. contributed to the conception of the study and designed the simulation model; J.L. helped to perform the analysis with constructive discussions.

Acknowledgments: This work is supported by the National Natural Science Foundation of China (Grant No. 51607179) and the Fundamental Research Funds for the Central Universities (2017QNB01).

Conflicts of Interest: No potential conflict of interest was reported by the authors.

References

1. Farokhian, F.M.; Roosta, A.; Gitizadeh, M. Stability analysis and decentralized control of inverter-based ac microgrid. *Prot. Control Mod. Power Syst.* **2019**, *4*, 65–86. [CrossRef]
2. Yang, H.; Zhang, Y.; Liang, J.; Gao, J.; Walker, P.; Zhang, N. Sliding-mode observer based voltage-sensorless model predictive power control of pwm rectifier under unbalanced grid condition. *IEEE Trans. Ind. Electron.* **2018**, *65*, 5550–5560. [CrossRef]
3. Xu, X.; Casale, E.; Bishop, M. Application guidelines for a new master controller model for statcom control in dynamic analysis. *IEEE Trans. Power Deliv.* **2017**, *32*, 2555–2564. [CrossRef]
4. Bosch, S.; Staiger, J.; Steinhart, H. Predictive current control for an active power filter with lcl-filter. *IEEE Trans. Ind. Electron.* **2018**, *65*, 4943–4952. [CrossRef]
5. Liu, B.; Wei, Q.; Zou, C.; Duan, S. Stability analysis of lcl-type grid-connected inverter under single-loop inverter-side current control with capacitor voltage feedforward. *IEEE Trans. Ind. Inform.* **2017**, *14*, 691–702. [CrossRef]
6. Jamshidifar, A.; Jovcic, D. Small-signal dynamic dq model of modular multilevel converter for system studies. *IEEE Trans. Power Deliv.* **2016**, *31*, 191–199. [CrossRef]
7. Bao, X.; Zhuo, F.; Tian, Y.; Tan, P. Simplified feedback linearization control of three-phase photovoltaic inverter with an lcl filter. *IEEE Trans. Power Electron.* **2013**, *28*, 2739–2752. [CrossRef]
8. Lamperski, A.; Ames, A.D. Lyapunov theory for zeno stability. *IEEE Trans. Autom. Control* **2012**, *58*, 100–112. [CrossRef]
9. Hwang, C.L.; Fang, W.L. Global fuzzy adaptive hierarchical path tracking control of a mobile robot with experimental validation. *IEEE Trans. Fuzzy Syst.* **2016**, *24*, 724–740. [CrossRef]
10. Kim, D.E.; Lee, D.C. Feedback linearization control of three-phase ups inverter systems. *IEEE Trans. Ind. Electron.* **2010**, *57*, 963–968.
11. Akhtar, A.; Nielsen, C.; Waslander, S.L. Path following using dynamic transverse feedback linearization for car-like robots. *IEEE Trans. Robot.* **2017**, *31*, 269–279. [CrossRef]
12. Choi, Y.S.; Han, H.C.; Jung, J.W. Feedback linearization direct torque control with reduced torque and flux ripples for ipmsm drives. *IEEE Trans. Power Electron.* **2015**, *31*, 3728–3737. [CrossRef]
13. Yang, L.Y.; Liu, J.H.; Wang, C.L.; Du, G.F. Sliding mode control of three-phase four-leg inverters via state feedback. *J. Power Electron.* **2014**, *14*, 1028–1037. [CrossRef]
14. Lascu, C.; Jafarzadeh, S.; Fadali, S.M.; Blaabjerg, F. Direct torque control with feedback linearization for induction motor drives. *IEEE Trans. Power Electron.* **2016**, *32*, 2072–2080. [CrossRef]
15. Yang, S.; Wang, P.; Tang, Y. Feedback linearization based current control strategy for modular multilevel converters. *IEEE Trans. Power Electron.* **2017**, *33*, 161–174.

© 2019 by the authors. Licensee MDPI, Basel, Switzerland. This article is an open access article distributed under the terms and conditions of the Creative Commons Attribution (CC BY) license (http://creativecommons.org/licenses/by/4.0/).

Article

A Novel RPWN Selective Harmonic Elimination Method for Single-Phase Inverter

Guohua Li [1,2,*], Chunwu Liu [1], Zhenfang Fu [1] and Yufeng Wang [1]

[1] Faculty of Electrical and Control Engineering, Liaoning Technical University, Huludao 125105, Liaoning, China; lcw19950204@163.com (C.L.); fuzhenfanglntu@163.com (Z.F.); wyf792@163.com (Y.W.)

[2] College of Mechanical Engineering, Liaoning Technical University, Fuxin 123000, China

* Correspondence: liguohua@lntu.edu.cn

Received: 30 December 2019; Accepted: 12 March 2020; Published: 16 March 2020

Abstract: In the existing random pule width modulation (RPWM) selective harmonic elimination methods, the formula of switching cycle T_{N+1} is complex, and the duty ratio D_{N+1} of the next switching cycle needs to be calculated in advance. However, in the case of unknown T_{N+1}, D_{N+1} is also difficult to calculate accurately, and the two parameters are based on each other. A novel selective harmonic elimination method in RPWM is proposed in this paper. The PWM voltage pulse is placed at the back of the switch cycle, which simplifies the formula of the switch cycle T_{N+1} and eliminates the need to calculate the duty ratio D_{N+1}. Two kinds of RPWM selective harmonic elimination ideas are summarized. The general formulas of the switch cycle, the effective random number k, and the upper and lower limits of switch frequency corresponding to k are derived. The spectrum shaping of inverter output voltage can be realized without using digital filter in this method. Simple algorithm, small calculation and easy implementation are characteristics of the proposed method. The simulation and experimental results confirm the ability of the proposed method for reducing harmonics at the specific frequency in power spectral density (PSD).

Keywords: harmonic; RPWM; selective voltage harmonic elimination; single-phase inverter

---

1. Introduction

Random pulse width modulation (RPWM) is an effective method to suppress the electromagnetic interference (EMI) and the electromagnetic vibration and noise of the load [1]. As shown in Figure 1, RPWM can mainly be classified as 1) random switching frequency pulse width modulation (PWM) [2], 2) random pulse position PWM, 3) random switching PWM [3], 4) or hybrid random PWM [4]. According to the randomness of the pulse position, it can be divided into random lead-lag [5], random zero vector [6], random pulse center displacement [7], random pulse position [8], random phase-shifted PWM [9], variable delay random PWM [10], asymmetric carrier random PWM [11], single random pulse position [12], and fractal space vector modulation [13]. However, the traditional RPWM cannot suppress specific harmonics selectively, such as the resonance frequency of motors and other loads. The commonly used selective harmonic elimination pulse width modulation (SHEPWM) [14,15] can eliminate specific harmonics, but it is mainly aimed at low-order harmonics such as $6k \pm 1$ (k is the harmonic order), and has little effect on eliminating the resonance frequency of the loads.

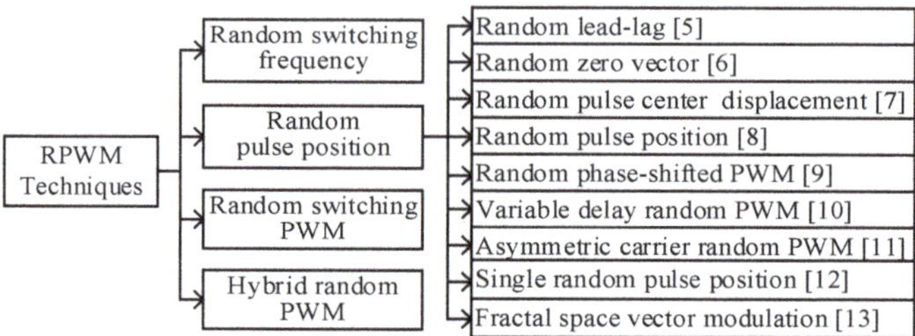

Figure 1. Random pulse width modulation (PWM) techniques.

In order to shape the noise spectrum of the inverter output voltage, the method of random modulation for selectively reducing the noise power at one or more frequencies was proposed in [16], but this method led to an increase in the peak of power spectral density. The literature [17] can also reduce the noise power at a specific frequency, but with this method, the switching frequency of the inverter must be less than the resonance frequency. When the resonance frequency is low, the switching frequency of the inverter will be over low. The low-pass filters and band-pass filters are used in [18–20], respectively, to reduce the harmonic power in the specific frequency range. However, the digital filter brings large computation costs in [18–20], and the harmonics in the specific frequency range are avoided, increasing rather than being completely eliminated in the random spread-spectrum. In other words, the harmonic content in the specific frequency range is just not increased compared with the original.

In RPWM selective harmonic elimination method, the specific harmonics can be eliminated by canceling each other with the preceding and succeeding terms in the Fourier series of the output voltage for the inverter. Theoretically, the specific harmonics can be completely eliminated by this method, which is mainly aimed at high-order harmonics such as 7 kHz, 9 kHz, etc. In [21], only harmonics whose frequencies are larger than 20 kHz can be eliminated. However, there is little practical value for the reason that the resonant frequency of the load is mostly lower than 20 kHz. The problem in [21] was solved by the method in [22], and the harmonics whose frequencies are lower than 10 kHz can be eliminated. However, the calculation of T_{n+1} in [22] is complicated, and D_{n+1} needs to be calculated in advance. Because D_{n+1} is calculated according to the midpoint of T_{n+1}, T_{n+1} and D_{n+1} are based on each other. In addition, the general formulas of the random number k and its corresponding switching frequency extreme value are not given in [22].

A novel RPWM selective harmonic elimination method for single-phase voltage source inverters (VSI) is proposed. In this method, the PWM pulse is placed at the back of the switching cycle. The calculation of switching cycle T_{n+1} is simplified and the contradiction between T_{n+1} and D_{n+1} is solved. The general formulas of switching cycle and the random number k and its corresponding switching frequency extreme value are also given. The noise of the specific frequency and its multiples can be selectively reduced in this method while realizing the function of traditional RPWM.

2. Strategy of Selective Harmonic Elimination in RPWM

The calculation of the switching cycle T_{n+1} is very significant in the RPWM selective harmonic elimination method. The formula for T_{n+1} and its corresponding pulse position in [22] are shown in Equation (1) and Figure 2a, where D_n and D_{n+1} are the duty ratios of the nth and $(n+1)$th switching cycles, k is the random number, f_0 is the frequency to be eliminated, and T_n is the nth switching cycle.

$$T_{n+1} = \frac{2k - f_0 T_n(1 + D_n)}{f_0(1 + D_{n+1})} \tag{1}$$

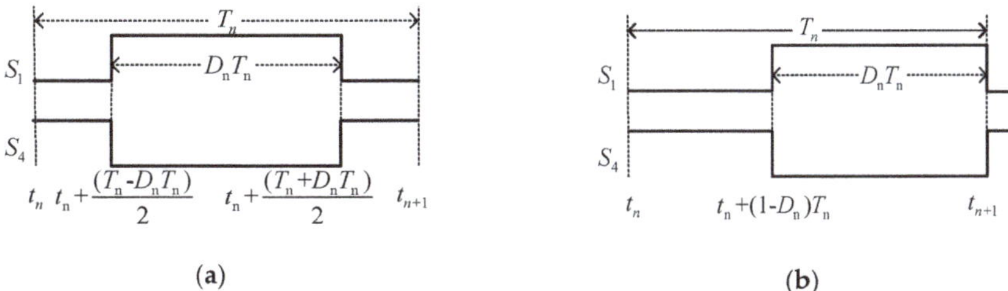

(**a**) (**b**)

Figure 2. The PWM pulse is located at (**a**) the center of switching cycle; (**b**) the back of switching cycle.

The D_{n+1} needs to be calculated in advance to calculate T_{n+1} in Equation (1). In the strategy of SPWM, the duty ratio D is calculated as $(1+M\sin(\omega t))/2$, where M is the modulation ratio. The value of ωt is calculated according to the time of the midpoint in each switching cycle. Namely, the time of the midpoint in T_{n+1} is used to calculate D_{n+1}. Therefore, when T_{n+1} is unknown, the duty ratio D_{n+1} cannot be calculated. Moreover, Equation (1) is complex and there are many parameters in it.

As shown in Figure 2b, the PWM pulse is located at the back of the switching cycle in this paper. This sequence pulse can be regarded as the sum of the output voltage u_{AB} and the DC-link voltage V_{dc} in single-phase VSI, as shown in Figure 3. If there are not the specific harmonics in the sequence pulse spectrum, there are not those harmonics in the output voltage u_{AB} spectrum of single-phase VSI, either.

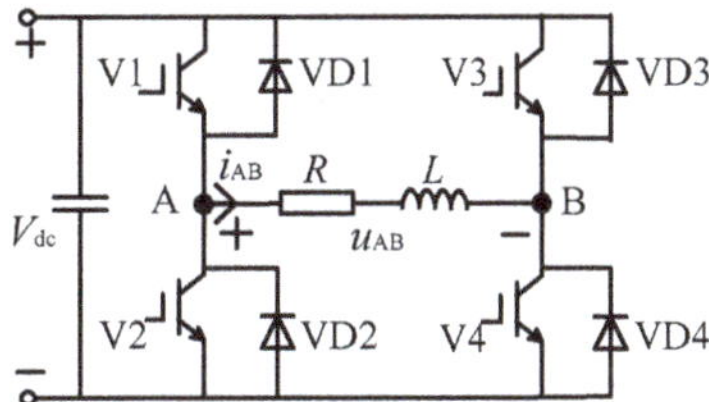

Figure 3. Topology of single-phase voltage source inverter.

The equation of the nth cycle of the sequence pulse is shown in Equation (2), where A is high value of output voltage and the equation of the sequence pulse is shown in Equation (3), where t_n and t_{n+1} are the starting and ending time of the nth switching cycle. The Fourier transform of Equation (3) is given at Equation (4). The real and imaginary parts in Equation (4) are special cases of Equation (5) [15,16]. If $c(f_0)$ at the frequency to be eliminated is zero for any φ, the result in Equation (4) will also be zero. That is, selective harmonic elimination is realized in the spectrum of the sequence pulse.

$$g_n(t) = \begin{cases} A & t_n+(1-D_n)T_n \leq t < t_{n+1} \\ 0 & otherwise \end{cases} \tag{2}$$

$$g(t) = \lim_{N\to\infty} \sum_{n=1}^{N} g_n(t) \tag{3}$$

$$G(f) = \int_{-\infty}^{\infty} g(t)e^{-j\omega t}dt = \int_{-\infty}^{\infty} g(t)\cos(\omega t)dt - j\int_{-\infty}^{\infty} g(t)\sin(\omega t)dt \tag{4}$$

$$c(f_0) = \int_{-\infty}^{\infty} g(t)\sin(2\pi f_0 t + \varphi)dt \tag{5}$$

3. Strategy for Novel RPWM Selective Harmonic Elimination

3.1. Calculation for Switching Cycle

Insert Equations (2) and (3) in Equation (5) for calculating Equation (6).

$$c(f_0) = \lim_{N \to \infty} \sum_{m=1}^{N} \left(\int_{t_m+(1-D_m)T_m}^{t_{m+1}} A \sin(2\pi f_0 t + \varphi) dt \right) = \frac{A}{2\pi f_0} \sum_{m=1}^{\infty} (\cos\{2\pi f_0 [t_m + (1 - D_m)T_m] + \varphi\} - \cos(2\pi f_0 t_{m+1} + \varphi)) \tag{6}$$

Two RPWM selective harmonic elimination ideas can be summarized on the basis of Equation (6). The first idea is that the first summation of the nth term is removed with the second summation of the $(n+e)$th term. The first summation of the $(n+1)$th term is removed with the second summation of the $(n+e+1)$th term, etc. The second idea is that the second summation of the nth term is removed with the first summation of the $(n+e)$th term. The second summation of the $(n+1)$th term is removed with the $(n+e+1)$th term of the first summation, etc. The following Equations (7)–(11) are given when e is equal to 1 and 2 according to the first idea, where e is a positive number.

$$\cos\{2\pi f_0[t_n+(1 - D_n)T_n] + \varphi\} - \cos(2\pi f_0 t_{n+e+1} + \varphi) = 0 \tag{7}$$

$$2\pi f_0 t_{n+e+1} + \varphi = 2\pi f_0[t_n+(1 - D_n)T_n] + \varphi + 2k\pi \tag{8}$$

$$T_n = \frac{1}{1 - D_n}\left(\sum_{m=n}^{n+e} T_m - \frac{k}{f_0}\right) \tag{9}$$

$$T_{n+1} = \frac{k}{f_0} - D_n T_n \ (e = 1) \tag{10}$$

$$T_{n+2} = \frac{k}{f_0} - D_n T_n - T_{n+1} \ (e = 2) \tag{11}$$

The comparison between Equation (10) and Equation (1) shows that the calculation of T_{n+1} with the method in this paper is simpler than the method in [22]. In addition, there is no need to calculate D_{n+1}. Thereby, the contradiction between T_{n+1} and D_{n+1} is solved and it is conducive to practical application.

3.2. Random Number k and Its Corresponding Extreme Value of Switching Frequency

The following Equations ((12) and (13)) for k_{max} and k_{min} are given by using Equation (10) if those conditions (f_0, D_{max}, D_{min}, f_{ma}, and f_{min}) are given, where k_{max} and k_{min} are the maximum and minimum values of random number k. D_{max} and D_{min} are the maximum and minimum values of the duty ratio. F_{max} and f_{min} are the maximum and minimum values of the instantaneous switching frequency of the inverter, which are usually preset.

$$k_{max} \leq \frac{f_0(1 + D_{max})}{f_{min}} \tag{12}$$

$$k_{min} \geq \frac{f_0(1 + D_{min})}{f_{max}} \tag{13}$$

Generally, there are many numbers for k that satisfy Equations (12) and (13), and the general formulas for f_{kmax} and f_{kmin} corresponding to each k are shown in Equations (14) and (15), where f_{kmax} and f_{kmin} are the maximum and minimum values of the switching frequencies corresponding to k. It can be seen from Equations (14) and (15) that the switching frequency of the inverter decreases with increasing k, or increases with decreasing k.

$$f_{kmax} = \frac{1}{\dfrac{k}{f_0} - \dfrac{D_{max}}{f_{min}}} \tag{14}$$

$$f_{k\min} = \cfrac{1}{\cfrac{k}{f_0} - \cfrac{D_{\min}}{f_{\max}}}$$

(15)

It should be noted that the calculation result of Equation (14) may be negative if the k is small. It shows that the denominator can be zero without taking the extreme value of f, and the D and the maximum value of the frequency is $+\infty$.

3.3. Calculation Process for Selective Harmonic Cancellation

As shown in Figure 4, the frequency to be eliminated (f_0) and the maximum value (f_{max}) and minimum value (f_{min}) of the instantaneous switching frequency are given in advance. The D_{max} and D_{min} are calculated according to the modulation ratio M. The range of the random number k can be acquired with the mentioned conditions inserted in Equations (12) and (13). T_{n+1} can be calculated with T_n and D_n inserted in Equation (10) when k is selected. Based on the above conditions, the PWM drive signals are generated by assigning a value to the comparison register in DSP TMS320F2812 to eliminate the harmonics at the specific frequency f_0 in RPWM.

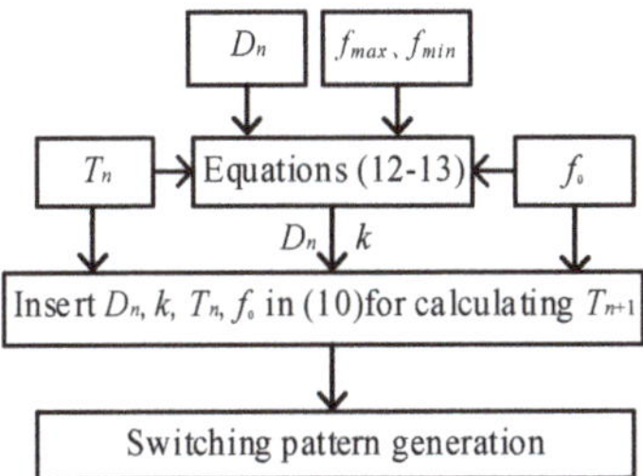

Figure 4. Flow chart of switching cycle calculation.

4. Simulation and Experiment

4.1. Parameters of System

The experimental system is shown in Figure 5. The parameters of the simulation and experimental system are shown in Table 1. The power electronic component in the inverter is IGBT. The driving circuit adopts IGBT-integrated driving module DA962D and the system main control chip adopts 32-bit DSP TMS320F2812. The dead time of inverter is 4.27 µs. In the experiment, the oscilloscope is DS1052E and the power quality analyzer is HIOKI PW3198.

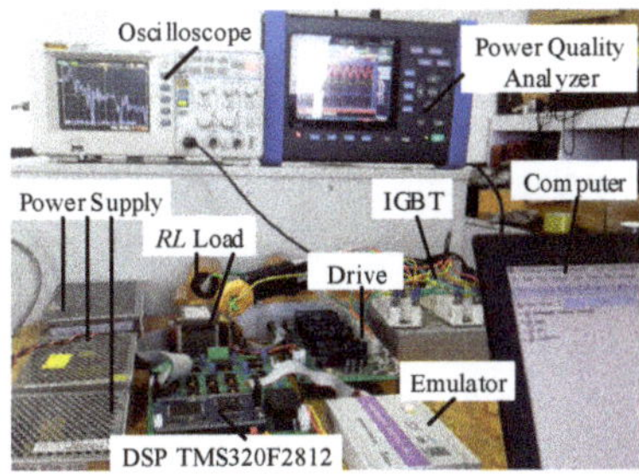

Figure 5. Experimental system.

Table 1. System parameters.

Parameter	Value	Parameter	Value
f_0(kHZ)	7 and 9	$R(\Omega)$	5
F_{max}(kHZ)	8	L(mH)	5
f_{min}(kHZ)	1.5	DC-link voltage(V)	24

4.2. Results and Analysis

The simulation waveforms of output voltage power spectral density (PSD) for single-phase VSI are shown in Figure 6. The traditional SPWM of fixed switching frequency (3 kHz) is adopted in Figure 6a. As seen from Figure 6a, the harmonics are mainly concentrated near 3 kHz and its multiples. Compared with the fixed switching frequency SPWM, it can be seen from Figure 6b with traditional RPWM that there is no outstanding peak in PSD. Figure 6c–e adopt the proposed method in this paper. The frequency to be eliminated (f_0) is 7 kHz. The modulation ratio M is 0.9, 0.7, and 0.5, respectively. Harmonics can be distributed randomly and can be reduced greatly at f_0 and its multiples.

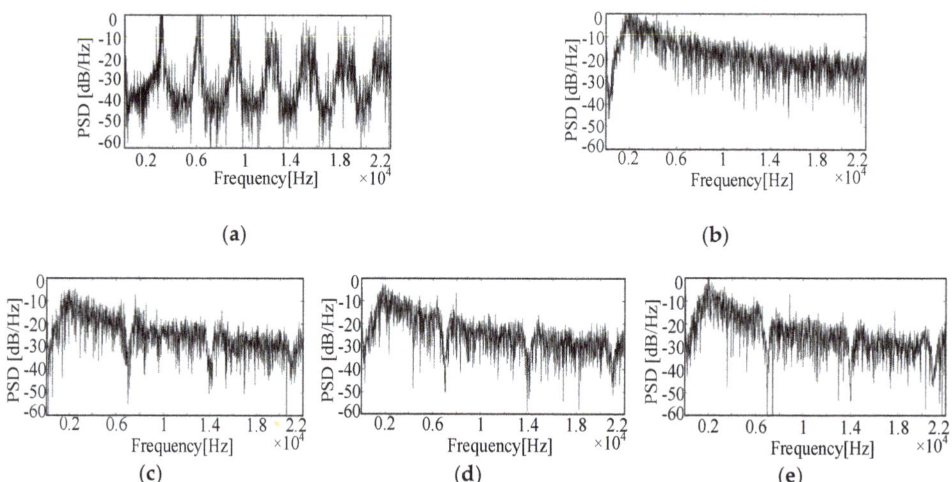

Figure 6. Simulated waveforms of voltage power spectral density (PSD). (**a**) Fixed switching frequency; (**b**) Traditional RPWM; and proposed method when $f_0 = 7$ kHz (**c**) $M = 0.9$; (**d**) $M = 0.7$; (**e**) $M = 0.5$.

Figure 7 is the simulation waveforms for the output voltage and current of the single-phase VSI when M is 0.9 and f_0 is 7 kHz. The enlarged waveforms of Figure 7 are shown in Figure 8.

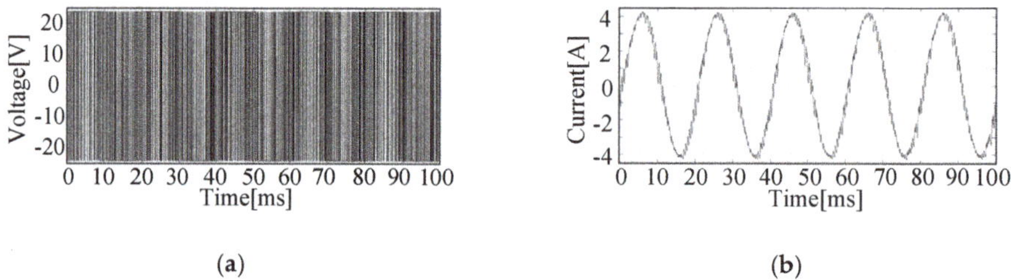

Figure 7. Simulated waveforms of single-phase inverter (**a**) voltage; (**b**) current.

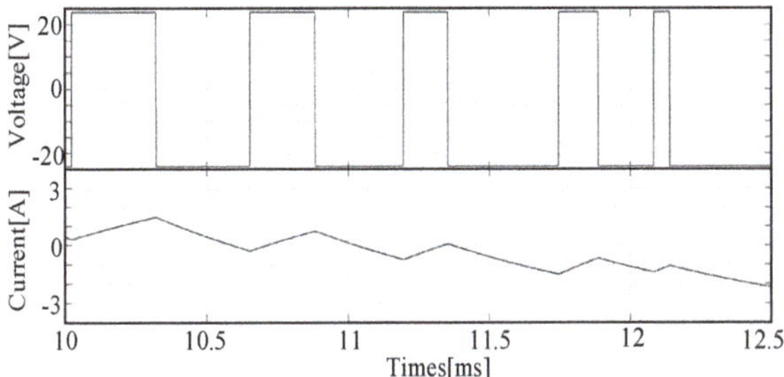

Figure 8. Enlarged waveform of Figure 7.

The experimental waveforms of output voltage and current PSD for single-phase VSI are shown in Figure 9. It can be seen that the harmonics near f_0 and its multiples are significantly reduced when f_0 is taken at 7 kHz and 9 kHz, respectively. The experimental results are basically consistent with the simulation results. As seen from the waveforms in Figures 6 and 9, there are obvious gaps in the range of several hundred hertz around f_0 and its multiples. The reason is that the frequencies close to f_0 will also be reduced when the harmonic at f_0 is completely eliminated. Moreover, the closer it is to f_0, the more it is reduced. The influence of system error can be overcome by this characteristic. Thus, the correctness of proposed method is proved by simulation and experimental results.

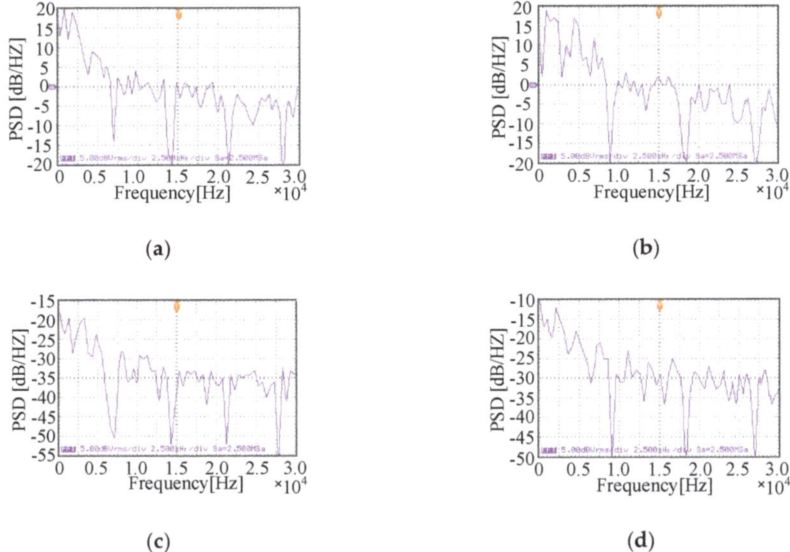

(a)

(b)

(c)

(d)

Figure 9. Experimental waveforms of PSD when M is 0.9. (**a**) The PSD of voltage when f_0 is 7 kHz; (**b**) the PSD of voltage when f_0 is 9 kHz; (**c**) the PSD of current when f_0 is 7 kHz; (**d**) the PSD of current when f_0 is 9 kHz.

The distributions of switching frequencies for two set numbers (k and k_1) when f_0 is 7 kHz and M is 0.9 are shown in Figure 10. All set random numbers $k = 1, 2, 3, 4, 5, 6, 7, 8, 9$ and smaller set numbers $k_1 = 1, 2, 3, 4$ for single-phase inverter are used for proposed method when switching frequencies are selected to eliminate harmonic at 7 kHz. In Figure 10, the switching frequencies have been selected randomly from 1.5 kHz to 8 kHz and the average switching frequency is equal to 2894 Hz (solid line). The distribution of switching frequencies gradually decreases and the randomness is good. By using k_1, the average switching frequency is increased to 3723 Hz (dot line). It can be known from Equations (14) and (15) that the random number k is inversely proportional to f_{kmax} and f_{kmin}. The average switching frequency can be increased by using smaller k when the average switching frequency is low to prevent the increase of current ripples. The average switching frequency can be reduced by using larger k when the average switching frequency is high. In addition, the larger f_0 is, the more k can be selected.

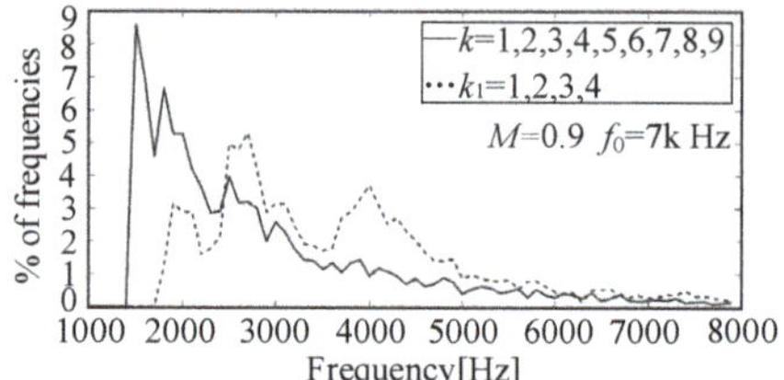

Figure 10. Distribution of switching frequencies by using the set numbers of k and k_1.

Figure 11 is the experimental waveforms for the output voltage and current of the single-phase VSI when M is 0.9 and f_0 is 7 kHz. As can be seen from Figure 11, the voltage pulse width varies randomly in each switching cycle affected by k and D. The current waveform of the inverter is sinusoidal.

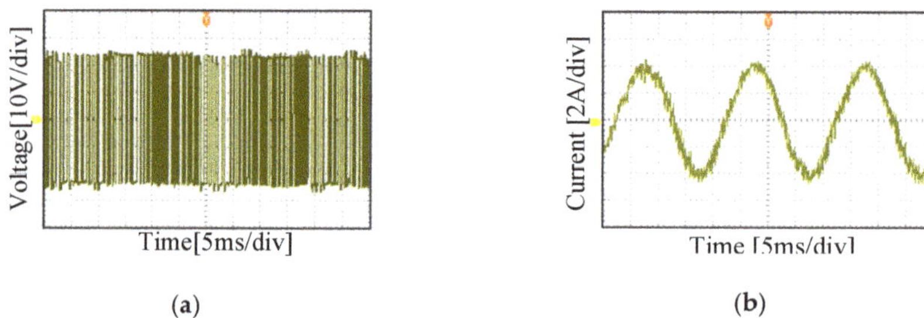

(a)   (b)

Figure 11. Experimental waveforms of single-phase inverter (**a**) voltage; (**b**) current.

The enlarged waveforms of Figure 11 are shown in Figure 12. With other cases unchanged, the larger the switching cycle is, the larger the current ripples are.

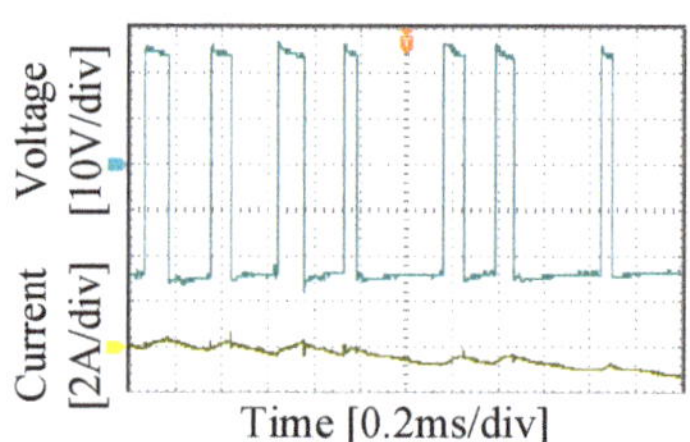

Figure 12. Enlarged waveform of Figure 11.

5. Conclusions

In this paper, a novel RPWM selective harmonic elimination method for single-phase VSI is proposed. A new pulse position which is placed at the back of the switching cycle is provided for RPWM selective harmonic elimination method. By using that, it can remove harmonics with certain frequency from output voltage and current. In fact, this has been done using switching cycle determination by switching cycles in the previous cycle and duty ratio. Compared with the fixed switching frequency SPWM and the traditional RPWM, the harmonics can be distributed uniformly in certain frequency range, and some unwanted harmonics can be eliminated successfully with the proposed method. A new pulse position for the RPWM selective harmonic elimination method is introduced in this paper, which is beneficial to improve the randomness of RPWM. When calculating T_{n+1}, D_{n+1} does not need to be calculated first. The random arrangement of the pulse position in switching cycle is worth studying in next steps.

Author Contributions: Conceptualization, G.L. and C.L.; methodology, Y.W.; software, Z.F.; validation, G.L. and C.L.; formal analysis, Y.W.; writing—original draft preparation, G.L.; writing—review and editing, C.L. All authors have read and agreed to the published version of the manuscript.

Funding: This research was funded by National Natural Science Foundation of China, grant number 51307076 and Natural Science Foundation of Liaoning Province, China, grant number 20180550268.

Acknowledgments: The authors thank Liaoning Technical University for providing the support to enable this research to be carried out.

Conflicts of Interest: The authors declare no conflict of interest.

References

1. Gamoudi, R.; Chariag, D.E.; Sbita, L. A review of spread-spectrum-based PWM techniques—A novel fast digital implementation. *IEEE Trans. Power Electron.* **2018**, *33*, 10292–10307. [CrossRef]

2. Bhattacharya, S.; Mascarella, D.; Joos, G.; Moschopoulos, G. Reduced switching random PWM technique for two-level inverters. In Proceedings of the 2015 IEEE Energy Conversion Congress and Exposition (ECCE), Montreal, QC, Canada, 20–24 September 2015; pp. 695–702.

3. Agelidis, V.G.; Vincenti, D. Optimum non-deterministic pulse-width modulation for three-phase inverters. In Proceedings of the IECON '93-19th Annual Conference of IEEE Industrial Electronics, Maui, HI, USA, 15–19 November 1993; pp. 1234–1239.

4. Huang, Y.; Xu, Y.; Zhang, W.; Zou, J. Hybrid RPWM Technique Based on Modified SVPWM to Reduce the PWM Acoustic Noise. *IEEE Trans. Power Electron.* **2019**, *34*, 5667–5674. [CrossRef]

5. Paramasivan, M.; Paulraj, M.M.; Balasubramanian, S. Assorted carrier-variable frequency-random PWM scheme for voltage source inverter. *IET Power Electron.* **2017**, *10*, 1993–2001. [CrossRef]

6. Liu, Y.; Wang, Q.; Zhao, J. Dual Randomized PWM Based on Vector Control System. *Proc. CSEE* **2010**, *30*, 98–102.

7. Trzynadlowski, A.M. Power spectra of a PWM inverter with randomized pulse position. *IEEE Power Electron. Spec. Conf.* **1993**, *9*, 1041–1047. [CrossRef]

8. Oh, S.; Jung, Y.; Lim, Y. A two-phase dual-zero vector RCD-PWM (DZRCD) technique. In Proceedings of the 30th Annual Conference of the IEEE Industrial Electronics Society, Busan, Korea, 2–6 November 2004.

9. Jadeja, R.; Ved, A.; Chauhan, S. An investigation on the performance of random pwm controlled converters. *Eng. Technol. Appl. Sci. Res.* **2015**, *5*, 86–876.

10. Shyu, J.L.; Liang, T.J.; Chen, J.F. Digitally-controlled PWM inverter modulated by multi-random technique with fixed switching frequency. *IEE Proc. Electr. Power Appl.* **2001**, *148*, 62–68. [CrossRef]

11. Mathe, L.; Lungeanu, F.; Sera, D.; Rasmussen, P.O.; Pedersen, J.K. Spread spectrum modulation by using asymmetric-carrier random PWM. *IEEE Trans. Ind. Electron.* **2012**, *59*, 3710–3718. [CrossRef]

12. Mathe, L.; Lungeanu, F.; Rasmussen, P.O.; Pedersen, J.K. Asymmetric carrier random PWM. In Proceedings of the 2010 IEEE International Symposium on Industrial Electronics, Bari, Italy, 4–7 July 2010; pp. 1218–1223.

13. Shiny, G.; Baiju, M.R. A space vector based pulse width modulation scheme for a 5-level induction motor drive. In Proceedings of the IEEE 9th International Conference on Power Electronics and Drive Systems (PEDS), Singapore, 5–8 December 2011.

14. Perez-Basante, A.; Ceballos, S.; Konstantinou, G.; Pou, J.; Kortabarria, I.; de Alegría, I.M. A universal formulation for multilevel selective- harmonic-eliminated PWM with half-wave symmetry. *IEEE Trans. Power Electron.* **2019**, *34*, 943–957. [CrossRef]

15. Haghdar, K. Optimal DC source influence on selective harmonic elimination in multilevel inverters using teaching–learning-based optimization. *IEEE Trans. Power Electron.* **2020**, *67*, 942–949. [CrossRef]

16. Pedersen, J.K.; Blaabjerg, F.; Frederiksen, P. Reduction of acoustic noise emission in AC-machines by intelligent distributed random modulation. In Proceedings of the 1993 Fifth European Conference on Power Electronics and Applications, Brighton, UK, 13–16 September 1993.

17. Wang, Z.; Chau, K.; Cheng, M. A chaotic PWM motor drive for electric propulsion. In Proceedings of the 2008 IEEE Vehicle Power and Propulsion Conference, Harbin, China, 3–5 September 2008.

18. Kang, B.J.; Liaw, C.M. Random hysteresis PWM inverter with robust spectrum shaping. *IEEE Trans. Aerosp. Electron. Syst.* **2001**, *37*, 619–629. [CrossRef]

19. Heping, L.; Qing, L.; Wei, Z.; Yiru, M.; Ping, L. Random PWM technique for acoustic noise and vibration reduction in induction motors used by electric vehicles. *Trans. China Electro Tech. Soc.* **2019**, *34*, 1488–1495.

20. Chai, J.Y.; Lin, Y.W.; Liaw, C.M. Comparative study of switching controls in vibration and acoustic noise reductions for switched reluctance motor. *IEE Proc. Electr. Power Appl.* **2006**, *153*, 348–360. [CrossRef]

21. Kirlin, R.L.; Lascu, C.; Trzynadlowski, A.M. Shaping the noise spectrum in power electronic converters. *IEEE Trans. Ind. Electron.* **2011**, *58*, 2780–2788. [CrossRef]

22. Peyghambari, A.; Dastfan, A.; Ahmadyfard, A. Selective voltage noise cancellation in three-phase inverter using random SVPWM. *IEEE Trans. Power Electron.* **2016**, *31*, 4604–4610. [CrossRef]

 © 2020 by the authors. Licensee MDPI, Basel, Switzerland. This article is an open access article distributed under the terms and conditions of the Creative Commons Attribution (CC BY) license (http://creativecommons.org/licenses/by/4.0/).

Article

Common Mode Voltage Elimination for Quasi-Switch Boost T-Type Inverter Based on SVM Technique

Duc-Tri Do [1], Minh-Khai Nguyen [1,2,*], Van-Thuyen Ngo [1], Thanh-Hai Quach [1] and Vinh-Thanh Tran [1]

[1] Faculty of Electrical and Electronics Engineering, Ho Chi Minh City University of Technology and Education, 1 Vo Van Ngan Street, Thu Duc district, Ho Chi Minh City 700000, Vietnam; tridd@hcmute.edu.vn (D.-T.D.); thuyennv@hcmute.edu.vn (V.-T.N.); haiqt@hcmute.edu.vn (T.-H.Q.); tranvinhthanh.tc@gmail.com (V.-T.T.)

[2] School of Electrical Engineering and Computer Science, Queensland University of Technology, Brisbane 4000, Australia

* Correspondence: nmkhai00@gmail.com; Tel.: +84-28-3722-1223

Received: 12 November 2019; Accepted: 19 December 2019; Published: 1 January 2020

Abstract: In this paper, the effect of common-mode voltage generated in the three-level quasi-switched boost T-type inverter is minimized by applying the proposed space-vector modulation technique, which uses only medium vectors and zero vector to synthesize the reference vector. The switching sequence is selected smoothly for inserting the shoot-through state for the inverter branch. The shoot-through vector is added within the zero vector in order to not affect the active vectors as well as the output voltage. In addition, the shoot-through control signal of active switches of the impedance network is generated to ensure that its phase is shifted 90 degrees compared to shoot through the signal of the inverter leg, which provides an improvement in reducing the inductor current ripple and enhancing the voltage gain. The effectiveness of the proposed method is verified through simulation and experimental results. In addition, the superiority of the proposed scheme is demonstrated by comparing it to the conventional pulse-width modulation technique.

Keywords: space-vector pulse-width modulation; common-mode voltage elimination; quasi-switched boost; three-level T-type inverter; impedance network

---

1. Introduction

Nowadays, applications of multilevel inverters (MLIs) are a significant part of the industry and the civilian population, such as renewable energy systems [1], uninterruptable power supplies [2], wind turbine systems [3], motor drives [4], etc. Instead of using additional diodes like conventional neutral-point clamped (NPC) topology, the three-level T-type inverter (3LT^{2}I) uses three bidirectional-switches blocks half of DC-link voltage, which provides superiority in medium and low power applications [5]. In addition, compared to the cascade topology [6] or flying capacitor topology [7], the 3LT^{2}I is superior in size and reliability because of not using additional isolated DC sources, as well as high power capacitors. Moreover, the 3LT^{2}I provides advantages for both MLIs and two-level inverter, such as high quality at output voltage, small output filter requirement, the low voltage stress on power elements of inverter, and it is easy to control [8]. However, the conventional MLIs only have capability for buck conversion [9,10]. This is because the DC-link voltage (input DC voltage of inverter) is considered as a constant power source, while the peak–peak output phase voltage is not higher than the DC-link voltage. Therefore, to achieve the desired amplitude of output voltage, the DC–DC converter is considered to be placed between the input power source and inverter leg to boost the DC-link voltage [11]. This system is known as a two-stage power converter. On the other hand, the low-frequency AC transformer is considered to be placed after the inverter to obtain the desired AC output voltage. These solutions lead to enhance the size, weight, cost, and reduce the efficiency of the power converter. One of the important issues that the traditional MLIs are facing is the shoot-through (ST) phenomenon, which is generated by switching any of the switches in any phase leg of the inverter [12]. This phenomenon can destroy switching devices used in the inverter, as well as cause a short circuit at input power supply [13]. This issue can be addressed by incorporating deadtime with a control signal fed to the inverter switching devices [13,14]. However, the effectiveness of the converter is not guaranteed, since it causes the distortion of an output waveform [11–15].

The Z-source (ZS) inverter topology, known as a single-stage power converter with a buck-boost capability and ST immune, is proposed in [16] to overcome the limitation of traditional MLIs. By adding the intermediate impedance network between the input power supply and inverter leg, the DC-link of the inverter is boost to the desired value instead of using an additional DC–DC converter before the inverter leg. Moreover, the ST state is used as the main mode during operation of the system, which provides the ST immunity without deadtime inclusion. As a result, the performance of the converter is significantly increased and has more reliability [17]. In order to incorporate the advantages of MLIs with the buck-boost capability of the ZS network, two ZS networks are connected in series with NPC structure [18]. This combination provides superior advantages in output quality compared with [16]. However, this structure also increases weight, size, and cost of the system because of using a large number of passive components. The new combination of a ZS network and NPC inverter by using two equal capacitors to split the input power supply into two equal sources is presented in [19]. In this method, the output of the ZS network and the midpoint of power supply provide three levels at output terminal without more ZS network requirement. Even so, this type of intermediate network also has drawbacks, such as high voltage stress on capacitors, as well as discontinuous input current because of using the input diode [20].

The quasi-Z-source (qZS) inverter topology derived from the ZS network was proposed to overcome the limitation of ZS topology [21]. Like the ZS inverter, this structure also behaves as a boost converter, which operates in shoot through (ST) and non-shoot through (NST) modes. The combination of the qZS inverter topology and three-level inverter has some advantages, such as low voltage stress on power elements and continuous input current [21]. The new topology of the qZS network was presented to enhance the boost capability [22]. This topology is implemented by adding one more active switch into the intermediate network to obtain the higher voltage gain. However, this structure also provides disadvantages, such as using a large number of passive components, such as inductors or capacitors that lead to increase the size of the power inverter, and the ripple of inductor current is quite large. Moreover, the boost capability of the qZS network just depends on the ST ratio of the inverter leg, therefore, the voltage gain is not flexible to control.

By using only one inductor and one capacitor in order to boost the DC input voltage, the quasi-switched boost (qSB) topology is superior in reducing a number of passive components compared to the qZS inverter, whereas the voltage gain is maintained as qZS inverter [23,24]. Moreover, this structure uses one more active switch, so it is very flexible to control. In [25], the qSB network is added to the NPC inverter, which connects two identical qSB networks in series to create three-level voltage at output of impedance network fed to the three-level inverter leg. In this structure, the qSB network uses two power supplies and two inductors resulting in unnecessary complexity. A single-stage active impedance source three-phase T-type inverter was proposed in [26] with a reduced component count and voltage stress on devices. In [27], a pulse-width modulation (PWM) strategy is proposed to improve the inductor current ripple and voltage gain of the converter based on the quasi-switched boost T-type inverter (qSBT^{2}I) topology. Applying the scheme proposed in [27], a PWM strategy is proposed in [28] to enhance the stabilization of the system by solving the open-circuit faults of switching devices in the inverter leg as well as the intermediate network.

Due to the superior quality and amplitude of output voltage, the space-vector modulation (SVM) is applied to qSBT^{2}I as presented in [29,30]. Because of not containing the zero vector in some of the region of the space vector diagram, these schemes use upper ST (UST) and lower ST (LST) modes, which are only included in small vectors, to ensure the voltage gain of the converter. As a result, the boost capability is enhanced while the quality of output voltage is improved significantly.

During operation, the inverter generates the common-mode voltage (CMV), which causes a lot of problems, such as bearing currents and shaft voltage in motor drives applications or electromagnetic interference [31]. Thus, it will reduce the life of the inductor motor or affect other electronic devices operating near the inverter. A PWM strategy used to reduce the magnitude of the CMV was discussed in [32]. In this technique, a reference vector is synthesized by using large vectors, medium vectors, and zero vector, which limits the amplitude of CMV from $-V_{DC\text{-link}}/6$ to $V_{DC\text{-link}}/6$. Based on [32], the switching sequence is then modified to insert the ST state, which guarantees the boost capability, whereas the magnitude of CMV is maintained and applied to the qZS inverter [33]. However, these methods do not eliminate CMV completely. A technique to eliminate CMV applied to five-level inverter based on SVM and sine PWM was founded in [34]. The switching loss for five-level NPC and cascade inverter was also analyzed. However, due to using conventional inverter topology, the voltage boosting capability cannot be achieved.

In this paper, the SVM strategy to eliminate CMV for 3L qST^{2}I is proposed. In this technique, the medium vectors and zero vector are adopted to generate the reference vector. As a result, the CMV is maintained zero during operation. To ensure the inverter operates under boost condition, the switching sequence and ST insertion are discussed carefully. Moreover, the control scheme for two active-switches of the impedance network proposed in [27] is adopted in this paper so the inductor current ripple is improved compared to the qZS inverter. In addition, the continuous input current is guaranteed by not using the input diode like the ZS network. Simulation and experimental results based on power simulation (PSIM) software and a practical prototype are presented to verify the effectiveness of the proposed scheme. The results are also compared with other conventional methods to demonstrate the effectiveness of this scheme. The advantages of the proposed space-vector pulse-width modulation (SVPWM) scheme over the conventional SVPWM scheme are as follows:

- The CMV is maintained zero during operation, and;
- The inductor current ripple is improved compared to the qZS inverter.

2. Three-Level Quasi-Switched Boost T-Type Inverter (qSBT^{2}I) Topology

The topology of qSBT^{2}I is illustrated in Figure 1, where the intermediate qSB network is placed in front of the 3LT^{2}I. The qSB network is constructed by one inductor (L_B), two capacitors (C_1 and C_2), two active switches (S_1 and S_2), and four diodes (D_1, D_2, D_3, and D_4). A three-phase low-pass filter which consists of L_f and C_f is used to reduce total harmonic distortion (THD) at the output. The connection of two identical qSB creates the neutral point "O" connected to load through three bi-directional switches, which consist of two active switches connected in series, shown in Figure 1. This topology provides superior advantages in medium and low application compared to conventional topology of MLIs because these bi-directional switches block only half of the DC-link voltage. Moreover, the high side switches (S_{1x}, $x = a, b, c$) and low side switches (S_{3x}) are used to connect the positive point "P" and negative point "N" of qSB network outputs to the load, while the middle switches (S_{2x}) is connected to point "0" of the qSB network to the load, as shown in Figure 1. By this example, the inverter leg can produce three-level voltages at the output terminal as "P", "O", and "N", which ensure the advantages of MLIs. Moreover, by using the qSB network, this topology can behave as a boost converter which is able to generate desired AC output voltage at the load.

Like [27], the inverter topology can operate in two main modes: ST and NST, which consist of four sub-NST modes as illustrated in Figure 2. As presented in Figure 2a, inductor L_B stores energy in NST mode 3 by switching both switches S_1 and S_2, where the inverter leg operates normally by producing three-level voltages at output. On the other hand, the inductor is also charged by turning on all switches of the inverter leg in ST mode, which produces "O" state at output terminal. Therefore, the ST state is considered to insert within "O" state in order to not affect the normal operation of the converter. Due to the use of ST state, the efficiency and reliability of the system are improved, which are not guaranteed in a conventional inverter when the deadtime is included in switching control signal. In NST mode, the inverter operates similar to conventional MLIs, which is able to achieve three-level at output terminal, the energy stored in inductor and power source is transferred to the load by triggering the corresponding switches, as detailed in Table 1.

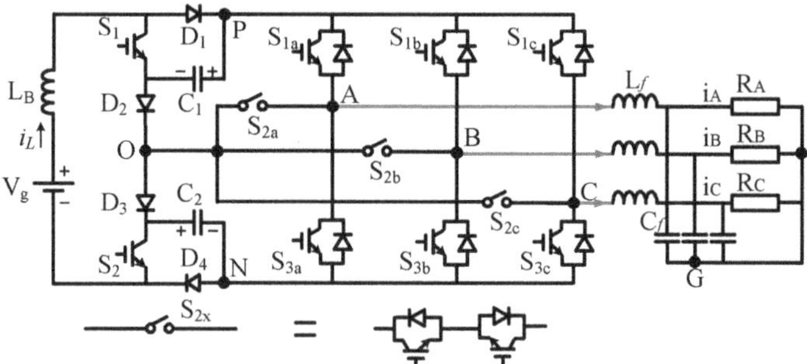

Figure 1. Topology of 3L qSBT^{2}I.

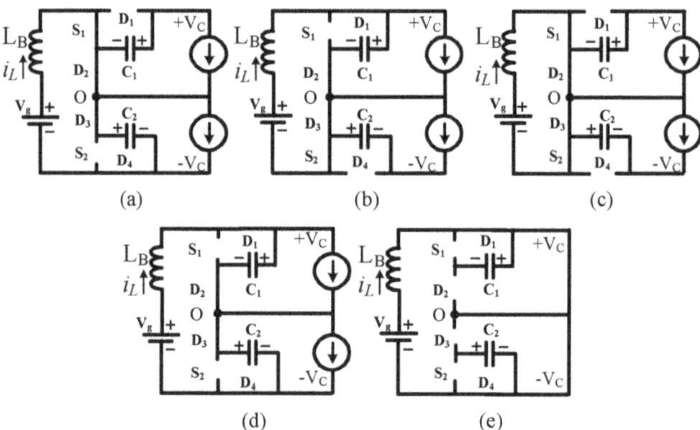

Figure 2. Modes of the quasi-switched boost T-type inverter (qSBT^{2}I). (**a**) NST 1, (**b**) NST 2, (**c**) NST 3, (**d**) NST 4, (**e**) shoot-through (ST).

Table 1. Switching states of qSBT^{2}I ($x = a, b,$ and c).

Mode	Triggered Switches	ON Diodes	V_X
NST 1	S_1	D_2, D_3, D_4	$+V_C, 0,$ or $-V_C$
NST 2	S_2	D_1, D_2, D_3	$+V_C, 0,$ or $-V_C$
NST 3	S_1, S_2	D_2, D_3	$+V_C, 0,$ or $-V_C$
NST 4	S_{1x}	D_1, D_2, D_3, D_4	$+V_C$
	S_{2x}		0
	S_{3x}		$-V_C$
ST	S_{1x}, S_{2x}, S_{3x}	D_1, D_4	0

The steady state is processed in the same condition compared to [27] in which the qSB network is considered symmetrical ($C_1 = C_2$) and the voltage across these capacitors is assumed constant ($V_{C1} = V_{C2}$). As a result, the capacitor voltage is expressed as (1).

$$V_C = V_{C1} = V_{C2} = \frac{V_g}{2 - 3D_0 - d}, \tag{1}$$

where,

V_{C1}, V_{C2}—capacitor voltage of C_1 and C_2;
V_g—input DC voltage of the inverter;
D_0—ST duty ratio;
d—the ratio identifying duty cycle of intermediate network switches (S_1 and S_2).

The relationship between ST duty ratio D_0 and the ratio d is shown in (2).

$$d + D_0 \leq 1. \tag{2}$$

When applying the ST algorithm presented in [27], the inductor current ripple of qSBT^{2}I can be decreased compared with the qZS network. As a result, the performance of the system is significantly improved. The value of inductor current ripple is expressed as follows:

$$\Delta I_L = \frac{1}{2L_B f_c} . V_g D_0, \tag{3}$$

where,

ΔI_L—the inductor current ripple;
L_B—the value of the boost inductor;
f_C—the frequency of the carrier.

The RMS value of output voltage can be identified through capacitor voltage and calculated as [27]

$$V_{x,RMS} = \frac{m.V_C}{\sqrt{2}} = \frac{m}{\sqrt{2}} . \frac{V_g}{2 - 3D_0 - d}, \tag{4}$$

where,

$V_{x,RMS}$—the RMS value of output voltage;
m—modulation index.

In order to avoid affecting the output voltage, the relationship between modulation index m and ST duty cycle D_0 must be

$$\begin{cases} m \leq 1 \\ m + D_0 \leq 1 \end{cases}. \tag{5}$$

3. The Proposed SVM Scheme to Eliminate CMV for 3L qSBT^{2}I

During operation, the MLIs generate CMV, which is identified as the voltage between load neutral point "G" and DC-link neutral point "O". It can be calculated through three-phase output voltage and the CMV, as presented in (6).

$$V_{CMV} = V_{GO} = \frac{V_{AO} + V_{BO} + V_{CO}}{3}, \tag{6}$$

where V_{AO}, V_{BO}, and V_{CO} are three-phase output pole voltages.

There are 27 vectors corresponding to 27 switching states of the $3LT^2I$. The CMV value is produced according to the vector adopted to generate the output voltage. Table 2 lists the CMV value of $3LT^2I$.

As illustrated in Table 2, the maximum magnitude of CMV is reached when the zero vectors [PPP] or [NNN] are adopted, its value is $\pm V_C$. When small vectors are adopted to synthesize the reference vector, the CMV value is changed from $-2V_C/3$ to $+2V_C/3$, while the large vectors just generate the CMV whose peak value is $V_C/3$. Along with 27 vectors listed in Table 2, the zero vector [OOO] and medium vectors generate the CMV with the minimum value, which is zero. Therefore, when zero vector [OOO] and medium vectors are adopted to create the output voltage, the CMV is eliminated.

Table 2. Common-mode voltage (CMV) value of the three-level T-type inverter ($3LT^2I$).

Vectors	State	V_{CMV}	State	V_{CMV}	State	V_{CMV}
Zero	[OOO]	0	[PPP]	$+V_C$	[NNN]	$-V_C$
P-Type	[POO]	$+V_C/2$	[PPO]	$+2V_C/3$	[OPO]	$+V_C/3$
Small	[OPP]	$+2V_C/3$	[OOP]	$+V_C/3$	[POP]	$+2V_C/3$
N-Type	[ONN]	$-2V_C/3$	[OON]	$-V_C/3$	[NON]	$-2V_C/3$
Small	[NOO]	$-V_C/3$	[NNO]	$-2V_C/3$	[ONO]	$-V_C/3$
Medium	[PON]	0	[OPN]	0	[NPO]	0
	[NOP]	0	[ONP]	0	[PNO]	0
Large	[PNN]	$-V_C/3$	[PPN]	$+V_C/3$	[NPN]	$-V_C/3$
	[NPP]	$+V_C/3$	[NNP]	$-V_C/3$	[PNP]	$+V_C/3$

This paper proposes a SVM strategy by using the medium vectors and zero vector [OOO], which synthesize the reference vector applied to $qSBT^2I$. As a result, CMV will be eliminated. The space vector diagram is illustrated in Figure 3, where the magnitude of medium vectors and zero vector [OOO] are $2V_C/\sqrt{3}$ and 0, respectively. The maximum magnitude of reference vector is V_C. The space vector diagram is divided into six sectors, which is used to analyze the operation principle of the inverter. During operation, the ST state is added into zero vector in order to avoid affecting the output voltage and provide a boost capability. The drew-time calculation, switching sequence, and ST insertion are presented in this section. Generally, the sector I is considered as a representative instant to analysis. The calculation for other sectors can be achieved in a way similar to the sector I.

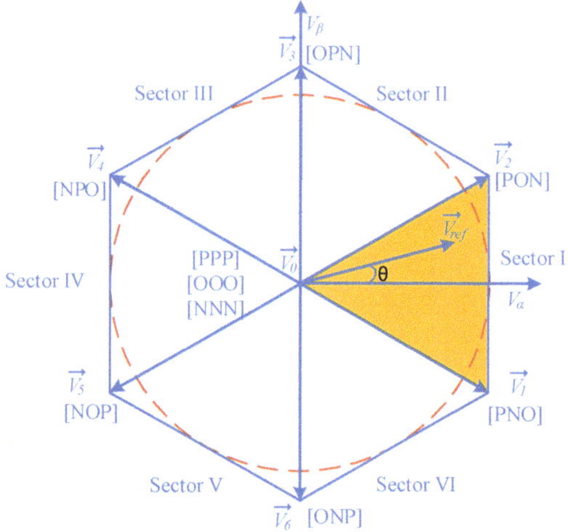

Figure 3. Space vector diagram of the proposed strategy.

Assume that the reference vector is located in the sector I, the medium vectors $(\vec{V_1}, \vec{V_2})$ and zero vector $\vec{V_0}$ are adopted to synthesize the reference vector. Their relationship will be presented in (7).

$$\begin{cases} \vec{V_{ref}}.T_s = \vec{V_0}.T_0 + \vec{V_1}.T_1 + \vec{V_2}.T_2 \\ T_s = T_0 + T_1 + T_2 \end{cases}, \tag{7}$$

where,

$\vec{V}_{ref}$—reference vector;

$\vec{V}_0$—zero vector;

$\vec{V}_1$, $\vec{V}_2$—medium vectors;

T_s—switching period of the inverter;

T_0, T_1, T_2—the on-times of $\vec{V}_0$, $\vec{V}_1$, and $\vec{V}_2$, respectively.

The reference vector $\vec{V}_{ref}$, zero vector $\vec{V}_0$, and medium vector $\vec{V}_1$, $\vec{V}_2$ are identified as follows:

$$\begin{cases} \vec{V}_{ref} = m.V_C/2.e^{j\theta} \\ \vec{V}_0 = \vec{0} \\ \vec{V}_1 = m.V_C/\sqrt{3}.e^{-j\pi/6} \\ \vec{V}_2 = m.V_C/\sqrt{3}.e^{j\pi/6} \end{cases}, \tag{8}$$

where m is the modulation index.

Based on (7) and (8), the drew time of these vectors can be calculated as follows:

$$\begin{cases} T_1 = T_s.m.\sin(\pi/6 - \theta) \\ T_2 = T_s.m.\sin(\pi/6 + \theta) \\ T_0 = T_s - T_1 - T_2 \end{cases}. \tag{9}$$

The switching sequence for sector I is [OOO]-[PON]-[PNO]-[OOO]-[PNO]-[PON]-[OOO].

When the ST state is obtained, all switches of the inverter leg are turned on at the same time. As a result, the output voltage is zero, which is similar to zero vector [OOO]. Therefore, to guarantee the boost capability of the qSB network and to not affect the output voltage, the ST vector is added into zero vector, so the zero vector is changed as

$$\begin{cases} \vec{V}_0'.T'_0 = \vec{V}_0 T_0 - \vec{V}_{ST}.T_{ST} \\ T'_0 = T_0 - T_{ST} = T_0 - D_0 T_S \end{cases}. \tag{10}$$

By incorporating the ST state with the switching sequence mentioned above, the designed switching sequence is changed to [FFF]-[OOO]-[PON]-[PNO]-[OOO]-[FFF]-[OOO]-[PNO]-[PON]-[OOO]-[FFF], where [FFF] is ST vector. For more details, ST insertion and control signals of intermediate network switches (S_1 and S_2) are presented in Figure 4. The ST control signal of impedance network, which has the phase shift of 90° compared to ST of the inverter leg, is generated to ensure the advantage in reducing the inductor current ripple. Moreover, by adding more duty cycle for switching S_1 and S_2 (which is represented by $dT_S/2$ in Figure 4), the voltage gain of the converter is further enhanced.

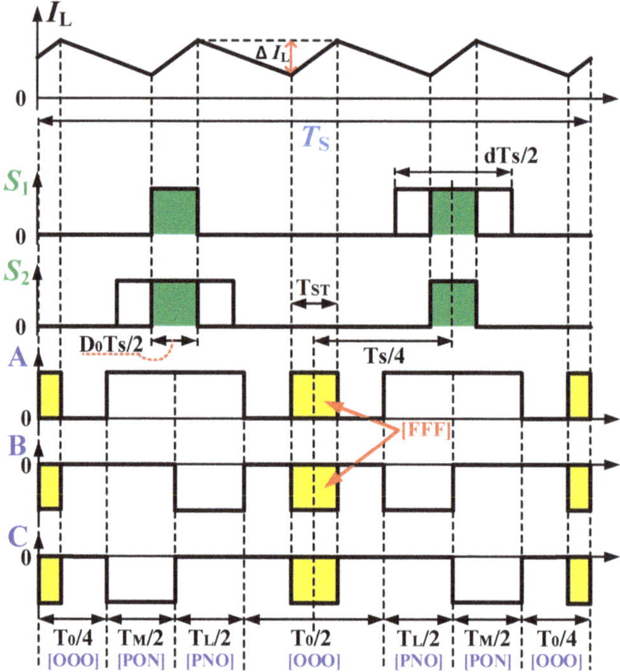

Figure 4. Switching sequence in sector I and control signals of S_1 and S_2.

Similar to the sector I, the switching sequence of other sectors and ST insertion can be achieved in the same way, as presented in Table 3.

Table 3. Switching sequence and ST insertion of the proposed method.

Sector	Switching Sequence
I	[FFF]-[OOO]-[PON]-[PNO]-[OOO]-[FFF]-[OOO]-[PNO]-[PON]-[OOO]-[FFF]
II	[FFF]-[OOO]-[OPN]-[PON]-[OOO]-[FFF]-[OOO]-[PON]-[OPN]-[OOO]-[FFF]
III	[FFF]-[OOO]-[NPO]-[OPN]-[OOO]-[FFF]-[OOO]-[OPN]-[NPO]-[OOO]-[FFF]
IV	[FFF]-[OOO]-[NOP]-[NPO]-[OOO]-[FFF]-[OOO]-[NPO]-[NOP]-[OOO]-[FFF]
V	[FFF]-[OOO]-[ONP]-[NOP]-[OOO]-[FFF]-[OOO]-[NOP]-[ONP]-[OOO]-[FFF]
VI	[FFF]-[OOO]-[PNO]-[ONP]-[OOO]-[FFF]-[OOO]-[ONP]-[PNO]-[OOO]-[FFF]

4. Simulation and Experimental Results

The effectiveness of the proposed SVPWM is verified with the help of PSIM and a practical prototype. The circuit parameters used in the simulation and the experiment are shown in Table 4.

Three PWM techniques are considered to use in the simulation and the experiment to validate the performance of this research: method 1—traditional phase disposition (PD) sinusoidal PWM scheme, method 2—traditional phase shift (PS) sinusoidal PWM scheme discussed in [30,31], method 3proposed scheme. The results of method 1 are achieved by applying the ST insertion method, which is used in UST and LST insertion explored in [29,30], to conventional PD sinusoidal PWM strategy.

When using modulation index as 0.8 and ST duty ratio as 0.2, as illustrated in Table 4, the *d* ratio is calculated as 0.63 to achieve 110 V_{RMS} at output phase voltage from 150 V at the input power supply.

Table 4. Simulation and experiment parameters.

Parameter/Components		Values
DC input voltage	V_g	150 V
Output voltage	$V_{o,RMS}$	110 V_{RMS}
Output frequency	f_o	50 Hz
Carrier frequency	f_s	5 kHz
ST duty cycle	D_0	0.2
Modulation index	m	0.8
Boost inductor	L_B	3 mH/20 A
Capacitors	$C_1 = C_2$	2200 µF/600 V
LC filter	L_f and C_f	3 mH and 10 µF
Resistor load	R	40 Ω

Figure 5 presents the simulation results about input voltage (V_g), capacitor voltages (V_{C1} and V_{C2}), and inductor current (I_L) of the proposed method. The simulation is conducted with the parameters listed in Table 4. As a result, the capacitor voltage is boosted to 194.8 V in theory, and the simulation measures 196 V and 193 V for C_1 and C_2, respectively. The peak value of the DC-link voltage is the sum of two voltages across capacitors. Therefore, it is 389 V in simulation. The average input current (also inductor current) in simulation is 6.1 A, as shown in Figure 5. In one period of output load voltage, the maximum and minimum values of the inductor current are 7 A and 5.2 A, respectively.

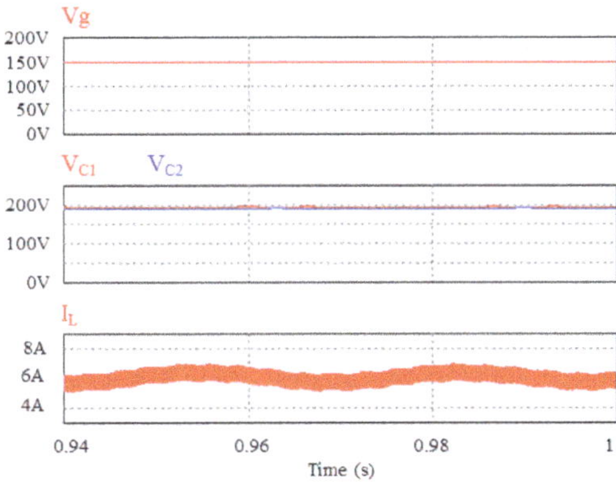

Figure 5. Simulation results of DC input voltage (V_g), capacitor voltages (V_{C1} and V_{C2}), and inductor current (I_L) of the proposed method.

Figure 6 illustrates the simulation results of the DC-link voltage, phase voltage (V_{AG}), and CMV in three PWM techniques mentioned above. Among these techniques, the peak-value of the DC-link voltage is similar, which is 389 V. Because of using UST and LST, the DC-link voltage of method 1 varies from $V_{DC-link}/2$ to $V_{DC-link}$, while the DC-link voltage of method 2 and the proposed method has two values, which are 0V and $V_{DC-link}$, since they use full ST to boost the DC power source. As shown in Figure 6, the THD value of the phase voltage is increasing from method 1 to the proposed method. The simulation produces results of 42.07%, 67.32%, and 77.08% for THD values of method 1, method 2, and the proposed method, respectively, while the CMV value of method 1 has the maximum magnitude, which is $2V_C/3$ and equals to 130 V. Method 2 has the medium magnitude of CMV, which varies from $+V_C/3$ (65 V) to $-V_C/3$ (–65 V). When the proposed method is adopted, the CMV generated by qSBT^{2}I is eliminated. The RMS values of CMV for the three methods are 71 V_{RMS}, 41.6 V_{RMS}, and 0

V_{RMS}, respectively. Since CMV is eliminated, the output phase voltage is similar to the pole voltage, which just has three-level $+V_C$, 0, and $-V_C$, as shown in Figure 6c.

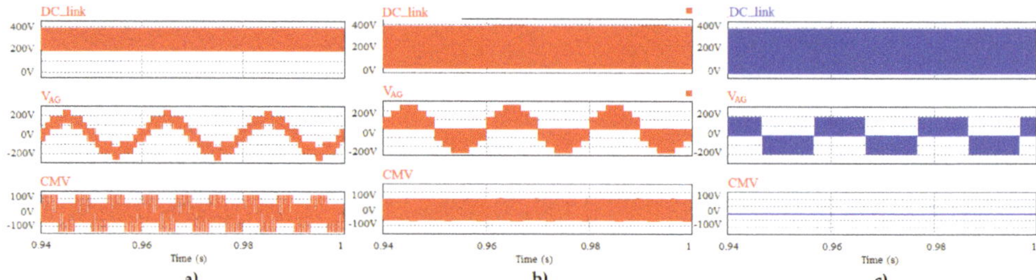

Figure 6. Simulation results of the DC-link voltage, phase voltage (V_{AG}), and CMV. (**a**) Method 1, (**b**) method 2, (**c**) proposed method.

Figure 7 shows the simulation results of output line–line voltage (V_{AB}), output load voltage (V_{RA}), and output load current I_A. When the proposed method is used to minimize the effect of CMV, the THD of the output voltage is increased, consequently. In simulation, the THD value of output line–line voltage (V_{AB}) for three methods are 42%, 67.3%, and 77.1%, respectively. The magnitude of line–line voltage is equal to peak DC-link voltage, the variable of V_{AB} varies from 0 to 390 V at the top part, as illustrated in Figure 7. Since this topology uses low pass filter (LC filter) at output, the THD value of output load current is very low, and they are 0.7%, 0.51%, and 1.28% for method 1, method 2, and the proposed method, respectively. The RMS value of output load voltage and output load current for three strategies are similar, which are 110 V_{RMS} and 2.77 A_{RMS}.

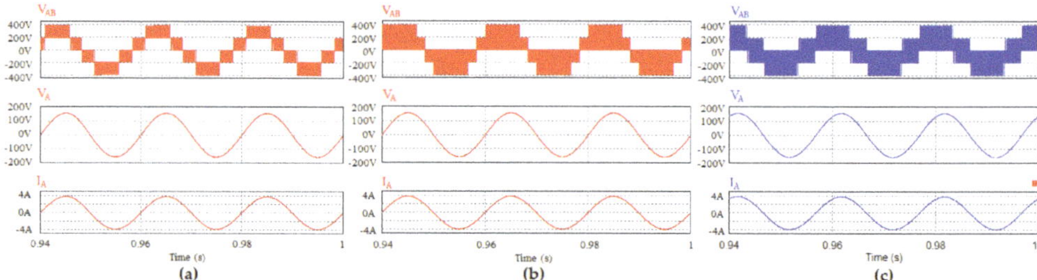

Figure 7. Simulation results of output line–line voltage (V_{AB}), load voltage (V_A), and output phase current (I_A). (**a**) Method 1, (**b**) method 2, (**c**) proposed method.

The THD value of output phase voltage (V_{AG}) and RMS value of the CMV investigated under difference modulation indexes are performed in the simulation. The value of the modulation index (m) varies from 0.25 to 0.8, whereas the ST duty ratio is kept constant. The results are shown in Figures 8 and 9, respectively. As illustrated in Figure 8, by increasing the value of the modulation index, the quality of output phase voltage is improved, which is illustrated through the decrease of the THD value. Figure 8 presents that method 1 is superior in THD, because of using small vectors [34], which generates $\pm 2V_C/3$ of CMV, while the other two techniques have higher value of THD, and the difference is not too large. Figure 9 shows the results of investigating the RMS value of CMV. Method 1 has the maximum value of CMV, which achieves the maximum value of RMS CMV at 0.6 of the modulation index. The RMS CMV value generated by the proposed method stays at zero when the value of modulation index changes, whereas in method 2, it is increased with the increase of modulation index.

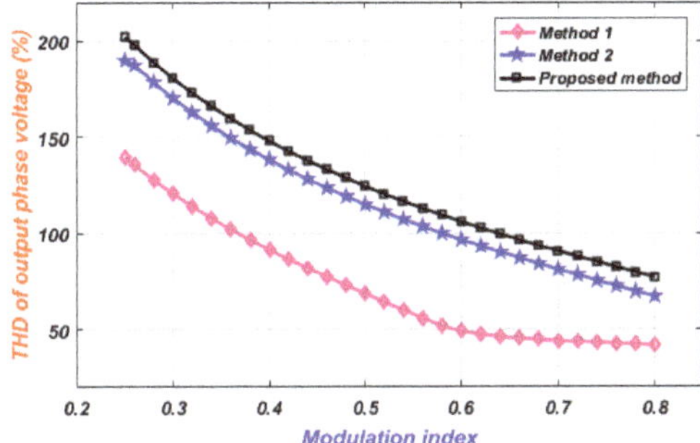

Figure 8. Total harmonic distortion (THD) of output phase voltage (V_{AG}) under different modulation indexes.

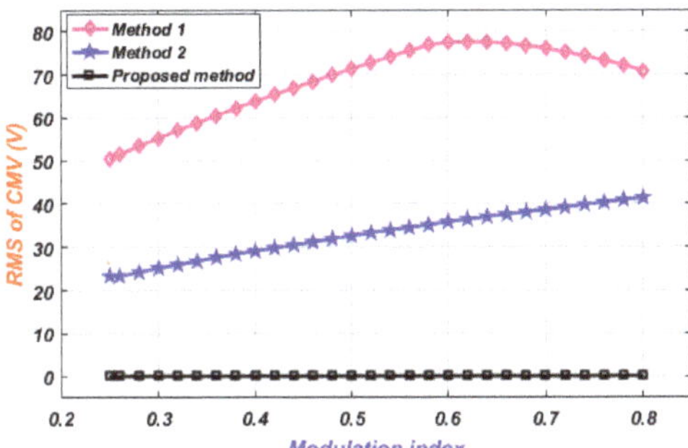

Figure 9. CMV RMS value under different modulation indexes.

To demonstrate the performance of the proposed SVM method, a 1kW 3L qSBT^{2}I prototype was created based on DSP TMSF28335. Figure 10 shows a photo of the hardware setup. Parameters of devices used in the experiment are listed in Table 4. Diodes and IGBTs used in the experiment are DSE160-12A and FGL40N150D, respectively. The modulation index, ST duty ratio, and d are selected similar to the simulation. With the 150 V of the DC input voltage, the capacitor voltages are boosted to 187 V and 190 V corresponding to voltages across C_1 and across C_2. They are maintained during the operation of the system. The average value of inductor current measured in the experiment is 6.02 A, as shown in Figure 11.

Figure 10. A photo of the hardware setup.

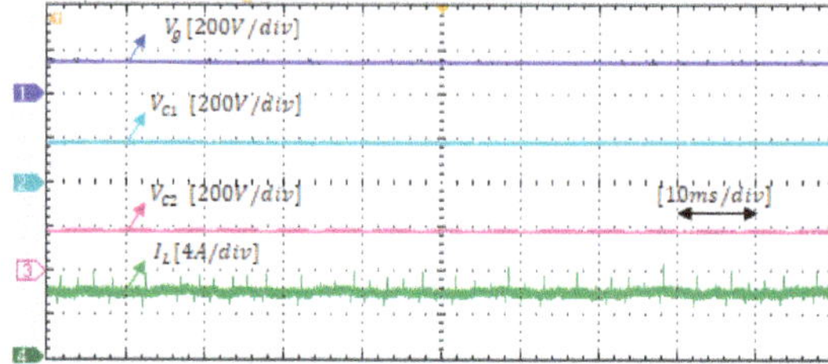

Figure 11. Experimental results of the DC input voltage (V_g), capacitor voltages (V_{C1} and V_{C2}), and inductor current (I_L) of the proposed method.

Figure 12 shows the experimental waveform results of the DC-link voltage, phase voltage (V_{AG}), and CMV of three methods. For method 1, presented in Figure 12a, the variation of the DC-link voltage is from 190 to 380 V, approximately, because of using UST and LST to ensure the boost capability of the qSB network, while the DC-link voltage of the other two methods varies from 0 to 380 V by using FST insertion in order to boost the DC input voltage. Since CMV of the proposed method is eliminated, the output phase voltage has just three levels: −190 V, 0 V, and +190 V, approximately, as presented in Figure 12c. Method 1 and method 2 have a higher number of level at output phase voltage, as illustrated in Figure 12a,b, so the quality of their output voltage is improved compared to the proposed method. However, among the three techniques, the RMS CMV of the proposed method measured in the experiment is 5.73 V_{RMS}, while it is 67.8 V_{RMS} and 40.2 V_{RMS} in method 1 and method 2, respectively. As a result, the CMV of the proposed method is reduced to 91.5% and 85.7% compared to method 1 and method 2, respectively.

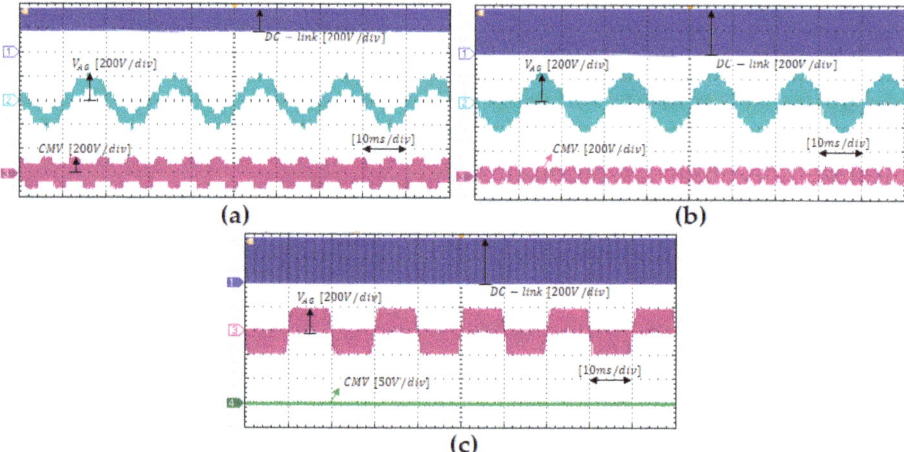

Figure 12. Experimental results of the DC-link voltage, phase voltage (V_{AG}), and CMV. (**a**) Method 1, (**b**) method 2, (**c**) proposed method.

Figure 13 presents the experimental results of the output line–line voltage (V_{AB}), load voltage (V_{RA}), and output phase current (I_A) of three methods. Similar to the output phase voltage mentioned above, the output line–line voltage generated by the proposed method has the lowest quality because of just using medium vectors and zero vector to produce output voltage. The maximum magnitude of the output line–line voltage is the same as the magnitude of the DC-link, which is 380 V, approximately. By using the low pass filter at output, the amplitude of high frequency harmonic is reduced significantly. As a result, the output load voltage and output load current have less THD value, as shown in Figure 13. The RMS value of output load voltage and current are the same when three methods are adopted, which are 105 V_{RMS}, 104 V_{RMS}, and 104 V_{RMS} corresponding to method 1, 2, and the proposed method, respectively. The load currents are 2.6 A_{RMS}, 2.58 A_{RMS}, and 2.55 A_{RMS} for the three methods.

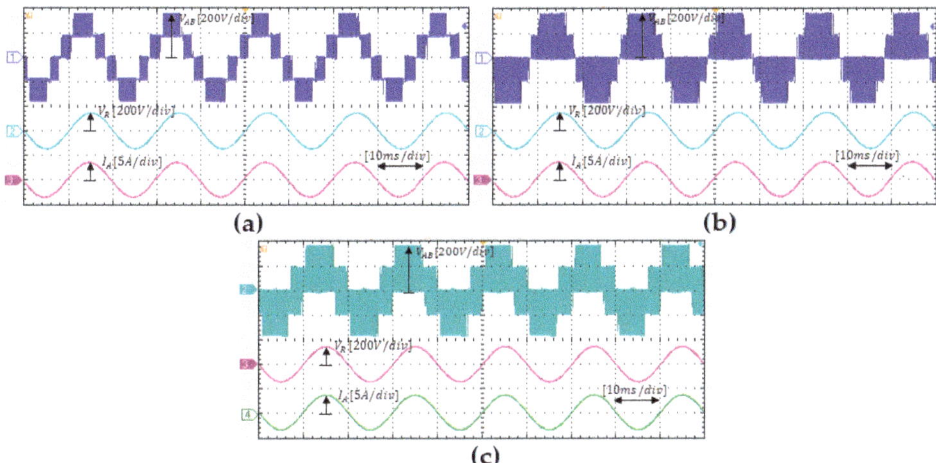

Figure 13. Experimental results of output line–line voltage (V_{AB}), load voltage (V_{RA}), and output phase current (I_A). (**a**) Method 1, (**b**) method 2, (**c**) proposed method.

In Figures 14 and 15, the FFT analyses of output phase voltage (V_{AG}) and output load current (I_A) are conducted accordingly for the three methods. Considering harmonics spectrums of V_{AG} and I_A, the magnitudes of first order harmonic of output load voltage and the output load current are the same for three methods, which are 104.5 V and 2.55 A. The THD values of V_{AG} and I_A are calculated by using results of harmonics spectrums, as shown in Table 5. Among the three methods, the THD value of the proposed method is maximum, which is 89.18%, while it is 57.62% and 81.25% in method 1 and method 2, respectively. Because of appearing three-phase low pass filter at output, the THD value of load current is quite smaller than output phase voltage. They are 2.62%, 2.34%, and 3.3% corresponding to method 1, method 2, and proposed method.

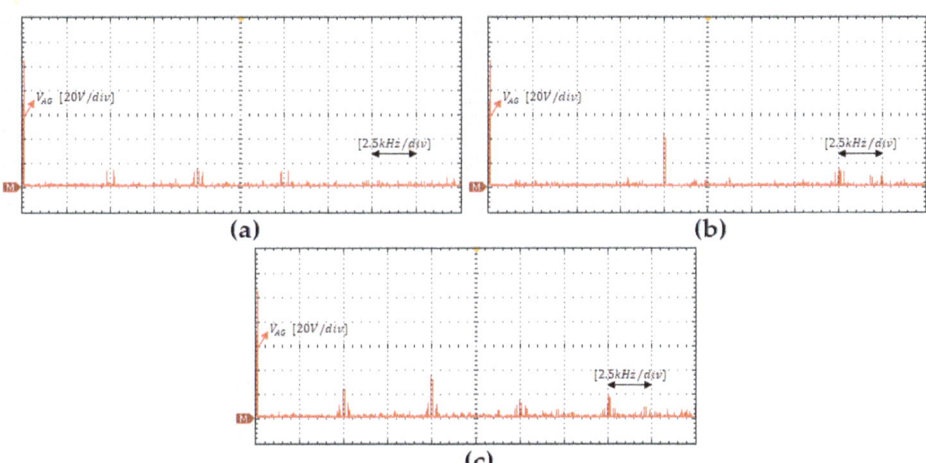

Figure 14. Experimental results of THD analysis of output phase voltage (V_{AG}). (**a**) Method 1, (**b**) method 2, (**c**) proposed method.

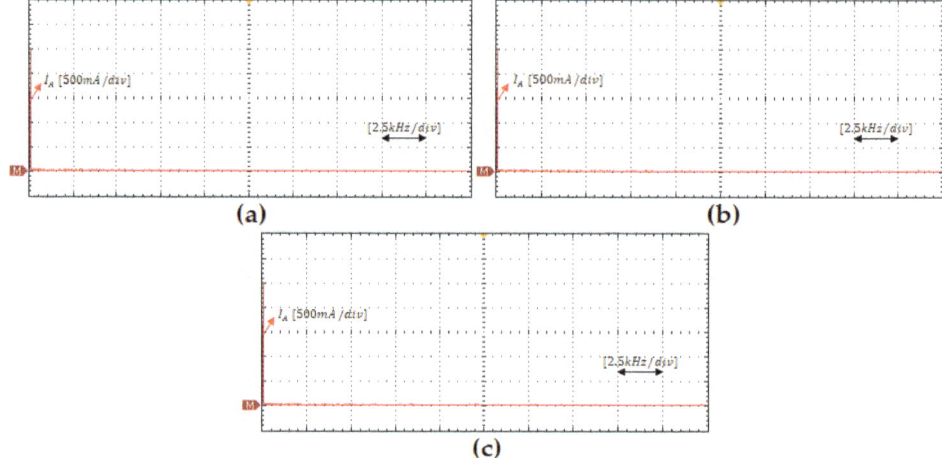

Figure 15. Experimental results of THD analysis of output load current (I_A). (**a**) Method 1, (**b**) method 2, (**c**) proposed method.

Table 5. THD analyses of output phase voltage (THD$_V$) and output load current (THD$_I$).

	Method 1	Method 2	Proposed Method
THD$_V$	57.62%	81.25%	89.18%
THD$_I$	2.62%	2.34%	3.3%

5. Conclusion

This paper has proposed the SVM technique to eliminate the CMV for the 3L qSBT^{2}I. In this scheme, the zero and medium vectors are adopted to synthesize the reference vector. The drew-time calculation of these vectors is also presented in detail. Switching sequence is selected to ensure the symmetry of output voltage and the ST vector is inserted into zero vector to ensure the boost capability, as well as not causing distortion at the output voltage. In addition, the control signals of switches S_1 and S_2 were presented to guarantee the advantages of the qSB network compared to ZS and qZS networks, such as high voltage gain, continuous input current, and low inductor current ripple. To validate the performance of the proposed method, the simulation and the experiment were conducted in PSIM simulation and laboratory. The comparison between the proposed method and other conventional schemes was carried out to confirm the effectiveness of the proposed technique. The results show that the proposed SVM method successfully eliminates the CMV with a little bit increment in THD of the load current. However, this THD value is still 5% smaller than the standard IEC61000-4-30 Edition 2 Class A [35].

Author Contributions: This paper was a collaborative effort among all authors. All authors conceived the methodology, conducted the performance tests and wrote the paper. All authors have read and agreed to the published version of the manuscript.

Funding: This research was funded by CT.2019.04.03 project.

Acknowledgments: This work was supported by the Advanced Power Electronics Laboratory, D405 at Ho Chi Minh City University of Technology and Education, Viet Nam.

Conflicts of Interest: The authors declare no conflict of interest.

References

1. Huang, Q.; Huang, A.Q.; Yu, R.; Liu, P.; Yu, W. High-Efficiency and High-Density Single-Phase Dual-Mode Cascaded Buck-Boost Multilevel Transformerless PV Inverter with GaN AC Switches. *IEEE Trans. Power Electron.* **2019**, *34*, 7474–7488. [CrossRef]
2. Elthokaby, Y.; Elshafei, L.; Abdel-Rahim, N.; Abdel-Rahim, E.S. Finite-Control Set Model-Predictive Control for SinglePhase Voltage-Source UPS Inverters. In Proceedings of the 2016 Eighteenth International Middle East Power Systems Conference (MEPCON), Cairo, Egypt, 27–29 December 2016.
3. Pena-Alzola, R.; Roldan-Perez, J.; Bueno, E.; Huerta, F.; Campos-Gaona, D.; Liserre, M.; Burt, G. Robust Active Damping in LCL-filter based Medium-Voltage Parallel Grid-Inverters for Wind Turbines. *IEEE Trans. Power Electron.* **2018**, *33*, 10846–10857. [CrossRef]
4. Yadav, A.K.; Gopakumar, K.; Raj, R.K.; Umanand, L.; Matsuse, K.; Kubota, H. Instantaneous Balancing of Neutral Point Voltages for Stacked DC-link Capacitors of Multilevel Inverter for Dual Inverter fed Induction Motor Drives. *IEEE Trans. Power Electron.* **2019**, *34*, 2505–2514. [CrossRef]
5. Xu, S.; Zhang, J.; Hang, J. Investigation of a Fault-tolerant Three-level T-type Inverter System. *IEEE Trans. Ind. Appl.* **2017**, *53*, 4613–4623. [CrossRef]
6. Hu, J.; Lin, J.; Chen, H. A Discontinuous Space Vector PWM Algorithm in abc Reference Frame for Multilevel Three-Phase Cascaded H-Bridge Voltage Source Inverters. *IEEE Trans. Ind. Electron.* **2017**, *64*, 8406–8414. [CrossRef]
7. Guo, X.; Wei, B.; Zhu, T.; Lu, Z.; Tan, L.; Sun, X.; Zhang, C. Leakage Current Suppression of Three-Phase Flying Capacitor PV Inverter with New Carrier Modulation and Logic Function. *IEEE Trans. Power Electron.* **2018**, *33*, 2127–2135. [CrossRef]
8. Schweizer, M.; Kolar, J.W. Design and implementation of a highly efficient three-level T-type converter for low-voltage applications. *IEEE Trans. Power Electron.* **2013**, *28*, 899–907. [CrossRef]
9. Nguyen, M.K.; Tran, T.T. A Single-Phase Single-Stage Switched-Boost Inverter with Four Switches. *IEEE Trans. Power Electron.* **2018**, *33*, 6769–6781. [CrossRef]
10. Jagan, V.; Kotturu, J.; Das, S. Enhanced-Boost Quasi-z-Source Inverters with Two Switched Impedance Network. *IEEE Trans. Ind. Electron.* **2017**, *64*, 6885–6897. [CrossRef]
11. Fang, X.; Ma, B.; Gao, G.; Gao, L. Three phase trans-Quasi-z-source inverter. *CPSS Trans. Power Electron. Appl.* **2018**, *3*, 223–231. [CrossRef]
12. Li, T.; Cheng, Q. A comparative study of Z-source inverter and enhanced topologies. *CES Trans. Electr. Mach. Syst.* **2018**, *2*, 284–288. [CrossRef]
13. Ahmed, H.F.; Cha, H.; Kim, S.; Kim, H. Switched-Coupled-Inductor Quasi-z-Source Inverter. *IEEE Trans. Power Electron.* **2016**, *31*, 1241–1254. [CrossRef]
14. Pires, V.F.; Cordeiro, A.; Foioto, D.; Martins, J.F. Quasi-z-Source Inverter with a T-Type Converter in Normal and Failure Mode. *IEEE Trans. Power Electron.* **2016**, *31*, 7462–7470. [CrossRef]
15. Zhu, X.; Zhang, B.; Qiu, D. A New Nonisolated Quasi-z-Source Inverter with High Voltage Gain. *IEEE J. Emerg. Sel. Top. Power Electron.* **2019**, *7*, 2012–2028. [CrossRef]
16. Peng, F.Z. Z-source inverter. *IEEE Trans. Ind. Appl.* **2003**, *39*, 504–510. [CrossRef]
17. Tang, Y.; Xie, S.; Zhang, C.; Xu, Z. Improved Z-Source Inverter with Reduced Z-Source Capacitor Voltage Stress and Soft-Start Capability. *IEEE Trans. Power Electron.* **2009**, *24*, 409–415. [CrossRef]
18. Loh, P.C.; Gao, F.; Blaabjerg, F. Topological and Modulation Design of Three-Level Z-Source Inverters. *IEEE Trans. Power Electron.* **2008**, *23*, 2268–2277. [CrossRef]
19. Xing, X.; Zhang, C.; Chen, A.; He, J.; Wang, W.; Du, C. Space-Vector-Modulated Method for Boosting and Neutral Voltage Balancing in Z-Source Three-Level T-Type Inverter. *IEEE Trans. Ind. Appl.* **2016**, *52*, 1621–1631. [CrossRef]
20. Yang, S.; Peng, F.Z.; Lei, Q.; Inoshita, R.; Qian, Z. Current-Fed Quasi-z-Source Inverter with Voltage Buck–Boost and Regeneration Capability. *IEEE Trans. Ind. Appl.* **2011**, *47*, 882–892. [CrossRef]
21. Anderson, J.; Peng, F.Z. Four quasi-z-Source inverters. In Proceedings of the 2008 IEEE Power Electronics Specialists Conference, Rhodes, Greece, 15–19 June 2008.
22. Gu, Y.; Chen, Y.; Zhang, B. Enhanced-Boost Quasi-z-Source Inverter with An Active Switched Z-Network. *IEEE Trans. Ind. Electron.* **2018**, *65*, 8372–8381. [CrossRef]

23. Nguyen, M.K.; Le, T.V.; Park, S.J.; Lim, Y.C. A Class of Quasi-Switched Boost Inverters. *IEEE Trans. Ind. Electron.* **2015**, *62*, 1526–1536. [CrossRef]

24. Nguyen, M.K.; Lim, Y.C.; Park, S.J. A Comparison Between Single-Phase Quasi-z-Source and Quasi-Switched Boost Inverters. *IEEE Trans. Ind. Electron.* **2015**, *62*, 6336–6344. [CrossRef]

25. Sahoo, M.; Keerthipati, S. A Three Level LC-Switching Based Voltage Boost NPC Inverter. *IEEE Trans. Ind. Electron.* **2017**, *64*, 2876–2883. [CrossRef]

26. Nguyen, M.K.; Tran, T.T.; Zare, F. An active impedance-source three-level T-type inverter with reduced device count. *IEEE J. Emerg. Sel. Top. Power Electron.* **2019**, in press. [CrossRef]

27. Do, D.T.; Nguyen, M.K. Three-Level Quasi-Switched Boost T-Type Inverter: Analysis, PWM Control, and Verification. *IEEE Trans. Ind. Electron.* **2018**, *65*, 8320–8329. [CrossRef]

28. Do, D.T.; Nguyen, M.K.; Quach, T.H.; Tran, V.T.; Blaabjerg, F.; Vilathgamuwa, M. A PWM Scheme for a Fault-Tolerant Three-Level Quasi-Switched Boost T-Type Inverter. *IEEE J. Emerg. Sel. Top. Power Electron.* **2019**, *1*. [CrossRef]

29. Do, D.T.; Nguyen, M.K.; Quach, T.H.; Tran, V.T.; Le, C.B.; Lee, K.W.; Cho, G.B. Space Vector Modulation Strategy for Three-Level Quasi-Switched Boost T-Type Inverter. In Proceedings of the 2018 IEEE 4th Southern Power Electronics Conference (SPEC), Singapore, 10–13 December 2018.

30. Tran, V.T.; Do, D.T.; Nguyen, M.K.; Nguyen, D.T. Space Vector Modulation Scheme for Three-Level T-Type Quasi-Switched Boost Inverter to Reduce Common Mode Voltage. In Proceedings of the 2019 10th International Conference on Power Electronics and ECCE Asia (ICPE 2019—ECCE Asia), Busan, Korea, 27–30 May 2019.

31. Takahashi, S.; Ogasawara, S.; Takemoto, M.; Orikawa, K.; Tamate, M. Common-Mode Voltage Attenuation of an Active Common-Mode Filter in a Motor Drive System Fed by a PWM Inverter. *IEEE Trans. Ind. Appl.* **2019**, *55*, 2721–2730. [CrossRef]

32. Lee, J.S.; Lee, K.B. New Modulation Techniques for a Leakage Current Reduction and a Neutral-Point Voltage Balance in Transformerless Photovoltaic Systems using a Three-Level Inverter. *IEEE Trans. Power Electron.* **2014**, *29*, 1720–1732. [CrossRef]

33. Qin, C.; Zhang, C.; Chen, A.; Xing, X.; Zhang, G. A Space Vector Modulation Scheme of the Quasi-z-Source Three-Level T-Type Inverter for Common-Mode Voltage Reduction. *IEEE Trans. Ind. Electron.* **2018**, *65*, 8340–8350. [CrossRef]

34. Nguyen, V.N.; Nguyen, T.K.T.; Lee, H.H. A Reduced Switching Loss PWM Strategy to Eliminate Common-Mode Voltage in Multilevel Inverters. *IEEE Trans. Power Electron.* **2015**, *30*, 5425–5438. [CrossRef]

35. IEC 61000-4-30: 2015. *Testing and Measuring Techniques—Power Quality Measurement Methods*; IEC: Geneva, Switzerland, 2015.

© 2020 by the authors. Licensee MDPI, Basel, Switzerland. This article is an open access article distributed under the terms and conditions of the Creative Commons Attribution (CC BY) license (http://creativecommons.org/licenses/by/4.0/).

 *electronics*

Article

Modulation Strategy with a Minimal Number of Commutations for a Five-Level H-Bridge NPC Inverter

Florent Becker [1], Ehsan Jamshidpour [2], Philippe Poure [3,*] and Shahrokh Saadate [4]

[1] Naval Academy Research Institute, Ecole Navale CC-600, F-29240 Brest CEDEX 9, France; florent.becker@ecole-navale.fr
[2] ECAM Strasbourg Europe—ICube laboratory (UMR7357), F-67400 Illkirch-Graffenstaden, France; ehsan.jamshidpour@ecam-strasbourg.eu
[3] Institut Jean Lamour (UMR7198), Université de Lorraine, 54000 Nancy, France
[4] GREEN Laboratory, Université de Lorraine, 54000 Nancy, France; shahrokh.saadate@univ-lorraine.fr
*   Correspondence: philippe.Poure@univ-lorraine.fr; Tel.: +33-(0)6-6087-8917

Received: 13 February 2019; Accepted: 18 April 2019; Published: 23 April 2019

Abstract: In this paper, a so-called OPTimized Pulse Width Modulation (OPT-PWM) strategy with a minimal number of commutations for a multilevel converter (MC) is proposed. The principle is based on the reduction of the number of switch commutations by removing the unnecessary ones for each voltage level transition. The OPT-PWM strategy is applied to a five-level H-Bridge Neutral Point Clamped (HB-5L-NPC) inverter. A specific block based on a state machine is added to conventional modulation techniques to perform the transitions from a given voltage level to another one via the best trajectory with a minimal number of commutations. The principle of this additional block can be applied to any modulation technique. In this paper, the proposed strategy is validated first by simulation and then through experimental tests.

Keywords: multilevel inverter; Pulse Width Modulation; minimal number of commutations; state machine; Neutral Point Clamped Converter

---

1. Introduction

In recent decades, the use of power converters in high-power and medium-voltage (MV) industrial applications such as the integration of renewable resources, distributed power generation systems, microgrids, energy storage systems, and motor drives has significantly increased. There are two major families for voltage source inverters (VSI): two-level and multilevel inverters (MLI) where the prefix "multi" means any integer greater than two. In MV applications, with respect to device rating limits, MLI has attracted increasing attention in the last decade. MLI provides significant advantages for these applications. Among these advantages, the THD (Total Harmonic Distortion) improvement in output signals at low switching frequencies is most discussed in the literature. Due to this improvement, the size of the output passive filters can be reduced. Moreover, a lower switching frequency allows performance with higher efficiency [1].

In the literature, four major multilevel inverters are commonly studied [2]: Neutral Point Clamped (NPC) [3–10], Flying Capacitor (FC) [11,12], Cascaded H-Bridge (CHB) [13,14], and Modular Multilevel Converter (MMC) [15,16]. They have major advantages compared to two-level inverters, mainly because they operate at lower switching frequencies. More, they provide a higher number of levels; thus, the staircase waveform is increased. In a recent review paper [17], these attractive features of MLI are highlighted and discussed.

In addition, modulation techniques are very often discussed in the literature [1,8,15]; they aim to perform harmonic reduction of the system for the same converter output power but at lower switching frequency [1,17]. The originality of this paper is to propose an additional advanced functionality for the modulation strategy that evaluates possible switching trajectories in multilevel inverters and implements the trajectory with a minimal number of commutations. The proposed advanced functionality can also be applied to any modulation strategy and provides a significant improvement compared with modulation approaches published in the literature.

In this paper, a five-level H-bridge NPC (HB-5L-NPC) inverter controlled by an OPTimized Pulse Width Modulation (OPT-PWM) strategy is proposed. The goal of the proposed OPT-PWM control is to perform the minimal number of commutations when the conventional Level Shift-PWM (LS-PWM) control is used. This means that this control method produces the same output voltage as the conventional LS-PWM but with a reduced number of switch commutations.

Applying the proposed OPT-PWM on MMCs can be useful to reduce switching losses and consequently improve the efficiency and the lifetime of the converters. In multilevel converters, even at low frequencies, the switching power losses of the applied IGBT (Insulated Gate Bipolar Transistor) devices are considerable. As an example, they can amount to roughly one-third of the total losses if 4.5 kV IGBT components are used [18]. Thus, there is a permanent requirement for further improvement of efficiency by power loss reduction, especially in the high-power range applications where energy cost and cooling equipment are mainly concerned [18]. A few works have studied some modulation strategies to minimize the number of commutation processes in matrix converters [19,20]. In reference [21], an advanced model predictive control technique was proposed for a back-to-back NPC converter; this control was applied to wind energy conversion systems (WECS). By applying the proposed predictive control, the authors considered the redundant switching states of a converter to reduce the number of commutations. The presented results show that the active power delivered to the grid was increased where the extracted power of the generator was constant [21]. This means that the reduction in the number of commutations decreases the power loss in the converter. In the same spirit, this paper presents a so-called OPT-PWM modulation strategy that removes unnecessary switching to reduce the number of commutations per switching cycles with respect to the classical modulation strategy.

According to surveys [22,23], one of the semiconductor aging processes is the accumulation of electrical switching cycles (ΔV and Δi cycles) that cause metallurgical defects on the semiconductor die and change the electrical characteristics as well [23]. Therefore, a reduction in the number of commutations can also postpone the aging of the semiconductor.

The paper is organized as follows. The next section presents the proposed HB-5L-NPC converter and all the possible output voltage levels. The proposed OPT-PWM control is detailed in Section 3. Simulation results for the HB-5L-NPC inverter are presented in Section 4, and experimental tests are presented and discussed in Section 6, followed by a general conclusion in the last section.

2. Studied System Description

The system studied in this paper is shown in Figure 1; it is based on a single-phase HB-5L-NPC. The converter supplies an inductive load (R, L series). The proposed HB-5L-NPC consists of two parallel connected 3-level NPC (Figure 2a). Each 3-level NPC consists of four switches and six diodes (Figure 2b). The DC bus consists of two identical capacitors C, connected in series, where the midpoint is denoted 'O'.

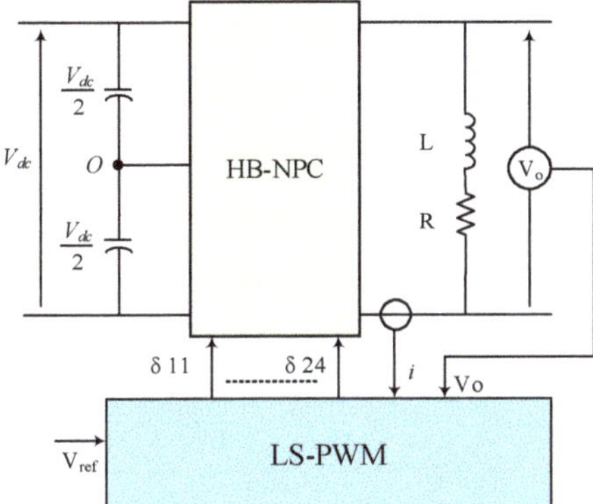

Figure 1. Studied system based on an five-level H-Bridge Neutral Point Clamped (HB-5L-NPC) inverter.

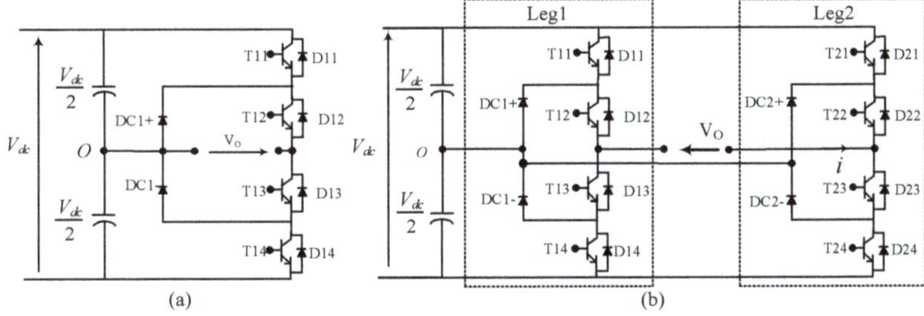

Figure 2. (**a**) Three-level NPC topology (for one phase); (**b**) HB-5L-NPC topology (for one phase).

2.1. Control of the HB-5L-NPC by Classical LS-PWM

To control the converter, a conventional modulation strategy was considered. The block LS-PWM in Figure 1 generates the switching patterns by comparing two triangular carriers with two sinusoidal references. As shown in Figure 3a, the positive triangular carrier signal (Car_Pos) and the sinusoidal reference (Ref_Pos) are used to generate the control signals for the switches T11 and T13 (leg1 for 3-level HB-NPC in Figure 3b). To generate the control signals of the components T12 and T14 (Leg1), the positive sinusoidal reference is compared with the negative triangular signal (Car_Neg). The controls for the four switches of the Leg2 (T21, T22, T23, and T24) are generated by comparing the negative sinusoidal reference (Ref_Neg) with the two carriers. The generated outputs of the LS-PWM block and the control signals are given in Figure 3b. Thus, as can be seen, the output voltage (at the bottom of Figure 3b) behavior has five levels ($-V_{dc}$, $-V_{dc}/2$, 0, $V_{dc}/2$, and V_{dc}).

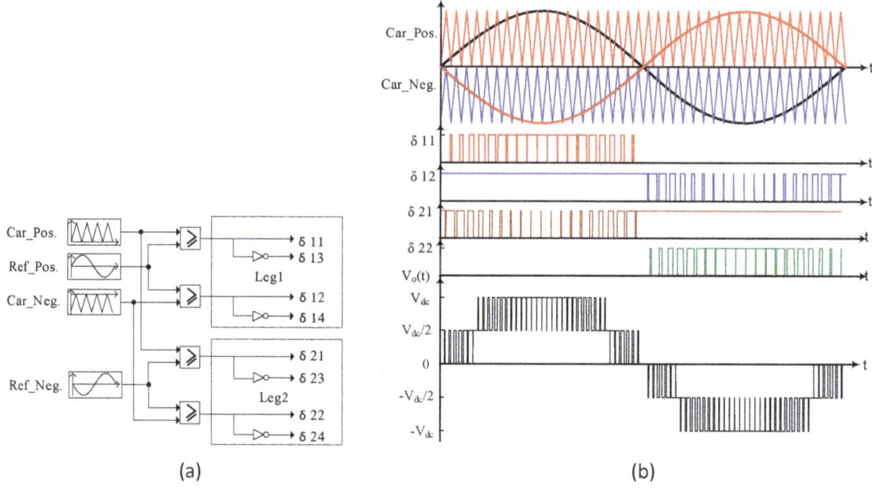

Figure 3. Switching pattern generation by classical Level Shift Pulse Width Modulation (LS-PWM): (**a**) block diagram; (**b**) switching orders and the output voltage $V_o(t)$.

2.2. Simulation Results for the HB-5L-NPC Converter Controlled by LS-PWM

Some simulations in the Matlab/Simulink environment using the SimPowerSystem library were performed to study the system depicted in Figure 2. The simulation results illustrate the possibility of reducing the number of commutations. The system parameters are introduced in Table 1, where f_{pwm} is the PWM switching frequency and m is the modulation index.

The simulation results for the output voltage are shown in Figure 4. As mentioned before, the goal of this paper is to propose an additional advanced functionality for the modulation strategy that provides the minimal number of switch commutations. For this, let us first analyze the output voltage generated by the classical LS-PWM, as given in Figure 4.

Table 1. Simulation parameters.

Symbol	Quantity
V_{dc}	100 V
R	27.7 Ω
L	9 mH
f_{pwm}	1 kHz
m	0.8

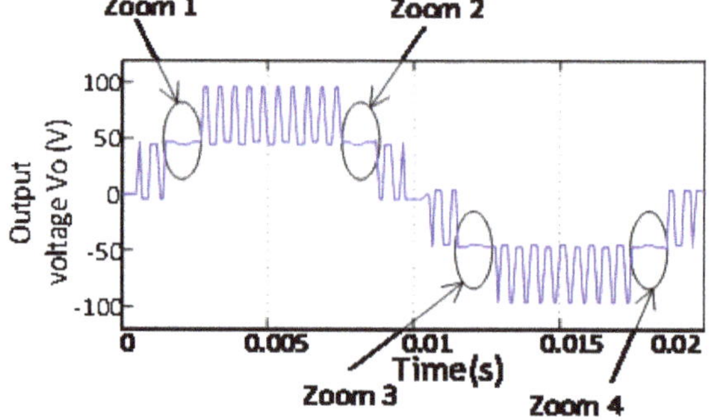

Figure 4. Zooms of the output voltage of the HB-5L-NPC.

2.3. The Switching Number Reduction

As indicated in Figure 4, some small variations in the output voltage $V_o(t)$ at the $V_{dc}/2$ and $-V_{dc}/2$ levels occur. To better understand our approach, these small variations are highlighted in Zooms 1 to 4 in Figure 4.

These small variations are not generated by changes in the level of the output voltage and are due to some switch commutations. It is clear that when the output voltage reference level remains the same (equal to $V_{dc}/2$ or $-V_{dc}/2$), it is not necessary to have any commutations. In order to clarify the origin of these voltage variations, the 18 possible combinations (states) of the eight switches to generate a given output voltage level (V_o) are illustrated in Figure 5. In this figure, a state S_X (with x from 1 to 18) corresponds to the realization of a given voltage level obtained by on-state switches of the converter (in red color).

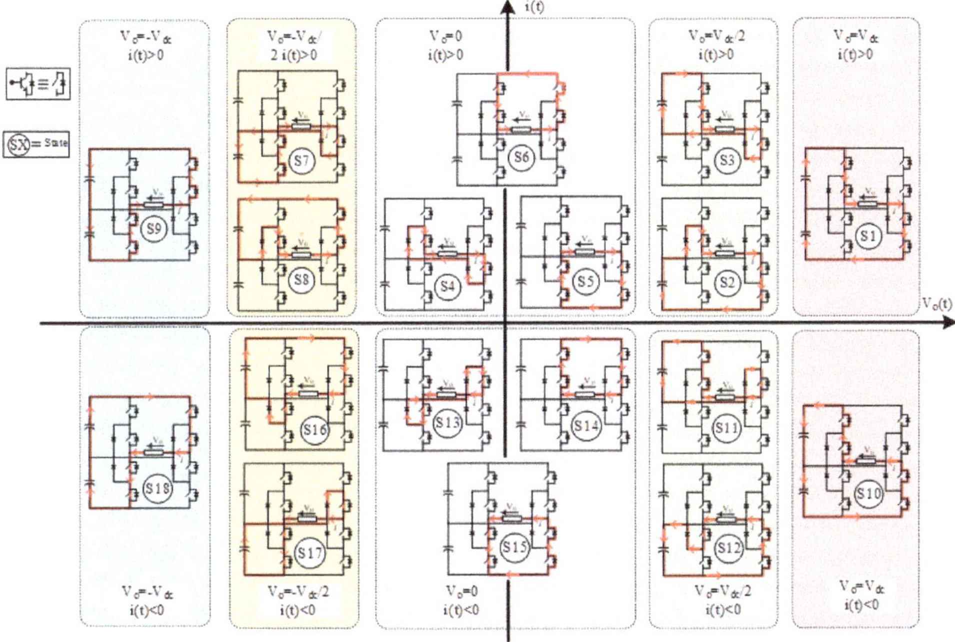

Figure 5. HB-5L-NPC converter configurations to generate the five voltage levels.

In Figure 5, when the requested output voltage level is equal to $V_{dc}/2$ (in the case where i(t) > 0), there are two possibilities (states S2 and S3). Therefore, only T11 and T24 can be commuted when the states change (but V_o remains the same). When the requested voltage level is equal to $-V_{dc}/2$ (with i(t) < 0, states S16 and S17) only T14 and T21 are switching. Thus, only the corresponding switching patterns are analyzed.

In order to understand what happens during these four intervals (Zooms 1 to 4), the simulation results of the output voltage were observed. To confirm the commutation of one switch, not only were the switching patterns (sent by the LS-PWM block to the switch) observed but the currents passing in the switches were also verified. Figures 6–9 give the switching patterns and the associated switch currents during the highlighted phases in Figure 4 (Zooms 1 to 4).

As illustrated in Figures 6–9, the output voltage $V_o(t)$ is not constant and small variations occur while it should be constant and equal to $V_{dc}/2$. The currents and switching patterns confirm that these oscillations are due to unnecessary commutations. For example, in Figure 6, the converter is in state S2 and then passed to state S3 (Figure 5). Both states generate $V_o = V_{dc}/2$; therefore, four unnecessary commutations for the same output voltage level are generated. These commutations inevitably decrease the efficiency of the converter. Therefore, to improve the efficiency of the converter, these unnecessary commutations should be avoided. After analyzing all switching patterns and currents over one period of V_o, it is confirmed that the unnecessary commutations occur only in the considered four zones (detailed before). In the next section, the proposed advanced modulation strategy to eliminate some 16 unnecessary commutations per period is developed.

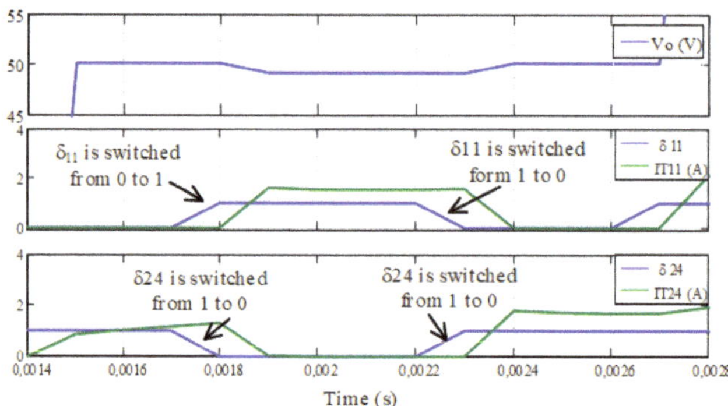

Figure 6. Output voltage $V_o(t)$, currents IT11 and IT24, and command of the switches T11 and T24 (Zoom 1).

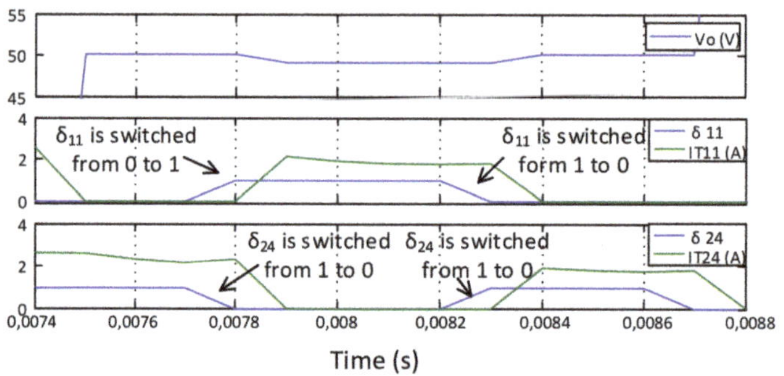

Figure 7. Output voltage $V_o(t)$, currents IT11 and IT24, and command of the switches T11 and T24 (Zoom 2).

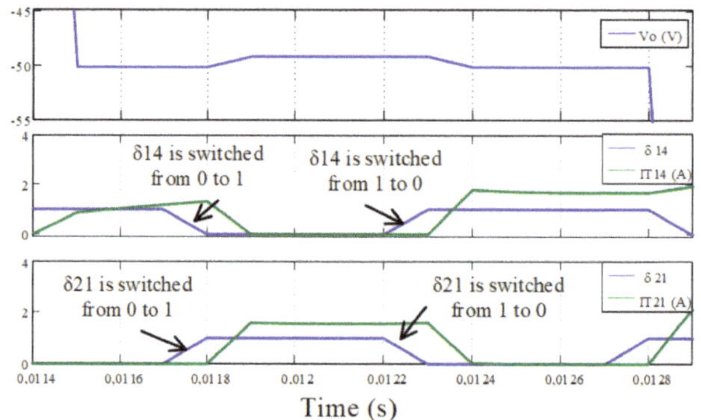

Figure 8. Output voltage $V_o(t)$, currents IT21 and IT14, and command of the switches T21 and T14 (Zoom 3).

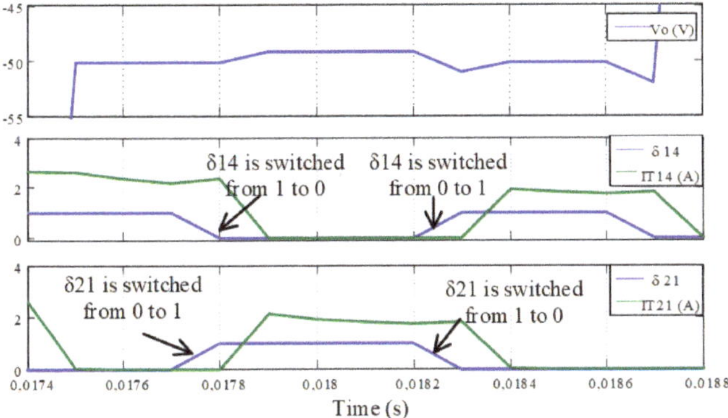

Figure 9. Output voltage $V_o(t)$, currents IT21 and IT14, and command of the switches T21 and T14 (Zoom 4).

3. The Principle of the Proposed OPT-PWM Strategy

As mentioned before, several modulation strategies have been conventionally proposed in the literature to control the output voltage of multilevel inverters [17]. In this paper, to illustrate our contribution, the classical LS-PWM is considered and an additional block is proposed to provide a minimal number of commutations. Nevertheless, any conventional PWM method can be used ([1,8,15] and [24–27]), and the same modification of the modulation strategy we propose can be applied in all cases.

Before discussing the OPT-PWM, it is necessary to define the following terms used in this paper:

- Transition: passing from one state to another one.
- Trajectory: A trajectory is made up of all the transitions that make it possible to pass from the initial output voltage level to the desired voltage level. A trajectory may consist of one or more transitions.
- NOT: number of transitions for a trajectory.
- NOC: number of commutations.
- NOC_{Sx-Sy}: number of commutations to switch from state S_X to state S_Y.
- NOC_{total}: total number of commutations made by the switches during a trajectory.

Figure 10 illustrates the principle of the OPT-PWM proposed in this paper. This method selects a trajectory with a minimal number of commutations after analyzing all possible trajectories to pass from the initial voltage level to the desired voltage level.

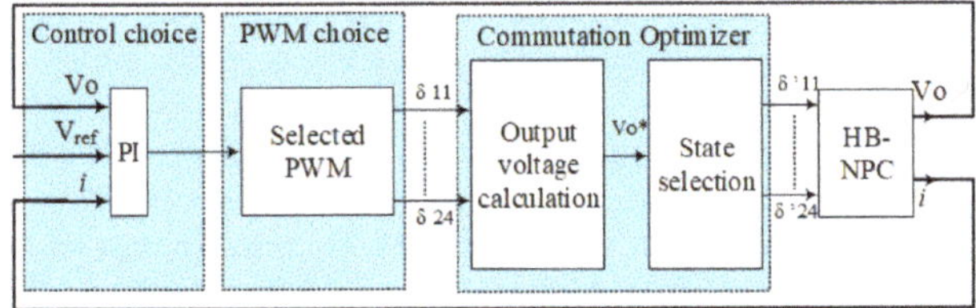

Figure 10. Scheme of the proposed OPT-PWM control strategy.

In the first step, the requested voltage level (V_o^*) will be determined by using the switching patterns generated by the conventional PWM block used in the classical modulation strategy. Then, the state selection algorithm will find the best trajectory and states to obtain the voltage level V_o^* with a minimal number of commutations. This is performed by the "Commutation Optimizer" block given in Figure 10. This optimization only depends on the initial and final voltage levels and consequently does not depend on the used modulation strategy.

3.1. Determination of the Voltage Level V_o*

For the commutation optimizer algorithm to find the best trajectory, the reference voltage level (V_o*) has to be determined. Based on the 18 possible states (Figure 5), suitable control was predefined and is recorded in Table 2. As an example for the state S1, the switches T11, T12, T22, and T24 should be "on" whereas the other switches are "off". Then, the block called "Output Voltage Calculation" compares the generated PWM switching patterns with the predefined table (Table 2) to find the required voltage level V_o*.

Table 2. Possible voltage levels and states for the single-phase HB-5L-NPC topology.

V_o*	$i(t) > 0$		$i(t) < 0$	
	States	**Passing Component**	**States**	**Passing Components**
V_{dc}	S1	T11, T12, T23, T24	S10	D11, D12, D23, D24
$V_{dc}/2$	S2	DC1+, T12, T23, T24	S11	D11, D12, T22, DC2+
	S3	T11, T12, T23, DC2−	S12	DC1−, T13, D23, D24
0	S4	DC1+, T12, T23, DC2−	S13	DC1−, T13, DC2+, T22
	S5	T23, T24, D14, D13	S14	D11, D12, T21, T22
	S6	T11, T12, D22, D21	S15	T13, T14, D23, D24
$-V_{dc}/2$	S7	D13, D14, DC2−, T23	S16	T13, T14, T22, DC2+
	S8	DC1+, T12, D21, D22	S17	DC1−, T13, T21, T22
$-V_{dc}$	S9	D13, D14, D21, D22	S18	T13, T14, T21, T22

3.2. Search for the Trajectory with a Minimal Number of Commutations

Once the output voltage V_o* (Figure 10) is fixed by the classical PWM control, a second block called "State Selection" selects the best state and trajectory with the minimum number of commutations to perform the required voltage level.

To pass from one level of voltage to another, several trajectories are possible. In addition, each trajectory can contain several transitions between different states. Each transition $(S_X \rightarrow S_Y)$ is associated with a given number of commutations $(NOC_{Sx\text{-}Sy})$, directly related to the number of switches changing their status (on $\leftrightarrow$ off) from S_X to S_Y (one voltage level to another). Thus, to achieve the goal of this method, the total number of commutations (NOC_{total}) during each change in the output voltage level has to be minimized. For this, all possible trajectories are observed. It is clear that NOC increases with the number of transitions (NOT). To accelerate the execution of the proposed algorithm, the trajectories with more than two transitions are ignored. Indeed, in all cases, the maximum value of NOT will be equal to 2, which allows selecting the optimal trajectories. After trajectory selection, the NOC of each transition and then NOC_{total} is calculated. Finally, the trajectory with the minimum NOC_{total} is selected.

The optimal trajectory search method is summarized in the flowchart presented in Figure 11. This algorithm is used to determine the optimal trajectory for one change in voltage level. Note that this algorithm (Figure 11) can lead to the choice of one or more trajectories. In the case where more than one trajectory is chosen, the trajectory using the switches which had the lowest number of commutations during the period before will be selected.

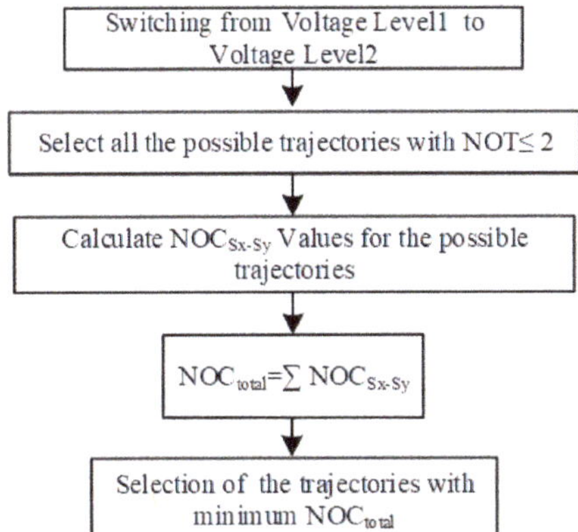

Figure 11. Search for the minimal number of commutations.

3.3. Selection of an Optimized Trajectory out of the Optimal Trajectories

As mentioned before, the "Output Voltage Calculation" block (Figure 10) compares the switching patterns generated by the PWM (δij) to the possible states in Table 2 to find the corresponding state. Once the concerned state is obtained, the value of the next output voltage level (V_o^*) is known. As illustrated in Figure 5, each voltage level may have more than one state. Now, the task of the "State Selection" block is the selection of a trajectory with the minimal NOC$_{total}$ to pass from S_X (actual state) to S_y (the state with V_o^*). Figure 12 illustrates all of the trajectories (based on Table 2 with i(t) > 0) that could be used to pass from S_X to S_Y.

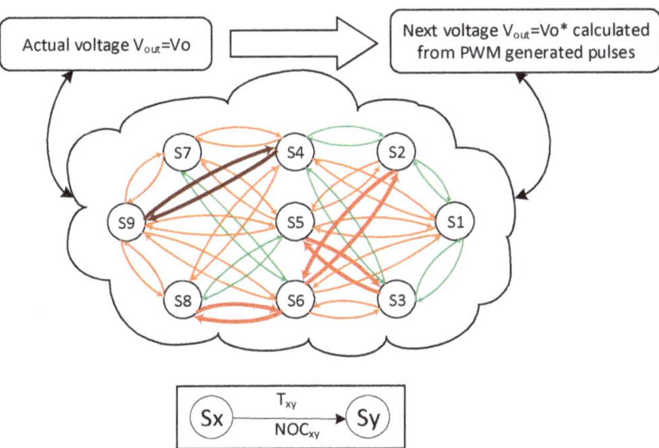

Figure 12. States of switching and associated trajectories when i(t) is positive.

As can be seen in Figure 12 (and Figure 5), in some cases, it may be necessary to perform several transitions to switch from one voltage level to another. As mentioned before, the trajectories with NOT > 2 will be ignored by the proposed algorithm (Figure 13).

The following color code (see Figure 12) is used for the transitions according to the number of communications:

Green (finest line): 2 commutations,

Orange: 4 commutations,

Red: 6 commutations,

Brown (thickest line): 8 commutations.

For instance, one of the possible trajectories to pass from S9 to S6 is S9 → S8 → S6. The NOC$_{S9\text{-}S6}$ of this trajectory is 10, as detailed below:

$$NOC_{S9-S6} = NOC_{S9-S8} + NOC_{S8-S6} = 4 + 6 = 10. \tag{1}$$

After calculation of the NOC for all possible trajectories, the results are presented in Table 3. The selected trajectories lead to a minimum NOC$_{total}$ with the minimum number of transitions.

In the case where several trajectories with the same NOC$_{total}$ are available, as mentioned before, a new criterion will be taken into account. This criterion is the number of uses (Nu) of each transition. For this, the Nu of each transition is stocked during a period, and then the trajectory with the minimal Nu will be selected. This new criterion makes it possible to better distribute the commutations of the switches over each cycle and increase the lifetime of the components. Figure 13 summarizes the different steps taken to perform the proposed OPT-PWM.

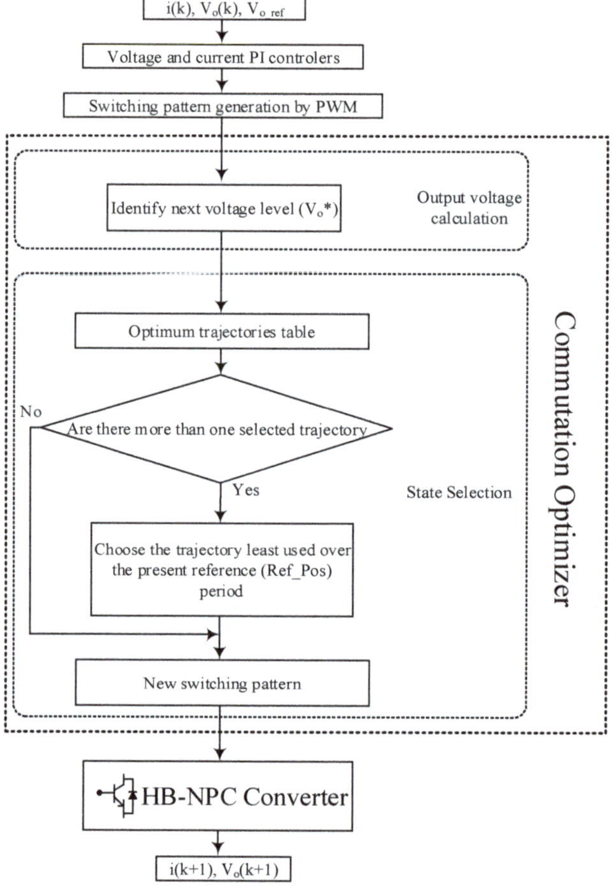

Figure 13. The final algorithm for OPTimized (OPT)-PWM.

Table 3. Transitions and possible trajectories when passing from one voltage level to another (Case where i(t) > 0).

Initial Output Voltage Level	Final Output Voltage Level	Selected Trajectories
$V_o = V_{dc}$	$V_o = V_{dc}/2$	S1 → S2 S1 → S3
$V_o = V_{dc}$	$V_o = 0$	S1 → S4 S1 → S5 S1 → S6
$V_o = V_{dc}$	$V_o = -V_{dc}/2$	S1 → S5 → S8, S1 → S6 → S7
$V_o = V_{dc}$	$V_o = -V_{dc}$	S1 → S5 → S9 S1 → S6 → S9
$V_o = V_{dc}/2$	$V_o = 0$	S2 → S4 S3 → S4
$V_o = V_{dc}/2$	$V_o = -V_{dc}/2$	S2 → S4 → S7 S2 → S4 → S8 S2 → S5 → S8 S3 → S4 → S7 S3 → S4 → S8 S3 → S6 → S7
$V_o = V_{dc}/2$	$V_o = -V_{dc}$	S2 → S5 → S9 S3 → S6 → S9
$V_o = -V_{dc}$	$V_o = 0$	S9 → S5 S9 → S6
$V_o = -V_{dc}$	$V_o = -V_{dc}/2$	S9 → S7 S9 → S8
$V_o = -V_{dc}/2$	$V_o = 0$	S7 → S6 S8 → S5

4. Simulation Results

To validate the performance of the proposed OPT-PWM applied to the HB-5L-NPC topology (Figure 2), some simulations were performed in the Matlab/Simulink environment using the SimPowerSystem library developed by MathWorks (Natick, MA, USA). The simulated system was the same as that illustrated in Figure 1 with an RL load and the OPT-PWM control. The parameters are presented in Table 1.

Figure 14 shows the output voltage when the converter is controlled by the proposed OPT-PWM. By comparing Figures 4 and 14, it can be seen that the output voltage waveforms are the same. To verify the elimination of the unnecessary commutations on the $V_{dc}/2$ (and also $-V_{dc}/2$) level, the same zooms as in Figure 4 were performed.

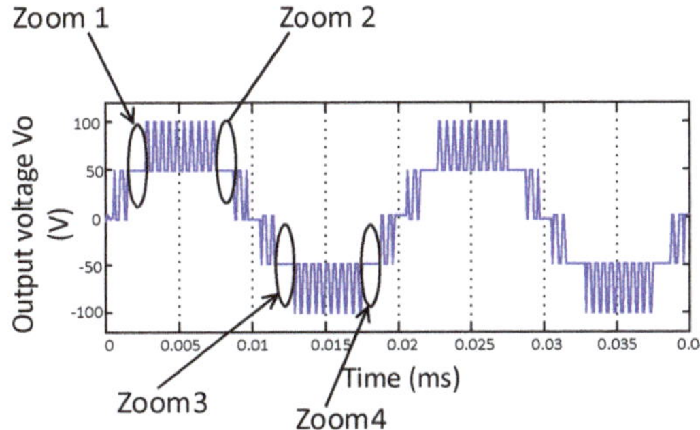

Figure 14. Output voltage of the converter with the proposed OPT-PWM control.

Figures 15–18 can be compared with Figures 6–9. As can be observed, any commutation when the voltage level remains constant (equal at $V_{dc}/2$ and $-V_{dc}/2$) is no longer performed. Thus, the total number of commutations is decreased by 16 over one period.

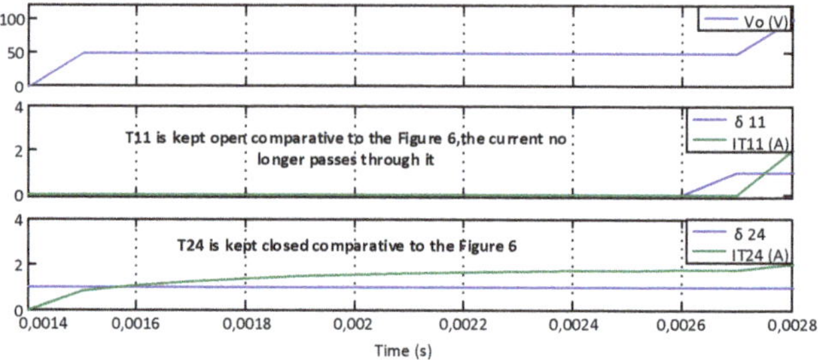

Figure 15. Voltage V_o, currents IT11 and IT24, and switching patterns δ11 and δ24 (Zoom 1).

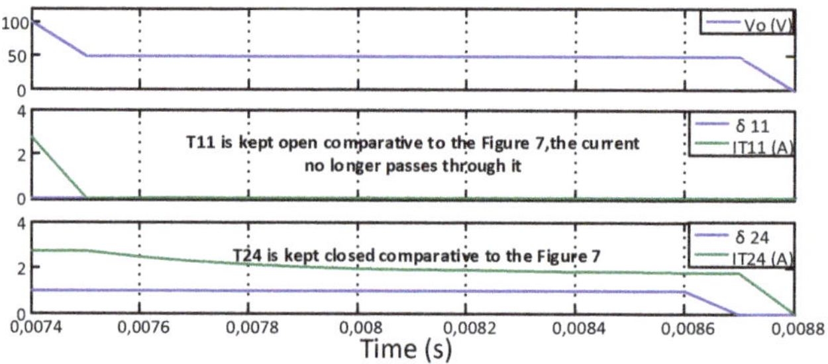

Figure 16. Voltage V_o, currents IT11 and IT24, and switching patterns δ11 and δ24 (Zoom 2).

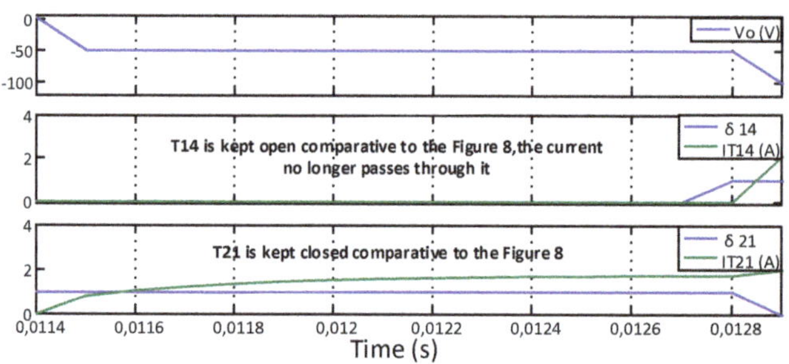

Figure 17. Voltage V_o, currents IT21 and IT14, and switching patterns δ21 and δ14 (Zoom 3).

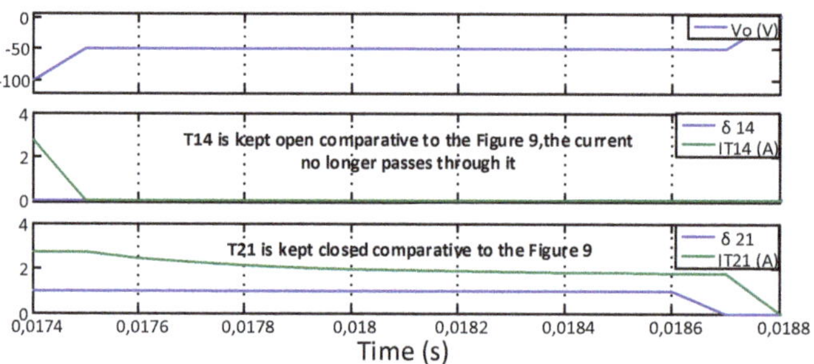

Figure 18. Voltage V_o, currents IT21 and IT14, and switching patterns δ21 and δ14 (Zoom 4).

To compare the proposed OPT-PWM and the classical LS-PWM, the operation of the system was tested for different switching frequencies while other parameters were kept unchanged. Table 4 shows the result when the system was operated under different switching frequencies. The THD of the injected current for the OPT-PWM is lower than for the LS-PWM at all frequencies. It should be mentioned that the proposed method always reduces the number of commutations, thus decreasing the switching losses in the system.

Table 4. Total Harmonic Distortion (THD) of the injected current at different frequencies for SPWM and OPT-PWM.

Frequency	Current THD OPT-PWM	Current THD LS-PWM
1 kHz	7.17%	7.50%
2 kHz	4.28%	4.48%
3 kHz	3.38%	4.11%
4 kHz	2.49%	3.2%

5. Experimental Results

To verify the validity of the proposed OPT-PWM control applied to the HB-5L-NPC, several experimental tests were carried out. The same parameters as those used in the simulations were considered. The test bench (Figure 19) was based on IGBT modules commercialized by the SEMIKRON Company (Nuremberg, Germany), reference SKM50GB123D. These IGBTs were driven by SKHI 22A drivers, distributed by the same company. The DC bus capacity value was 2200 µF.

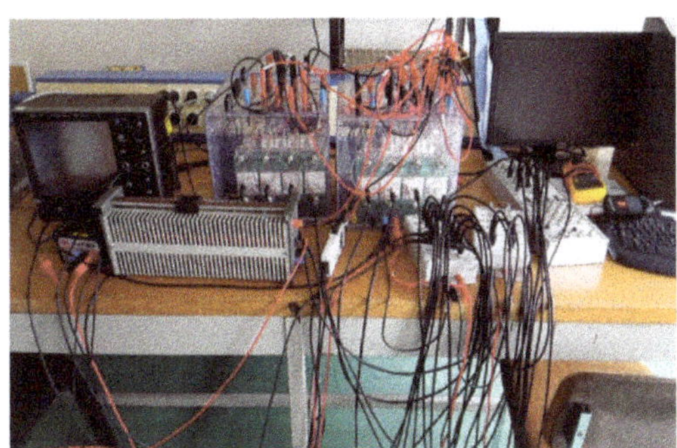

Figure 19. Experimental test bench.

The control method was applied by using a dSPACE system (Paderborn, Germany) containing a DS1005 control card as well as a DS2004 for high-resolution analog conversion (16 bit–0.8 µs) and a DS5101 PWM card with 12 outputs. The hardware implementation on the DS1005 was based on the modeling of the control algorithm carried out in the Matlab environment with classical blocks from the Simulink toolbox.

Some tests with a LS-PWM control were first performed and then the proposed OPT-PWM was applied to control the system. The results obtained from the two modulation strategies are compared in the following sections.

5.1. Experimental Results with LS-PWM Control

Figure 20 shows the output voltage of the converter with LS-PWM control. These results show clearly that there are the same small variations (peaks) at the same intervals as in the simulations. To confirm this, two zooms, one when the voltage remains a long time at $V_{dc}/2$ and the other at $-V_{dc}/2$, were made and are presented in Figures 21 and 22.

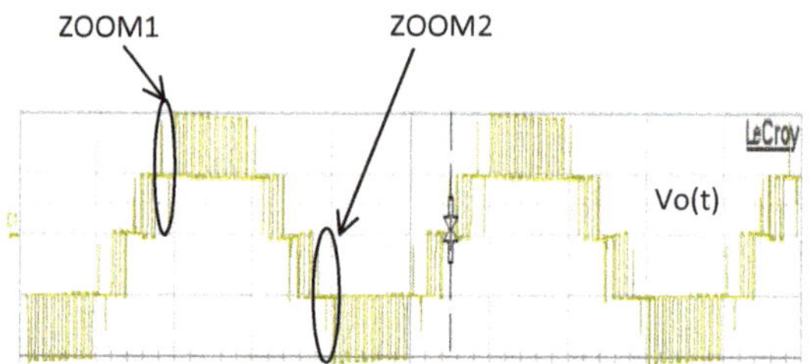

Figure 20. Output voltage of the converter with LS-PWM control.

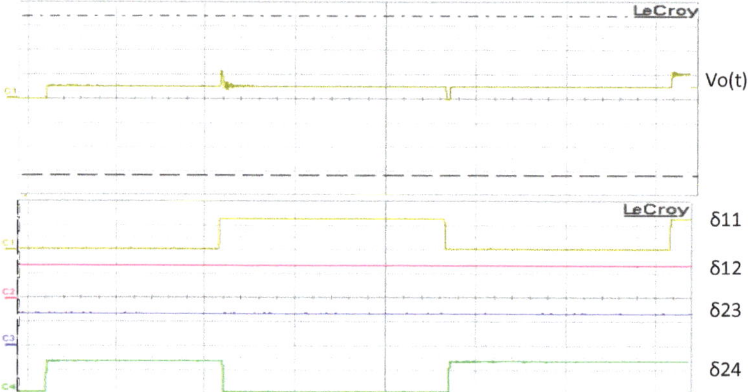

Figure 21. Output voltage $V_o(t)$ and switching patterns sent to T11, T12, T23, and T24 (ZOOM 1 in Figure 20).

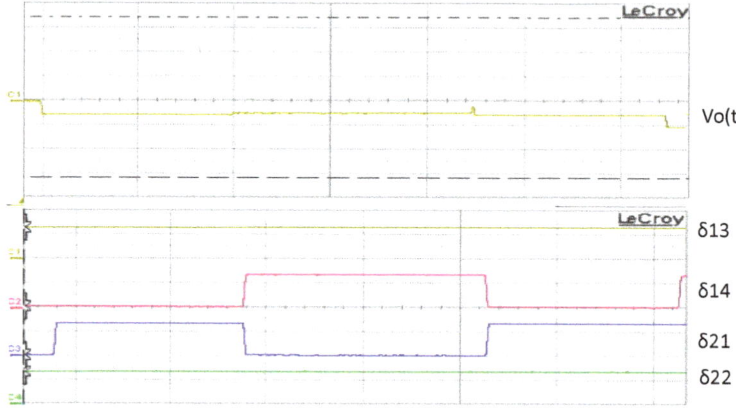

Figure 22. Output voltage $V_o(t)$ and switching patterns sent to T11, T12, T23, and T24 (ZOOM 2 in Figure 20).

5.2. Experimental Results with the Proposed OPT-PWM Control

Figure 23 shows the output voltage of the HB-5L-NPC converter with the proposed OPT-PWM control.

Figures 24 and 25 show the experimental results for the OPT-PWM. As can be seen, there are no unnecessary commutations. This means that the goal of the proposed control was met: the advanced functionality for the modulation strategy that evaluates possible switching state trajectories and implements the trajectory that provides the minimal number of commutations is efficient.

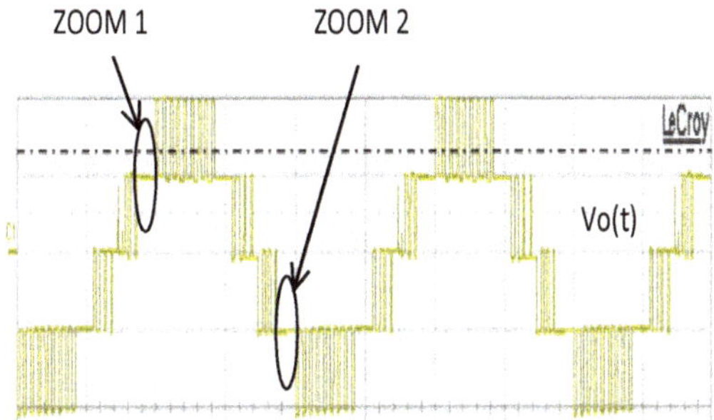

Figure 23. Output voltage during OPT-PWM control.

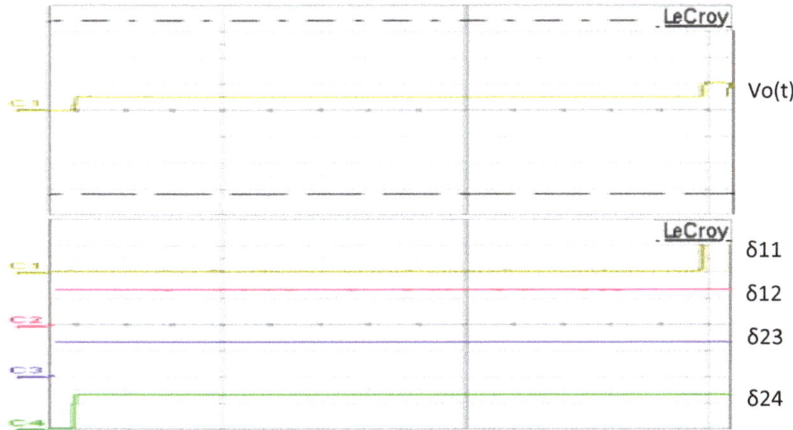

Figure 24. Output voltage $V_o(t)$ and switching patterns sent to T11, T12, T23, and T24 (ZOOM 1 of Figure 23).

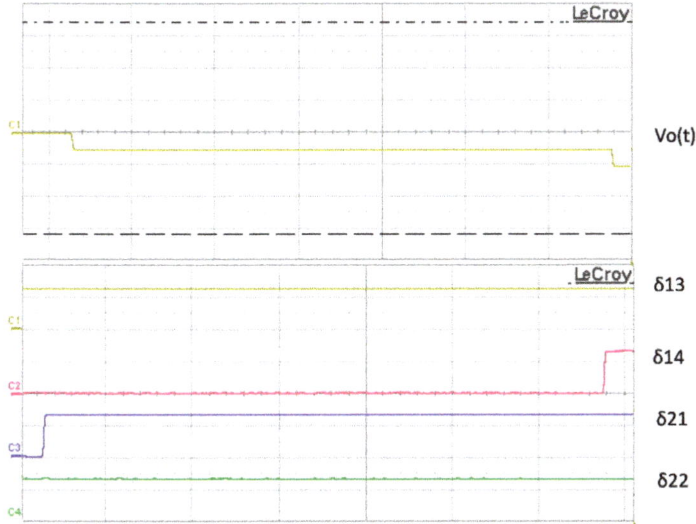

Figure 25. Output voltage $V_o(t)$ and switching patterns sent to T13, T14, T21, and T22 (ZOOM 2 of Figure 23).

6. Conclusions

This paper proposed an advanced PWM strategy applied to a five-level H-Bridge Neutral Point Clamped inverter. This modulation strategy optimizes the number of commutations without any degradation of the conventional PWM control performance. Both simulations and experimental tests were performed to confirm the validity of the proposed method. The results obtained with conventional PWM control (LS-PWM) and the proposed OPT-PWM control present similar output voltages for the HB-5L-NPC. The OPT-PWM results confirm that unnecessary commutations were removed.

By considering the number of commutations for each switch in simulations and experimental tests for two cases, thanks to the OPT-PWM control, the efficiency of the converter was improved. We note again that this advanced modulation strategy with reduced complexity can be applied to any modulation technique because its implementation uses an additional block.

Author Contributions: Investigation, F.B., P.P.; Resources, F.B., E.J. and P.P.; Writing—review and editing, F.B., E.J., P.P. and S.S.

Funding: This research received no external funding.

Conflicts of Interest: The authors declare no conflict of interest.

References

1. Bouhali, O.; Rizoug, N.; Mesbahi, T.; Francois, B. Modeling and control of the three-phase NPC multilevel converter using an equivalent matrix structure. In Proceedings of the 7th IET International Conference on Power Electronics, Machines and Drives (PEMD 2014), Manchester, UK, 8–10 April 2014; pp. 1–6.
2. Akagi, H. Multilevel Converters: Fundamental Circuits and Systems. *Proc. IEEE* **2017**, *105*, 2048–2065. [CrossRef]
3. Nabae, A.; Takahashi, I.; Akagi, H. A New Neutral-Point-Clamped PWM Inverter. *IEEE Trans. Ind. Appl.* **1981**, *IA-17*, 518–523. [CrossRef]
4. Konstantinou, G.; Pou, J.; Capella, G.J.; Song, K.; Ceballos, S.; Agelidis, V.G. Interleaved Operation of Three-Level Neutral Point Clamped Converter Legs and Reduction of Circulating Currents Under SHE-PWM. *IEEE Trans. Ind. Electron.* **2016**, *63*, 3323–3332. [CrossRef]
5. Barros, J.D.; Silva, J.F.A.; Jesus, É.G. Fast-predictive optimal control of NPC multilevel converters. *IEEE Trans. Ind. Electron.* **2013**, *60*, 619–627. [CrossRef]
6. Behera, R.K.; Das, S.P. A forced switching technique for current controlled three-level NPC ac-dc converter. In Proceedings of the Joint International Conference on Power Electronics, Drives and Energy Systems (PEDES), New Delhi, India, 20–23 December 2010; pp. 1–6.
7. Choudhury, A.; Pillay, P.; Williamson, S.S. DC-Bus Voltage Balancing Algorithm for Three-Level Neutral-Point-Clamped (NPC) Traction Inverter Drive With Modified Virtual Space Vector. *IEEE Trans. Ind. Appl.* **2016**, *52*, 3958–3967. [CrossRef]
8. Kuo, C.C.; Tzou, Y.Y. FPGA predictive control for single-phase active NPC grid inverters with multi-sampling technique. In Proceedings of the Annual Conference of the IEEE Industrial Electronics Society (IECON 2016), Florence, Italy, 23–26 October 2016; pp. 2295–2300.
9. Campanhol, L.B.G.; da Silva, S.A.O.; de Oliveira, A.A.; Bacon, V.D. Dynamic Performance Improvement of a Grid-Tied PV System Using a Feed-Forward Control Loop Acting on the NPC Inverter Currents. *IEEE Trans. Ind. Electron.* **2017**, *64*, 2092–2101. [CrossRef]
10. Sebaaly, F.; Vahedi, H.; Kanaan, H.Y.; Moubayed, N.; Al-Haddad, K. Design and Implementation of Space Vector Modulation-Based Sliding Mode Control for Grid-Connected 3L-NPC Inverter. *IEEE Trans. Ind. Electron.* **2016**, *63*, 7854–7863. [CrossRef]
11. Ebrahimi, J.; Karshenas, H. A New Single dc Source Six-Level Flying Capacitor Based Converter with Wide Operating Range. *IEEE Trans. Power Electron.* **2019**, *34*, 34–2158. [CrossRef]
12. Stillwell, A.; Pilawa-Podgurski, R.C.N. A 5-Level Flying Capacitor Multi-Level Converter with Integrated Auxiliary Power Supply and Start-Up. In Proceedings of the IEEE Applied Power Electronics Conference and Exposition (APEC), Tampa, FL, USA, 26–30 March 2017; pp. 2932–2938. [CrossRef]

13. Alamri, B.; Darwish, M. Power loss investigation in HVDC for cascaded h-bridge multilevel inverters (CHB-MLI). In Proceedings of the IEEE Eindhoven PowerTech, Eindhoven, The Netherlands, 29 June–2 July 2015; pp. 1–7. [CrossRef]

14. Rojas, C.A.; Kouro, S.; Edwards, D.; Bin, W.; Rivera, S. Five-level H-bridge NPC central photovoltaic inverter with open-end winding grid connection. In Proceedings of the Conference of the IEEE Industrial Electronics Society (IECON 2014), Dallas, TX, USA, 29 October–1 November 2014; pp. 4622–4627. [CrossRef]

15. Fan, S.; Zhang, K.; Xiong Jand Xue, Y. An Improved Control System for Modular Multilevel Converters with New Modulation Strategy and Voltage Balancing Control. *IEEE Trans. Power Electron.* **2015**, *30*, 358–371. [CrossRef]

16. Rodriguez, J.; Bernet, S.; Steimer, P.K.; Lizama, I.E. A Survey on Neutral-Point-Clamped Inverters. *IEEE Trans. Ind. Electron.* **2010**, *57*, 2219–2230. [CrossRef]

17. Hasan, N.S.; Rosmin, N.; Osman DA, A.; Musta'amal, A.H. Reviews on multilevel converter and modulation techniques. *Renew. Sustain. Energy Rev.* **2017**, *80*, 163–174. [CrossRef]

18. Marquardt, R. Modular Multilevel Converters—Sate of the art and future progress. *IEEE Power Electron. Mag.* **2018**, *5*, 24–31. [CrossRef]

19. Kobravi, K.; Iravani, R. A novel modulation strategy to minimize the number of commutation processes in the matrix converter. In Proceedings of the 2010 IEEE Energy Conversion Congress and Exposition, Atlanta, GA, USA, 12–16 September 2010.

20. Yusuke, A.; Takeshita, T. PWM control of matrix converter for reducing a number of commutations and output voltage harmonics. In Proceedings of the 2007 Power Conversion Conference, Nagoya, Japan, 2–5 April 2007; pp. 769–775.

21. Calle-Prado, A.; Alepuz, S.; Bordonnau, J.; Cortes, P.; Rodriguez, J. Predictive control of a back-to-back NPC converter-based wind power system. *IEEE Trans. Ind. Electron.* **2016**, *63*, 4615–4627. [CrossRef]

22. Durand, C.; Klinger, M.; Coutellier, D.; Naceur, H. Power cycling reliability of power module: A survey. *IEEE Trans. Device Mater. Reliab.* **2016**, *16*, 80–97. [CrossRef]

23. Niu, H. A review of power cycle driven fatigue, aging and failure modes for semiconductors power module. In Proceedings of the 2017 IEEE International Electrical Machines and Drives Conference, Miami, FL, USA, 21–24 May 2017; pp. 1–8.

24. Ounejjar, Y.; AL-Haddad, K. Current control of the three phase five-level PUC-NPC converter. In Proceedings of the 38th Annual Conference on IEEE Industrial Electronics Society (IECON 2012), Montreal, QC, USA, 25–28 October 2012; pp. 4949–4954. [CrossRef]

25. Song, W.; Feng, X.; Xiong, C. A neutral point voltage regulation method with SVPWM control for single-phase three-level NPC converters. In Proceedings of the IEEE Vehicle Power and Propulsion Conference (VPPC2008), Harbin, China, 3–5 September 2008; pp. 1–4. [CrossRef]

26. Li, J.; Liu, Y.; Bhattacharya, S.; Huang, A.Q. An optimum PWM Strategy for 5-level active NPC (ANPC) converter based on real-time solution for THD minimization. In Proceedings of the IEEE Energy Conversion Congress and Exposition, San Jose, CA, USA, 20–24 September 2009; pp. 1976–1982. [CrossRef]

27. Song, K.; Konstantinou, G.; Mingli, W.; Acuna, P.; Aguilera, R.P.; Agelidis, V.G. Windowed SHE-PWM of Interleaved Four-Quadrant Converters for Resonance Suppression in Traction Power Supply Systems. *IEEE Trans. Power Electron.* **2017**, *32*, 7870–7881. [CrossRef]

© 2019 by the authors. Licensee MDPI, Basel, Switzerland. This article is an open access article distributed under the terms and conditions of the Creative Commons Attribution (CC BY) license (http://creativecommons.org/licenses/by/4.0/).

Article

A PWM Scheme for Five-Level H-Bridge T-Type Inverter with Switching Loss Reduction

Thanh-Hai Quach [1], Duc-Tri Do [1] and Minh-Khai Nguyen [2,*]

[1] Faculty of Electrical and Electronics Engineering, Ho Chi Minh City University of Technology and Education, Ho Chi Minh City 700000, Vietnam; haiqt@hcmute.edu.vn (T.-H.Q.); tridd@hcmute.edu.vn (D.-T.D.)

[2] Department of Power Engineering, School of Electrical Engineering and Computer Science, Queensland University of Technology, Brisbane 4000, Australia

[*] Correspondence: minhkhai.nguyen@qut.edu.au; Tel.: +61-7-31382830

Received: 21 May 2019; Accepted: 17 June 2019; Published: 21 June 2019

Abstract: In this paper, a new pulse width modulation (PWM) scheme using an offset function to reduce switching loss in the five-level H-bridge T-type inverter (5L-HBT^{2}I) is proposed. The proposed modulation technique is implemented with a third harmonic offset voltage function. A new control voltage, that is adding the offset voltage into the initial control, is shifted to the top or bottom position of the carrier, simultaneously—where the absolute value of its load current is high or medium in comparison to other phase load currents. Due to reducing the intersection between a control voltage and the carriers, the number of switch commutations of the inverter is reduced. As a result of reducing the number of commutation count with a high current at the non-switching position, the switching losses of the inverter are decreased. Analysis and comparison of switching losses on the two-level and three-level inverters, which are components of 5L-HBT^{2}I are presented. The power loss analysis on the 5L-HBT^{2}I is performed. The proposed technique implements the switching loss reduction strategy based on setting the operation of the two-level inverter in six-step mode. PSIM software is used to clarify the proposed technique. The simulation results show that the total switching losses of the proposed technique in 5L-HBT2I reduce in comparison to the conventional sine PWM technique. A prototype is built to validate the proposed scheme. Simulation and experimental results match the analysis.

Keywords: carrier-based pulse width modulation; offset function; switching loss reduction; H-bridge five-level inverter; multilevel inverter

---

1. Introduction

Five-level inverters are power electronic converters that play important roles in applications of mechanical-electrical systems, transportation, power quality management, renewable energy conversion system, and motor drives [1–4]. In the design of five-level inverters, different topologies have been evolved, including the diode-clamped [5–9], flying capacitor [10–13], cascaded H bridge [14–16], and T type topologies [17–20]. A new single-phase five-level T-type inverter can reduce the harmonic components compared with that of the traditional full-bridge three-level PWM inverter under the same conditions of DC voltage source and switching frequency [18]. In [19] the DC-DC boost converter is connected to a five-level inverter and PI control is used, to control the inverter with much less total harmonic distortion (THD) and near unity power factor. To enhance the input voltage, three PV strings in cascade and parallel configurations with the five-level inverter are found in [19]. To maintain the power factor at near unity, the inverters in [19] and [20] use a proportional–integral (PI) current control scheme. The main solutions for the small-scale rooftop PV applications are proposed in [20]. That paper presents a comparison of four multilevel converters based on the T-type topology.

The switching loss is very important to evaluate the performance of the inverter, especially when multilevel inverter operates at a high frequency. Several studies on switching loss reduction have been introduced in recent years [21–26]. The complete analytical calculation of the switching loss is proposed in [23] for the basic inverters. The switching losses of the IGBT depend on switching frequency, load current, input DC voltage, and characteristics of switches are presented in [24]. The paper [24] shows that the total switching losses of the inverter on a control voltage period are proportional to switching frequency, load current, and DC voltage supply. The switching frequency depends on the carrier frequency and the modulation method. Note that the switching frequency is opposite to the total harmonic distortion (THD) of the output voltage. So that, lower carrier frequencies based switching loss reduction can increase THD value. Therefore, a new technique for switching loss reduction without THD increment is needed to be solved. As a result, the switching energy of the switches when the switches operate is turned on (Ec(on)) or turned off (Ec(off)) mode, are determined by current and voltage across the switch (I_C and V_{CE}) [25].

The average switching loss in the switch (P_S) during each period T_s is calculated as Equation (1) [25].

$$P_S = \frac{1}{6}\left(V_{CE}I_s \frac{t_{c(on)} + t_{c(off)}}{T_s}\right) - \frac{1}{3}\left(V_{on}I_s \frac{t_{c(on)} + t_{c(off)}}{T_s}\right) \tag{1}$$

where:

f_s—switching frequency

V_{CE}—voltage across the switch when the switch is turned off

V_{on}—voltage across the switch when the switch is turned on

I_s—current through the switch when the switch is turned on

$t_{c(on)}$—turn-on cross-over interval

$t_{c(off)}$—turn-off cross-over interval

From Equation (1), it can be seen that to reduce switching losses it needs to reduce switching frequency (f_s). Or for reduction switching loss, it needs to choose a switching state based on I_C and V_{CE} of phase to have the smallest value in the three-phase. As a result, f_s, I_C and V_{CE} are three basic elements that affect to reduce switching loss. A new space vector modulation strategy with two PI regulators is used in both DC voltage and load voltage sides for controlling the PWM parameters. So, the non-switching state can be controlled by these parameters to generate pulses as found in [26]. The switching losses are reduced by reducing either switching frequency or instantaneous current and voltage values at the time of switching in [26]. Since using two PI regulators, the control technique is very complex and hard to control for the system.

A reduced switching loss PWM strategy to eliminate common mode voltage in multilevel inverters is presented in [27]. This PWM method has increased switching times for the phase with the smallest load current by comparing the three-phase load current to eliminate common mode voltage. Simulation and experimental results in [27] are performed based on the standard models with the collector-emitter voltage of the IGBTs equaled. Therefore, for application to other models, which the voltage crossed IGBTs is not the same, as if the 5L-HBT^{2}I, its effect of loss reduction may not be very high. The two modified space vector modulation strategies were proposed in [28] for the reduction of the qZSI number of switch commutations at high current level with shorter periods during the fundamental cycle, i.e., reducing the switching loss, simplifying the generation of the gate signals by utilizing only three reference signals, and achieving a single switch commutation at a time were also presented in [28].

This paper proposes a new modulation technique to reduce the number of switch commutations for switching losses reduction. There is no switching on the phase which the control voltage has the smallest displacement to top or bottom of the carrier, and the absolute of the load current being the first or second large. In addition, reducing turn on and off the switch that has a large across voltage, will be done to reduce switching losses. The paper is organized as follows: in Section 2, the topology, operating principles, and circuit analysis of the three-phase 5L-HBT^{2}I are presented. Section 3 presents switching loss calculation for 5L-HBT^{2}I. Section 4 presents the proposed PWM scheme. Simulation and experimental results are shown in Section 5. Finally, the result is summarized in Section 6.

2. Three-Phase Five-Level H-Bridge T-Type Inverter Topology

Figure 1 shows the three-phase 5L-HBT^{2}I. As shown in Figure 1a, the 5L-HBT^{2}I consists of three DC sources (U_a, U_b, U_c), six capacitors (C_{a1}, C_{b1}, C_{c1}, C_{a2}, C_{b2}, and C_{c2}), and three T-type H-bridge circuits.

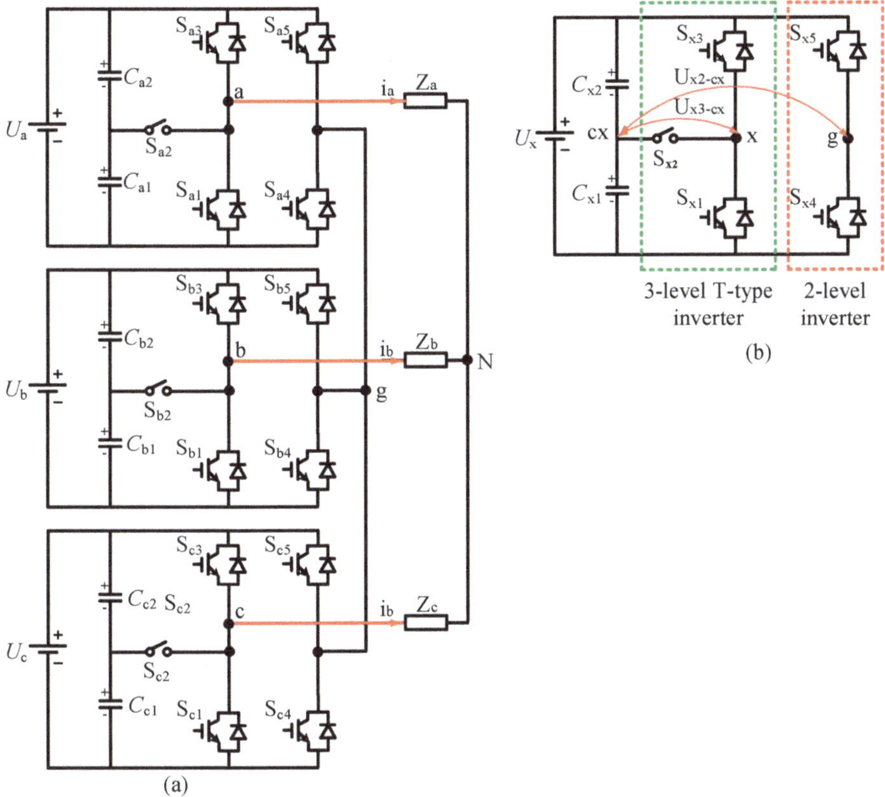

Figure 1. Five-level T-type H-bridge inverter topologies. (a) Three-phase inverter and (b) One phase module.

Define T_{Sxi} is status of the ith switch on phase x (x = a, b, c); where i is the index of switches (1 to 5), with conditions:

$$T_{Sxi} = \begin{cases} 0 \text{ if } Sxi \text{ off} \\ 1 \text{ if } Sxi \text{ on} \end{cases} \tag{2}$$

$$T_{Sx5} + T_{Sx4} = 1 \tag{3}$$

$$T_{Sx1} + T_{Sx2} + T_{Sx3} = 1 \tag{4}$$

Figure 1b shows a structure of phase x, where a leg of 5L-HBT^{2}I is built from two single-phase inverters: two-level and three-level T-type.

From Equations (3) and (4), the pole voltage of two-level ($U_{x2\text{-}cx}$) and three-level T type inverter ($U_{x3\text{-}cx}$) are determined:

$$U_{x2-cx} = (2T_{Sx5} - 1)\frac{V_{dc}}{2} \tag{5}$$

$$U_{x3-cx} = (2T_{Sx3} + T_{Sx2} - 1)\frac{V_{dc}}{2} \tag{6}$$

Hence, the voltage from phase to pole (U_{xg}) is determined as

$$V_{xg} = U_{x3-cx} - U_{x2-cx} = (2T_{Sx3} + T_{Sx2} - 2T_{Sx5})\frac{V_{dc}}{2} \tag{7}$$

Set T_{Sx} is status of phase x:

$$T_{Sx} = 2T_{Sx3} + T_{Sx2} - 2T_{Sx5} \tag{8}$$

So, the phase to pole voltages is given as Equation (9).

$$\begin{bmatrix} V_{ag} \\ V_{bg} \\ V_{cg} \end{bmatrix} = \frac{V_{dc}}{2} \begin{bmatrix} T_{Sa} \\ T_{Sb} \\ T_{Sc} \end{bmatrix} \tag{9}$$

The output phase and line-to-line voltages of this inverter is given as Equations (10) and (11).

$$\begin{bmatrix} V_{an} \\ V_{bn} \\ V_{cn} \end{bmatrix} = \frac{1}{3} \begin{bmatrix} 2 & -1 & -1 \\ -1 & 2 & -1 \\ -1 & -1 & 2 \end{bmatrix} \begin{bmatrix} V_{ag} \\ V_{bg} \\ V_{cg} \end{bmatrix} \tag{10}$$

$$\begin{bmatrix} V_{ab} \\ V_{bc} \\ V_{ca} \end{bmatrix} = \begin{bmatrix} 1 & -1 & 0 \\ 0 & 1 & -1 \\ -1 & 0 & 1 \end{bmatrix} \begin{bmatrix} V_{an} \\ V_{bn} \\ V_{cn} \end{bmatrix} \tag{11}$$

As a result, the third order harmonic does not appear in the output phase voltage (V_{xn}), while it appears in both pole voltage (V_{xg}) and line-to-line voltage. Therefore, the proposed algorithm has offset the function at the third order harmonic, which will not affect the third order harmonic magnitude of the load. From Equation (8) to Equation (11), the values of V_{xg} can be determined in Table 1. As shown in Table 1, the pole voltage V_{xg} has five levels including two positive, two negative, and one zero.

Table 1. Switching state of five-level H-bridge T-type inverter (5L-HBT^{2}I) for phase-x index.

State	T_{Sx3}	T_{Sx2}	T_{Sx1}	T_{Sx5}	T_{Sx}	V_{xg}
1	0	0	1	0	0	0
2	0	1	0	0	1	$\frac{V_{dc}}{2}$
3	1	0	0	0	2	$2\frac{V_{dc}}{2}$
4	0	0	1	1	-2	$-2\frac{V_{dc}}{2}$
5	0	1	0	1	-1	$-\frac{V_{dc}}{2}$
6	1	0	0	1	0	0

3. Switching Loss Calculation for 5L-HBT^{2}I

The switching loss depends on the load current and collector-emitter voltage of the power switches and the number of commutations during the entire fundamental cycle [24,25]. Therefore, the switching losses of the 5L-HBT^{2}I will be calculated based on the switching losses of the power switches of each phase, where phase x consists of five switches: S_{x1}–S_{x5}, x = a, b, c.

Figure 1b and Table 1 can be seen that

$$V_{CE1} = V_{CE2} = V_{CE3} = V_{CE4}/2 = V_{CE5}/2 = V_{dc}/2 \tag{12}$$

where $V_{CE,i}$ is the voltage across switch S_{xi} when S_{xi} is turned-off ($i = 1, 2, 3, 4, 5$).

And the current through switches, i_s is determined as

$$i_{si} = |i_{Lx}(\theta)| \qquad \text{where } \theta \text{ is phase angle, } x = a, b, c \tag{13}$$

In Equation (13), i_{Si} and i_{Lx} are current through switch S_{xi} when S_{xi} is turned-on ($i = 1, 2, 3, 4, 5$) and the load current over phase x ($x = a, b, c$), respectively.

Hence, the average value of the local (per control voltage cycle) switching loss over switches S_{x1}–S_{x5} (for instance, for phase-a) can be calculated as [27]

$$P_{S1} = \frac{1}{2\pi}(t_{c(on)} + t_{c(off)}) \int_0^{2\pi} \left(\frac{V_{CE1}}{6} - \frac{V_{on}}{3}\right) i_{s1} d\theta = \frac{1}{2\pi}(t_{c(on)} + t_{c(off)}) \left(\frac{V_{dc}}{12} - \frac{V_{on}}{3}\right) \int_0^{2\pi} i_{s1} d\theta \tag{14}$$

Similarly, we have

$$P_{S2} = \frac{1}{2\pi}(t_{c(on)} + t_{c(off)}) \left(\frac{V_{dc}}{12} - \frac{V_{on}}{3}\right) \int_0^{2\pi} i_{s2} d\theta \tag{15}$$

$$P_{S3} = \frac{1}{2\pi}(t_{c(on)} + t_{c(off)}) \left(\frac{V_{dc}}{12} - \frac{V_{on}}{3}\right) \int_0^{2\pi} i_{s3} d\theta \tag{16}$$

$$P_{S4} = \frac{1}{2\pi}(t_{c(on)} + t_{c(off)}) \left(\frac{V_{dc}}{6} - \frac{V_{on}}{3}\right) \int_0^{2\pi} i_{s4} d\theta \tag{17}$$

$$P_{S5} = \frac{1}{2\pi}(t_{c(on)} + t_{c(off)}) \left(\frac{V_{dc}}{6} - \frac{V_{on}}{3}\right) \int_0^{2\pi} i_{s5} d\theta \tag{18}$$

where P_{Si} is an average value of the local (per carrier cycle) switching loss over the switch ($i = 1, 2, 3, 4, 5$). $V_{on,i}$ is voltage across switch S_{xi} in turned-on state.

Equations (14)–(18) can show that switching loss decreases if the switching is on the phase, which has a small load current. Due to the decrement of the number of commutations on two-level inverter switches (S_{x5} or S_{x4}), the switching loss will be smaller when compared with three-level T-type inverter switches (S_{x1} or S_{x2} or S_{x3}).

As a result, the proposed technique helps to reduce the switching loss by reducing the number of commutations on the phase, which has the absolute of the load current are first or second large in three-phase and by reducing the number of commutations on two level inverter.

4. Proposed Algorithm

4.1. Principle of the Proposed Algorithm

The switching loss depends on the current through power switches, the voltage across the switch, and the number of commutations in the period of control voltage is shown in Equations (14)–(18). Since the switching loss can be decreased by reducing switching frequency of S_{x5} and S_{x4} and reducing the number of commutations on the phase, which has the absolute of the load current is high or medium in comparison to another phase. Therefore, in the first stage, the new control voltages are determined with a non-switching state on the phase, which has the smallest displacement to top or bottom peak (of the carrier) and the absolute of the load current being the first or second largest. In addition, in the second stage, the control voltages that have been determined in the previous stage, are divided into the control voltage for the two-level inverter and three-level T-type inverter for reducing switch turn-on/turn-off on the two-level inverter.

The first stage:

It is defined that v_x is the initial control voltage phase x (x = a, b, c) and v_{rx} is control voltage that it is determined as the new control voltage by adding v_{offset} into v_x from first stage of proposed algorithm. The maximum ampplitude of carrier is selected by 1. Due to the 5-level inverter, threshold comparison of the carrier is 0, 1, 2, 3 and 4. The initial control voltage of phase v_x is determined as Equation (19).

$$v_x = m \frac{4}{\sqrt{3}} \cos(\omega t + \theta_{0x}) + 2 \tag{19}$$

where v_x, m, ω and θ_{0x} are the initial control voltage of phase x, modulation index, angular velocity, and initial phase angle, respectively.

The error of v_x and L_x are determined:

$$L_x = \left[\begin{array}{ll} int(v_x) & if\ int(v_x) < 4 \\ int(v_x - 1) & else \end{array} \right. ;\ H_x = L_x + 1 \tag{20}$$

$$e_x = v_x - L_x \tag{21}$$

Call I_{xABS} is the absolute of current across phase x, then

$$i_{xABS} = \left| i_{Lx}(\theta) \right| \tag{22}$$

where θ is phase angle, I_{Lx} is the load current of phase x (x = a, b, c).

The maximum, minimum, medium of error (e_{max}, e_{min}, e_{med}) and the maximum and medium of absolute of load current (I_{max}, I_{med}) are defined:

$$e_{max} = \max(e_a, e_b, e_c),\ e_{min} = \min(e_a, e_b, e_c),\ e_{med} = \mathrm{med}(e_a, e_b, e_c) \tag{23}$$

$$i_{max} = \max(i_{aABS}, i_{bABS}, i_{cABS}),\ i_{med} = \mathrm{med}(i_{aABS}, i_{bABS}, i_{cABS}) \tag{24}$$

Case 1: When the i_{xABS} is the largest and ($e_x = e_{min}$) or ($e_x = e_{max}$), the none switching phase is x and the offset voltage (v_{offset}) is determined through the value e_x as Equation (25)

$$v_{offset} = \left[\begin{array}{l} -e_x\ if\ e_x = e_{min} \\ 1 - e_x\ if\ e_x = e_{max} \end{array} \right. . \tag{25}$$

Case 2: When i_{aABS} is the largest and e_a equal the e_{med}, for reducing error of output voltages, the offset voltage will be not equal to e_a. In this case, another phase has the absolute of the load current as medium (assuming it is phase b) and the error (e_b) of v_b and L_b fix the maximum or minimum. Then, the offset voltage is calculated following i_{bABS} and e_b as Equation (26)

$$v_{offset} = \left[\begin{array}{l} 1 - e_b\ if\ (e_b = e_{max})\ and\ (i_{bABS} = i_{med}) \\ -e_b\ if\ (e_b = e_{min})\ and\ (i_{bABS} = i_{med}) \end{array} \right. \tag{26}$$

Thus, the offset values can be determined as i_{xABS} and e_x in Table 2.

Table 2. The offset voltage of the 5L-HBT^{2}I.

i_{aABS}	i_{bABS}	i_{cABS}	e_a	e_b	e_c	v_{offset}	Figure
i_{max}	NA	NA	e_{min}	NA	NA	$-e_{min}$	2a
i_{max}	NA	NA	e_{max}	NA	NA	$1 - e_{max}$	2b
i_{max}	i_{med}	NA	e_{med}	e_{min}	NA	$-e_{min}$	2c
i_{max}	i_{med}	NA	e_{med}	e_{max}	NA	$1 - e_{max}$	2d
i_{max}	NA	i_{med}	e_{med}	NA	e_{min}	$-e_{min}$	-
i_{max}	NA	i_{med}	e_{med}	NA	e_{max}	$1 - e_{max}$	-

NA: Not applicable.

If the matrixes from Equation (27) to Equation (31) are defined, the offset voltage will be calculated as Equation (32).

$$\begin{cases} [I_{max}] = \begin{bmatrix} a_{max}, & b_{max}, & c_{max} \end{bmatrix} \\ x_{max} = \begin{bmatrix} 1 \ if \ i_{xABS} = i_{max} \\ 0 \ else \end{bmatrix} \end{cases} \tag{27}$$

$$\begin{cases} [I_{med}] = \begin{bmatrix} a_{med}, & b_{med}, & c_{med} \end{bmatrix} \\ x_{med} = \begin{bmatrix} 1 \ if \ i_{xABS} = i_{med} \\ 0 \ else \end{bmatrix} \end{cases} \tag{28}$$

$$\begin{cases} [E_{max}] = \begin{bmatrix} e_{amax}, & e_{bmax}, & e_{cmax} \end{bmatrix} \\ e_{xmax} = \begin{bmatrix} 1 \ if \ e_x = e_{max} \\ 0 \ else \end{bmatrix} \end{cases} \tag{29}$$

$$\begin{cases} [E_{med}] = \begin{bmatrix} e_{amed}, & e_{bmed}, & e_{cmed} \end{bmatrix} \\ e_{xmed} = \begin{bmatrix} 1 \ if \ e_x = e_{med} \\ 0 \ else \end{bmatrix} \end{cases} \tag{30}$$

$$\begin{cases} [E_{min}] = [e_{amin}, e_{bmin}, e_{cmin}] \\ e_{xmin} = \begin{bmatrix} 1 \ if \ e_x = e_{min} \\ 0 \ else \end{bmatrix} \end{cases} \tag{31}$$

$$v_{offset} = \begin{bmatrix} [I_{max}][E_{max}]^T(1 - e_{max}) - [I_{max}][E_{min}]^T e_{min} \ if \ [I_{max}][E_{med}]^T = 0 \\ [I_{med}][E_{max}]^T(1 - e_{max}) - [I_{med}][E_{min}]^T e_{min} \ else \end{bmatrix} \tag{32}$$

The new control voltage v_{rx} is determined

$$v_{rx} = v_x + v_{offset} \tag{33}$$

Since matrixes $[I_{max}]$, $[I_{med}]$, $[E_{max}]$, $[E_{med}]$ and $[E_{min}]$ have "0" and "1" values, which can be determined by simple comparison commands lead to the calculation of the offset voltage easy and quick. The new control voltage (v_{rx}) is the old control voltage (v_x) add the offset voltage. The new control voltages (v_{rx}) on the phase which the absolute of the load current are first or the second large in three-phase and e_x equal e_{max} or e_{min} will be shifted to the top or bottom peak of the carrier.

Figure 2 shows the shift of the control voltages according to the conditions of the proposed algorithm. As shown in Figure 2a, when $(i_{aABS} = i_{max})$ and $(e_a = e_{min})$, the offset voltage is $-e_a$. This offset voltage is added to the original control voltages so the new control voltages will shift down to the new positions. The new position of the control voltage of phase-a will be L_a. Therefore, there will be no switching on the phase-a when $(i_{aABS} = i_{max})$ and $(e_a = e_{min})$ as shown in Figure 2b. Similarly, the switching state on the phase-b is off if $(I_{aABS} \geq I_{bABS} \geq I_{cABS})$, $(e_a = e_{med})$ and $(e_b = e_{min}$ or $e_{max})$, as shown in Figure 2c or 2d.

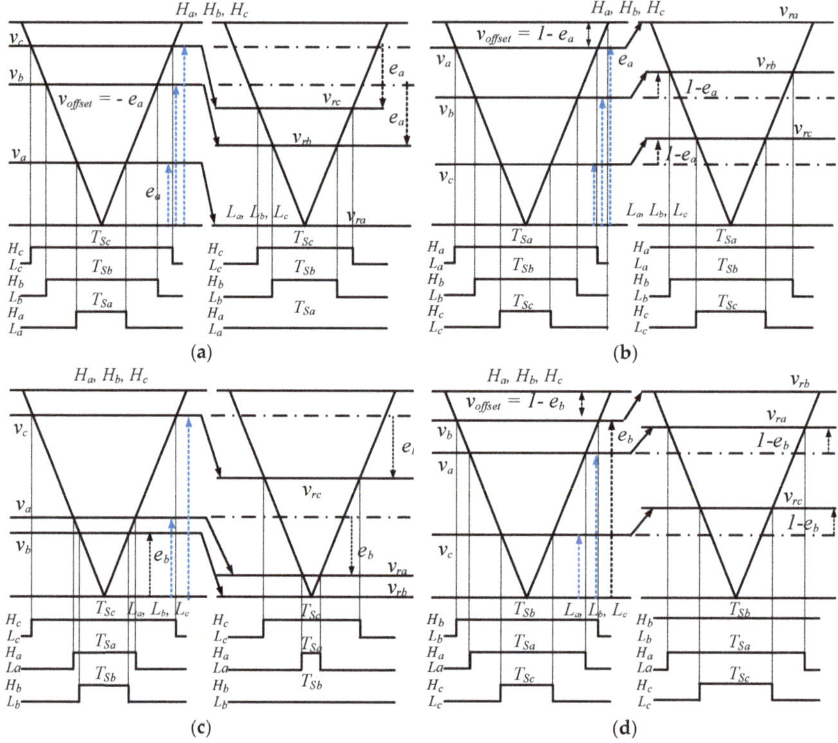

Figure 2. Offset voltages under the proposed algorithm. (**a**) $(i_{aABS} = i_{max})$ *and* $(e_a = e_{min})$; (**b**) $(i_{aABS} = i_{max})$ *and* $(e_a = e_{max})$; (**c**) $(i_{aABS} \geq i_{bABS} \geq i_{cABS})$, $(e_a = e_{med})$ *and* $(e_b = e_{min})$; (**d**) $(i_{aABS} \geq i_{bABS} \geq i_{cABS})$, $(e_a = e_{med})$ *and* $(e_b = e_{max})$.

The second stage:

In the second stage, the control voltages that were created in the previous stage will be divided into the control voltage for the two-level inverter and three-level T-type inverter. Since the two-level inverter is operated in six-step mode, its control voltage can be calculated as

$$v_{x,2l} = \left[\begin{array}{l} 1 \; if \; v_{rx} \geq 2 \\ 0 \; else \end{array} \right. \tag{34}$$

In addition, from Equation (7), it is easy to determine the control voltages for three-level T type, which are:

$$v_{x,\,3l} = v_{rx} - 2v_{x,2l} \tag{35}$$

where v_{rx} is the control voltage, created form first stage; $v_{x,2l}$ and $v_{x,3l}$ are the control voltages for two-level inverter and T-type inverter on phase x.

4.2. Flow Chart

Figure 3 shows a flow chart of the proposed algorithm using simple commands such as subtraction, and comparison on the program. The comparison of the phase currents can be done by comparing hardware circuits that do not require the use of expensive sensors. Thus, calculation time of the algorithm is low and suitable for closed-loop control or other control methods.

For example, assuming that control voltages v_a, v_b, and v_c are 1.12 V, 0.64 V and 3.24 V, respectively, from Equation (22), the error of v_x and L_x are $e_a = 0.12$, $e_b = 0.64$, and $e_c = 0.24$. From Equation (24), $e_{min} = 0.12$, $e_{max} = 0.64$, $e_{mid} = 0.24$.

Assuming that $I_{aABS} > I_{bABS} > I_{cABS}$, from Equations (27) and (28), $[I_{max}] = [1\ 0\ 0]$, $[I_{mid}] = [0\ 1\ 0]$. From Equations (30)–(32), $[E_{max}] = [0\ 1\ 0]$, $[E_{med}] = [0\ 0\ 1]$, $[E_{min}] = [1\ 0\ 0]$. From Equations (22)–(32), the offset voltage is $v_{offdet} = -e_{min} = -0.12$. Then, new control voltages are $v_{ra} = v_a - e_{min} = 1$ V, $v_{rb} = v_b - e_{min} = 0.52$ V, $v_{rc} = v_c - e_{min} = 3.12$ V.

When $v_{ra} = 1$ V, switches S_1, S_3, and S_5 are turned off while switches S_2 and S_4 are turned on. Thus, the phase-a switches will not be switched in one cycle T. Similarly, when $v_{rc} = 3.12$ V, S_1, S_3, and S_4 are turned off while S_2 and S_5 are switching. As a result, the phase-c switches will be switched in one cycle T. Therefore, the proposed algorithm can reduce the switching times of the switches.

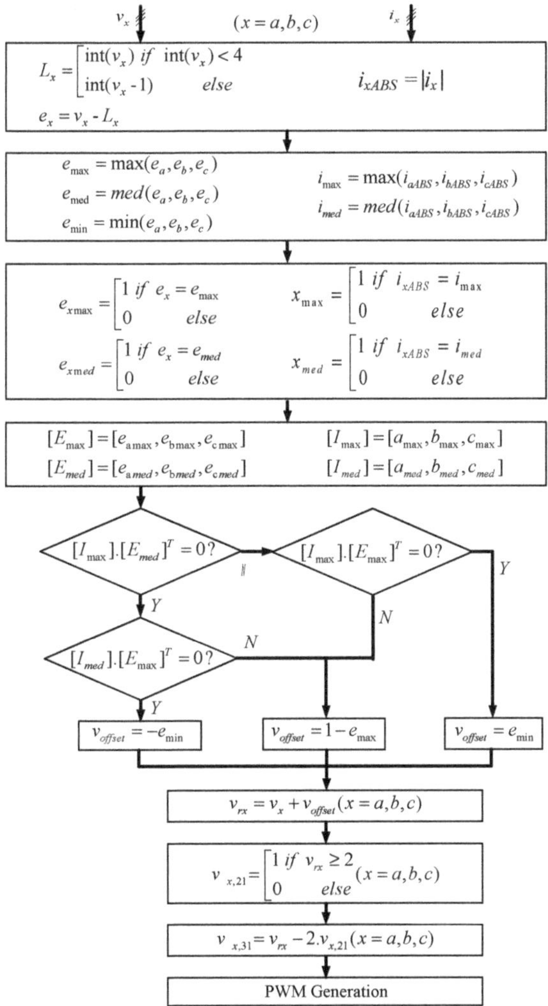

Figure 3. The flow chart of the proposed scheme.

5. Simulation and Experimental Results

5.1. Simulation Results

Parameters used in simulation is shown in Table 3. Figure 4 shows the simulation results of the proposed algorithm when modulation index are 0.4 and 0.9, respectively. The waveforms from top to bottom in Figure 4 are three-phase absolute current, three-phase error (e_a, e_b, e_c), maximum and minimum errors (e_{max} and e_{min}), the initial voltage (v_x) and new control voltage (V_{rx}), control voltage of two-level inverter (v_{a2L}), and control voltage of three-level T-type inverter (v_{a3L}), the next waveforms are gating signals of S_{x1} to S_{x5} and the bottom is the pole voltage output. It can be seen that from t_1 to t_2, i_{aABS} is maximum and e_a hit maximum (e_{max}), then v_{offset} will be $1 - e_a$ lead to the control voltage of phase a move up to $3V$; see Figure 4a and obtain 4V; see Figure 4b. As shown in Figure 4a, during a time interval of [t_1 to t_2], the control phase-a voltage is 4 V. As a result, the switching states on phase-a are ($S_{5a} = 0$, $S_{4a} = 1$, $S_{3a} = 1$, $S_{2a} = 0$ and $S_{1a} = 0$) and its phase pole voltage is $V_{ag} = V_{dc} = 100$ V. Similarly, in Figure 4b, the phase-a to pole voltage is $V_{ag} = \frac{V_{dc}}{2} = 50$ V when ($S_{5a} = 0$, $S_{4a} = 1$, $S_{3a} = 0$, $S_{2a} = 1$ and $S_{1a} = 0$). From t_3 to t_4, i_{bABS} is maximum, the e_b is medium. Simultaneously, the absolute of phase a current load (i_{aABS}) is medium and e_a is e_{min} as analyzed in the above section, the offset voltage is $-e_a$ and the new control voltage of phase a shift down 0 V ($m = 0.9$) and 1 V ($m = 0.4$), so there is no switching in phase-a. Similarly, the phase-b and phase-c in the situation have no switching.

Table 3. Parameters used in simulation.

Parameter/Component		Attributes
Input voltage	V_{dc}	100 V
Output frequency	f_o	50 Hz
Carrier frequency	f_s	5 kHz, 10 kHz, 15 kHz and 20 kHz
Capacitors	$C_1 = C_2$	47,000 µF/300 V
Three-phase RL load	R_{load} and L_{load}	40 Ω and 10 mH

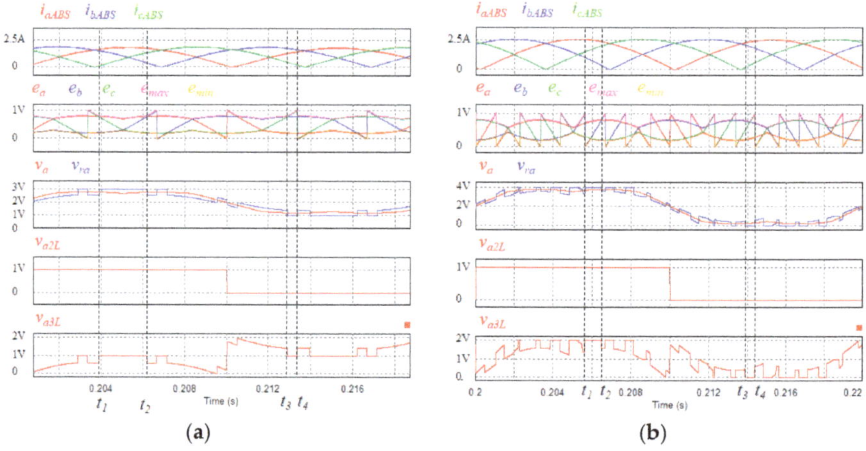

(a)

(b)

Figure 4. *Cont.*

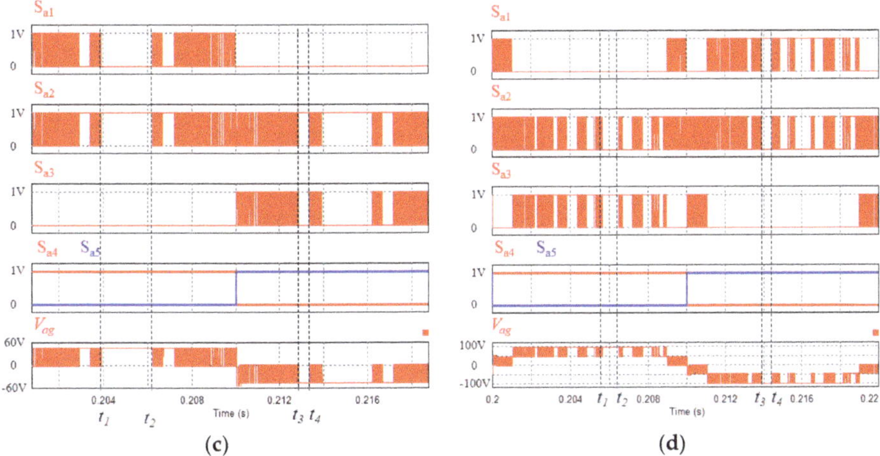

Figure 4. Control voltage generation, gating signals of S_{a1} to S_{a5} and the phase pole voltage in the proposed technique with (**a,c**) $m = 0.4$ and (**b,d**) $m = 0.9$.

To evaluate the efficiency of switching loss reduction, IGBT CM1000HA-24H module (Mitsubishi Electric., Tokyo, Japan) in PSIM's device database is used in the simulation. Figure 5 shows the conduction loss (*Pc-sinPWM*), switching loss (*Ps-sinPWM*) using sine PWM algorithm and conduction loss (*Pc-proposed*), switching loss (P_S-*proposed*) using the proposed algorithm on a phase on changing carrier frequency. As a result, the proposed technique can reduce switching losses. The switching loss reduction value is maximized when $m = 0.5$, equivalent to 78% reduction. There is no significant difference in conduction loss (P_c) between the proposed algorithm and sinPWM algorithm, especially at low carrier frequencies. The difference in conduction loss is due to the difference in the high harmonic amplitude between the proposed algorithm and sinPWM algorithm. When the carrier frequency is not high enough (e.g., at $f_c = 5$ kHz), the load phase voltage that applies the proposed technique has the high harmonic amplitude at some modulation index values. Therefore, the conduction loss in this case is greater than that in the sinPWM method as in Figure 5c,d. The harmonics spectrum of the proposed and sinPWM algorithms are presented in Figure 6. The harmonics with a large amplitude in the proposed technique focus around the carrier frequency, while they are around twice the carrier frequency in the sinPWM algorithm. The THD of the phase current in the two algorithms is slightly different, as seen in Figure 7.

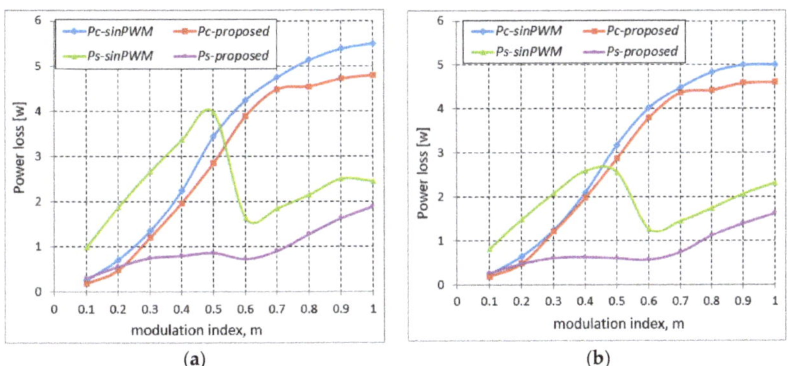

Figure 5. *Cont.*

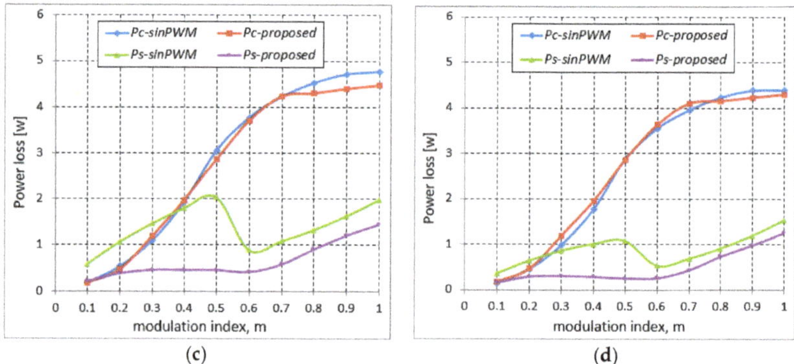

Figure 5. Power loss on a phase under sinPWM method and proposed algorithm at carrier frequency of (**a**) 20 kHz, (**b**) 15 kHz, (**c**) 10 kHz, and (**d**) 5 kHz.

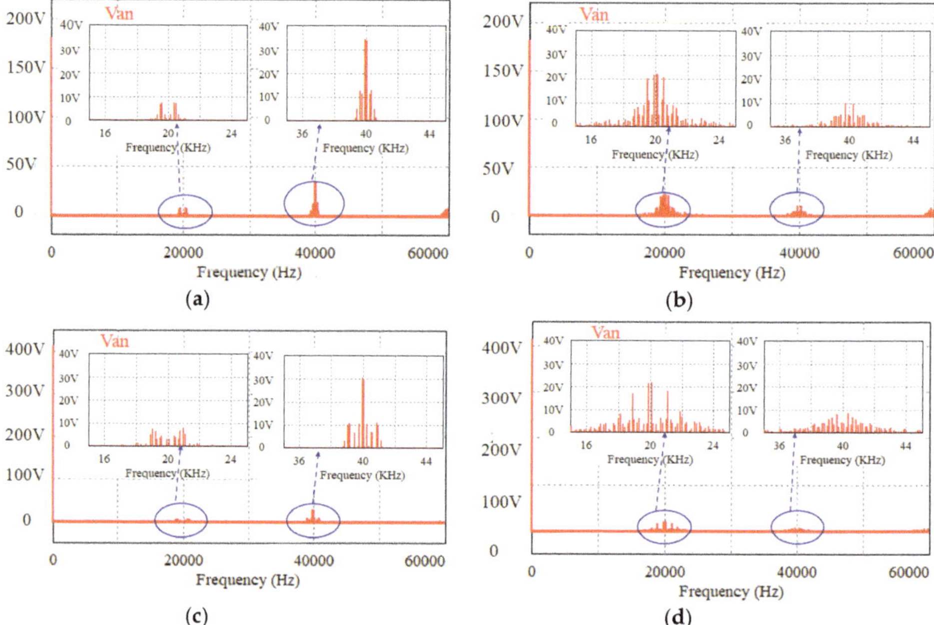

Figure 6. Harmonics spectrum of phase voltage at f_s = 20 kHz using (**a**) sinPWM method with m = 0.4, (**b**) proposed method with m = 0.4, (**c**) sinPWM method with m = 0.9, and (**d**) proposed method with m = 0.9.

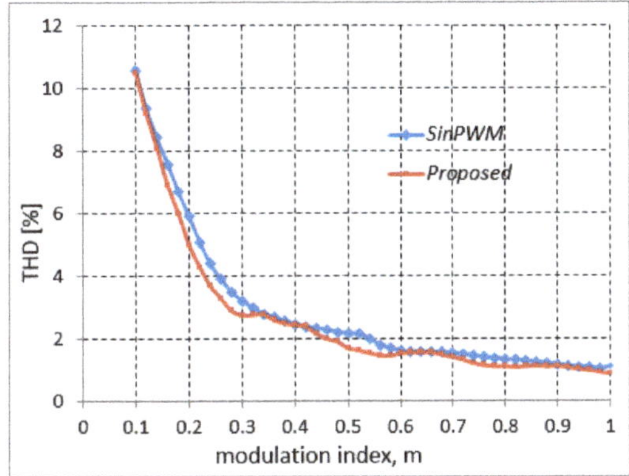

Figure 7. Total harmonic distortion (THD) comparison between sinPWM method and proposed technique at f_s = 20 kHz.

Figure 7 shows a comparison of THD values of the phase current between sinPWM method and the proposed technique. As shown in Figure 7, THD value of the phase voltage of 5L-HBT^{2}I under the proposed algorithm is smaller than that with the sinPWM method at f_c = 20 kHz. This can explain that the two-level inverter only operates in a six-step mode with rarely rapid changes of phase voltage on this inverter. As a result, the output harmonics have a small amplitude and conductive losses are also lower than sinPWM techniques, as shown in Figure 5.

Under the load of 40 Ω and 10 mH at the output frequency of 50 Hz, the calculated load power factor is 0.997. Table 4 shows the power factor (PF) and the power factor displacement (PFD) in simulation. As shown in Table 4, the simulated power factor of the inverter is lower than the calculated load power factor of 0.997. This is due to the harmonics distortion of load current and voltage.

Table 4. Simulated power factor and power factor displacement.

m	THDi	THDu	PFDi	PFDu	PF
0.4	2.45%	40%	0.997	0.928	0.9264
0.9	1.12%	17%	0.999	0.986	0.9826

5.2. Experimental Results

An experimental model based on the DSP TMS320F28335 microcontroller (Texas Instruments, Dallas, TX, US) is built in the laboratory to verify the effectiveness of the proposed control technique with the elements in Table 5. Figure 8 shows a laboratory prototype. The input voltage is 100 V. The output frequency is 50 Hz, while the switching frequency of the inverter circuit is 5 kHz. A three-phase inductive load of 40 Ω and 10 mH is used in the experiment.

Table 5. Elements used in experiment.

Elements	Type
IGBT	FGL40N150D
Current sensor	LEM-LA 25-P
Microcontroller	DSP TMS320F28335
Tektronix oscilloscope	MSO 2024B

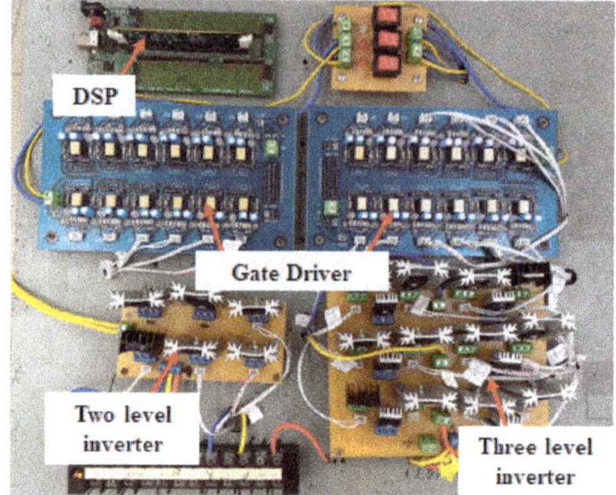

Figure 8. Experimental setup of the prototype.

Figure 9 shows the experimental result of the proposed algorithm when $m = 0.9$ and $m = 0.4$. In Figure 9, the waveforms from the top to the bottom are the gating control signals of the power switches S_{A5}, S_{A3}, S_{A2} and S_{A1}, the pole voltage (V_{ag}), the phase voltage that its harmonic spectrum is in Figure 10. The experimental result of the proposed algorithm in Figure 9 are under modulation index $m = 0.4$ (Figure 9a) and $m = 0.9$ (Figure 9b).

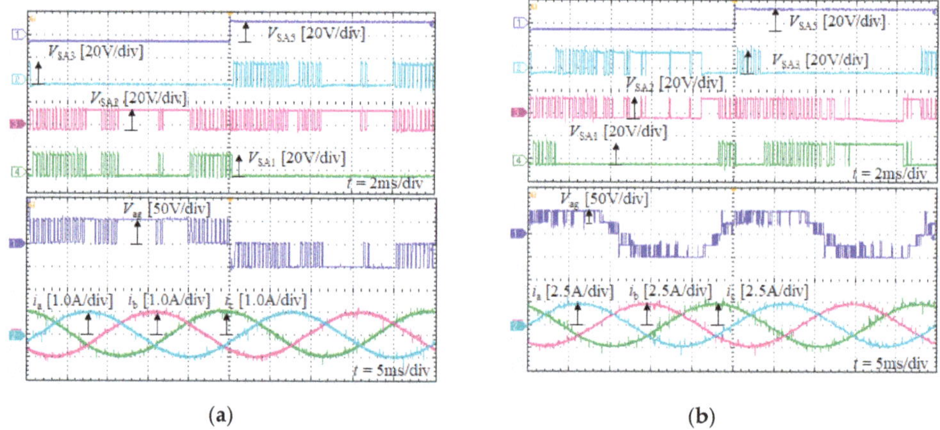

(a)   (b)

Figure 9. Experimental results at (**a**) $m = 0.4$ and (**b**) $m = 0.9$.

The experimental results are close to the simulation results. Figure 9a,b show the position "no switching" in the phase, which has the absolute of its load current hitting maximum or medium in three phases. The gating signal of S_{A5} is similar to that of S_{A4}, that is the switch of the two-level inverter. This is similar to its waveform in the six-steps mode.

Figure 10 shows the experimental results of the load current fast Fourier transform (FFT) and its THD at $m = 0.4$ and 0.9. The experimental results are close to the simulation results. As shown in Figure 10, the THD of the load current with the proposed algorithm at $m = 0.4$ is bigger it at $m = 0.9$. Since the THD of the load current under the proposed algorithm is smaller than 5%, this complies in the standard IEC61000-4-30 Edition 2 Class A. The experimental results in Figure 10 show that the amplitude of lower harmonics is very small. That is one of the advantages of the proposed technique. The measured THD values in Figure 10 are close to the simulation results shown in Figure 7.

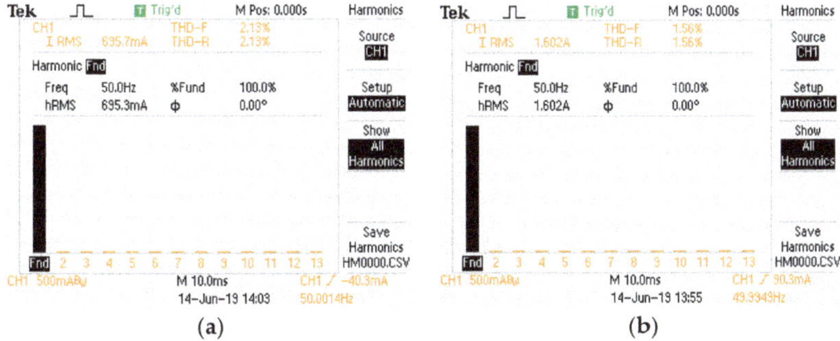

Figure 10. The load current fast Fourier transform (FFT) and its THD. (**a**) $m = 0.4$ and (**b**) $m = 0.9$.

6. Conclusions

The paper presents the carrier based PWM algorithm for 5L-HBT^{2}I by using the offset function to reduce the number of commutations on the power switches. Since the phase leg switches are reduced, the number of commutations at maximum phase current are absolute, the two-level inverter in 5L-HBT^{2}I is on six-step mode, and the switching losses of the inverter 5L-HBT^{2}I are reduced under the proposed algorithm. The switching loss reduction value, in the proposed algorithm for 5L-HBT^{2}I, is maximized when $m = 0.5$, that is equivalent to a reduction of 78%. Since there is no increase conductive loss when applying the proposed algorithm on 5L-HBT2I, the power loss of the inverter is also reduced. Due to no increase of THD, this algorithm not only reduces the number of commutations but also hits ME standards as IEC61000-4-30 Edition 2 Class A at small modulation indexes [28].

Author Contributions: This paper was a collaborative effort among all authors. All authors conceived the methodology, conducted the performance tests and wrote the paper.

Funding: This research was funded by KC 186 project.

Acknowledgments: This work was supported by the Advanced Power Electronics Laboratory, D405 at Ho Chi Minh City University of Technology and Education, Viet Nam.

Conflicts of Interest: The authors declare no conflict of interest.

Nomenclature

5L-HBT^{2}I	Five-level H-bridge T-type inverter
f_s	Switching frequency
NA	Not applicable
Pc-sinPWM	Conduction loss using sine PWM
Pc-proposed	Conduction loss using proposed
PI	Proportional integral
Ps-proposed	Switching loss using proposed
Ps-sinPWM	Switching loss using sine PWM
PV	Photovoltaic
PWM	Pulse width modulation

References

1. Bin, W.; Mehdi, N. *High-Power Converters and Ac Drives*; IEEE Press/Wiley: Hoboken, NJ, USA, 2016.
2. Zhang, J. Modified cascaded multilevel grid-connected inverter to enhance european efficiency and several extended topologies. *IEEE Trans. Ind. Inform.* **2015**, *11*, 1358–1365.
3. Baiju, M.R. Five-level cascaded multilevel motor driver for electrical vehicle with battery charge management. In Proceedings of the Power Engineering Conference, Sydney, NSW, Australia, 14–17 December 2008.

4. Nallamekala, K.K.; Kalyan, U.M.; Sivakumar, K. Harmonic reduction technique with a five-level inverter for four pole induction motor drive. In Proceedings of the International Future Energy Electronics Conference (IFEEC), Tainan, Taiwan, 3–6 November 2013; pp. 482–487.

5. Akagi, H. Multilevel converters: Fundamental circuits and systems. *Proc. IEEE* **2017**, *105*, 2048–2065. [CrossRef]

6. Ye, Z.B.; Xu, Y.M.; Wu, X.; Tan, G.; Deng, X.M.; Wang, Z.C. A simplified PWM strategy for a neutral-point-clamped (NPC) three-level converter with unbalanced dc links. *IEEE Trans. Power Electron.* **2016**, *31*, 3227–3238. [CrossRef]

7. Do, D.T.; Nguyen, M.K. Three-level quasi-switched boost T-type inverter: Analysis, PWM control, and verification. *IEEE Trans. Ind. Electron.* **2018**, *65*, 8320–8329. [CrossRef]

8. Barros, J.D.; Silva, J.F.; Jesus, É.G. Fast-predictive optimal control of NPC multilevel converters. *IEEE Trans. Ind. Electron.* **2013**, *60*, 619–627. [CrossRef]

9. Choudhury, A.; Pillay, P.; Williamson, S. DC-bus voltage balancing algorithm for three-level neutral-point-clamped (NPC) traction inverter drive with modified virtual space vector. *IEEE Trans. Ind. Appl.* **2016**, *52*, 3958–3967. [CrossRef]

10. Abdelhakim, A.; Mattavelli, P.; Spiazzi, G. Three-phase three-level flying capacitors split-source inverters: Analysis and modulation. *IEEE Trans. Ind. Electron.* **2017**, *64*, 4571–4580. [CrossRef]

11. He, L.; Cheng, C. A flying-capacitor-clamped five-level inverter based on bridge modular switched-capacitor topology. *IEEE Trans. Ind. Electron.* **2016**, *63*, 7814–7822. [CrossRef]

12. Du, S.; Wu, B.; Zargari, N. Common-mode voltage elimination for variable-speed motor drive based on flying-capacitor modular multilevel converter. *IEEE Trans. Power Electron.* **2018**, *33*, 5621–5628. [CrossRef]

13. Sayli, K.; Rohini, M.; Amarjeet, P. A 5-Level single phase flying capacitor multilevel inverter. *Int. Res. J. Eng. Technol. (IRJET)* **2017**, *4*, 348–352.

14. Gadalla, A.S.; Yan, X.; Altahir, S.Y.; Hasabelrasul, H. Evaluating the capacity of power and energy balance for cascaded H-bridge multilevel inverter using different PWM techniques. *J. Eng.* **2017**, *13*, 1713–1718. [CrossRef]

15. Malinowski, M.; Gopakumar, K.; Rodriguez, J.; Pe, X.; Rez, M.A. A Survey on cascaded multilevel inverters. *IEEE Trans. Ind. Electron.* **2010**, *57*, 2197–2206. [CrossRef]

16. Nordin, N.M.; Idris, N.R.N.; Azli, N.A. Direct Torque Control with 5-level cascaded H-bridge multilevel inverter for induction machines. In Proceedings of the IECON 2011—37th Annual Conference of the IEEE Industrial Electronics Society, Melbourne, VIC, Australia, 7–10 November 2011; pp. 4691–4697.

17. Shin, H.; Lee, K.; Choi, J.; Seo, S.; Lee, J. Power loss comparison with different PWM methods for 3L-NPC inverter and 3L-T type inverter. In Proceedings of the IEEE International Power Electronics and Application Conference and Exposition (APEC), Shanghai, China, 5–8 November 2014; pp. 1322–1327.

18. Rahim, N.A.; Selvaraj, J. Multistring five-level inverter with novel PWM control scheme for PV application. *IEEE Trans. Ind. Electron.* **2010**, *57*, 2111–2123. [CrossRef]

19. Selvaraj, J.; Rahim, N.A. Multilevel inverter for grid-connected PV system employing digital PI controller. *IEEE Trans. Ind. Electron.* **2010**, *56*, 149–158. [CrossRef]

20. Verdugo, C.C.; Kouro, S.; Rojas, C.; Meynard, T. Comparison of single-phase T-type multilevel converters for grid-connected PV systems. In Proceedings of the IEEE Energy Conversion Congress and Exposition (ECCE), Montreal, QC, Canada, 20–24 September 2015; pp. 3319–3325.

21. Vijaybabu, S.; Naveen Kumar, A.; Rama Krishna, A. Reducing switching losses in cascaded multilevel inverters using hybrid-modulation techniques. *Int. J. Eng. Sci. Invent.* **2013**, *2*, 26–36.

22. Sri Matha, S.; Thirumala, P. Switching losses and harmonic investigations in multi-level inverter. *Int. J. Innov. Res. Technol. IJIRT* **2017**, *4*, 172–178.

23. Zhao, D.; Narayanan, G.; Ayyanar, R. Switching loss characteristics of sequences involving active state division in space vector based PWM. In Proceedings of the IEEE Applied Power Electronics Conference and Exposition (APEC), Anaheim, CA, USA, 22–26 February 2004; pp. 479–485.

24. Bierhoff, M.H.; Fuchs, F.W. Semiconductor losses in voltage source and current source IGBT converters based on analytical derivation. In Proceedings of the IEEE Power Electronics Specialists Conference (SPEC), Aachen, Germany, 20–25 June 2004; Volume 4, pp. 2836–2842.

25. Chaturvedi, P.K.; Jain, S.; Agrawal, P.; Nema, R.K.; Sao, K.K. Switching losses and harmonic investigations in multilevel inverters. *IETE J. Res.* **2008**, *54*, 295–305. [CrossRef]

26. Wu, B.; Cheng, Z. A novel switching sequence design for five-level H NPC-bridge inverters with improved output voltage spectrum and minimized device switching frequency. *IEEE Trans. Power Electron.* **2007**, *22*, 2138–2145.

27. Nguyen, V.N.; Nguyen, K.T.T.; Quach, T.H.; Lee, H.H. A reduced switching loss PWM strategy to eliminate common mode voltage in multilevel inverters. In Proceedings of the IEEE Energy Conversion Congress and Exposition, Pittsburgh, PA, USA, 14–18 September 2014; pp. 219–226.

28. Abdelhakim, A.; Davari, P.; Blaabjerg, F.; Mattavelli, P. Switching loss reduction in the three-phase quasi-Z-source inverters utilizing modified space vector modulation strategies. *IEEE Trans. Ind. Electron.* **2018**, *33*, 4045–4060. [CrossRef]

 © 2019 by the authors. Licensee MDPI, Basel, Switzerland. This article is an open access article distributed under the terms and conditions of the Creative Commons Attribution (CC BY) license (http://creativecommons.org/licenses/by/4.0/).

 *electronics*

Article

New "Full-Bridge Buck Inverter–DC Motor" System: Steady-State and Dynamic Analysis and Experimental Validation

Eduardo Hernández-Márquez [1], Carlos Alejandro Avila-Rea [2], José Rafael García-Sánchez [3], Ramón Silva-Ortigoza [2,*], Magdalena Marciano-Melchor [2], Mariana Marcelino-Aranda [4], Alfredo Roldán-Caballero [2] and Celso Márquez-Sánchez [5]

[1]   Departamento de Mecatrónica, Instituto Tecnológico Superior de Poza Rica, Veracruz 93230, Mexico; eduardo.hernandez@itspozarica.edu.mx

[2]   Laboratorio de Mecatrónica & Energía Renovable, Centro de Innovación y Desarrollo Tecnológico en Cómputo, Instituto Politécnico Nacional, Ciudad de México 07700, Mexico; alejandro8@gmail.com (C.A.A.-R.); mmarciano@ipn.mx (M.M.-M.); alfredo_rol@outlook.com (A.R.-C.)

[3]   Departamento de Procesos Productivos, Unidad Lerma, Universidad Autónoma Metropolitana, Estado de México 52005, Mexico; j.garcia@correo.ler.uam.mx

[4]   Sección de Estudios de Posgrado e Investigación, Unidad Profesional Interdisciplinaria de Ingeniería y Ciencias Sociales y Administrativas, Instituto Politécnico Nacional, Ciudad de México 08400, Mexico; mmarcelino@ipn.mx

[5]   Departamento de Sistemas de Información y Comunicaciones, Unidad Lerma, Universidad Autónoma Metropolitana, Estado de México 52005, Mexico; c.marquez@correo.ler.uam.mx

*   Correspondence: rsilvao@ipn.mx; Tel.: +52-55-5729-6000 (ext. 52530)

Received: 18 September 2019; Accepted: 20 October 2019; Published: 24 October 2019

Abstract: A mathematical model of a new "full-bridge Buck inverter–DC motor" system is developed and experimentally validated. First, using circuit theory and the mathematical model of a DC motor, the dynamic behavior of the system under study is deduced. Later, the steady-state, stability, controllability, and flatness properties of the deduced model are described. The flatness property, associated with the mathematical model, is then exploited so that all system variables and the input can be differentially parameterized in terms of the flat output, which is determined by the angular velocity. Then, when a desired trajectory is proposed for the flat output, the input signal is calculated offline and is introduced into the system. In consequence, the validation of the mathematical model for constant and time-varying duty cycles is possible. Such a validation of this mathematical model is tackled from two directions: (1) by circuit simulation through the SimPowerSystems toolbox of Matlab-Simulink and (2) via a prototype of the system built by using Matlab-Simulink and a DS1104 board. The good similarities between the circuit simulation and the experimental results allow satisfactorily validating the mathematical model.

Keywords: motor drives; power converters; full-bridge Buck inverter; DC motor; mathematical model; differential flatness; time-varying duty cycle; circuit simulation; experimental validation

---

1. Introduction

Electronic power converters as drivers for DC motors have been recently studied [1–30]. According to the literature on power converters, the Buck [1–23], the Boost [24–26], and the Buck-Boost [27–30] topologies are the most used. The Buck topology received the most attention. This is, in part, due to the fact that the mathematical model of the Buck topology is linear and, compared with the Boost and the Buck-Boost topologies, the Buck topology does not have a nonminimum phase output variable [31]. As the present paper focuses on the Buck power converter as a driver for a DC motor, a review of state-of-the-art of this topic is presented below.

1.1. Related Work

The literature reviewed has been divided into two approaches: (1) DC/DC Buck converter–DC motor systems [1–20] and (2) DC/DC Buck converter–inverter–DC motor systems [21–23].

Regarding a DC motor when it is fed by a DC/DC Buck converter, the most relevant literature is the following; Lyshevski, in [1], designed a nonlinear PI control that regulates the velocity of the motor shaft. Ahmad et al., in [2], presented a performance assessment of the PI, fuzzy PI, and LQR controls for the tracking problem. Bingöl and Paçaci reported, in [3], a virtual laboratory based on neural networks to control the angular velocity. Sira-Ramírez and Oliver-Salazar [4] used the concepts of active disturbance rejection and differential flatness to design a tracking control for two configurations of a DC/DC Buck converter connected to a DC motor. In [5], Silva-Ortigoza et al. introduced a two-stage sensorless tracking control based on flatness, whose implementation was executed via a Σ–Δ-modulator. In [6], Hoyos et al. designed a robust adaptive quasi-sliding mode regulation control, which is generated through the zero average dynamics (ZAD) technique and a fixed point inducting control (FPIC). Later, Silva-Ortigoza et al. proposed a robust hierarchical control approach based on differential flatness in [7]. Wei et al. in [8] reported a robust adaptive controller based on dynamic surface and sliding mode. A two-stage controller based on sliding mode plus PI control and flatness was reported by Silva-Ortigoza et al. in [9]. Hernández-Guzmán et al., in [10], proposed a simple control scheme by using sliding mode, to regulate the converter current, and three PI controls. These latter to regulate the converter voltage, the motor current, and the angular velocity. Moreover, via sensorless load torque estimation schemes, a passive tracking control based on the exact tracking error dynamics was proposed by Kumar and Thilagar in [11]. Khubalkar et al., in [12], presented the design and realization of standalone digital fractional order PID control for the Buck converter–DC motor system, where the dynamic particle swarm optimization (dPSO) technique is used to tune the gains and the order of the control. By using the concept of differential flatness and a derivative-free nonlinear Kalman filter, Rigatos et al., in [13], designed a control to solve the trajectory tracking problem. Another solution was proposed by Nizami et al. in [14], where a neuroadaptive backstepping tracking control was developed for the system. The dynamic analysis of the Buck converter that uses the combined ZAD-FPIC technique to control the speed of the DC motor, when different reference values are considered, was developed in [15] by Hoyos et al. Khubalkar et al., in [16], presented for the DC/DC Buck converter driving a DC motor a digital implementation of a fractional order PID control, whose parameters are tuned through the improved inertia weight dPSO technique. A flatness-based tracking control implemented in successive loops was presented by Rigatos et al. in [17]. More recently, the speed regulation problem was addressed by Yang et al. in [18], by using a robust predictive control via a discrete-time reduced-order GPI observer. Additionally, other important contributions related to the connection of a DC/DC Buck converter and a DC motor have been reported in [19,20].

On the other hand, regarding a DC motor when it is fed by a DC/DC Buck converter–inverter, the literature is as follows. In [21], Silva-Ortigoza et al. developed and experimentally validated a mathematical model associated with the DC/DC Buck converter–inverter–DC motor system. Silva-Ortigoza et al. reported, in [22], a passive tracking control based on the exact tracking error dynamics. Robust tracking controls were proposed by Hernández-Márquez et al. in [23].

1.2. Discussion of Related Work and Contribution

In accordance with the aforementioned, different approaches have been proposed for a DC motor fed by a DC/DC Buck converter when the unidirectional rotation of the motor shaft is considered [1–19]. This unidirectional rotation emerges because the Buck converter only delivers unipolar voltages. In this regard, when an inverter is integrated between the converter and the DC motor, bidirectional rotation of the motor shaft is achieved, giving rise to the "DC/DC Buck converter–inverter–DC motor" system [21].

Related to this system, the trajectory tracking problem has been addressed in [22,23]. Note that as such a system includes an inverter connected to the DC motor, an abrupt behavior of the voltages and currents is generated because of the hard switching of the transistors composing the inverter; consequently, the useful life of the DC motor could be reduced. One manner to attenuate the abruptness of the voltages and currents and at the same time to drive a bipolar voltage to the DC motor is through the full-bridge Buck inverter, giving rise to the new "full-bridge Buck inverter–DC motor" system [32]. Thus, compared with [32], the contribution of the present paper is fourfold:

1. The steady-state, stability, controllability, and flatness properties associated with the dynamic behavior of the "full-bridge Buck inverter–DC motor" system are presented.
2. The differential flatness property [33] linked to the mathematical model of the system under study is exploited, with the aim of obtaining the reference trajectories of the system offline so that the mathematical model can be validated when time-varying duty cycles are considered.
3. Obtaining circuit simulation results when the input and the reference signals are introduced into the system via the SimPowerSystems toolbox of Matlab-Simulink.
4. Obtaining experimental results when the input and the reference signals are introduced to a prototype of the system built through Matlab-Simulink along with a DS1104 board.

The remainder of this paper is organized as follows. In Section 2, the generalities of the full-bridge inverter DC–motor system and some static/dynamic properties and the generation of the reference trajectories, via the flatness concept, are presented. To validate the proposed mathematical model, in Section 3, circuit simulation and experimental results are shown. Later, a discussion of the obtained results is introduced in Section 4. Finally, concluding remarks are given in Section 5.

2. Materials and Methods

This section describes the key concepts of the full-bridge Buck inverter–DC motor system. First, the mathematical model of the system is obtained by using the circuit theory and the mathematical model of a DC motor. Second, some static and dynamic properties related to the deduced model are listed. Last, the generation of the reference trajectories is introduced.

2.1. Model of the "Full-Bridge Buck Inverter–DC Motor" System

In the following, the generalities of the full-bridge Buck inverter–DC motor system and the deduction of its corresponding mathematical model are presented.

2.1.1. Generalities of the Full-Bridge Buck Inverter–DC Motor System

The electronic diagram of the full-bridge Buck inverter–DC motor topology is drawn in Figure 1a. Such a circuit can be divided into two parts: (1) the full-bridge Buck inverter and (2) the DC motor. The full-bridge Buck inverter modulates and supplies the bipolar voltage v to the DC motor via the input signal u. This part also contains the following; a power supply E; an array of four MOSFET transistors—Q_1, $\overline{Q}_1$, Q_2, and $\overline{Q}_2$; and a low-pass filter composed of R, C, and L (where the current i flows through). In addition, the DC motor is the actuator of the system and is made up of the elements L_a, R_a and a variable i_a, corresponding to the inductance, resistance, and armature current. Here, ω is the angular velocity of the shaft. Other important values of the DC motor are b, k_m, J, and k_e, which correspond to the viscous friction coefficient of the motor, the motor torque constant, the moment of inertia of the rotor, and the counterelectromotive force constant, respectively. Additionally, Figure 1b depicts the ideal configuration of the system, which will be described in the following section.

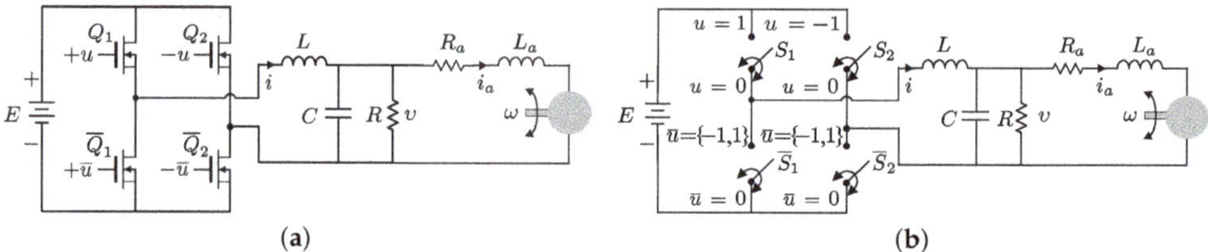

Figure 1. The full-bridge Buck inverter–DC motor topology: (**a**) proposed structure and (**b**) ideal configuration.

2.1.2. Mathematical Model of the Proposed "Full-Bridge Buck Inverter–DC Motor" Topology

In Figure 1b, the ideal structure of the proposed "full-bridge Buck inverter–DC motor" topology is depicted. In this figure, the transistors Q_1, Q_2, $\overline{Q}_1$, and $\overline{Q}_2$ are replaced by switches S_1, S_2, $\overline{S}_1$, and $\overline{S}_2$, respectively. When switches S_1 or S_2 are on, the possible values of the input signal u are 1 or -1. These values depend on the voltage polarity, or operating cycle, that is desired to be generated in load R. That is, for a positive voltage, or positive cycle, the switch S_1 will be on and the input signal u will be 1. For a negative voltage, or negative cycle, the switch S_2 will be on and the output signal u will be -1. On the other hand, when switches S_1 and S_2 are off, the input signal u is 0. During this commutation process, the switches $\overline{S}_1$ and $\overline{S}_2$ are activated complementarily to S_1 and S_2. With the aim of easing the deduction of the mathematical model of the full-bridge Buck inverter–DC motor topology, different structures (associated with the positions of the inverter switches) will be analyzed. Thus, in accordance with the bipolarity of the voltage v, the deduction of the mathematical model will be divided into positive and negative cycles.

Positive Cycle

The generation of a positive voltage, i.e., the clockwise movement of the motor shaft, is executed when the ideal circuit of Figure 1b is simplified to the circuit shown in Figure 2a. It is noteworthy that switches S_2 and $\overline{S}_2$ are in a fixed position, whereas switches S_1 and $\overline{S}_1$ work complementarily, switching their position according to the input signal u. Thus, similar to a Buck converter, the energy charging in the LC filter occurs when the input signal $u = 1$, whereas energy discharging occurs when $u = 0$. This behavior is summarized in Figure 2b.

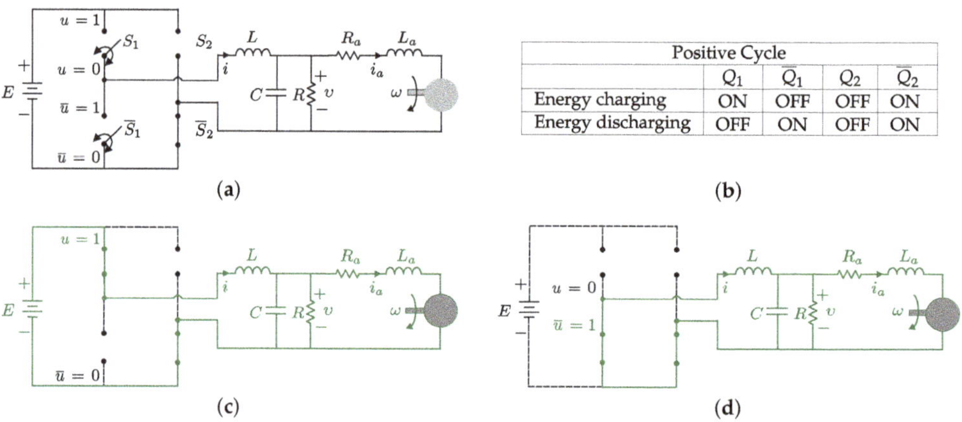

	Positive Cycle			
	Q_1	$\overline{Q}_1$	Q_2	$\overline{Q}_2$
Energy charging	ON	OFF	OFF	ON
Energy discharging	OFF	ON	OFF	ON

(a) (b) (c) (d)

Figure 2. The full-bridge Buck inverter–DC motor system: (**a**) ideal circuit for positive cycle; (**b**) in positive cycle Q_1 and $\overline{Q}_1$ of Figure 1 operate as a transistor and as a diode, respectively; (**c**) energy charging mode of operation; and (**d**) energy discharging mode of operation.

(i) Energy charging. In this mode of operation, a part of the energy supplied by the power supply is stored in the LC filter. Figure 2c depicts the structure of this operating mode. By using the mathematical model of a DC motor [34,35] and applying Kirchhoff's laws to the electric circuit of Figure 2c, the following system of differential equations is obtained:

$$L\frac{di}{dt} = -v + E, \tag{1}$$

$$C\frac{dv}{dt} = i - \frac{v}{R} - i_a, \tag{2}$$

$$L_a\frac{di_a}{dt} = v - R_a i_a - k_e \omega, \tag{3}$$

$$J\frac{d\omega}{dt} = k_m i_a - b\omega. \tag{4}$$

(ii) Energy discharging. Here, as the LC filter is no longer connected to the power supply, the energy stored in the filter is released directly to the resistance R and to the DC motor. Figure 2d shows the connection of this operating mode. By using the mathematical model of a DC motor and applying Kirchhoff's laws, the following system associated with the circuit of Figure 2d is obtained:

$$L\frac{di}{dt} = -v, \tag{5}$$

$$C\frac{dv}{dt} = i - \frac{v}{R} - i_a, \tag{6}$$

$$L_a\frac{di_a}{dt} = v - R_a i_a - k_e \omega, \tag{7}$$

$$J\frac{d\omega}{dt} = k_m i_a - b\omega. \tag{8}$$

Negative Cycle

In generating the negative voltage, the ideal circuit of Figure 1b reduces to the one shown in Figure 3a. Here, switches S_1 and $\overline{S}_1$ are in a fixed position, whereas switches S_2 and $\overline{S}_2$ work complementarily, switching their position according to the input u. Similarly to the positive cycle, there is energy charging or discharging in the LC filter. This behavior is summarized in Figure 3b.

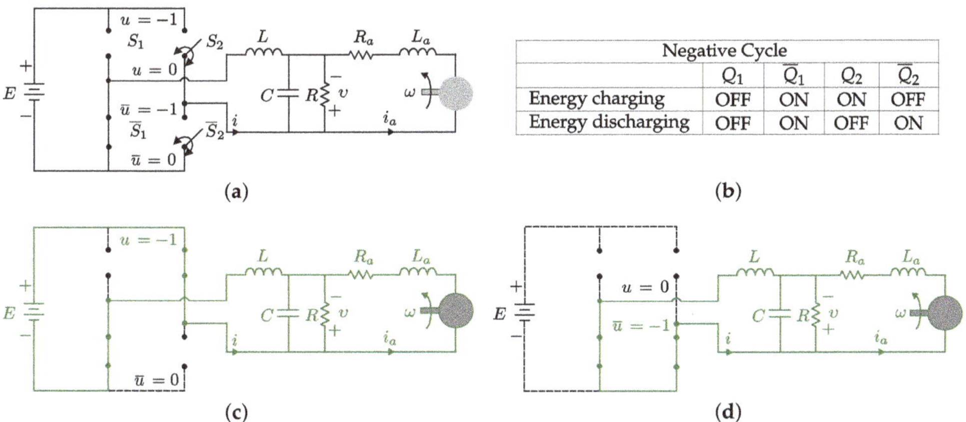

Negative Cycle				
	Q_1	$\overline{Q}_1$	Q_2	$\overline{Q}_2$
Energy charging	OFF	ON	ON	OFF
Energy discharging	OFF	ON	OFF	ON

(a)   **(b)**

(c)   **(d)**

Figure 3. The full-bridge Buck inverter–DC motor system: (**a**) ideal circuit for negative cycle; (**b**) in negative cycle Q_2 and $\overline{Q}_2$ of Figure 1 operate as a transistor and as a diode, respectively; (**c**) energy charging mode of operation; and (**d**) energy discharging mode of operation.

(i) Energy charging. The circuit allowing energy charging for negative cycles is shown in Figure 3c. The model related to the circuit of Figure 3c, after applying Kirchhoff's laws and considering the mathematical model of a DC motor, is given by

$$L\frac{di}{dt} = -v - E, \tag{9}$$

$$C\frac{dv}{dt} = i - \frac{v}{R} - i_a, \tag{10}$$

$$L_a\frac{di_a}{dt} = v - R_a i_a - k_e \omega, \tag{11}$$

$$J\frac{d\omega}{dt} = k_m i_a - b\omega. \tag{12}$$

(ii) Energy discharging. Lastly, the energy discharging in this mode of operation is presented in Figure 3d. For this mode (see Figure 3d), the mathematical model is determined by the following system of differential equations:

$$L\frac{di}{dt} = -v, \tag{13}$$

$$C\frac{dv}{dt} = i - \frac{v}{R} - i_a, \tag{14}$$

$$L_a\frac{di_a}{dt} = v - R_a i_a - k_e \omega, \tag{15}$$

$$J\frac{d\omega}{dt} = k_m i_a - b\omega. \tag{16}$$

By unifying the four modes (see Figures 2c,d, and 3c,d), represented by Equations (1)–(16), the model of the full-bridge Buck inverter–DC motor topology is given by

$$L\frac{di}{dt} = -v + Eu, \tag{17}$$

$$C\frac{dv}{dt} = i - \frac{v}{R} - i_a, \tag{18}$$

$$L_a\frac{di_a}{dt} = v - R_a i_a - k_e \omega, \tag{19}$$

$$J\frac{d\omega}{dt} = k_m i_a - b\omega, \tag{20}$$

where $u \in \{-1, 0, 1\}$ are the positions of the switches. Due to the discrete nature of the system modeled by Equations (17)–(20), it is usual to call it a "switched model". In contrast, the continuous model or "average model" associated with the full-bridge Buck inverter–DC motor system is described by

$$L\frac{di}{dt} = -v + Eu_{av}, \tag{21}$$

$$C\frac{dv}{dt} = i - \frac{v}{R} - i_a, \tag{22}$$

$$L_a\frac{di_a}{dt} = v - R_a i_a - k_e \omega, \tag{23}$$

$$J\frac{d\omega}{dt} = k_m i_a - b\omega, \tag{24}$$

with $u_{av} \in [-1, 1]$ the average input.

2.2. Properties of the "Full-Bridge Buck Inverter–DC Motor" System

This section presents the most relevant static and dynamic properties of the full-bridge Buck inverter–DC motor system. These properties bring qualitative information about the behavior of such a system. Particularly, the steady-state, stability, controllability, and flatness properties of the system are described.

2.2.1. Steady-State

The steady-state analysis predicts the behavior of the full-bridge Buck inverter–DC motor system, given by (21)–(24), when its variables and input are in equilibrium. This is,

$$0 = -\bar{v} + E\bar{u}_{av}, \tag{25}$$

$$0 = i - \frac{\bar{v}}{R} - \bar{i}_a, \tag{26}$$

$$0 = \bar{v} - R_a\bar{i}_a - k_e\bar{\omega}, \tag{27}$$

$$0 = k_m\bar{i}_a - b\bar{\omega}, \tag{28}$$

where the overline means the nominal or constant value of such variables and input. After performing some algebraic manipulations, the equilibrium point (25)–(28) can be expressed in terms of the variable of interest $\bar{\omega}$ as follows,

$$\bar{i}_a = \frac{b}{k_m}\bar{\omega}, \tag{29}$$

$$\bar{v} = \left(\frac{bR_a}{k_m} + k_e\right)\bar{\omega}, \tag{30}$$

$$\bar{i} = \left(\frac{bR_a + k_ek_m + bR}{k_mR}\right)\bar{\omega}, \tag{31}$$

$$\bar{u}_{av} = \left(\frac{bR_a + k_ek_m}{Ek_m}\right)\bar{\omega}. \tag{32}$$

2.2.2. Stability

When analyzing the stability of a dynamic linear system two cases arise. The first is related to the zero state-response, where the output is expected to be bounded if the input is also bounded and the initial condition is equal to zero, meaning that the system is BIBO stable. The second case is about the zero-input response, where the system has no input and has nonzero initial condition. In this way, the system will be stable in the sense of Lyapunov if the output response is bounded and will be asymptotically stable if the output response approaches zero as $t \to \infty$. Both cases can be assessed through the roots of the characteristic polynomial associated with matrix $\mathbf{A}$ of the state space model representation. Thus, if the roots of such a characteristic polynomial have negative real part then the system is completely stable.

Regarding the full-bridge Buck inverter–DC motor system, the state space representation of its model (21)–(24) is given by

$$\dot{x} = \mathbf{A}x + \mathbf{B}u_{av},$$
$$y = \mathbf{C}x, \tag{33}$$

where

$$x = \begin{bmatrix} i \\ v \\ i_a \\ \omega \end{bmatrix}, \qquad A = \begin{bmatrix} 0 & -\frac{1}{L} & 0 & 0 \\ \frac{1}{C} & -\frac{1}{RC} & -\frac{1}{C} & 0 \\ 0 & \frac{1}{L_a} & -\frac{R_a}{L_a} & -\frac{k_e}{L_a} \\ 0 & 0 & \frac{k_m}{J} & -\frac{b}{J} \end{bmatrix}, \qquad B = \begin{bmatrix} \frac{E}{L} \\ 0 \\ 0 \\ 0 \end{bmatrix}, \qquad C = [0\ 0\ 0\ 1]. \tag{34}$$

While the characteristic polynomial associated with **A** is

$$P(s) = a_0 s^4 + a_1 s^3 + a_2 s^2 + a_3 s + a_4, \tag{35}$$

with

$$a_0 = 1,$$
$$a_1 = \frac{bL_a RC + JR_a RC + JL_a}{JL_a RC},$$
$$a_2 = \frac{JL_a R + JRL + bR_a RCL + k_e k_m RCL + bL_a L + JR_a L}{JL_a RCL},$$
$$a_3 = \frac{bL_a R + bRL + JR_a R + bR_a L + k_e k_m L}{JL_a RCL},$$
$$a_4 = \frac{bR_a + k_e k_m}{JL_a CL}.$$

By using the following Routh array,

$$\begin{array}{c|ccc} s^4 & a_0 & a_2 & a_4 \\ s^3 & a_1 & a_3 & \\ s^2 & b_1 & b_2 & \\ s^1 & c_1 & & \\ s^0 & d_1 & & \end{array} \tag{36}$$

where

$$b_1 = \frac{a_1 a_2 - a_0 a_3}{a_1},$$
$$b_2 = \frac{a_1 a_4 - a_0 a_5}{a_1} = a_4,$$
$$c_1 = \frac{b_1 a_3 - a_1 b_2}{b_1} = \frac{a_1 a_2 a_3 - a_0 a_3^2 - a_1^2 a_4}{a_1 a_2 - a_0 a_3},$$
$$d_1 = \frac{c_1 b_2 - b_1 b_3}{c_1} = b_2 = a_4,$$

It can be demonstrated that the roots of (35) have negative real part if a_0, a_1, a_2, a_3, a_4, b_1, c_1, and d_1 are positive. As all system parameters associated with (21)–(24) are positive and after computing (36), it is concluded that the full-bridge Buck inverter–DC motor system is completely stable.

2.2.3. Controllability

The controllability property of a dynamic system is crucial in control theory. This property states that if an input to a system can be found such that it takes the vector state from a desired initial state to a desired final state, the system is controllable; in other case, the system is uncontrollable. With the aim of determining whether a system is controllable or not, a controllability matrix $\mathcal{C}$ can be constructed. If matrix C is of rank n, being n the dimension of the vector state, then the system is completely controllable.

Regarding the full-bridge Buck inverter–DC motor system, represented by Equation (33), the associated controllability matrix is given by

$$\mathcal{C} = [\mathbf{B} \ \mathbf{AB} \ \mathbf{A}^2\mathbf{B} \ \mathbf{A}^3\mathbf{B}]$$

$$= \begin{bmatrix} \frac{E}{L} & 0 & -\frac{E}{L^2C} & \frac{E}{RL^2} \\ 0 & \frac{E}{LC} & -\frac{E}{RL} & -\frac{E(R^2LC+R^2L_aC-LL_aC^4)}{R^2L^2L_aC^3} \\ 0 & 0 & \frac{E}{LL_aC} & -\frac{E(RR_a+L_aC)}{RLL_a^2C} \\ 0 & 0 & 0 & \frac{Ek_m}{JLL_aC} \end{bmatrix}, \tag{37}$$

with matrices $\mathbf{A}$ and $\mathbf{B}$ defined in (34). In this way, after calculating the determinant of matrix $\mathcal{C}$, one obtains

$$det\, \mathcal{C} = \frac{E^4 k_m}{JL^4 L_a^2 C^3} \neq 0, \tag{38}$$

meaning that the system is controllable.

On the other hand, an important property directly linked with controllability is that of differential flatness. This latter states that if a system is differentially flat [33], then it is controllable. This, in turn, means that the vector state and the input can be differentially parameterized in terms of the flat output and a finite number of its derivatives with respect to time. Moreover, there is a relation between the differential parametrization and the steady-state behavior, as the latter can be also obtained when the time derivatives of the flat output are equating to zero.

Note that the differential flatness property has been exploited during the past few years in DC/DC power converters-DC motor and DC/DC power converters-inverter-DC motor systems for different purposes. The most common ones are: (a) as a generator of time-varying reference trajectories to be used in validating mathematical models [21,25] and in passive controls [22,24,28] and (b) for control design purposes [4,5,7,9,23,29]. This paper exploits the flatness property with the intention of generating the reference trajectories for validating the obtained mathematical model, as will be presented in the following section.

2.3. Generation of Reference Trajectories via Differential Flatness

After finding that $det\, \mathcal{C} \neq 0$, i.e., the full-bridge Buck inverter–DC motor system is differentially flat, the flat output of the overall system is found through the following mathematical statement:

$$[0\,0\,0\,1]\mathcal{C}^{-1}x = \frac{JLL_aC}{Ek_m}\omega, \tag{39}$$

and, without loss of generality, the flat output of the system described by Equations (21)–(24) can be taken as

$$\mathcal{S} = \omega, \tag{40}$$

which corresponds to the angular velocity of the full-bridge Buck inverter–DC motor system. Therefore, the variables i_a, v, i, and the input u_{av} of the system can be expressed in terms of $\mathcal{S}$ and its successive derivatives with respect to time as follows,

$$\omega = \mathcal{S}, \tag{41}$$

$$i_a = \frac{J}{k_m}\dot{\mathcal{S}} + \frac{b}{k_m}\mathcal{S}, \tag{42}$$

$$v = \frac{JL_a}{k_m}\ddot{\mathcal{S}} + \left(\frac{bL_a + JR_a}{k_m}\right)\dot{\mathcal{S}} + \left(\frac{bR_a}{k_m} + k_e\right)\mathcal{S}, \tag{43}$$

$$i = \left(\frac{JL_aC}{k_m}\right)\mathcal{S}^{(3)} + \left(\frac{bRL_aC + JRR_aC + JL_a}{Rk_m}\right)\ddot{\mathcal{S}}$$
$$+ \left(\frac{bL_a + JR_a + JR + bRR_aC + RK_eK_mC}{k_mR}\right)\dot{\mathcal{S}} + \left(\frac{bR_a + k_ek_m + bR}{k_mR}\right)\mathcal{S}, \tag{44}$$

$$u_{av} = \left(\frac{JL_aLC}{Ek_m}\right)\mathcal{S}^{(4)} + \left(\frac{bRLL_aC + JRR_aLC + JLL_a}{Ek_mR}\right)\mathcal{S}^{(3)}$$
$$+ \left(\frac{bLL_a + JR_aL + JRL + bRR_aLC + K_eK_mRLC + JRL_a}{Ek_mR}\right)\ddot{\mathcal{S}}$$
$$+ \left(\frac{bR_aL + k_ek_mL + bRL + bRL_a + JRR_a}{Ek_mR}\right)\dot{\mathcal{S}} + \left(\frac{bR_a + k_ek_m}{Ek_m}\right)\mathcal{S}. \tag{45}$$

From the previous results, if a desired trajectory $\mathcal{S}^*$ is proposed, i.e., ω^*, then from Equation (45), the input to be introduced into the full-bridge Buck inverter–DC motor system is

$$u_{av}^* = \left(\frac{JL_aLC}{Ek_m}\right)\mathcal{S}^{*(4)} + \left(\frac{bRLL_aC + JRR_aLC + JLL_a}{Ek_mR}\right)\mathcal{S}^{*(3)}$$
$$+ \left(\frac{bLL_a + JR_aL + JRL + bRR_aLC + K_eK_mRLC + JRL_a}{Ek_mR}\right)\ddot{\mathcal{S}}^*$$
$$+ \left(\frac{bR_aL + k_ek_mL + bRL + bRL_a + JRR_a}{Ek_mR}\right)\dot{\mathcal{S}}^* + \left(\frac{bR_a + k_ek_m}{Ek_m}\right)\mathcal{S}^*, \tag{46}$$

achieving that ω be similar to ω^*, meaning that the mathematical model is sufficiently accurate. In addition, when $\mathcal{S}^*$ is replaced in Equations (42)–(44), the reference trajectories of the system, i.e., i_a^*, v^*, and i^*, are obtained offline.

3. Results

The mathematical model of the full-bridge Buck inverter–DC motor system will be validated here. This validation is carried out in two directions: (1) by circuit simulation through the SimPowerSystems toolbox of Matlab-Simulink and (2) via a built prototype of the system by using Matlab-Simulink and a DS1104 board. The results of the circuit simulation of the full-bridge Buck inverter–DC motor system will be presented first. Later, the corresponding experimental results associated with the system will be presented.

3.1. Circuit Simulation Results

The connection diagram of the system, built on Matlab-Simulink, along with some simulation results are presented here.

3.1.1. Connection Diagram of the System

The circuit simulation results are obtained through the diagram of the system shown in Figure 4, whose implementation has been executed via the SimPowerSystems toolbox of Matlab-Simulink. The blocks composing the diagram of this figure are detailed below:

- Desired and reference trajectories. In this block, the desired trajectory ω^* and the differential parametrization, see Equations (31)–(33), are programmed to obtain the reference trajectories i_a^*, v^*, and i^*.
- Signals to be plotted. The variables to be plotted are defined in this block. These variables are related to the full-bridge Buck inverter–DC motor system and to the differential parametrization.

- Input signal and PWM. Here, the input u^*_{av}, given by Equation (37), is programmed. Also, through this block, the full-bridge transistors are driven when the switched inputs are generated through the PWM signal.
- Full-bridge Buck inverter–DC motor circuit. This block corresponds to the overall system. The parameters of the full-bridge Buck inverter are given by the following values:

$$R = 48 \,\Omega, \quad C = 4.7 \,\mu\text{F}, \quad L = 4.94 \,\text{mH}, \quad E = 32 \,\text{V}. \tag{47}$$

The sampling frequency of the four transistors, associated with the full-bridge is 50 kHz. The DC motor was manufactured by ENGEL with a 3.1 gearbox with reduction ratio of 14.5:1. Such a motor is a GNM5440E-G3.1 $(24 \,\text{V}, 95 \,\text{W})$, whose parameters are

$$
\begin{aligned}
L_a &= 2.22 \,\text{mH}, & k_m &= 120.1 \times 10^{-3} \,\tfrac{\text{N·m}}{\text{A}}, \\
R_a &= 0.965 \,\Omega, & k_e &= 120.1 \times 10^{-3} \,\tfrac{\text{V·s}}{\text{rad}}, \\
J &= 118.2 \times 10^{-3} \,\text{kg·m}^2, & b &= 129.6 \times 10^{-3} \,\tfrac{\text{N·m·s}}{\text{rad}}.
\end{aligned}
\tag{48}
$$

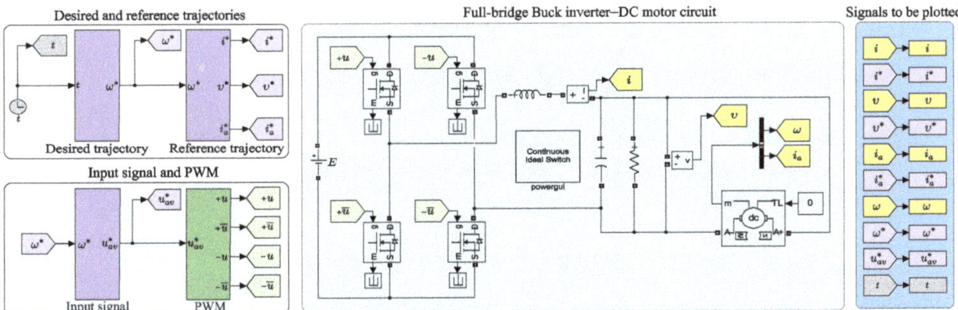

Figure 4. Circuit of the "full-bridge Buck inverter–DC motor" system designed via the SimPowerSystems toolbox of Matlab-Simulink. This circuit is used for validating the deduced mathematical model of such a system.

3.1.2. Circuit Simulation Results

With the intention of validating the obtained mathematical model of the full-bridge Buck inverter–DC motor system, this section presents the circuit simulation results for different desired trajectories of the angular velocity.

Circuit Simulation 1

In this simulation, the desired trajectory ω^* is generated via the following Bézier polynomial:

$$\omega^*(t) = \overline{\omega}_i(t_i) + [\overline{\omega}_f(t_f) - \overline{\omega}_i(t_i)]\varphi(t, t_i, t_f), \tag{49}$$

where $\varphi(t, t_i, t_f)$ is given by

$$
\varphi\left(t, t_i, t_f\right) =
\begin{cases}
0 & \text{for } t \leq t_i, \\[2mm]
\left(\frac{t-t_i}{t_f-t_i}\right)^5 \times \left[r_1 + r_2\left(\frac{t-t_i}{t_f-t_i}\right) + r_3\left(\frac{t-t_i}{t_f-t_i}\right)^2 \right. \\[2mm]
\left. + r_4\left(\frac{t-t_i}{t_f-t_i}\right)^3 + r_5\left(\frac{t-t_i}{t_f-t_i}\right)^4 + r_6\left(\frac{t-t_i}{t_f-t_i}\right)^5 \right] & \text{for } t \in (t_i, t_f), \\[2mm]
1 & \text{for } t \geq t_f,
\end{cases}
\tag{50}
$$

and

$$r_1 = 252, \quad r_2 = -1050, \quad r_3 = 1800, \quad r_4 = -1575, \quad r_5 = 700, \quad r_6 = -126. \tag{51}$$

With proposal (50) and coefficients (51), ω^* smoothly interpolates between $\overline{\omega}_i = -10\frac{\text{rad}}{\text{s}}$ and $\overline{\omega}_f = 10\frac{\text{rad}}{\text{s}}$ over the time interval $[t_i, t_f] = [4\text{ s}, 6\text{ s}]$. Note that if (50) were of different order, then coefficients (51) would be also different. The corresponding results are presented in Figure 5.

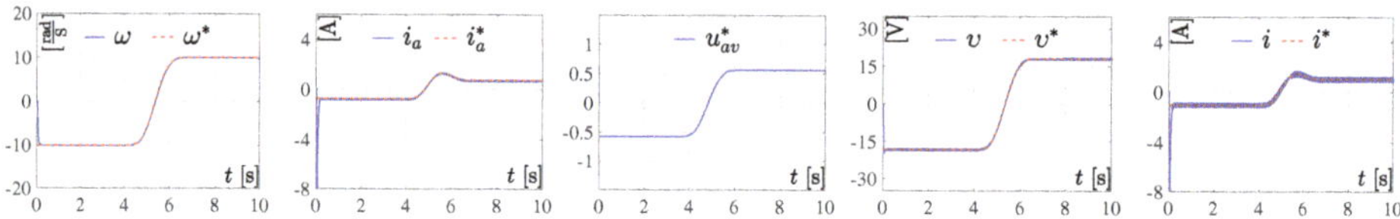

Figure 5. Circuit simulation results for a Bézier polynomial-type trajectory. The results related to the mathematical model are denoted by ω, i_a, v, i, and the results associated with the reference variables are labeled as ω^*, i_a^*, u_{av}^*, v^*, i^*.

Circuit Simulation 2

Here, ω^* is defined by the following sinusoidal function:

$$\omega^*(t) = 10\sin(0.8\pi t). \tag{52}$$

Figure 6 depicts the corresponding simulation results.

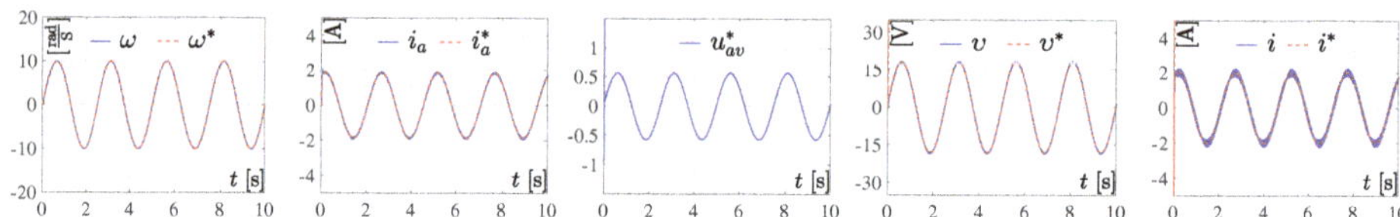

Figure 6. Circuit simulation results for a sinusoidal trajectory. The results of the mathematical model correspond to variables ω, i_a, v, i, whereas the results related to the reference variables are ω^*, i_a^*, u_{av}^*, v^*, i^*.

Circuit Simulation 3

In this case, the trajectory to be tracked has been proposed as

$$\omega^*(t) = 10\left(1 - e^{-2t^2}\right)\sin(0.8\pi t), \tag{53}$$

and the results are shown in Figure 7.

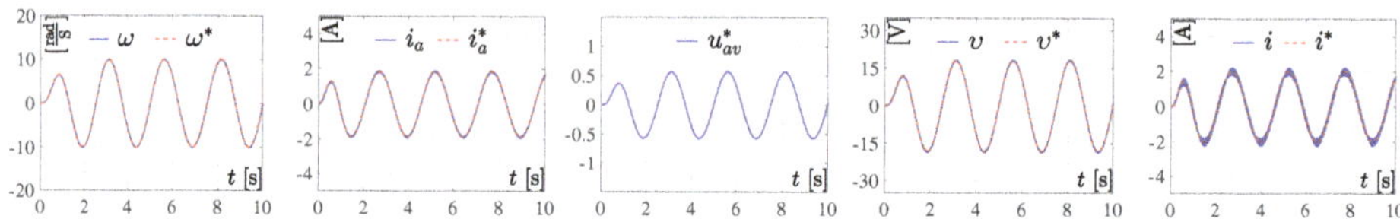

Figure 7. Circuit simulation results for a sinusoidal trajectory with exponential amplitude. The results associated with the mathematical model are ω, i_a, v, i, and the results of the reference variables correspond to ω^*, i_a^*, u_{av}^*, v^*, i^*.

Circuit Simulation 4

Lastly, the trajectory to be tracked in this simulation is given by Equation (54) and the corresponding results are presented in Figure 8.

$$\omega^*(t) = 10 \sin\left(0.125\pi t^{\frac{3}{2}}\right). \tag{54}$$

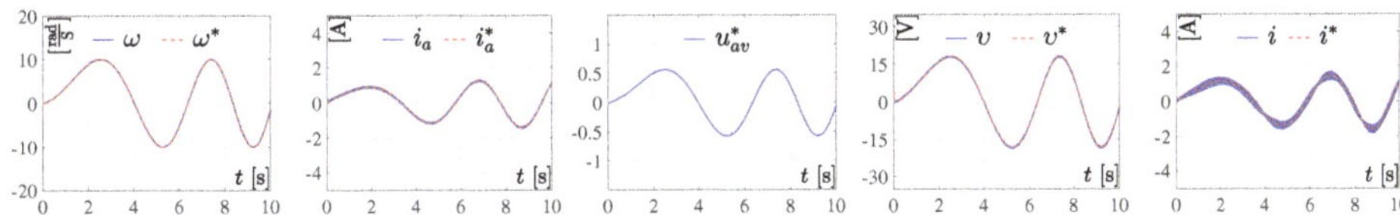

Figure 8. Circuit simulation results for a sinusoidal trajectory with time-varying frequency. The results related to the mathematical model correspond to the signals denoted by ω, i_a, v, i, and the results associated with the reference variables are labeled as ω^*, i_a^*, u_{av}^*, v^*, i^*.

3.2. Results from the Experimental Prototype

This section describes the connection diagram that allows the implementation of the input signal u_{av}^* on the built prototype of the full-bridge Buck inverter–DC motor system shown in Figure 9. Also, the corresponding experimental results are presented.

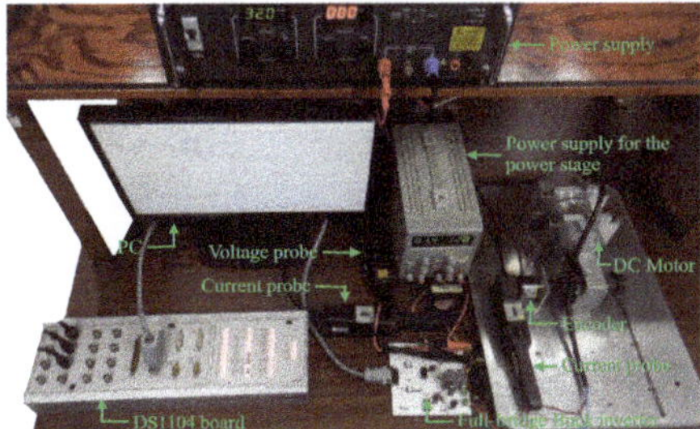

Figure 9. Experimental prototype of the full-bridge Buck inverter–DC motor system.

3.2.1. Experimental Diagram of the System

The experimental results were obtained by using the connection diagram depicted in Figure 10. The blocks composing this figure are described below.

- Desired and reference trajectories. The desired trajectory ω^* and the Equations (31)–(33) to obtain the reference trajectories i_a^*, v^*, and i^* are programmed in this block.
- Input signal. The input u_{av}^*, see Equation (37), is programmed here.
- Signals to be plotted. The variables to be plotted are specified in this block.
- Board and signal processing. This block shows the connections between the DS1104 board and the system. As can be seen, signal conditioning (SC) is executed over the angular position θ, the voltage v, and the currents i_a and i. Also, the input signal u_{av}^* is introduced into the DS1104 board so that the PWM signal can be generated. This latter is processed through the sub-block conditioning circuit (CC) allowing the correct activation of the transistors.
- Full-bridge Buck inverter–DC motor circuit. This block corresponds to the system under study and the values of its parameters were defined in (47). The sampling frequency of the four IRF640 MOSFET transistors associated with the full-bridge is 50 kHz. Additionally, the parameters of the DC motor are the same of those considered in (48).

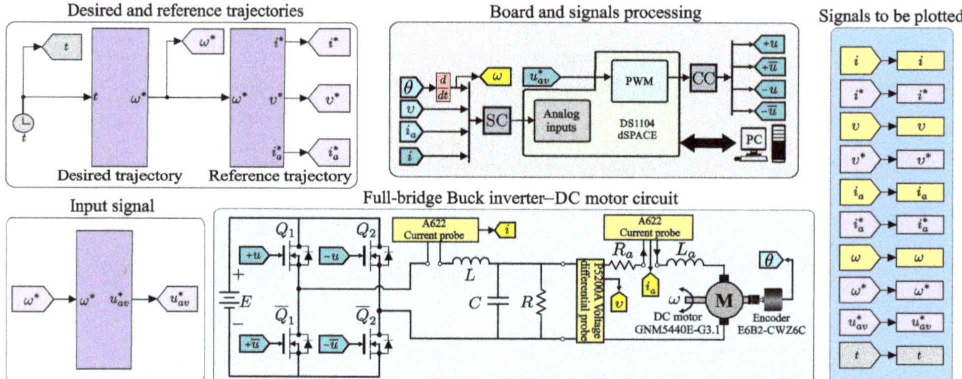

Figure 10. Experimental diagram of the "full-bridge Buck inverter–DC motor" system.

3.2.2. Experimental Results

With the purpose of validating the obtained mathematical model (21)–(24), this section presents the experimental results for the system. In these experiments, and with the purpose of making a fair comparison with the simulation results, the desired trajectories for ω are the same as those considered in the simulation results.

Experiment 1

In this experiment, ω^* is the Bézier-type trajectory defined in (49). The experimental results of the system are presented in Figure 11.

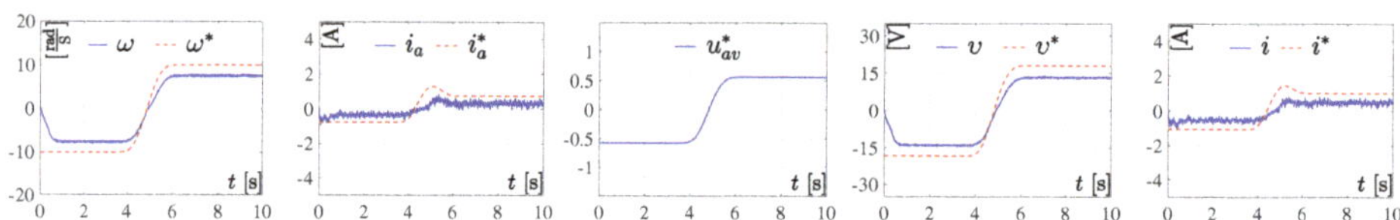

Figure 11. Experimental results for a Bézier-type trajectory. The results of the measured variables are ω, i_a, v, i, and the results of the reference variables are ω^*, i_a^*, u_{av}^*, v^*, i^*.

Experiment 2

Here, ω^* is the sinusoidal trajectory (52). In Figure 12, the corresponding experimental results are depicted.

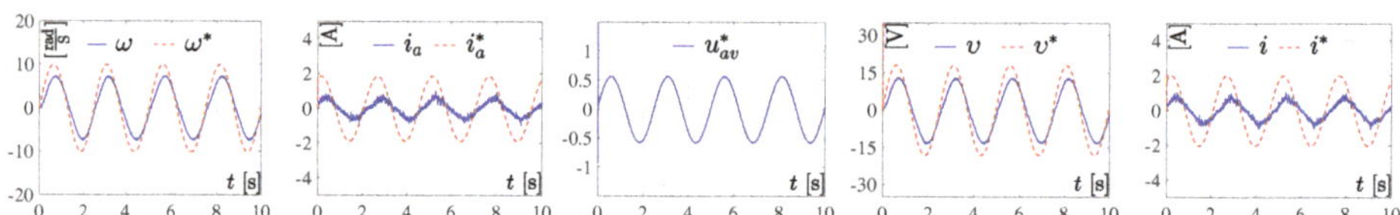

Figure 12. Experimental results for a sinusoidal trajectory. The results related to the measured variables correspond to the signals denoted by ω, i_a, v, i, whereas the results associated with the reference variables are labeled as ω^*, i_a^*, u_{av}^*, v^*, i^*.

Experiment 3

In this experiment, the desired angular velocity is defined as in (53). The corresponding experimental results are shown in Figure 13.

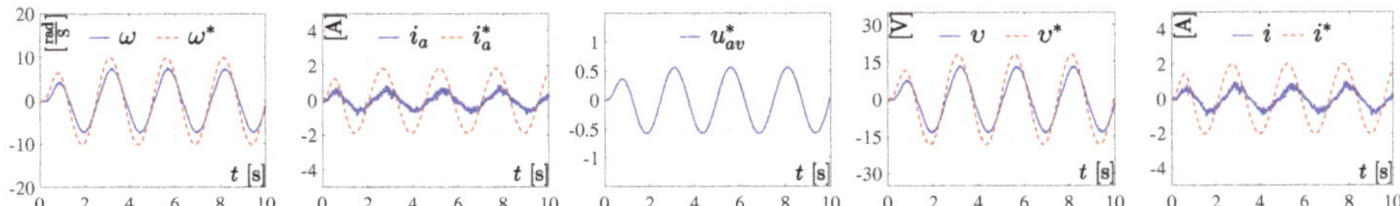

Figure 13. Experimental results for a sinusoidal trajectory with exponential amplitude. The results of the measured variables correspond to ω, i_a, v, i, and the results of the reference variables are ω^*, i_a^*, u_{av}^*, v^*, i^*.

Experiment 4

Lastly, Figure 14 presents the experimental result associated with Equation (54).

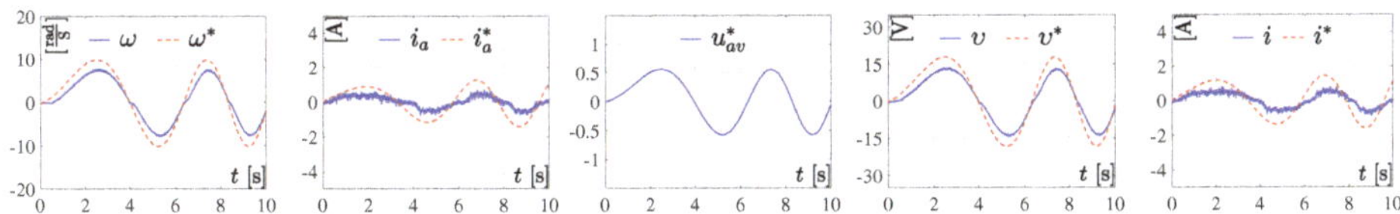

Figure 14. Experimental results for a sinusoidal trajectory with time-varying frequency. The results associated with the measured variables are denoted as ω, i_a, v, i, whereas the results related to the reference variables are labeled as ω^*, i_a^*, u_{av}^*, v^*, i^*.

4. Discussion

As the mathematical model developed herein for the full-bridge Buck inverter–DC motor system was differentially flat, all system variables were parameterized in terms of the flat output. Later, by proposing and replacing ω^* into the differential parametrization of the system, the reference variables (i_a^*, i^*, v^*) and the input signal (u_{av}^*) were found offline. This was made to compare the results of the circuit simulation with the differential parametrization results, both shown in Figures 5–8. The same thing was done in order to compare the experimental results and the differential parametrization results, both shown in Figures 11–14. Regarding these latter, a tracking error between system variables ω, i_a, v, i and reference variables ω^*, i_a^*, v^*, i^* can be observed. Such an error appears because some dynamics were not included into the mathematical model, i.e., energy losses associated with semiconductors and parasitic resistances related to capacitor and inductors. In this sense, note that, due to these omitted dynamics and the existence of the load R, the efficiency of the Buck converter is 89.14 %. On the other hand, if all the neglected dynamics were taken into account, then the mathematical model would be more complex and this is far beyond the scope of this paper. In brief, the obtained results depicted in Figures 5–8 and 11–14 validate the good accuracy of the proposed mathematical model.

5. Conclusions

With the aim of validating, through circuit simulations and experimental results, the proposed mathematical model of the new "full-bridge Buck inverter–DC motor" system, the flatness property has been exploited. Likewise, the steady-state, stability, and controllability properties associated with the dynamic behavior of such a system have been developed.

In the development of the mathematical model of the new "full-bridge Buck inverter–DC motor" system, all components were considered as ideal. This was done with the intention of obtaining a non-complex mathematical model that would still enjoy relatively good accuracy. On the other hand, by applying the flatness concept to the proposed mathematical model, it was found that the flat output of the system is given by ω. Therefore, the vector state and the input signal were differentially parameterized, in terms of the flat output and its successive derivatives with respect to time. Thus, with the help of such a parametrization, the validation of the deduced mathematical model was carried out through circuit simulation and a built prototype of the system. The circuit simulation results (presented in Figures 5–8) and the experimental results (depicted in Figures 11–14) validates the proposed mathematical model despite the small tracking error between system variables and reference variables. It is worth mentioning that such an error could be minimized if the electronic and electric elements were considered nonideal, meaning that energy losses and parasitic resistances should be considered into the mathematical model. However, the advantages of using the model presented in this paper is its simplicity and its accuracy.

Future research will be devoted to the design of feedback tracking controls and their experimental implementation on the prototype of the system that has been built.

Author Contributions: Conceptualization, E.H.-M. and R.S.-O.; Data curation, C.A.A.-R., J.R.G.-S., M.M.-M., A.R.-C., and C.M.-S.; Funding acquisition, E.H.-M., R.S.-O., and M.M.-A.; Investigation, E.H.-M., C.A.A.-R., J.R.G.-S., R.S.-O., A.R.-C., and C.M.-S.; Project administration, M.M.-A.; Resources, E.H.-M., R.S.-O., and M.M.-M.; Software, C.A.A.-R., J.R.G.-S., A.R.-C., and C.M.-S.; Supervision, E.H.-M. and R.S.-O.; Validation, C.A.A.-R., J.R.G.-S., M.M.-M., A.R.-C., and C.M.-S.; Visualization, C.A.A.-R., J.R.G.-S., M.M.-M., M.M.-A., A.R.-C., and C.M.-S.; Writing—original draft, E.H.-M., C.A.A.-R., J.R.G.-S., R.S.-O., A.R.-C., and C.M.-S.; Writing—review & editing, E.H.-M., C.A.A.-R., J.R.G.-S., R.S.-O., M.M.-M., M.M.-A., A.R.-C., and C.M.-S.

Funding: This research was funded by the Comisión de Operación y Fomento de Actividades Académicas (COFAA) and the Secretaría de Investigación y Posgrado (SIP), both from the Instituto Politécnico Nacional, México.

Acknowledgments: This work has been supported by the Secretaría de Investigación y Posgrado del Instituto Politécnico Nacional (SIP-IPN), México. J. R. García-Sánchez and C. Márquez-Sánchez acknowledge the financial support received from the SNI-México. R. Silva-Ortigoza, M. Marciano-Melchor, and M. Marcelino-Aranda acknowledge the financial support received from the IPN programs EDI and COFAA and from the SNI-México. Finally, the work of A. Roldán-Caballero has been supported by the CONACYT-México and BEIFI scholarships.

Conflicts of Interest: The authors declare no conflicts of interest.

Abbreviations

DC	Direct current
PI	Proportional-integral
PID	Proportional-integral-derivative
LQR	Linear quadratic regulator
ZAD	Zero average dynamics
FPIC	Fixed point inducting control
dPSO	Dynamic particle swarm optimization
GPI	Generalized proportional-integral
MOSFET	Metal-oxide-semiconductor field-effect transistor
BIBO	Bounded-input bounded-output
PWM	Pulse width modulation
SC	Signal conditioning
CC	Conditioning circuit

Notation

i	Converter inductor current
v	Converter capacitor voltage
i_a	DC motor armature circuit current
ω	DC motor angular velocity
θ	DC motor angular position
u	Switch position function
$\bar{u}$	Complementary switch position function
u_{av}	Duty cycle or Average input signal
E	Power supply

R	Converter output resistance
C	Converter capacitance
L	Converter inductance
R_a	DC motor armature circuit resistance
L_a	DC motor armature inductance
J	Moment of inertia of the rotor
b	DC motor viscous friction coefficient
k_e	Counterelectromotive force constant
k_m	DC motor torque constant
Q_1, Q_2	MOSFET transistors
$\overline{Q}_1, \overline{Q}_2$	Complementary MOSFET transistors
S_1, S_2	Ideal switches
$\overline{S}_1, \overline{S}_2$	Complementary ideal switches
LC	Analog filter conformed by L and C
x	State vector
$\bar{i}, \bar{v}, \bar{i}_a, \bar{\omega}, \bar{u}_{av}$	Nominal variables and nominal input
$\overline{\omega}_i, \overline{\omega}_f$	Constant angular velocities for interpolating the Bézier polynomial
$\mathbf{A}, \mathbf{B}, \mathbf{C}$	Constant matrices
y	System output
$\mathcal{C}$	Controllability matrix
$\mathcal{S}$	Flat output
$\mathcal{S}^*$	Flat output reference trajectory
$i^*, v^*, i_a^*, \omega^*, u_{av}^*$	System reference trajectories and reference input

References

1. Lyshevski, S.E. *Electromechanical Systems, Electric Machines, and Applied Mechatronics*; CRC Press: Boca Raton, FL, USA, 2000; ISBN 0-8493-2275-8.
2. Ahmad, M.A.; Ismail, R.M.T.R.; Ramli, M.S. Control strategy of Buck converter driven DC motor: A comparative assessment. *Austral. J. Basic Appl. Sci.* **2010**, *4*, 4893–4903.
3. Bingöl, O.; Paçaci, S. A virtual laboratory for neural network controlled DC motors based on a DC-DC Buck converter. *Int. J. Eng. Educ.* **2012**, *28*, 713–723.
4. Sira-Ramírez, H.; Oliver-Salazar, M.A. On the robust control of Buck-converter DC-motor combinations. *IEEE Trans. Power Electron.* **2013**, *28*, 3912–3922. [CrossRef]
5. Silva-Ortigoza, R.; García-Sánchez, J.R.; Alba-Martínez, J.M.; Hernández-Guzmán, V.M.; Marcelino-Aranda, M.; Taud, H.; Bautista-Quintero, R. Two-stage control design of a Buck converter/DC motor system without velocity measurements via a $\Sigma - \Delta$-modulator. *Math. Probl. Eng.* **2013**, *2013*, 929316. [CrossRef]
6. Hoyos, F.E.; Rincón, A.; Taborda, J.A.; Toro, N.; Angulo, F. Adaptive quasi-sliding mode control for permanent magnet DC motor. *Math. Probl. Eng.* **2013**, *2013*, 693685. [CrossRef]
7. Silva-Ortigoza, R.; Márquez-Sánchez, C.; Carrizosa-Corral, F.; Antonio-Cruz, M.; Alba-Martínez, J.M.; Saldaña-González, G. Hierarchical velocity control based on differential flatness for a DC/DC Buck converter-DC motor system. *Math. Probl. Eng.* **2014**, *2014*, 912815. [CrossRef]
8. Wei, F.; Yang, P.; Li, W. Robust adaptive control of DC motor system fed by Buck converter. *Int. J. Control Autom.* **2014**, *7*, 179–190. [CrossRef]
9. Silva-Ortigoza, R.; Hernández-Guzmán, V.M.; Antonio-Cruz, M.; Muñoz-Carrillo, D. DC/DC Buck power converter as a smooth starter for a DC motor based on a hierarchical control. *IEEE Trans. Power Electron.* **2015**, *30*, 1076–1084. [CrossRef]
10. Hernández-Guzmán, V.M.; Silva-Ortigoza, R.; Muñoz-Carrillo, D. Velocity control of a brushed DC-motor driven by a DC to DC Buck power converter. *Int. J. Innov. Comp. Inf. Control* **2015**, *11*, 509–521.
11. Kumar, S.G.; Thilagar, S.H. Sensorless load torque estimation and passivity based control of Buck converter fed DC motor. *Sci. World J.* **2015**, *2015*, 132843. [CrossRef]

12. Khubalkar, S.; Chopade, A.; Junghare, A.; Aware, M.; Das, S. Design and realization of stand-alone digital fractional order PID controller for Buck converter fed DC motor. *Circuits Syst. Signal Process.* **2016**, *35*, 2189–2211. [CrossRef]

13. Rigatos, G.; Siano, P.; Wira, P.; Sayed-Mouchaweh, M. Control of DC–DC converter and DC motor dynamics using differential flatness theory. *Intell. Ind. Syst.* **2016**, *2*, 371–380. [CrossRef]

14. Nizami, T.K.; Chakravarty, A.; Mahanta, C. Design and implementation of a neuro-adaptive backstepping controller for Buck converter fed PMDC-motor. *Control Eng. Pract.* **2017**, *58*, 78–87. [CrossRef]

15. Hoyos Velasco, F.E.; Candelo-Becerra, J.E; Rincón Santamaria, A. Dynamic analysis of a permanent magnet DC motor using a Buck converter controlled by ZAD-FPIC. *Energies* **2018**, *11*, 3388. [CrossRef]

16. Khubalkar, S.W.; Junghare, A.S.; Aware, M.V.; Chopade, A.S.; Das, S. Demonstrative fractional order-PID controller based DC motor drive on digital platform. *ISA Trans.* **2018**, *82*, 79–93. [CrossRef]

17. Rigatos, G.; Siano, P.; Ademi, S.; Wira, P. Flatness-based control of DC-DC converters implemented in successive loops. *Electr. Power Compon. Syst.* **2018**, *46*, 673–687. [CrossRef]

18. Yang, J.; Wu, H.; Hu, L.; Li, S. Robust predictive speed regulation of converter-driven DC motors via a discrete-time reduced-order GPIO. *IEEE Trans. Ind. Electron.* **2019**, *66*, 7893–7903. [CrossRef]

19. Rigatos, G.; Siano, P.; Sayed-Mouchaweh, M. Adaptive neurofuzzy H-infinity control of DC-DC voltage converters. *Neural Comput. Appl.* **2019**, 1–14. [CrossRef]

20. Ismail, A.A.A.; Elnady, A. Advanced drive system for DC motor using multilevel DC/DC Buck converter circuit. *IEEE Access* **2019**, *7*, 54167–54178. [CrossRef]

21. Silva-Ortigoza, R.; Alba-Juárez, J.N.; García-Sánchez, J.R.; Antonio-Cruz, M.; Hernández-Guzmán, V.M.; Taud, H. Modeling and experimental validation of a bidirectional DC/DC Buck power electronic converter–DC motor system. *IEEE Latin Am. Trans.* **2017**, *15*, 1043–1051. [CrossRef]

22. Silva-Ortigoza, R.; Alba-Juárez, J.N.; García-Sánchez, J.R.; Hernández-Guzmán, V.M.; Sosa-Cervantes, C.Y.; Taud, H. A sensorless passivity-based control for the DC/DC Buck converter–inverter–DC motor system, *IEEE Latin Am. Trans.* **2016**, *14*, 4227–4234. [CrossRef]

23. Hernández-Márquez, E.; García-Sánchez, J.R.; Silva Ortigoza, R.; Antonio-Cruz, M.; Hernández-Guzmán, V.M.; Taud, H.; Marcelino-Aranda, M. Bidirectional tracking robust controls for a DC/DC Buck converter-DC motor system. *Complexity* **2018**, *2018*, 1260743. [CrossRef]

24. Linares-Flores, J.; Reger, J.; Sira-Ramírez, H. Load torque estimation and passivity-based control of a Boost-converter/DC-motor combination. *IEEE Trans. Control Syst. Technol.* **2010**, *18*, 1398–1405. [CrossRef]

25. García-Rodríguez, V.H.; Silva-Ortigoza, R.; Hernández-Márquez, E.; García-Sánchez, J.R.; Taud, H. DC/DC Boost converter–inverter as driver for a DC motor: Modeling and experimental verification. *Energies* **2018**, *11*, 2044. [CrossRef]

26. García-Sánchez, J.R.; Hernández-Márquez, E.; Ramírez-Morales, J.; Marciano-Melchor, M.; Marcelino-Aranda, M.; Taud, H.; Silva-Ortigoza, R. A robust differential flatness-based tracking control for the "MIMO DC/DC Boost converter–inverter–DC motor" system: Experimental results. *IEEE Access* **2019**, *7*, 84497–84505. [CrossRef]

27. Linares-Flores, J.; Barahona-Avalos, J.L.; Sira-Ramírez, H.; Contreras-Ordaz, M.A. Robust passivity-based control of a Buck–Boost-converter/DC-motor system: An active disturbance rejection approach. *IEEE Trans. Ind. Appl.* **2012**, *48*, 2362–2371. [CrossRef]

28. Hernández-Márquez, E.; Silva-Ortigoza, R.; García-Sánchez, J.R.; Marcelino-Aranda, M.; Saldaña-González, G. A DC/DC Buck-Boost converter–inverter–DC motor system: Sensorless passivity-based control. *IEEE Access* **2018**, *6*, 31486–31492. [CrossRef]

29. Hernández-Márquez, E.; Avila-Rea C.A.; García-Sánchez, J.R.; Silva-Ortigoza, R.; Silva-Ortigoza, G.; Taud, H.; Marcelino-Aranda, M. Robust tracking controller for a DC/DC Buck-Boost converter–inverter–DC motor system. *Energies* **2018**, *11*, 2500. [CrossRef]

30. Ghazali, M.R.; Ahmad, M.A.; Ismail, R.M.T.R. Adaptive safe experimentation dynamics for data-driven neuroendocrine-PID control of MIMO systems. *IETE J. Res.* **2019**, 1–14. [CrossRef]

31. Sira-Ramírez, H.; Pérez-Moreno, R.A.; Ortega, R.; García-Esteban, M. Passivity-based controllers for the stabilization of DC-to-DC power converters. *Automatica* **1997**, *33*, 499–513. [CrossRef]

32. Hernández-Márquez, E.; Avila-Rea, C.A.; Silva-Ortigoza, R.; García-Sánchez, J.R.; Antonio-Cruz, M.; Taud, H.; Marcelino-Aranda, M. Modeling and simulation of a DC motor fed by a full-bridge Buck inverter. In Proceedings of the 2017 International Conference on Mechatronics, Electronics and Automotive Engineering, ICMEAE 2017, Cuernavaca, Mexico, 21–24 November 2017; pp. 77–81.

33. Sira-Ramírez, H.; Agrawal, S.K. *Differentially Flat Systems*; Marcel Dekker: New York, NY, USA, 2004; ISBN 0-8247-5470-0.

34. Hernández-Guzmán, V.M.; Silva-Ortigoza, R. Velocity Control of a Permanent Magnet Brushed Direct Current Motor. In *Automatic Control with Experiments. Advanced Textbooks in Control and Signal Processing*; Springer: Cham, Switzerland, 2019; pp. 605–644. ISBN 978-3-319-75804-6._10. [CrossRef]

35. Usman, H.M.; Mukhopadhyay, S.; Rehman, H. Permanent magnet DC motor parameters estimation via universal adaptive stabilization. *Control Eng. Pract.* **2019**, *90*, 50–62. [CrossRef]

© 2019 by the authors. Licensee MDPI, Basel, Switzerland. This article is an open access article distributed under the terms and conditions of the Creative Commons Attribution (CC BY) license (http://creativecommons.org/licenses/by/4.0/).

Article

Comparative Evaluation of Wide-Range Soft-Switching PWM Full-Bridge Modular Multilevel DC–DC Converters

Jingwen Chen, Xiaofei Li, Hongshe Dang * and Yong Shi

School of Electrical and Control Engineering, Shaanxi University of Science and Technology, Xi'an 710021, China; chenjw@sust.edu.cn (J.C.); lxiaofei1214@163.com (X.L); shiyong@sust.edu.cn (Y.S)
* Correspondence: danghs@sust.edu.cn

Received: 26 December 2019; Accepted: 27 January 2020; Published: 31 January 2020

Abstract: This paper discusses some wide-range soft-switching full-bridge (FB) modular multilevel dc–dc converters (MMDCs), and a comparative evaluation of these FB MMDCs is also presented. The discussed topologies have all merits belonging to conventional FB MMDCs, e.g., smaller voltage stress on the primary switches, no added primary clamping devices and modular primary structure. In addition, the primary switches in each converter can obtain zero-voltage switching (ZVS) or zero-current switching (ZCS) in a wide load range. Two presented topologies are selected as examples to discuss in detail. The proposed FB MMDCs are assessed and evaluated based on performance, components and topology structure indices, such as soft switching characteristics, current stress, power loss distribution, number of added devices, complexity of structure and added cost. Experimental results are also included to verify the feasibility and advantages of the new topologies.

Keywords: full bridge (FB); modular multilevel dc-dc converters (MMDCs); zero-voltage switching (ZVS); zero-current switching (ZCS)

---

1. Introduction

With the rapid development of smart grid systems, dc-based distributed power systems and micro grids attract more and more attention due to some clearly good features, e.g., high power transfer efficiency, low system cost, high system stability and easy system control [1,2]. Commonly, the input voltage of these dc interfaces is very high to obtain optimum system performance, which makes high input voltage and high frequency isolated dc-dc converters become hot and causes challenging issues in power electronics. In the high voltage applications, how to reduce the voltage stress on the primary switches is a key point, and several methods can be used to solve this problem. The first way is to connect the power switches in series directly, but this method is seldom used in the high frequency applications due to the serious static and dynamic voltage balance problems [3]. Second, three-level (TL) dc-dc converters (TLDCs) are suitable topologies for high voltage applications due to only half the input voltage stress on the primary switches [4]. The first diode-clamped TLDC was proposed in 1992 [4], and then, many other efforts have been made on this topic, e.g., novel topologies and corresponding control strategies [4–11], wide range soft switching solutions [12–16] and reduced filter size solutions [16–19]. Finally, modular multilevel dc-dc converters (MMDCs) can also be used in high voltage applications [20]. MMDCs are built of modular cells with input series and output series or parallel connected, and can be extended to higher voltage levels easily due to their modular structure. Many TLDCs, e.g., diode-clamped TLDCs, can also be extended to a higher voltage level. But, as mentioned in [20], the number of achievable voltage levels is limited due to not only the dynamic voltage unbalance problem but also the complexity of the primary circuit structure and modulation strategy. Therefore, MMDCs may be a better choice for applications with super high input voltage, i.e., 1400 V or higher. A half bridge (HB) MMDC was proposed and discussed in [4], which is built of two-level HB modules. Several HB MMDCs for higher input voltage ratings have been proposed [21,22]. In [23], an FB MMDC was proposed, which is composed of two input series and output series connected two-level FB cells. Compared to HB MMDCs, FB MMDCs are more suitable for high input and large power industry applications due to the lower VA rating of the primary components, more modular structure and easy control. A number of new FB MMDCs were proposed and discussed in [20,23–26]. A new FB MMDC with auto-balance ability among series connected primary modules was proposed in [20]. Controlling strategies for output parallel-connected FB MMDCs were analyzed in [27–29]. However, improvements are still required. In the high voltage applications, soft switching performance of the primary switches is a key point to ensure higher efficiency due to the switching loss increases squared with the input voltage. Fortunately, the switching scheme of FB MMDCs is quite similar to that of the two-level phase shift (PS) FB dc-dc converter. Hence, common wide-range soft-switching solutions for conventional two-level PS FB dc-dc converters can be directly used in FB MMDCs. But, the system performance, structure complexity and added cost of different solutions for FB MMDCs are quite different because more primary cells are involved. Therefore, wide-range soft-switching FB MMDCs and a comparative evaluation of these solutions are still interesting subjects.

In this paper, some new wide-range soft-switching FB MMDCs are discussed and compared. The main contributions of this paper are:

(1)   Based on conventional FB MMDC, this paper proposes eight novel soft-switching solutions;
(2)   Through comparative study of them, the two most promising switching solutions are found;
(3)   Converter losses, efficiency and costs of the two solutions are analyzed and tested for verification.

This paper is organized as follows. In Section 2, the circuits of the presented converters are described. The operation principles and relevant analysis of an improved zero-voltage switching (IZVS) FB MMDC are provided in Section 3. The operation principles and relevant analysis of a zero-voltage and zero-current switching (ZVZCS) FB MMDC are discussed in Section 4. A comparative evaluation of the presented converters is provided in Section 5. In Section 6, experimental results are presented and analyzed. Finally, some brief conclusions are given.

2. Wide-Range Soft-Switching PWM FB MMDCs

Figure 1 illustrates a conventional FB MMDC, which is built of two two-level FB cells. The switches in each cell are switched in PS mode. Thus, eight switches in Figure 1 can also be divided into two groups, i.e., the leading and lagging switches. The lagging switches will face more difficulty to obtain ZVS because only the energy stored in the leakage inductances can be used.

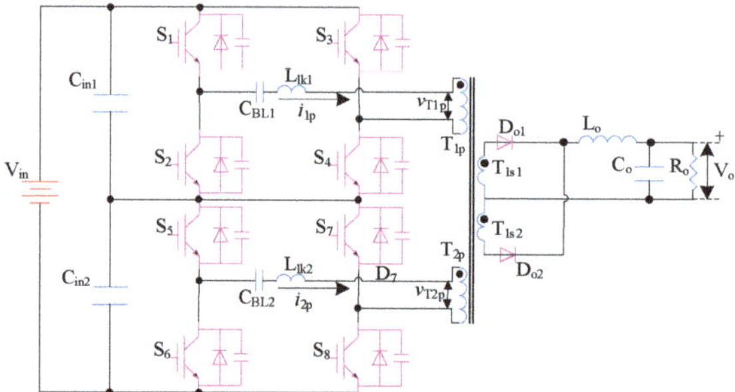

Figure 1. Conventional full-bridge (FB) modular multilevel dc–dc converter (MMDC).

As the switching scheme of Figure 1 is quite similar to that of a two-level PS FB dc-dc converter, common wide-range soft-switching solutions for two-level PS FB dc-dc converters can also be used [30–37]. These solutions can be concluded into two types: IZVS and ZVZCS. IZVS converters extend the ZVS range of the lagging switches by increasing either the value of the primary equivalent inductance or the currents of the switches [30–34]. In the ZVZCS converters, the lagging switches can realize ZCS by resetting the primary current during the free-wheeling mode. According to the different reset voltage generated methods, the ZVZCS converters can be further divided into two kinds, which are primary reset and secondary reset ZVZCS converters [35–37]. Generally, ZVZCS solutions are more suitable for the converters with IGBTs because of the large tailing current during the switching commutation.

2.1. IZVS FB MMDCs

Figure 2 shows four IZVS FB MMDCs. In IZVS_CD, the upper FB primary cell is built of the switches S1 to S4, the primary coil T1p, Lr1, Dcl1, Dcl2 and the blocking capacitor CBL1; another primary cell is comprised of the switches S5 to S8, the primary coil T2p, Lr2, Dcl3, Dcl4 and the blocking capacitor CBL2. Lr1 and Lr2 are added to enlarge ZVS range of the lagging switches. Dcl1 to Dcl4 are used to eliminate the oscillation and clamp the secondary rectified voltage reflected to the primary side. The power transformer is built of two independent magnetic cores, two primary coils and two common secondary coils. Each primary coil is wired around an independent magnetic core; while the common secondary coils enclose both cores. Cin1 and Cin2 are the input capacitors with the same value, and these capacitors share the input voltage evenly during the operation stages, i.e., $V_{Cin1} = V_{Cin2} = V_{in}/2$. Llk1 and Llk2 are leakage inductances of T1p and T2p. Do1 and Do2 are the rectifier diodes, and the output filter is built of Lo and Co. Ro is the load resistor.

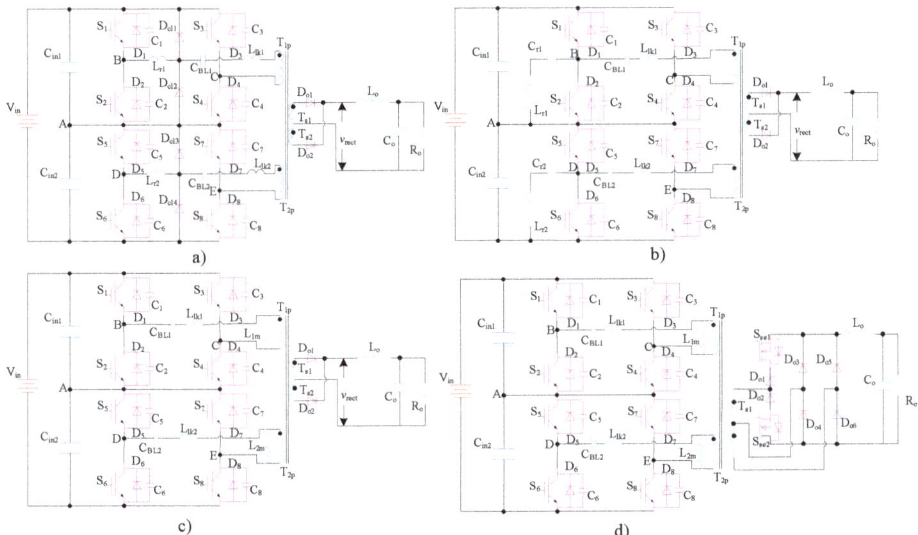

Figure 2. Improved zero-voltage switching (IZVS) FB MMDCs: (**a**) IZVS with clamped diode (IZVS_CD); (**b**) IZVS with commutation auxiliary circuit (IZVS_CAC); (**c**) IZVS with small magnetizing inductance (IZVS_SMI); (**d**) IZVS with secondary modulation method (IZVS_SMM).

In IZVS_CAC, two commutation auxiliary circuits (CACs) are added to enlarge the ZVS range of the lagging switches. The secondary rectifier of IZVS_CAC is identical to that of IZVS_CD. The configuration of IZVS_SMI is quite similar to that of Figure 1. However, the magnetic currents of the transformer in IZVS_SMI are increased to provide more resonant energy. The structure of the power transformers in IZVS_CAC and IZVS_SMI is identical to that of IZVS_CD. IZVS_SMM shows a secondary modulated FB MMDC. The primary structure of IZVS_SMM is identical to that of

IZVS_SMI, and the magnetizing currents are also increased to help the ZVS of the primary switches. The transformer in IZVS_SMM has two primary coils and two secondary coils. Do3 to Do6 are the rectifier diodes. Sse1 and Sse2 are two added secondary switches. The output filter is built of Lo and Co. Ro is the load resistor.

2.2. ZVZCS FB MMDCs

Figure 3 depicts four ZVZCS FB MMDCs. IZVZCS_DCF is a primary current reset ZVZCS converter with two cut-off diodes in each FB cell. CBL1 and CBL2 are designed to a specific value to reset the primary currents. D3, D4, D7 and D8 are used to block the reverse primary currents. The secondary circuit of IZVZCS_DCF equals that of IZVS_CD. IZVZCS_SAC&CAC and IZVZCS_SRC&CAC illustrate two secondary reset ZVZCS FB MMDCs. In IZVZCS_SAC&CAC, Sse and Cse are added to conventional FB MMDC to reset the primary currents; while the secondary reset circuit in IZVZCS_SRC&CAC has the same function, which is composed of Dse1, Dse2 and Cse. In IZVZCS_SI, CBL1 and CBL2 are designed to a specific value to reset the primary currents, and two saturable inductors, i.e., Lr1 and Lr2, are used to limit the reverse primary currents. The structure of the power transformers in Figure 3 is identical to that of IZVS_CD.

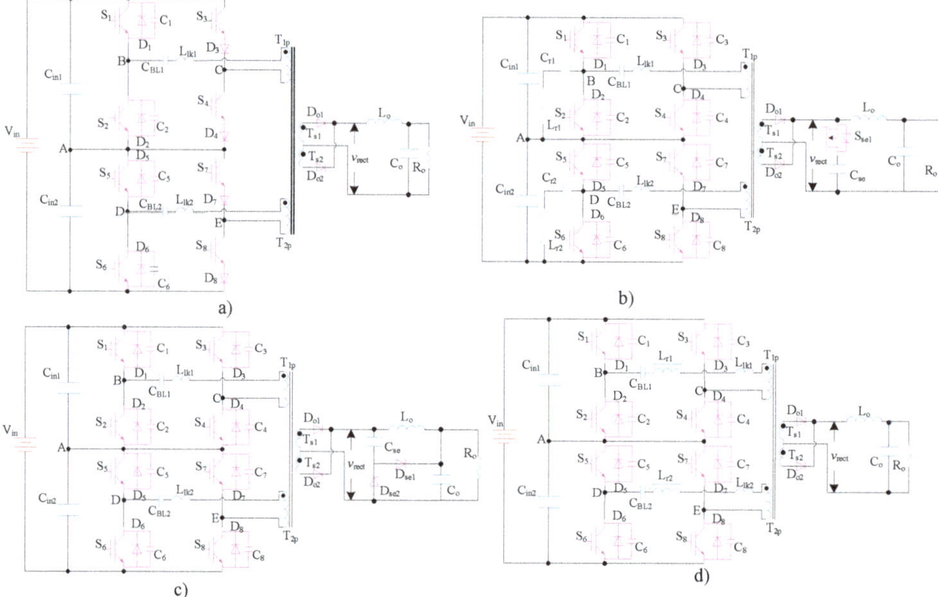

Figure 3. Improved zero-voltage and zero-current switching (IZVZCS) FB MMDCs: (**a**) IZVZCS with diode cutting-off (IZVZCS_DCF); (**b**) IZVZCS with secondary active clamping and commutation auxiliary circuit (IZVZCS_SAC&CAC); (**c**) IZVZCS with secondary reactive clamping and commutation auxiliary circuit (IZVZCS_SRC&CAC); (**d**) IZVZCS with saturable inductance (IZVZCS_SI).

3. IZVS FB MMDC with Secondary Modulated

In order to simplify the description, the operation principle and characteristics of IZVS_CD, IZVS_CAC and IZVS_SMI is not presented here, and corresponding information can be found in [30–33]. The IZVS_SMM is selected as an example to analyze in detail in this part.

3.1. Operation Principle

Key waveforms of IZVS_SMM are shown in Figure 4. There are six operation stages during the whole switching cycle, and the operation stages in the first half switching cycle are illustrated in Figure 5. Before the analysis, some assumptions are set to simplify the explanation: all the components in the topology are ideal; Cin1 and Cin2 are large enough to be considered as voltage sources valued $V_{in}/2$, and the voltage ripple on them can be neglected; CBL1 and CBL2 are large enough, and the voltage ripple on them can be neglected; $L_{m1} = L_{m2} = L_m$; $L_{L1K1} = L_{L1K2} = L_{1K}$; Im is the peak value of the magnetizing currents; output filter and load are replaced by a constant current source Io; k_{T1} and k_{T2} are the turn ratios $k_T' = (k_{T1} \times k_{T2})/(k_{T1} + k_{T2})$. The output capacitance of each switch is identical and represented as Cos in the following equations.

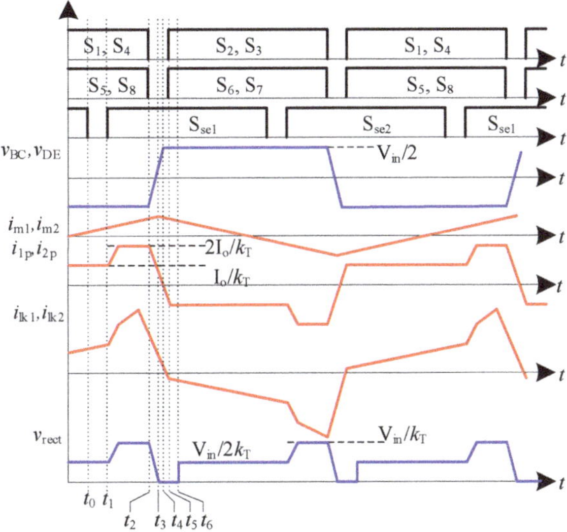

Figure 4. Key waveforms of IZVS_SMM.

Stage 1 [Figure 5a]: before $t0$, the circuit is operated in steady condition. Input source powers the load. S1, S4, S5 and S8 are on; Do2 is conducted; Sse2 is also on, and the current flowing through Sse2 is zero due to Do4 is off. $v_{BC} = v_{DE} = V_{in}/2$; $v_{rect} = V_{in}/k_{T2}$; $i_{1p} = i_{2p} = I_o/k_{T2}$; $i_{L1k1} = i_{1p} + i_{m1}$; $i_{L1k2} = i_{2p} + i_{m2}$; im1 and im2 increase with time linearly, and the slope is

$$\frac{di_{m1}}{dt} = \frac{di_{m2}}{dt} = \frac{V_{in}}{2L_m} \tag{1}$$

Stage 2 [Figure 5b, $t0$-$t1$]: At $t0$, Sse2 is turned off at zero-current. Primary side powers the load. $v_{BC} = v_{DE} = V_{in}/2$; $v_{rect} = V_{in}/k_{T2}$; $i_{1p} = i_{2p} = I_o/k_{T2}$; $i_{L1k1} = i_{1p} + i_{m1}$; $i_{L1k2} = i_{2p} + i_{m2}$; im1 and im2 keep increasing.

Stage 3 [Figure 5c, $t1$-$t2$]: At $t1$, Sse1 is turned on; Do1 is conducted and Do2 is off. Primary powers the load. $v_{BC} = v_{DE} = V_{in}/2$; $v_{rect} = V_{in}/2k_T'$; $i_{1p} = i_{2p} = I_o/k_T'$; $i_{L1k1} = i_{1p} + i_{m1}$; $i_{L1k2} = i_{2p} + i_{m2}$; im1 and im2 increase with time linearly.

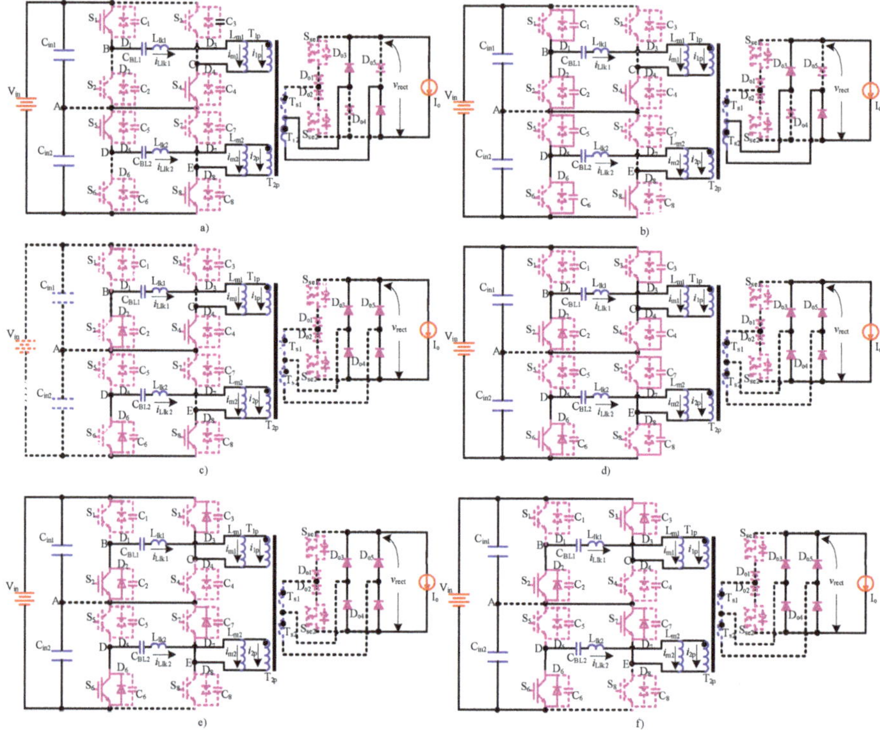

Figure 5. Operation stages of IZVS_SMM: (**a**) stage 1; (**b**) stage 2; (**c**) stage 3; (**d**) stage 4; (**e**) stage 5; (**f**) stage 6.

Stage 4 [Figure 5d, *t2-t4*]: At *t2*, S1, S4, S5 and S8 are simultaneously turned off at zero-voltage due to the existence of C1, C4, C5 and C8; *i*Llk1 charges C1 and C4, and discharges C2 and C3 linearly with time; *i*Llk2 charges C5 and C8, and discharges C6 and C7 linearly with time. During this interval, *i*m1 and *i*m2 reach their peak value *Im* and keep constant. Before $v_{\text{rect}} > 0$, the energy stored in the output inductor can still be used to charge or discharge the output capacitance of each primary switch. When *v*rect is zero, one half of the final voltage on each primary switch has been charged or discharged [34]. Thus, less resonant energy is required to obtain ZVS for the primary switches, and this is the advantage of IZVS_SMM. After *t3*, the circuit will be operated into the free-wheeling mode. *i*Llk1 charges C1 and C4, and discharges C2 and C3 linearly with time; *i*Llk2 charges C5 and C8, and discharges C6 and C7 linearly with time. This stage ends until $v_{\text{C1}} = v_{\text{C4}} = v_{\text{C5}} = v_{\text{C8}} = V_{\text{in}}/2$ and $v_{\text{C2}} = v_{\text{C3}} = v_{\text{C6}} = v_{\text{C7}} = 0$.

Stage 5 [Figure 5e, *t4-t5*]: At *t4*, D2, D3, D6 and D7 conduct naturally. The circuit operates in the free-wheeling mode; *i*Llk1 and *i*Llk2 decrease due to negative voltage applied to the terminals of Llk1 and Llk2; during this stage, S2, S3, S6 and S7 must be turned on to achieve ZVS. According to Figure 4, S2, S3, S6 and S7 are turned on at *t5*.

Stage 6 [Figure 5f, *t5-t6*]: At *t5*, S2, S3, S6 and S7 are switched on; *i*1p and *i*2p increase in the inverse direction. When these currents reach $-I_{\text{o}}/k_{\text{T2}}$, the free-wheeling mode is over. Primary powers load. After *t6*, $v_{\text{BC}} = v_{\text{DE}} = -V_{\text{in}}/2$; $v_{\text{rect}} = -V_{\text{in}}/k_{\text{T2}}$; $i_{\text{1p}} = i_{\text{2p}} = -I_{\text{o}}/k_{\text{T2}}$; *i*Llk1 equals the sum of *i*1p and *i*m1; *i*Llk2 equals the sum of *i*2p and *i*m2; *i*m1 and *i*m2 decrease with time linearly, and the slope is determined by (1). The current flowing through Sse1 is zero due to Do1 is off. After stage 6, the circuit will be operated in the second half switching cycle.

3.2. Soft Start

During the soft-start operation, the IZVS_SMM can be treated as a conventional FB MMDC in Figure 1. Two secondary switches are off, and four primary switches in each FB module are switched in PS mode. The output voltage can be regulated down to zero by increasing the phase angle among the primary switches. Detailed operation principle about this procedure can be found in [23,24].

3.3. ZVS Condition of The Primary Switches

With proper design of $i1m$ and $i2m$, all primary switches can obtain ZVS down to 0 load currents. S2 and S3 in the upper module are selected to describe as an example. Figure 5d shows the equivalent circuit of this procedure. Before $v$rect decays to zero, the load current can still be used to charge or discharge corresponding capacitors. As discussed above, 50% of the final voltage across C1 to C4 has been discharged or charged before $v$rect decays to zero. Thus, following equation should be fitted to obtain ZVS.

$$\frac{1}{2}L_{lk}I_m^2 \geq 2 \times C_{os}\left(\frac{V_{in}}{4}\right)^2 \tag{2}$$

The required Im is

$$I_m \geq \frac{V_{in}}{2}\sqrt{\frac{C_{os}}{L_{lk}}} \tag{3}$$

The peak to peak value of im is

$$\Delta i_m = \frac{V_{in}T_s}{2L_m} = 2I_m \tag{4}$$

Thus, Im is

$$I_m = \frac{V_{in}T_s}{4L_m} \tag{5}$$

Substituting (5) into (3) yields

$$L_m \leq \frac{T_s}{2}\sqrt{\frac{L_{lk}}{C_{os}}} \tag{6}$$

Therefore, S2 and S3 can obtain ZVS down to zero load current with a specific value of Lm decided by (6).

3.4. ZCS Condition of The Secondary Switches

As proved in Figures 4 and 5, all secondary switches can obtain ZCS independent of the load current. Sse2 is selected to describe as an example. As shown in Figure 5a, Sse2 is on in this stage. But, the current flowing through Sse2 is zero due to the reverse voltage applied to Do4. As shown in Figure 5b, Sse2 is switched off at zero current. Therefore, the switching loss of the secondary switches can be minimized.

3.5. Turn Ratios

The output is regulated by the phase angle among the primary and secondary switches. The turn ratios of IZVS_SMM should be designed according to the input voltage range. At maximum input voltage, the phase angle between S1 and Sse1 is zero; primary powers the load through Ts2. With decreasing of the input voltage, the phase angle between S1 and Sse1 is increased, and primary powers the load through both Ts1 and Ts2. Hence, $kT2$ is

$$k_{T2} \leq \frac{V_{inmax}}{V_o} \tag{7}$$

And $kT1$ can be computed by

$$\frac{k_{T1}k_{T2}}{k_{T1} + k_{T2}} = \frac{V_{inmin}}{V_o} \tag{8}$$

As for a converter with 100 to 300 V input and 50 V output (used in the prototype), $kT2$ can be decided by (7) and the value is 16; according to (8), $k_{T1} = 48$.

3.6. Duty Ratio Loss

The time between $t5$ and $t6$ is defined as duty ratio loss time, and corresponding states are plotted in Figures 5e and 5f. The primary side currents are

$$i_{kp} = \frac{I_o}{k'_T} - \frac{V_{in}}{2L_{lk}}\Delta t_{56}, k = 1, 2 \tag{9}$$

when $i_{kp} = -I_o/k_{T2}, k = 1, 2$, the free-wheeling mode is finished. The time of this interval is

$$\Delta t_{56} = \frac{2I_o L_{lk}}{V_{in}}\left(\frac{k'_T + k_{T2}}{k'_T k_{T2}}\right) \tag{10}$$

The duty ratio loss is

$$\Delta D = \frac{\Delta t_{56}}{T_s/2} = \frac{4I_o L_{lk} f_s}{V_{in}}\left(\frac{k'_T + k_{T2}}{k'_T k_{T2}}\right) \tag{11}$$

3.7. Reduced Filter Size

The reduction of the output inductance with TL secondary rectified voltage waveform has been discussed in [16–19]. According to these references, the required output inductance of the converters with TL secondary rectified voltage waveform is about one-third of that of conventional two-level converters. Therefore, the volume of the output filter in the IZVS_SMM can be significantly reduced.

4. ZVZCS FB MMDC with Secondary Active Reset

The IZVZCS_SAC&CAC. is selected as an example to analyze in detail in this part. In order to simplify the description, the operation principle and characteristics of IZVZCS_DCF, IZVZCS_SRC&CAC and IZVZCS_SI are not presented here, and detailed information can be found in [35,36] and [37].

4.1. Operation Principle

Key waveforms of IZVZCS_SAC&CAC are provided in Figure 6.

There are 12 operation stages in each switching cycle, and eight switching stages in the first half switching cycle are provided in Figure 7. Before the analysis, some assumptions are set to simplify the explanation: all the components in the topology are ideal; the voltage ripple on Cin1 and Cin2 can be neglected; $L_{L1K1} = L_{L1K2} = L_{1K}$; kT is the turn ratio; the output filter and load are replaced by a constant current source Io. The output capacitance of each primary switch is identical and represented as Cos in the following equations.

Stage 1 [Figure 7a, $t0$-$t1$]: At $t0$, primary powers the load. S1, S4, S5 and S8 are on; Do1 is conducted while Do2 is off. The secondary rectified voltage is clamped by Cse through the anti-parallel diode of Sse. $iLlk1$ and $iLlk2$ are

$$i_{L1k1} = i_{L1k2} = \frac{1}{L_{lk}}\left(\frac{V_{in}}{2} - k_T v_{Cse}\right)t \tag{12}$$

The current of the clamping capacitor Cse is

$$i_{Cse} = k_T i_{L1k1} - I_o \tag{13}$$

This stage ends until i_{Cse} is 0.

Stage 2 [Figure 7b, *t1-t2*]: At *t1*, the anti-parallel diode of Sse is turned off; primary powers the load. S1, S4, S5 and S8 are on; Do1 is conducted while Do2 is off. $i_{L1k1} = i_{L1k2} = I_o/k_T$; $v_{rect} = V_{in}/k_T$.

Stage 3 [Figure 7c, *t2-t3*]: At *t2*, S1 and S5 are turned off at zero-voltage due to the existence of D1 and D5. *i*Llk1 charges C1 and discharges C2 linearly with time, and *i*Llk2 charges C5 and discharges C6 linearly with time, the voltage of B is

$$v_B(t) = V_{in} - \frac{I_o}{k_T}\frac{t}{2C_{os}} \tag{14}$$

And the voltage of D is

$$v_D(t) = \frac{V_{in}}{2} - \frac{I_o}{k_T}\frac{t}{2C_{os}} \tag{15}$$

The time of this stage is

$$T_{32} = \frac{V_{in}C_{os}k_T}{I_o} \tag{16}$$

After *t3*, D2 and D6 are turned on. Then, S2 and S6 should be gated to achieve ZVS, and according to Fig.6, S2 and S6 are switched at *t3*.

Stage 4 [Figure 7d, *t3-t4*]: At *t3*, S2 and S6 are turned on at zero-voltage. Sse is also turned on at this time. *v*rect is forced to *V*Cse, and this voltage is applied to the primary leakage inductances. Hence, *i*Llk1 and *i*Llk2 decrease, and the slope is

$$\frac{di_{L1k1}}{dt} = \frac{di_{L1k2}}{dt} = -\frac{k_T v_{Cse}}{L_{lk}} \tag{17}$$

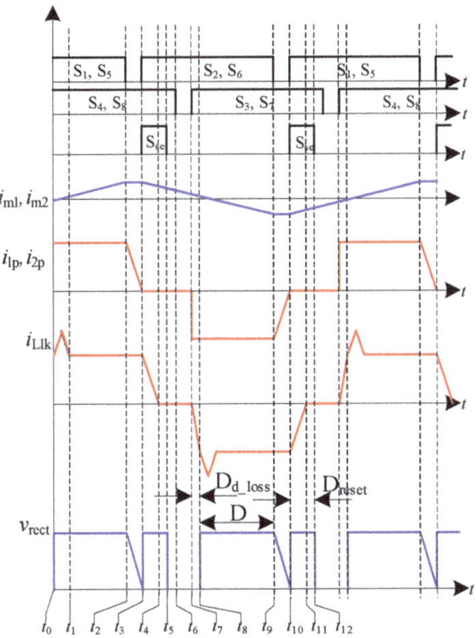

Figure 6. Key waveforms of IZVZCS_SAC&CAC.

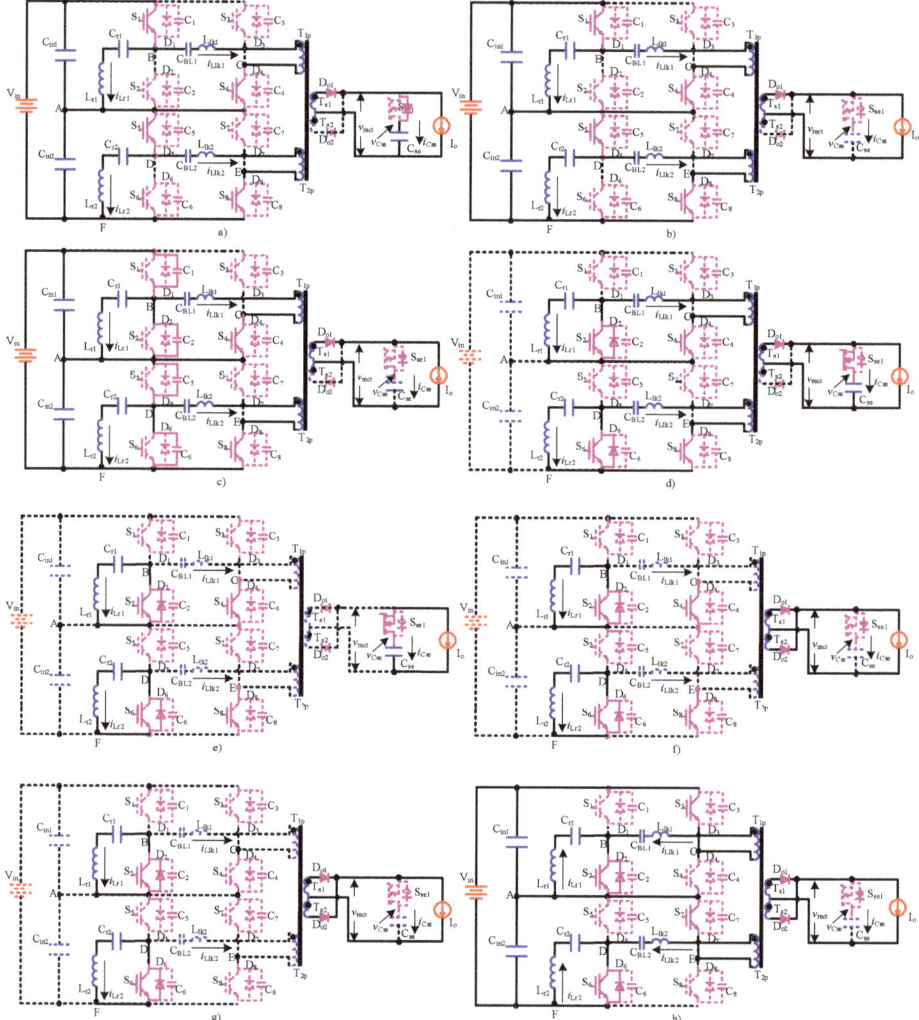

Figure 7. Operation principle of IZVZCS_SAC&CAC: (**a**) stage 1; (**b**) stage 2; (**c**) stage 3; (**d**) stage 4; (**e**) stage 5; (**f**) stage 6; (**g**) stage 7; (**h**) stage 8.

Stage 5 [Figure 7e, *t*4-*t*5]: At *t*4, *i*Llk1 and *i*Llk2 are 0, all rectifier diodes are off. Io is free-wheeled through Sse and Cse.

Stage 6 [Figure 7f, *t*5-*t*6]: At *t*5, Sse is off, and all rectified diodes are conducted. *i*Llk1 and *i*Llk2 keep zero.

Stage 7 [Figure 7g, *t*6-*t*7]: At *t*6, S4 and S8 are turned off at zero current.

Stage 8 [Figure 7h, *t*7-*t*8]: At *t*7, S3 and S7 are turned on at zero current due to the existence of Llk1 and Llk2. *i*Llk1 and *i*Llk2 decrease due to negative value applied to Llk1 and Llk2. When these currents reach -Io/*k*T, the free-wheeling mode is over. After stage 8, the circuit is operated in the second half period.

4.2. Duty Ratio Loss

The time between $t7$ and $t8$ is defined as the duty loss caused by the leakage inductances, and corresponding state is plotted in Figure 7 h. The primary side currents are

$$i_{L1ki} = 0 - \frac{V_{in}}{L_{lk}}\Delta t_{78}, i = 1,2 \tag{18}$$

when $i_{L1ki} = -I_o/k_T$, the free-wheeling mode is accomplished, and the time of this interval is

$$\Delta t_{78} = \frac{2I_o L_{lk}}{V_{in}} \tag{19}$$

$$D_{L\_loss} = \frac{\Delta t_{78}}{T_s/2} = \frac{4I_o L_{lk} f_s}{k_T V_{in}} \tag{20}$$

where D_{L_loss} is the duty ratio loss caused by the leakage inductances.

As shown in Figure 6, the primary currents are reset to zero during the interval of [$t4$, $t5$]. In order to ensure safe ZCS of the lagging switches, the maximum primary duty ratio should be limited as

$$D_{p\_max} = 1 - D_{reset} \tag{21}$$

where D_{p_max} is the maximum primary duty ratio of IZVZCS_SAC&CAC.

However, as depicted in Figure 6, the primary reset time can be well compensated by the boost effect of secondary clamping, which is defined as D_{Boost} in Figure 6. D_{Boost} is identical to D_{reset}, thus, total duty ratio loss of IZVZCS_SAC&CAC is

$$D_{loss} = D_{L\_loss} + D_{reset} - D_{Boost} = \frac{4I_o L_{lk} f_s}{k_T V_{in}} \tag{22}$$

where D_{loss} is total duty ratio loss of IZVZCS_SAC&CAC.

4.3. ZVS Condition of the Leading Switches

S2 is chosen as an example, the switching instant is provided in Figure 7c. ZVS criteria for S2 is

$$\frac{1}{2}L_p \left(\frac{I_o}{k_T}\right)^2 \geq 2C_{os}\left(\frac{V_{in}}{2}\right)^2 \tag{23}$$

where L_p is equal to $L_{1k1} + k_{T2}L_o$.

The minimum load current to keep safe ZVS is

$$I_{o\_min} = k_{T1}V_{in}\sqrt{\frac{C_{os}}{L_{1k1} + k_{T1}^2 L_o}} \tag{24}$$

The stored energy in the output inductance is large enough to conduct D2, so S2 can obtain ZVS in wide load range.

4.4. ZCS Condition of the Lagging Switches

As illustrated in Figure 6, the lagging switches should be turned off after $iLlk1$ and $iLlk2$ decay to zero. The reset time is T43, and its value is

$$T_{43} = \frac{I_o L_{lk}}{k_T^2 V_{Cse}} \tag{25}$$

When IGBTs are used as the lagging switches, the minority carriers in the component can be combined in a specific time, and this interval is determined by the component itself and defined as TCOM. Therefore, the following equation should be confirmed to ensure ZCS operation:

$$T_{43} + T_{com} \leq T_{reset} \tag{26}$$

5. Comparative Evaluation

In this section, a comparative evaluation of the proposed converters with regard to soft-switching characteristics, duty ratio loss, the current rating of the primary components, power loss distribution, number of added components and added cost is provided to highlight the advantages and disadvantages of each converter and to help the selection of a candidate for a given application.

5.1. Specifications

The presented converters are compared based on the specifications listed as follows. The input voltage is varied from 100 to 300 V. The rated output voltage is 50 V, and the output current is 20 A. The switching frequency is 20 kHz. 1000 V/60 A IGBTs are used as the primary switches, and the output capacitance of IGBTs is estimated as 1nF. The equivalent leakage inductances of the transformer in each converter are set to be 10 μH. The ideal value of $kT1$ in IZVS_SMM is 48 regardless of the effect of the leakage inductances, while $k_{T2} = 16$. The ideal value of kT in other converters is 12. The peak value of the magnetic currents of IZVS_CD and IZVS_SMM are set as 0.6 times of the primary rate current, and the magnetic currents of other converters can be omitted.

5.2. Duty Ratio Loss

The duty ratio loss caused by the leakage inductances is a disadvantage of PS-controlled dc-dc converters. Large duty ratio loss requires the transformer turn ratio to comprise, which degrades the performance of the converter. Table 1 shows the duty ratio loss comparison. As an additional inductor is series-connected with each primary coil, the duty ratio loss of IZVS_CD is highest among all the converters. The primary currents of IZVS_SMM are TL waveforms, thus, the duty ratio loss of IZVS_SMM is smallest among all IZVS converters. The primary currents in ZVZCS converters are reset to zero during freewheeling stages, and the duty ratio loss caused by leakage inductances of ZVZCS converters is smaller than that of IZVS converters. However, as a specific time for primary current resetting is required, thus, the duty ratio loss of IZVZCS_DCF, IZVZCS_SRC&CAC and IZVZCS_SI is a little higher. As proved in Figure 6, the primary reset time of IZVZCS_SAC&CAC can be compensated by the duty ratio boost effect, thus, the duty ratio loss of this converter is smallest. Figure 8 shows the duty ratio loss at rated load current. As shown in Figure 8, when $I_o = 20A$, the duty ratio loss of IZVS_CD is about 0.089. Considering the duty ratio loss, the turn ratios of the proposed converters should be revised, and detailed information is listed in Table 2.

Table 1. Duty ratio loss comparison [30–37].

Converter	Duty ratio loss
IZVS_CD	$D_{loss} = \dfrac{8I_o(L_{1k}+L_r)f_s}{k_T V_{in}} = \dfrac{0.67I_o(L_{1k}+L_r)f_s}{V_{in}}$
IZVS_CAC and IZVS_SMI	$D_{loss} = \dfrac{8I_o L_{1k}f_s}{k_T V_{in}} = \dfrac{0.67I_o L_{1k}f_s}{V_{in}}$
IZVS_SMM	$D_{loss} = \dfrac{4I_o L_{1k}f_s}{V_{in}}\dfrac{k_T+k_{T2}}{k'_T k_{T2}} = \dfrac{0.58I_o L_{1k}f_s}{V_{in}}$
IZVZCS_DCF, IZVZCS_SRC&CAC and IZVZCS_SI	$D_{loss} = \dfrac{4I_o L_{1k}f_s}{k_T V_{in}} + D_{reset} = \dfrac{0.33I_o L_{1k}f_s}{V_{in}} + 0.05$
IZVZCS_SAC&CAC	$D_{loss} = \dfrac{4I_o L_{1k}f_s}{k_T V_{in}} = \dfrac{0.33I_o L_{1k}f_s}{V_{in}}$

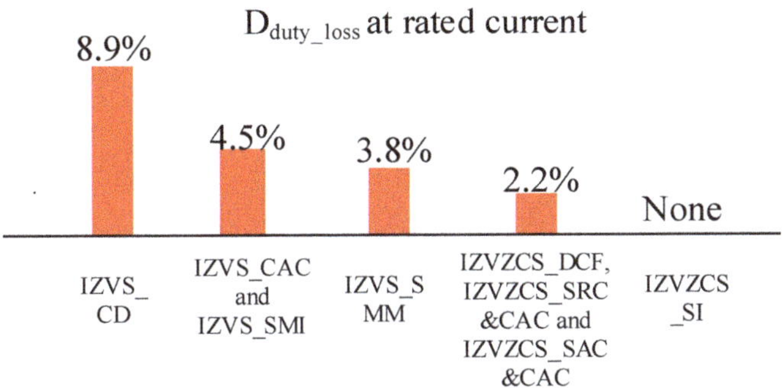

Figure 8. Duty ratio loss at rated output current.

Table 2. Turn ratios after considering the duty ratio loss.

Item	IZVS_CD	IZVS_CAC and IZVS_SMI	IZVS_SMM
Turn ratios	$k_T = 10.9$	$k_T = 11.4$	$k_{T1} = 41.4, k_{T2} = 16$
Item		IZVZCS_DCF, IZVZCS_SRC&CAC and IZVZCS_SAC&CAC	IZVZCS_SI
Turn ratios		$k_T = 11.1$	$k_T = 11.7$

5.3. Soft Switching Load Range

The soft switching load range is defined as

$$\eta_{LR} = \frac{I_{o\_rate} - I_{o\_min}}{I_{o\_rate}} \tag{27}$$

where I_{o_rate} represents the rated output current, and its value is 20A in this paper; I_{o_min} is the minimum load current to ensure ZVS of the switches.

(1) Leading switches:

As the energy stored in the output inductance can be used, all leading switches can obtain ZVS in wide load range. As proved in [31], the minimum load of the leading switches in IZVS_CD is

$$I_{o\_min} = k_T V_{in} \sqrt{\frac{C_{os}}{L_r + L_{1k} + k_T^2 L_o}} \tag{28}$$

As proved in [32] and [35–37], the minimum load of the leading switches in IZVS_CAC and Figure 3 is

$$I_{o\_min} = k_T V_{in} \sqrt{\frac{C_{os}}{L_{1k} + k_T^2 L_o}} \tag{29}$$

As the magnetizing current of the leading switches is increased in IZVS_SMI and IZVS_SMM, the leading switches in IZVS_SMI and IZVS_SMM can obtain ZVS down to 0 load current. Thus, the η_{LR} of the leading switches is concluded in Table 3.

Table 3. η_{LR} (leading-switches, 300 V).

Item	IZVS_CD	IZVS_CAC	IZVS_SMI	IZVS_SMM
η_{LR}	0.967	0.996	1	1
Item	IZVZCS_DCF	IZVZCS_SRC&CAC	IZVZCS_SAC&CAC	IZVZCS_SI
η_{LR}	0.996	0.996	0.996	0.996

(2) Lagging switches:

A resonate inductance is added to enlarge the ZVS load range of the lagging switches in IZVS_CD, and, as shown in [31], the minimum load of the lagging switches in IZVS_CD is

$$I_{o\_min} = k_T V_{in} \sqrt{\frac{C_{os}}{L_r + L_{1k}}} \tag{30}$$

With proper design, the lagging switches of IZVS_CAC, IZVS_SMI and IZVS_SMM can obtain ZVS down to zero loads. The lagging switches in Figure 3 are operated in ZCS mode, and the ZCS operation can be ensured under D_{p_max} at rated load current. Thus, the η_{LR} of the lagging switches is listed in Table 4.

Table 4. η_{LR} (lagging-switches, 300 V).

Item	IZVS_CD	IZVS_CAC	IZVS SMI	IZVS_SMM
η_{LR}	0.76	1	1	1
Item	IZVZCS_DCF	IZVZCS_SRC&CAC	IZVZCS_SAC&CAC	IZVZCS_SI
η_{LR}	1	1	1	1

5.4. Relative Current Rating of the Primary Components

The rate primary current is defined as

$$I_{p\_rate} = I_{o\_rate} / k_{T\_ideal} = 20/12 = 1.667 (A) \tag{31}$$

where I_{p_rate} is the primary rate current; k_{T_ideal} is 12.

The primary relative average absolute current rating is

$$\tau_{C\_AV} = \frac{I_{p\_AV}}{I_{p\_rate}} \tag{32}$$

where I_{p_AV} is the primary average absolute current.

The primary relative RMS current rating is

$$\tau_{C\_RMS} = \frac{I_{p\_RMS}}{I_{p\_rate}} \tag{33}$$

where I_{p_RMS} is the primary RMS current.

Table 5 illustrates τ_{C_AV} and τ_{C_RMS} of the primary components. The magnetizing currents of IZVS_SMI are increased to help ZVS of the lagging switches, and these currents keep their peak value during the whole free-wheeling time. Thus, τ_{C_AV} and τ_{C_RMS} of IZVS_SMI is highest, which results highest primary side conduction loss. The magnetizing currents of IZVS_SMM are also enlarged, but, the average value in the half switching cycle of these currents is zero, and these currents are not in phase with the load current. Hence, τ_{C_AV} and τ_{C_RMS} of IZVS_SMM are much smaller than that of IZVS_SMI. The IZVZCS_SAC&CAC has the smallest τ_{C_AV} and τ_{C_RMS}.

Table 5. τ_{C_AV} and τ_{C_RMS} of the primary components.

Item	τ_{C_AV}		τ_{C_RMS}	
	Switches	**Added diodes**	**Transformer**	**Added inductor**
IZVS_CD	1.1	0.55	1.1	1.1
IZVS_CAC	1.05	None	1.05	0.21
IZVS_SMI	1.89	None	1.3	None
IZVS_SMM	1.04	None	1.09	None
IZVZCS_DCF	1.08	0.54	1.08	None
IZVZCS_SAC&CAC	1.02	None	1.02	None
IZVZCS_SRC&CAC	1.08	None	1.08	None
IZVZCS_SI	1.08	None	1.08	1.08

5.5. Power Loss Distribution

The losses of soft-switching FB converters mainly include switching losses, transformer losses, output rectifier diode losses, resonance inductance losses, and other losses. The switching loss model mainly covers light-load and heavy-load senarios. The losses of a MOSFET can be calculated by (34), where P_{Driver} is the driving loss, P_{SW} is the switching loss of the MOSFET, P_{MOS_Lead} is the conduction loss of the leading MOSFET, P_{MOS_Lag} is the conduction loss of the lagging MOSFET.

$$\begin{cases} P_{MOSFET} = P_{Drive} + P_{SW} \\ P_{MOSFET} = P_{MOS\_Lead} + P_{MOS\_Lag} \end{cases} \tag{34}$$

Transformer loss mainly includes copper loss($P_{Winding}$) and iron loss, which includes eddy-current loss(P_e) and hysteresis loss(P_h). Then, the total transformer loss, defined as P_T, is

$$P_T = P_{Winding} + P_h + P_e \tag{35}$$

The output rectifier diode loss includes three parts: turn-off loss(P_{D_off}), turn-on loss(P_{D_on}), and on-state loss(P_{Con}). The total loss of a diode, defined as P_{D_loss}, is

$$P_{D\_loss} = P_{D\_off} + P_{D\_on} + P_{Con} \tag{36}$$

Inductor losses mainly include copper loss,P_{Cu_lo} and iron loss P_{Fe_lo}. Then the total inductor loss of L_o and L_r are

$$P_{Lo\_loss} = P_{Fe\_lo} + P_{Cu\_lo} \tag{37}$$

$$P_{Lr\_loss} = P_{Fe\_lr} + P_{Cu\_lr} \tag{38}$$

Relative switching loss is

$$\delta_{S\_loss} = \frac{\sum_n P_{S\_Loss}}{P_o} \tag{39}$$

where P_o is the output power and δ_{S_loss} is the corresponding switching loss.

Relative conduction loss is

$$\delta_{C\_loss} = \frac{\sum_n P_{C\_Loss}}{P_o} \tag{40}$$

where P_o is the output power and P_{C_Loss} is corresponding conduction loss.

The power loss distribution of the proposed converters is estimated in Table 6. The magnetizing currents of IZVS_SMM are enlarged to help the ZVS of the primary switches and the peak value of these currents can provide more resonant energy with increasing of the input voltage, thus, the switching loss of the primary switches of IZVS_SMM is smallest among all IZVS converters. When $V_{in}= 300V$, the phase angle between the primary switches and secondary switches is zero, and the primary currents of IZVS_SMM are much smaller than that of other IZVS converters. Therefore, the primary conduction loss of IZVS_SMM is also smallest among all IZVS converters. As depicted in Table 6, some secondary conduction loss and switching loss is added to the converter in IZVS_SMM because two secondary switches are required. However, the efficiency of IZVS_SMM is still highest among all IZVS MMDCs. As shown in Table 6, the IZVZCS_SAC&CAC has the smallest conduction loss among all ZVZCS MMDCs due to smaller duty ratio loss. In addition, the turn-off loss of the leading switches of IZVZCS_SAC&CAC is also smaller, which results in smaller switching loss of the primary switches. Hence, the efficiency of IZVZCS_SAC&CAC is highest among all ZVZCS converter.

Table 6. δS_loss and δC_loss ($V_{in} = 300$V).

Item	$\delta_{\text{S\_loss}} \times 10^{-3}$				$\delta_{\text{C\_loss}} \times 10^{-3}$			
	Primary		Secondary		Primary		Secondary	
	$I_o=5$A	$I_o=20$A	$I_o=5$A	$I_o=20$A	$I_o=5$A	$I_o=20$A	$I_o=5$A	$I_o=20$A
IZVS_CD	73.5	49.3	23.3	15.5	18.1	20.1	17.8	19.8
IZVS_CAC	57.6	38.6	23.3	15.5	14.5	16.1	17.8	19.8
IZVS_SMI	79.1	53.1	23.3	15.5	21.2	23.6	17.8	19.8
IZVS_SMM	40.4	28	24.6	16.5	9.6	10.7	19.6	21.8
IZVZCS_DCF	71.1	47.4	23.3	15.5	13.9	15.4	17.8	19.8
IZVZCS_SAC&CAC	47.3	36.4	24.6	16.5	9.6	10.7	18.1	20.1
IZVZCS_SRC&CAC	67.5	45	23.3	15.5	10.4	11.6	18.5	20.5
IZVZCS_SI	74.3	49.5	23.3	15.5	13.0	14.4	17.8	19.8

5.6. Structure and Cost Comparison

Table 7 shows a comparison of added components of the proposed converters. The IZVS_SMI, IZVS_SMM, IZVZCS_SAC&CAC and IZVZCS_SRC&CAC have no added primary components, which means the primary circuits of these converters are simpler and more compact compared to that of other converters. A smaller number of primary components means not only cheaper BOM cost but also simpler and more compact primary structure. In addition, less area in the primary side of these converters is required to ensure safe electrical clearance due to the smaller number of primary components and simpler connection between these components. Therefore, the primary circuit volume of the IZVS_SMI, IZVS_SMM, IZVZCS_SAC&CAC and IZVZCS_SI is smaller and more compact, which are attractive features for high input voltage industry applications. As depicted in Table 7, the IZVS_CD, IZVS_CAC, IZVS_SMI, IZVZCS_DCF and IZVZCS_SI require no added secondary components. In some industrial applications, the input voltage may be 1400 V or higher. Due to the modular structure, the proposed converters can be extended to higher voltage levels easily. As shown in Table 7, the added primary components of the extension topologies of the IZVS_CD, IZVS_CAC, IZVZCS_DCF and IZVZCS_SI increase linearly with the number of the primary modules. However, as depicted in Table 7, the added secondary components of the extension topologies of the IZVS_SMI, IZVS_SMM and IZVZCS_SRC&CAC are not increased with primary cells. Thus, the IZVS_SMI, IZVS_SMM, IZVZCS_SAC&CAC and IZVZCS_SI are more suitable for super-high-voltage applications due to smaller added components and a simpler primary structure. The added cost of the proposed converters is listed in Table 8.

From above analyses, some brief conclusions can be drawn as follows. The IZVS_SMM and IZVZCS_SRC&CAC show some clear advantages compared to other solutions, e.g., smaller duty ratio loss, wide range soft switching operation, less conduction loss, simpler and more compact primary circuit, and lower added cost. Thus, these two converters are more suitable for high input voltage applications with more primary modules. The current stress of Sse and Cse in IZVZCS_SRC&CAC increases with the output current, which may arise several implementation problems. Hence, the IZVS_SMM is a better choice for large output current applications. The IZVS_SMM has some special characteristics. The secondary rectified voltage is a TL waveform, which results lower input and output filter requirements. The IZVS_SMM is the only topology, which can be used in the high input applications with controllable multi-output ports.

Table 7. Added component comparison. (N is series-connected primary modules number).

Converter	Added primary components	Added secondary components
IZVS_CD	N × inductors and 2N × diodes	None
IZVS_CAC	N × inductors and N × capacitors	None
IZVS_SMI	None	None
IZVS_SMM	None	2 × MOSFETs and 2 × diodes
IZVZCS_DCF	2N × diodes	None
IZVZCS_SAC&CAC	None	1 × MOSFET and 1 × capacitors
IZVZCS_SRC&CAC	None	2 × diodes and 1 × capacitors
IZVZCS_SI	N × inductors	None

Table 8. Added cost comparison. (Two series-connected primary modules).

Converter	Added components	Added cost	Ratio of the total cost
IZVS_CD	2 × inductors and 4 × diodes	$150	8.8%
IZVS_CAC	2 × inductors and 2 × capacitors	$60	3.3%
IZVS_SMI	None	None	None
IZVS_SMM	2 × MOSFETs, 2 × diodes and corresponding drive circuits	$80	4.4%
IZVZCS_DCF	4 × diodes	$120	6.7%
IZVZCS_SAC&CAC	1 × MOSFETs, 1 × diodes and corresponding drive circuits	$50	2.8%
IZVZCS_SRC&CAC	2 × diodes and 1 × capacitors	$45	2.6%
IZVZCS_SI	2 × inductors	$30	1.7%

6. Experimental Results

The IZVS_SMM and IZVZCS_SAC&CAC are selected as examples to test in this section, and the conventional FB MMDC in Figure 1 is also tested in the efficiency comparison. The main parameters of the prototype are listed in Table 9. The waveforms of IZVS_SMM and IZVZCS_SAC&CAC are provided in Figure 9.

Table 9. Main parameters of the prototypes.

Item	Parameters
Input	100–300 V
Output	50 V/20 A
Switching frequency	20 kHz
I GBT	G60N100
C_{BL1} and C_{BL2}	100 μF
k_{T1} and k_{T2} (IZVS_SMM)	$k_{T1} = 41$, $k_{T2} = 16$
k_T (IZVZCS_SAC&CAC)	11.7
k_T (Figure 1)	11.4
S_{se1} and S_{se2} (IZVS_SMM))	IPB180N04S4 ×4
S_{se} (IZVZCS_SAC&CAC)	IPB180N04S4 ×1
C_{se} (IZVZCS_SAC&CAC)	3 μF/200 A
Rectifier diodes	MBR30100
L_o	7 μH
C_o	1000 μF

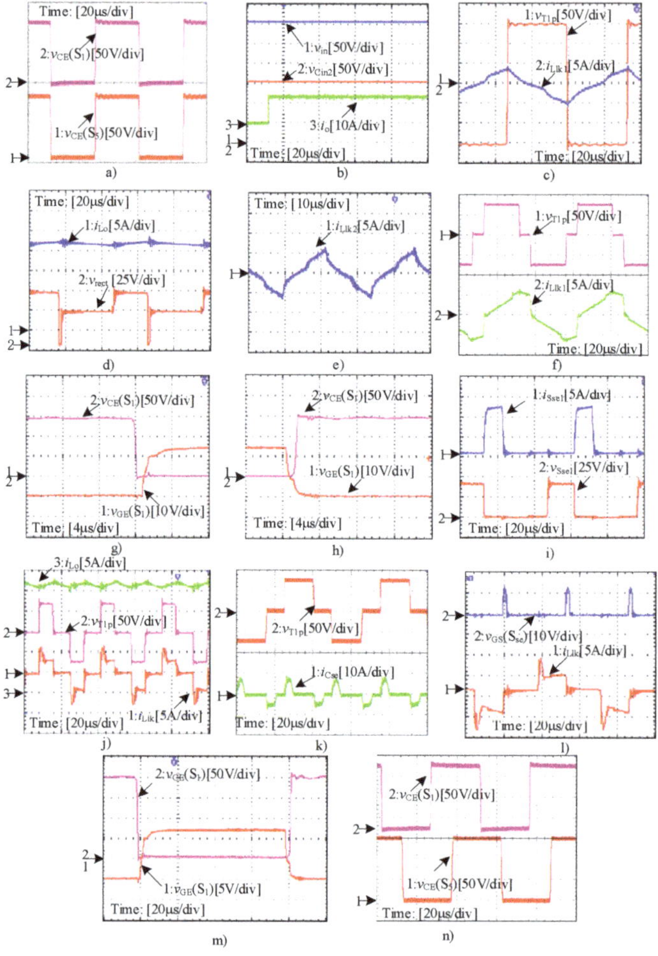

Figure 9. Experimental waveforms: (**a**) vCE(S1) and vCE(S5) of IZVS_SMM; (**b**) $v$cin1, $v$in2 and io of IZVS_SMM; (**c**) vBC and iLlk1 of IZVS_SMM; (**d**) iLo and $v$rect of IZVS_SMM; (**e**) iLlk2 of IZVS_SMM; (**f**) vBC and iLlk1 of IZVS_SMM (soft start); (**g**) vGE(S1) and vCE(S1) of IZVS_SMM (turn-on); (**h**) vGE(S1) and vCE(S1) of IZVS_SMM (turn-off); (**i**) vDS(Sse1) and iDS(Sse1) of IZVS_SMM; (**j**) vBC, iLlk1and iLo of IZVZCS_SAC&CAC; (**k**) vBC and $i$se of IZVZCS_SAC&CAC; (**l**) vGE(Sse) and iLlk1 of IZVZCS_SAC&CAC; (**m**) vGE(S1) and vCE(S1) of IZVZCS_SAC&CAC; (**n**) vCE(S1) and vCE(S5) of IZVZCS_SAC&CAC.

A conventional circuit of a phase-shifted full-bridge includes IZVS and ZVZCS operation modes. IZVS implement ZVS on both the leading switches and the lagging switches. Due to the existence of the transformer leakage inductance and the output inductance, the current does not change suddenly when the leading switches are turned off, and only ZVS. IZVS mode has good switching characteristics and high on-state loss. For the ZVZCS mode, it achieves ZVS of the leading switches and ZCS on the lagging switches. ZVZCS mode has low on-state loss and current overshoot. Some of the proposed FB-MMDCs discussed in this paper have clear advantages compared with conventional FB MMDC. For example, IZVZCS-SAC&CAC (see Figure 3b) has less voltage stress on the primary switch no additional primary clamping device and modular primary structure. In addition, the primary switch in all converters can achieve ZVS or ZCS over a wide load range.

As shown in Figure 9a, the off-state voltage of the primary switches in IZVS_SMM is even during normal operation stages, and the midpoint voltage of the input capacitors is stable and equals Vin/2. $v$in, vCin2 and Io are depicted in Figure 9b, and the mid-point voltage of the input capacitors is balanced even during the output dynamic instant. As proved in Figure 9c, iLlk1 is not a constant value because i1m is enlarged to help the ZVS of the primary switches. As i1m is not in phase with the load current, the added primary RMS current is smaller. Thus, added conduction loss is also smaller. As depicted in Figure 9 c, vBC does not have free-wheeling time. Therefore, the input current ripple is smaller. As proved in Figure 9d, the secondary rectified voltage is a TL waveform, which can significantly reduce the volume of output filter. The iLlk2 is provided in Figure 9e, and vBC and iLlk1 during the soft-start operation are shown in Figure 9f. The voltage waveforms of the gate-emitter and the collector-emitter of S1 are depicted in Figure 9g and 9h. In Figure 9g, the gate-emitter voltage of S1 is much lower than gate-emitter threshold voltage when the collector-emitter voltage of S1 decreases to zero, thus, S1 can obtain ZVS. According to Figure 9i, Sse1 can obtain ZCS.

The waveforms of IZVZCS_SAC&CAC are depicted in Figure 9j to 9n. As proved in Figure 9j, iLlk1 is reset by the secondary clamping capacitor and keeps zero during the whole free-wheeling stage. Thus, the lagging switches can obtain ZCS. In addition, the primary circling energy is zero. $i$se is provided in Figure 9k. The gate signal of Sse and the primary current is depicted in Figure 9l, the secondary switches are turned on at the beginning of the free-wheeling stages to reset the primary currents. In Figure 9m, the gate-emitter voltage of S1 is much lower than the gate-emitter threshold voltage when the collector-emitter voltage of S1 decreases to zero; thus, S1 can obtain ZVS. As shown in Figure 9n, the off-state voltage of the primary switches in IZVZCS_SAC&CAC is even during normal operation stages, and the mid-point voltage of the input capacitors is stable and equals Vin/2.

Figure 10 shows the efficiency comparison between the converters in the conventional FB MMDC (Figure 1), the IZVS with secondary modulation method (IZVS_SMM, Figure 2d) and the IZVZCS with secondary active clamping and commutation auxiliary circuit (IZVZCS_SAC&CAC, Figure 3b). During the efficiency test, the auxiliary power of the controller and driver are taken into account, and the power of the force air cooling fan is also included. As depicted in Figure 10a, the efficiency of IZVS_SMM and IZVZCS_SAC&CAC is higher than that of Figure 2 because of wide-range soft-switching for all primary switches. As illustrated in Figure 10a, the light-load efficiency of IZVZCS_SAC&CAC is a bit lower than that of IZVS_SMM because the switching loss of the leading switches is a little higher, and the high load efficiency of IZVZCS_SAC&CAC is a bit higher than that of IZVS_SMM due to the smaller turned off loss of the lagging switches and lower conduction loss. Figure 10b shows the efficiency curves with larger output capacitance of the primary switches. Turn-off loss of the switches decreases with the increasing of the output capacitance, but, turn-on loss may increase due to narrow ZVS load range. As all primary switches in IZVS_SMM can still obtain ZVS turned-on in wide load range with less added conduction loss, the efficiency of IZVS_SMM is higher than others. Hence, the converter in IZVS_SMM can gain optimum efficiency performance by more flexible selecting of the trade-off among turn-on loss, turn-off loss and conduction loss. Figures 10c and 10d show the efficiency comparison with variable input voltage. The efficiency curves decrease with the increasing of the input voltage. As the magnetizing inductances can provide more resonant energy, the converter in IZVS_SMM has higher efficiency under high input voltage condition.

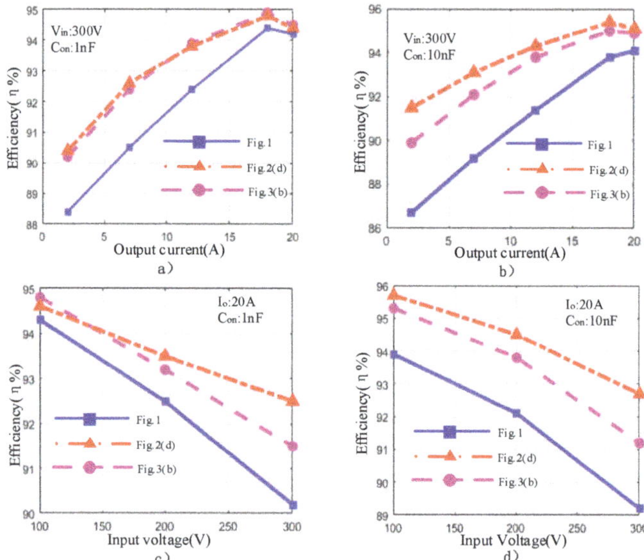

Figure 10. Efficiency comparison: (**a**) variable output current (Vin = 300 V, Cos = 1 nF); (**b**) variable output current (Vin = 300 V, Cos = 10 nF);(**c**) variable input voltage (Io = 20 A, Cos = 1 nF); (**d**) variable input voltage (Io = 20 A, Cos = 10 nF).

7. Conclusions

Eight wide-range soft-switching FB MMDCs are proposed and discussed. Two of the most promising converters are found and explained in detail, and a comparative evaluation among the proposed converters is also provided. The experimental results agree with the theoretical prediction. The IZVS_SMM and IZVZCS_SAC&CAC show clear advantages compared to other solutions, e.g., smaller duty ratio loss, wide-range soft switching, less conduction loss, a simpler and more compact primary circuit and lower added cost. Thus, these two converters are more suitable for high input voltage applications with more primary modules. Furthermore, the IZVS_SMM has some special characteristics. its secondary rectified voltage is a TL waveform, which results in lower input and output filter requirements. The IZVS_SMM is the only topology which can be used in the high input voltage applications with controllable multi-output ports.

Author Contributions: Software, X.L.; formal analysis, J.C. and Y.S.; data curation, X.L.; writing—original draft preparation, J.C.; writing—review and editing, Y.S.; project administration, H.D. and Y.S.; All authors have read and agreed to the published version of the manuscript.

Funding: This research received no external funding.

Conflicts of Interest: The authors declare no conflict of interest.

References

1. Kakigano, H.; Miura, Y.; Ise, T. Low-voltage bipolar-type DC micro grid for super high quality distribution. *IEEE Trans. Power Electron.* **2010**, *25*, 3066–3075. [CrossRef]
2. Anand, S.; Fernandes, B.G. Reduced-order model and stability analysis of low-voltage DC micro grid. *IEEE Trans. Ind. Electron.* **2013**, *60*, 5040–5049. [CrossRef]
3. Lai, J.; Peng, F.Z. Multi Level converters—A new breed of power converters. *IEEE Trans. Ind. Appl.* **1996**, *32*, 509–517.

4. Pinheiro, J.R.; Barbi, I. The three-level ZVS PWM converter—A new concept in high-voltage DC-to-DC conversion. In Proceedings of the 1992 International Conference on Industrial Electronics, Control, Instrumentation, and Automation, San Diego, CA, USA, 13 November 1992; pp. 173–178.

5. Shi, Y.; Gui, X.; Wang, X.; Xi, J.; Yang, X. Large Power Hybrid Soft Switching Mode PWM Full Bridge DC–DC Converter With Minimized Turn-ON and Turn-OFF Switching Loss. *IEEE Trans. Power Electron.* **2019**, *34*, 11629–11644. [CrossRef]

6. Duarte, J.L.; Lokos, J.; vanHorck, F.B.M. Phase-Shift-Controlled Three-Level Converter With Reduced Voltage Stress Featuring ZVS Over the Full Operation Range. *IEEE Trans. Power Electron.* **2013**, *28*, 2140–2150. [CrossRef]

7. Zhang, Y.; Shi, J.; Zhou, L. Wide input-voltage range boost three-level DC–DC converter with quasi-Z source for fuel cell vehicles. *IEEE Trans. Power Electron.* **2017**, *32*, 6728–6738. [CrossRef]

8. Shi, Y.; Xi, J.; Wang, X.; Gui, X.; Yang, X. Large Power ZVZCS Full Bridge Three-Level DC-DC Converter with Wide Operation Range and Its Application in Sapphire Crystal Furnace Power Supply. *IEEE J. Emerg. Sel. Top. Power Electron.* **2019**. [CrossRef]

9. Liu, F.; Hu, G.; Ruan, X. Three-Phase Three-Level DC/DC Converter for High Input Voltage and High-Power Applications Adopting Symmetrical Duty Cycle Control. *IEEE Trans. Power Electron.* **2014**, *29*, 56–65. [CrossRef]

10. Liu, D.; Deng, F.; Zhang, Q.; Chen, Z. Zero-voltage switching PWM strategy based capacitor current-balancing control for half-bridge three-level DC/DC converter. *IEEE Trans. Power Electron.* **2018**, *33*, 357–369. [CrossRef]

11. Yu, X.; Jin, K.; Liu, Z. Capacitor Voltage Control Strategy for Half-Bridge Three-Level DC/DC Converter. *IEEE Trans. Power Electron.* **2014**, *29*, 1557–1561. [CrossRef]

12. Ruan, X.; Zhou, L.; Yan, Y. Soft-switching PWM three-level converters. *IEEE Trans. Power Electron.* **2001**, *16*, 612–622. [CrossRef]

13. Shi, Y.; Yang, X. Wide Range Soft Switching PWM Three-Level DC–DC Converters Suitable for Industrial Applications. *IEEE Trans. Power Electron.* **2014**, *29*, 603–616. [CrossRef]

14. Shi, Y. Wide Load Range Capacitor Clamped ZVZCS Half Bridge Three-level DC-DC Converter with Two Unsymmetrical Bi-directional Switches. *Energies* **2019**, *12*, 2362. [CrossRef]

15. Lee, I.O.; Moon, G.W. Analysis and Design of a Three-Level LLC Series Resonant Converter for High- and Wide-Input-Voltage Applications. *IEEE Trans. Power Electron.* **2012**, *27*, 2966–2979. [CrossRef]

16. Shi, Y.; Yang, X. Zero-Voltage Switching PWM Three-Level Full-Bridge DC-DC Converter with Wide ZVS Load Range. *IEEE Trans. Power Electron.* **2013**, *28*, 4511–4524. [CrossRef]

17. Liu, F.; Yan, J.; Ruan, X. Zero-Voltage and Zero-Current-Switching PWM Combined Three-Level DC/DC Converter. *IEEE Trans. Ind. Electron.* **2010**, *57*, 1644–1654.

18. Shi, Y.; Yang, X. Wide-Range Soft-Switching PWM Three-Level Combined DC–DC Converter Without Added Primary Clamping Devices. *IEEE Trans. Power Electron.* **2014**, *29*, 4511–4524. [CrossRef]

19. Kim, D.Y.; Kim, J.K.; Moon, G.W. A Three-Level Converter with Reduced Filter Size Using Two Transformers and Flying Capacitors. *IEEE Trans. Power Electron.* **2013**, *28*, 46–53. [CrossRef]

20. Li, W.; Jiang, Q.; Mei, Y.; Li, C.; Deng, Y.; He, X. Modular Multilevel DC/DC Converters With Phase-Shift Control Scheme for High-Voltage DC-Based Systems. *IEEE Trans. Power Electron.* **2015**, *30*, 99–107. [CrossRef]

21. Li, W.; He, Y.; He, X.; Sun, Y.; Wang, F.; Ma, L. Series asymmetrical half-bridge converters with voltage auto balance for high input-voltage applications. *IEEE Trans. Power Electron. Aug.* **2013**, *28*, 3665–3674. [CrossRef]

22. Sun, T.T.; Chung, H.S.H.; Ioinovici, A. A high-voltage DC-DC converter with Vin/3—Voltage stress on the primary switches. *IEEE Trans. Power Electron.* **2007**, *22*, 2124–2137.

23. Lin, B. Hybrid DC/DC converter based on dual three-level circuit and half-bridge circuit. *IET Power Electron.* **2016**, *9*, 817–824. [CrossRef]

24. Miller, M. New technologies on telecom power conversion. In Proceedings of the INTELEC 95. 17th International Telecommunications Energy Conference, The Hague, Netherlands, 29 October–1 November 1995.

25. Shi, Y.; Xu, Z. Wide Load Range ZVS Three-level DC-DC Converter: Modular Structure, Redundancy Ability, and Reduced Filters Size. *Energies* **2019**, *12*, 3537. [CrossRef]

26. Ralph, P.; Vladimir, B.; Todd, J.K. ELECTRICAL POWER CONVERTER POWER SUPPLY AND INVERTER WITH SERIES CONNECTED SWITCHING CIRCUITS. United States Patent No. 5546295, 13 August 1996.

27. Ruan, X.; Lulu, C.; Tao, Z. Control strategy for input-series output paralleled converter. In Proceedings of the 37th IEEE Power Electronics Specialists Conference, Jeju, Korea, 18–22 June 2006.
28. Kim, J.W.; You, J.S.; Cho, B.H. Modeling, control, and design of input-series-output-parallel-connected converter for high-speed-train power system. *IEEE Trans. Ind. Electron.* **2001**, *48*, 536–544.
29. Shi, J.; Luo, J.; He, X. Common-duty-ratio control of input-series output-parallel connected phase-shift full-bridge DC-DC converter modules. *IEEE Trans. Power Electron.* **2011**, *26*, 3318–3329. [CrossRef]
30. Jin, L.; Duan, S. Comparative analysis of three-level dual active bridge DC–DC converter between reflux-power-optimised and current-stress-optimised phase shift control. *IET Power Electron.* **2018**, *11*, 1681–1688. [CrossRef]
31. Hong, F.; Li, L.; Wu, Y.; Ji, B.; Zhou, Y. 1500 V three-level forward converter with phase-shifted control. *IET Power Electron.* **2018**, *11*, 1547–1555. [CrossRef]
32. Liu, D.; Deng, F.; Gong, Z.; Chen, Z. Input-parallel output-parallel three-level DC/DC converters with interleaving control strategy for minimizing and balancing capacitor ripple currents. *IEEE J. Emerg. Sel. Top. Power Electron.* **2017**, *5*, 1122–1132. [CrossRef]
33. Jain, P.K.; Kang, W.; Soin, H.; Xi, Y. Analysis and Design Considerations of a Load and Line Independent Zero Voltage Switching Full Bridge DC/DC Converter Topology. *IEEE Trans. Power Electron.* **2002**, *17*, 649–657. [CrossRef]
34. Ayyanar, R.; Mohan, N. A Novel Full-Bridge DC–DC Converter for Battery Charging Using Secondary-Side Control Combines Soft Switching Over the Full Load Range and Low Magnetics Requirement. *IEEE Trans. Ind. Appl.* **2001**, *37*, 559–565. [CrossRef]
35. Jung-Goo, C.; Ju-Won, B.; Chang-Yong, J.; Geun-Hie, R. Novel Zero-Voltage and Zero-Current-Switching Full-Bridge PWM Converter Using a Simple Auxiliary Circuit. *IEEE Trans. Ind. Appl.* **1999**, *35*, 15–20. [CrossRef]
36. Seok, K.; Kwon, B. An improved zero-voltage and zero-current switching full-bridge PWM converter using a simple resonant circuit. *IEEE Trans. Ind. Electron.* **2001**, *48*, 1205–1209. [CrossRef]
37. Ruan, X.; Yan, Y. A novel zero-voltage and zero-current-switching PWM full bridge converters using two diodes in series with the lagging leg. *IEEE Trans. Ind. Electron.* **2001**, *48*, 777–785. [CrossRef]

© 2020 by the authors. Licensee MDPI, Basel, Switzerland. This article is an open access article distributed under the terms and conditions of the Creative Commons Attribution (CC BY) license (http://creativecommons.org/licenses/by/4.0/).

 *electronics*

Article

Analysis of a DC Converter with Low Primary Current Loss and Balance Voltage and Current

Bor-Ren Lin

Depart. of Electrical Engineering, National Yunlin University of Science and Technology (NYUST), Yunlin 640, Taiwan; linbr@yuntech.edu.tw; Tel.: +886-912312281

Received: 6 April 2019; Accepted: 15 April 2019; Published: 17 April 2019

Abstract: A dc/dc pulse width modulation (PWM) circuit was investigated to realize the functions of reduced primary current loss and balanced voltage and current distribution. In the presented dc/dc converter, two full bridge pulse width modulation circuits were used with the series/parallel connection on the high-voltage/low-voltage side. The flying capacitor was adopted on the input side to achieve voltage balance on input split capacitors. The magnetic coupling element was employed to achieve current sharing between two parallel circuits. A capacitor-diode passive circuit was adopted to lessen the primary current at the commutated interval. The phase-shifted duty cycle control approach was employed to regulate load voltage and implement soft switching characteristics of power metal-oxide-semiconductor field-effect transistors (MOSFETs). Finally, the experimental results using a 1.68 kW prototype converter were obtained to confirm the performance and feasibility of the studied circuit topology.

Keywords: dc converter; full bridge converter; zero voltage operation

---

1. Introduction

High power density/efficiency dc/dc converters [1,2] have been proposed to reduce unnecessary power loss in order to reduce environmental pollution and met global energy demands. Clean and renewable energy sources meet these requirements to provide available ac or dc power to ac utilities, dc micro-grids, residential houses and commercial buildings by using power electronic based converters or inverters. High-voltage pulse width modulation (PWM) converters have been presented for industry power units [3,4], dc micro-grid systems [5–7] and dc traction or dc light rail transportation systems [8,9]. For these applications, the dc bus voltage is regulated between 750 V and 800 V. Full bridge circuits with a high switching frequency high-voltage rating SiC MOSFET or high-voltage rating insulated gate bipolar transistor (IGBT) devices can be adopted for high dc voltage applications. However, 1200 V SiC MOSFET devices are expensive and 1200 V IGBT devices have low switching frequency problems. Cost and performance are always the two main issues in the development of the available and high reliability power converters. Therefore, MOSFET power devices have been widely adopted in high efficiency power electronics. To improve the low-voltage drawback of MOSFETs in high-voltage applications, zero voltage switching, multi-level circuit topologies [10–14] have been developed and proposed to decrease conduction and switching losses. The phase shift pulse width modulation (PSPWM) and frequency modulation have normally been adopted to regulate duty cycle or switching frequency. However, the main disadvantage of conventional PSPWM is high circulating current losses at low effective duty cycle. In order to reduce circulating current, an active or passive snubber used on the low-voltage side has been proposed in [15–17]. A modular dc/dc converter [18] using low power rating modular circuits in series or parallel connection has been proposed to increase circuit efficiency and power rating. However, the current between each modular circuit may be unbalanced. To accomplish current balance issue, several current balancing approaches have been presented in [19,20].

A series/parallel-connected dc/dc converter is proposed to accomplish reduced primary circulating current, a soft switching operation, and balanced input voltages and output currents. Two full bridge circuits with series/parallel connection on input/output side are used in the studied circuit to achieve voltage and current sharing for the high voltage input and current output. Therefore, power devices on each full bridge circuit have $V_{in}/2$ voltage stress. To prevent input split voltages imbalance, a voltage balance capacitor was used on the input side to automatically achieve split voltages balance. Passive snubber circuits were employed on output side to create a positive voltage on the secondary rectified terminal under the commutated state so that the primary current at the commutated state can be reduced. To realize current sharing of two full bridge circuits, a current-balance magnetic component was adopted on the high-voltage side. If two currents are unbalanced, then one induced voltage of MC component is decreased to lessen the larger converter current. Therefore, two converter currents can be compensated automatically. PSPWM was adopted to control power devices and accomplish soft switching operation of active switches. The organization of this paper is as follows: The circuit structure and the principle of operation are discussed in Section 2. The steady state operation of the presented converter is shown in Section 3. Test results with a laboratory prototype are presented and investigated in Section 4. The conclusions of the studied converter are discussed in Section 5.

2. Circuit Structure and Principle of Operation

High-voltage converters were researched and presented for industry power supplies, dc traction vehicle and dc micro-grid systems. Ac/dc power converters with low harmonic currents, high power factor (PFC) and a stable dc voltage shown in Figure 1a are required for industry power converters. Normally, the dc voltage V_H is controlled at 760 V. Then, a high-frequency link converter is adopted to provide low-voltage output. Figure 1b illustrates the basic power distributed diagrams on dc light rail vehicle. A high-voltage dc converter can convert a 760 V input to supply a low-voltage output for battery charger, control, and communication demands. Figure 1c demonstrates the blocks of a bipolar dc micro-grid system to integrate ac utility system, clean energy generators and industry load and residential load into a common dc bus voltage.

Figure 2 gives the converter configuration of the developed circuit with series/parallel connection of full bridge circuits to achieve low circulating current loss, soft switching for power MOSFETs, and balanced input split voltages and output current sharing. The first full bridge circuit has power MOSFETs S_1-S_4 with the output capacitors C_1-C_4, leakage or external inductor L_{r1}, transformer T_1, magnetic core MC, filter inductor L_{o1}, secondary-side rectifier diodes D_1 and D_2, and passive circuit including C_{a1}, D_{a1} and D_{b1}. Likewise, the second full bridge circuit includes the components S_5-S_8, C_5-C_8, L_{r2}, T_2, MC, L_{o2}, D_3, D_4, C_{a2}, D_{a2} and D_{b2}. $C_{in,1}$ and $C_{in,2}$ are input split capacitors and C_b is the balance capacitor. Each circuit can transfer one-half rated power to the secondary side. The magnetic core [20] is used to balance i_{Lr1} and i_{Lr2}. Under the balanced primary currents ($|i_{Lr1}| = |i_{Lr2}|$), the induced voltages V_{L1} and V_{L2} are zero. If i_{Lr1} and i_{Lr2} are unbalanced (such as $|i_{Lr1}| > |i_{Lr2}|$), the induced voltage V_{L1} of MC cell is decreased in order to decrease i_{Lr1} and V_{L2} is increased to increase i_{Lr2}. Thus, i_{Lr1} will equal i_{Lr2} and $V_{L1} = V_{L2} = 0$. PSPWM is used to control S_1~S_8 and have zero voltage switching on the MOSFETs. Power switches S_1~S_4 and S_5~S_8 have identical PWM waveforms. The balance capacitor C_b is linked between points a and c. If S_1 and S_5 are on and S_2 and S_6 are off, then $V_{Cb} = V_{Cin,1}$. Likewise, $V_{Cb} = V_{Cin,2}$ if S_2 and S_6 are on and S_1 and S_5 are in the off. Since the duty cycle $d_{S1} = d_{S2} = 0.5$, it is obvious that $V_{Cb} = V_{Cin,1} = V_{Cin,2} = V_{in}/2$. Thus, the drain voltages of S_1~S_8 are $V_{in}/2$. To decrease the primary-side circulating currents, passive snubbers, C_{a1}, D_{a1}, D_{b1}, C_{a2}, D_{a2}, D_{b2}, are used on the presented circuit. At the same time, the filter inductor voltages $v_{Lo1} = v_{Ca1} - V_o$ and $v_{Lo2} = v_{Ca2} - V_o$ rather than $-V_o$ in the traditional full bridge duty cycle converters.

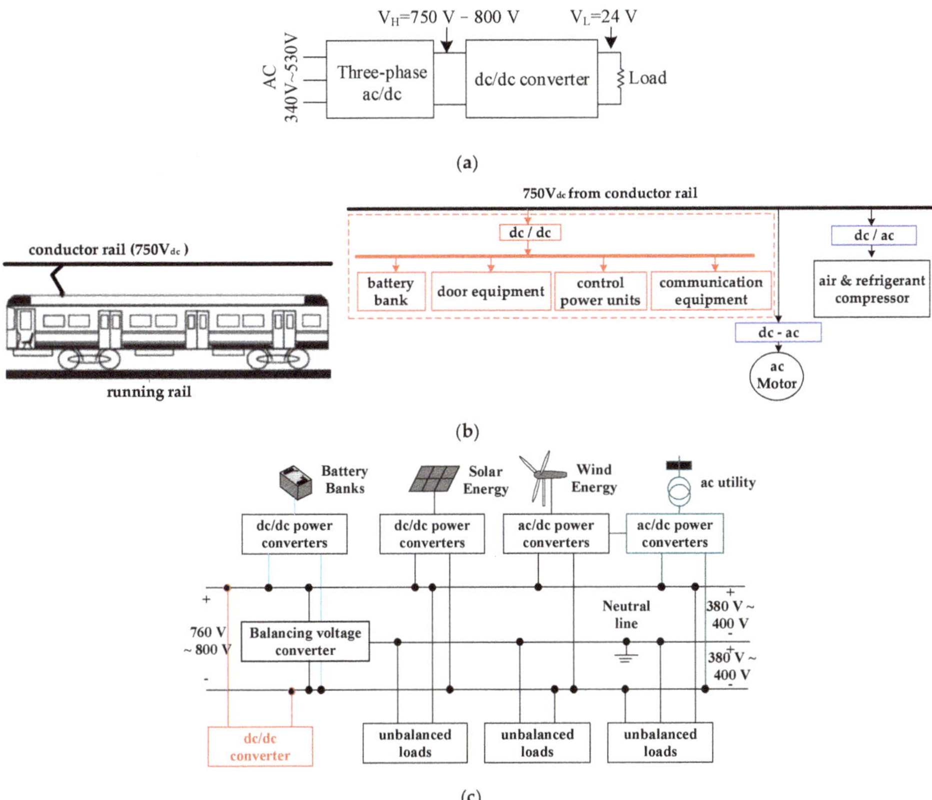

Figure 1. High-voltage dc/dc converters in (**a**) ac/dc converters, (**b**) dc light rail transportation vehicle, (**c**) dc micro-grid system with bipolar voltages.

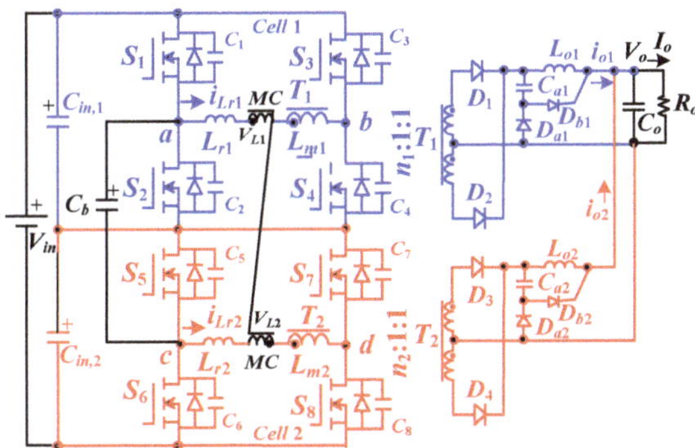

Figure 2. Circuit configuration of the studied converter.

The operation principle of the presented circuit is discussed with the assumptions: (1) $C_{in,1} = C_{in,2}$; (2) $C_1 = \ldots = C_8 = C_{oss}$; (3) $C_{a1} = C_{a2} = C_a$; (4) $L_{m1} = L_{m2} = L_m$; (5) $L_{o1} = L_{o2} = L_o$; (6) $L_{r1} = L_{r2} = L_r$; (7) $V_{Cb} = V_{Cin,1} = V_{Cin,2} = V_{in}/2$; and (8) $n_1 = n_2 = n$. Figure 3 shows the PWM waveforms of the studied converter. The duty cycle of $S_1 \sim S_8$ is 0.5. The gating signals of S_4 (S_3) is shifted to S_1 (S_2), respectively. Due to the switching states of power devices, the converter has fourteen steps in a switching period. The PWM waveforms are symmetrical for every half cycle. Therefore, only the first seven steps are discussed and the circuits of first seven operating steps are illustrated in Figure 4.

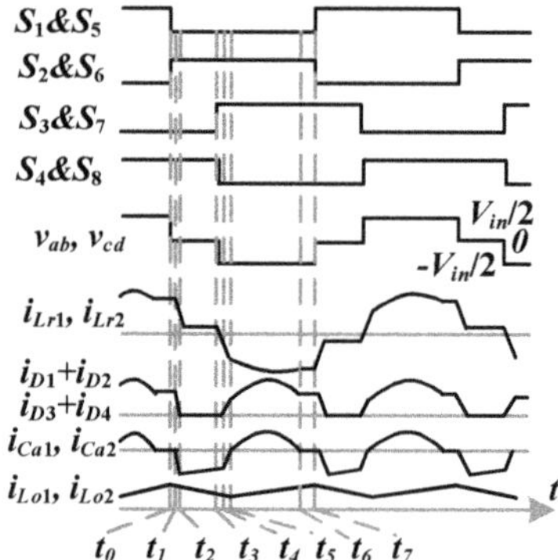

Figure 3. Pulse width modulation (PWM) waveforms of the presented converter.

Step 1 [t_0, t_1]: Before t_0, S_1, S_4, S_5, S_8, D_1 and D_3 conduct, $i_{Lr1} > 0$ and $i_{Lr2} > 0$. After t_0, power MOSFETs S_1 and S_5 turn off. C_1 and C_5 are charged and C_2 and C_6 are discharged. The energy on L_{r1}, L_{r2}, L_{o1} and L_{o2} can discharge C_2 and C_6. Equations (1) and (2) give the soft switching conditions of S_2 and S_6.

$$(L_{r1} + n_1^2 L_{o1})i_{Lr1}^2(t_0) \geq C_{oss}V_{in}^2/2 \tag{1}$$

$$(L_{r2} + n_2^2 L_{o2})i_{Lr2}^2(t_0) \geq C_{oss}V_{in}^2/2 \tag{2}$$

If these conditions are satisfied, then $v_{C1} = v_{C5} = V_{in}/2$ and $v_{C2} = v_{C6} = 0$ at t_1. The time duration in step 1 is obtained in Equation (3).

$$\Delta t_{01} = t_1 - t_0 = \frac{C_{oss}V_{in}}{i_{Lr1}(t_0)} \approx \frac{n_1 C_{oss}V_{in}}{i_{Lo,peak}} \approx \frac{2n_1 C_{oss}V_{in}}{I_o} \tag{3}$$

The time duration Δt_{01} is related to the load current and input voltage.

Step 2 [t_1, t_2]: At t_1, $v_{C2} = v_{C6} = 0$. The positive primary currents i_{Lr1} and i_{Lr2} will flow through the body diode of S_2 and S_6. Thus, power MOSFETs S_2 and S_6 can achieve zero-voltage switching after t_1. The ac side voltages $v_{ab} = v_{cd} = 0$ and i_{Lr1} and i_{Lr2} decrease. Therefore, D_{a1} and D_{a2} conduct at this freewheeling state. The magnetizing voltages v_{Lm1} and v_{Lm2} are clamped at v_{Ca1} and v_{Ca2}, respectively. The primary inductor voltages $v_{Lr1} = -n_1 v_{Ca1}$ and $v_{Lr2} = -n_2 v_{Ca2}$ and the filter inductor voltages $v_{Lo1} = v_{Ca1} - V_o < 0$ and $v_{Lo2} = v_{Ca2} - V_o < 0$. Therefore, i_{Lr1}, i_{Lr2}, i_{Lo1} and i_{Lo2} are all decreased in this step.

$$i_{Lr1}(t) \approx i_{Lr1}(t_1) - \frac{n_1 v_{Ca1}}{L_{r1}}(t - t_1) \tag{4}$$

$$i_{Lr2}(t) \approx i_{Lr2}(t_1) - \frac{n_2 v_{Ca2}}{L_{r2}}(t - t_1) \tag{5}$$

$$i_{Lo1}(t) \approx i_{Lo1}(t_1) + \frac{v_{Ca1} - V_o}{L_{o1}}(t - t_1) \tag{6}$$

$$i_{Lo2}(t) \approx i_{Lo2}(t_1) + \frac{v_{Ca2} - V_o}{L_{o2}}(t - t_1) \tag{7}$$

Thus, the circulating current is decreased under the commutated state. In this step, the balance capacitor voltage $V_{Cb} = V_{Cin,2}$.

Step 3 [t_2, t_3]: At time t_2, $i_{D1} = i_{D3} = 0$ and D_1 and D_3 are off. Then, $i_{Ca1} = -i_{Lo1}$, $i_{Ca2} = -i_{Lo2}$, $i_{Lr1} = i_{Lm1}(t_2)$ and $i_{Lr2} = i_{Lm2}(t_2)$. Since $i_{Lm1}(t_2)$ and $i_{Lm2}(t_2)$ are close to zero, the circulating current is removed. The filter inductor voltages $v_{Lo1} = v_{Ca1} - V_o < 0$ and $v_{Lo2} = v_{Ca2} - V_o < 0$. The secondary-side currents i_{Lo1} and i_{Lo2} are lessened.

Step 4 [t_3, t_4]: At t_3, power MOSFETs S_4 and S_8 turn off. C_3 and C_7 are discharged and C_4 and C_8 are charged. The energy on L_{r1} and L_{r2} can discharge C_3 and C_7 and the soft switching conditions of S_3 and S_7 are expressed in Equations (8) and (9).

$$L_{r1}i_{Lr1}^2(t_3) \geq C_{oss}V_{in}^2/2 \tag{8}$$

$$L_{r2}i_{Lr2}^2(t_3) \geq C_{oss}V_{in}^2/2 \tag{9}$$

Step 5 [t_4, t_5]: This step starts at t_4 when $v_{C3} = v_{C7} = 0$. Due to $i_{Lr1}(t_4) > 0$ and $i_{Lr2}(t_4) > 0$, the body diodes of S_3 and S_7 conduct. Then, S_3 and S_7 naturally conduct at zero-voltage switching after t_4. In this step, $v_{ab} = -V_{Cin,1}$, $v_{cd} = -V_{Cin,2}$, $V_{Cb} = V_{Cin,2}$, $v_{Lm1} = -n_1 v_{Ca1}$, $v_{Lm2} = -n_2 v_{Ca2}$, $v_{Lo1} = v_{Ca1} - V_o$ and $v_{Lo2} = v_{Ca2} - V_o$. In order to balance i_{Lr1} and i_{Lr2}, the magnetic core MC is employed on the high-voltage side. Under current balance condition, $V_{L1} = V_{L2} = 0$. Therefore, $v_{Lr1} \approx n_1 v_{Ca1} - V_{in}/2$ and $v_{Lr2} \approx n_2 v_{Ca2} - V_{in}/2$. It can be observed that i_{Lr1}, i_{Lr1}, i_{Lo1} and i_{Lo2} all decrease in this step. When the secondary-side diode currents $i_{D2} = i_{Lo1}$ and $i_{D4} = i_{Lo2}$ at time t_5, diodes D_{a1} and D_{a2} are reverse biased. In this step, i_{D2} and i_{D4} increase from zero to i_{Lo1} and i_{Lo2}, respectively and the time duration in step 5 can be obtained as:

$$\Delta t_{45} = \frac{L_{r1}I_{Lo1,min}}{n_1 V_{in}/2 - n_1^2 v_{Ca1}} \approx \frac{L_{r1}I_o}{n_1 V_{in} - 2n_1^2 v_{Ca1}} \tag{10}$$

Since D_{a1} and D_{a2} are still conducting in this step, the duty loss is calculated in Equation (11).

$$d_{5,loss} \approx \frac{L_{r1}I_o f_{sw}}{n_1 V_{in} - 2n_1^2 v_{Ca1}} \tag{11}$$

Step 6 [t_5, t_6]: At time t_5, $i_{D2} = i_{Lo1}$ and $i_{D4} = i_{Lo2}$. Then, the secondary-side diodes D_{a1} and D_{a2} are off and D_{b1} and D_{b2} are on in this step. The inductances $L_{r1}/(n_1)^2$ ($L_{r2}/(n_2)^2$) and C_{a1} (C_{a2}) are resonant with frequency $f_R = n_1/(2\pi \sqrt{L_{r1}C_{ca1}})$. Since $v_{Lo1} = v_{Ca1}$ and $v_{Lo2} = v_{Ca2}$, i_{Lo1} and i_{Lo2} both increase in this step. In order to force $i_{Db1} = i_{Db2} = 0$ at t_6, the half resonant cycle $1/(2f_R)$ must be less than $d_{eff,min}T_{sw}/2$. The primary magnetizing voltages $v_{Lm1} = n_1(v_{Ca1} + V_o)$ and $v_{Lm2} = n_2(v_{Ca2} + V_o)$ and the primary inductor current $i_{Lr1} \approx -(i_{Lo1} + i_{Ca1})/n_1$ and $i_{Lr2} \approx -(i_{Lo2} + i_{Ca2})/n_2$.

Step 7 [t_6, t_7]: At time t_6, $i_{D2} = i_{Lo1}$ and $i_{D4} = i_{Lo2}$. Then diodes D_{b1} and D_{b2} are off. The filter inductor voltages $v_{Lo1} \approx V_{in}/(2n_1) - V_o$ and $v_{Lo2} \approx V_{in}/(2n_2) - V_o$ so that i_{Lo1} and i_{Lo2} increase. At t_7, S_2 and S_6 turn off.

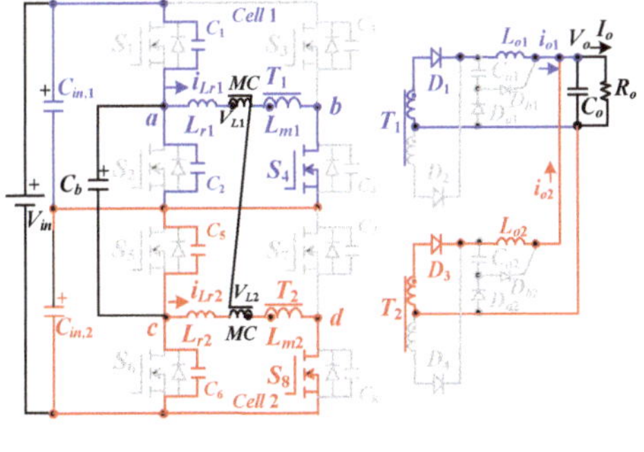

(a)

Figure 4. *Cont.*

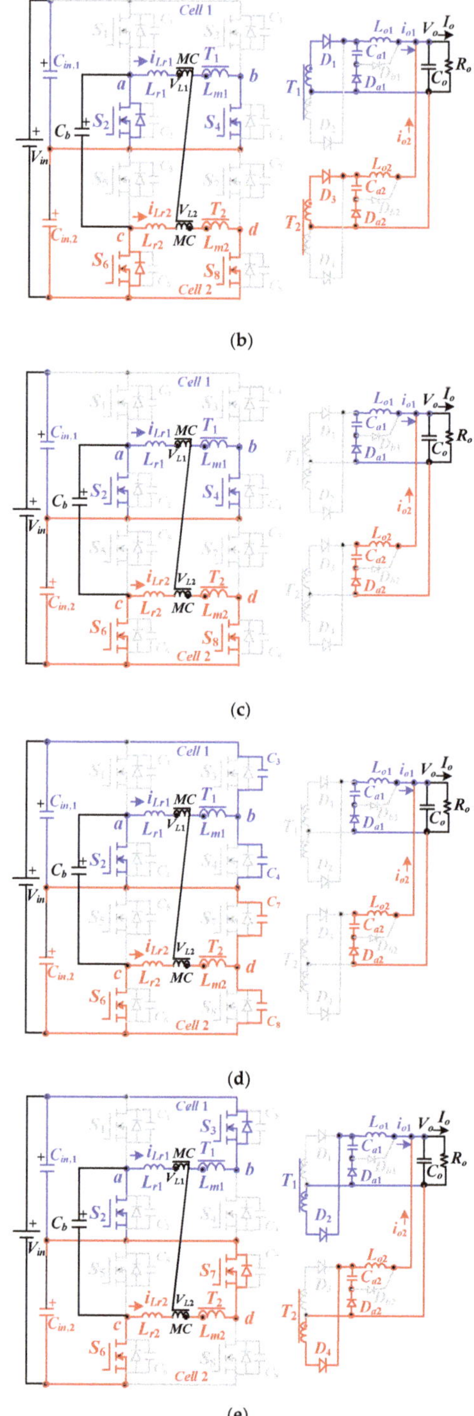

(b)

(c)

(d)

(e)

Figure 4. *Cont.*

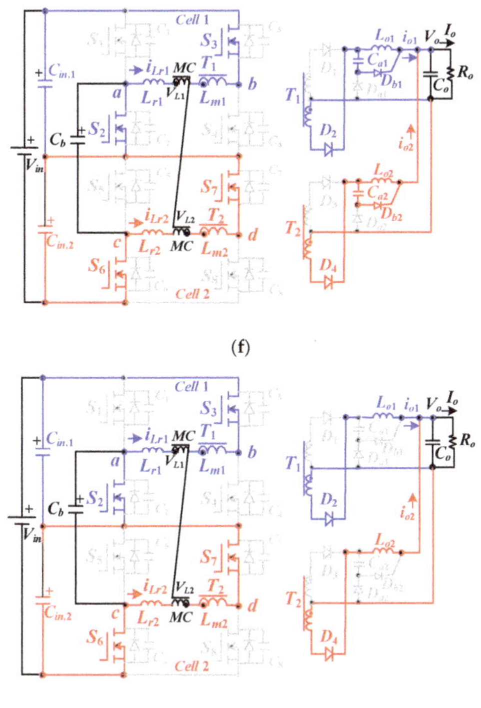

(f)

(g)

Figure 4. Equivalent circuits in first seven steps (**a**) step 1, (**b**) step 2, (**c**) step 3, (**d**) step 4, (**e**) step 5, (**f**) step 6, (**g**) step 7.

3. Steady State Analysis

Each full bridge converter in the presented converter supplies one-half of load power to low-voltage side. To balance i_{Lr1} and i_{Lr2}, the magnetic component MC is employed in the studied converter. Under current balance condition, $V_{L1} = V_{L2} = 0$. It can be observed that the average voltages $V_{Ca1} = V_{Ca2} = V_{in}/(2n_1) - V_o$ in step 6. Under steady state operation, i_{Lo1} and i_{Lo2} at t_0 and $t_0 + T_{sw}$ in every switching period are identical, $i_{Lo1}(t_0) = i_{Lo1}(t_0 + T_{sw})$ and $i_{Lo2}(t_0) = i_{Lo2}(t_0 + T_{sw})$. Thus, Equation (12) is derived according to voltage-second balance condition.

$$(d - d_5 - d_6)\left(\frac{V_{in}}{2n_1} - V_o\right) + d_6 V_{Ca1} = \left(\frac{1}{2} - d + d_5\right)(V_o - V_{Ca1}) \tag{12}$$

where d_5 and d_6 are duty cycles in steps 5 and 6, respectively. Based on $V_{Ca1} = V_{in}/(2n_1) - V_o$, V_o can be derived in Equation (13) under steady state.

$$V_o = \frac{V_{in}}{4n_1(1 - d + d_5)} = \frac{V_{in}}{4n_1(1 - d_{eff})} \tag{13}$$

where $d_{eff} = d - d_5$ is an effective duty ratio and δ is duty ratio when S_1 (S_2) and S_4 (S_3) are conducting. From Equation (13), the voltage gain is expressed in Equation (14).

$$G_{dc} = V_o/V_{in} = \frac{1}{4n_1(1 - d_{eff})} \tag{14}$$

From the given input and output voltages, the turns ratio n is derived as:

$$n = n_1 = n_2 = \frac{V_{in}}{4V_o(1 - d_{eff})} \tag{15}$$

Therefore, the minimum primary turns n_p and secondary turns n_s are derived in Equation (16).

$$n_p \geq \frac{V_{Lm}dT_{sw}}{\Delta B_{max}A_e}, \; n_s = \frac{n_p}{n} \tag{16}$$

where ΔB_{max}: maximum flux density range, A_e: effective cross area, and V_{Lm}: primary voltage. In steady state, $I_{Lo1} = I_{Lo2} = I_o/2$, $V_{Cin,1} = V_{Cin,2} = V_{Cb} = V_{in}/2$ and $v_{S1,ds} = \ldots = v_{S8,ds} = V_{in}/2$. If the effective duty cycle is defined, then the ripple currents ΔL_{o1} and ΔL_{o2} can be obtained in Equation (17).

$$\Delta i_{Lo1} = \Delta i_{Lo2} - (V_o - V_{Ca1})(0.5 - d_{eff})T_{sw}/L_o \approx (2V_o - \frac{V_{in}}{2n_1})(0.5 - d_{eff})T_{sw}/L_o \tag{17}$$

If the ripple currents $\Delta i_{Lo1} = \Delta i_{Lo2} = \Delta i_{Lo}$ are given or selected, then output inductances are achieved in Equation (18).

$$L_{o1} = L_{o2} = L_o \geq (2V_o - \frac{V_{in}}{2n_1})(0.5 - d_{eff})T_{sw}/\Delta i_{Lo} \tag{18}$$

The winding turns of filter inductors L_{o1} and L_{o6} are expressed as:

$$n_{Lo1} = n_{Lo2} \geq \frac{L_{o1}i_{Lo1,peak}}{B_{max}A_e} \tag{19}$$

The voltage ratings and average currents on $D_1 \sim D_4$ are expressed in Equations (20)–(23).

$$V_{D1,rating} = \ldots = V_{D4,rating} \approx V_{in}/n_1 \tag{20}$$

$$V_{Da1,rating} = V_{Da2,rating} = V_{Db1,rating} = V_{Db2,rating} \approx V_o \tag{21}$$

$$I_{D1,av} = I_{D2,av} = I_{D3,av} = I_{D4,av} \approx I_o/4 \tag{22}$$

$$I_{Da1,av} = I_{Da2,av} = I_{Db1,av} = I_{Db2,av} \approx dI_o/2 \tag{23}$$

Based on Equation (11), the necessary inductances L_{r1} and L_{r2} are obtained as:

$$L_{r1} = L_{r2} \approx \frac{d_{5,loss}(n_1 V_{in} - 2n_1^2 v_{Ca1})}{I_o f_{sw}} \tag{24}$$

4. Test Results

The studied circuit was verified through a prototype. In the prototype circuit, V_{in} was between 750 V and 800 V, V_o was 24 V, $I_{o,rated}$ was 70 A, f_{sw} is 60 kHz, the effective duty cycle d_{eff} was 0.35, and the duty loss in step 5 was 0.01. Therefore, the turn-ratio and the primary-side inductances were obtained as:

$$n = n_1 = n_2 = \frac{V_{in,min}}{4V_o(1 - d_{eff})} \approx 12 \tag{25}$$

$$L_{r1} = L_{r2} \approx \frac{d_{5,loss}(n_1 V_{inmin} - 2n_1^2 v_{Ca1})}{I_o f_{sw}} \approx 16.5 \; \mu H \tag{26}$$

In the prototype circuit, the actual magnetizing inductance, $L_{m1} = L_{m2} = 2$ mH, the primary turns $n_{1,p}$ and $n_{2,p}$ were 48 and the secondary turns $n_{1,s}$ and $n_{2,s}$ were 4 with TDK EER-42 magnetic core. The maximum ripple currents Δi_{Lo} was assumed as 4 A. The L_{o1} and L_{o2} can be obtained in Equation (27).

$$L_{o1} = L_{o2} = \left(2V_o - \frac{V_{in,min}}{2n_1}\right)(0.5 - d_{eff})T_{sw}/\Delta i_{Lo} \approx 10.5\ \mu\text{H} \tag{27}$$

MOSFETs SiHG20N50C with 500 V/20 A ratings were selected for $S_1 \sim S_8$. MBR40100PT with 100 V/40 A ratings were selected for $D_1 \sim D_4$ and $D_{a1} \sim D_{b2}$. The magnetic coupling (MC) transformer $n_p{:}n_s =$ 24 turns:24 turns. The adopted capacitors were $C_{in,1} = C_{in,2} = 240\ \mu\text{F}/450$ V, $C_b = 1\ \mu\text{F}$, $C_{a1} = C_{a2} =$ 8.8 μF and $C_o = 2200\ \mu\text{F}/100$ V. The PSPWM integrated circuit UCC3895 was selected as a controller to control $S_1 \sim S_8$.

The experimental test bench is given in Figure 5. The dc power source was using a two Chroma 62016P-600-8 programmable dc power supply connected in series to supply 800 V at the input side of the proposed circuit. The dc electronic load was a Chroma 63112A programmable dc load. The digital oscilloscope Tektronix TDS3014B was adopted to measure the test waveforms. Figure 6 illustrates the test waveforms of the gate voltages of switches $S_1 \sim S_4$ in first full bridge circuit at rated power. The phase-shift angle between S_1 and S_4 depended on the input voltage. The gating voltages of $S_5 \sim S_8$ in second full bridge circuit were identical to $S_1 \sim S_4$, respectively. The gating voltages $v_{S1,g}$ and $v_{S4,g}$ and ac voltages v_{ab} and v_{cd} at rated power are demonstrated in Figure 7. Three-level voltages were observed on v_{ab} and v_{cd}. Figure 8 illustrates the test waveforms v_{ab}, v_{cd}, i_{Lr1} and i_{Lr2} under 20% power and rated power. Two primary-side currents were well balanced with each other due to the current balance magnetic core is used to achieve current sharing. Figure 9 gives the test results of V_{Cin1}, V_{Cin2} and V_{Cb} at rated power. Split capacitor voltages were well balanced with $V_{Cin1} = 401.6$ V and V_{Cin2} = 398.4 V. Figure 10 gives the measured waveforms of output side currents. Figure 11 provides the test waveforms of i_{Lo1}, i_{Db1}, i_{o1} and i_{o2} under different load conditions. It is clear that i_{o1} and i_{o2} were balanced. Figure 12 provides the measured results of S_1 under 20% power and rated power. The soft switching of S_1 was succeeded from 20% power to rated power. The other switches such as S_2, S_5 and S_6 had the same characteristics as S_1. Therefore, S_2, S_5 and S_6 were also turned on with soft switching turn-on from 20% power. Figure 13 demonstrates the measured waveforms of S_4 under half and full loads. The soft switching of S_4 are realized from 50% load due to the energy on L_{m1} and L_{m2}. Likewise, S_3, S_7 and S_8 have same characteristics as S_4 and S_3, S_7 and S_8 with soft switching turn-on from 50% power. The test efficiencies of the presented circuit are 91.7% at 20% power, 93.5% at 50% power and 92.9% at 100% power and shown in Figure 14.

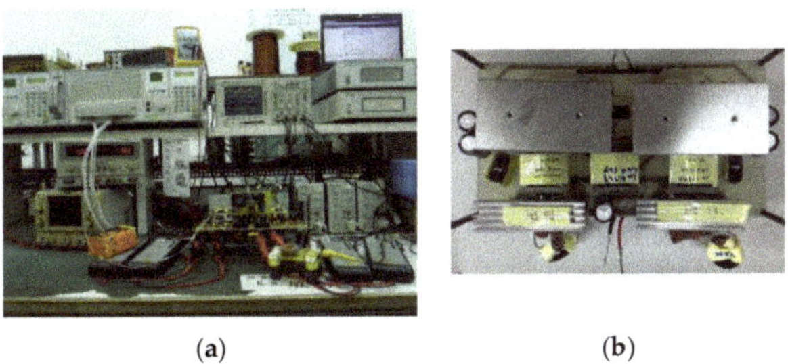

(a)   (b)

Figure 5. Pictures of the presented circuit in the laboratory: (**a**) experimental setup and (**b**) prototype circuit.

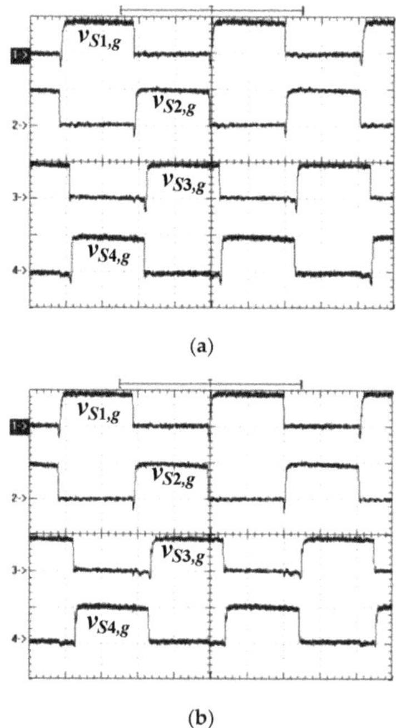

(a)

(b)

Figure 6. Measured gate voltages of $S_1 \sim S_4$ in first full bridge circuit at full load and (**a**) V_{in} = 750 V and (**b**) V_{in} = 800 V ($v_{S1,g} \sim v_{S4,g}$: 10 V/div; time: 4 µs/div).

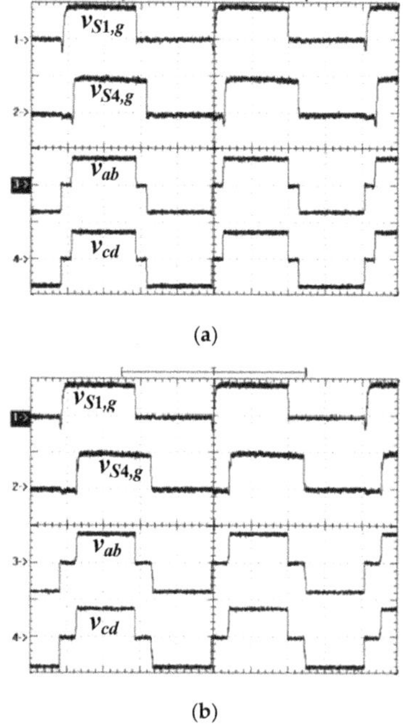

(a)

(b)

Figure 7. Experimental results of $v_{S1,g}$, $v_{S4,g}$, v_{ab} and v_{cd} at full load (**a**) V_{in} = 750 V and (**b**) V_{in} = 800 V ($v_{S1,g}$, $v_{S4,g}$: 10 V/div; v_{ab}, v_{cd}: 500 V/div; time: 4 µs/div).

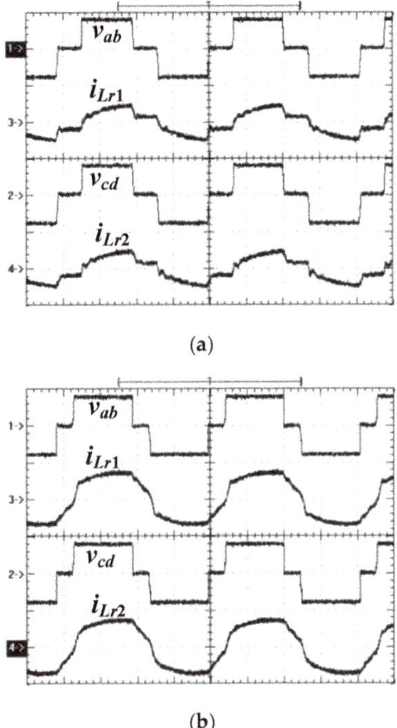

Figure 8. Measured waveforms of v_{ab}, v_{cd}, i_{Lr1} and i_{Lr2} at 800 V input (**a**) 20% load (v_{ab}, v_{cd}: 500 V/div; i_{Lr1}, i_{Lr2}: 2 A/div; time: 4 μs/div) and (**b**) full load (v_{ab}, v_{cd}: 500 V/div; i_{Lr1}, i_{Lr2}: 5 A/div; time: 4 μs/div).

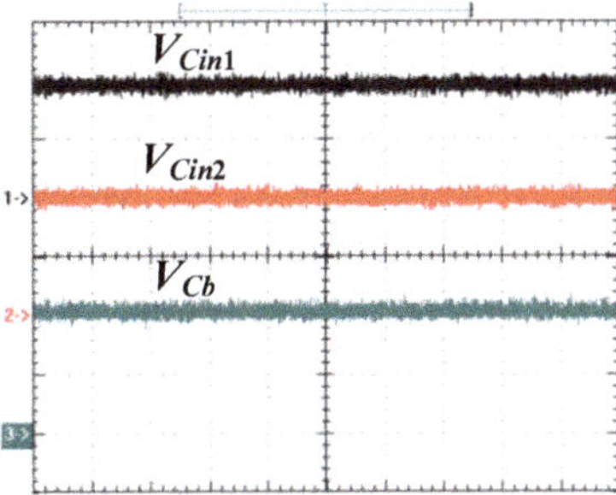

Figure 9. Measured waveforms of split voltages V_{Cin1} and V_{Cin2} and balance capacitor voltage V_{Cb} at 800 V input and full load (V_{Cin1}, V_{Cin2}, V_{Cb}: 200 V/div; time: 4 μs/div).

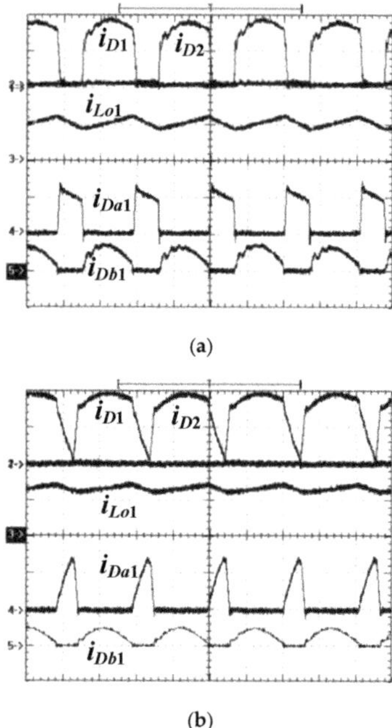

(a)

(b)

Figure 10. Measured waveforms of the secondary-side currents in first full bridge circuit (**a**) 20% load (i_{D1}, i_{D2}, i_{Lo1}, i_{Da1}, i_{Db1}: 5 A/div; time: 4 μs/div) and (**b**) full load (i_{D1}, i_{D2}, i_{Lo1}, i_{Da1}, i_{Db1}: 20 A/div; time: 4 μs/div).

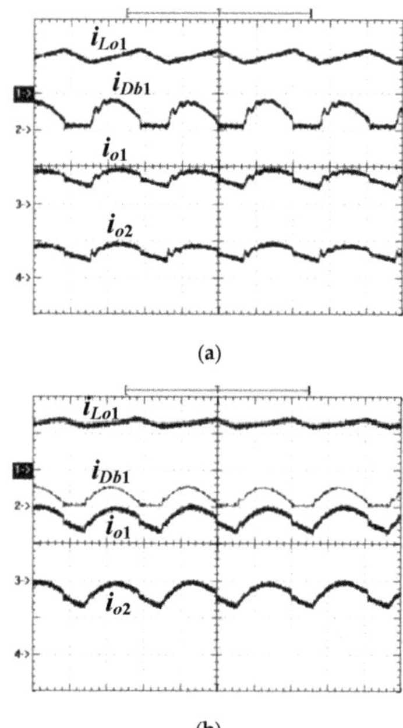

(a)

(b)

Figure 11. Measured waveforms of the secondary-side currents i_{Lo1}, i_{Db1}, i_{o1} and i_{o2} (**a**) 20% load (i_{Lo1}, i_{Db1}: 5 A/div; i_{o1}, i_{o2}: 10 A/div; time: 4 μs/div) and (**b**) full load (i_{Lo1}, i_{Db1}, i_{o1}, i_{o2}: 20 A/div; time: 4 μs/div).

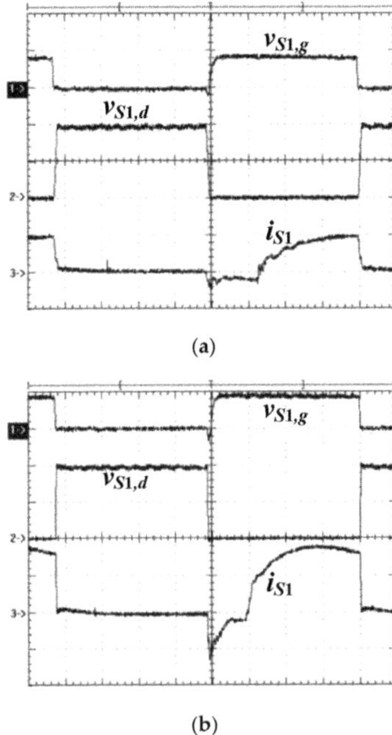

Figure 12. Measured waveforms of the leading-leg switch S_1 at 800 V input (**a**) 20% load ($v_{S1,g}$: 10 V/div; $v_{S1,d}$: 200 V/div; i_{S1}: 1 A/div; time: 2 μs/div) and (**b**) full load ($v_{S1,g}$: 10 V/div; $v_{S1,d}$: 200 V/div; i_{S1}: 2 A/div; time: 2 μs/div).

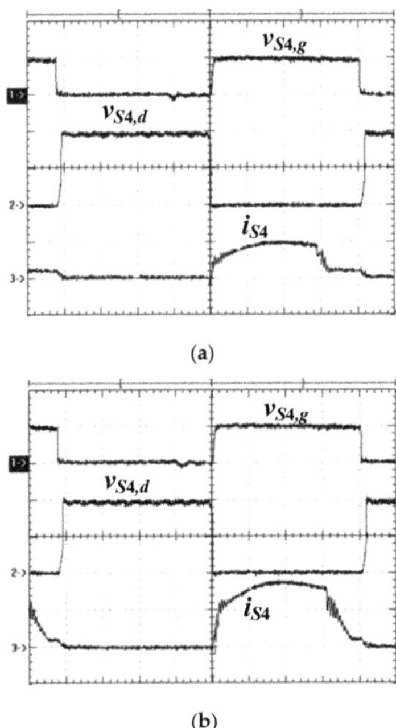

Figure 13. Measured waveforms of the lagging-leg switch S_4 at 800 V input (**a**) 50% load and (**b**) full load ($v_{S4,g}$: 10 V/div; $v_{S4,d}$: 200 V/div; i_{S4}: 2 A/div; time: 2 μs/div).

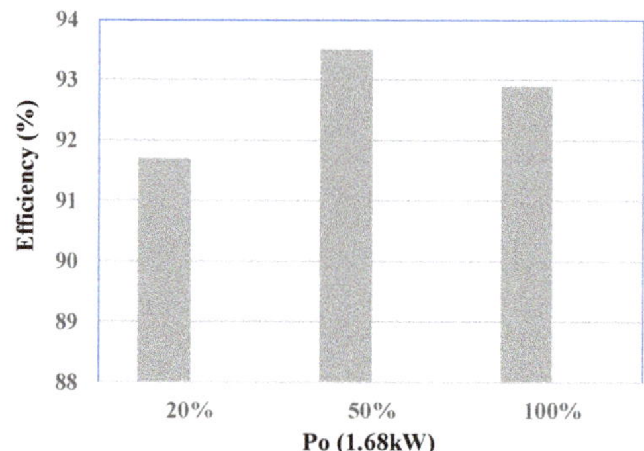

Figure 14. Measured circuit efficiencies of the proposed circuit.

5. Conclusions

A parallel dc/dc converter was proposed and investigated to achieve the main benefits of balanced input split voltages, load current sharing and reduced circulating current loss. Input split voltage balance was realized using a balance capacitor to automatically charge/discharge split capacitors in every switching cycle. The current sharing of two series full bridge circuits was achieved using a magnetic core. Passive circuits were employed on output side to decrease primary circulating current at commutated state. However, the proposed converter was out of work under power switch failure. Therefore, some bypass circuits may be added in the circuit to protect the converter from damage. This protection procedure is the next study case to further improve the circuit reliability. The theoretical analysis was well supported by the test waveforms of a prototype.

Author Contributions: B.-R.L. designed the project and was responsible for writing the paper.

Funding: This research is funded by the Ministry of Science and Technology, Taiwan, grant number MOST 107-2221-E-224-013.

Acknowledgments: This research is supported by the Ministry of Science and Technology, Taiwan, under contract MOST 107-2221-E-224-013. The author would like to thank Mr. Wei-Po Liu for his help to measure the circuit waveforms in the experiment.

Conflicts of Interest: The author declares no potential conflict of interest.

References

1. Cho, I.-H.; Cho, K.-M.; Kim, J.-W.; Moon, G.-W. A New Phase-Shifted Full-Bridge Converter With Maximum Duty Operation for Server Power System. *IEEE Trans. Power Electron.* **2011**, *26*, 3491–3500. [CrossRef]
2. Sun, B.; Burgos, R.; Boroyevich, D. Ultralow Input–Output Capacitance PCB-Embedded Dual-Output Gate-Drive Power Supply for 650 V GaN-Based Half-Bridges. *IEEE Trans. Power Electron.* **2019**, *34*, 1382–1393. [CrossRef]
3. Fu, D.; Lee, F.C.; Qiu, Y.; Wang, F. A Novel High-Power-Density Three-Level LCC Resonant Converter with Constant-Power-Factor-Control for Charging Applications. *IEEE Trans. Power Electron.* **2008**, *23*, 2411–2420. [CrossRef]
4. Song, B.-M.; McDowell, R.; Bushnell, A.; Ennis, J. A Three-Level DC–DC Converter with Wide-Input Voltage Operations for Ship-Electric-Power-Distribution Systems. *IEEE Trans. Plasma Sci.* **2004**, *32*, 1856–1863. [CrossRef]
5. Nejabatkhah, F.; Li, Y.W. Overview of Power Management Strategies of Hybrid AC/DC Microgrid. *IEEE Trans. Power Electron.* **2015**, *30*, 7072–7089. [CrossRef]
6. Dragicevic, T.; Lu, X.; Vasquez, J.C.; Guerrero, J. DC Microgrids—Part I: A Review of Control Strategies and Stabilization Techniques. *IEEE Trans. Power Electron.* **2016**, *31*, 4876–4891. [CrossRef]
7. Dragicevic, T.; Lu, X.; Vasquez, J.C.; Guerrero, J. DC Microgrids—Part II: A Review of Power Architectures, Applications, and Standardization Issues. *IEEE Trans. Power Electron.* **2016**, *31*, 3528–3549. [CrossRef]

8. Chua, T.Z.Y.; Ong, Y.T.; Toh, C.L. Transformerless DC traction power conversion system design for light-rail-transit (LRT). In Proceedings of the IEEE Energy Conversion Conf. (CENCON), Kuala Lumpur, Malaysia, 30–31 November 2017; pp. 38–43.

9. Lin, B.-R. Investigation of a Resonant dc–dc Converter for Light Rail Transportation Applications. *Energies* **2018**, *11*, 1078. [CrossRef]

10. Lin, B.R. Hybrid dc/dc converter based on dual three-level circuit and full circuit. *IET Power Electron.* **2016**, *9*, 817–824. [CrossRef]

11. Lin, B.-R. Soft Switching DC Converter for Medium Voltage Applications. *Electronics* **2018**, *7*, 449. [CrossRef]

12. Ren, R.; Liu, B.; Jones, E.A.; Wang, F.F.; Zhang, Z.; Costinett, D. Capacitor-Clamped, Three-level GaN-Based DC–DC Converter with Dual Voltage Outputs for Battery Charger Applications. *IEEE J. Emerg. Sel. Top. Electron.* **2016**, *4*, 841–853. [CrossRef]

13. Haga, H.; Kurokawa, F. A novel modulation method of the full bridge three-level LLC resonant converter for battery charger of electrical vehicles. In Proceedings of the IEEE ECCE, Montreal, QC, Canada, 20–24 September 2015; pp. 5498–5504.

14. Dusmez, S.; Li, X.; Akin, B. A New Multi-Input Three-Level DC/DC Converter. *IEEE Trans. Power Electron.* **2016**, *31*, 1230–1240. [CrossRef]

15. Choi, H.-S.; Kim, J.-W.; Cho, B.H. Novel zero-voltage and zero-current-switching (ZVZCS) full-bridge PWM converter using coupled output inductor. *IEEE Trans. Power Electron.* **2002**, *17*, 641–648. [CrossRef]

16. Pahlevaninezhad, M.; Das, P.; Drobnik, J.; Jain, P.K.; Bakhshai, A. A Novel ZVZCS Full-Bridge DC/DC Converter Used for Electric Vehicles. *IEEE Trans. Power Electron.* **2012**, *27*, 2752–2769. [CrossRef]

17. Mishima, T.; Akamatsu, K.; Nakaoka, M. A high frequency-link secondary-side phase-shifted full-bridge soft-switching PWM DC-DC converter with ZCS active rectifier for EV battery charger. *IEEE Trans. Power Electron.* **2013**, *28*, 5758–5773. [CrossRef]

18. You, H.; Xu, C. A family of un-isolated modular dc/dc converters. In Proceedings of the IEEE IPEMC, Hefei, China, 22–26 May 2016; pp. 696–702.

19. Shi, J.J.; Liu, T.J.; Cheng, J.; He, X.N. Automatic current sharing of an input-parallel output-parallel (IPOP)-connected DC-DC converter system with chain-connected rectifiers. *IEEE Trans. Power Electron.* **2015**, *30*, 2997–3016. [CrossRef]

20. Liu, C.; Xu, X.; He, D.; Liu, H.; Tian, X.; Guo, Y.; Cai, G.; Ma, C.; Mu, G. Magnetic-coupling current-balancing cells based input-parallel output-parallel LLC resonant converter modules for high-frequency isolation of DC distribution systems. *IEEE Trans. Power Electron.* **2016**, *31*, 6968–6979. [CrossRef]

© 2019 by the author. Licensee MDPI, Basel, Switzerland. This article is an open access article distributed under the terms and conditions of the Creative Commons Attribution (CC BY) license (http://creativecommons.org/licenses/by/4.0/).

 *electronics*

Article

Multiple Modulation Strategy of Flying Capacitor DC/DC Converter

Pengcheng Li [1,*]**, Chunjiang Zhang** [2]**, Sanjeevikumar Padmanaban** [3,*] **and Leonowicz Zbigniew** [4]

[1]   Department of Electrical Engineering, Hebei University of Science and technology, Shijiazhuang 050018, China
[2]   Department of Electrical Engineering, Yanshan University, Qinhuangdao 066004, China
[3]   Department of Energy Technology, Aalborg University, 6700 Esbjerg, Denmark
[4]   Faculty of Electrical Engineering, Wroclaw University of Science and Technology, Wyb. Wyspianskiego 27, 50370 Wroclaw, Poland
*   Correspondence: lipengcheng@hebust.edu.cn (P.L.); san@et.aau.dk (S.P.); Tel.: +86-18630322865 (P.L.)

Received: 24 May 2019; Accepted: 5 July 2019; Published: 11 July 2019

Abstract: Flying-capacitor multiplexed modulation technology is suitable for bipolar DC microgrids with higher voltage levels and higher current levels. The module combination and corresponding modulation method can be flexibly selected according to the voltage level and capacity level. This paper proposes a bipolar bidirectional DC/DC converter and its interleaved-complementary modulation strategy that is suitable for bipolar DC microgrids. The converter consists of two flying-capacitor three-level bidirectional DC/DC converters that are interleaved in parallel 90°, and then cascaded with another module to form a symmetrical structure of the upper and lower arms; the complementary modulation of the upper and lower half bridges constitutes an interleaved complementary multilevel bidirectional DC/DC converter. If the bidirectional converter needs to provide a stronger overcurrent capability, more bridge arms can be interleaved in parallel. Once n bridge arms are connected in parallel, the bridge arms should be interleaved 180°/n in parallel. In bipolar DC microgrids, the upper and lower arms should be complementarily modulated, and the input and output are isolated by the inductance. To solve the current difference, caused by the inconsistent parasitic, the voltage-current double closed-loop-control is used, and the dynamic response is faster during bidirectional operation. This paper proposes theoretical analysis and experiments that verify bipolar bidirectional DC/DC converter for high-power energy storage.

Keywords: bidirectional DC/DC converter (BDC); dual mode operation; current sharing; multiplexed modulation

---

1. Introduction

With the high penetration of intermittent energy, such as solar and wind [1–3], a power electronic interface for distributed energy storage is becoming increasingly attractive. The bidirectional DC/DC converter (BDC) is an important piece of equipment for distributed energy storage in DC microgrids, which helps to promote intermittent energy scale applications. The BDCs are widely used in DC microgrids, due to their simple structure, easy expansion, and transmission power being independent of transformers [4–7].

In particular, it plays a large irreplaceable role in the distributed energy storage of high voltage and high power. For BDCs, current research focuses on buck/boost two-level converters and control strategies for suppressing load disturbances [8,9]; however, switches are subject to low-voltages applications. The voltage and current stresses of the converter are relatively high, so a multilevel converter is required. A multiplexed multiphase and multilevel BDC is used for a wide range of voltage variations (voltage conversion level less than 10 times); different from multilevel converters required for a high-voltage DC transmission (voltage conversion level more than 10 times), relying on transformer boosting to achieve a higher level of voltage conversion [10–12]. The n-level structure of the multiplexed multilevel BDC reduce voltage stress only 1/n of the high-side voltage; the m-phase of the multiplexed multiphase BDC reduce current stress only 1/m of the low-side current [13]. The multiphase and multilevel BDC adopts the interleaved phase modulation technology to improve the output current ripple frequency, reduce the filter capacitor ripple value in a DC microgrid [14–17].

When a battery is connected to a DC microgrid by a BDC, the BDC needs to have strong input and output impedance matching capability to keep the system stable. In particular, when the BDC operates in buck mode, the input impedance is large; when operating in boost mode, the input impedance is small; when operating in bidirectional mode, the impedance adjustment range is wider, and the response speed is faster, which is beneficial to the system stability [18].

H. L. Do. [19] proposed a soft-switching DC/DC converter with high voltage gain by a boost cell and a coupled inductor cell. Soft-switching characteristic reduces the switching loss of active power switches and increase the converter efficiency. However, the converter can only work in boost mode, and only S_2 and D_4 can achieve ZVS turn-on. The S_1-D_4 current stress is uneven, and S_1 has a higher current stress. In high-voltage and high-power distributed energy storage, it is necessary to consider the equalization and current sharing problems. X. S. Zhang et al. [20] proposed the idea of battery energy storage systems (BESSs) with integrated wind farms to stabilize the grid power. High-power BDCs are required to meet high power requirements.

R. Naderi et al. [21] proposed a dual flying capacitor active-neutral-point-clamped (DFC-ANPC) DC/AC inverter with a five-level modulation method that achieves soft switching and neutral point voltage balancing. More importantly, the five-level modulation method eliminates the transient voltage balancing issue by series-connected switches of S_5 and S_6 and decreases the switching loss. F. Mohammadi [22] discussed the configuration, operation and decoupled control mode of a VSC-HVDC. Compared to the pulse width modulation (PWM) strategies, the vector control method generates fewer voltage harmonics and allows to control the active and reactive power independently. The vector control method is used to control the VSC-HVDC system, which is based on transforming a 3-φ system into a 2-φ system by d-q frame. Droop control is easily affected by the line impedance and the frequency fluctuation of the power grid, which reduces the distribution accuracy of the active and reactive power. In the future, the bidirectional DC/DC converter (BDC) will be connected to a bipolar DC microgrid by droop control.

Reference [23] used a DC bus capacitor to provide a three-level state (buck/boost synchronous PWM modulation). However, the converter operates at high gain, and S1 is subjected to high current stress, which exacerbates the switching losses. In Reference [24], a two-level buck/boost converter is cascaded to form a three-level bidirectional DC/DC converter. However, the modulation method easily causes inconsistent duty ratios of the upper and lower half bridges to be inconsistent, thereby affecting the voltage equalization effect of the DC bus, and requires an additional voltage equalization control loop at boost mode, which increases the complexity of the control structure.

This paper proposes a multiplexed modulation technique of the flying capacitor DC/DC converter to meet the high-voltage and high-power requirements. The BDC has many advantages: (1) the bus capacitor voltage is easily stabilized by the midpoint potential balance control of the rear inverter circuit; (2) the BDC is easy to combine by the PWM sequence to achieve multiple modulations; (3) it is easy to design the control loop and suppress phase-to-phase circulation; and (4) the BDC has a strong fault tolerance ability, and failure of any one of the arms does not affect the operation of other arms. The BDC is suitable for battery energy storage systems in bipolar DC microgrids.

This paper organized as follows. The Flying capacitor type three-level DC/DC basic unit in Section 2. Multiple modulation techniques are presented in Section 3. Controller design in Section 4. Experimental verifications in Section 5. Some conclusions are given in Section 6.

2. Flying Capacitor Type Three-level DC-DC Basic Unit

This paper proposes a bidirectional DC/DC converter (BDC) topology with multiplexed modulation strategy for a high-power system, as shown in Figure 1. The parallel operation improves the current capability of the BDC; the voltage level is increased in series operation; and the high-voltage and large-capacity characteristics are realized in series-parallel operation. The symbols and reference directions are indicated in the figure. The basic unit of the flying-capacitor-type three-level bidirectional DC/DC converter (3L_BDC) easily forms a bipolar BDC of high-voltage and high-current systems.

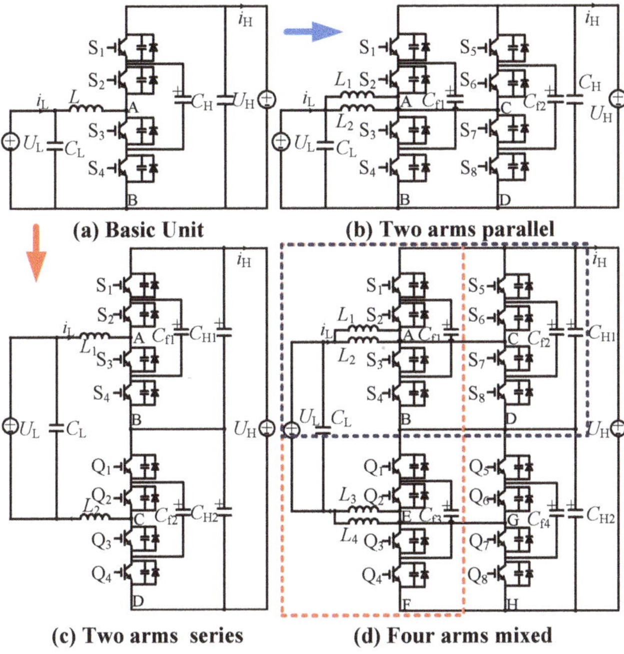

Figure 1. The proposed bidirectional DC/DC converter. (**a**) Flying capacitor type three-level DC-DC basic unit. (**b**) Two basic units parallel. (**c**) Two basic units parallel. (**d**) Four basic units mixed.

In Figure 1a, U_L is the input-side voltage, U_H is the DC bus side voltage, C_L is the battery-side capacitance, and C_H is the DC bus side capacitance. The conduction time of S_3 and S_4 is defined as T_{on} in the switching period T_s, and the duty ratio is $D = T_{on}/T_s$. To facilitate the analysis, define the switch function as follows:

$$M_k = \begin{cases} 00 \ (S_1, S_2 \ \text{ON}, \ S_3, S_4 \ \text{OFF}) \\ 01 \ (S_2, S_4 \ \text{ON}, \ S_1, S_3 \ \text{OFF}) \\ 10 \ (S_1, S_3 \ \text{ON}, \ S_2, S_4 \ \text{OFF}) \\ 11 \ (S_3, S_4 \ \text{ON}, \ S_1, S_2 \ \text{OFF}) \end{cases} . \tag{1}$$

The modal analysis is performed on the basic unit of the flying-capacitor-type three-level bidirectional DC/DC converter. The key waveform is shown in Figure 2. When the operating mode at $D > 0.5$, the switching mode of S_3 and S_4 is only 01, 10, 11, and not 00. The voltage at the two points of AB is $0.5U_H$ or 0; at $D < 0.5$, the switching mode of S_3 and S_4 is only 00, 01, 10, and not 11. The voltage of AB is U_H or $0.5U_H$. The inductive current flowing from the low-voltage side to the high-voltage side is defined as the positive direction.

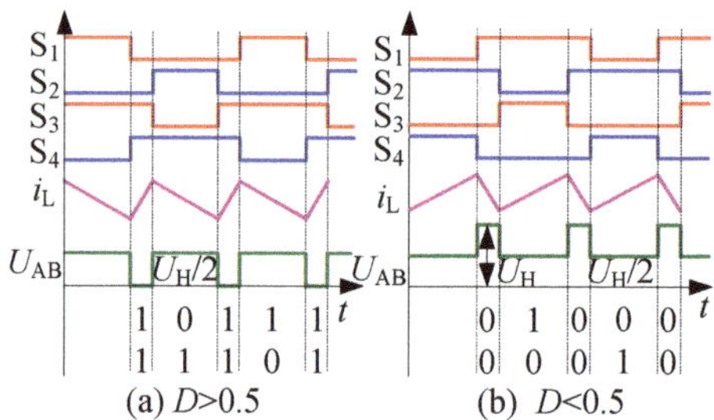

Figure 2. The main waveform of the basic unit. (**a**) Mode $D > 0.5$. (**b**) Mode $D < 0.5$.

2.1. Operational Modal Analysis $D > 0.5$

During the switching cycle, there are three modes of 11, 01, and 10 for each arm.

1. Mode $M_3 = 11$, S_1 and S_2 are turned off, S_3 and S_4 are turned on, S_1 and S_2 voltage stress are $U_H/2$, the voltage of AB two-point is $U_{AB} = 0$, the inductor voltage across L_1 is U_L, and i_L flows to the high-voltage side and linearly increases:

$$\begin{cases} \dot{i}_L = -\frac{r_L}{L}i_L + \frac{1}{L}U_L \\ \dot{U}_H = -\frac{1}{R_H C_H}U_H \\ \dot{U}_f = 0 \end{cases} \quad . \tag{2}$$

2. Mode $M_1 = 01$, S_1 and S_3 turn off, S_2 and S_4 turn on, the S_1 and S_3 voltage stresses are $U_H/2$, the flying capacitor C_{f1} charge according to the differential equation $C_{f1}\, dU_{f1}/dt = I_c$, that is $\Delta U_{f1} = t_2 I_c/C_{f1} = Q_c/C_{f1}$, $U_{AB} = U_{f1} = U_H/2$, the inductor L_1 is $U_L - U_{AB} < 0$, the inductor current i_L decreases linearly, and the average inductor current is $I_L = I_c$.

$$\begin{cases} \dot{i}_L = -\frac{r_L}{L}i_L + \frac{1}{L}U_L - \frac{1}{L}U_f \\ \dot{U}_H = -\frac{1}{R_H C_H}U_H \\ \dot{U}_f = \frac{1}{C_f}i_L \end{cases} \quad . \tag{3}$$

3. Mode $M_3 = 11$, S_1 and S_2 are turned off, S_3 and S_4 are turned on, the S_1 and S_2 voltage stresses are $U_H/2$, the voltage of the two-point of AB is $U_{AB}=0$, the voltage of the inductance L_1 is U_L, and i_L flows to high voltage side and linearly increase, as shown in Equation (2).

4. Mode $M_2 = 10$, S_2 and S_4 are turned off, S_1 and S_3 are turned on, the S_2 and S_4 voltage stresses are $U_H/2$, flying capacitor C_{f1} is discharged, $U_{AB} = U_H - U_{f1} = U_H/2$, the voltage across inductor L_1 is $U_L - U_{AB} < 0$, and i_L flows to the high-pressure side and decreases linearly. The column differential equation can be obtained as:

$$\begin{cases} \dot{i}_L = -\frac{r_L}{L}i_L + \frac{1}{L}U_L + \frac{1}{L}U_f - \frac{1}{L}U_H \\ \dot{U}_H = \frac{1}{C_H}i_L - \frac{1}{R_H C_H}U_H \\ \dot{U}_f = -\frac{1}{C_f}i_L \end{cases} \quad . \tag{4}$$

According to the duty cycle definition, each equation group for the modal action time can be listed during the switching period:

$$\begin{cases} t_1 + t_2 + t_3 = DT_s \\ t_1 + t_3 + t_4 = DT_s \\ t_1 + t_2 + t_3 + t_4 = T_s \end{cases} \Rightarrow \begin{cases} t_2 = (1-D)T_s \\ t_4 = (1-D)T_s \\ t_1 + t_3 = (2D-1)T_s \end{cases} \quad . \tag{5}$$

The inductance satisfies the volt-second balance condition:

$$U_L(2D-1)T_s + (U_L - U_H/2)(2-2D)T_s = 0 \Rightarrow \frac{U_H}{U_L} = \frac{1}{1-D}. \tag{6}$$

2.2. Operational Modal Analysis D < 0.5

During the switching cycle, each arm has three modes of 00, 10, and 01.

1. Mode $M_0 = 00$, S_3 and S_4 are turned off, S_1 and S_2 are turned on, the voltage of AB is $U_{AB} = U_H$, S_3 and S_4 voltage stress are $U_H/2$, the voltage of inductance L_1 is $U_L - U_H$, i_L flows to the high-voltage side and decreases linearly, and the corresponding differential equation can be expressed as:

$$\begin{cases} \dot{i}_L = -\frac{r_L}{L}i_L + \frac{1}{L}(U_L - U_H) \\ \dot{U}_H = -\frac{1}{R_H C_H}U_H \\ \dot{U}_f = 0 \end{cases} \tag{7}$$

2. Mode $M_2 = 10$, S_2 and S_4 are turned off, S_1 and S_3 are turned on, S_2 and S_4 voltage stress are $U_H/2$, C_{f1} is discharging, $U_{AB} = U_H - U_{f1} = U_H/2$, the voltage across inductor L_1 is $U_L - U_{AB}>0$, i_L flows to the high-voltage side and increases linearly, and the differential equation can be expressed as Equation (4).

3. Mode $M_0 = 00$, S_3 and S_4 are turned off, S_1 and S_2 are turned on, the voltage of AB is $U_{AB} = U_H$, S_3 and S_4 voltage stress are $U_H/2$, inductance L_1 voltage is $U_L - U_H$, i_L flow to high voltage side and the linearity is reduced, and the differential equation can be expressed as Equation (7).

4. Mode $M_1 = 01$, S_1 and S_3 are turned off, S_2 and S_4 are turned on, S_1 and S_3 voltage stresses are $U_H/2$, flying capacitor C_{f1} is charging, $U_{AB} = U_{f1} = U_H/2$, the inductor voltage across L_1 is $U_L - U_{AB} > 0$, i_L flows to the high-voltage side and increases linearly, and the differential equation can be expressed as Equation (3). According to the duty cycle definition, each mode action time can be expressed as:

$$\begin{cases} t_2 = DT_s \\ t_4 = DT_s \\ t_1 + t_3 = (1-2D)T_s \end{cases} \tag{8}$$

The inductance satisfies the volt-second balance:

$$2DT_s(U_L - U_H/2) + (1-2D)T_s(U_L - U_H) = 0 \Rightarrow \frac{U_H}{U_L} = \frac{1}{1-D} \tag{9}$$

3. Multiple Modulation Technique

3.1. Two Arms Interleaved Parallel Modulation

Two arms are paralleled, as shown in in Figure 1b, to increase the overcurrent capability and reduce the input side ripple. Arm 1 is composed of S_1-S_4, the inductor L_1 and the flying capacitor C_{f1}; and arm 2 is composed of S_5, S_6, S_7, S_8, the inductor L_2 and the flying capacitor C_{f2}. Among them, S_1 and S_4 are turned on complementarily, S_2 and S_3 are turned on complementarily, the modulated waves of S_1 and S_2 are interleaved 180°, and the S_3 and S_4 modulated waves are interleaved 180°, as shown in Figure 3. Arm 2 is modulated in the same manner as arm 1 with a phase lag of 90°.

Two arms are interleaved 90° in parallel; eight modes are used at $0.5 < D < 0.75$, and other eight modes are used at $0.25 < D < 0.5$. The working mode of the space ratio is shown in Table 1. In the forward power flow (Boost mode), the inductor currents i_{L1} and i_{L2} are positive and flow from the low voltage side to the high voltage side. When the negative power flows (Buck mode), the inductor currents of i_{L1} and i_{L2} are negative, and the high voltage side flows to the low voltage side. The driving signal between the two arms is interleaved 90°, and the other side bridge arm switch maintains the original state when one side bridge arm is operated. After the inductor current is superimposed, the low-voltage side current is $i_L = i_{L1} + i_{L2}$ doubling the pulsation frequency, and the ripple of the i_L is reduced.

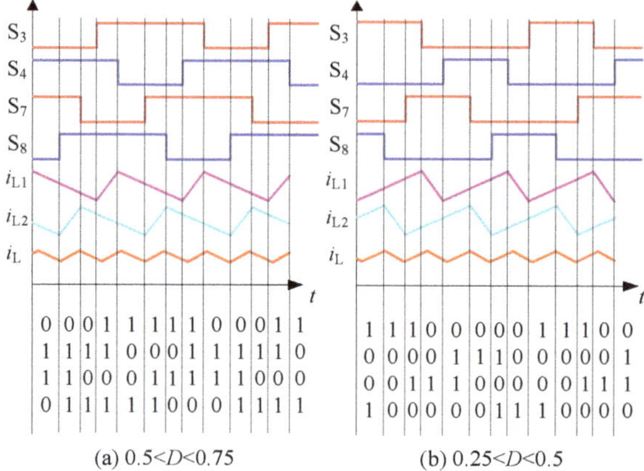

(a) 0.5<*D*<0.75 (b) 0.25<*D*<0.5

Figure 3. The main waveform of two parallel arms. (**a**) Mode 0.5 < *D* < 0.75, (**b**) Mode 0.25 < *D* < 0.5.

Table 1. Two arms interleaved parallel coding.

0~0.25	*D* = 0.25	0.25~0.5	*D* = 0.5	0.5~0.75	*D* = 0.75	0.75~1
0000	0010	1010	1010	0101	1101	1111
0010	0010	0010	1010	1101	1101	1101
0000	0100	0110	0110	1001	1011	1111
0100	0100	0100	0110	1011	1011	1011
0000	0001	0101	0101	1010	1110	1111
0001	0001	0001	0101	1110	1110	1110
0000	1000	1001	1001	0110	0111	1111
1000	1000	1000	1001	0111	0111	0111

Then the inductor current is superimposed, the low-voltage side current is $i_L = i_{L1} + i_{L2}$ doubling the pulsation frequency, and the ripple of the i_L is reduced. According to Equations (2)–(9), the ripple current of three-level bi-directional DC/DC(3L-BDC) in Figure 1b, $\Delta I_{L1\_3L\_LBDC}$, can be calculated as

$$\Delta I_{L1\_3L\_BDC} = \begin{cases} \dfrac{(U_H - 2U_L)(1-D)}{2L_1 f_s}, & (D > 0.5) \\ \dfrac{(2U_L - U_H)D}{2L_1 f_s}, & (D \le 0.5) \end{cases} \tag{10}$$

Correspondingly, the ripple current of the inductor of two-level bidirectional DC/DC (2L_ BDC) can be calculated as

$$\Delta I_{L1\_2L\_BDC} = \frac{(U_H - U_L)D}{L_1 f_s}. \tag{11}$$

To reduce currents ripple, *n* arms can be cascaded, and the drive signals between the arms are interleaved 180°/*n*. The more arms that participate in interleaved parallel connection, the more obvious the ripple reduction effect, and the more characteristic points of zero ripple appear at the same time—these zero ripple points show a uniform distribution law. The aforementioned analysis shows that the voltage stress on the switches and the flying capacitors of the 3L_BDC is half of U_H, which is just half of the traditional 2L_BDC. To reduce the voltage stress, it is necessary to employ a series connection.

3.2. Two Arms Complementary Series Modulation

The topology is connected in series with two inductors to reduce the inductor current ripple amplitude; the low voltage side is isolated from the output side by two inductors, which can improve the energy storage unit safety; the series structure can reduce the voltage stress of the switches and increase the voltage level of the DBC, as shown in Figure 1c. The inductor current waveform during complementary modulation is shown in Figure 4. S_4 and Q_1 are the same drive signal, S_3 and Q_2 are the same drive signal, S_2 and Q_3 are the same drive signal, and S_1 and Q_4 are the same drive signal; that is, the switch modulation is based on the topology. The switching period inductance fluctuation frequency is equal to twice of the switching frequency, the inductance fluctuation amplitude is half of that of the single submodule, and the remaining mode complementary series modulation is shown in Table 2.

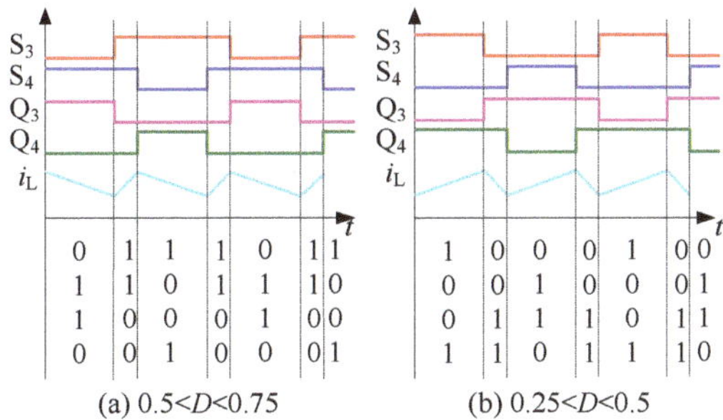

(a) 0.5<*D*<0.75　　　　(b) 0.25<*D*<0.5

Figure 4. The key waveform of two arms series. (**a**) Mode 0.5 < *D* < 0.75, (**b**) Mode 0.25 < *D* < 0.5.

Table 2. Two arms complementary series coding.

0~0.25	0.25	0.25~0.5	0.5	0.5~0.75	0.75	0.75~1
00/11	00/11	00/11	01/10	11/00	11/00	11/00
01/10	01/10	01/10	10/01	10/01	10/01	10/01
00/11	00/11	00/11	01/10	11/00	11/00	11/00
10/01	10/01	10/01	10/01	01/10	01/10	01/10

The voltage stress on the switches and the flying capacitors is 0.25 U_H in Figure 1c. In order to reduce the voltage and current stress, it is necessary to increase the series and parallel bridge arms simultaneously. The ripple current of L_1 in Figure 1c, $\Delta I_{L1\_3L\_LBDC}$, can be calculated as

$$\Delta I_{L1\_3L\_BDC} = \begin{cases} \dfrac{(U_H-2U_L)(1-D)}{2(L_1+L_3)f_s}, & (D > 0.5) \\ \dfrac{(2U_L-U_H)D}{2(L_1+L_3)f_s}, & (D \leq 0.5) \end{cases}. \tag{12}$$

3.3. Four Arms Mixed Modulation

To meet the large-capacity requirements in the bipolar DC bus, the voltage and current stress of the switching tube should be reduced, so a four-arms mixed converter is proposed, that is, the cascaded form of the interleaved parallel flying-capacitor type three-level converter, as shown in Figure 1d. The left and right parallel arms are interleaved 90° parallel modulation, and the upper and lower series arms are complementarily connected in series. For example, when 0.5 < *D* < 0.75, in mode 0101/1010, it means S_3 is off, S_4 is on, S_7 is off, S_8 is on; while, Q_3 is on, Q_4 is off, Q_7 is on, and Q_8 is off. The modulation rule of the BDC is shown in Figure 5.

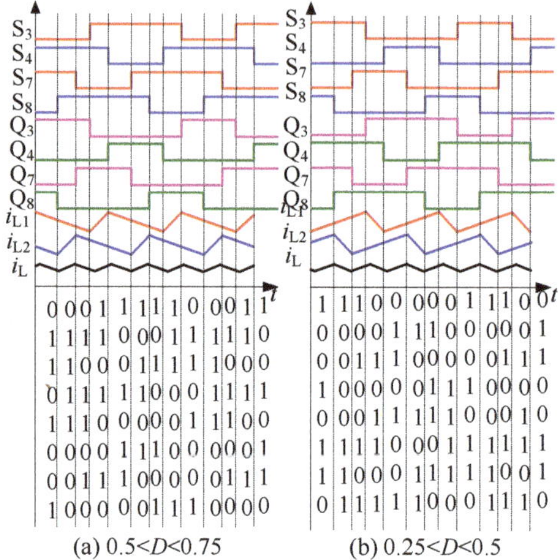

(a) 0.5<*D*<0.75   (b) 0.25<*D*<0.5

Figure 5. The key waveform of the four arms mixed. (**a**) Mode 0.5 < *D* < 0.75, (**b**) Mode 0.25 < *D* < 0.5.

By mixed modulation, the average inductor current is only half of the single-arm current, the ripple frequency is doubled, the flying capacitor voltage is 0.25 U_H, and the voltage stress of the switch is only 0.25 U_H compared to the 2L_BDC. The modulation is shown in Table 3. To compare the characteristics of the structure, shown in Figure 1, the voltage and current stress conditions are listed in Table 3, and the four arms mixed modulation is more suitable for bipolar high-power applications. This modulation strategy helps to control and protect the design of the circuit. A topological comparison of the proposed BDC in this paper with others are shown in Table 4.

Table 3. Four arms mixed modulation coding.

0~0.25	*D* = 0.25	0.25~0.5	*D* = 0.5	0.5~0.75	*D* = 0.75	0.75~1
0000/1111	0010/1101	1010/0101	1010/0101	0101/1010	1101/0010	1111/0000
0010/1101	0010/1101	0010/1101	1010/0101	1101/0010	1101/0010	1101/0010
0000/1111	0100/1011	0110/1001	0110/1001	1001/0110	1011/0100	1111/0000
0100/1011	0100/1011	0100/1011	0110/1001	1011/0100	1011/0100	1011/0100
0000/1111	0001/1110	0101/1010	0101/1010	1010/0101	1110/0001	1111/0000
0001/1110	0001/1110	0001/1110	0101/1010	1110/0001	1110/0001	1110/0001
0000/1111	1000/0111	1001/0110	1001/0110	0110/1001	0111/1000	1111/0000
1000/0111	1000/0111	1000/0111	1001/0110	0111/1000	0111/1000	0111/1000

4. Controller design

Define the state variable as $x = [\, i_L,\ U_H,\ U_f\,]^T$, the input variable as $u = U_L$, the transfer function as $G_{id}(s)$ from the duty cycle d to the inductor current i_L and the transfer function $G_{ud}(s)$ from the duty cycle d to the output voltage U_H, according to Equations (2)–(9).

$$G_{id}(s) = \frac{\frac{U_L C_H R_H}{1-D}s + \frac{2U_L}{1-D}}{C_H L R_H s^2 + Ls + R_H(1-D)^2 + r_L}, \tag{13}$$

$$G_{ud}(s) = \frac{-\frac{L U_L}{(1-D)^2}s + R_H U_L - \frac{r_L U_L}{(1-D)^2}}{C_H L R_H s^2 + Ls + R_H(1-D)^2 + r_L}, \tag{14}$$

where, R_H is R_{load}, and r_L is the inductance parasitic resistance.

The main circuit *parameters* of the converter are shown in Table 5. Due to the inconsistent impedance of the IGBT parasitic parameters and the inductor winding process, the main loop has a certain degree of impedance difference. The controller's PI regulator is $G_c(s) = k_1 + k_2/s$, where $k_1 = 0.1$, $k_2 = 50$, and the control block diagram is shown in Figure 6.

Table 4. Topological comparison.

(a) Comparison in terms of passive component and out gain, inductor ripple current and switching frequency.

Proposed	Number of Elements	Gain of Voltage	Δi_L		$f_{\Delta i_L}$
SC [23]	$C = 3\ S = 4\ L = 1$	$\frac{2U_L}{1-D}$	$\frac{(U_H - 2U_L)D}{2Lf_s}$		f_s
Double Buck/Boost [24]	$C = 3\ S = 4\ L = 2$	$\frac{U_L}{1-D}$	$\frac{U_H(1-D)D}{Lf_s}$		$2f_s$
Basic FC [5]	$C = 3\ S = 4\ L = 1$	$\frac{U_L}{1-D}$	$\frac{(U_H - 2U_L)(1-D)}{2L_1 f_s}$, $\frac{(2U_L - U_H)D}{2L_1 f_s}$,	$(D > 0.5)$ $(D \le 0.5)$	$2f_s$
Interleaved FC [7]	$C = 4\ S = 8\ L = 2$	$\frac{U_L}{1-D}$	$\frac{(U_H - 2U_L)(1-D)}{2L_1 f_s}$, $\frac{(2U_L - U_H)D}{2L_1 f_s}$,	$(D > 0.5)$ $(D \le 0.5)$	$4f_s$
Complementary FC [21]	$C = 5\ S = 8\ L = 2$	$\frac{U_L}{1-D}$	$\frac{(U_H - 2U_L)(1-D)}{2(L_1 + L_3)f_s}$, $\frac{(2U_L - U_H)D}{2(L_1 + L_3)f_s}$,	$(D > 0.5)$ $(D \le 0.5)$	$2f_s$
This paper	$C = 7\ S = 16\ L = 4$	$\frac{U_L}{1-D}$	$\frac{(U_H - 2U_L)(1-D)}{2(L_1 + L_3)f_s}$, $\frac{(2U_L - U_H)D}{2(L_1 + L_3)f_s}$,	$(D > 0.5)$ $(D \le 0.5)$	$4f_s$

(b) Comparison in terms of out capacitor, flying capacitor voltage, voltage across switch, inductor current and fault tolerant capabilities

Proposed.	Output Capacitor Voltage	Flying capacitor voltage	Switch Voltage	Inductor Current	Fault Tolerance
SC [23]	$2\ U_L/(1-D)$	$0.5\ U_H$	$0.5\ U_H$	I_L	Weak
Double Buck/Boost [24]	$U_L/(1-D)$	No	$0.5\ U_H$	I_L	Weak
Basic FC [5]	$U_L/(1-D)$	$0.5\ U_H$	$0.5\ U_H$	I_L	Weak
Interleaved FC [7]	$U_L/(1-D)$	$0.5\ U_H$	$0.5\ U_H$	$0.5\ I_L$	Average
Complementary FC [21]	$U_L/(2-2D)$	$0.25\ U_H$	$0.25\ U_H$	I_L	Average
This paper	$U_L/(2-2D)$	$0.25\ U_H$	$0.25\ U_H$	$0.5\ I_L$	Strong

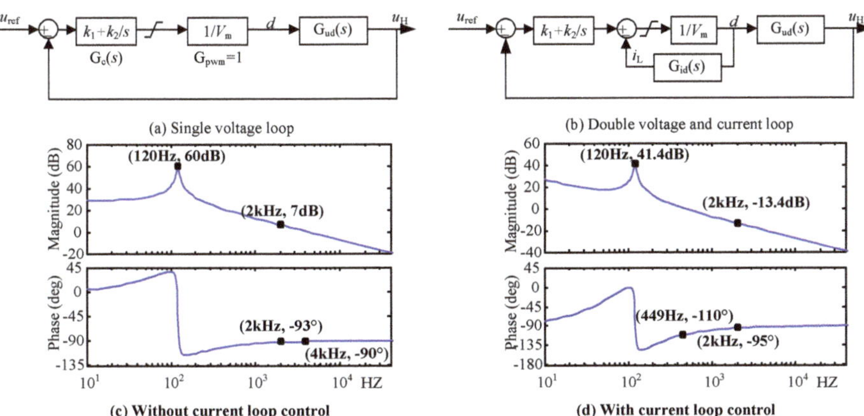

Figure 6. The control strategy. (**a**) & (**c**) are single voltage loop and its Bode without current loop control. (**b**) & (**d**) are double voltage-current loop and its Bode with current loop control.

The Bode diagram of the proposed control strategy by a single voltage loop, as shown in Figure 6c. The low frequency range is 30 dB, the high frequency band traverses 0 dB with a slope of −20 dB/dec, and the corner frequency is 120 Hz. The gain crossover frequency is 4 kHz, the phase margin is 90°, and the system is stable; however, the gain is higher than 0dB at 0.1 times of switching frequency. The current inner loop uses inductor current feedback, as shown in Figure 6d. The PI regulator is $G_c(s) = 0.1 + 50/s$, the crossing frequency of the open-loop transfer function is 449 Hz when crossing the 0 dB line, and the phase margin is 70° at the crossing frequency. It can be judged that the closed loop system is stable. A first-order low-pass filter is added to the loop due to the influence of high frequency noise. The cutoff frequency of the first-order low-pass filter ($\omega_c = 2\pi fc$) is set at 0.1 times the switching frequency, and the gain margin kg ($f_c = 2$ kHz) is 13.4 dB, which is ideal. The pole introduced by the first-order LPF is far from the real axis, and has little effect on the bode diagram and can be ignored.

Table 5. The Parameters of BDC.

Parameters	Value	Parameters	Value
U_L/V	150~220	C_L/µF	220
U_H/V	400	C_H/µF	110
P_o/kW	1	C_{f1}/µF	110
L_1~L_2/mH	2	C_{f2}/µF	110
r_L/Ω	0.2	f_s/kHz	20

5. Experimental result

The experimental platform is shown in Figure 7. In the platform, the DC power supply E and resistance *R* are used to simulate the power generation change of the renewable energy source. The rated voltage of the DC power supply *E* is 450 V, and the DC bus voltage rating is 400 V. Super capacitor rated voltage is 250 V, rated capacity is 10 F, maximum discharge current is 15 A, charging current is 10 A, super capacitor voltage U_L range is 150~220 V; switching frequency f_s is 20 kHz. DC power supply Chroma 62050H-600 (Chroma Systems Solutions, Inc., Foothill Ranch, CA, USA), DC probe YOKOGAWA 701934 (Yokogawa Electric, Inc., Tokyo, Japan), oscilloscope Tek DPO2024B (Tektronix, Inc., Beaverton, OR, USA). The control chip uses DSP (TMS28335) combined with FPGA (EP3C25Q240); DSP is used for signal sampling and control signal generation, and FPGA is used to generate the modulated wave. To test the feasibility of multiple applications, super capacitor U_L = 200V, high side load R_H = 200 Ω, adjustable power supply *E* = 450 V, resistance *R* = 10 Ω; during switch *S* disconnection, super capacitor discharge, converter operates on boost 0.5 < *D* < 0.75 mode, the high side capacitor voltage is stable to U_H = 400 V. Super capacitor U_L = 250 V, high-voltage-side load R_H = 200 Ω, adjustable power supply *E* = 450 V, resistance *R* = 10 Ω; during switch *S* closing, super capacitor charging.

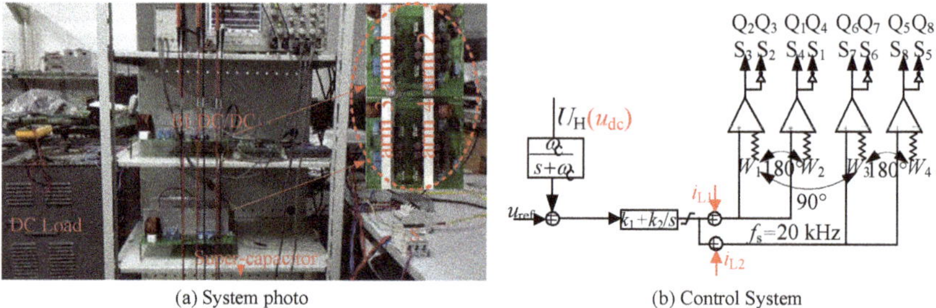

(a) System photo   (b) Control System

Figure 7. Experimental setup. (a) The experimental system, (b) Control system logic.

In the project, due to the inconsistent IGBT parasitic parameters and the inconsistent impedance caused by the inductor winding process, the main loop objectively has impedance differences.

The voltage and current double closed-loop PI regulators are used for impedance matching. The two control loops share the voltage outer loop, and the current inner loop uses respective inductor current feedback. The control loop is shown in Figure 7b. The difference in the modulation signals generated by the control loop adjusts the respective output impedances to achieve current sharing control. The left and right arms are interleaved 90° parallel, the two inductor current ripples cancel each other, and the output voltage is stable, as shown in Figures 8 and 9. The current i_L fluctuating frequency is twice of the switching frequency, and the inductor current fluctuation amplitude is reduced, due to i_L through two inductors evenly. Two arms series, the flying capacitor voltage is 0.25 U_H, as shown in Figure 10. The inductor current ripple is small, and the flying capacitor voltage is equal to 0.25 U_H.

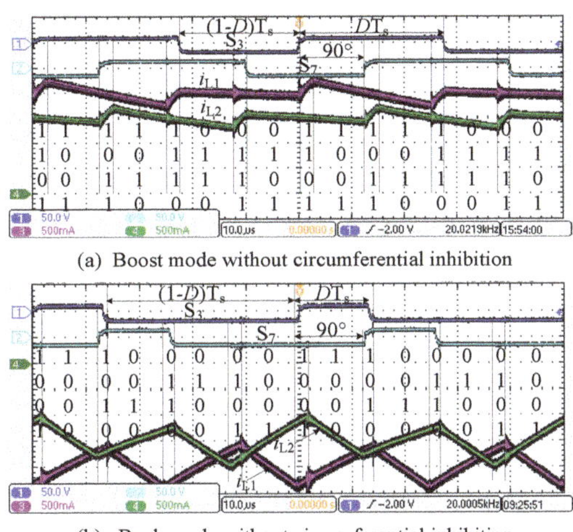

(a) Boost mode without circumferential inhibition

(b) Buck mode without circumferential inhibition

Figure 8. Left and right arms interleaved 90° parallel without circumferential inhibition. (**a**) Boost mode. (**b**) Buck mode.

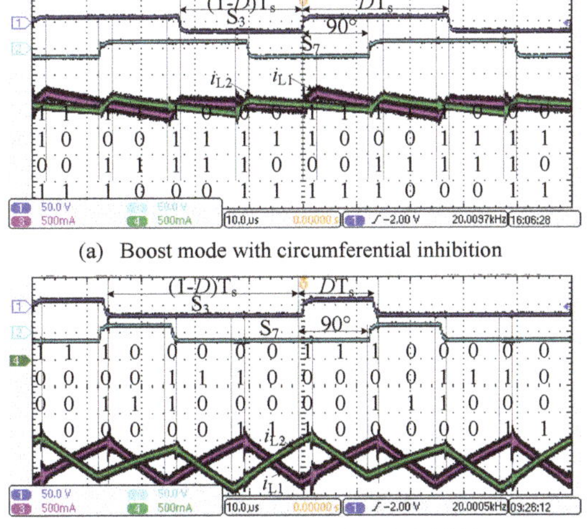

(a) Boost mode with circumferential inhibition

(b) Buck mode with circumferential inhibition

Figure 9. Left and right arms interleaved 90° parallel with circumferential inhibition. (**a**) Boost mode, (**b**) Buck mode.

Switch S is turned on at t_{ON}, and the supercapacitor charging power is 800 W in buck mode. Switch S is turned off at t_{OFF}, and the supercapacitor discharging power is 1200 W in boost mode, as shown in Figure 11. The dynamic response time is less than 20 ms from charging to discharging. The dynamic response time is

400 ms from discharging to charging in the four-arms mixed modes. The flying capacitor voltage is always stable, and the DC bus voltage fluctuation is less than 20 V. The input voltage varies between 150–220 V, the output voltage is stable at 400 V, and the change range of D is 0.63–0.45. From the experimental results in Figures 8–11, it can be seen that the input and output side voltage ripple is less than 1%, and the current ripple frequency is relatively small, which is beneficial to the stable operation of the energy storage unit. When Q_7 shorted, i_{L4} ripple is only once per cycle, losing the advantage of three levels. However, the overall performance of the converter remains stable, and the input and output voltage ripples are low, as shown in Figure 12. Therefore, the BDC has a strong fault tolerance ability, and failure of any one of the arms does not affect the operation of other arms.

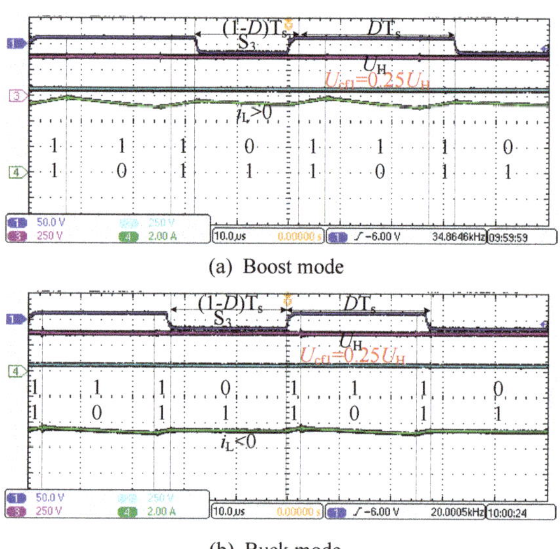

(a) Boost mode

(b) Buck mode

Figure 10. Two arms complementary series. (**a**) Boost mode, (**b**) Buck mode.

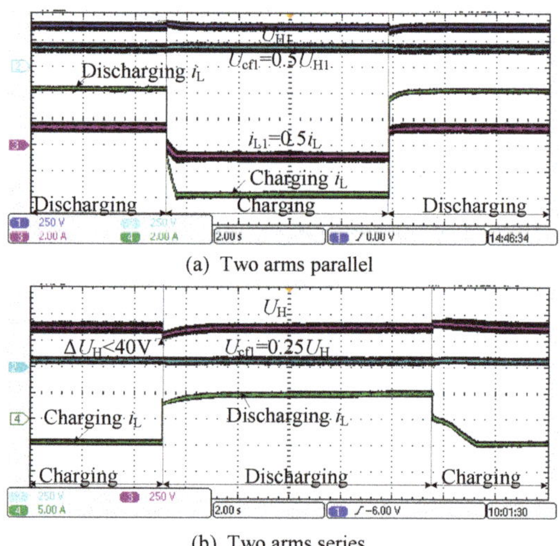

(a) Two arms parallel

(b) Two arms series

Figure 11. Transient response for bidirectional operation. (**a**) Two arms interleaved parallel transient response, (**b**) Two arms complementary series transient response.

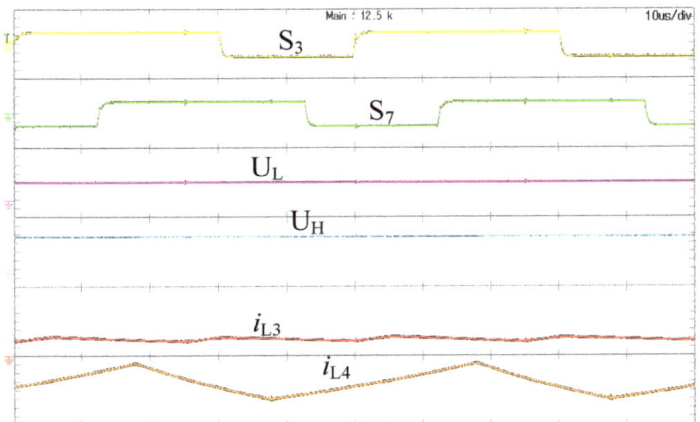

Figure 12. Switch Q7 short circuit experiment.

The efficiency is reduced, due to the inherent loss increase with a light load. Since there is a dead time of 1.5 μs when the converter is actually running, energy can be transferred from the low voltage side to the high voltage side through the IGBT body diode during dead time, as shown in Figure 13. Thus, the boost mode efficiency is higher than the buck mode. The efficiency curve is more than 90% from light load to heavy load, meeting design requirements.

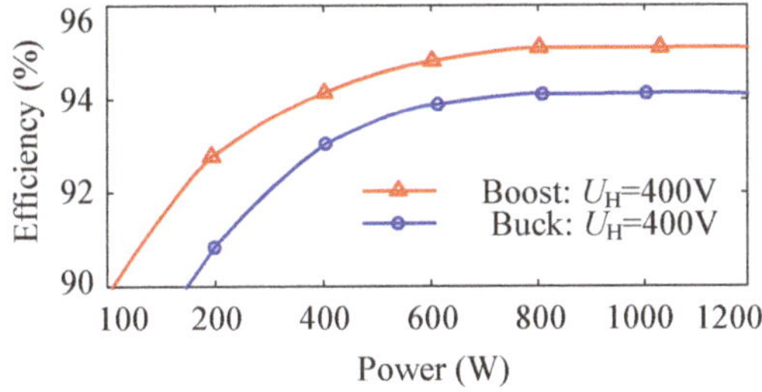

Figure 13. The efficiency curves.

6. Conclusions

This paper proposes a BDC topology with a multiplexed modulation strategy for high-power energy storage in bipolar DC microgrids. The parallel arms divide the input side current, which can effectively overcome the current difference caused by the inconsistent parasitic parameters of the parallel arms. The series arms divide the voltage of the high voltage side, which can effectively reduce the voltage stress of the switch and the flying capacitor. The bidirectional transient response is milliseconds, which ensures the dynamic performance and operating efficiency of the converter. The BDC has a strong fault tolerance ability, and the failure of any one of the arms does not affect the operation of other arms. The proposed BDC topology and its modulation strategy can effectively solve the issue of high-power energy storage in bipolar DC microgrids.

Author Contributions: P.L. designed the prototype and was responsible for writing the paper. C.Z., S.P. and L.Z. were responsible for guidance in the experiment and thesis writing process.

Funding: This research was supported by the National Nature Science Foundation of China under Grants 51477148 and the Science Foundation of Hebei University of Science and technology Grants PYB2019011. This research also received funding from EEEIC International, Poland.

Conflicts of Interest: The authors declare no potential conflict of interest.

References

1. Ji, Y.; Yang, Y.; Zhou, J.; Ding, H.; Guo, X.; Padmanaban, S. Control Strategies of Mitigating Dead-time Effect on Power Converters: An Overview. *Electronics* **2019**, *8*, 196. [CrossRef]
2. Guo, X.; Jia, X. Hardware-based cascaded topology and modulation strategy with leakage current reduction for transformerless PV systems. *IEEE Trans. Ind. Electron.* **2016**, *62*, 7823–7832. [CrossRef]
3. Guo, X.; Yang, R.Y.; He, R.; Wang, B.; Blaabjerg, F. Transformerless Z-source four-leg PV inverter with leakage current reduction. *IEEE Trans. Power Electron.* **2019**, *34*, 4343–4352. [CrossRef]
4. Mahmoudi, M.A.H.; Ahmadi, R. Modulated model predictive control of three level flying capacitor buck converter. In Proceedings of the IEEE Power and Energy Conference at Illinois (PECI), Champaign, IL USA, 23–24 February 2017; pp. 1–5.
5. Jin, K.; Yang, M.; Ruan, X.; Xu, M. Three-Level Bidirectional Converter for Fuel-Cell/Battery Hybrid Power System. *IEEE Trans. Ind. Electron.* **2010**, *57*, 1976–1986. [CrossRef]
6. Zhang, C.; Li, P.; Kan, Z.; Chai, X.; Guo, X. Integrated Half Bridge CLLC Bidirectional Converter for Energy Storage Systems. *IEEE Trans. Ind. Electron.* **2018**, *65*, 3879–3889. [CrossRef]
7. Li, P.; Zhang, C.; Kan, Z.; Fu, Y. An Interleaving 90° Three-Level DC-DC Converter and Current Sharing Control. In Proceedings of the 2nd IEEE International Power Electronics and Application Conference and Exposition, Shenzhen, China, 4–7 November 2018.
8. Dragičević, T.; Guerrero, J.M.; Vasquez, J.C.; Škrlec, D. Supervisory Control of an Adaptive-Droop Regulated DC Microgrid with Battery Management Capability. *IEEE Trans. Power Electron.* **2014**, *29*, 695–706. [CrossRef]
9. Huang, X.; Lee, F.C.; Li, Q.; Du, W. High-Frequency High-Efficiency GaN-Based Interleaved CRM Bidirectional Buck/Boost Converter with Inverse Coupled Inductor. *IEEE Trans. Power Electron.* **2016**, *31*, 4343–4352. [CrossRef]
10. Xiao, G.; Zhou, J.; He, R.; Jia, X.; Rojas, C.A. Leakage current attenuation of a three-phase cascaded inverter for transformerless grid-connected pv systems. *IEEE Trans. Ind. Electron.* **2018**, *65*, 676–686.
11. Chen, W.; Ruan, X.; Yan, H.; Tse, C.K. DC/DC Conversion Systems Consisting of Multiple Converter Modules: Stability, Control, and Experimental Verifications. *IEEE Trans. Power Electron.* **2009**, *24*, 1463–1474. [CrossRef]
12. Jin, L.; Liu, B.Y.; Duan, S.X. ZVS Operation Range Analysis of Three-Level Dual Active Bridge DC-DC Converter with Phase-Shift Control. In Proceedings of the 2017 Thirty Second Annual IEEE Applied Power Electronics Conference and Exposition (Apec), Tampa, FL, USA, 26–30 March 2017.
13. Wang, Y.; Yuan, Z.; Fu, J.; Li, Y.; Zhao, Y. A feasible coordination protection strategy for MMC-MTDC systems under DC faults. *Int. J. Electr. Power Energy Syst.* **2017**, *90*, 103–111. [CrossRef]
14. Kakigano, H.; Miura, Y.; Ise, T. Low-Voltage Bipolar-Type DC Microgrid for Super High-Quality Distribution. *IEEE Trans. Power Electron.* **2010**, *25*, 3066–3075. [CrossRef]
15. Xiao, G.; Yang, Y.; Wang, B.; Blaabjerg, F. Leakage Current Reduction of Three-phase Z-source three-level four-leg inverter for transformerless PV system. *IEEE Trans. Power Electron.* **2019**, *34*, 6299–6308.
16. Dusmez, S.; Khaligh, A.; Hasanzadeh, A. A Zero-Voltage-Transition Bidirectional DC/DC Converter. *IEEE Trans. Ind. Electron.* **2015**, *62*, 3152–3162. [CrossRef]
17. Guo, Z.; Chai, X.; Li, P.; Zhang, C. A new balance control method of three-level converter NPP. In Proceedings of the IEEE 8th International Power Electronics and Motion Control Conference (IPEMC-ECCE Asia), Piscataway, NJ, USA, 22–26 May 2016; pp. 2356–2361.
18. Singh, S.N. Selection of non-isolated DC-DC converters for solar photovoltaic system. *Renew. Sustain. Energy Rev.* **2017**, *76*, 1230–1247.
19. Do, H.L. A Soft-Switching DC/DC Converter with High Voltage Gain. *IEEE Trans. Power Electron.* **2010**, *25*, 1193–1200.
20. Zhang, X.S.; Yuan, Y.; Hua, L.; Cao, Y.; Qian, K.J. On Generation Schedule Tracking of Wind Farms with Battery Energy Storage Systems. *IEEE Trans. Sustain. Energy* **2017**, *8*, 341–353. [CrossRef]
21. Naderi, R.; Sadigh, A.K.; Smedley, K.M. Dual Flying Capacitor Active-Neutral-Point-Clamped Multilevel Converter. *IEEE Trans. Power Electron.* **2016**, *31*, 6476–6484. [CrossRef]
22. Mohammadi, F. Power Management Strategy in Multi-Terminal VSC-HVDC System. In Proceedings of the 4th National Conference on Applied Research in Electrical, Mechanical, Computer and IT Engineering, Tehran, Iran, 5 January 2018.

23. Uno, M. High Step-Down Converter Integrating Switched Capacitor Converter and PWM Synchronous Buck Converter. In Proceedings of the 35th International Telecommunications Energy Conference, Hamburg, Germany, 13–17 October 2013.

24. Li, X.; Zhang, W.; Li, H.; Xie, R.; Xu, D. Design and control of bi-directional DC/DC converter for 30kW fuel cell power system. In Proceedings of the 8th International Conference on Power Electronics-ECCE Asia, Jeju, Korea, 30 May–3 June 2011; pp. 1024–1030.

© 2019 by the authors. Licensee MDPI, Basel, Switzerland. This article is an open access article distributed under the terms and conditions of the Creative Commons Attribution (CC BY) license (http://creativecommons.org/licenses/by/4.0/).

 *electronics*

Article

Application of the Lyapunov Algorithm to Optimize the Control Strategy of Low-Voltage and High-Current Synchronous DC Generator Systems

Jinfeng Liu [1,*], Xiaohai Tan [1], Xudong Wang [1] and Herbert Ho-Ching IU [2]

[1] Ministry of Education, Engineering Research Center of Automotive Electronics Drive Control and System Integration, Harbin University of Science and Technology, Harbin 150080, China
[2] School of Electrical, Electronic and Computer Engineering, The University of Western Australia, Crawley, WA 6009, Australia
* Correspondence: liujinfeng@hrbust.edu.cn; Tel.: +86-181-0368-8668

Received: 24 June 2019; Accepted: 4 August 2019; Published: 6 August 2019

Abstract: In the present study, a novel multiple three-phase low-voltage and high-current permanent magnet synchronous generation system is proposed, which has only half-turn coils per phase. The proposed system is composed of a generator and two confluence plates with 108 rectifier modules. The output can reach up to 10,000 A continuous DC power supply, which is suitable for the outdoors and non-commercial power supply. The application of the Lyapunov algorithm in the synchronous rectification control was optimized. A current sharing loop control was added to the closed-loop control to ensure a stable output voltage and the output current sharing of each rectifier module. Since the two control variables solved by the Lyapunov algorithm were coupled and the negative definite function of the Lyapunov algorithm could not be guaranteed in this system, a simple decoupling method was used to decouple the control variables. Compared to the conventional control, the proposed strategy highly improved the dynamic performance of the system. The effectiveness of the proposed strategy was verified by the simulation. The 5 V/10,000 A hardware experiment platform was built, which proved the feasibility and validity of the proposed strategy for a high-power generation system.

Keywords: low-voltage and high-current; Lyapunov algorithm; current sharing control; confluence plate

---

1. Introduction

Low-voltage and high-current generation systems with currents higher than 10,000 A are widely used in ships, electrolytic plating, and industrial fields. Considering the integration of motor and power electronics technology, the synchronous generation system has developed rapidly from high power density, high reliability, and high fault tolerance points of view [1]. In the present study, a three-phase rectifier module is used to generate direct current (DC) through a confluence plate, based on a permanent magnet synchronous motor with only half-turn coil per phase. In order to ensure that the integrated DC generation system generates high quality electrical power, the present study focuses on the control strategy of this system

The control strategy of three-phase rectifier circuits has been extensively studied by many researchers. Thomas A.F. first proposed the hysteresis current control; in [2,3] this control method was adopted in the current inner loop. It is also proven that the disadvantage of this method is that the switching frequency changes as the load current changes. In other words, this method cannot guarantee a fixed switching frequency in a power cycle. This causes additional stress on the switching device and reduces the service life of the device. Kalman proposed the dead-beat control theory, which is a control method based on circuit equations. In [4], an improved implementation of the deadbeat current controller was considered that is aimed at two purposes: The minimization of the small-signal response delay and the optimization of the large-signal step response. This method cancels the zero with the pole by using state feedback and configures another pole at the origin [5]. However, studies showed that this method has major drawbacks, including a high real-time requirement for calculating the pulse width and the sensitivity of system stability to the circuit parameters and weaknesses. Although the aforementioned drawbacks can be resolved by employing a complicated algorithm and a current observer, a large error will be generated. Zadeh L.A. proposed the fuzzy control based on fuzzy reasoning, which mimics the human mindset and controls a model where it is difficult to establish an accurate mathematical expression. In [6], a modified zero-voltage switching pulse width-modulation inverter with a digital signal processor-based proportional integral derivative-like fuzzy controller was implemented. The switches of the inverter achieved a soft-switching feature that largely reduced switching losses and improved the converting efficiency. The disadvantage of this control strategy is that it only relies on the experience and attempt to design the controller. There is no systematic method of analyzing and designing the controller. The cycle is long and the precision is low. Once the adaptive limit is exceeded, this method will no longer be applicable [7]. Tokuo Ohnishi proposed the direct power control method. The proposed control method indirectly controls the output current through the direct control of active and reactive power. Moreover, this method establishes the voltage and the current double closed-loop model, according to the AC voltage and instantaneous power to select the output state corresponding to the switch table. In [8], a direct power control using the natural switching surface was proposed; the proposed control considered the output voltage when selecting the switching states. Therefore, the proposed control does not need an outer voltage control loop and can highly improve the dynamic performance of the DC output voltage. The advantages of this method are the high-power factor, the high efficiency factor, and the ability to be used in a wide variety of applications [9,10]. However, due to the limitation of the current loop, it easily generates harmonic distortion in the input current, thereby reducing the power factor.

Lyapunov proposed the stability criterion control theory, which initially constructs a scalar energy-like function for the system and then designs the controller under the premise that the change in time of this function is negative [11]. This method was introduced into the rectifier control of a three-phase pulse with modulation (PWM) by Hasan K. [12]. He utilized the Lyapunov algorithm to control the three-phase rectifier system. This algorithm has the advantage that the system's stability is not interfered with by large signals and is independent of circuit parameters. At present, the application of the Lyapunov algorithm to control PWM rectifiers is concentrated on grid voltage rectification [13,14] where there is a mutual coupling phenomenon between the obtained control variables, which cannot guarantee the long-term stable operation of the system. This study first proposes the application of the Lyapunov algorithm to a multiple three-phase permanent magnet in the synchronous generator system. It is expected that the system with the proposed control scheme can provide a reasonable transient response, unified power factor control, reduced harmonics on the AC side and an increase in the reliability of the generation system. In order to ensure the stable operation of the system, the proposed method decouples the control variables in a simple way.

2. Methods

2.1. Parallel Scheme of the Multiple Rectification Module

The studied high-current generation system is a new DC power generation system that realizes electromechanical energy conversion by using a multiple three-phase permanent magnet synchronous motor. The system directly generates low-ripple and high-intensity DC power through the combination of the modular rectification component and the high current confluence plate [15]. The advantages of this system are as the following:

1. The phase voltage and the phase current can be reduced to generate low voltage and high power.
2. The harmonic loss can be reduced to improve generator efficiency.

3. Each phase is driven separately to improve the fault tolerance and the reliability of the system.
4. Considering the special design of the confluence plate, the ripple of the output voltage is low when the system generates the high current [16]. Therefore, the output current can be used with no filtering for the majority of applications.

Figure 1 illustrates the designed model of the multiple three-phase permanent magnet synchronous high-current generation system. It indicates that the designed system is mainly composed of a permanent magnet synchronous generator with a half-turn coil for each phase and two confluence plates. Figure 1a describes the minimum cell which is composed of each winding and the leg of the bridge, and Figure 1b describes the five MOSFETs on each leg which are fixed on the heat sink. Moreover, Figure 1c shows a confluence plate that fixes the heat sink to be used as the output bus. In order to improve the heat-dissipation efficiency, the inside of the confluence plate is machined with a water-cooled passage. Figure 1d shows the overall model of the system containing the permanent magnet in the synchronous generator.

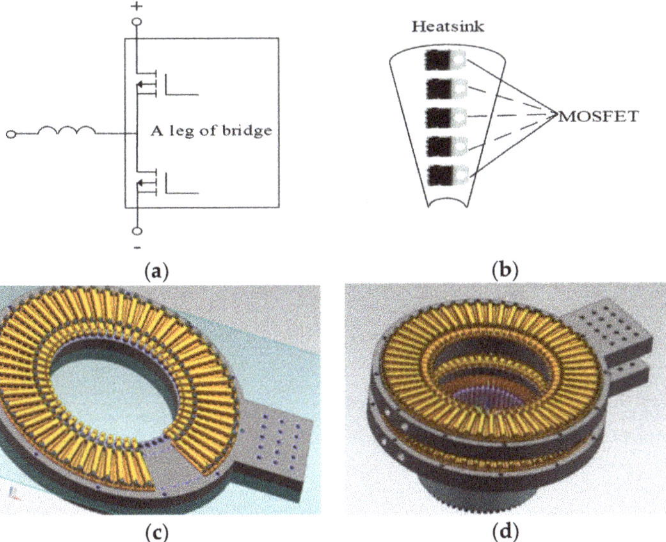

Figure 1. Schematic models of the rectifier module and the power generation system: (**a**) A leg of bridge; (**b**) heat sink; (**c**) schematic of the confluence plate; (**d**) schematic model of a large DC power generation system.

The generator rotor contains 6 pairs of magnetic poles and the stator core is internally opened with 54 slots. Each slot is embedded with a half-turn coil, containing the insulation as one phase, while the three-phase stator windings are placed at an electrical angle of 120°. It should be indicated that the two adjacent three-phase winding units differ from each other by an electrical angle of $2\pi/9$. Moreover, each positive and negative confluence plate is composed of 54 MOSFET rectifier modules. The rectifier modules can be distributed along the circumference surface of the confluence plate as a stationary part of the synchronous generator. In order to simplify the analysis, only three rectifier modules were considered in the design and analysis. These modules correspond to the spaces of nine stator windings of the synchronous generator, which can form a 360° electrical angular space. The system is managed hierarchically, where the top layer is the management layer and the controller area network (CAN) bus transmits the current sharing information. The second layer of each control module analyzes the collected module voltage and compares it with current signals to generate the PWM control information to drive the switching device of the rectifier. The bottom layer is a three-phase PWM rectifier module, which is fixed at the bottom of the generator together with the confluence plate to generate high currents. Figure 2 shows the block diagram of the synchronous DC power generation system with the multi-phase permanent magnet.

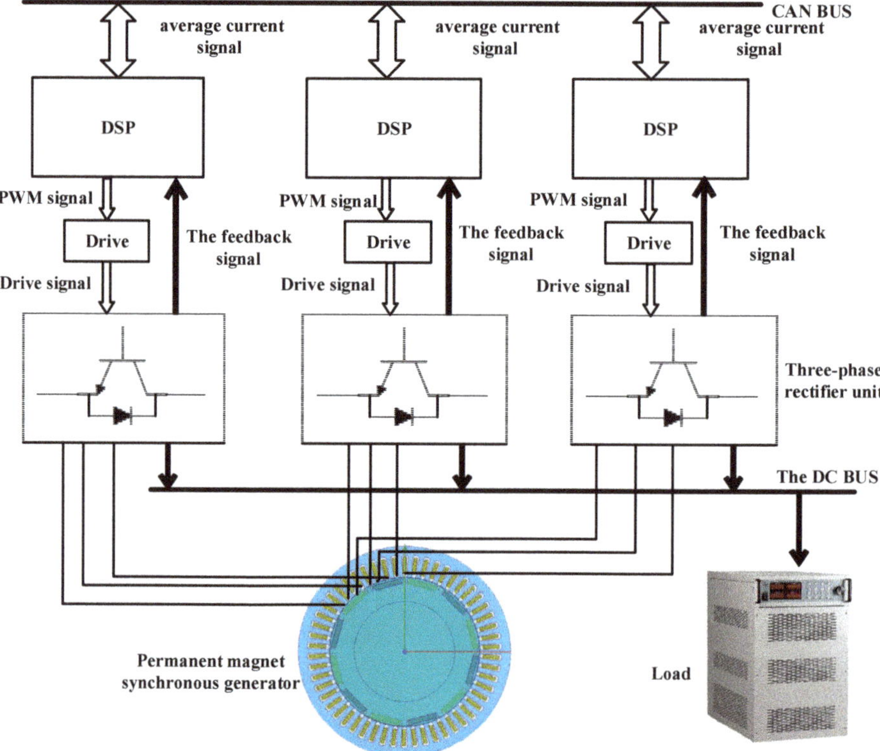

Figure 2. The block diagram of the synchronous DC power generation system with the multi-phase permanent magnet.

The digital current sharing by the CAN bus can avoid the defects of the conventional master–slave control mode. This reduces the number of controllers and avoids the failure of the main control module not achieving the appropriate current sharing [17]. The CAN bus realizes the current sharing by means of competition where the module which generates the maximum current is determined as the main module and the corresponding maximum current is used as the reference current of the submodule. Even if the main module fails and shuts down, the other modules can continue to qualify as the main module in a competitive manner.

A modular rectifier improves system performance. First of all, the modular parallel connection is adopted in this rectification system, and the ripple of output is small, so that a rapid voltage adjustment method can be adopted without causing distortion of the input current, and the dynamic performance is good. Secondly, when the modules are connected in parallel to supply power to the load, even if a module fails, as long as it is removed from the parallel system in time, the load power supply can be ensured without interruption, further improving the reliability and flexibility of the power supply system. Finally, when powered by a parallel system, the power of each parallel module will be reduced, while the low power module can operate at a higher switching frequency, which can reduce the volume of the filter capacitor and increase the power density of the parallel system, which is very suitable for low- voltage and high-power systems.

2.2. Design of the Three-Phase Rectifier Controller, Based on the Lyapunov Algorithm

2.2.1. Mathematical Model of the Three-Phase Voltage Source PWM Rectifier in the d-q Synchronous Rotating Coordinate System

The state equation of the three-phase voltage source PWM rectifier is described by the unipolar binary logic switching function. When the coordinate transformation of the constant power is carried out, the mathematical models in the three-phase rotating coordinate system can be written as the following [18]:

$$L\frac{di_d}{dt} = \omega L i_q - R i_d + e_d - v_o p_d \tag{1}$$

$$L\frac{di_q}{dt} = -\omega L i_d - R i_q + e_q - v_o p_q \tag{2}$$

$$C\frac{dv_o}{dt} = \frac{3}{2}\left(P_d i_d + P_q i_q\right) - i_o \tag{3}$$

where v_o, i_o, L and C denote the output voltage, module current, AC side inductance, and the DC side filter capacitance, respectively. On the other hand, ω is the angular frequency. Furthermore, p_d and p_q are voltage modulation ratios on d and q axes, respectively. Finally, i_d, i_q, e_d, e_q denote the currents and voltages of the network side in the synchronous rotating coordinate system, respectively.

2.2.2. Lyapunov Control Algorithm

Studies showed that the hysteresis current control, the deadbeat control and the fuzzy control methods yield various advantages and disadvantages [19–21] from different points of view, including the circuit complexity, switching frequency, and the transient state. A common disadvantage of these methods is that system stability cannot be guaranteed under large signal interference. Since the system generates high current, the voltage produces large fluctuations when subjected to large signal interference, which affects the normal operation of the system. In [9], the controller is derived based on direct Lyapunov stability theory in order to calculate proper switching functions, Setting up the controller with this switch function, the results show the appropriate performance of the proposed controllers during both steady-state and transient dynamic conditions. Considering the abovementioned challenge, a nonlinear control strategy based on the Lyapunov direct method [22,23] is proposed to strictly guarantee the global asymptotic stability of the rectifier. The purpose of this control strategy is to make the output voltage close to the reference voltage (V_r) and provide a unity power factor with a nearly sinusoidal input current. When the system is controlled by the Lyapunov algorithm, it is necessary to construct an "energy-like" function for the system, and then design the controller under the premise of ensuring that the function is always negative. Finally, establishing the Lyapunov function based on the quantitative correlation of the energy storage of the inductor and the capacitor. Suppose that the Lyapunov function is a positive definite function, while its derivative is negative definite function. When x tends to infinity in any direction, $V(x)$ also approaches infinity. The equilibrium point at the origin is globally asymptotically stable.

Letting the operating point of the system energy stability be the equilibrium point, we define a positive definite Lyapunov function as the following:

$$V(\overline{x}) = \frac{3}{2}Lx_1^2 + \frac{3}{2}Lx_2^2 + Cx_3^2 \tag{4}$$

where x_1, x_2 and x_3 are system state variables defined as the following:

$$\begin{cases} x_1 = i_d - i_{d0} \\ x_2 = i_q \\ x_3 = V_r - v_o + k(I_r - i_o) \end{cases} \tag{5}$$

where V_r, I_r, k and i_{d0} denote the reference voltage, reference current, proportional coefficient, and the steady value of i_d, respectively. It should be indicated that the reference current is maximum current in the multiple rectification. The derivative of the Lyapunov function is

$$V'(\overline{x}) = 3x_1 Lx_1' + 3x_2 Lx_2' + 2x_3 Cx_3' \tag{6}$$

According to the first Lyapunov stability theorem, when the Lyapunov derivative is negative, the system is stable in the equilibrium point. Let $i_q = 0$ to ensure unit power factor, when the system is stable, the parameter values of the equilibrium point are as follows:

$$\begin{cases} v_o = V_r \\ i_o = I_r \\ i_d = i_{d0} \\ e_d = E_m \\ i_q = 0 \\ p_d = p_{d0} \\ p_q = p_{q0} \end{cases} \tag{7}$$

where p_{d0} and p_{q0} are the control variables when the system is stable and E_m is the amplitude of the phase voltage.

After substituting Equation (7) into Equations (1)–(3) and a simple manipulation, the following expressions are obtained:

$$p_{d0} = (E_m - Ri_{d0})/V_r \tag{8}$$

$$p_{q0} = -\omega L i_{d0}/V_r \tag{9}$$

$$i_o - \frac{3}{2V_r}(E_m i_{d0} - Ri_{d0}^2) \tag{10}$$

$$i_{d0} = \frac{1}{2}\left\{ \frac{E_m}{R} \pm \left[\left(\frac{E_m}{R}\right)^2 - \left(\frac{8V_r i_o}{3R}\right) \right]^{\frac{1}{2}} \right\} \tag{11}$$

Assuming that the system is disturbed, the variations of voltage-space vector modulation ratio on d and q axes are Δp_d and Δp_q, respectively. Then the actual modulation ratios of output voltage can be obtained as the following:

$$p_d = p_{d0} + \Delta p_d \tag{12}$$

$$p_q = p_{q0} + \Delta p_q \tag{13}$$

Substitute Equations (5), (8) and (12) into Equation (1) results in the following equation:

$$Lx_1' = \omega Lx_2 - V_r \Delta p_d - \frac{(E_m - Ri_{d0})[x_3 + k(I_r - i_o)]}{V_r} \\ -[x_3 + k(I_r - i_o)]\Delta p_d - Rx_1 \tag{14}$$

Similarly, Equations (5), (9) and (13) are applied to Equation (2) so that the following equation is obtained:

$$Lx_2' = -\omega Lx_1 - V_r \Delta p_q - \frac{\omega L i_{d0}[x_3 + k(I_r - i_o)]}{V_r} \\ -[x_3 + k(I_r - i_o)]\Delta p_q - Rx_2 \tag{15}$$

Moreover, Equations (5), (10), (12) and (13) are implemented into Equation (3) to obtain the following equation:

$$Cx_3' = \frac{3}{2}\left[\frac{(E_m - Ri_{d0})}{V_r}x_1 + i_{d0}\Delta p_d + \Delta p_d x_1 \\ -\frac{\omega L i_{d0}}{V_r}x_2 + \Delta p_q x_2 \right] \tag{16}$$

Substituting Equations (10) and (14)–(16) into Equation (6) yields the following equation:

$$V'(\bar{x}) = \begin{array}{l} -3\{[V_r + k(I_r - i_o)]x_1 - i_{d0}x_3\}\Delta p_d \\ -3x_2[V_r + k(I_r - i_o)]\Delta p_q - 3R\left(x_1^2 + x_2^2\right) \end{array} \tag{17}$$

When the following conditions are met, the Lyapunov derivative along any of the trajectories of the system is negative:

$$\Delta p_d = \gamma\{[V_r + k(I_r - i_o)]x_1 - i_{d0}x_3\}, \gamma > 0 \tag{18}$$

$$\Delta p_q = \beta x_2[V_r + k(I_r - i_o)], \beta > 0 \tag{19}$$

where β and γ are arbitrary real constants.

2.2.3. Saturation Constraint and Decoupling Control Variables

The switching state of the rectifier is determined by the space vector pulse width modulation method (SVPWM) [24]. In order to ensure that the rectifier is in sinusoidal steady-state operation and the switching function is not saturated, the following conditions must be met:

$$\left(p_{d0} + \Delta p_d\right)^2 + \left(p_{q0} + \Delta p_q\right)^2 \le \frac{4}{3} \tag{20}$$

The control variables, which satisfy the SVPWM are obtained from Equation (20) as the following:

$$\left(\Delta p_d\right)_{m1} = \frac{2(p_{d0} + \Delta p_d)}{\sqrt{3\left[(p_{d0} + \Delta p_d)^2 + \left(p_{q0} + \Delta p_q\right)^2\right]}} - p_{d0} \tag{21}$$

$$\left(\Delta p_q\right)_{m1} = \frac{2\left(p_{q0} + \Delta p_q\right)}{\sqrt{3\left[(p_{d0} + \Delta p_d)^2 + \left(p_{q0} + \Delta p_q\right)^2\right]}} - p_{q0} \tag{22}$$

The two control variables $(\Delta p_d)_{m1}$ and $(\Delta p_q)_{m1}$ are mutual coupling and non-linear variables, which cannot guarantee a negative Lyapunov derivative. Therefore, it is necessary to decouple the control variables. Assuming that the system is controlled by p_q, (i.e., $p_d = 0$), then the range of Δp_q can be written as

$$-\left(p_{q0m} + p_{q0}\right) \le \Delta p_q \le p_{q0m} - p_{q0} \tag{23}$$

P_{q0m} represents the maximum possible steady-state value that p_q can take for the maximum DC load current. Similarly, the range of Δp_d can be written as

$$-\left(p_{d0m} + p_{d0}\right) \le \Delta p_d \le p_{d0m} - p_{d0} \tag{24}$$

P_{d0m} is the maximum steady-state value of p_d, which can be calculated by Equation (25) as the following:

$$p_{d0m} = \sqrt{\frac{4}{3} - p^2_{q0m}} \tag{25}$$

Therefore, the control variables are rewritten as the following:

$$(\Delta p_d)_{m2} = \begin{cases} -(p_{d0m} + p_{d0}) & \gamma\{[V_r + k(I_r - i_o)]x_1 - i_{d0}x_3\} < -(p_{d0m} + p_{d0}) \\ \gamma\{[V_r + k(I_r - i_o)x_1 - i_{d0}x_3]\} & -(p_{d0m} + p_{d0}) \le \gamma\{[V_r + k(I_r - i_o)]x_1 - i_{d0}x_3\} \le (p_{d0m} - p_{d0}) \\ p_{d0m} - p_{d0} & \gamma\{[V_r + k(I_r - i_o)x_1 - i_{d0}x_3]\} > (p_{d0m} - p_{d0}) \end{cases} \tag{26}$$

$$\left(\Delta p_q\right)_{m_2} = \begin{cases} -\left(p_{q0m} + p_{q0}\right) & \beta x_2[V_r + k(I_r - i_o)] < -\left(p_{q0m} + p_{q0}\right) \\ \beta x_2[V_r + k(I_r - i_o)] & -\left(p_{q0m} + p_{q0}\right) \le \beta x_2[V_r + k(I_r - i_o)] \le p_{q0m} - p_{q0} \\ p_{q0m} - p_{q0} & \beta x_2[V_r + k(I_r - i_o)] > p_{q0m} - p_{q0} \end{cases} \tag{27}$$

Figure 3 illustrates that the voltage vector of the rotation is in the rectangular area of the dotted round line. Under the control rules of Equations (26) and (27), the Lyapunov derivative is a negative definite function. Meanwhile, the stability of the system is independent of the circuit parameters.

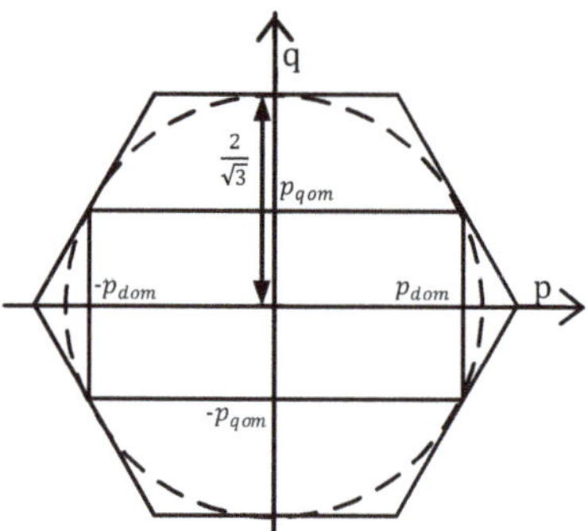

Figure 3. Modified control vector area.

2.2.4. SVPWM Algorithm

The final voltage modulation ratio p_d, p_q is as follows:

$$p_d = p_{d0} + (\Delta p_d)_{m2} \tag{28}$$

$$p_d = p_{d0} + (\Delta p_d)_{m2} \tag{29}$$

Then the inverse Park transformation on p_d, p_q, is as shown in Figure 4.

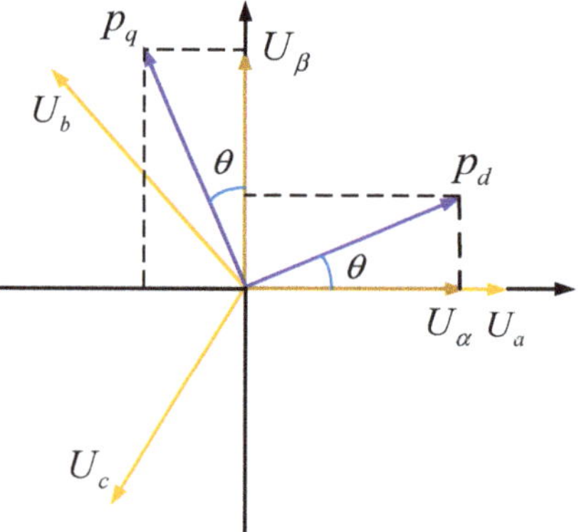

Figure 4. Coordinate transformation.

Here, p_d and p_q are inversely transformed into U_α and U_β, and the electrical angle θ is taken as the angle. Then, p_d and p_q are respectively vector-decomposed, and the voltage components on the α-axis and the β-axis are calculated as follows:

$$U_\alpha = p_d * \cos\theta - p_q * \sin\theta \tag{30}$$

$$U_\beta = p_q * \cos\theta - p_d * \sin\theta \tag{31}$$

Then, we apply the inverse Park transformation on U_α, U_β, converting the target voltage vector U_{out} into a balanced three-phase voltage vector U_a, U_b, U_c, as follows:

$$\begin{cases} U_a = U_\beta \\ U_b = \dfrac{-U_\beta + \sqrt{3}U_\alpha}{2} \\ U_c = \dfrac{-U_\beta - \sqrt{3}U_\alpha}{2} \end{cases} \tag{32}$$

We can determine the sector of U_{out} by defining variables a, b, and c according to Equation (33).

$$N = 4a + 2b + c \tag{33}$$

The correspondence between the value of N and the sector is as Table 1:

Table 1. The sector.

N	3	1	5	4	6	2
Sector	I	II	III	IV	V	VI

In Figure 5, the target voltage vector U_{out} is analyzed in the I sector, TS is the PWM carrier period, T_0, T_7 is zero voltage vector time, T_4 is the U_4 time, T_6 is the U_6 time, and θ is electrical angle of the U_{out}. The following relationships can be obtained from Figure 5.

$$\begin{bmatrix} U_\alpha \\ U_\beta \end{bmatrix} T_s = U_{out} \begin{bmatrix} \cos\theta \\ \sin\theta \end{bmatrix} T_s = \frac{2}{3}U_{dc} \begin{bmatrix} 1 \\ 0 \end{bmatrix} T_4 + \frac{2}{3}U_{dc} \begin{bmatrix} \cos\frac{\pi}{3} \\ \sin\frac{\pi}{3} \end{bmatrix} T_6 \tag{34}$$

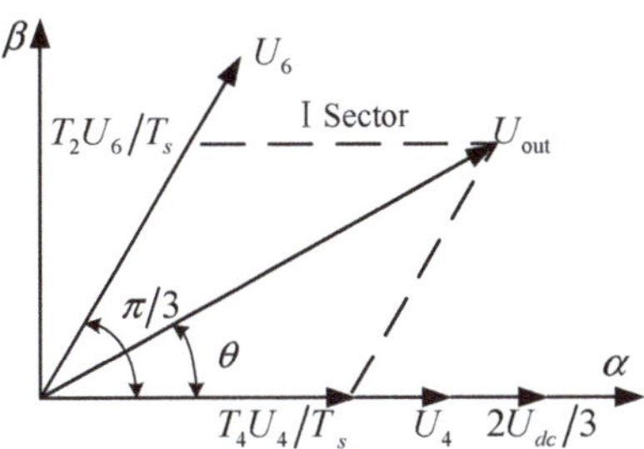

Figure 5. I sector.

From Equation (34), there are

$$\begin{cases} T_4 = \frac{\sqrt{3}T_s}{U_{dc}}\left(\frac{\sqrt{3}}{2}U_\alpha - \frac{U_\beta}{2}\right) \\ T_6 = \frac{\sqrt{3}T_s}{U_{dc}}U_\beta \\ T_0 = T_7 = \frac{1}{2}(T_s - T_4 - T_6) \end{cases} \tag{35}$$

In the same way, the time of each vector of U_{out} in other sectors is obtained. The result is shown in Table 2, where the $K = \sqrt{3}T_s/U_{dc}$.

Table 2. The time of the basic space vector of each sector.

Sector I	$T_4 = \frac{\sqrt{3}T_s}{U_{dc}}(\frac{\sqrt{3}}{2}U_\alpha - \frac{U_\beta}{2}) = KU_b$ $T_6 = \frac{\sqrt{3}T_s}{U_{dc}}U_\beta = KU_a$ $T_0 = T_7 = (T_s - T_4 - T_6)/2$	Sector II	$T_6 = \frac{\sqrt{3}T_s}{U_{dc}}(\frac{\sqrt{3}}{2}U_\alpha + \frac{U_\beta}{2}) = KU_c$ $T_2 = -\frac{\sqrt{3}T_s}{U_{dc}}(\frac{\sqrt{3}}{2}U_\alpha - \frac{U_\beta}{2}) = -KU_b$ $T_0 = T_7 = (T_s - T_2 - T_6)/2$
Sector III	$T_2 = \frac{\sqrt{3}T_s}{U_{dc}}U_\beta = KU_a$ $T_3 = \frac{\sqrt{3}T_s}{U_{dc}}(-\frac{\sqrt{3}}{2}U_\alpha - \frac{U_\beta}{2}) = KU_c$ $T_0 = T_7 = (T_s - T_2 - T_3)/2$	Sector IV	$T_3 = \frac{\sqrt{3}T_s}{U_{dc}}(-\frac{\sqrt{3}}{2}U_\alpha + \frac{U_\beta}{2}) = -KU_b$ $T_1 = -\frac{\sqrt{3}T_s}{U_{dc}}U_\beta = -KU_a$ $T_0 = T_7 = (T_s - T_1 - T_3)/2$
Sector V	$T_1 = \frac{\sqrt{3}T_s}{U_{dc}}(-\frac{\sqrt{3}}{2}U_\alpha - \frac{U_\beta}{2}) = KU_c$ $T_5 = \frac{\sqrt{3}T_s}{U_{dc}}(\frac{\sqrt{3}}{2}U_\alpha - \frac{U_\beta}{2}) = KU_b$ $T_0 = T_7 = (T_s - T_1 - T_5)/2$	Sector VI	$T_5 = -\frac{\sqrt{3}T_s}{U_{dc}}U_\beta = -KU_a$ $T_4 = \frac{\sqrt{3}T_s}{U_{dc}}(\frac{\sqrt{3}}{2}U_\alpha + \frac{U_\beta}{2}) = -KU_c$ $T_0 = T_7 = (T_s - T_4 - T_5)/2$

3. Results and Discussion

In order to verify the efficiency of the proposed control strategy compared to that of conventional methods, the direct power control algorithm [25,26], dead-Beat control algorithm, which are widely used in three-phase PWM rectifier systems, were selected in the present study. The studied parameters included the dynamic performance, reliability and independency of the large signal interference for the control of the new synchronous generation system with a multiple three-phase permanent magnet. Figure 6 shows the block diagram of the control system. Moreover, Figure 7 illustrates the three-module parallel simulation model. The simulation parameters of the three-phase PWM rectifier are as Table 3:

Table 3. Electrical parameters power circuit.

Parameter	Value
Line to line ac voltage, E	4.2 V
Source voltage frequency, f	50 Hz
Switching frequency	20 kHz
dc-bus capacitor, C	2200 uF
dc-bus voltage, vdc	5 V
Sample frequency	50 μs
Resistance of smoothing inductor, R	0.56 Ω
Inductance of smoothing inductor, L	1.5 mH
Load resistance RL	32 Ω

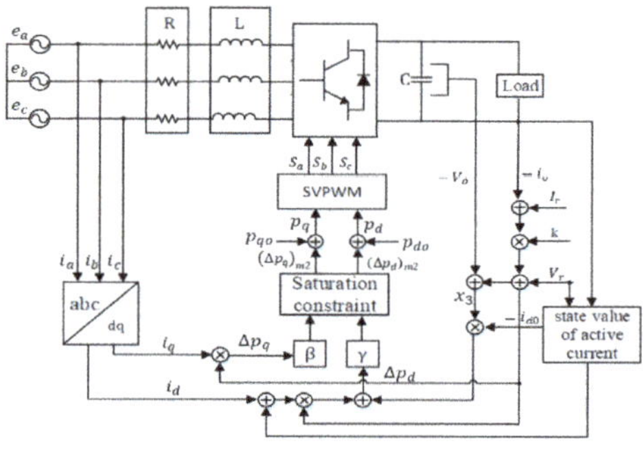

Figure 6. System diagram of the single module control of the space vector pulse width modulation method (SVPWM) rectifier.

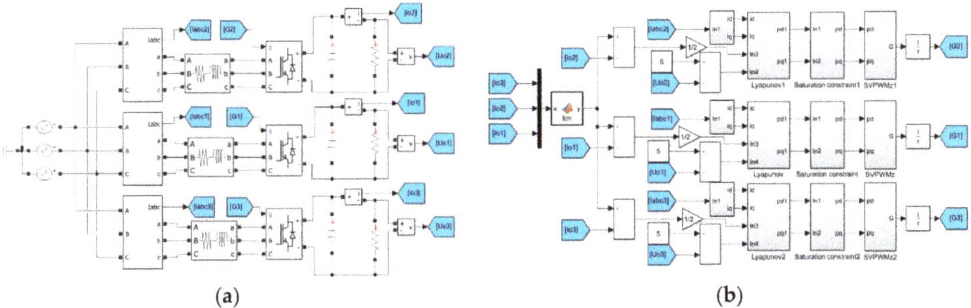

(a)   (b)

Figure 7. The three-module parallel simulation model: (**a**) The main circuit model of a three-module system during the parallel simulation; (**b**) control loop model of a three-module system during the parallel simulation.

3.1. The Control Based on the Lyapunov Algorithm

Figure 8 shows the output voltage, current and AC side A-phase voltage and current response curves of the system during the load is 32 Ω. Figure 8a indicates that during the power-on process, the system voltage overshoot is 10%, the system reaches the steady state in 0.02 s and the output voltage is 5 V. Moreover, Figure 8b shows that the output current is 150A. Figure 8c shows the waves of the input. It is observed that the current fluctuates before 0.02 s and there is a short overshoot. Meanwhile, the system is stable after 0.02 s and achieves unit power factor. Figure 8d shows the spectrum of the A-phase current. It is found that the A-phase current contains low harmonics, where the total harmonic distortion (THD) rate is THD = 1.02%.

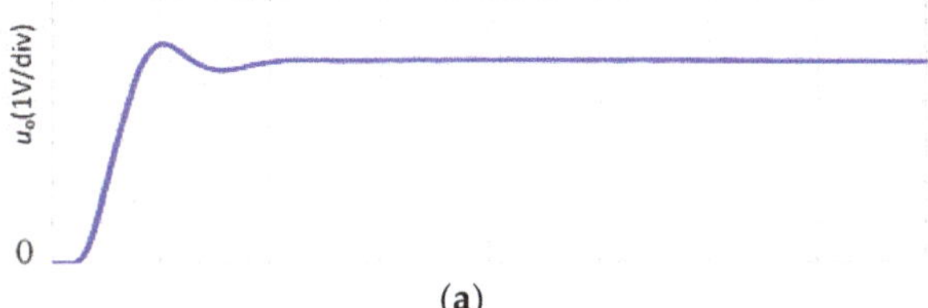

(a)

Figure 8. *Cont.*

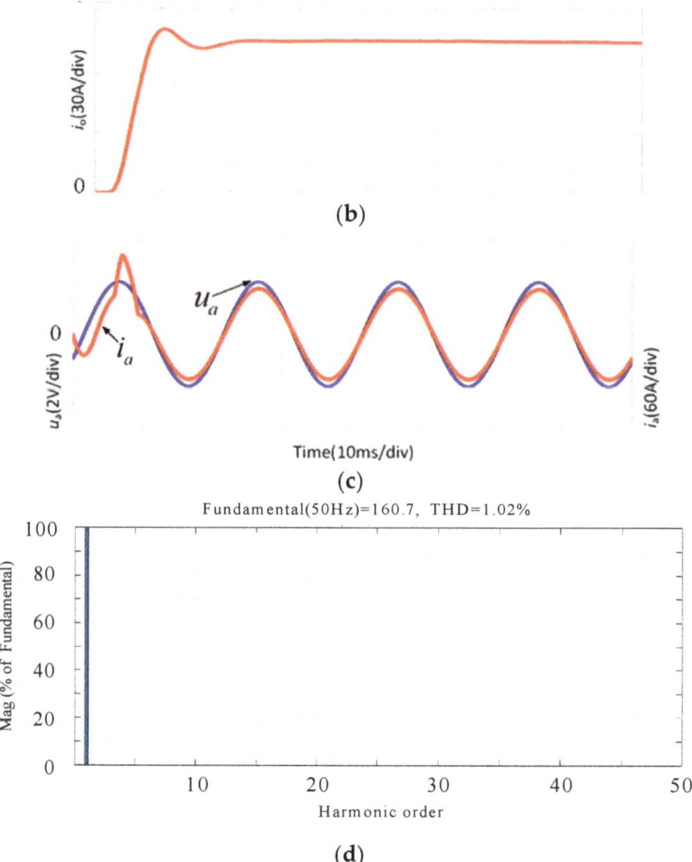

Figure 8. Simulation waves of the Lyapunov control: (**a**) Steady-state voltage wave; (**b**) steady-state current wave; (**c**) A-phase voltage and current waves; (**d**) A-phase current spectrum.

Figure 9 shows the output voltage, current and AC side of the A-phase voltage and current response curves of the system when the load suddenly changes from 32 to 16 Ω. Figure 9a illustrates that the system is stable in the 0.02 s, the load is halved in 0.03 s and the voltage changes to about 0.2 V. Moreover, Figure 9b shows that the system reaches the steady state after 0.02 s and the output current is doubled to 300 A. Figure 9c shows that the phase voltage and current briefly fluctuate and they re-implement the unit power factor.

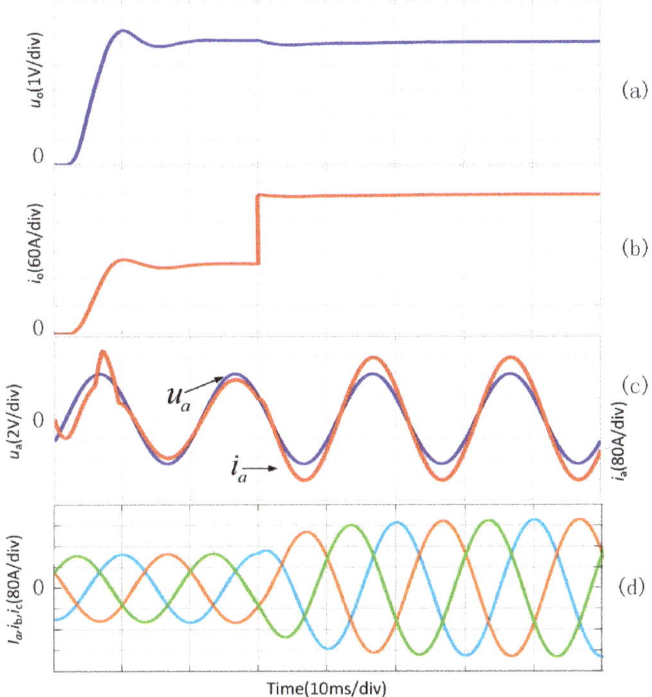

Figure 9. Simulation waves of the disturbed Lyapunov control: (**a**) Voltage wave; (**b**) Current wave; (**c**) A-phase voltage and current waves. (**d**) three-phase AC current waveform.

3.2. The Control Based on the Direct Power Algorithm

Figure 10 shows the voltage, current and AC side of the A-phase voltage and current response curves of the system during the control of the direct power algorithm. Figure 10a shows that the system is stable at 0.02 s, while the voltage drops when the load is halved at 0.04 s. The system adjusts to a new steady state after 0.03 s and the output voltage drops to 4.8 V, which is less than the reference voltage. Figure 10b shows the output current wave. It is observed that the current increases rapidly and reaches 148 A. Furthermore, Figure 10c shows the AC side of the A-phase voltage and current waveforms. It is found that the current increases after 0.04 s, while the unit power factor is still not guaranteed.

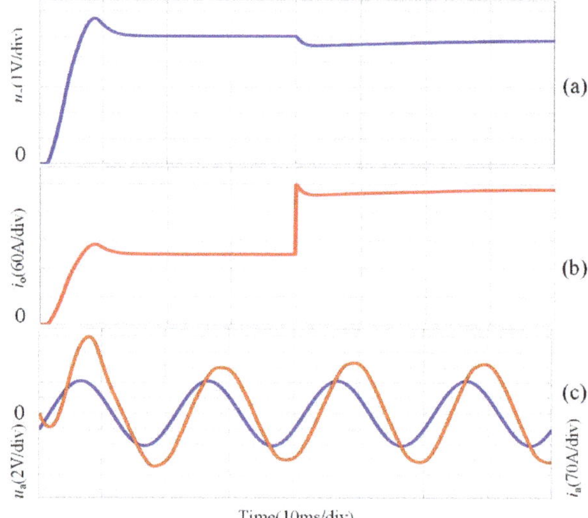

Figure 10. Simulation waves of the disturbed direct power control: (**a**) Voltage wave; (**b**) current wave; (**c**) A-phase voltage and current waves.

3.3. The Control Based on the No Beat Control Algorithm

Figure 11 is the waveform with no beat control. It can be seen that the voltage overshoot is 20%, the system reaches the steady state in 0.03 s when the load resistance suddenly changes from 32 Ω to 16 Ω at 0.06 s, the voltage change is about 0.8 V, and the steady state is restored after 0.03 s. The output current is doubled to 300 A, and the phase A voltage and current achieve a unit power factor after a short period of fluctuation.

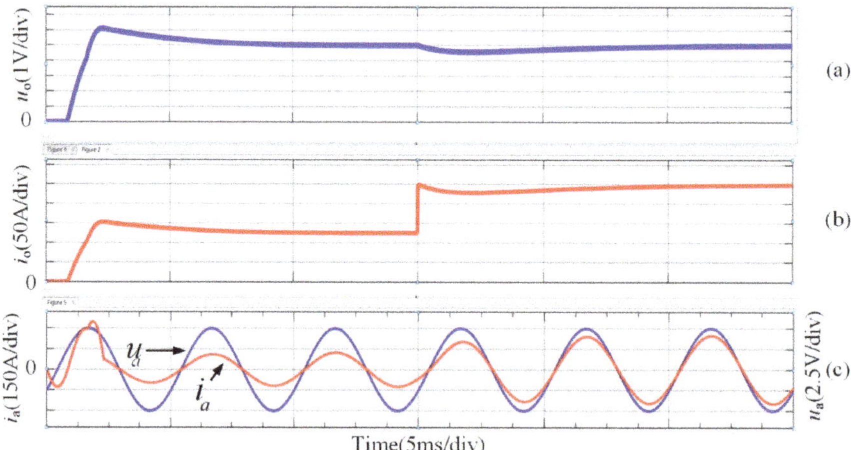

Figure 11. Simulation waves of the no beat control algorithm: (**a**) Voltage wave; (**b**) current wave; (**c**) A-phase voltage and current waves.

Figure 12 shows output current waveforms of the three-module system in parallel mode. Each rectifier module is controlled by the autonomous current sharing. In all of the parallel modules, the module with the highest output current automatically becomes the main module through the current sampling and the current sharing control bus sends reference current information to other modules to achieve the current sharing. The simulation results indicate that after the current sharing control, each module can generate a stable current of 300 A.

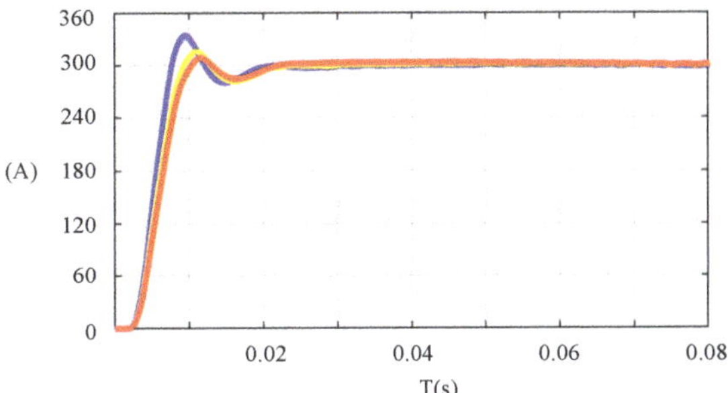

Figure 12. Output current waveform of the three modules in parallel mode.

It is concluded from the abovementioned comparisons that the voltage overshoot is 10%, 14%, and 20% when the system is controlled by the Lyapunov algorithm, direct power algorithm, and dead-beat algorithm, respectively. Moreover, when the load varies, the Lyapunov algorithm restores the steady state after 0.02 s adjustment, while the output voltage is 5 V. On the other hand, the direct power algorithm reaches a new steady state after 0.03 s, while the output voltage is 4.8 V. In other words, the steady state voltage of the Lyapunov algorithm is slightly higher than that of the direct power algorithm. Meanwhile, the voltage drops and produces 4% error. When the dead-beat algorithm is used as control, the load changes and the system reaches a new steady state after 0.03 s. Here, the restoring time of steady state is longer than that of Lyapunov algorithm. On the AC side, the Lyapunov control algorithm can realize the unit power factor and the overshoot of the current before and after the load mutation is low. However, when the direct power control is adopted, the phase difference of the single-phase voltage and current exists all the time. The unit power factor cannot be realized by the direct power control algorithm because it needs to estimate the instantaneous reactive power and then estimate the voltage vector of the AC side. Since the AC current is large, when the current transient tracking index is satisfied, the AC side inductance is limited. The inductance affects the accuracy of the estimation of instantaneous reactive power, which leads to the phase difference between the voltage and current on the AC side. Thus, the unit power factor cannot be realized.

3.4. Verification of the Experimental Result

The main electrical parameters of the power circuit and control data, used in the implementation tests, are given in Table 3. The development of control algorithms was performed and simulated with Matlab/Simulink and the real-time implementation with a Texas Instrument digital signal processor (DSP) board (TMS320F28335). Each three-phase winding was used as an independent module. Each module, using a 32-bit DSP28335 processor, operated at a frequency of 150 MHz, with a bandwidth of 600 Mbps and a single precision floating-point-unit (FPU). The current sharing control was realized by CAN bus among the controllers, which improved the independence and reduced the interference of the modules, so the traditional fixed-point MCU would affect the result of algorithm processing. The control strategy can be used in the mainstream 32-bit floating-point microprocessor and 10-bit or more AD sampling. Figure 13 shows the experimental platform, which is the setup for experimental verification. This setup was used to verify the validity of the proposed Lyapunov control algorithm. When carrying out a load test, the setup can adjust the resistor box terminal to provide 10–1000 mΩ.

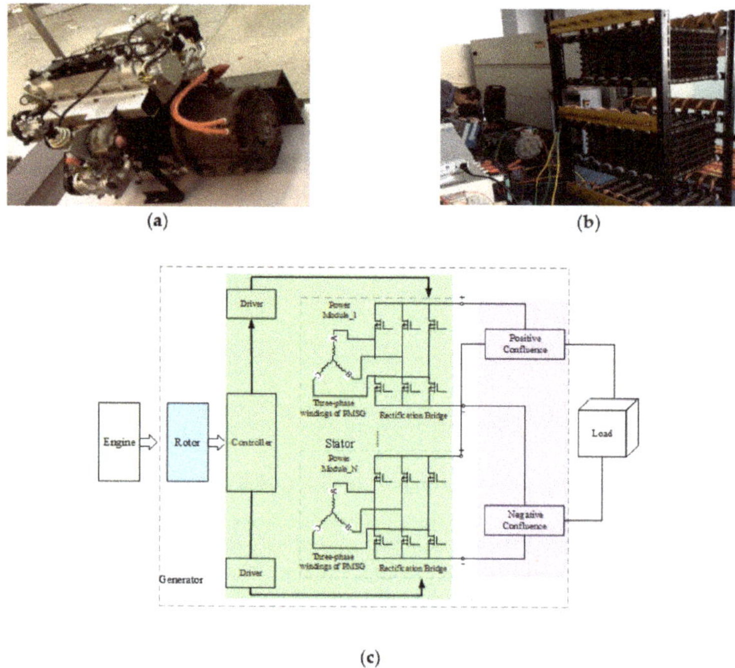

Figure 13. Large current load experiment of the integrated DC output system: (**a**) Experimental platform; (**b**) load experiment; (**c**) the schematic of the synchronous system.

Figure 14 shows the experiment waves of the output current and the three-phase AC current waveforms when a single rectifier module is in operation. When the load suddenly changed from 32 to 16 Ω at 0.04 s, as can be seen, the system reached the steady state in less than 0.02 s, their steady-state performances were both quite good. The proposed control can highly improve the dynamic performance of the DC output current with a shorter transition time. It is consistent with the simulation results.

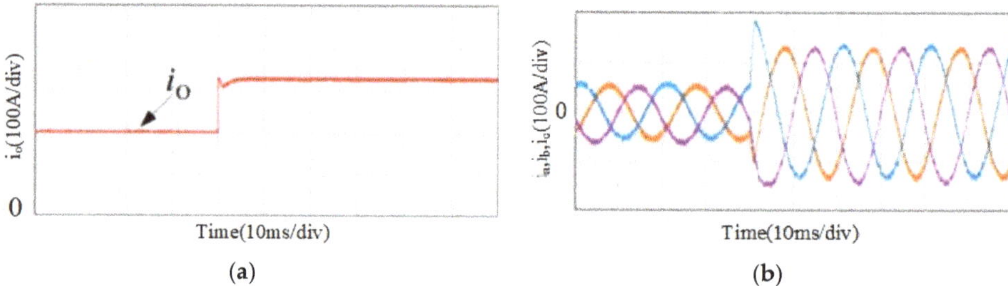

Figure 14. The experiment waves of a single rectifier module: (**a**) Output current wave; (**b**) three-phase AC current waveform.

Figure 15 shows the experiment waves of the output voltage and the corresponding frequency spectrum when a single rectifier module is in operation. The logarithmic uniform distribution is used to facilitate the comparison of the longitudinal axis. The diagram shows that the output voltage of the rectifier module is stable. The amplitude of the voltage is 5 V when the main frequency is 0 Hz, and the maximum value of the ripple of each frequency does not exceed 0.06 V, which meets the requirements of the DC output.

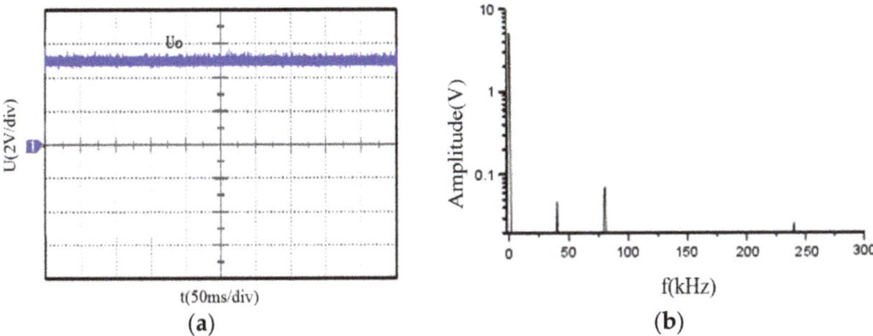

Figure 15. The experiment voltage waves of the DC power generation: (**a**) Output voltage wave; (**b**) output voltage spectrum.

Figure 16 shows the experiment waves of the output current and the corresponding spectrum when a single rectifier module is in operation. The vertical axis is logarithmically and uniformly distributed. Moreover, the unit ratio is 100 A. When the main wave of the output current is stable, the main frequency is still 0 Hz, which conforms to DC characteristics. Although there is a certain amplitude in other frequency bands, none of them exceeds 3 A.

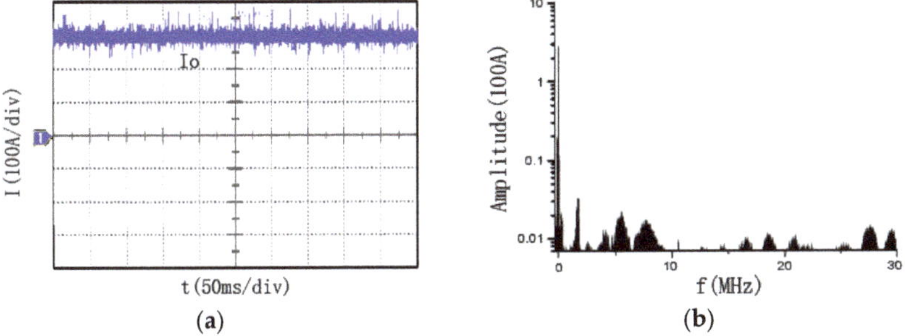

Figure 16. The experiment current waves of the DC power generation: (**a**) Output current wave; (**b**) output current spectrum.

Figure 17 shows the waveforms' output voltage when the system load increases and decreases suddenly. The diagram illustrates that when the load changes, the output external characteristics of the rectifier module of the generation system changes, under the condition of the same duty cycle. This may be attributed to the internal resistance inside the device. The output voltage in the diagram can be quickly restored to a stable state without overshooting, which proves that the dynamic control performance of the Lyapunov algorithm is superior over the conventional methods.

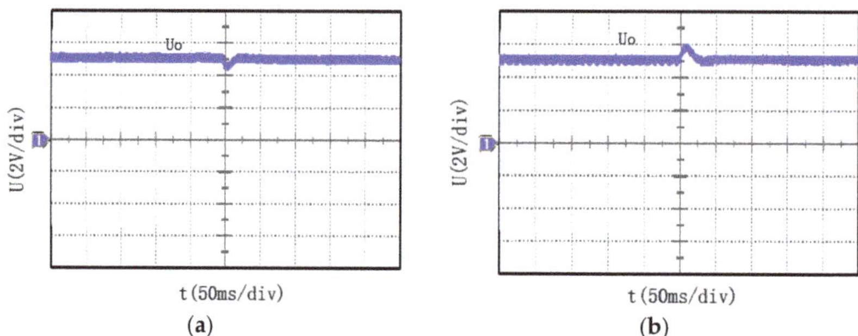

t(50ms/div)

(a)   t(50ms/div)

(b)

Figure 17. Output voltage waves of resistance mutation: (**a**) Load increases; (**b**) load decreases.

Figures 18 and 19 present the current waveforms of the paralleling system when the light load switches to the heavy load, and vice versa. They indicate that the current variation and adjustment times are different between module 1 and 2. This originates from the two modules within the inherent relationship between different parameters and their corresponding master–slave structure, in which the regulating speed of the submodule is slower than that of the main module. Meanwhile, the control process contains the vibration. However, the two modules can quickly achieve stable output and current sharing, which proves the better dynamic response of digital current sharing, compared to the conventional methods.

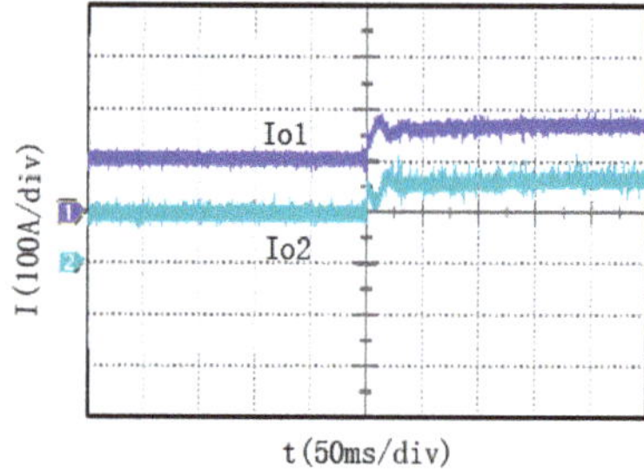

t(50ms/div)

Figure 18. Switching the light load to the heavy load.

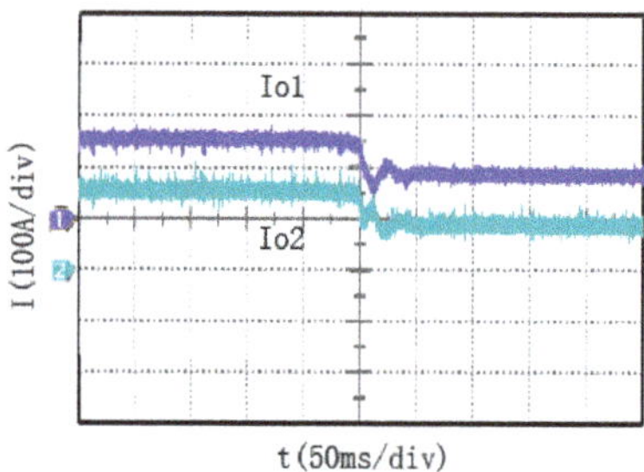

t(50ms/div)

Figure 19. Switching the heavy load to the light load.

4. Conclusions

Based on the current situation that a low-voltage high-current generation system focuses mainly on rectifying the voltage of the power grid, the present study proposes a synchronous generator system with a multiple three-phase permanent magnet, which is controlled by the Lyapunov algorithm. Through analysis, there is a problem of mutual coupling between the control variables. To solve this problem, a simple decoupling method was adopted. By decoupling, the Lyapunov derivative of the system is always negative, thus ensuring the stability of the system. Since the direct power control method and dead-beat algorithm are widely used, in this paper, these two methods were used to compare the performance. The simulation results show that when the Lyapunov algorithm is utilized for the synchronous generator system with a multiple three-phase permanent magnet, the system has low voltage overshoot, and reasonable dynamic and static performance. It is found that system stability is not disturbed by large signals and has reasonable performance for the unit power factor. The experimental results are consistent with the simulation, which proves that the control strategy is an effective and reliable control scheme.

Author Contributions: Conceptualization, J.L. and X.T.; methodology, J.L. and X.T.; software, X.T.; validation, J.L. and X.T.; formal analysis, J.L. and X.T.; investigation, J.L and X.W.; writing—original draft preparation, J.L. and X.T.; writing—review and editing, J.L and X.T.; visualization, H.H.-C.I.; supervision, H.H.-C.I.; project administration, X.W.; funding acquisition, X.W. and J.L.

Funding: This research was funded by National Natural Science Foundation of China, grant number 51177031.

Conflicts of Interest: The authors declare no conflict of interest.

References

1. Kassakian, J.G.; Jahns, T.M. Evolving and Emerging Applications of Power Electronics in Systems. *IEEE J. Emerg. Sel. Top. Power Electron.* **2013**, *1*, 47–58. [CrossRef]
2. Mohseni, M.; Islam, S.M. A New Vector-Based Hysteresis Current Control Scheme for Three-Phase PWM Voltage-Source Inverters. *IEEE Trans. Power Electron.* **2010**, *25*, 2299–2309. [CrossRef]
3. Du, J.; Zhao, H.; Zheng, T.Q. Study on a switching-pattern hysteresis current control method based on zero voltage vector used in three-phase PWM converter. *IET Power Electron.* **2017**, *10*, 2190–2198. [CrossRef]
4. Buso, S.; Tommaso, C.G.; Danilo, I.B. Dead-Beat Current Controller for Voltage-Source Converters with Improved Large-Signal Response. *IEEE Trans. Ind. Appl.* **2016**, *52*, 1588–1596. [CrossRef]
5. Cheng, C.W.; Nian, H.; Wang, X.H.; Sun, D. Dead-beat predictive direct power control of voltage source inverters with optimised switching patterns. *IET Power Electron.* **2017**, *10*, 1438–1451. [CrossRef]
6. Tao, C.-W.; Wang, C.-M.; Chang, C.-W. A Design of a DC–AC Inverter Using a Modified ZVS-PWM Auxiliary Commutation Pole and a DSP-Based PID-Like Fuzzy Control. *IEEE Trans. Ind. Electron.* **2016**, *63*, 397–405. [CrossRef]
7. Cecati, C.; Dell'Aquila, A.; Liserre, M.; Ometto, A. A fuzzy-logic-based controller for active rectifier. *IEEE Trans. Ind. Appl.* **2003**, *39*, 105–112. [CrossRef]
8. Ge, J.; Zhao, Z.; Yuan, L.; Lü, T.; He, F. Direct Power Control Based on Natural Switching Surface for Three-Phase PWM Rectifiers. *IEEE Trans. Power Electron.* **2015**, *30*, 2918–2922. [CrossRef]
9. Bouafia, A.; Krim, F.; Gaubert, J.-P. Fuzzy-Logic-Based Switching State Selection for Direct Power Control of Three-Phase PWM Rectifier. *IEEE Trans. Ind. Electron.* **2009**, *56*, 1984–1992. [CrossRef]
10. Malinowski, M.; Kaźmierkowski, M.; Hansen, S.; Blaabjerg, F.; Marques, G. Virtual-flux-based direct power control of three-phase PWM rectifiers. *IEEE Trans. Ind. Appl.* **2001**, *37*, 1019–1027. [CrossRef]
11. Li, H.; Wang, Y. Lyapunov-Based Stability and Construction of Lyapunov Functions for Boolean Networks. *SIAM J. Control Optim.* **2017**, *55*, 3437–3457. [CrossRef]
12. Hasan, K.; Osman, K. Lyapunov based control for three-phase PWM AC/DC voltage source converters. *IEEE Trans. Power Electron.* **1998**, *13*, 801–813.
13. Hadi, H.K.; Mahdi, B.; Ali, A.K. Direct-Lyapunov-based control scheme for voltage regulation in a three-phase islanded microgrid with renewable energy sources. *Energies* **2018**, *11*, 1161.
14. Arkadiusz, L.; Zbigniew, K.; Haitham, A.R. Space-vector pulse width modulation for three-level NPC converter with the neutral point voltage control. *IEEE Trans. Ind. Electron.* **2011**, *58*, 5076–5086.
15. Gu, Y.L.; Huang, G.S.; Zhang, J.F. A synchronous rectifier driving circuit for modules in parallel. *Proc. CSEE* **2005**, *4*, 25–29.
16. Mohamed, A. Harmonic elimination in three-phase voltage source inverters by particle swarm optimization. *JEET* **2011**, *6*, 334–341.
17. Lin, B.-R. Analysis of a DC Converter with Low Primary Current Loss and Balance Voltage and Current. *Electronics* **2019**, *8*, 439. [CrossRef]

18. Song, Z.F.; Tian, Y.J.; Yan, Z.; Chen, Z. Direct Power Control for Three-Phase Two-Level Voltage-Source Rectifiers Based on Extended-State Observation. *IEEE Trans. Ind. Electron.* **2016**, *63*, 4593–4603. [CrossRef]

19. Hang, L.; Liu, S.; Yan, G.; Qu, B.; Lu, Z.-Y. An Improved Deadbeat Scheme with Fuzzy Controller for the Grid-side Three-Phase PWM Boost Rectifier. *IEEE Trans. Power Electron.* **2011**, *26*, 1184–1191. [CrossRef]

20. Cheng, Q.M.; Cheng, Y.M.; Xue, Y. A summary of current control methods for three-phase voltage-source PWM rectifiers. *Power Syst. Prot. Control* **2012**, *40*, 145–155.

21. Qiu, X.-D.; Jung, Y.-G.; Lim, Y.-C. Three-Phase Z-Source PWM Rectifier Based on the DC Voltage Fuzzy Control. *Trans. Korean Inst. Power Electron.* **2013**, *18*, 466–476. [CrossRef]

22. Kwak, S.; Yoo, S.-J.; Park, J. Finite control set predictive control based on Lyapunov function for three-phase voltage source inverters. *IET Power Electron.* **2014**, *7*, 2726–2732. [CrossRef]

23. Abdul, M.D.; Tey, K.S.; Saad, M.; Mutsuo, N. Lyapunov law based on model predictive control scheme for grid connected three phase three level neutral point clamped inverter. In Proceedings of the 3rd International Future Energy Electronics Conference and ECCE Asia, Kaohsiung, Taiwan, 4–7 June 2017; pp. 512–516.

24. Feng, Y.L.; Zhang, C.N. Core loss analysis of interior permanent magnet synchronous machines under SVPWM excitation with considering saturation. *Energies* **2017**, *10*, 1716. [CrossRef]

25. Vazquez, S.; Sanchez, J.; Carrasco, J.; Leon, J.I.; Galvan, E. A Model-Based Direct Power Control for Three-Phase Power Converters. *IEEE Trans. Ind. Electron.* **2008**, *55*, 1647–1657. [CrossRef]

26. Wang, J.H.; Li, H.D.; Wang, L.M. Direct power control system of three phase boost type PWM rectifiers. *Proc. CSEE* **2006**, *26*, 54–60.

 © 2019 by the authors. Licensee MDPI, Basel, Switzerland. This article is an open access article distributed under the terms and conditions of the Creative Commons Attribution (CC BY) license (http://creativecommons.org/licenses/by/4.0/).

 *electronics*

Article

Improvement of Stability in a PCM-Controlled Boost Converter with the Target Period Orbit-Tracking Method

Yanfeng Chen [1], Fan Xie [1,*], Bo Zhang [1], Dongyuan Qiu [1], Xi Chen [1], Zi Li [1] and Guidong Zhang [2]

[1] School of Electric Power, South China University of Technology, Guangzhou 510641, China; eeyfchen@scut.edu.cn (Y.C.); epbzhang@scut.edu.cn (B.Z.); epdyqiu@scut.edu.cn (D.Q.); xichen_1021@hotmail.com (X.C.); zili100097@gmal.com (Z.L.)
[2] School of Automation, Guangdong University of Technology, Guangzhou 510640, China; guidong.zhang@gdut.edu.cn
*   Correspondence: epfxie@scut.edu.cn

Received: 24 October 2019; Accepted: 27 November 2019; Published: 30 November 2019

Abstract: Thee peak-current-mode (PCM) control strategy is widely adopted in pulse width-modulated (PWM) DC-DC converters. However, the converters always involve a sub-harmonic oscillating state or chaotic state if the active duty ratio is beyond a certain range. Hence, an extra slope signal in the inductor-current loop is used to stabilize the operation of the converter. This paper presents a new technique for enlarging the stable range of PCM-controlled DC-DC converters, in which the concept of utilizing unstable period-1 orbit (UPO-1) of DC-DC converters is proposed and an implementation scenario based on the parameter-perturbation method is presented. With the proposed technique, perturbations are introduced to the reference current of the control loop, and the converters operating in a chaotic state can be tracked, and thus be stabilized to the target UPO-1. Therefore, the stable operating range of the converters is extended. Based on an example of a PCM-controlled boost converter, simulations are presented as a guide to a detailed implementation process of the proposed technique, and comparisons between the proposed technique and techniques in terms of ramp compensation are provided to show the differentiation in the performance of the converter. Experimental results in the work confirm the effectiveness of the proposed technique.

Keywords: peak-current-mode (PCM) control; boost converter; stability; parameter perturbation; target period orbit tracking

---

1. Introduction

Pulse width-modulated (PWM) DC-DC converters are typical piecewise-smooth dynamical systems [1], and an external control loop is required for them to produce a precise and stable output voltage. Because of some desirable features, such as fast dynamic response, automatic overload protection, good current sharing, current limiting, and so on [2], the peak-current-mode (PCM) control has been widely applied [3].

However, it is commonly accepted that converters with PCM control are confronted with instability issues [4,5]. A wide variety of nonlinear dynamic phenomena, such as bifurcations and chaos, have been observed in these converters [6–13], which could deteriorate the performances of the converters, and are undesired in practice. As a result, the active duty ratio of PCM control in DC-DC converters is usually restricted in the range of (0, 1/2) in continuous-conduction-mode (CCM) and (0, 2/3) in discontinuous-conduction-mode (DCM) [14,15]. Hence, these limitations in turn lead to a restrained application of DC-DC converters. In renewable energy grid-connected power systems, for example, DC-DC converters with high step-up gain are needed, yet the classical boost topology cannot service the request in these scenarios [16]. An approach to overcoming this drawback is to build new topologies [17–21]. Unfortunately, some unavoidable features of these new proposed topologies, especially a complicated structure, have brought difficulties in design and burdened costs in scalable applications.

Building new topologies is not the only solution that can be found; in fact, improving control strategies is another option. For example, hybrid predictive control is designed to improve the features of a boost converter operating in both CCM and DCM operation [22,23]. Additionally, slope compensation is the conventional strategy preferred by engineers [24]. According to analysis works on PCM-controlled DC-DC converters with CCM operation, when the active duty ratio is beyond the range of (0, 1/2), the quality factor, Q, of the control-to-inductor transfer function is negative [24,25]. It means that the eigenvalues of the converters distribute on the right half of the S plane, and the converters are in an unstable state. However, by adding a sufficient compensation ramp signal to the current loop, the factor can turn into a positive value, which causes the converters to return to a stable state.

Unfortunately, for the conventional ramp compensation technique, there is an over-compensation (low Q) or under-compensation (high Q) problem that would limit the control bandwidth. Therefore, a dynamic ramp compensation scheme was put forward in [26]. Another drawback associated with the conventional ramp compensation technique is that the use of constant slope compensation reduces some of the benefits of PCM control. For example, the peak value of the inductor current deviates from the desired reference, which is not desirable in applications where accurate tracking of the reference signal is needed [27]. Hence, a self-compensation technique was proposed in [27], which can provide a more accurate current limiting capacity and does not require an external slope generator. In [28], a time-varying ramp compensation technique was presented to eliminate the fast-scale instability of a PCM-controlled power factor correction (PFC) boost converter. In [29], adaptive ramp compensation was proposed to improve the robustness of the conventional ramp compensation technique. Yet, the inherent over-compensation or under-compensation problem has still not been solved well.

In this paper, a new technique based on the parameter-perturbation method [30] is proposed, which can make PCM control work in a wider duty ratio range without using ramp compensation. The parameter-perturbation method is a recently proposed chaos control strategy, which aims to stabilize a chaotic system to a desired unstable period orbit (UPO). This method applies only very small perturbations to a carefully chosen system parameter once per control period. Compared with some classical chaos control methods [31], the parameter-perturbation method does not require the unstable fixed point to be a saddle node, and thus there is no limitation on the types of targeting UPOs. Hence, this method is applied to enhance the performance of a PCM-controlled DC-DC converter. In some works on nonlinear dynamic analysis of PCM-controlled DC-DC converters [8,10,13], the duty cycle of PCM control can be simply approximated by a function of the reference current. Therefore, the reference current of PCM control is used as a parameter to be perturbed in this paper. Then, the UPO-1 of the converters can be calculated by the real-time sampling data of inductor current and capacitor voltage, and the stabilizer can be designed.

The rest of the paper begins with a quick glimpse of the typical limitations of a boost converter with classical PCM control in Section 2. The principles of the parameter-perturbation method and the detailed procedures of implementing this method are presented in Section 3. Section 4 analyzes the performances of the boost converter with the proposed control scheme. In Section 5, experiments are performed for a further verification of the proposed scheme. Finally, the conclusion is outlined in Section 6.

2. Typical Limitation of Boost Converter with Classical PCM Control Scheme

The circuit diagram of the boost converter is shown in Figure 1a, where r_L, r_C, and r_T denote the parasitic resistances of the inductor, L, the capacitor, C, and the switch, S_T, respectively. The control loop consists of a comparator, and an R-S flip-flop. The operation can be briefly described as follows. The flip-flop is set periodically by the clock signal, turning on the switch, S_T. Then, the inductor current, i_L, goes up linearly, and is compared with the reference level, I_{ref}. When the peak value of i_L reaches the level, I_{ref}, the output of the comparator resets the flip-flop, thereby turning off S_T. When S_T is off, the inductor current falls almost linearly.

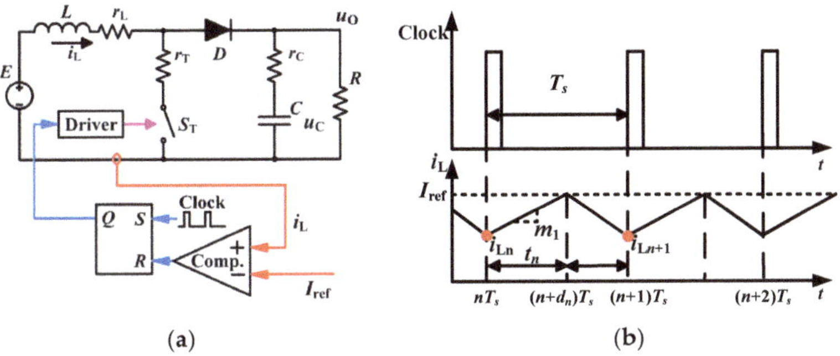

(a)   (b)

Figure 1. Boost converter with a classical PCM control scheme: (**a**) Basic circuit diagram; (**b**) typical steady-state waveforms.

Assuming that the converter operates in CCM, there are two operating modes depending on whether or not S_T is on, and typical waveforms of i_L and the clock signal are shown in Figure 1b. The PCM-controlled boost converter can be described by:

$$\begin{cases} \dot{x} = \mathbf{A}_1 x + \mathbf{B}_1 E, & nT_s \leq t < (n+d_n)T_s \\ \dot{x} = \mathbf{A}_2 x + \mathbf{B}_2 E, & (n+d_n)T_s \leq t < (n+1)T_s \end{cases} \tag{1}$$

where E, T_s, and d_n denote the input voltage, switching period, and duty ratio in the nth cycle, respectively. The state vector was set to be $x = [i_L, u_O]^{\text{Tr}}$, where the superscript 'Tr' means the transposition of a matrix. The system matrices, $\mathbf{A}_i$ ($i = 1, 2$) and $\mathbf{B}_i$ ($i = 1, 2$), are given by:

$$\mathbf{A}_1 = \begin{bmatrix} -\frac{r_T+r_L}{L} & 0 \\ 0 & -\frac{1}{C(R+r_C)} \end{bmatrix}, \quad \mathbf{A}_2 = \begin{bmatrix} -\frac{r_L}{L} & -\frac{1}{L} \\ \frac{R(L-r_Cr_LC)}{C(R+r_C)} & -\frac{L+RCr_C}{LC(R+r_C)} \end{bmatrix}, \tag{2}$$

and:

$$\mathbf{B}_1 = \begin{bmatrix} \frac{1}{L} \\ 0 \end{bmatrix}, \quad \mathbf{B}_2 = \begin{bmatrix} \frac{1}{L} \\ \frac{Rr_C}{L(R+r_C)} \end{bmatrix}. \tag{3}$$

A stroboscopic map is a discrete map or iterative map obtained by sampling a continuous system periodically, which describes the dynamics of a discrete variable in terms of a difference equation. According to (1), the stroboscopic map can be obtained as:

$$\begin{aligned} x_{n+1} &= \mathbf{F}(x_n, I_{\text{ref}}) \\ &= \mathbf{\Phi}_2((1-d_n)T_s)[\mathbf{\Phi}_1(d_nT_s)x_n + \mathbf{\Psi}_1(d_nT_s)E] + \mathbf{\Psi}_2((1-d_n)T_s)E, \end{aligned} \tag{4}$$

in which x_n and x_{n+1} denote the state vector at the instant of $t = nT_s$ and $t = (n+1)T_s$, respectively. $\mathbf{\Phi}_i(\xi)$ and $\mathbf{\Psi}_i(\xi)$ are calculated by the following equations as:

$$\mathbf{\Phi}_i(\xi) = e^{\mathbf{A}_i(\xi)} = \mathbf{I} + \sum_{k=1}^{\infty} \frac{1}{k!}\mathbf{A}_i^k\xi^k, \quad \mathbf{\Psi}_i(\xi) = \int_0^{\xi} \mathbf{\Phi}_i(\tau)\mathbf{B}_i d\tau, \tag{5}$$

where 'I' is a symbol of the unit matrix, and the subscript $i = 1, 2$. The switching function is defined as:

$$s(x_n, d_n) = d_n - (I_{\text{ref}} - i_{Ln})/m_1T_s, \tag{6}$$

in which $m_1 = E/L$ is the rising slope of i_L, and d_n can be determined by setting the switching function to be zero, i.e., $s(\mathbf{x}_n, d_n) = 0$.

Parameters of the converter are listed in Table 1, which were chosen to be the same as those in [32]. When I_{ref} is changed from 0.5 to 5.5 A, typical bifurcation diagrams of i_L and u_C versus I_{ref} can be obtained by numerical simulation. Bifurcation is the sudden change of the qualitative behavior of a system when one or more parameters are varied. Bifurcation literally means splitting into two parts. In nonlinear dynamics, the term has been used to mean the splitting of the behavior of a system at a threshold parameter value into two qualitatively different behaviors, corresponding to parameter values below and above the threshold. So, a bifurcation diagram is a summary chart of the behavioral changes as some selected parameters are varied. DC-DC converters are typically nonlinear, and the bifurcation diagram has become a very common tool for analysis of the dynamic behaviors of converters. It can be seen in Figure 2 that the period doubling bifurcation occurs at $I_{ref} = 1.52$ A with $d = 0.39$ and the voltage set-up ratio $M = 1.64$. Along with the increase of I_{ref}, the converter undergoes several period doubling bifurcations, and evolves into a chaotic state eventually. Obviously, when the system operates in a chaotic state, there are usually different kinds of values of u_C corresponding to the same I_{ref}. Then, it is desirable to stabilize the system from a chaotic state to the target period-1 orbit with larger u_C, by some kind of chaotic control method, such as the method based on parameter perturbation.

Table 1. Circuit parameters of the PCM-controlled boost converter.

Parameters	Values	Units
Input voltage E	10	V
Switching frequency f	10	kHz
Inductance L	1	mH
Capacitance C	10	μF
Load resistance R	20	Ω
On resistance r_T	50	mΩ
Parasitic resistance r_C	30	mΩ
Parasitic resistance r_L	40	mΩ
Reference current I_{ref}	0.5 to 5.5	A

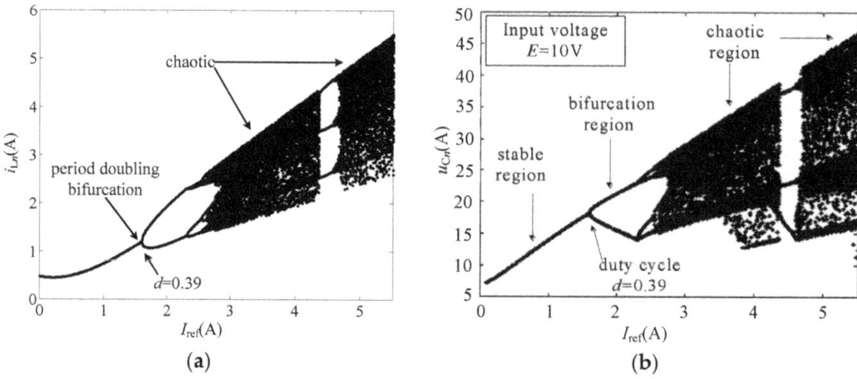

Figure 2. Bifurcation diagram of a classical PCM-controlled boost converter: (**a**) For the inductor current i_L versus I_{ref}; (**b**) For the output capacitor voltage u_C versus I_{ref}.

3. An Improving Scheme Based on the Target UPO Tracking Method

3.1. Principle of the Parameter-Perturbation Method

The parameter-perturbation method is designed to stabilize a discrete chaotic system on a desired unstable period-1 orbit (UPO-1) [30]. For a two-dimensional discrete system, it can be described by the following equation as:

$$x_{n+1} = \mathbf{F}(x_n, \, p_n), \tag{7}$$

where $\mathbf{F}(\cdot)$ is a smooth vector function, $x \in R^2$ is state vector of the system, P is a parameter that can be changed in a neighborhood of a nominal value of P [30], and the subscript n denotes the nth iteration. Assuming that the system is in chaotic state, and has an unstable fixed point, $\mathbf{X}_P$, the following equation can be obtained:

$$\mathbf{X}_P = \mathbf{F}(\mathbf{X}_P, \, P). \tag{8}$$

Then, in a sufficiently small neighborhood of $\mathbf{X}_P$, the system (Equation (7)) can be approximated by a linear map as:

$$x_{n+1} = \mathbf{J}_x(x_n - \mathbf{X}_P) + \mathbf{J}_p(p_n - P) + \mathbf{X}_P, \tag{9}$$

where the coefficient matrices, $\mathbf{J}_x$ and $\mathbf{J}_p$, are defined as:

$$\mathbf{J}_x = \left. \frac{\partial \mathbf{F}(x_n, \, p_n)}{\partial x} \right|_{(\mathbf{X}_p, \, P)}, \, \mathbf{J}_p = \left. \frac{\partial \mathbf{F}(x_n, \, p_n)}{\partial p} \right|_{(\mathbf{X}_p, \, P)}. \tag{10}$$

According to Equation (9), two steps of perturbations are needed for the selected parameter, the iterative function for x_{n+2} can be obtained, that is:

$$x_{n+2} = \mathbf{J}_x^2(x_n - \mathbf{X}_p) + \begin{bmatrix} \mathbf{J}_x\mathbf{J}_p & \mathbf{J}_p \end{bmatrix} \begin{bmatrix} p_n - P \\ p_{n+1} - P \end{bmatrix} + \mathbf{X}_p, \tag{11}$$

and it is the expression for the UPO-1 that corresponds to $\mathbf{X}_P$.

When the system is stabilized to the desired UPO-1, then $x_{n+2} = x_{n+1} = \mathbf{X}_P$, and the following perturbation increments for P_n can be acquired by:

$$\begin{bmatrix} \Delta p_n \\ \Delta p_{n+1} \end{bmatrix} = \begin{bmatrix} p_n - P \\ p_{n+1} - P \end{bmatrix} = \begin{bmatrix} \mathbf{J}_x\mathbf{J}_p & \mathbf{J}_p \end{bmatrix}^{-1} \mathbf{J}_x^2(\mathbf{X}_p - x_n) = \mathbf{M}(\mathbf{X}_p - x_n). \tag{12}$$

3.2. Implementation Scenario

This subsection provides a scenario to illustrate the common functionality that can be implemented in a PCM-controlled boost converter. Based on dynamic analysis of the boost converter with a classical PCM control scheme, one ensures that the converter operates in a chaotic state when the reference level is set to be $I_{\mathrm{ref}} = 3$ A, and the other parameters are shown in Table 1. According to the basic principle of the parameter-perturbation method above, let $x_{n+1} = x_n$ in Equation (4), one can obtain the unstable fixed point as:

$$\mathbf{X}_p = \begin{bmatrix} I_{\mathrm{L}p}, & U_{\mathrm{O}p} \end{bmatrix}^{\mathrm{Tr}} = [2.4344, \, 26.2895]^{\mathrm{Tr}}. \tag{13}$$

The perturbing values for the reference current can be calculated according to Equation (12), and the reference current becomes:

$$I_{\mathrm{refn}}^{(p)} = \Delta I_{\mathrm{refn}} + I_{\mathrm{ref}}. \tag{14}$$

By using the above computed perturbations, the UPO-1 of the converter can be found, which will be in the form of Equation (11). The schematic diagram of the stabilizer can be designed. As shown in Figure 3, in addition to the conventional peak current mode control scheme, a perturbation module is added to generate the disturbance increments for the controlled parameter. Additionally, the reference level in each period is the sum of the nominal value, I_{ref}, and the disturbance increments, ΔI_{refn}, which is calculated by the perturbation module at the beginning of the nth period according to Equation (14). The additional switch, S, in the control loop is used to enable and disable this target orbit control scheme.

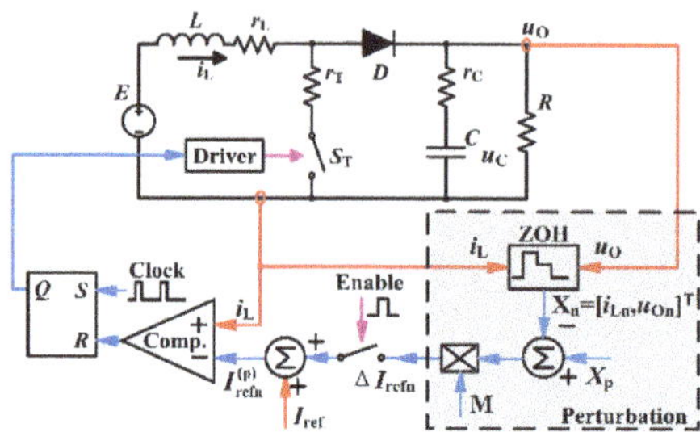

Figure 3. Circuit diagram of a PCM-controlled boost converter with a parameter-perturbation module.

Figure 4a,b respectively show the simulated time-domain waveforms and the phase trajectories of the state variables at $I_{ref} = 3$ A. Obviously, it can be seen from Figure 4a that before $t = 0.02$ s, the conventional PCM-controlled boost converter operates in a chaotic state, with larger ripples accompanying the inductor current and output voltage. When the switch, S, is enabled at the instant of $t = 0.02$ s, the proposed control scheme comes into force. Additionally, the system can be stabilized quickly from the chaotic state to the desired target period-1 orbit (UPO-1), as shown in Figure 4b. Moreover, the peak-peak values of the inductor current and output voltage are reduced to 0.57 A and 5.5 V from 1.45 A and 15 V, respectively.

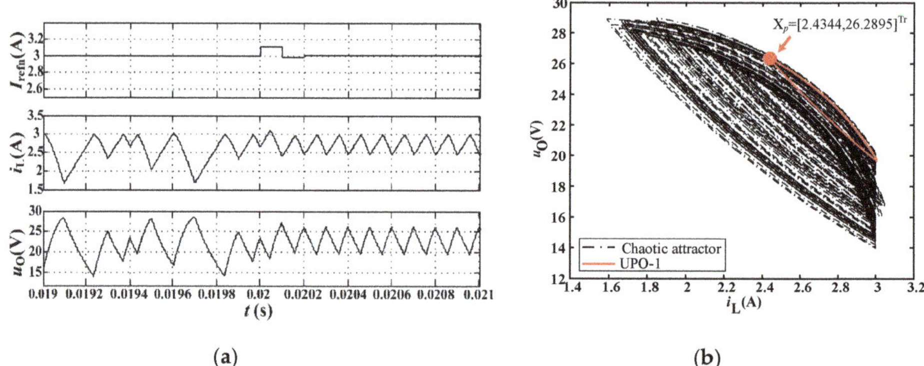

(a)

(b)

Figure 4. Simulations of the boost converter with the proposed control scheme: (**a**) Time-domain waveforms of state variables; (**b**) chaotic attractor and the stabilized UPO-1.

3.3. Performance Assessment

The performances of the converter can be evaluated by checking the movement of the eigenvalues when some chosen circuit parameters are varied. Any crossing from the interior of the unit circle to the exterior indicates a bifurcation. Particularly, if a real eigenvalue goes through -1 as it moves out of the unit circle, a period-doubling occurs.

According to the stroboscopic map of the boost converter, the Jacobian matrix evaluated in the neighborhood of $\mathbf{X}_p$ (the equilibrium point) is defined as:

$$\mathbf{J}(\mathbf{X}_p) = \frac{\partial \mathbf{F}}{\partial \mathbf{x}_n} - \frac{\partial \mathbf{F}}{\partial d_n}\left(\frac{\partial s}{\partial d_n}\right)^{-1}\frac{\partial s}{\partial d_n}\Bigg|_{\mathbf{x}_n=\mathbf{X}_p}, \tag{15}$$

one can get the following derivatives by using Equations (4) and (6):

$$\frac{\partial \mathbf{F}}{\partial \mathbf{x}_n} = \boldsymbol{\Phi}_2((1-d_n)T_s)\boldsymbol{\Phi}_1(d_nT_s), \tag{16}$$

$$\begin{aligned}\frac{\partial \mathbf{F}}{\partial d_n} &= T_s\boldsymbol{\Phi}_2((1-d_n)T_s)(\mathbf{A}_1 - \mathbf{A}_2)\boldsymbol{\Phi}_1(d_nT_s)\mathbf{x}_n \\ &\quad -T_s\mathbf{A}_2\boldsymbol{\Phi}_2((1-d_n)T_s)\mathbf{A}_2^{-1}\mathbf{B}_2E + \boldsymbol{\Phi}_2((1-d_n)T_s)\mathbf{A}_1^{-1}\mathbf{B}_1, \\ &\quad \cdot E[-T_s\mathbf{A}_2\boldsymbol{\Phi}_2((1-d_n)T_s) + T_s\mathbf{A}_1\boldsymbol{\Phi}_1(d_nT_s) + T_s\mathbf{A}_2]\end{aligned} \tag{17}$$

$$\frac{\partial s}{\partial d_n} = \frac{(r_\mathbf{L} + r_\mathbf{T})T_s}{L}\left(i_\mathbf{L} - \frac{E}{r_\mathbf{L} + r_\mathbf{T}}\right)e^{-\frac{r_\mathbf{L}+r_\mathbf{T}}{L}d_nT_s}, \tag{18}$$

$$\frac{\partial s}{\partial d_n} = \left[0.5042 - e^{-\frac{r_\mathbf{L}+r_\mathbf{T}}{L}d_nT_s}, \; -0.0002\right]. \tag{19}$$

Additionally, by introducing the above derivatives into Equation (15), it leads to the following characteristic equation as:

$$det\left(\lambda\mathbf{I} - J(\mathbf{X}_p)\right) = 0. \tag{20}$$

Then, eigenvalues can be obtained by solving Equation (20).

When the reference level is set to be $I_{ref} = 3$ A, the loci of eigenvalues of the converter with the proposed control technique is provided in Figure 5, in which the load and source disturbances are considered. For the variation of the load resistance, R, from 1 to 90 Ω, the loci of the eigenvalues can be obtained as shown in Figure 5a. From Figure 5a, the boost converter is stable when the load resistance is in the range of 1 to 57.4 Ω. The first point of period doubling bifurcation occurs at $R = 57.4$ Ω when one of the eigenvalues equals −1. After that, an eigenvalue moves out of the unit circle with the increase of the load resistance, R. For the variation of the input voltage, E, from 1 to 20 V, the loci of eigenvalues is depicted in Figure 5b. One can see that the converter is stable when the input voltage, E, is in the range of 4.95 to 20 V. Otherwise, the converter will be in an unstable state.

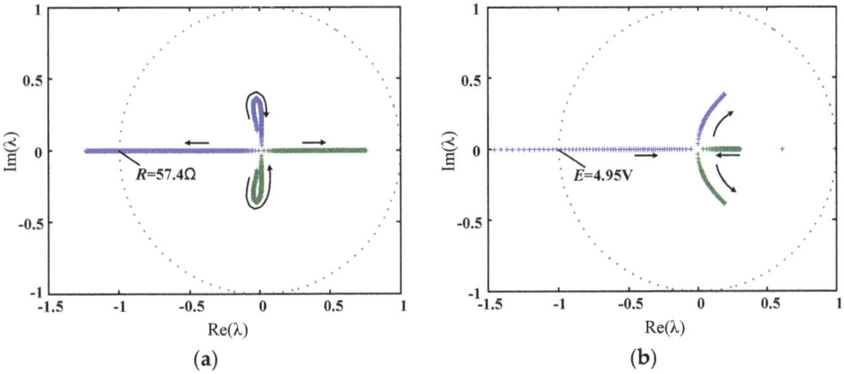

(a)   (b)

Figure 5. Loci of eigenvalues of a PCM-controlled boost converter with the proposed technique: (**a**) Arrows indicate the direction of movement of the eigenvalues with R increasing, (**b**) Arrows indicate the direction of movement of the eigenvalues with E increasing.

As a comparison, the loci of eigenvalues of the converter with the conventional PCM control technique is provided in Figure 6. It can be seen from Figure 6 that with the conventional PCM control scheme, the boost converter is stable for the load resistance being in the range of 1 to 3.06 Ω, and the source voltage, E, being in the range of 18.78 to 20 V.

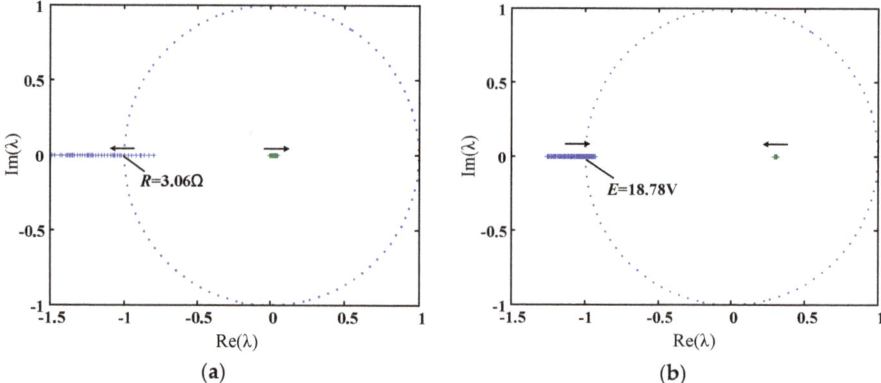

(a)

(b)

Figure 6. Loci of eigenvalues of the classical PCM-controlled boost converter: (**a**) Arrows indicate the direction of movement of the eigenvalues with R increasing; (**b**) Arrows indicate the direction of movement of the eigenvalues with E increasing.

From Figures 5 and 6, the stable operating range for input disturbance is expanded remarkably from 18.78 to 20 V to 4.95 to 20 V, and similarly, the stable range for the load variation is also enlarged from 1 to 3.06 Ω to 1 to 57.4 Ω. Obviously, the performances of the classical PCM control scheme can be improved with the proposed stabilizer.

4. Comparison and Discussion

The general remedy to avoid the chaotic operation state of PCM-controlled converters is to introduce a compensating ramp to I_{ref}. The basic circuit diagram of a PCM-controlled boost converter with ramp compensation and the typical steady-state waveforms are shown in Figure 7, where m_c denotes the slope of the compensating signal.

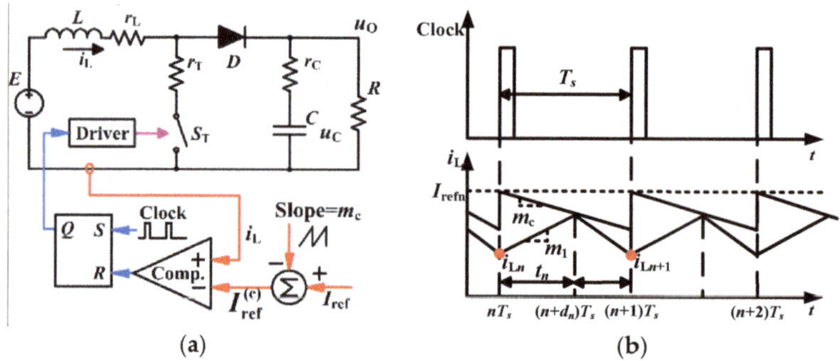

(a)

(b)

Figure 7. PCM-controlled boost converter with ramp compensation: (**a**) Basic circuit diagram; (**b**) typical steady-state waveforms.

According to the ramp compensation technique as shown in Figure 7, the reference current is given by the following equation:

$$I_{\mathrm{ref}}^{(c)} = I_{\mathrm{ref}} - m_c(t \bmod T), \tag{21}$$

where the term '$t \bmod T$' means a modulo operation, which is equivalent with '$t\text{-}kT$', and k is the integer quotient of t/T. Additionally, if the self-compensation technique in [25] is adopted, the reference can be described by:

$$I_{\text{ref}}^{(s)} = I_{\text{ref}} - \frac{1}{T} \int_0^t \left[I_{\text{ref}} - \frac{m_1 DT}{2} - i_{\text{L}}(\tau) \right] d\tau. \tag{22}$$

In Equations (21) and (22), m_1 is the rising slope of the Boost converter. The reference current of the converter is set to be $I_{\text{ref}} = 3$ A, and all the other circuit parameters are chosen as the same as those listed in Table 1. Additionally, to confirm that the converter with a constant slope of m_c operates in a stable state, bifurcation analysis is performed. According to the results in Figure 8, period-doubling bifurcation occurs when $m_c = 3220$ V/s. Then, in the following simulations, the value of m_c is set to be 3250 V/s, and the converter can attain a higher voltage output than that in the case of $m_c > 3250$ V/s.

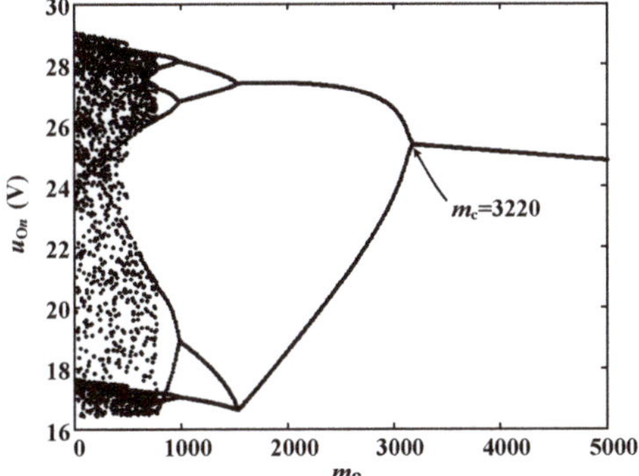

Figure 8. Bifurcation diagram of a PCM-controlled boost converter with constant ramp compensation.

To perform a comparison, both the conventional ramp compensation technique and the self-compensation technique are adopted in simulations. The behaviors and characteristics of the converter are put together with the proposed technique. The waveforms of applying different techniques to stabilize the PCM-controlled boost converter can be seen in Figure 9, where a transition from a chaotic state to a stable state is depicted.

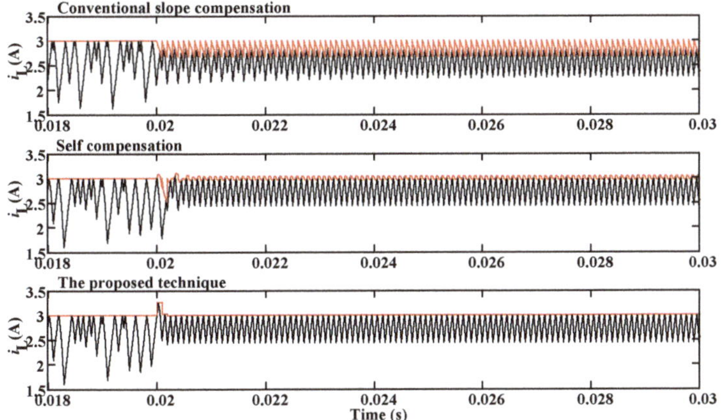

Figure 9. Stabilizing PCM-controlled boost converter by different techniques.

In Figure 9, the red and black solid lines represent the reference current and the inductor current of the converter, respectively. It can be seen when stabilizers come into effect at $t = 0.02$ s, the converter with both the self-compensation technique and the proposed technique can be stabilized quickly. Yet, the converter with the conventional ramp compensation technique takes more than 20 switching cycles to enter a stable state.

Furthermore, the simulated stable current and voltage waveforms of the boost converter with different control strategies are provided in Figure 10. The black solid lines represent the results with the conventional ramp compensation technique, the blue dotted lines are obtained by using the self-compensation technique, and the red solid lines result from the proposed technique. It is seen that the three control techniques can be used to stabilize the converter from a chaotic state. Additionally, by setting UPO-1 as the control objective, the proposed technique can achieve a performance very close to that of using the self-compensation technique. However, the output voltage of the converter with the conventional ramp compensation technique is a little lower than those with the self-compensation technique and the proposed technique. Moreover, both the self-compensation technique and the proposed technique can ensure that the peak value of the inductor current does not deviate from the desired reference, whereas the conventional slope compensation technique cannot.

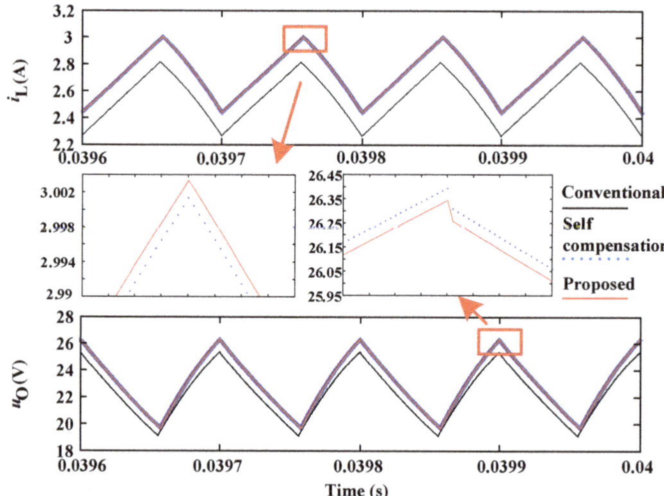

Figure 10. Stable waveforms of the PCM-controlled boost converter with different techniques.

Additionally, with variations of the load from 2 to 80 Ω, and the input voltage from 0.5 to 20 V, respectively, the performances of the boost converter with three different kinds of control technique are compared, as listed in in Table 2. According to the simulation results summarized in Table 2, the boost converter has almost the same voltage gain under the proposed control technique and the self-compensation method. Both methods provide an accurate current limiting capacity, and their robustness is better than the conventional ramp compensation technique.

Table 2. Comparison of performance quotas.

	The Proposed Technique	Self-Compensation	Conventional Ramp Compensation
Duty cycle d	0.5675	0.5661	0.5527
Voltage gain M	2.3121	2.3047	2.2356
Peak values of i_L	3.0033 A	3.0016 A	2.8170 A
Transition time	3 switching cycles	6 switching cycles	more than 20 switching cycles
Stable range for R	(2 to 57.4) Ω	(12.2 to 70) Ω	(2 to 20.6) Ω
Stable range for E	(4.95 to 20) V	(5.96 to 20) V	(10.1 to 20) V

5. Experiments

In this section, experiments are performed, and the schematic diagram is shown in Figure 11. The experimental system includes the main circuit and the control loop, where the main circuit parameters are listed in Table 1. In the control loop, a hall sensor CSM050LX with the transfer ratio of 0.8 Ω is adopted to acquire the inductor current signal, so the value obtained from the oscilloscope should be divided by this coefficient. The amplifier AD620 is used to collect the output voltage signal. The conversion of A/D, and calculation of the disturbance increment of the controlled parameter are processed by an ARM-based 32-bit MCU (Microcontroller Unit) STM32F103C8T6. The 12-bit Digital-to-Analog Converter TLV5618 is adopted to perform the conversion of D/A. The voltage comparison and the RS flip-flop function are completed by LM339 and HD74LS02P, respectively. An optical coupler is used to drive the MOSFET, and it can also achieve the isolation between the main circuit and the control loop.

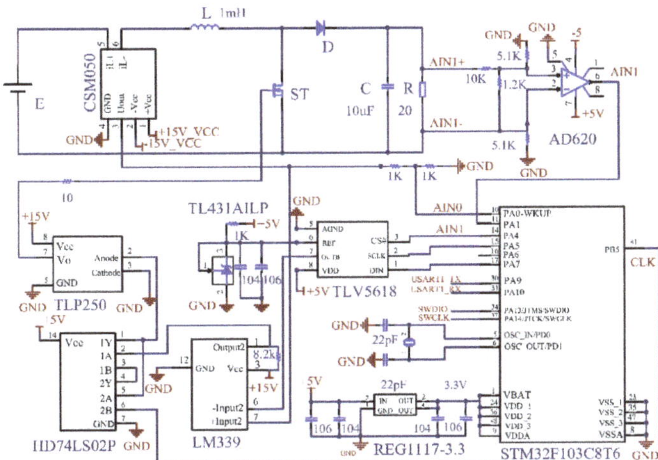

Figure 11. Schematic diagram.

The experimental setup and the experimental results are shown in Figure 12. As shown in Figure 12b,c, the upper curves represent the inductor current and the lower curves represent the output voltage. The waveforms of the converter operating from the chaotic state to a stable state are shown in Figure 12a. The close-up views of the state variables in the chaotic and stable period-1 states are shown in Figure 12b,c respectively. It can be seen from Figure 12a that when the converter operates in the stable state, the sampling values for the inductor current at the bottom and peak point are 1.915 and 2.42 V, corresponding to the real inductor current values of 2.394 and 3.025 A, respectively. Additionally, the transfer ratio of the current sensor is 0.8 Ω. Thus, the peak-peak value of the inductor current is about 0.63 A. For the output voltage, as shown in Figure 12c, the valley and peak values are 18.78 and 25.01 V, respectively. It can also be seen from Figure 12 that when the converter returns to a stable state from a chaotic state, the peak-peak value of the output voltage is reduced to 6.228 from 13.56 V, and that of the inductor current is almost cut in half, i.e., from about 1.25 to 0.63 A.

Moreover, for the boost converter operating in a stable state with the proposed control scheme, a comparison of the data from both the simulation and experiment is summarized in Table 3. It should be noticed that the experimental value for the output voltage is about 1.3 V less than the simulated value, which is caused by the output diode. Thus, considering the conductance voltage drop of the Schottky diode, the experimental results agree quite well with the simulations. Another noteworthy thing is that in order to observe the effect of the proposed control method, the parameters of the example circuit, which result in the converter having larger ripples with the state variables, were chosen to be the same as those in [32].

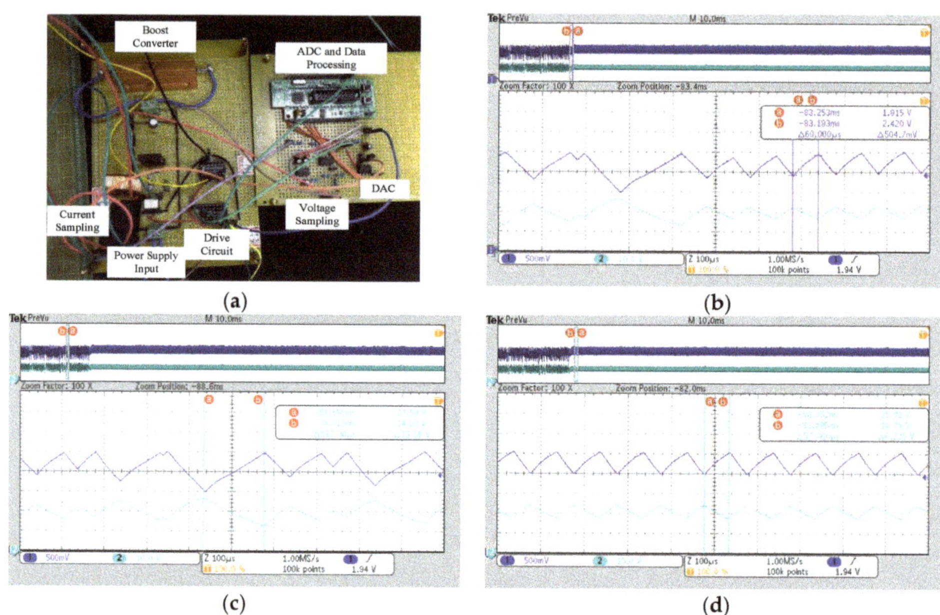

Figure 12. (**a**) Experimental setup and experimental waveforms (channel 1: Inductor current i_L; Channel 2: Output voltage u_O): (**b**) From chaotic state to period-1 state, (**c**) chaotic operation, (**d**) period-1 operation.

Table 3. Results from simulation and experiment of the PCM boost converter with the proposed control.

Parameters	Simulation	Experiment
Duty cycle d	0.5675	0.59
Voltage gain M	2.31	2.19
Variation range of i_L	(2.43 to 3) A	(2.394 to 3.025) A
Peak-peak value of inductor current i_L	0.57 A	0.6312 A
Variation range of u_O	(19.5 to26.3) V	(18.78 to 25.01) V
Peak-peak value of output voltage u_O	6.8 V	6.228 V

6. Conclusions

A novel scheme for improving the classical PCM control method was presented in this paper, in which the principle of the parameter-perturbation method was introduced and adopted. The converter can be stabilized to operate in the target period orbit from the chaotic state. By applying this scheme to the DC-DC boost converter, the performances of the converter were improved, such as the converter's stability range, the accuracy of current limiting, and so on. Moreover, compared to the conventional ramp compensation method, the proposed scheme demonstrated some advantages as follows. The transition time of the converter from the chaotic state to the steady state was greatly shortened. The capacity of the resisting load and source disturbances was expanded significantly, which means a better robustness was accomplished with the proposed control scheme. In addition to those mentioned above, the proposed method can achieve the same effect as the self-compensation method yet does not require an external signal generator for the ramp as in both the conventional ramp compensation and self-compensation scheme. The results from both simulations and experiments support the theoretical analysis, which indicates that the proposed technique can be used as an alternative to improve the performances of PCM-controlled DC-DC converters.

Author Contributions: Conceptualization, B.Z., and Y.C.; methodology, Y.C.; software, X.C., and Z.L.; validation, F.X.; formal analysis, X.C., and Z.L.; investigation, Z.L.; resources, B.Z., and D.Q.; data curation, F.X.; writing—original draft preparation, X.C., and Z.L.; writing—review and editing, Y.C., D.Q., and G.Z.; visualization, Y.C.; supervision, B.Z., and Y.C.; project administration, B.Z., and F.X.; funding acquisition, B.Z.

Funding: This research was funded by the Key Program of Natural Science Foundation of China, under Grant No.2018YFB0905804.

Conflicts of Interest: The authors declare no conflict of interest.

References

1. Bernardo, M.; Budd, C.; Champneys, A.R.; Kowalczyk, P. *Piecewise Smooth Dynamical Systems: Theory and Applications*; Springer: London, UK, 2008.
2. Leppäaho, J.; Suntio, T. Characterizing the Dynamics of the Peak-Current-Mode-Controlled Buck-Power-Stage Converter in Photovoltaic Applications. *IEEE Trans. Power Electron.* **2014**, *29*, 3840–3847. [CrossRef]
3. Lu, W.; Lang, S.; Zhou, L.; Iu, H.H.; Fernando, T. Improvement of Stability and Power Factor in PCM Controlled Boost PFC Converter with Hybrid Dynamic Compensation. *IEEE Trans. Circuits Syst. I* **2015**, *62*, 320–328. [CrossRef]
4. Tse, C.K. *Complex Behavior of Switching Power Converters*; CRC Press: New York, NY, USA, 2004.
5. Tse, C.K.; Li, M. Design-oriented Bifurcation Analysis of Power Electronics Systems. *Int. J. Bifurc. Chaos* **2011**, *21*, 175–187. [CrossRef]
6. Chen, Y.; Chi, K.T.; Qiu, S.S.; Lindenmuller, L.; Schwarz, W. Coexisting fast-scale and slow-scale instability in current-mode controlled DC/DC converters: Analysis, Simulation and Experimental Results. *IEEE Trans. Circuits Syst. I* **2008**, *55*, 3335–3348. [CrossRef]
7. Yang, R.; Zhang, B.; Xie, F.; Iu HH, C.; Hu, W. Detecting bifurcation types in DC-DC switching converter by duplicate symbolic sequence and weight complexity. *IEEE Trans. Ind. Electron.* **2013**, *60*, 3145–3156. [CrossRef]
8. Huang, L.; Qiu, D.; Xie, F.; Chen, Y.; Zhang, B. Modeling and Stability Analysis of a Single-Phase Two-Stage Grid-Connected Photovoltaic System. *Energies* **2017**, *10*, 2176. [CrossRef]
9. Luo, Z.; Xie, F.; Zhang, B.; Qiu, D. Quantifying the Nonlinear Dynamic Behavior of the DC-DC Converter via Permutation Entropy. *Energies* **2018**, *11*, 2747. [CrossRef]
10. Bao, B.; Zhou, G.; Xu, J.; Liu, Z. Unified Classification of Operation-State Regions for Switching Converters with Ramp Compensation. *IEEE Trans. Power Electron.* **2011**, *26*, 1968–1975. [CrossRef]
11. Ma, W.; Wang, M.; Liu, S.; Li, S.; Yu, P. Stabilizing the average current-mode-controlled boost PFC converter via washout-filter-aided method. *IEEE Trans. Circuits Syst. Express Briefs* **2011**, *58*, 595–599. [CrossRef]
12. Fan, J.W.T.; Chung, H.S.H. Bifurcation Phenomena and Stabilization with Compensation Ramp in Converter with Power Semiconductor Filter. *IEEE Trans. Power Electron* **2017**, *32*, 9424–9434. [CrossRef]
13. Wang, Y.; Yang, R.; Zhang, B.; Hu, W. Smale horseshoes and symbolic dynamics in buck-boost DC-DC converter. *IEEE Trans. Ind. Electron.* **2018**, *65*, 800–809. [CrossRef]
14. Suntio, T. Analysis and Modeling of Peak-Current-Mode-Controlled Buck Converter in DICM. *IEEE Trans. Ind. Electron.* **2001**, *48*, 127–135. [CrossRef]
15. Suntio, T. *Dynamic Profile of Switched-Mode Converter: Modeling, Analysis and Control*; Wiley-VCH: Weinheim, Germany, 2009.
16. Cha, H.; Peng, F.Z.; Yoo, D.W. Distributed Impedance Network (Z-Network) DC-DC Converter. *IEEE Trans. Power Electron.* **2010**, *25*, 2722–2733. [CrossRef]
17. Hu, X.; Gong, C. A High Gain Input-Parallel Output-Series DC/DC Converter with Dual Coupled Inductors. *IEEE Trans. Power Electron.* **2015**, *30*, 1306–1317. [CrossRef]
18. Zhang, G.; Zhang, B.; Li, Z.; Qiu, D.; Yang, L.; Halang, W.A. A 3-Z-Network Boost Converter. *IEEE Trans. Ind. Electron.* **2015**, *62*, 278–288. [CrossRef]
19. Shen, H.; Zhang, B.; Qiu, D.; Zhou, L. A Common Grounded Z-Source DC-DC Converter with High Voltage Gain. *IEEE Trans. Ind. Electron.* **2016**, *63*, 2925–2935. [CrossRef]

20. Shen, H.; Zhang, B.; Qiu, D. Hybrid Z-Source Boost DC-DC Converters. *IEEE Trans. Ind. Electron.* **2017**, *64*, 310–319. [CrossRef]
21. Zhang, Y.; Liu, Q.; Li, J.; Sumner, M. A Common Ground Switched-Quasi-Z-Source Bidirectional DC-DC Converter with Wide-Voltage-Gain Range for EVs With Hybrid Energy Sources. *IEEE Trans. Ind. Electron.* **2018**, *65*, 5188–5200. [CrossRef]
22. Shabestari, P.M.; Gharehpetian, G.B.; Riahy, G.H.; Mortazavian, S. Voltage controllers for DC-DC boost converters in discontinuous current mode. In Proceedings of the 2015 International Energy and Sustainability Conference (IESC), Farmingdale, NY, USA, 12–13 November 2015. [CrossRef]
23. Hejri, M.; Mokhtari, H. Hybrid predictive control of a DC–DC boost converter in both continuous and discontinuous current modes of operation. *Optim. Control Appl. Methods* **2011**, *32*, 270–284. [CrossRef]
24. Qian, T.; Lehman, B. An adaptive ramp compensation scheme to improve stability for DC-DC converters with ripple-based constant on-time control. In Proceedings of the 2014 IEEE Energy Conversion Congress and Exposition (ECCE), Pittsburgh, PA, USA, 14–18 September 2014; pp. 14–18. [CrossRef]
25. Liu, P.H.; Yan, Y.; Lee, F.C.; Mattavelli, P. Universal Compensation Ramp Auto-tuning Technique for Current Mode Controls of Switching Converters. *IEEE Trans. Power Electron.* **2018**, *33*, 970–974. [CrossRef]
26. Chen, W.W.; Chen, J.F.; Liang, T.J.; Wei, L.C.; Ting, W.Y. Designing a Dynamic Ramp with an Invariant Inductor in Current-Mode Control for an On-Chip Buck Converter. *IEEE Trans. Power Electron.* **2014**, *29*, 750–758. [CrossRef]
27. El Aroudi, A.; Mandal, K.; Giaouris, D.; Banerjee, S. Self compensation of DC-DC converters under peak current mode control. *Electron. Lett.* **2017**, *53*, 345–347. [CrossRef]
28. Cheng, W.; Song, J.; Li, H.; Guo, Y. Time-Varying Compensation for Peak Current-Controlled PFC Boost Converter. *IEEE Trans. Power Electron.* **2015**, *30*, 3431–3437. [CrossRef]
29. Morcillo, J.D.; Burbano, D.; Angulo, F. Adaptive ramp technique for controlling chaos and subharmonic oscillations in DC-DC power converters. *IEEE Trans. Power Electron.* **2016**, *31*, 5330–5343. [CrossRef]
30. Jimenez-Triana, A.; Chen, G.; Gauthier, A. A parameter-perturbation method for chaos control to stabilizing UPOs. *IEEE Trans. Circuits Syst. II Express Briefs* **2015**, *62*, 407–411. [CrossRef]
31. Schöll, E.; Schuster, H.G. *Handbook of Chaos Control*, 2nd ed.; Wiley-VCH: Weinheim, Germany, 2008.
32. Ayati, M. Chaos control of Boost converter via the fuzzy delayed-feedback method. In Proceedings of the 2016 4th International Conference on Control, Instrumentation, and Automation (ICCIA), Qazvin, Iran, 27–28 January 2016; pp. 324–328. [CrossRef]

 © 2019 by the authors. Licensee MDPI, Basel, Switzerland. This article is an open access article distributed under the terms and conditions of the Creative Commons Attribution (CC BY) license (http://creativecommons.org/licenses/by/4.0/).

Article

Enhancement of System Stability Based on PWFM

K. I. Hwu [1], C. W. Wang [1] and Y. T. Yau [2],*

[1] Department of Electrical Engineering, National Taipei University of Technology, Taipei 10608, Taiwan; eaglehwu@ntut.edu.tw (K.I.H.); terrywang@gmail.com (C.W.W.)
[2] Asian Power Devices Inc., Taoyuan City, Taoyuan County 330, Taiwan
* Correspondence: tsmc35@yahoo.com.tw; Tel.: +886-3-799078

Received: 9 March 2019; Accepted: 27 March 2019; Published: 3 April 2019

Abstract: In this paper, a pulse width and frequency modulation (PWFM) control strategy is presented, which combines the one-comparator counter-based pulse width modulation (PWM) control with pulse frequency modulation (PFM) control to increase pseudo-1-bit resolution under constant-frequency operation. Accordingly, system stability will be enhanced significantly. As compared with the traditional counter-based PWM control, there is no difference in off-chip circuit complexity except a slight change in on-chip hardware. Finally, a prototype circuit is used to verify the proposed control concept by some experimental results with no limit cycle oscillation.

Keywords: counter-based; one-comparator; PWFM; PWM; PFM

---

1. Introduction

Up to now, much research on the accuracy of the pulse width modulation (PWM) has been conducted. The reason why the development of the high-resolution PWM is needed is described below. One reason is that the PWM resolution should be higher than the analog-to-digital converter (ADC) resolution to avoid limit cycle oscillation [1–3]. The other reason is that under the fixed system clock, the PWM accuracy is inversely proportional to the switching frequency. However, PWM accuracy should be not too low. Consequently, under this constraint, the more the switching frequency is, the more the system clock, which is proportional to the switching frequency. Accordingly, due to limitations on the integrated circuit (IC) process, the dissipation power will be increased abruptly, including the charging/discharging of the complementary metal-oxide-semiconductor (CMOS) gate and the leakage current due to the miniature process. Based on the above two reasons, the PWM accuracy and the switching frequency are limited to some extent. It is possible that a special process, such as silicon-on-isolator (SOI) [4,5], may reduce the leakage current, and may keep low power dissipation and low temperature under the high-speed system clock of the miniature process. However, the corresponding cost is high, and this special process is usually used in the manufacture of the central processing unit (CPU) or the graphic processing unit (GPU). There is a lot of research on high-accuracy PWM. For example, the literature [6–21] focus on how to reduce the number of digital pulse width modulation (DPWM) steps, where the high-switching clock, e.g., counter-based DPWM, is achieved based on special structures [7,19–21]. Most of these structures are multiple interleaved to achieve high-switching clock or use very short delay elements, e.g., hybrid DPWM [1,22]. Although the delay line based DPWM can achieve high accuracy, the corresponding silicon area is relatively large compared with the traditional counter-based DPWM. In addition, the delay line based DPWM is sensitive to the operating temperature, process, and power interference [12].

On the other hand, some researches increase the effective duty cycle to achieve high-accuracy resolution, for example, digital dither [23], sigma-data [8,10,18], special modulation [24], and PFM [25]. The method taken by the literature [23] may cause the output voltage ripple to be small or large, thereby influencing the controller performance. The method followed in [24] tends to vary the turn-on and turn-off periods of the switch so the ADC sampling is difficult. As for the method shown in [25], it is restricted to PFM operation.

Based on aforementioned, this paper is an extension of the paper [26]. The latter takes the one-comparator counter-based PWM control, whereas the former takes the one-comparator counter-based PWM control with PFM control. By doing so, the former can increase pseudo-1-bit resolution under constant-frequency operation, so that system stability will be improved greatly. In addition, the difference in on-chip hardware between the two control strategies is slightly small, whereas there is no difference in off-chip circuit complexity between the two control strategies.

2. Problem Description

In many papers, the problem of limit cycle oscillation has been discussed as shown in Figure 1a. Most people say that this is because the resolution of PWM is larger than that of ADC. In fact, the answer to this problem is too brief. This is because the gain of the control loop should be taken into account. The solution of the limit cycle oscillation that is only based on high-resolution DPWM is not enough. Therefore, the detailed overall calculations and the corresponding program flow are discussed. From Figure 1b, it can be seen that although the output voltage V_O keeps up with the voltage reference V_{ref}, V_O swings up and down around V_{ref} due to the DPWM resolution being lower than ADC resolution. From the point of view of control, the difference in oscillation between V_O and V_{ref} is quite small, but V_O cannot keep up with V_{ref}. As the DPWM resolution is higher than the ADC resolution, the feedback error can be kept as small as possible, and hence no limit cycle oscillation occurs. From Figure 1, it gives us a hint that the traditional controller needs an integral gain so as to make V_O approach to V_{ref} as near as possible.

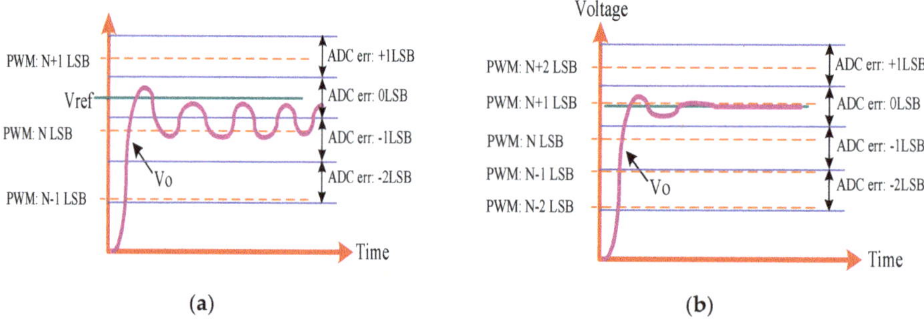

Figure 1. Quantization resolution in a digitally controlled pulse width modulation (PWM): (**a**) with limit cycle oscillation; (**b**) without limit cycle oscillation.

3. Discussion of Compensator Gain

Figure 2 shows the digital closed-loop system block diagram. There are three block diagrams. One is an analog-to-digital converter (ADC) block, another is a compensator block, and the other is a digital-to-analog converter (DAC) block. The last block includes the controlled plant. The gains for the ADC, compensator, and DAC blocks are described as (1), (2) and (3), respectively. In order to avoid limit cycle oscillation, Equation (4) must hold:

$$ADC_{Gain} = \frac{LSB_{ADC}}{Vo} \tag{1}$$

$$DAC_{Gain} = \frac{Vo}{LSB_{DAC}} \tag{2}$$

$$Comp_{Gain} = \frac{LSB_{DAC}}{LSB_{ADC}} \tag{3}$$

$$ADC_{Gain} \times Comp_{Gain} \cdot DAC_{Gain} \leq 1 \tag{4}$$

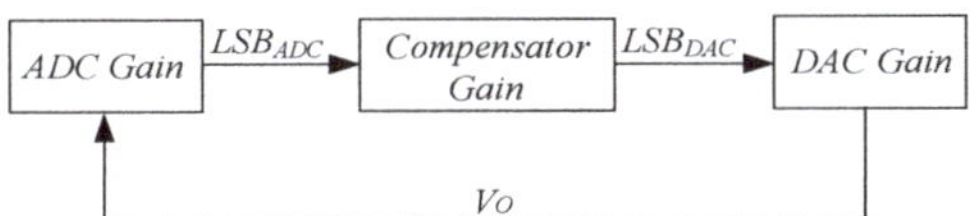

Figure 2. Digital closed-loop system block diagram.

4. ADC Strategy

Figure 3 shows the proposed system configuration, constructed by one synchronously-rectified (SR) buck converter and one feedback control circuit. The latter is built up by the field-programmable gate array (FPGA). Inside the FPGA, there are one proportional-integral-derivative (PID) control block, one DPWM controller, and one feedback control block. Outside the FPGA, there is one voltage divider, one saw-tooth generator, one analog circuit, and one comparator. The saw-tooth generator is constructed by one charging switch Q_3, one constant current source, one capacitor C_{ramp}, one DC-blocking capacitor C_b and one operational amplifier (OPA) with a voltage gain of −1. Furthermore, the feedback counter is a digital counter, which is inside the FPGA. As the output signal from the comparator, named *VFB*, is "1", the counter counts one, whereas the *VFB* signal is reset to zero as synchronized with the *PWM* signal. In addition, the sensed output signal v'_O is obtained by one feedback voltage divider built up by two resistors.

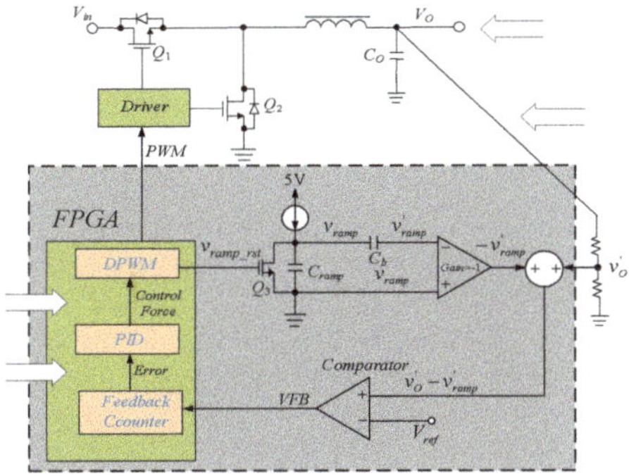

Figure 3. System configuration.

5. Basic Operating Principles

Prior to this section, there are some assumptions and symbol definitions described. It is assumed that the voltage ripple of the output voltage is quite small so the output voltage v_O can be regarded as an average value V_O. The triggering signal for Q_3 is signified by v_{ramp_rst}. The sawtooth waveform is represented by v_{ramp} which has a minimum value of zero and a peak value of V_{ramp_pp}. The signal v_{ramp} after the AC coupling capacitor C_b is signified by v'_{ramp}. Therefore, the signal $-\frac{1}{2}V_{ramp\_pp}$ is the minimum value of v'_{ramp} whereas the signal $\frac{1}{2}V_{ramp\_pp}$ is the maximum value of v'_{ramp}. The signal v'_{ramp} is changed to $-v'_{ramp}$ after the OPA with gain= −1. Afterwards, the sum of $-v'_{ramp}$ and v'_O has a minimum value of $\left(v'_O - \frac{1}{2}V_{ramp\_pp}\right)$ and a maximum value of $\left(v'_O + \frac{1}{2}V_{ramp\_pp}\right)$. Finally, the output of the comparator is a digital signal, determined by $\left(v'_O - v'_{ramp}\right)$ and V_{ref}. If the $\frac{v'_O}{v_O} = G_{fb}$, then the voltage range of V_O can locate between $\left(V_{ref} - \frac{1}{2}V_{ramp\_pp}\right)/G_{fb}$ and $\left(V_{ref} + \frac{1}{2}V_{ramp\_pp}\right)/G_{fb}$. Figure 4 shows the associated circuit waveforms, which are described based on the time sequence, that is, the operating states. There are four operating states in this circuit, to be shown below:

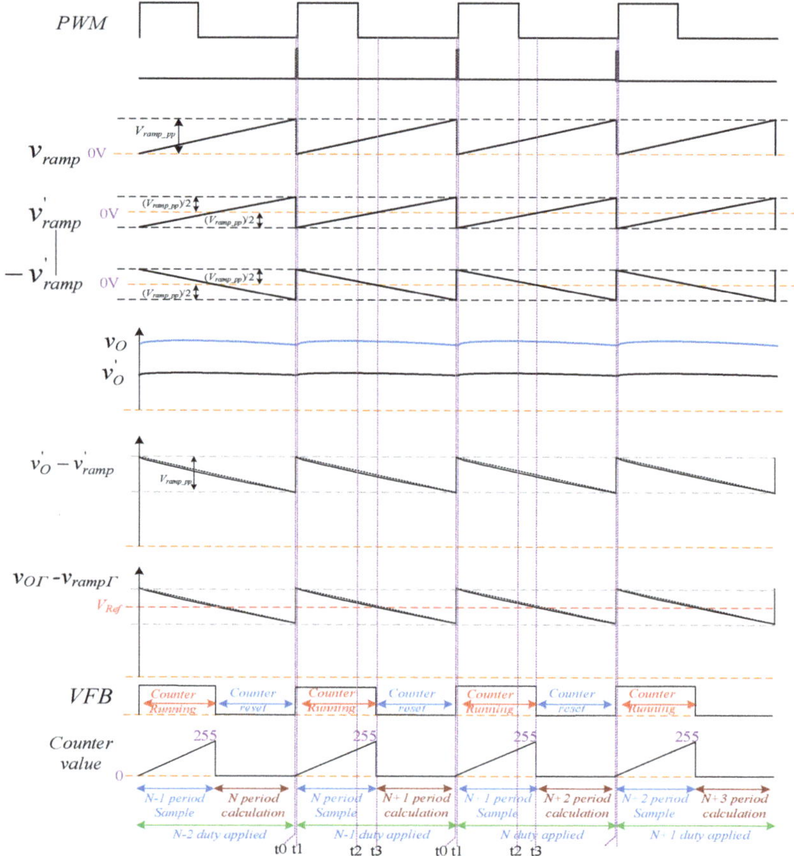

Figure 4. Waveforms relevant to sampling sequence.

(1) State 1 [t0~t1]: Let the voltage V_{ramp_rst} be high, thereby making the switch Q_3 turned on and the voltage across C_{ramp} zero. At the same time, the PWM is high, causing Q_1 to be turned on and Q_2 to be turned off. This time interval is short, smaller than 1% of the PWM switching period. The signal v_{ramp} is blocked by the DC-blocking capacitor C_b, and after the OPA with a gain of -1, the resulting signal $-v'_{ramp}$, which is added to v'_O so as to obtain $\left(v'_O - v'_{ramp}\right)$, which is subtracted from V_{ref} to obtain the signal VFB.

(2) State 2 [t1~t2]: The signal PWM keeps high. Let V_{ramp_rst} be low, thereby making Q_3 turned off and C_{ramp} is linearly charged. If $\left(v'_O - v'_{ramp}\right) > V_{ref}$, then VFB = "high", thereby making the counter value increased by one. As the preset turn-on time in the N period is reached, the operating state proceeds to state 3.

(3) State 3 [t2~t3]: At the time instant of t2, let the signal PWM be "low." If $(v'_O - v'_{ramp}) > V_{ref}$, then VFB = "high", thus rendering the counter value keeping increased by one. If $v'_O - v'_{ramp} < V_{ref}$, then the operating state goes to state 4.

(4) State 4 [t3~t0]: At the time instant of t3, VFB = "low," the corresponding counter value will be saved in the feedback register, and then the counter value is set to zero. During this state, the PID calculates the control force for the next period, according to the information in the feedback register. At the same time, the duty cycle information is downloaded to the DPWM. At the time of t0, the next

cycle begins. In addition, from Figure 4, it can be seen that as $v'_O = V_{ref}$, the counter value is 50%, which is a full scale.

Under the ideal condition, if the counter value locates within the sampling range of the comparator, VFB is linearly proportional to v'_O. For example, in Figure 4, if the range of the counter value is n-bit, the sampling resolution can be represented by $\frac{V_{ramp\_pp}}{counter\ value}$, that is, $\frac{V_{ramp\_pp}}{n\ bits}$ is the sampling resolution of the least significant bit (LSB). The smaller the V_{ramp_pp} is or the larger the bit number of the counter value is, the more the resolution. It is suggested that in order to enhance the linearity, $V_{ramp_{pp}} > 10 \times v_{O\_ripple}$ should hold in design.

If the sensed output voltage v'_O is out of the sampling range of the comparator, the sampled data will be saturated. For example, as $v'_O \geq V_{ref} + \frac{1}{2}V_{ramp\_pp}$, the comparator keeps VFB = "high", and hence the counter value keeps the maximum error. By the same way, as $v'_O \leq V_{ref} - \frac{1}{2}V_{ramp\_pp}$, the comparator keeps VFB = "low", and hence the counter value keeps the minimum error.

6. Resolution Design

6.1. Requirements of Resolution of DPWM and ADC

In the traditional buck converter with digital control, if the DPWM has N_{DPWM} bits, then the resolution of DPWM can be expressed by

$$Resolution_{DPWM} = \frac{V_{in}}{2^{N_{DPWM}}} \tag{5}$$

Each bit in DPWM causes a voltage variation to be $\Delta V_{DPWM} = \frac{V_{in}}{2^{N_{DPWM}}}$.. By the same way, if ADC has N_{ADC} bits, then ADC resolution can be represented by

$$Resolution_{ADC} = \frac{V_{ramp\_pp}}{2^{N_{ADC}}} \tag{6}$$

Each LSB in ADC causes a voltage variation to be $\Delta V_{ADC} = \frac{V_{ramp\_pp}}{2^{N_{ADC}}}$.

According to the literature [20], the resolution of DPWM and ADC is mainly influenced by the limit cycle oscillation. This is because when the DPWM resolution is lower than the ADC resolution, the DPWM cannot satisfy the sampled value of the ADC for any operating point such that the feedback error is not zero. Therefore, the controller outputs a keeping-jumping control force to the DPWM, to force the average output voltage to satisfy the voltage reference. However, such a keeping-jumping PWM control force will cause the output voltage to oscillate. When the DPWM resolution is higher than the ADC resolution, the DPWM can find some operating point to satisfy the sampled value of the ADC and to make the feedback error zero. By doing so, the limit cycle oscillation phenomenon can be avoided.

Therefore, the minimum requirement for the limit cycle oscillation is shown in (7):

$$Resolution_{DPWM} = \frac{V_{in}}{2^{N_{DPWM}}} = \frac{V_{ramp_{pp}}}{2^{N_{ADC}}} \tag{7}$$

Accordingly, the minimum value of V_{ramp_pp} is $V_{ramp\_pp\_min} = V_{in} \times \frac{2^{N_{ADC}}}{2^{N_{DPWM}}}$. Based on $I_{ramp} \times T_s = C_{ramp} \times v_{ramp}$, $V_{ramp\_pp\_min}$ can be signified by

$$V_{ramp\_pp\_min} = V_{in} \times \frac{2^{N_{ADC}}}{2^{N_{DPWM}}} = \frac{I_{ramp} \times T_s}{C_{ramp}} \tag{8}$$

6.2. Calculation of VFB Duty

From Figure 5, it can be seen that the relationship between the comparator output signal and the feedback counter value are as shown in (9):

$$Duty_{(VFB)} = \frac{T_{VFB\_Hi}}{T_S} = \frac{ADC_{VFB}}{ADC_{fullscale}} \tag{9}$$

where

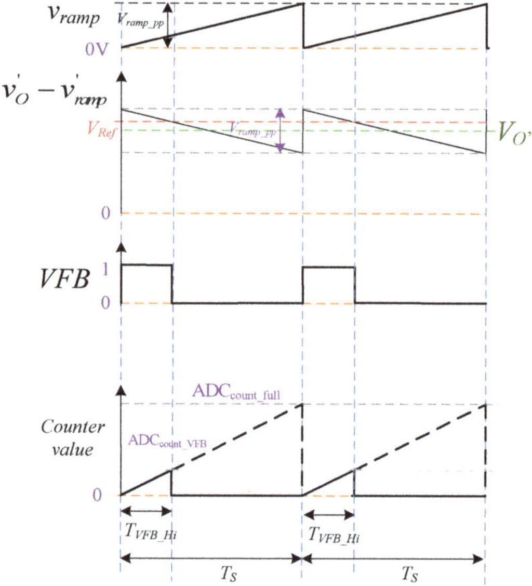

Figure 5. $\left(V_{ref} - \frac{V_{ramppp}}{2}\right) < v'_O < \left(V_{ref} + \frac{V_{ramp\_pp}}{2}\right).$

$ADC_{fullscale}$: Full-scale value of feedback counter.
ADC_{VFB}: Feedback value of N period
T_{VFB_Hi}: High level of the *VFB* signal
$Duty_{(VFB)}$: Duty cycle of the *VFB* signal

On the other hand, Figure 6a is under the condition that $Duty_{(VFB)}$ is 100%. In this case, $V_{O'_{max}} = V_{ref} + \frac{V_{ramp\_pp}}{2}$. Figure 6b is under the condition that $Duty_{(VFB)}$ is 0%. In this case, $v'_{Omin} = V_{ref} - \frac{V_{ramppp}}{2}$. Thus, based on the geometry theory, it can be found that any value of $Duty_{(VFB)}$ within the interval of $\left(V_{ref} - \frac{V_{ramp\_pp}}{2}\right) < v'_O < \left(V_{ref} + \frac{V_{ramp\_pp}}{2}\right)$ can be expressed as

$$Duty_{(VFB)} = \frac{v'_O - V_{ref} + \frac{V_{ramppp}}{2}}{v'_{Omax} - v'_{Omin}} = \frac{\left(v'_O - V_{ref} + \frac{V_{ramp\_pp}}{2}\right)}{V_{ramp\_pp}} = \frac{\left(v'_O - V_{ref}\right)}{V_{ramp\_pp}} + \frac{1}{2} \tag{10}$$

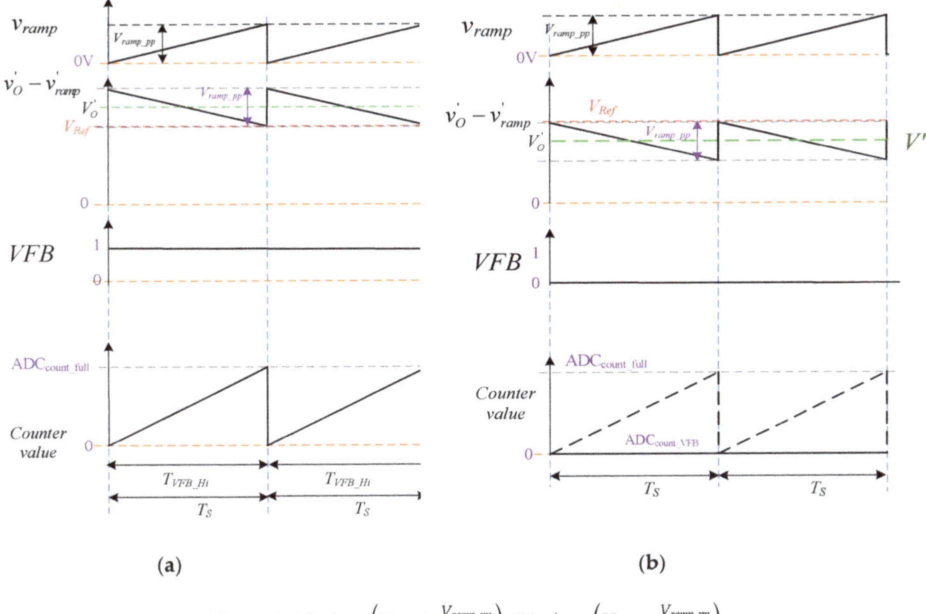

Figure 6. (a) $v'_O = \left(V_{ref} + \frac{V_{ramp\_pp}}{2}\right)$; (b) $v'_O = \left(V_{ref} - \frac{V_{ramp\_pp}}{2}\right)$.

Combining (9) and (10) yields

$$Duty_{(VFB)} = \frac{ADC_{VFB}}{ADC_{fullscale}} = \frac{\left(v'_O - V_{ref}\right)}{V_{ramp\_pp}} + \frac{1}{2} \tag{11}$$

Taking the output voltage divider transfer function into account, (11) can be rewritten to be

$$Duty_{(VFB)} = \frac{ADC_{VFB}}{ADC_{fullscale}} = \frac{\left(V_O \times G_{fb} - V_{ref}\right)}{V_{ramp\_pp}} + \frac{1}{2} \tag{12}$$

Substituting $V_{ramp\_pp\_min}$ shown in (8) into (12) yields

$$Duty_{(VFB)} = \frac{ADC_{VFB}}{ADC_{fullscale}} = \frac{\left(V_O \times G_{fb} - V_{ref}\right)}{V_{ramp\_pp\_min}} + \frac{1}{2} = \frac{\left(V_O \times G_{fb} - V_{ref}\right)}{\left(\frac{I_{ramp} \times T_s}{C_{ramp}}\right)} + \frac{1}{2} \tag{13}$$

7. Gain Analysis of Digital Compensator

Figure 7 shows the block diagram of the PID compensator. Inside this compensator, there is one rounded number block, one anti-saturation block, and one right z shift block beside the PID calculation.

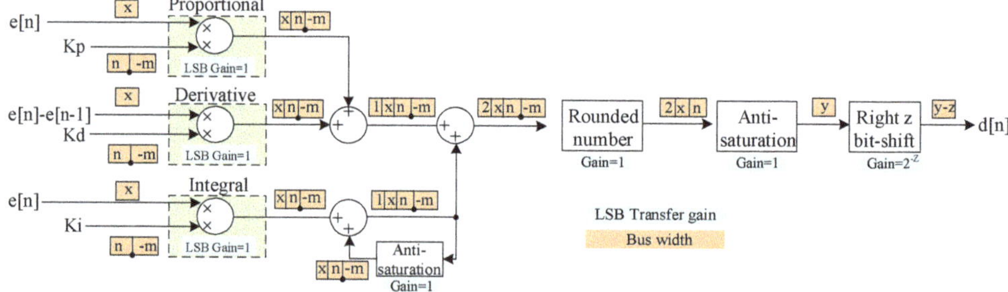

Figure 7. Proportional–integral–derivative (PID) compensator block diagram.

7.1. PID Calculation

The proportional gain Kp, the integral gain Ki, and the derivative gain Kd all have the data register form. This form implies an n-bit integer and an m-bit decimal fraction. For example, if $n = 5$, $m = 3$ and $Kp = 1$, then the corresponding digital value is "00001.000". If the bus width of e[n] is (x) and the bus widths of Kp, Ki, and Kd are all the same ($n.m$), then after PID calculation, the corresponding bus width is ($x + n.m$).

7.2. Rounded Number Block

The decimal bits are removed without any conditions, but there is no change of LSB gain, implying that the corresponding output bus width is still ($x + n.m$).

7.3. Anti-Saturation Block

This block is used as a protection of the arithmetic overflow of the register. There is no change of LSB gain and the corresponding output bus width is ($2 + x + n$).

7.4. Right z-Bit Shift Block

As the division is operated, redundant bits are removed. This is because the DPWM does not need so many bits. The right shift number is expressed by the symbol z. For each shift, the corresponding LSB gain is divided by two. The output bus width of this block is (y-z) and the corresponding LSB gain is 2^{-z}.

Accordingly, if $Kp = 2$ and $z = 1$, then the total LSB gain is one. If $Kp = 4$ and $z = 0$, then the total LSB gain is four. The LSB gain of the PID calculation is determined by the minimum value of {Kp, Ki, Kd}. For example, if $Kp = 2$, $Ki = 0.01$ and $Kd = 3$, then the LSB gain of the PID block is two. This is because the decimal fraction, created from the integrator, will be removed directly.

8. PWFM Control Concept

The PWFM strategy, combining the pulse width modulation (PWM) and the pulse frequency modulation (PFM), can improve 1-bit resolution. This strategy does not need to change the original control structure and circuit. In the following, one example, together with Table 1, is given. The first two columns are associated with traditional duty cycles and periods, respectively. There are three cases with duty cycles of 30%, 50%, and 80%. According to the traditional PWM, the jumping interval is about 0.195%. The last two columns are associated with the proposed duty cycles and periods, respectively. Under the condition of the period of 512 clocks (CLK), both the corresponding duty cycle for the PWM and PFWM are the same, whereas, under the condition of the period of 511 CLK, the corresponding duty cycle for the PFWM is larger than that for the previous PWM but smaller than that for the next PWM. By doing so, the resolution of the DPWM strategy is larger than that of the PWM strategy. In other words, a higher resolution can be achieved and expressed by ($n + 0.5$) CLK, which is different from the PWM strategy with a resolution expressed by n CLK. From Table 1, it can be seen that there are no errors in the duty cycle of around 50%, but there are some errors in the duty cycle of 30% and 80%. Whether these errors are useful or not depends on actual applications. For an example of a buck converter, it normally works with the duty cycle locating between 15% and 85%, so a little large error in duty cycle calculation is OK. Furthermore, the extreme duty cycle is used in a large transient response so the duty cycle linearity is not so important. From Figure 8, both the curves of duty cycle versus control force for the PWFM and PWM are almost the same. In Figure 9, it shows that the errors in the duty cycle locating between zero and 100%, where not all points have errors. For n CLK, there are no errors in duty cycle calculation if n CLK is activated, whereas, for $(n + 0.5)$ CLK, there are some errors in duty cycle calculation if $(n + 0.5)$ CLK is activated. From Figure 9, it can be seen that the maximum error is within 0.1%, showing that the proposed strategy possesses industrial applications to some extent.

Table 1. Duty cycle comparison between (PWM) and pulse width and frequency modulation (PWFM).

Traditional PWM	PWM Period	Duty (%)	Proposed PWFM	PWFM Period	Duty (%)
154	512	30.08	154	512	30.08
			154	511	30.14
155	512	30.27	155	512	30.27
			155	511	30.33
156	512	30.47	156	512	30.47
			156	511	30.53
157	512	30.66	157	512	30.66
			157	511	30.72
158	512	30.86	158	512	30.86
			158	511	30.92
159	512	31.05	159	512	31.05
			159	511	31.12
160	512	31.25	160	512	31.25
256	512	50.00	256	512	50.00
			256	511	50.10
257	512	50.20	257	512	50.20
			257	511	50.29
258	512	50.39	258	512	50.39
			258	511	50.49
259	512	50.59	259	512	50.59
			259	511	50.68
260	512	50.78	260	512	50.78
410	512	80.08	410	512	80.08
			410	511	80.23
411	512	80.27	411	512	80.27
			411	511	80.43
412	512	80.47	412	512	80.47
			412	511	80.63
413	512	80.66	413	512	80.66
			413	511	80.82
414	512	80.86	414	512	80.86
			414	511	81.02
415	512	81.05	415	512	81.05
			415	511	81.21
416	512	81.25	416	512	81.25

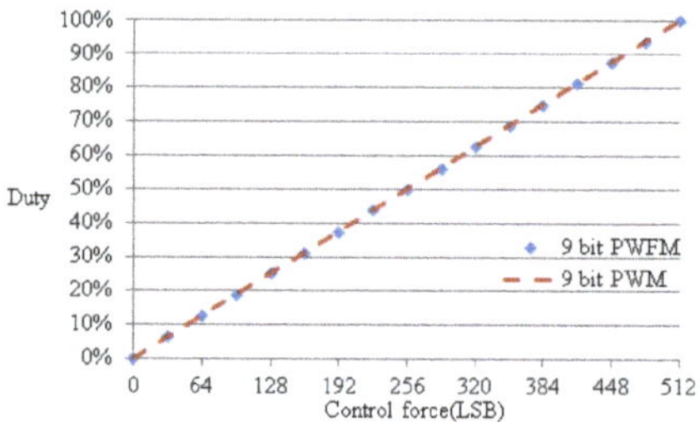

Figure 8. Curves of duty cycle versus control force for the PWFM and PWM strategies.

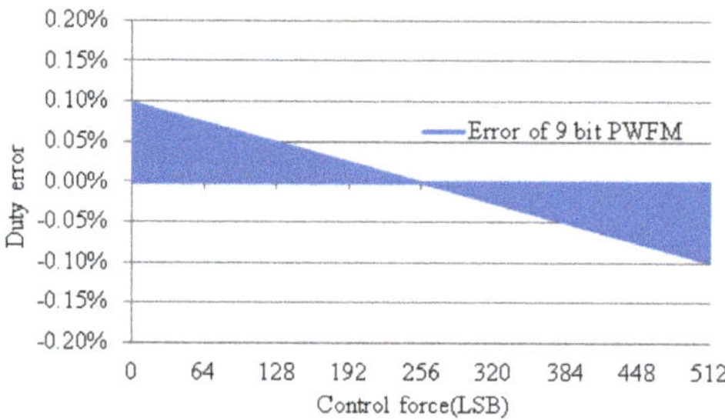

Figure 9. Curves of duty cycle error versus control force for the PWFM strategy.

9. PWFM Procedure

Figures 10 and 11 show the program flow charts for the traditional PWM strategy and the proposed PWFM strategy, respectively. From Figure 10, since the traditional 9-bit PWM has a period of 512 CLK, the 11-bit control force will be saved in a register with the last two bits cut off. At the same time, there is an up counter to be activated as PWM is equal to one, counting from zero. As soon as the counter value is equal to the duty cycle value, this counter will be set to zero. As for the PWFM shown in Figure 11, the 11-bit control force will be saved in a register with the last bit cut off. The first nine bits are integral values, similar to 9-bit PWM but the last bit is not an integral value, called 0.1bit, which will be finely modulated according to the PFM. The first nine bits will be put into a register and compared with the value of the up counter, and at the same time, the last bit will be checked. If the 0.1bit is equal to one, the accompanying period is 511 CLK, leading to the resolution of $(n + 0.5)$; otherwise, the integral bit information is obtained and the duty cycle is 512 CLK.

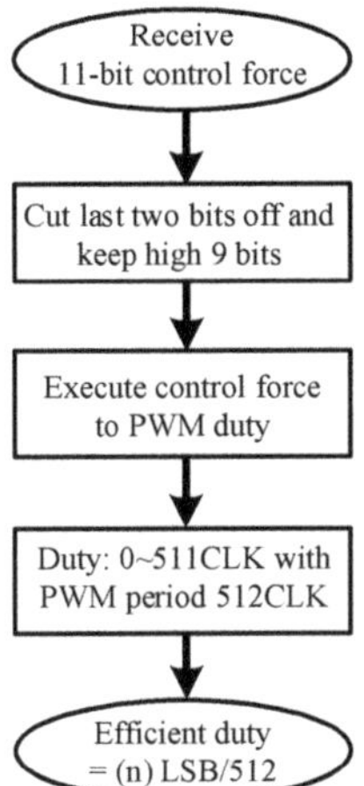

Figure 10. Program flow chart for the traditional PWM strategy.

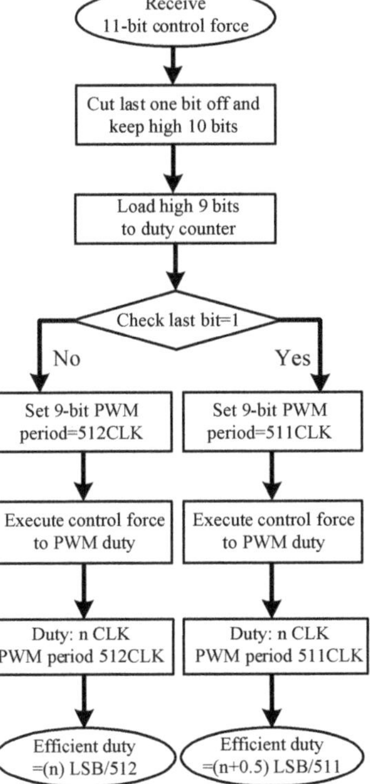

Figure 11. Program flow chart for the proposed PWFM strategy.

10. Experimental Results

Prior to this section, some specifications for a buck converter are given as follows: (i) The input voltage is 12 V; (ii) The output voltage is 5 V; (iii) The rated output current is 8 A; (iv) The switching frequency is 195 kHz; (v) The ADC is 9-bit with a peak-to-peak voltage of 1.7 V; (vi) The PWM is 9-bit and PWFM is 9-bit plus 1-bit; (vii) The value of the output inductor is 5 μH; (viii) The output capacitor is constructed by two 470 μF electrolytic capacitors and two 10 μF multilayer ceramic capacitors (MLCC), with all capacitors paralleled together; (x) The part names of the main switch Q_1 and synchronous rectifier Q_2 are the same, called IRL8113; (xi) The FPGA, belonging to Altera Cyclone 3 with operating clock of 100 MHz, has the part name of EP34C5T44.

Figures 12 and 13 show the waveforms relevant to the traditional PWM control strategy and the proposed PWFM control strategy under different loads, respectively. The controller is not well designed herein. The purpose of the controller is to show the limit cycle oscillation under the traditional PWM control strategy. Therefore, the transient part is not so important. As for the proportional gain k_p, it may affect the experimental results but may not be needed. As for the integral gain k_i, it must be needed to make the DC output voltage stable at a given value. In the following experiments, the value of k_i is 0.0625. From Figures 14 and 15, the limit cycle oscillation is removed, and the transient parts are almost the same as those shown in Figures 12 and 13. The features of the proposed PWFM control strategy are the same as those of the traditional PWM control strategy except that both the resolutions are different. Figures 16–19 show the zoom-in waveforms for Figures 12–15, respectively. From these figures, it can be seen that due to the switching frequency, the output voltages have high-frequency ripples, and the inductor currents also have high-frequency ripples.

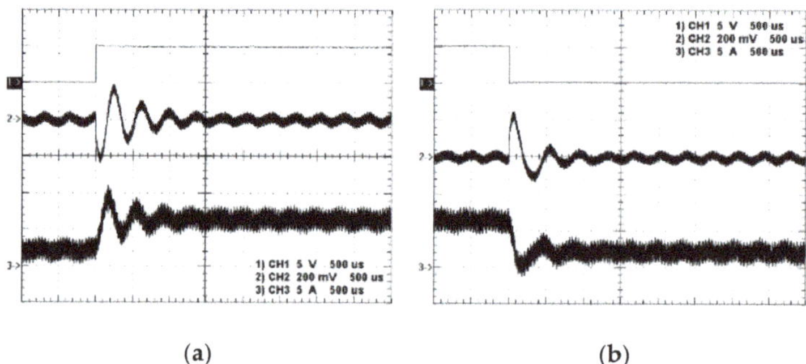

(a)   (b)

Figure 12. Waveforms based on the traditional PWM control strategy: (1) load enable; (2) AC output voltage; (3) inductor current, due to (**a**) from 25% to 75%; (**b**) from 75% to 25%.

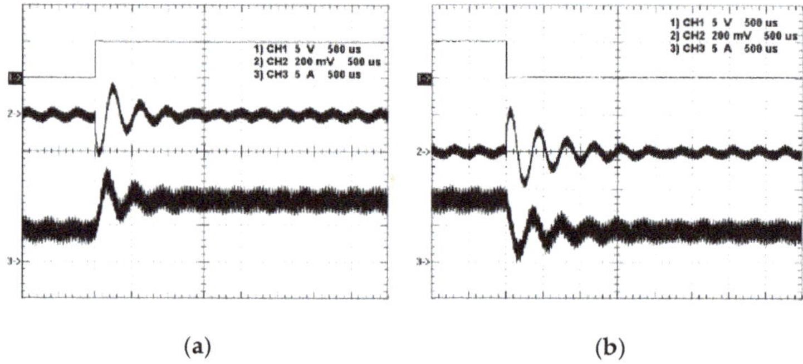

(a)   (b)

Figure 13. Waveforms based on the traditional PWM control strategy: (1) load enable; (2) AC output voltage; (3) inductor current, due to (**a**) from 50% to 100%; (**b**) from 100% to 50%.

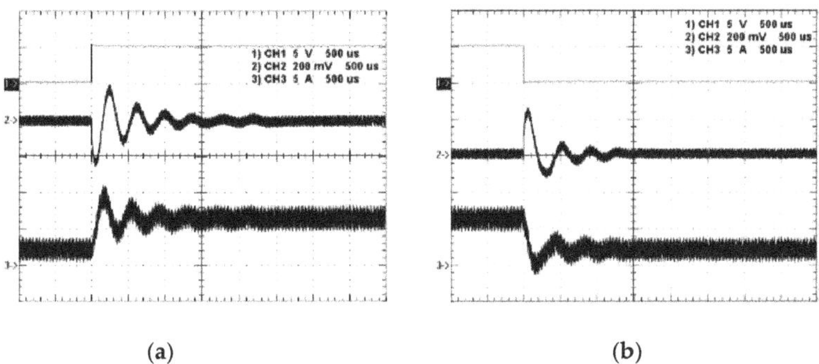

(a)　　　　　　　　　　　　(b)

Figure 14. Waveforms based on the proposed PWFM control strategy: (1) load enable; (2) AC output voltage; (3) inductor current, due to (**a**) from 25% to 75%; (**b**) from 75% to 25%.

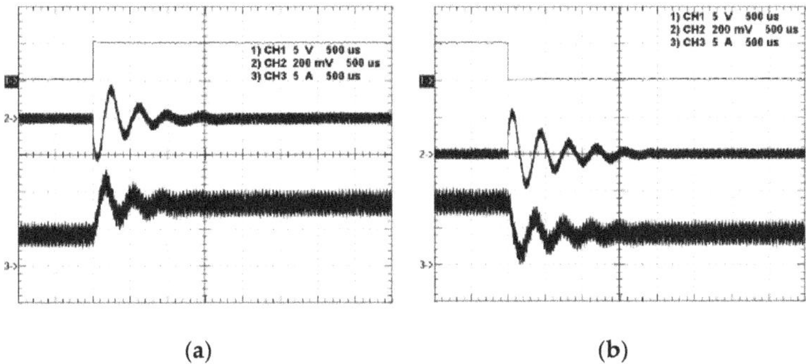

(a)　　　　　　　　　　　　(b)

Figure 15. Waveforms based on the proposed PWFM control strategy: (1) load enable; (2) AC output voltage; (3) inductor current, due to (**a**) from 50% to 100%; (**b**) from 100% to 50%.

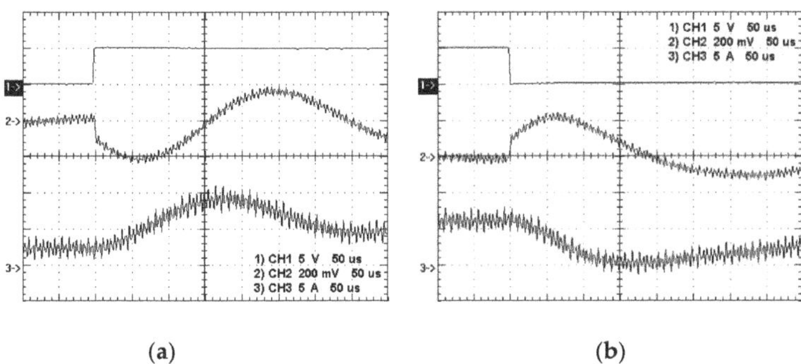

(a)　　　　　　　　　　　　(b)

Figure 16. Zoom-in waveforms based on the traditional PWM control strategy: (1) load enable; (2) AC output voltage; (3) inductor current, due to (**a**) from 25% to 75%; (**b**) from 75% to 25%.

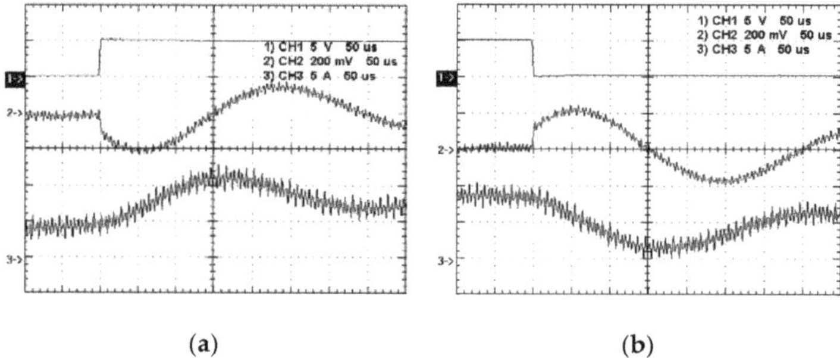

(a)   (b)

Figure 17. Zoom-in waveforms based on the traditional PWM control strategy: (1) load enable; (2) AC output voltage; (3) inductor current, due to (**a**) from 50% to 100%; (**b**) from 100% to 50%.

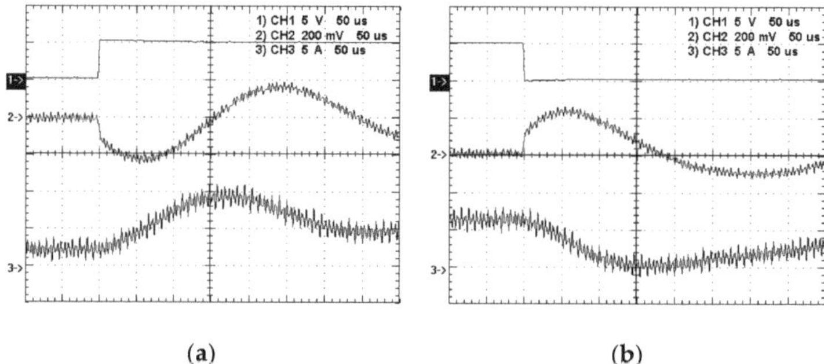

(a)   (b)

Figure 18. Zoom-in waveforms based on the proposed PWFM control strategy: (1) load enable; (2) AC output voltage; (3) inductor current, due to (**a**) from 25% to 75%; (**b**) from 75% to 25%.

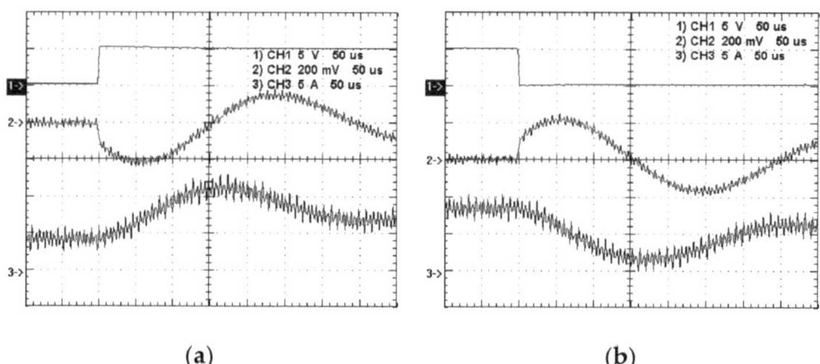

(a)   (b)

Figure 19. Zoom-in waveforms based on the proposed PWFM control strategy: (1) load enable; (2) AC output voltage; (3) inductor current, due to (**a**) from 50% to 100%; (**b**) from 100% to 50%.

11. Conclusions

A PWFM control strategy is presented herein by combining the one-comparator counter-based DPWM control with PFM control to increase pseudo-1-bit resolution under constant-frequency operation. By doing so, as compared with the traditional PWM control strategy, system stability will be enhanced, including no limit cycle oscillation although both have almost the same transient responses. Above all, the difference in internal structure between the two is quite small, and the circuit complexity and chip area are not altered. In the future, the number of bits for DPWM and PFM will be investigated to shorten the transient time.

Author Contributions: The conception was presented by K.I.H., who also was responsible for editing this paper. C.W.W. carried out experimental setup and verification. Y.T.Y. surveyed the existing papers and wrote the software program. K.I.H. was in charge of project administration.

Funding: This research was funded by the Ministry of Science and Technology, Taiwan, under the Grant Number MOST 107-2221-E-027-023.

Acknowledgments: The authors gratefully acknowledge the support of the Ministry of Science and Technology, Taiwan, under the Grant Number MOST 107-2221-E-027-023.

Conflicts of Interest: The authors declare no conflict of interest with commerce.

References

1. Peterchev, A.V.; Sanders, S.R. Quantization resolution and limit cycling in digitally controlled PWM converters. *IEEE Trans. Power Electron.* **2003**, *18*, 301–308. [CrossRef]
2. Peng, H.; Prodic, A.; Alarcon, E.; Maksimovic, D. Modeling of quantization effects in digitally controlled DC-DC converters. *IEEE Trans. Power Electron.* **2007**, *22*, 208–215. [CrossRef]
3. Prodic, A.; Maksimovic, D.; Erickson, R.W. Design and Implementation of a digital PWM controller for a high-frequency switching DC-DC power converter. In Proceedings of the 27th Annual Conference of the IEEE Industrial Electronics Society, Denver, CO, USA, 29 November–2 December 2001.
4. Bernstein, K.; Rohrer, N.J. *SOI circuit design concepts*; Springer: New York, NY, USA, 2007; pp. 6–9.
5. Curran, B.; Fluhr, E.; Paredes, J.; Sigal, L.; Friedrich, J.; Chan, Y.H.; Hwang, C. Power-constrained high-frequency circuits for the IBM POWER6 microprocessor. *IBM J. Res. Dev.* **2007**, *51*, 715–731. [CrossRef]
6. Dancy, A.P.; Chandrakasan, A.P. Ultra low power control circuits for PWM converters. In Proceedings of the 28th Annual IEEE Power Electronics Specialists Conference, Saint Louis, MO, USA, 27 June 1997.
7. Foley, R.; Kavanagh, R.; Marnane, W.; Egan, M. Multiphase digital pulsewidth modulator. *IEEE Trans. Power Electron.* **2006**, *21*, 842–846. [CrossRef]
8. Lukic, Z.; Wang, K.; Prodic, A. High-frequency digital controller for DC-DC converters based on multi-bit sigma-delta pulse-width modulation. In Proceedings of the Twentieth Annual IEEE Applied Power Electronics Conference and Exposition, Austin, TX, USA, 6–10 March 2005.
9. Foley, R.F.; Kavanagh, R.C.; Marnane, W.P.; Egan, M.G. An area efficient digital pulsewidth modulation architecture suitable for FPGA implementation. In Proceedings of the Twentieth Annual IEEE Applied Power Electronics Conference and Exposition, Austin, TX, USA, 6–10 March 2005.
10. Malley, E.O.; Rinne, K. A programmable digital pulse width modulator providing versatile pulse patterns and supporting switching frequencies beyond 15 MHz. In Proceedings of the Nineteenth Annual IEEE Applied Power Electronics Conference and Exposition, Anaheim, CA, USA, 22–26 February 2004.
11. Wang, K.; Rahman, N.; Lukic, Z.; Prodic, A. All-digital DPWM/DPFM controller for low-power DC-DC converters. In Proceedings of the Twenty-First Annual IEEE Applied Power Electronics Conference and Exposition, Dallas, TX, USA, 19–23 March 2006.
12. Yousefzadeh, V.; Takayama, T.; Maksimovic, D. Hybrid DPWM with digital delay-locked loop. In Proceedings of the 2006 IEEE Workshops on Computers in Power Electronics, Troy, NY, USA, 16–19 July 2006.
13. Peterchev, A.V.; Xiao, J.; Sanders, S.R. Architecture and IC implementation of a digital VRM controller. In Proceedings of the 2001 IEEE 32nd Annual Power Electronics Specialists Conference, Vancouver, BC, Canada, 17–21 June 2001.
14. Lukic, Z.; Blake, C.; Huerta, S.C.; Prodic, A. Universal and fault tolerant multiphase digital PWM controller IC for high-frequency DC-DC converters. In Proceedings of the Twenty-Second Annual IEEE Applied Power Electronics Conference and Exposition, Anaheim, CA, USA, 25 February–1 March 2007.
15. Zhang, J.; Sanders, S.R. A digital multi-mode multi-phase IC controller for voltage regulator application. In Proceedings of the Twenty-Second Annual IEEE Applied Power Electronics Conference and Exposition, Anaheim, CA, USA, 25 February–1 March 2007.
16. Huerta, S.C.; de Castro, A.; Garcia, O.; Cobos, J.A. FPGA based digital pulse width modulator with time resolution under 2 ns. *IEEE Trans. Power Electron.* **2008**, *23*, 3135–3141. [CrossRef]
17. Syed, A.; Ahmed, E.; Maksimovic, D.; Alarcon, E. Digital pulse width modulator architectures. In Proceedings of the 2004 IEEE 35th Annual Power Electronics Specialists Conference, Aachen, Germany, 20–25 June 2004.

18. Kelly, A.; Rinne, K. High resolution DPWM in a DC-DC converter application using digital sigma-delta techniques. In Proceedings of the 2005 IEEE 36th Power Electronics Specialists Conference, Recife, Brazil, 16 June 2005.

19. de Castro, A.; Todorovich, E. DPWM based on FPGA clock phase shifting with time resolution under 100 ps. In Proceedings of the 2008 IEEE Power Electronics Specialists Conference, Rhodes, Greece, 15–19 June 2008.

20. Carosa, T.; Zane, R.; Maksimovic, D. Scalable digital multiphase modulator. *IEEE Trans. Power Electron.* **2008**, *23*, 2201–2205. [CrossRef]

21. Batarseh, M.G.; Al-Hoor, W.; Huang, L.; Iannello, C.; Batarseh, I. Segmented digital clock manager-FPGA based digital pulse width modulator technique. In Proceedings of the 2008 IEEE Power Electronics Specialists Conference, Rhodes, Greece, 15–19 June 2008.

22. Foley, R.F.; Kavanagh, R.C.; Marnane, W.P.; Egan, M.G. A versatile digital pulsewidth modulation architecture with area-efficient FPGA implementation. In Proceedings of the 2005 IEEE 36th Power Electronics Specialists Conference, Recife, Brazil, 16 June 2005.

23. Li, P.; Kang, Y.; Pei, X.; Chen, J. A novel PWM technique in digital control. *IEEE Ind. Electron.* **2007**, *54*, 338–346. [CrossRef]

24. Qiu, Y.; Li, J.; Xu, M.; Ha, D.S.; Lee, F.C. Proposed DPWM scheme with improved resolution for switching power converters. In Proceedings of the Twenty-Second Annual IEEE Applied Power Electronics Conference and Exposition, Anaheim, CA, USA, 25 February–1 March 2007.

25. Li, J.; Qiu, Y.; Sun, Y.; Huang, B.; Xu, M.; Ha, D.S.; Lee, F.C. High resolution digital duty cycle modulation schemes for voltage regulators. In Proceedings of the Twenty-Second Annual IEEE Applied Power Electronics Conference and Exposition, Anaheim, CA, USA, 25 February–1 March 2007.

26. Yau, Y.T.; Hwu, K.I. One-comparator sampling design for digital power converters. In Proceedings of the 2018 7th International Symposium on Next Generation Electronics, Taipei, Taiwan, 7–9 May 2018.

© 2019 by the authors. Licensee MDPI, Basel, Switzerland. This article is an open access article distributed under the terms and conditions of the Creative Commons Attribution (CC BY) license (http://creativecommons.org/licenses/by/4.0/).

 *electronics*

Article

Direct Power-Based Three-Phase Matrix Rectifier Control with Input Power Factor Adjustment

Jae-Chang Kim, Dongyeon Kim and Sang-Shin Kwak *

School of Electrical and Electronic Engineering, Chung-ang University, Seoul 06974, Korea;
wockddld@naver.com (J.-C.K.); alexandros47@naver.com (D.K.)
* Correspondence: sskwak@cau.ac.kr; Tel.: +82-2-820-5346

Received: 22 October 2019; Accepted: 27 November 2019; Published: 29 November 2019

Abstract: In a current source rectifier such as a matrix rectifier, input voltage and current cannot be in phase unless an additional input power factor control technique is implemented. This paper proposes such a technique for a matrix rectifier using power-based space vector modulation (SVM). In the proposed method, the modulation index and phase required in order to apply the SVM are calculated based on the active and reactive power of the rectifier for intuitive power factor control. The active power that the rectifier should generate for the regulation of the output inductor current is obtained by the PI (proportional-integral) controller. The reactive power, which is supplied by the rectifier for adjustment of the power factor, is assigned differently depending on the output condition: for the output condition capable of unity power factor, it is set to a negative value of reactive power of the input capacitor, and when the unity power factor is not achievable, it is set with the maximum reactive power the rectifier can generate under the given condition to attain the maximum possible input power factor. It is determined whether the given condition is the light load condition by comparing the absolute value of the reactive power supplied by the input capacitor with the maximum rectifier reactive power that can be produced under the given condition. The SVM based on the active and reactive power of the rectifier in this technique allows the input power factor control to be intuitive and simple. The performance and feasibility of the technique were proved by simulation and experimentation.

Keywords: power factor adjustment; space vector modulation; current source converter; matrix rectifier

---

1. Introduction

When the charger of an electric vehicle is manufactured with a voltage source rectifier with a boost type characteristic, an additional DC-to-DC converter is needed. This DC-to-DC converter drops the voltage, because the battery voltage used in an electric vehicle is lower than the grid voltage [1–5]. However, because a matrix rectifier derived from an AC-to-AC matrix converter is a rectifier with a buck-type current source characteristic, an extra DC-to-DC converter is not needed when designing the charger using a matrix rectifier [6–12]. Therefore, an electric charger with a matrix rectifier has the advantage of reducing power conversion efficiency and volume compared to voltage source-based chargers.

The matrix rectifier is a type of current source rectifier that uses DC current flow in the output inductor to produce a pulse wave of the same size as the output inductor current at the rectifier input [13]. The duty value of the pulse wave is regulated to control the fundamental component of the rectifier input current. On the input side of the current source rectifier, an LC filter is installed; it removes the harmonic content of the rectifier input current and provides a path for the input current to flow when the rectifier input current is zero. The current source rectifier controls the phase of the rectifier input current in synchronization with the phase of the input voltage or input capacitor voltage.

However, this control method causes a phase difference between the input voltage and input current because of the current flowing in the input capacitor. Therefore, the input voltage and input current cannot be in phase unless an additional input power factor control technique is applied.

To regulate the input power factor of the current source rectifier, several techniques have been proposed [14–18]. In Choi et al. and Zargari et al. [14,15], a power factor adjustment delaying the phase of the rectifier input current was proposed. However, because the delay angle for unity power factor control was calculated without considering the voltage drop of the input inductor, there was a limitation in that the power factor was not precisely unity. In Zargari et al. and Zhang et al. [16,17], the rectifier input current, which puts the input voltage and input current in phase with each other, was obtained by mathematical calculations using the relationship between voltage and current at the input side. However, the calculation process used was complicated and not intuitive. Another limitation of these studies [14–17] is that there was no description of the power factor control technique where unity power factor control was not achievable. In Kim et al. [18], a power factor regulation technique using a virtual capacitor was proposed. To compensate for the current flowing through the input capacitor, which causes the nonunity power factor, a virtual capacitor was applied in parallel to the input capacitor. In order to implement the virtual capacitor, the current flowing through it was additionally generated in the rectifier input current. Additionally, a maximum achievable power factor (MAPF) control method for the light load condition was developed. However, for the use of the differentiator required to calculate the current of the virtual capacitor, a noise cancellation technique with dq conversion was used, which complicates the control technique. In addition, it is not intuitive to generate additional current realizing the virtual capacitor.

This paper proposes a power factor adjustment technique for a matrix rectifier using a power-based space vector modulation (SVM). In the proposed technique, the modulation index and phase for the SVM are calculated based on the active and reactive power of the rectifier for intuitive power factor control. The active power that the rectifier should generate for the regulation of the output inductor current is obtained by the PI (proportional-integral) controller. The reactive power supplied by the rectifier for adjustment of the power factor is assigned differently depending on the output condition. First, in the output condition capable of unity power factor control, the rectifier generates a negative value of reactive power of the input capacitor. In the light load condition, where MAPF is possible, the reactive power reference of the rectifier is set to the maximum reactive power the rectifier can produce under the given condition to attain the maximum possible input power factor. The determination of whether the given condition is the light load condition is judged by comparing both the absolute value of the reactive power supplied by the input capacitor and the maximum rectifier reactive power that can be produced under the given condition. The advantage of the proposed technique is that the input power factor adjustment technique is intuitive and simple, owing to the SVM based on the active and reactive power of the rectifier. The performance and feasibility of the proposed technique were proved by simulation and experimentation.

2. Conventional Space Vector Modulation

Figure 1 shows the matrix rectifier; S_1–S_6 indicate its bidirectional switch, L_i and C_i represent the inductor and capacitor of the input LC filter, L_o and C_o refer to the inductor and capacitor of the output LC filter, I_r denotes the rectifier input current, v_s denotes the input voltage, i_s represents the input current, and v_c is the input capacitor voltage. The voltage and current of the load are represented by I_{load} and V_{load}, respectively. I_{dc} is the current through the output inductor.

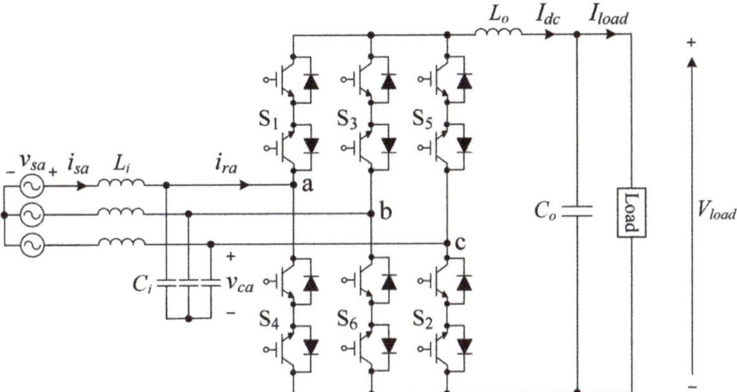

Figure 1. Matrix rectifier.

The matrix rectifier is the current source type that controls I_{dc} to a constant reference value. The matrix rectifier generates the pulse wave in the rectifier input side by I_{dc} and controls the fundamental component of the rectifier input current by regulating the duty of the pulse wave. A general technique for controlling the current source type rectifier is SVM, which uses a space vector obtained by *abc* to $\alpha\beta$ conversion of the three-phase rectifier input current. This is determined based on the switching state of the rectifier. Table 1 shows the space vectors based on the switching states, and Figure 2 plots the magnitude and phase of the space vector with respect to the $\alpha\beta$ axis.

Table 1. Space vector based on switching states.

	Space Vector	On-State Switches
	I_1	(S_1, S_6)
	I_2	(S_1, S_2)
	I_3	(S_3, S_2)
Active vectors	I_4	(S_3, S_4)
	I_5	(S_5, S_4)
	I_6	(S_5, S_6)
Zero vectors	I_0	$(S_1, S_4), (S_3, S_6), (S_5, S_2)$

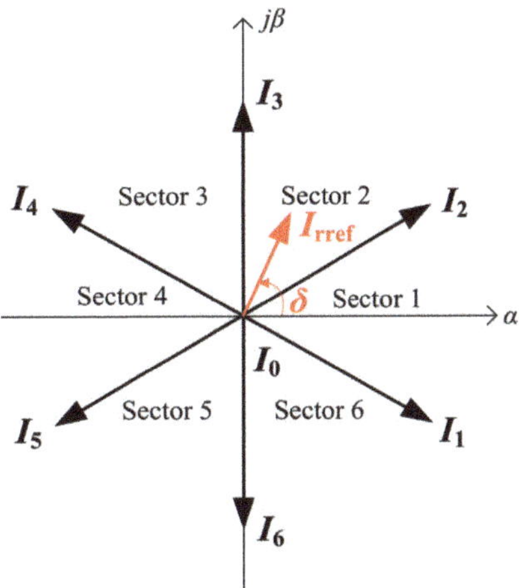

Figure 2. Space vector diagram in $\alpha\beta$ plane.

In Figure 2, I_{rref} and δ represent the reference of the rectifier input current and the phase of I_{rref}, respectively. As shown in Figure 2, the sector can be divided according to the phase of the reference of the rectifier input current. Each sector has different space vectors adjacent to the reference value. Two active vectors and one zero vector adjacent to the reference value in one sampling are used to make the rectifier input current follow the reference value. When I_{rref} is in Sector 1, the duty value of I_1, I_2, and I_0 is obtained by Equation (1) to track the reference of the rectifier input current.

$$
\begin{aligned}
d_1 &= m\sin\left(-\delta + \tfrac{\pi}{6}\right) \\
d_2 &= m\sin\left(\delta + \tfrac{\pi}{6}\right) \\
d_o &= 1 - (d_1 + d_2)
\end{aligned}
\tag{1}
$$

In Equation (1), d_1, d_2, and d_0 indicate the duty values of I_1, I_2, and I_0, respectively, and m represents the modulation index defined by Equation (2).

$$
m = \frac{|I_{rref}|}{I_{dc}}
\tag{2}
$$

Figure 3 shows a block diagram of a conventional SVM. As shown in Equation (1), m and δ are needed for the SVM. In conventional SVM control, the modulation index is obtained from the output inductor current and the PI controller. In addition, the phase of the rectifier input current is used by the phase of the input voltage and obtained from using a phase locked loop (PLL).

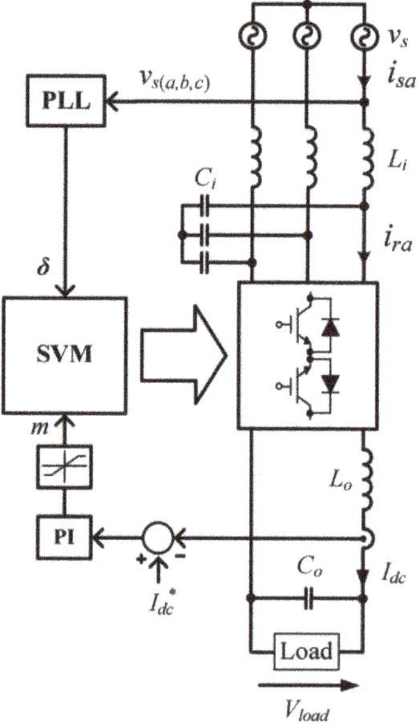

Figure 3. Block diagram of the conventional space vector modulation (SVM).

Figure 4 shows the phase relationship between i_s, i_{ci}, i_r, and input voltage v_s. In Figure 4, i_{ci} means input capacitor current, and θ denotes the phase difference between the input voltage and input current. When the SVM is implemented in the same manner as shown in Figure 3, the input voltage and rectifier input current i_r are in phase, as in Figure 4. As a result, i_{ci} causes the input current i_s to be out of phase with the input voltage. To remove this phase difference, the new rectifier input current that makes input power factor unity can be obtained from mathematical calculations using the voltage and current at the input side [17]. The virtual capacitor can also be applied in parallel to the input capacitor to compensate for the leading current [18]. However, because these techniques are not direct methods of controlling power, power factor control is not intuitive. In this paper, a new power-based SVM is proposed for intuitive power factor control.

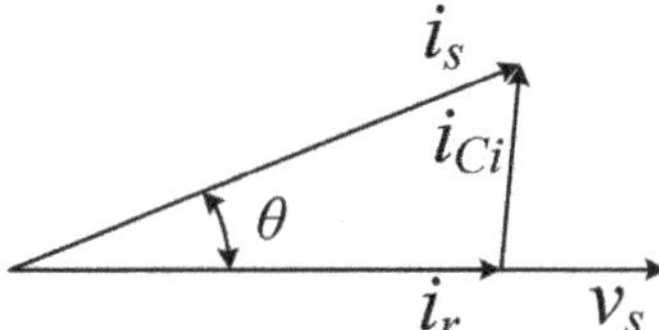

Figure 4. Phase relationship between i_s, i_{ci}, i_r, and v_s.

3. Proposed Method

Equation (3) is used to derive a new power-based SVM.

$$P_r = 1.5\left(v_{s\alpha}i_{r\alpha1} + v_{s\beta}i_{r\beta1}\right)$$
$$Q_r = 1.5\left(v_{s\beta}i_{r\alpha1} - v_{s\alpha}i_{r\beta1}\right) \tag{3}$$

In Equation (3), P_r and Q_r represent the active and reactive power produced by the fundamental component of the rectifier input current, and $v_{s\alpha}$ and $v_{s\beta}$ represent the result of the *abc* to $\alpha\beta$ conversion of the three-phase input voltage. In addition, $i_{r\alpha1}$ and $i_{r\beta1}$ denote fundamental components of the result of the *abc* to $\alpha\beta$ conversion of the three-phase rectifier input current. Equation (3) is arranged as Equation (4) for $i_{r\alpha1}$ and $i_{r\beta1}$.

$$i_{i\alpha1} = \frac{2}{3}\left(\frac{v_{s\alpha}}{v_{s\alpha}^2+v_{s\beta}^2}P_r + \frac{v_{s\beta}}{v_{s\alpha}^2+v_{s\beta}^2}Q_r\right)$$
$$i_{i\beta1} = \frac{2}{3}\left(\frac{v_{s\beta}}{v_{s\alpha}^2+v_{s\beta}^2}P_r - \frac{v_{s\alpha}}{v_{s\alpha}^2+v_{s\beta}^2}Q_r\right) \tag{4}$$

Equation (4) can be used to find the fundamental components of the rectifier input current using the active and reactive power of the rectifier. The modulation index and phase for SVM operation can be obtained as follows:

$$m = \frac{\sqrt{i_{i\alpha1}^2+i_{i\beta1}^2}}{I_{dc}} = \frac{2}{3I_{dc}}\sqrt{\frac{P_r^2+Q_r^2}{v_{s\alpha}^2+v_{s\beta}^2}}$$
$$\delta = \tan^{-1}\frac{i_{i\beta1}}{i_{i\alpha1}} = \frac{v_{s\beta}P_r - v_{s\alpha}Q_r}{v_{s\alpha}P_r + v_{s\beta}Q_r}. \tag{5}$$

Neglecting the losses in the rectifier, the respective active powers of the output and the rectifier are equal. Thus, the active power that must be produced by the rectifier to generate the desired output inductor current can be set as the reference value for the active power of the rectifier.

On the other hand, the reactive powers of the input, input capacitor, and rectifier have the following relationship:

$$Q_s = Q_c + Q_r. \tag{6}$$

In Equation (6), Q_s, Q_c, and Q_r represent the reactive powers of the input, input capacitor, and rectifier, respectively. In order for the input voltage and input current to be in phase, the input reactive power must be zero. Therefore, the rectifier should compensate for the reactive power generated by the input capacitor for the unity power factor control. The reactive power of the input capacitor can be attained as follows:

$$Q_c = -1.5\omega C_i V_s. \tag{7}$$

In Equation (7), ω is the angular frequency of the input voltage, and V_s is the peak magnitude of the input voltage. The input reactive power can be controlled to zero if the rectifier compensates the reactive power supplied by the input capacitor by generating the negative value of it obtained in (7).

However, the unity power factor cannot be accomplished in an all-output condition. The range of the modulation index is 0–1. As can be seen from Equation (5), once the active power to be produced by the rectifier is determined for the output inductor current control, the maximum reactive power that can be supplied under such output conditions is decided. Therefore, in order to determine the reactive power reference of the rectifier, it is first identified whether the unity power factor control is achievable under the given load condition. The maximum reactive power that the rectifier can generate at the given output conditions (Q_{rmax}) can be obtained by setting the modulation index to 1, as shown in Equation (8).

$$Q_{rmax} = \sqrt{(1.5I_{dc})^2\left(v_{ci\alpha}^2 + v_{ci\beta}^2\right) - P_r^2} \tag{8}$$

If the magnitude of Q_{rmax} calculated by Equation (8) is larger than or equal to the absolute value of the reactive power supplied by input capacitor, the reactive power reference of the rectifier is set to the negative reactive power value of the input capacitor. Otherwise, the reactive power reference of the rectifier is assigned to be the maximum reactive power Q_{rmax}. The method of determining the reference of the reactive power of the rectifier is summarized in Equation (9).

$$Q_r^* = \begin{cases} |Q_c| & Q_{rmax} \geq |Q_c| \\ Q_{rmax} & Q_{rmax} < |Q_c| \end{cases} \tag{9}$$

In Equation (9), the case where Q_{rmax} is greater or equal to $|Q_c|$ is called the normal condition in which the unity power factor can be achieved. On the other hand, the case where Q_{rmax} is lower than $|Q_c|$ is called the light load condition where MAPF can be attained.

A block diagram of the proposed technique is shown in Figure 5. Here, the superscript * marks a reference value. Because the proposed method performs SVM based on power, the power factor control is more intuitive than the conventional power factor control technique. With this intuitive power control, the reference value of the reactive power changes organically based on the load condition. As a result, the maximum power factor can be controlled even in the output condition where the unity power factor control is not achievable. Table 2 shows comparisons between conventional method and proposed method.

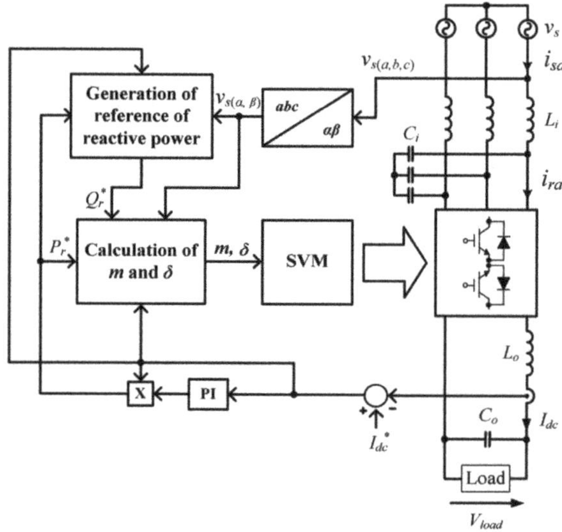

Figure 5. Block diagram of the proposed method.

Table 2. Comparisons between the conventional method and proposed method.

	Conventional Method	Proposed Method
Output current control	Possible	Possible
Unity power factor control	Impossible	Possible
Maximum achievable power factor control in light load condition	Impossible	Possible

4. Simulation Results

A simulation was carried out to confirm the performance of the proposed technique and to verify the power factor control performance by comparing the conventional SVM described in Figure 3 with the proposed scheme. Table 3 shows the parameters used for the simulation.

Table 3. Parameters for simulation and experiment.

Parameter	Value
Input phase peak voltage v_s	100 V
Input voltage frequency f	60 Hz
Input inductance L_i	1 mH
Input capacitance C_i	60 μF
Output inductance L_o	2.5 mH
Output capacitance C_o	40 μF
Load resistance R	20 Ω
Sampling and switching frequency	5 kHz

Figure 6 shows the simulation results of the conventional SVM and the proposed technique in the normal condition; Figure 6a shows the simulation result conducted by conventional SVM, and Figure 6b shows the simulation result obtained by proposed method. As shown in Figure 6a, the conventional SVM sets the phase of the rectifier input current to the phase of the input voltage, so that there is a phase difference between the input current and the input voltage. However, as shown in Figure 6b, in the proposed method, the input voltage and input current are set in phase by setting the reactive power of the rectifier to compensate the reactive power of the input capacitor through the power-based SVM.

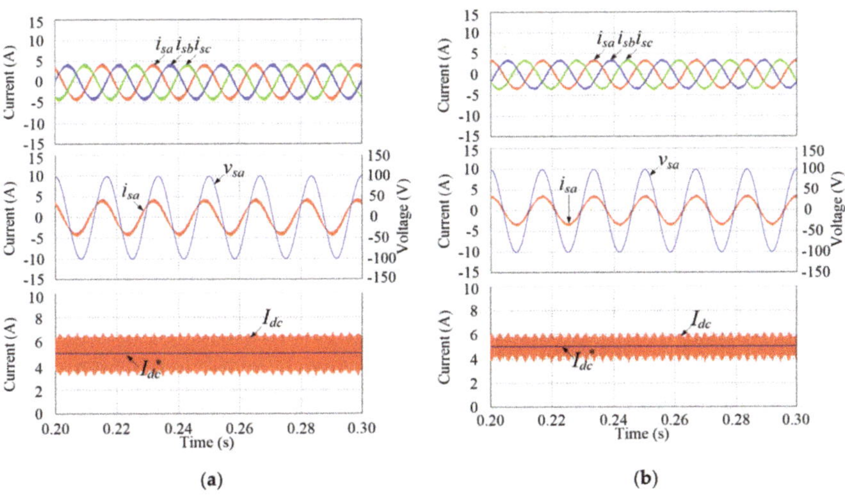

Figure 6. Simulation in normal conditions: (**a**) conventional SVM, (**b**) proposed method.

Figure 7 is the simulation result under the light load condition where the output power is about 20% of the output power of Figure 6. As shown in Figure 7a, the conventional SVM follows the reference value of the output current even under a light load, but there is the phase difference between the input voltage and the input current as in the normal condition. However, in Figure 7b, which is the simulation result of the proposed method under light load conditions, the input voltage and input current have the smallest phase difference even under the light load condition.

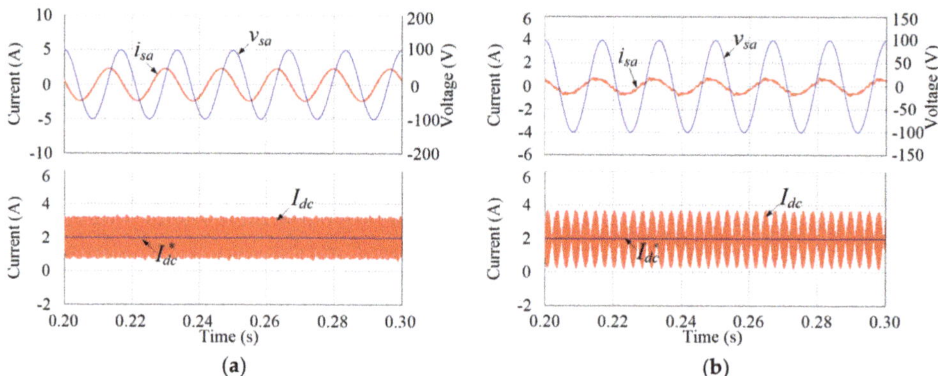

Figure 7. Simulation under light load condition: (**a**) conventional SVM, (**b**) proposed method.

Figure 8 shows the simulation result of a sudden change in the reference current of the output inductor. The reference current changes from 3 to 5 A in 0.2 s to evaluate the dynamic performance between the conventional SVM method and the proposed method. From Figure 8a,b, it can be seen that there is no difference in the dynamic performance between conventional SVM and the proposed scheme. In addition, Figure 8b shows not only that the proposed method can accurately follow the rapidly changing reference current, but that the unity power factor control can also be performed quickly.

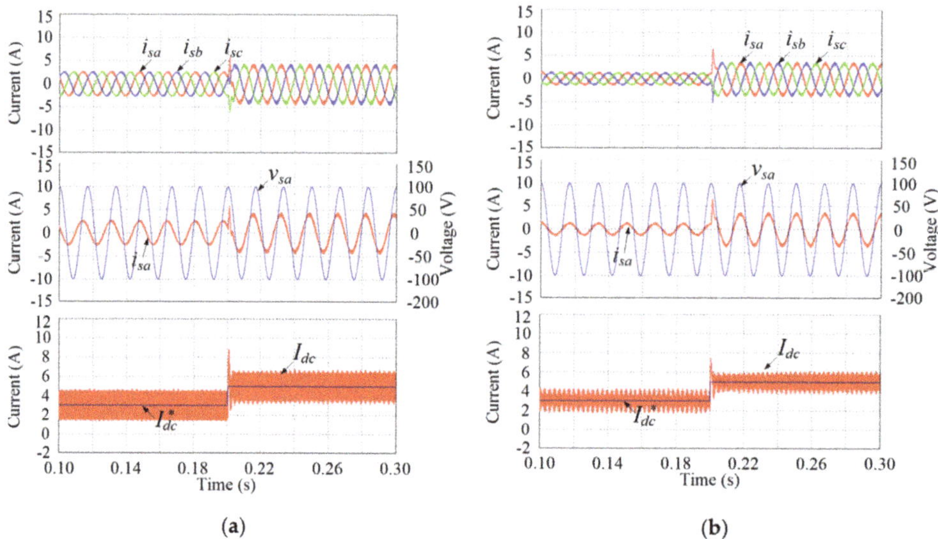

Figure 8. Simulation with load step change of 3–5 A: (**a**) conventional SVM, (**b**) proposed method.

5. Experimental Results

In order to evaluate the performance of the proposed method, experiments were performed using the conventional SVM and the proposed method in the same manner as the simulation. The experiment was conducted using the parameters shown in Table 3. The bidirectional switch was implemented using an insulated gate bipolar transistor (IGBT), where the part number is IXA37IF1200HJ and the rectifier was controlled by a TMS320F28335-based digital signal processor (DSP) board. Figure 9 shows the experiment setup used.

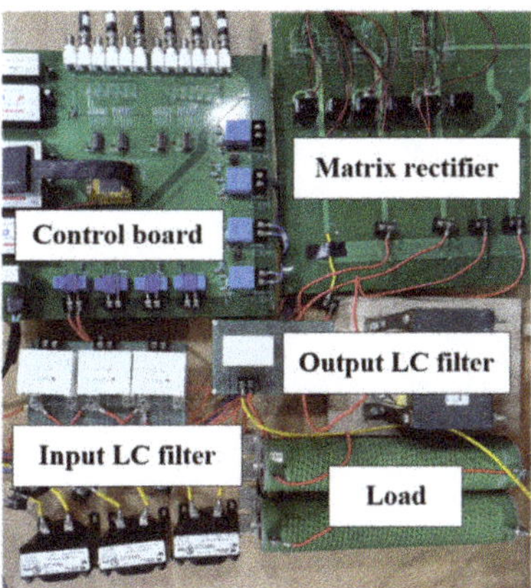

Figure 9. Experiment setup.

Figure 10 shows the experimental results of the conventional SVM and the proposed technique in the normal condition. The angle between the input voltage and input current is obtained by zero crossing points of the input voltage and input current using the oscilloscope. Figure 10a shows the experimental results of the conventional SVM. As in the simulation, the current of the output inductor follows the reference value of 5 A, but there is a phase difference between the input voltage and the input current due to the reactive power of the input capacitor. As a result, the input power factor of the conventional SVM is 0.90, in which the angle between the input voltage and input current is 26°. However, from Figure 10b, it can be seen that the output inductor current is well-controlled, and the unity power factor control is achieved by the reactive power reference value of the rectifier to compensate for the reactive power of input capacitor. The input power factor of the proposed method is 0.99, in which the angle between the input voltage and input current is almost 0°. In addition, the frequency analysis shows that the total harmonic distortion (THD) between the two techniques is almost identical.

Figure 11 shows the experimental results of the conventional SVM and the proposed power-based SVM under a light load. As in the simulation, in the conventional SVM, as shown in Figure 11a, the reference value of the output current is followed, but the input voltage and input current are not in phase. In the light load case, the power factors of the conventional SVM and the proposed method were 0.32 and 0.85, respectively. The phase between the input voltage and input current in the conventional method is 71°—a value that is too large. However, in the proposed method, the output inductor current follows the reference value of 2 A, and the maximum power factor is simultaneously realized. As a result, the phase difference between the input voltage and input current is reduced to 32°. Additionally, the THD of the proposed technique under the light load was 16.1%, which is much higher than that of the conventional SVM. This result is due to the reduction of the fundamental components of the input current to minimize the phase of the input voltage and input current [18,19].

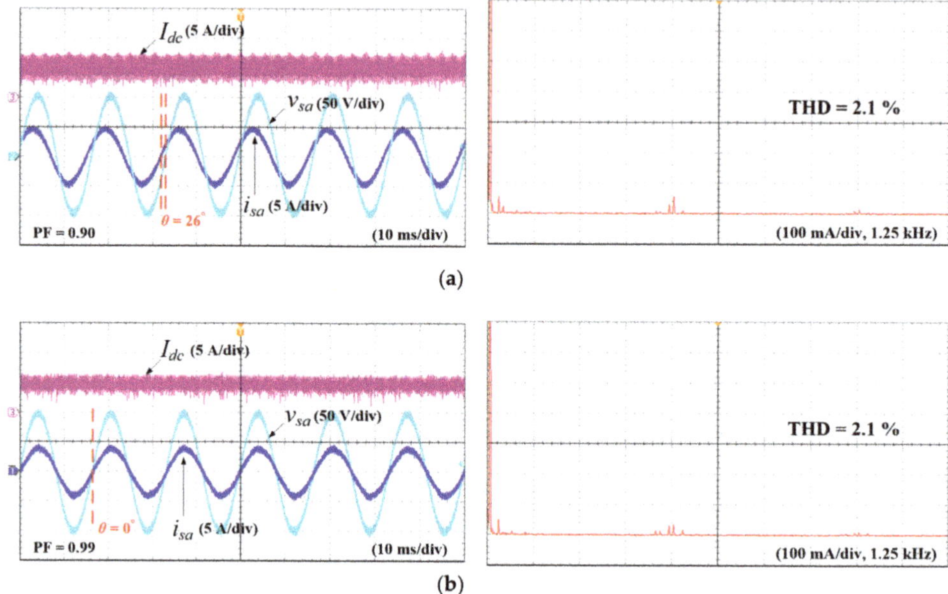

(**a**)

(**b**)

Figure 10. Experimental results of time domain and frequency domain in normal condition: (**a**) conventional SVM, (**b**) proposed method. THD: total harmonic distortion.

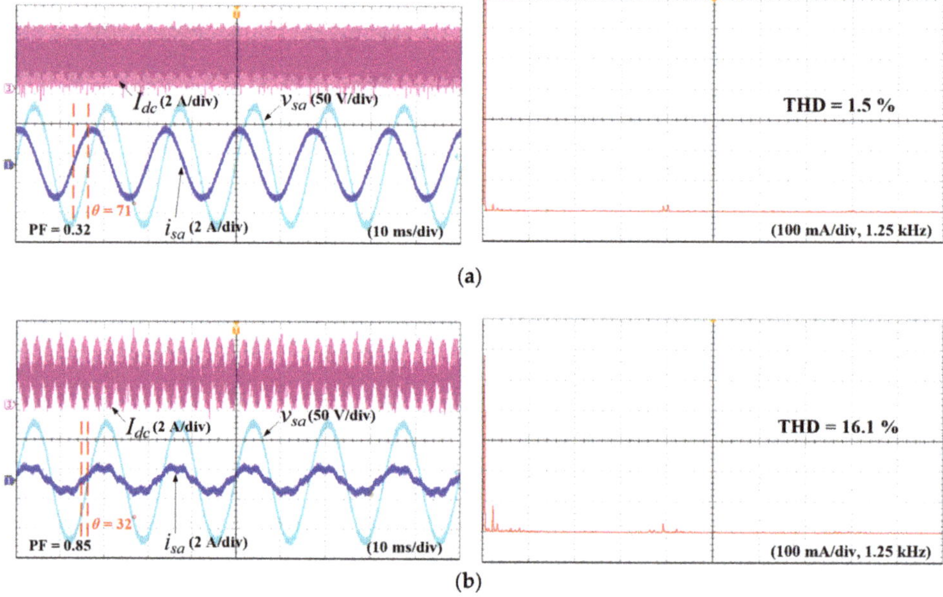

(**a**)

(**b**)

Figure 11. Experimental results of time domain and frequency domain in light load condition: (**a**) conventional SVM, (**b**) proposed method.

Figure 12 displays the experimental results that show the dynamic performance of the conventional SVM and the proposed method during the sudden change of the reference value of the output inductor current. Figure 12 shows that, as in simulation, there is no difference in dynamic performance between the conventional SVM and the proposed technique. Moreover, the proposed technique shows that the output inductor current rapidly follows the reference value and maintains the unity power factor in the step change of the inductor output current.

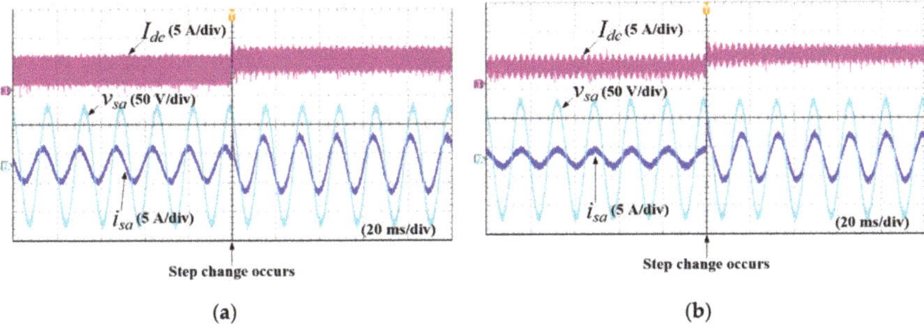

Figure 12. Experiment results in load step change: (**a**) conventional SVM, (**b**) proposed method.

Figure 13 shows the simulation results and the experiment results under the normal and light load conditions of the proposed method. Figure 13a compares the simulation results with the experiment results of the proposed method under the normal load condition. As the simulation results, it can be confirmed from Figure 13a that the input voltage and input current are controlled in phase and that the current of the output inductor follows the reference value well in the experiment results. Figure 13b shows the simulation results and the experiment results of the proposed method under light load condition. From Figure 13b, it can be seen that MAPF is obtained by minimizing the phase difference between the input voltage and the input current in both simulation and experiment.

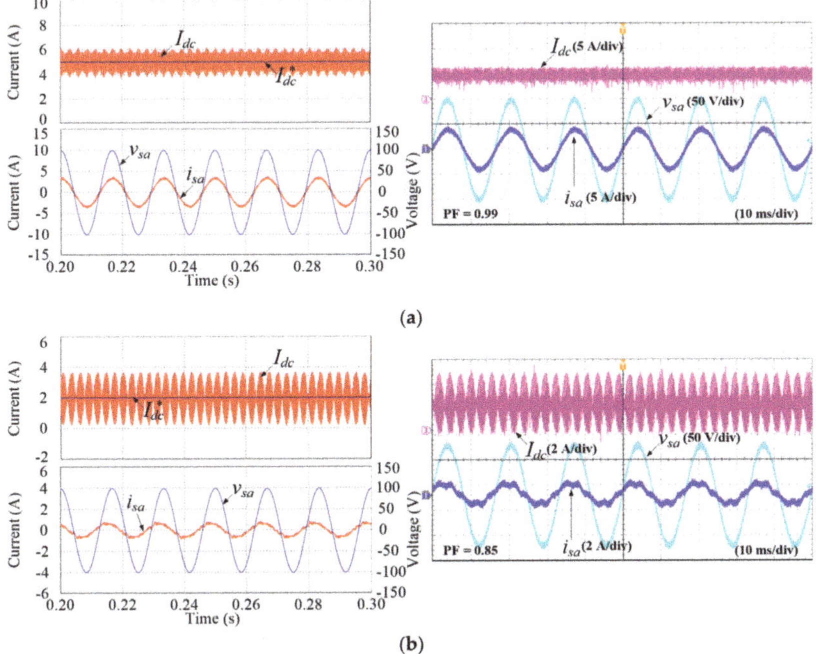

Figure 13. Comparisons between simulation results and experiment results of the proposed method in (**a**) the normal condition and (**b**) the light load condition.

6. Conclusions

In this paper, a novel SVM method based on the active and reactive power of a matrix rectifier was proposed. In the conventional SVM, the modulation index and phase required to drive the SVM were obtained from the PI controller to control the output inductor current and phase of the input voltage. However, this approach makes input power factor control unintuitive. In the proposed scheme, the modulation index and phase were calculated using the active and reactive power of the rectifier for intuitive power factor control. The active power reference for the rectifier was obtained from the PI controller, the output inductor current, and the reference of the output inductor current. In order to compensate for the reactive power of the input capacitor and to achieve the unity power factor, the reference of the rectifier reactive power was set to the negative value of the reactive power of the input capacitor. Under light load conditions, where unity power factor control was not possible, the MAPF was obtained by setting the reference value of the rectifier reactive power to the maximum reactive power that the rectifier can produce. The advantage of the proposed technique is that it can directly achieve the unity power factor using power-based SVM, and in the light load condition, MAPF can be accomplished by supplying the maximum reactive power of the rectifier. The simulation and experimental results confirm the performance of the proposed method.

Author Contributions: Conceptualization, S.-S.K.; methodology, S.-S.K., J.-C.K.; software, D.K.; validation, J.-C.K.; formal analysis, J.-C.K.; investigation, J.-C.K.; resources, S.-S.K.; data curation, D.K.; writing—original draft preparation, J.-C.K.; writing—review and editing, S.-S.K.; visualization, J.-C.K.; supervision, S.-S.K.; project administration, S.-S.K.; funding acquisition, S.-S.K.

Funding: This research was supported by the National Research Foundation of Korea (NRF) grant funded by the Korean government (MSIP) (2017R1A2B4011444).

Acknowledgments: This research was supported by the National Research Foundation of Korea (NRF) grant funded by the Korean government (MSIP) (2017R1A2B4011444).

Conflicts of Interest: The authors declare no conflict of interest.

References

1. Rodríguez, J.R.; Dixon, J.W.; Espinoza, J.R.; Pontt, J.; Lezana, P. PWM regenerative rectifiers: State of the art. *IEEE Trans. Ind. Electron.* **2005**, *52*, 5–22. [CrossRef]
2. Erb, D.C.; Onar, O.C.; Khaligh, A. Bi-directional charging topologies for plug-in hybrid electric vehicles. In Proceedings of the Twenty-Fifth Annual IEEE Applied Power Electronics Conference and Exposition (APEC), Palm Springs, CA, USA, 21–25 February 2010; pp. 2066–2072.
3. Gallardo-Lozano, J.; Milanés-Montero, M.I.; Guerrero-Martinez, M.A.; Romero-Cadaval, E. Three-phase bidirectional battery charger for smart electric vehicles. In Proceedings of the 7th International Conference-Workshop Compatibility and Power Electronics (CPE), Tallinn, Estonia, 1–3 June 2011; pp. 371–376.
4. Tan, N.M.L.; Abe, T.; Akagi, H. Design and performance of a bidirectional isolated dc–dc converter for a battery energy storage system. *IEEE Trans. Power Electron.* **2012**, *27*, 1237–1248. [CrossRef]
5. Yilmaz, M.; Krein, P. Review of battery charger topologies, charging power levels and infrastructure for plug-in electric and hybrid vehicles. *IEEE Trans. Power Electron.* **2013**, *28*, 2151–2169. [CrossRef]
6. Holmes, D.G.; Lipo, T.A. Implementation of a controlled rectifier using AC–AC matrix converter theory. *IEEE Trans. Power Electron.* **1992**, *7*, 240–250. [CrossRef]
7. Su, M.; Wang, H.; Sun, Y.; Yang, J.; Xiong, W.; Liu, Y. AC/DC matrix converter with an optimized modulation strategy for V2G applications. *IEEE Trans. Power Electron.* **2013**, *28*, 5736–5745. [CrossRef]
8. You, K.; Xiao, D.; Rahman, M.F.; Uddin, M.N. Applying reduced general direct space vector modulation approach of AC–AC matrix converter theory to achieve direct power factor controlled three-phase AC–DC matrix rectifier. *IEEE Trans. Ind. Appl.* **2014**, *50*, 2243–2257. [CrossRef]
9. Feng, B.; Lin, H.; Wang, X. Modulation and control of ac/dc matrix converter for battery energy storage application. *IET Power Electron.* **2015**, *8*, 1583–1594. [CrossRef]

10. Varajao, D.; Araujo, R.E.; Miranda, L.M.; Lopes, J.A.P. Modulation strategy for a single-stage bidirectional and isolated AC–DC matrix converter for energy storage systems. *IEEE Trans. Ind. Electron.* **2018**, *65*, 3458–3468. [CrossRef]
11. Afsharian, J.; Xu, D.; Wu, B.; Gong, B.; Yang, Z. A NEW PWM and Commutation Scheme for One Phase Loss Operation of Three-Phase Isolated Buck Matrix-Type Rectifier. *IEEE Trans. Power Electron.* **2018**, *33*, 9854–9865. [CrossRef]
12. Afsharian, J.; Xu, D.D.; Wu, B.; Gong, B.; Yang, Z. The Optimal PWM Modulation and Commutation Scheme for Three-Phase Isolated Buck Matrix Type Rectifier. *IEEE Trans. Power Electron.* **2018**, *33*, 110–124. [CrossRef]
13. Wu, B.; Narimani, M. *High-Power Converters and AC Drives*; John Wiley & Sons: Piscataway, NJ, USA, 2006.
14. Choi, J.H.; Kojori, H.A.; Dewan, S.B. High power GTO-CSC based power supply utilizing SHE-PWM and operating at unity power factor. In Proceedings of the Canadian Conference on Electrical and Computer Engineering, Vancouver, BC, Canada, 14–17 September 1993.
15. Zargari, N.R.; Xiao, Y.; Wu, B. Near unity input displacement factor for current source PWM drives. *IEEE Ind. Appl. Mag.* **1999**, *5*, 19–25. [CrossRef]
16. Zargari, N.R.; Joos, G. A three-phase current-source type PWM rectifier with feed-forward compensation of input displacement factor. In Proceedings of the Power Electronics Specialist Conference—PESC'94, Taipei, Taiwan, 20–25 June 1994; Volume 1, pp. 363–368.
17. Zhang, Y.; Yi, Y.; Dong, P.; Liu, F.; Kang, Y. Simplified model and control strategy of three-phase PWM current source rectifiers for dc voltage power supply applications. *IEEE J. Emerg. Sel. Topics Power Electron.* **2015**, *3*, 1090–1099. [CrossRef]
18. Kim, J.C.; Kwak, S.; Kim, T. Power factor control method based on virtual capacitors for three-phase matrix rectifiers. *IEEE Access* **2019**, *7*, 12484–124949. [CrossRef]
19. Kim, J.C.; Kwak, S. Direct power control method with minimum reactive power reference for three-phase AC-to-DC matrix rectifiers using space vector modulation. *IEEE Access* **2019**, *7*, 67515–67525. [CrossRef]

© 2019 by the authors. Licensee MDPI, Basel, Switzerland. This article is an open access article distributed under the terms and conditions of the Creative Commons Attribution (CC BY) license (http://creativecommons.org/licenses/by/4.0/).

Article

A Parametric Conducted Emission Modeling Method of a Switching Model Power Supply (SMPS) Chip by a Developed Vector Fitting Algorithm

Xuchun Hao, Shuguo Xie * and Ziyao Chen

School of Electronic and Information Engineering, Beihang University, Beijing 100083, China
* Correspondence: xieshuguo@buaa.edu.cn; Tel.: +86-10-8233-9409

Received: 21 May 2019; Accepted: 23 June 2019; Published: 26 June 2019

Abstract: This paper proposes a modeling method to establish a parametric-conducted emission model of a switching model power supply (SMPS) chip through a developed vector fitting algorithm. A common SMPS chip LTM8025 was taken as an example to explain the modeling process. According to the integrated circuit (IC) electromagnetic modeling (ICEM) standard, the parametric conducted emission model is divided into two parts: IC internal activity (ICIA) and IC passive distribution network (ICPDN). The parameters of ICIA are identified by measured data and correlated with key components; an improved vector-fitting algorithm is proposed to solve the fitting problem of ICPDN without phase information. This parametric model can be used with commercial simulation software together to achieve predictions of conducted emissions from power modules. The experiment results show that the maximum and 90% confidence interval of the forecast errors are 9.677 dB and (−4.56 dB, 6.52 dB) respectively, which achieve the international standard requirements and have sufficient accuracy and effectiveness.

Keywords: electromagnetic compatibility (EMC); switching model power supply (SMPS); conducted emission; parametric modeling method; vector fitting algorithm

---

1. Introduction

Electromagnetic compatibility (EMC) is one of the important conditions for measuring the electromagnetic strength of a device or a system. With an increasing number of electronic and electric devices integrated into complex electronic information systems, the electromagnetic environments, including the circuit principle and the coupling relationship are increasingly complicated. The inside or outside electromagnetic interference (EMI) of the system leads to an increase in cases where the sensitive devices may become degraded or even unable to work properly.

In order to solve this problem, in the actual development process, the iterative process of 'design-test-redesign' has been carried out to ensure electromagnetic compatibility of the electronic and electric devices and complex information systems. To reduce the development period and cost, it is particularly important to make a reasonable prediction of its EMC before the prototype is produced. The importance of EMC design work has become increasingly prominent.

As a large number of devices are integrated into complex systems, switching model power supply (SMPS) becomes essential for its role in improving power efficiency and reducing costs. However, its rapid on-off and parasitic effects may lead to serious electromagnetic emission problems. This makes it difficult to pass the appropriate industrial EMC/EMI control standards and may affect the functional ability of itself or other equipment. To ensure electromagnetic compatibility of an equipment or system, electromagnetic (EM) emission prediction is required [1–3]. EM emission can be further categorized into conducted emission (CE) and radiated emission (RE). Since the switching frequency of SMPS is between several tens of kHz to serval MHz, the emission of SMPS is mainly transmitted by conduction.

For the conducted emission modeling of SMPS, the earlier research starts from the physical parameters of the device and constructs the CE model [4–7]. However, this method requires detailed information of the device, the modeling process is complicated, and the established model has poor precision. Another mainstream method is behavioral level modeling. This method directly establishes the CE model of SMPS for analysis without considering the specific parameters of the switching power supply. In terms of behavioral model construction, Norton or Thevenin equivalent can generally be used to give an equivalent circuit of a CE model, and its parameters are determined by analysis or test [8,9]. In addition to these two aspects, there are other methods for modeling SMPS, such as using a concept-named fast reconstruction method (FRM) to deal with the time domain simulation of common mode conducted disturbances [10].

The SMPS modeling methods mentioned above need to measure the properties of internal components so as to ensure the accuracy of the model. With the miniaturization of devices, an increasing number of SMPS chips have emerged. For SMPS chips, traditional CE modeling methods are no longer applicable because their components are packaged inside the chip and cannot be measured. To calculate the EM emission accurately, it is necessary to model SMPS at the chip-level through simulation and external measurement.

As for these matters, various research has been reported in the past years [11–13]. Reference [14] gives an overview of the EMC research focusing on chip-level. IBIS (I/O buffer information specification) model is the standard for electronic behavioral specifications for integrated circuit input/output analog characteristics [15]. However, it is not sufficient to directly apply EM emission modeling because the interference associated with the internal activities of the component is not considered. IMIC (I/O interface model for integrated circuits) model can effectively solve the modeling accuracy problem [16]. Considering the modeling process may require confidential information to build a netlist and read it through dedicated simulation software, the portability is not relatively high. ICEM (integrated circuit electromagnetic modeling) is a simple and intuitive method that uses RLC (resistance, inductance, and capacitance) circuits to fit the measured port characteristics, which seems accurate enough for EMC predictions [17,18].

In recent years, the ICEM method was widely used to model different integrated circuits (ICs) and predict their electromagnetic compatibility [19–21]. Jean-Luc Levant, et al. used the ICEM model to simulate and predict the jitter of the integrated phase-locked loop and provide a correction solution [22]. Hyun Ho Park, et al. generated a macro model of the timing controller chip for running pseudo H pattern data from transistor level simulation and is used to estimate the power switching current on the printed circuit board (PCB) [23]. Abhishek Ramanujan, et al. designed a common interchange format for ICEM based on the well-known extensible markup language format and applied to Atmega88 microcontrollers [13]. Modeling methods for PCB and components with both chips and packages were published in [24] and [25], in order to calculate PCB power noise generated by the switching chip.

Although the ICEM model has been successfully applied in many chip-level EM emission predictions, it still has certain limitations. In the modeling process, the ICEM model divides the IC model into the passive distribution network (PDN) component and the internal activity (IA) component [26]. The establishment of IA parameter extraction depends on the extracted PDN parameters and the measured external current, thus the parameters have no specific physical meaning and cannot be parametrically modeled based on different practical scenarios. Therefore, when application parameters are different from the test board, the accuracy of the predicted results will be limited as well. Su, D. et al. proposed a new theory named 'basic emission waveform theory', which characterizes emission with four basic waveforms by its physical characteristics of the equipment. The identification and analysis of an emission source can be realized by analyzing the basic waveforms from the emission of a complex system [27].

Based on the basic emission waveform theory, this paper proposes a parametric modeling method to establish the conducted emission model of an SMPS chip through a developed vector fitting algorithm.

The organization of this paper is as follows. Section 2 describes the measurement configuration required for modeling. Section 3 illustrates the model division and parametric modeling methods of each part. In Section 4, a set of comparative experiments and an application example are given to verify the effectiveness of the modeling method. Conclusions are drawn in Section 5 to summarize the work proposed.

2. Measurement Configuration

This study takes a commercial SMPS chip called LTM8025 [28] as an example to illustrate the modeling method. LTM8025 is a step down micro module converter chip and widely used in consumer electronics as power supply modules.

It usually requires a dedicated chip test board to predict its conducted emission by the ICEM method at chip-level modeling. However, since the output voltage and switching frequency of the LTM8025 are closely related to the selection of component parameters in the peripheral circuits, the user needs to make a particular measurement board according to the used parameters to ensure the modeling accuracy. Furthermore, when a power supply module needs to convert one input voltage into multiple different output voltages, the user needs to integrate multiple ICs on one PCB. Under this case, in order to accurately predict the conducted emission of the PCB, multiple measurement boards are required to build a variety of chip models under each working condition. This will undoubtedly lengthen the design process time and increase costs.

Instead of the particular measurement board in ICEM method, an official demonstration circuit DC1379B [29] of LTM8025 is used as the test board in this paper. Figure 1 shows the PCB and schematic of DC1379B. The authors measured the conducted emission currents of the port VIN (Power Input Port) and VOUT (Power Output Port) to extract the parametric model. As can be seen from Figure 1b, the port VIN provides a voltage input to the DC1379B, which is stepped down by the LTM8025 and filtered by the peripheral circuit, and is output from the port VOUT to the downstream load.

The experimental configuration during the measurement were as follows: The port VIN was connected to a stabilized DC voltage supply with an internal resistance of $R_{PCB\_VIN} = 0.2\ \Omega$ to provide a 30 V input voltage to the board; the port VOUT was connected with a $R_{PCB\_VOUT} = 1.4\ \Omega$ high-power resistor as the downstream load; current monitor probe F-33-2 from the FCC company was used to connect to the spectrum analyzer to measure the conducted emission on the power lines (as shown in Figure 2). The measured results $U_{measured} = \begin{bmatrix} U_{VIN} & U_{VOUT} \end{bmatrix}^T$ are shown in Figure 3.

(a)

Figure 1. *Cont.*

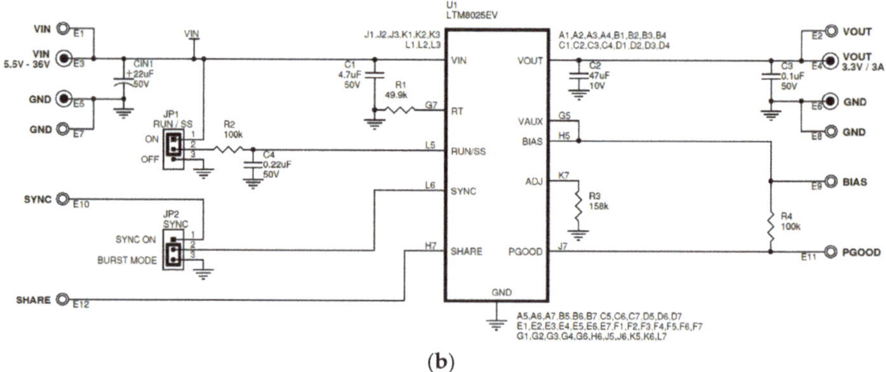

(b)

Figure 1. Board diagram and schematic of DC1379B [29]. (**a**) Printed circuit board (PCB) and its layouts; (**b**) Schematic.

(a)　　　　　　　　　　　　(b)

Figure 2. Experimental scenario of the test board at (**a**) VIN port; (**b**) VOUT port conduction emissions.

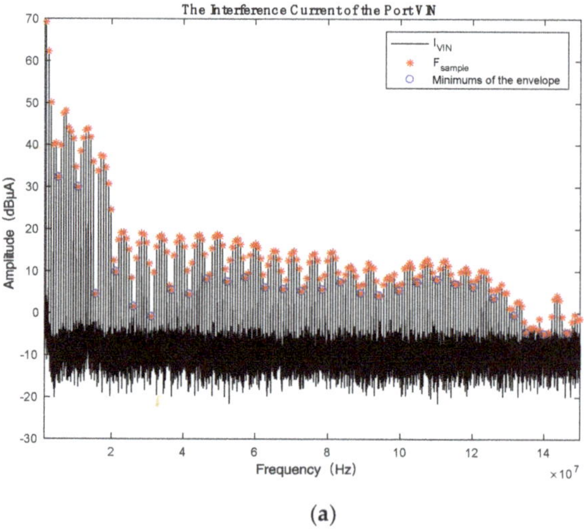

(a)

Figure 3. *Cont.*

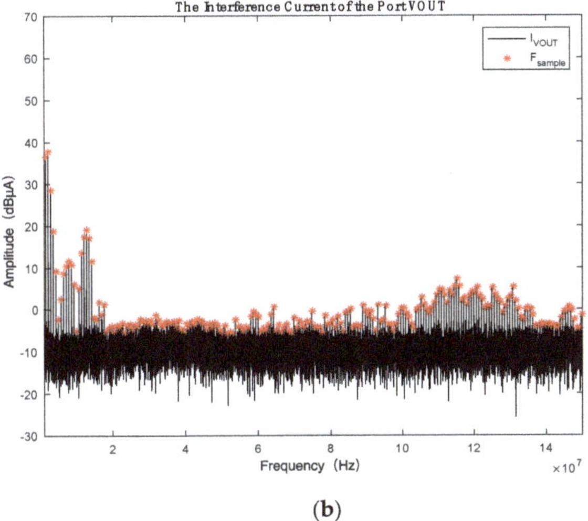

(b)

Figure 3. The measured results $U_{measured}$ (**a**) the measured conducted emission U_{VIN} at the port VIN; (**b**) the measured conducted emission U_{VOUT} at the port VOUT.

3. Parametric Modeling Method

Referring to the ICEM partitioning method, the SMPS chip model is also divided into ICIA and ICPDN parts. Different from ICEM, based on the 'basic emission waveform theory', the authors treat ICIA as a square waveform which parameters are related to the chip usage parameter settings and define the peripheral circuits as ICPDN, as Figure 4 shows.

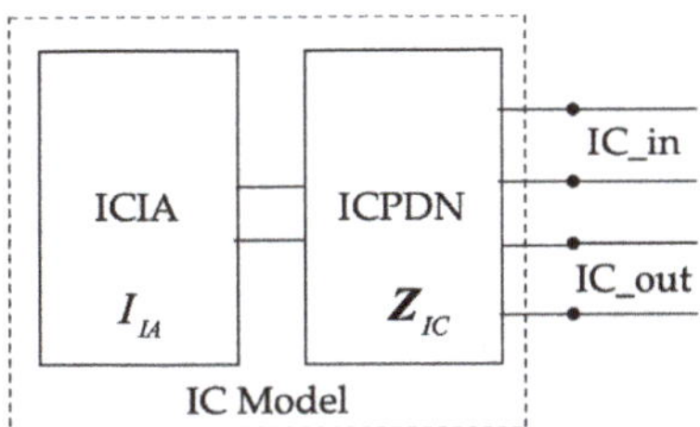

Figure 4. IC conducted emission model partitioning. ICIA and ICPDN.

The following sections will introduce the modeling methods of IA and PDN separately.

3.1. ICIA Parameters Extraction

The time-domain expression of an ideal square waveform can be represented by

$$i_{IA}(t) = \begin{cases} 1, \dfrac{n}{f_0} < t \le \dfrac{n+dc_0}{f_0} \\ 0, \dfrac{n+dc_0}{f_0} < t \le \dfrac{n+1}{f_0} \end{cases}, n = 1,2,\cdots \tag{1}$$

where f_0 represents the repetition frequency, dc_0 represents the duty cycle, n is a positive integer.

After Fourier transform to (1), the magnitude-frequency characteristics is given by

$$I_{IA}(f) = \sum_{n=-\infty}^{+\infty} \frac{dc_0}{f_0} \mathrm{Sa}(n\pi dc_0)\delta(f - nf_0) = \sum_{n=-\infty}^{+\infty} I_{IAn}\delta(f - nf_0), n \in \mathbf{N} \tag{2}$$

where N is the integer set, n represents an arbitrary integer and $I_{IAn} = \frac{dc_0}{f_0} \mathrm{Sa}(n\pi dc_0)$.

It can be found from (2) that the magnitude-frequency characteristics of a square wave can be determined by two key parameters which are repetition frequency f_0 and duty cycle dc_0. The following discussion will focus on the estimation process of these two parameters.

3.1.1. Repetition Frequency f_0 Estimation

It is easy to understand that the frequency spectra of the SMPS conducted emission are a bunch of discrete spectrum lines. The discrete frequency points set $\{F_{sample}\}$ could be described as

$$\{F_{sample}\} = nf_0, n = 0,1,2,\cdots. \tag{3}$$

It shows that elements in the set $\{F_{sample}\}$ are only related to f_0, and all of them are integral multiples of f_0. Therefore, theoretically speaking, we can obtain f_0 by reading the frequency intervals between adjacent spectrum lines Δf from the measured data $\boldsymbol{U}_{measured}$. Unfortunately, considering the RBW settings of spectrum analyzer, the measured Δf can hardly be equaled with f_0. Moreover, due to the spectrum analyzer's own algorithm, the frequency sampling intervals of the test data is non-uniform. The above-mentioned problems make it impossible to read f_0 from $\boldsymbol{U}_{measured}$ accurately and intuitively. Figure 5 shows an example of a spectrum analyzer display frequency interval data. In this case, the frequency range is from 100 kHz to 200 MHz, the number of sampling points is 32001, and RBW = 1 kHz. It can be seen that the frequency intervals are non-uniform.

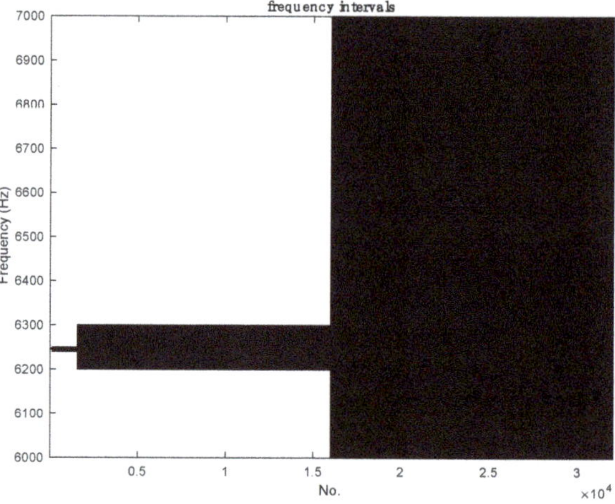

Figure 5. Frequency interval changes in one test.

The paper presents an engineering method to obtain accurate f_0 from measured data. Firstly, find all frequency points $\{F_{measured}\}$ 10 dB higher than the noise (as the red stars in Figure 3a) and record intervals between adjacent two frequency points as a set $\{\Delta f|_{measured}\}$ (as the black line in Figure 6). Secondly, find the mode $\Delta f|_{mode}$ in the set $\{\Delta f|_{measured}\}$ as the initial value. Finally, find all the elements in the set $\{\Delta f|_{measured}\}$ that satisfy $\left|\Delta f_i - \Delta f|_{mode}\right| < 0.1 * \Delta f|_{mode}$, $\forall \Delta f_i \in \{\Delta f|_{measured}\}$, and take the average as the final result of f_0, as shown in Equation (4).

$$\hat{f}_0 = \mathrm{average}(\{\Delta f_i\}|_{\forall \Delta f_i \in \{\Delta f|_{measured}\}, \text{s.t. } |\Delta f_i - \Delta f|_{mode}| < 0.1 * \Delta f|_{mode}}), i = 1,2,\ldots,N \tag{4}$$

In this case, $\hat{f}_0 = 706.39$ kHz.

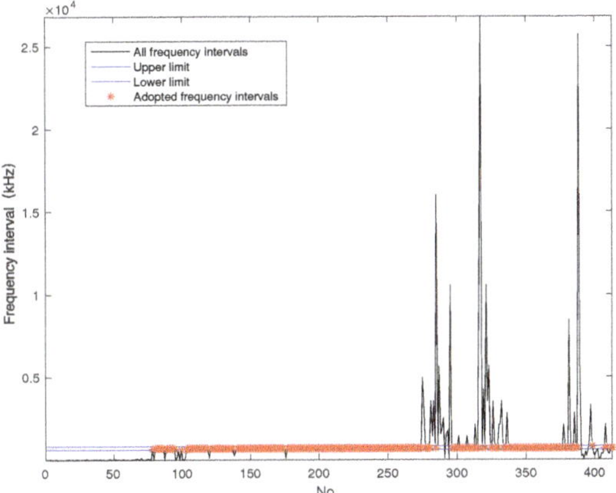

Figure 6. Δf_i selecting process.

3.1.2. Duty Cycle dc_0 Preliminary Estimating

The estimation process of dc_0 is divided into two steps. Firstly, the measured data $U_{measured}$ was used for preliminary estimation; secondly, a more accurate value will be estimated in the ICPDN modeling process. The process to estimate dc_0 preliminarily is explained hereafter.

We know that the envelope of the square waveform spectrum $I_{IA}(f)$ can be described by the envelope function

$$E(f) = \frac{dc_0}{f_0}\text{Sa}\left(\frac{\pi dc_0 f}{f_0}\right). \tag{5}$$

It can be seen from Equation (5) that when the frequency f satisfies $f = mf_0/dc_0, m \in \mathbf{N}$, there is $E(f) = 0$. According to the characteristic of the trigonometric function, the frequency interval between two adjacent zeros Δf_{zeros} is

$$\Delta f_{zeros} = f_0/dc_0. \tag{6}$$

Since it is usually impossible to exist a series of resonance points with multiple relations in the IC_PDN part, we can infer that most of the minimum values in the envelope of the measured data $U_{measured}$ are near the zeros of the envelope function $E(f)$.

Therefore, we use the following method to estimate dc_0 initially. Firstly, find the envelope of the measured data $U_{measured}$, which can be described by the amplitude value curve corresponding to the frequency set $\{F_{measured}\}$. Secondly, find all the minimums of this envelope (as the blue circles in Figure 3a), calculate the frequency interval between two adjacent minimums, and find their average, denoted as $\Delta \overline{f}_{min}$. Finally, estimate the initial value of dc_0 as Equation (7) shows.

$$\hat{dc}_0^0 = \hat{f}_0/\Delta \overline{f}_{min}. \tag{7}$$

The estimated result in the case is $\hat{dc}_0^0 = 13.09\%$.

3.1.3. Parameterization of ICIA

Considering the working principle of the SMPS, the repetition frequency f_0 is related to the resonance resistance, and the duty cycle dc_0 is related to the ratio of output voltage and input one.

Since the components on the test board DC1379B are difficult to be replaced, the curve of repetition frequency f_0 under different values of resonance resistance R_T are taken from the LTM8025 data sheet and showed in Figure 7. Using an inverse function to fit the data, the fitted curve equation is:

$$f_0(R_T) = \frac{4.489 \times 10^{10}}{R_T + 1.036 \times 10^4} - 9.630 \times 10^3. \tag{8}$$

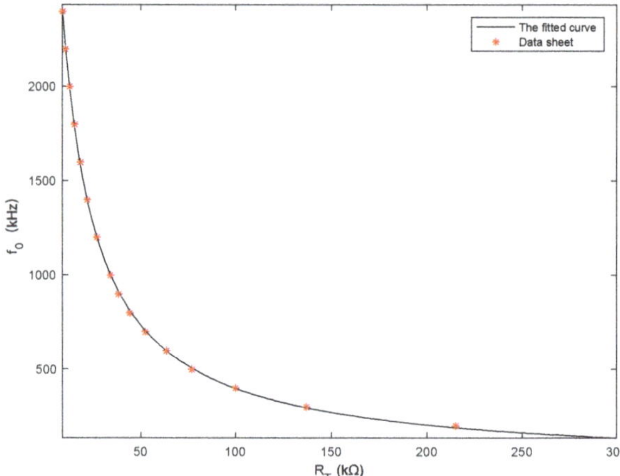

Figure 7. The curve of repetition frequency f_0 under different values of resonance resistance R_T.

Since the output voltage is related to the value of bias resistor R_{adj}, which is difficult to adjust. Therefore, the author obtained different input voltages by adjusting the regulated power supply and calculated the corresponding dc_0 by the method proposed in the Section 3.1.2. The relationship between dc_0 and the ratio of output voltage and input one is shown in Figure 8. Using a linear function to fit the data, the fitted curve equation is:

$$dc_0\left(\frac{V_{out}}{V_{in}}\right) = 120.93 \times \frac{V_{out}}{V_{in}} + 0.5379. \tag{9}$$

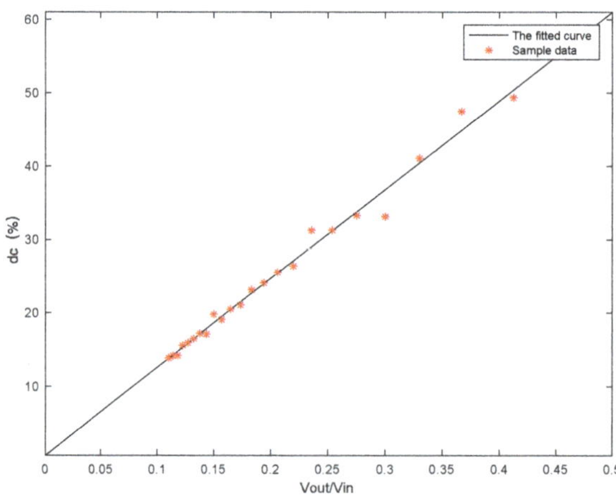

Figure 8. The curve of duty cycle dc_0 under different value of V_{out}/V_{in}.

Till now, the paper has completed the parametric modeling of ICIA. For practical use, according to designed parameters, the user could estimate f_0 and dc_0 by Equations (8) and (9); and obtain the model of I_{IA} according to Equation (2).

3.2. ICPND Modeling

In the ICEM method [26], the hardware set-up used to extract the PDN parameters consists of measurement equipment (usually the vector network analyzer), a measurement probe and a measurement board. Therefore, before extracting the PDN parameters, a de-embedding process is needed so as to remove all the parasitic elements of this set-up. Figure 9 shows its de-embedding principle.

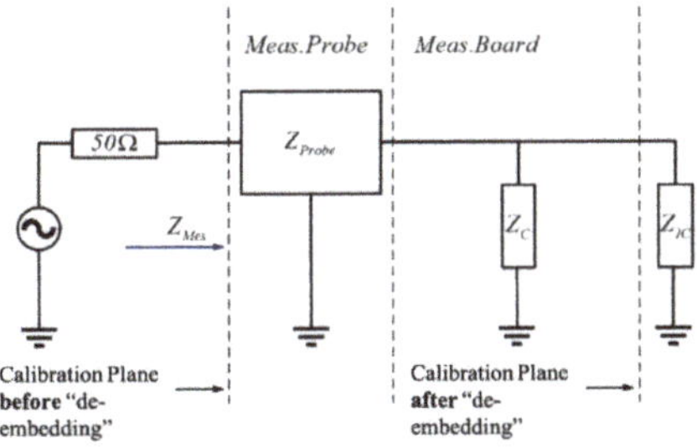

Figure 9. De-embedding principle [26].

The block diagram of the measurement setup of proposed method in this paper is shown in Figure 10. The coupling relationship can be expressed as Equation (10).

$$|Y_{PCB} \cdot Z_{IC} \cdot I_{IA}| / |Z_{probe}| = U_{measured} \tag{10}$$

where Y_{PCB} and Z_{probe} represent the influence of the measured board and the current probe respectively. And Z_{IC} represents the ICPDN model.

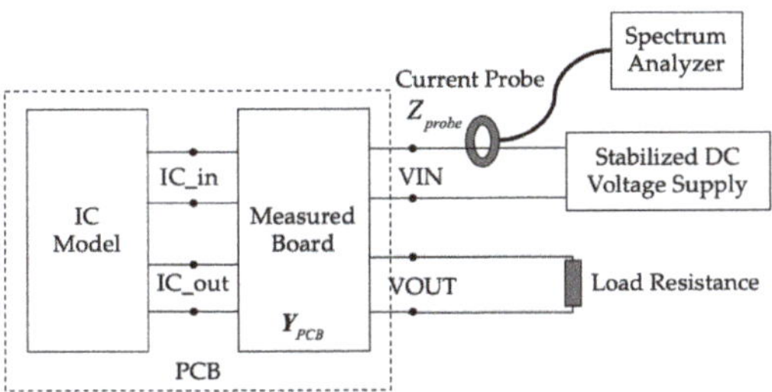

Figure 10. The block diagram of the measurement setup.

One of the most important differences is that the study uses spectrum analyzer to measure the spectra of the input and output ports respectively, instead of using VNA to gain the impedance between two ports in the ICEM method. This leads to a lack of phase information for the measured data. Therefore, it is impossible to use the same de-embedding and Z_{IC} fitting method with ICEM.

3.2.1. De-Embedding Process

- Current monitor probe F-33-2

In this paper, the authors used F-33-2 current monitor probe to measure the conducted emission on the power lines. The F-33-2 is for laboratory and field testing. The useable frequency range of this probe is 1–250 MHz. A typical calibration curve $|Z_{probe}|$ is shown below [30].

From the curve shown in Figure 11, the measured results $U_{measured}$ can be converted into the spectrum of the interference current on the power line, which is $I_{PCB} = \begin{bmatrix} I_{VIN} & I_{VOUT} \end{bmatrix}^T$.

$$I_{PCB} = U_{measured} / |Z_{probe}| \tag{11}$$

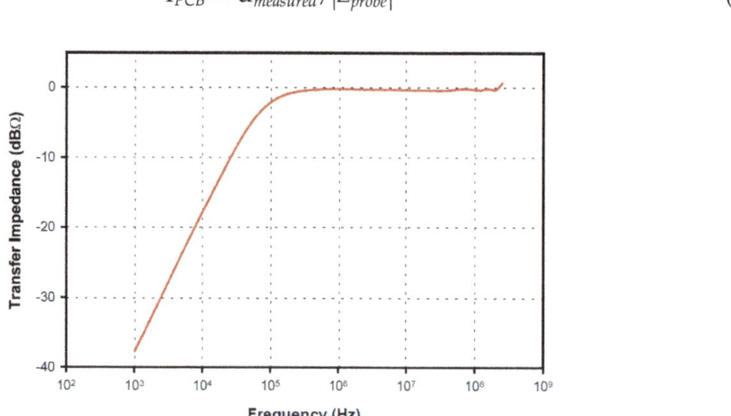

Figure 11. The calibration curve of F-33-2 [30].

- Test board DC1379B

As for the modeling of the in-band characteristics of passive linear components, the extant methods are relatively mature, especially when the parasitic parameters of the main components are known. Therefore, a commercial simulation software Ansys SIwave [31] is used to model the test board DC1379B. ANSYS SIwave is a specialized design platform for modeling, analyzing and simulating of IC packages and PCBs. The test board consisted of four layers; the relative permittivity of the dielectric substrate is $\varepsilon_r = 4.4$; all conductors are made by copper material with conductivity $\sigma = 5.8 \times 10^7 \ S/m$. The given layout and geometry of the DC1379B were imported into the software. The locations and basic descriptions of the four defined ports are shown in Figure 12 and Table 1.

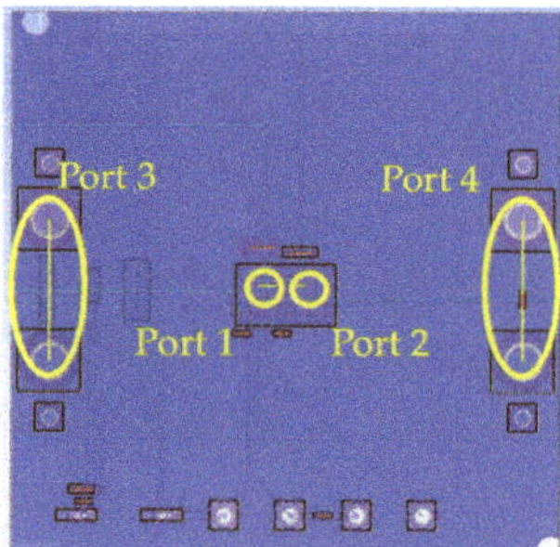

Figure 12. Port definitions of DC1379B.

Table 1. Parameter settings of each port.

No.	Name	Description
1	IC_in	Internal port: IC input voltage port
2	IC_out	Internal port: IC output voltage port
3	VIN	External port: PCB input voltage port
4	VOUT	External port: PCB output voltage port

Therefore, the PCB can be regarded as a four-port network, and the effect of DC1379B on the measured results can be represented by a Y-parameter matrix $\mathbf{Y}_{PCB}$.

$$\mathbf{Y}_{PCB} = \begin{bmatrix} Y_{13} & Y_{23} \\ Y_{14} & Y_{24} \end{bmatrix}. \tag{12}$$

Using the full-wave analysis method, the Y-parameters were calculated. The results are shown in Figure 13.

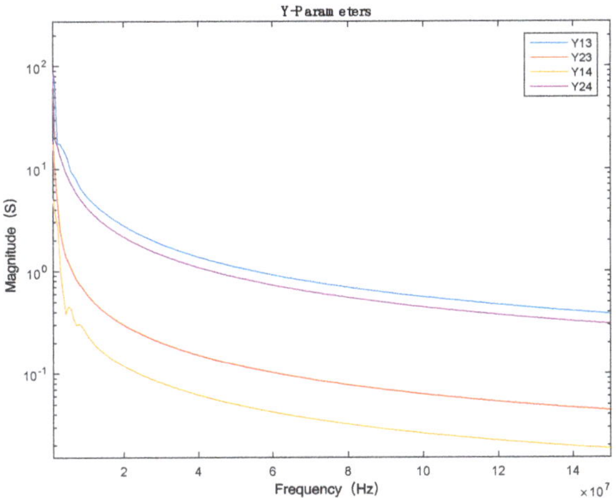

Figure 13. Y parameter simulation results.

Theoretically, the output interference signals $\mathbf{U}_{IC} = \begin{bmatrix} U_{ICin} & U_{ICout} \end{bmatrix}^{T}$ from the chip pins can be obtained by solving the linear equations of Equation (13) when $\mathbf{I}_{PCB}$ and $\mathbf{Y}_{PCB}$ are known.

$$|\mathbf{Y}_{PCB} \cdot \mathbf{U}_{IC}| = \mathbf{I}_{PCB}. \tag{13}$$

However, since $\mathbf{U}_{measured}$ has no phase information, $\mathbf{I}_{PCB}$ has no phase information either. Therefore, the solution of the complex coefficient Equation (13) may not be unique. Under the test configuration described in this paper, $\mathbf{Z}_{IC}$ cannot be directly obtained referring the de-embedding process in the ICEM standard.

To solve this problem, the paper proposes a developed vector fitting algorithm as follows.

3.2.2. A Developed Vector Fitting Algorithm

Let $f_0 = \hat{f}_0$ and $dc_0 = \hat{dc}_0^0$, Equation (2) could be represented as below:

$$I_{IA}(\hat{f}_0, \hat{dc}_0^0) = \frac{\hat{dc}_0^0}{\hat{f}_0} \sum_{n=-\infty}^{+\infty} \mathrm{Sa}(n\pi \hat{dc}_0^0)\delta(f - n\hat{f}_0), n \in \mathbf{N}. \tag{14}$$

Bring (14) into (10), there is:

$$\left| \begin{bmatrix} Y_{13} & Y_{23} \\ Y_{14} & Y_{24} \end{bmatrix} \cdot \begin{bmatrix} Z_{ICin} \\ Z_{ICout} \end{bmatrix} \cdot I_{IA} \right| = \begin{bmatrix} I_{VIN} \\ I_{VOUT} \end{bmatrix} \tag{15}$$

or:

$$\begin{cases} |Y_{13}Z_{ICin} + Y_{23}Z_{ICout}| = I_{VIN}/|I_{IA}| \\ |Y_{14}Z_{ICin} + Y_{24}Z_{ICout}| = I_{VOUT}/|I_{IA}| \end{cases}. \tag{16}$$

Let the auxiliary functions Z_{in} and Z_{out} as

$$\begin{cases} Z_{in} = Y_{13}Z_{ICin} + Y_{23}Z_{ICout} \\ Z_{out} = Y_{14}Z_{ICin} + Y_{24}Z_{ICout} \end{cases}. \tag{17}$$

Since Z_{in} and Z_{out} were fitted in the same the process, researchers took Z_{in} as an example to illustrate the algorithm. The traditional vector fitting [32] algorithm makes a clever use of matrix transformation to provide a feasible numerical solution method for solving frequency domain responses rational approximation problems. Contrast with vector fitting algorithm, considering the rational function approximations of the impedance parameter Z_{in} as:

$$Z_{in}(s) = \sum_{n=1}^{N} \frac{c_n}{s - a_n} + d + sh, \tag{18}$$

let auxiliary function $\sigma(f)$ as

$$\begin{bmatrix} \sigma Z_{in}(s) \\ \sigma(s) \end{bmatrix} = \begin{bmatrix} \sum_{n=1}^{N} \frac{c_n}{s-\bar{a}_n} + d + sh \\ \sum_{n=1}^{N} \frac{\bar{c}_n}{s-\bar{a}_n} + 1 \end{bmatrix}. \tag{19}$$

Multiplying the second row in (20) with Z_{in} yields the following relation

$$\sum_{n=1}^{N} \frac{c_n}{s - \bar{a}_n} + d + sh - Z_{in} \cdot \sum_{n=1}^{N} \frac{\bar{c}_n}{s - \bar{a}_n} = Z_{in}. \tag{20}$$

However, since $U_{measured}$ were measured by the spectrum analyzer, it contains no phase information. Therefore, it is unable to obtain Z_{in} with accurate phase information. This can greatly affect the accuracy of vector fitting and even lead to serious errors. Therefore, the paper proposed a developed algorithm to reduce the impact of the uncertain phase information. When only the amplitude information is considered, the Equation (20) turns into:

$$\left| \sum_{n=1}^{N} \frac{c_n}{s - \bar{a}_n} + d + sh \right| - |Z_{in}| \cdot \left| \sum_{n=1}^{N} \frac{\bar{c}_n}{s - \bar{a}_n} \right| = |Z_{in}| \tag{21}$$

where $c_n, \bar{c}_n, d, h$ are unknowns.

It can be seen that Equation (21) is no longer a linear problem. Therefore, the linear least squares optimization process in the original method is changed to a nonlinear problem, and a margin control parameter is added into the iterative process to reduce the impact of the uncertain phase information. From Equation (21), record the residual function $r(c_n, \bar{c}_n, d, h)$ as:

$$r(c_n, \bar{c}_n, d, h) = \left| \sum_{n=1}^{N} \frac{c_n}{s - \bar{a}_n} + d + sh \right| - |Z_{in}| \cdot \left| \sum_{n=1}^{N} \frac{\bar{c}_n}{s - \bar{a}_n} \right| - |Z_{in}|. \tag{22}$$

Therefore, the fitting problem on the sample data set $\{F_{measured}\}$ can be described by a nonlinear least squares problem as Equation (23).

$$\begin{aligned} \underset{\{F_{measured}\}}{\text{minimize}} \quad & f(\{F_{measured}\}) = \sum_{i=1}^{M} r_i^2 \\ \text{subject to} \quad & r_i = \left| \sum_{n=1}^{N} \frac{c_n}{jf_i - \bar{a}_n} + d + jf_i h \right| - |Z_{in}^i(f_i)| \cdot \left| \sum_{n=1}^{N} \frac{\bar{c}_n}{jf_i - \bar{a}_n} \right| - |Z_{in}^i(f_i)| \\ & f_i \in \{F_{measured}\} \end{aligned} \tag{23}$$

This nonlinear problem can be solved by the Levenberg-Marquardt algorithm. The pseudo-code algorithm is shown in the reference [33].

Therefore, Z_{in} and Z_{out} could be calculated. According to Equation (17), the values of Z_{ICin} and Z_{ICout} can be obtained, and $\hat{I}_{PCB}$ can be obtained from (10) and (11).

As an example, Figure 14 shows the comparison results between the calculated I_{VIN} and the predicted $\hat{I}_{VIN}$ which is under the condition that $\hat{dc}_0^0 = 13.57\%$. I_{VIN} is represented by the black line in the figure which is as same as the red star in the Figure 3a. Since it was directly converted from the test result $U_{measured}$, the values of I_{VIN} were used as a reference to measure the accuracy of the fitting algorithm. The blue line in the figure shows the predicted result calculated by the traditional vector fitting algorithm, while the red line is estimated by the developed algorithm proposed in this paper. Comparing the two curves, it can be seen that the proposed algorithm can effectively solve the problem that the traditional vector fitting algorithm has low fitting precision when there is no phase information in the fitted data.

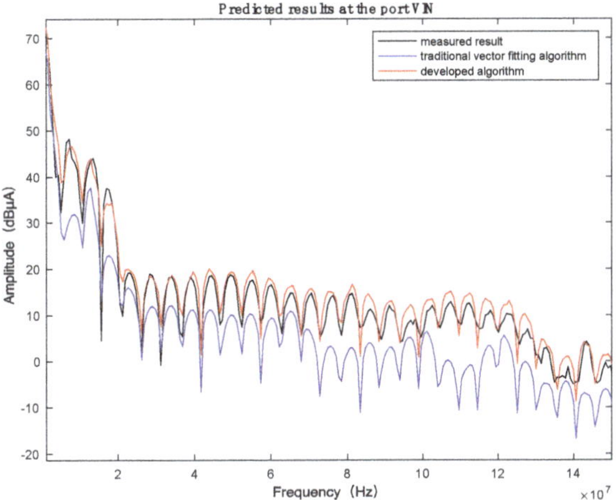

Figure 14. The predicted results at the port VIN.

3.2.3. Further Estimation of dc_0 and ICPND

Due to the influence of measurement accuracy, etc., the accuracy of the estimated $\hat{dc}_0^0$ in Section 3.1.2 is limited. Therefore, this subsection describes a process establishing the model of Z_{IC} while estimating a more accurate value of dc_0.

It can be seen from Equations (10) and (14), Z_{IC} is related to the value of dc_0. Therefore, the problem can be converted into finding an optimal dc_0 which minimizes the error between the predicted output I_{PCB}^i and the calculated data I_{PCB}, as is shown in (24).

$$\begin{aligned} \underset{dc_0}{\text{minimize}} \quad & \|I_{PCB}^i - I_{PCB}\|_2 \\ \text{subject to} \quad & I_{PCB}^i = Y_{PCB} \cdot Z_{IC}(dc_0) \cdot I_{IA}(f_0, dc_0) \\ & f_0 = \hat{f}_0 \end{aligned} \tag{24}$$

To solve the above mentioned optimization problem (24), errors function $errors^i$ was built as the 2-norm of the difference between I_{PCB}^i and I_{PCB} at the sample set $\{F_{measured}\}$, which can be represented as:

$$errors^i = \|I_{PCB}^i(\{F_{measured}\}) - I_{PCB}(\{F_{measured}\})\|_2. \tag{25}$$

At this point, (24) can be transformed to minimize (25).

Repeat the above process under different dc_0 values in the neighborhood of $\hat{dc}_0^0$. Then the final duty cycle $\hat{dc}_0$ is estimated as the one which made (25) achieve its minimum. And let the corresponding $\mathbf{Z}_{IC}^i = [\ Z_{ICin}^i \quad Z_{ICout}^i\ ]^{\mathrm{T}}$ be the impedance parameters of ICPDN part $\hat{\mathbf{Z}}_{IC}$.

Figure 15 shows the variation of *errors* under different dc_0, and the red star represents the final estimated result $\hat{dc}_0 = 13.57\%$.

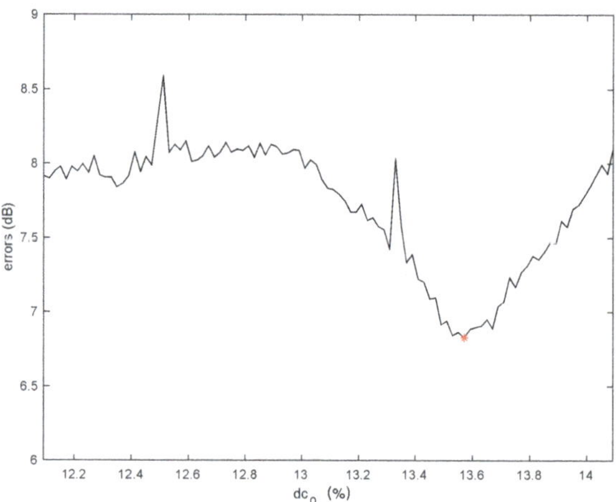

Figure 15. The *errors* curve under different dc_0.

4. Experimental Results and Discussion

In this section, an application example and a set of comparative experiments are given to verify the effectiveness of the modeling method.

4.1. Application Example

An application example was taken to verify the accuracy of the above-mentioned modeling method. In this example, the 12 V input voltage were convert to 5.4 V, 5.4 V, and 3.8 V output voltages respectively by three LTM8025 chips. The PCB board adopts a four-layer board structure. Its photo and topology diagram are shown in Figures 16 and 17 respectively.

(a)   (b)

Figure 16. Photos of the power supply module instance. (**a**) PCB and its layouts; (**b**) Experimental scenario.

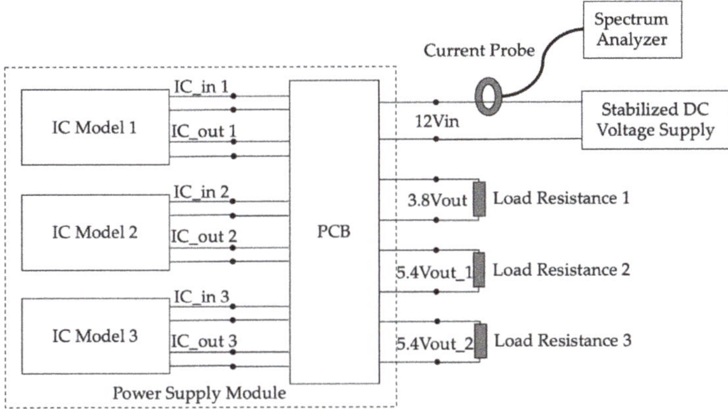

Figure 17. The topology of the power supply module.

The author treated the PCB sub-module in the instance as a 10-port network and simulated its Y-parameters by Ansys SIwave. The port definitions were shown in Figure 18 and Table 2. And the calculation results were shown in Figure 19.

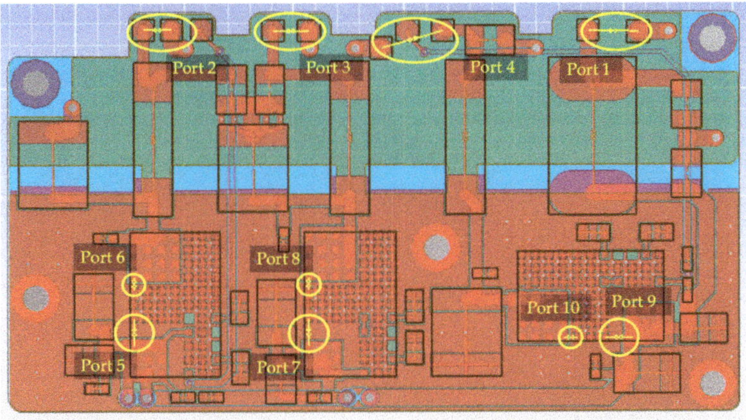

Figure 18. Port definitions of the power supply module instance.

Table 2. Parameter settings of each port.

No.	Name	Description
1	12Vin	External port: PCB input voltage port
2	5.4Vout_1	External port: PCB output voltage port (comes from IC3)
3	5.4Vout_2	External port: PCB output voltage port (comes from IC 2)
4	3.8Vout	External port: PCB output voltage port (comes from IC 1)
5	IC1_in	Internal port: IC1 input voltage port
6	IC1_out	Internal port: IC1 output voltage port
7	IC2_in	Internal port: IC2 input voltage port
8	IC2_out	Internal port: IC2 output voltage port
9	IC3_in	Internal port: IC3 input voltage port
10	IC3_out	Internal port: IC3 output voltage port

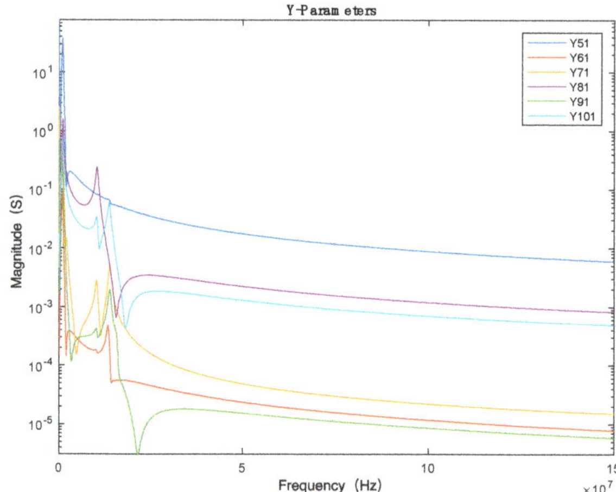

Figure 19. Y parameter simulation results between the Port 1 and other internal ports.

As for IC sub-modules modeling, according to the design parameters, f_0 and dc_0 of the three chips are as shown in Table 3. Therefore, the functions of the three ICIAs can be calculated according to Equation (10). Furthermore, the CE model of each chip could be obtained according to Equation (6). As an example, the estimated results of IC sub-module 1 are shown in Figure 20.

Table 3. Design parameters of the three chips.

	R_1 (kΩ)	f_0 (kHz)	V_{in} (V)	V_{out} (V)	dc_0 (%)
IC 1	39.5	890.7	12	5.4	38.83
IC 2	22	1377.5	12	5.4	54.96
IC 3	20	1468.9	12	3.8	54.96

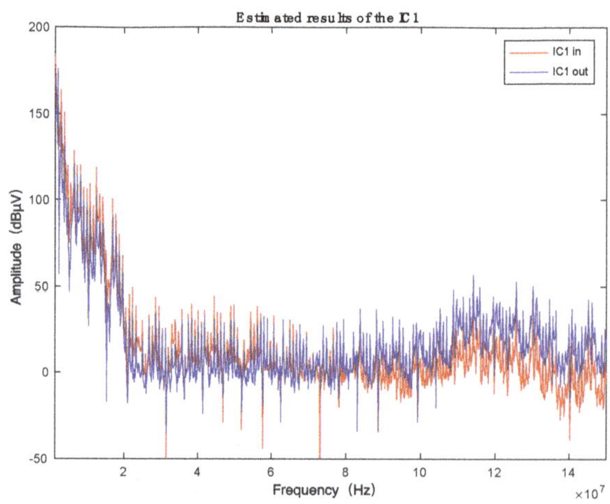

Figure 20. The estimated results of the IC sub-module 1.

However, in practical applications, the true value of a resistor often deviates from its nominal value, which may cause the estimated interference spectrum to offset from the measured value. Furthermore, for a regular electromagnetic interference such as switching signals, we tend to pay more attention to the characteristics of its envelope rather than a single frequency point. Therefore, the authors used its envelope to evaluate the accuracy of the estimated result in the following part. At this point, the conducted emission measured result at the port 12Vin could be estimated according to Equation (10).

Figure 21a shows the comparison between the envelope of the measured result (as the red line) and the estimated one (as the blue line). And the forecast errors and its 90% confidence interval are shown in Figure 21b. It can be found that the maximum error is 9.677 dB @23.79 MHz, and its 90% confidence interval is (−4.56 dB, 6.52 dB).

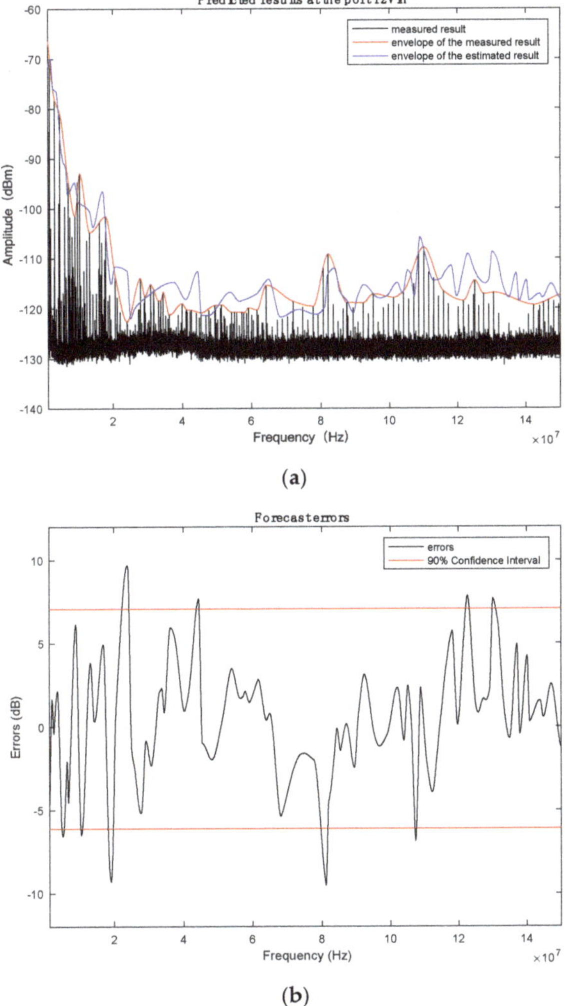

(a)

(b)

Figure 21. Comparison between the measured result and the estimated one. (**a**) Predicted results; (**b**) forecast errors.

As an authoritative international standard for IC conduction emission modeling, the examples given in the reference [26] suggest that 'the agreement is very good' when the forecast error is less than ±10 dB. In contrast, it can be seen that the modeling method proposed in this paper could achieve the standard requirements and has sufficient accuracy and effectiveness.

4.2. Comparative Experiments

According to ICEM of test configurations described in the reference [26], a dedicated chip test board for LTM8025 is made. In order to ensure the normal operation of the tested chip, the resonance resistance $R_T = 15\,k\Omega$ and the bias resistance $R_{adj} = 200\,k\Omega$ are set. Using the impedance analyzer and the oscilloscope to measure the impedance curve and output waveform of the chip. The ICEM model is obtained, and the following two comparative experiments are used to illustrate the practicality of the algorithm. With reference to DC1379B, the authors created three demo boards with different parameter settings as models to be used for comparison experiments.

Table 4 shows the parameter settings for the three demo boards. It can be seen that the parameter settings of Board 1 are same with the ICEM test board while the other two boards are different.

Table 4. Design parameters of the three boards.

	R_T (kΩ)	f_0 (kHz)	R_{adj} (kΩ)	V_{in} (V)	V_{out} (V)	dc_0 (%)
Board 1	15	1745.4	200	5.8	2.75	57.88
Board 2	27	1180.5	75	10	5.87	71.52
Board 3	42	847.6	115	5.8	4.21	88.36

It should be noted that since the impedance analyzer used in ICEM modelling process only covers a frequency band from 20 kHz to 100 MHz, the CE predicted results were only considered in this range in the comparison experiments.

- **Board 1:** The usage parameters are same as the ICEM test board.

When the actual board parameter settings are exactly same as the ICEM test board, the two methods are used to predict the conducted emissions separately. The comparison results are shown in Figure 22.

It can be seen that in this case, the 90% confidence intervals of the two methods are (−6.82 dB, 7.32 dB) and (−7.04 dB, 7.54 dB), respectively. The accuracy of the two modeling methods is not much different. Both the proposed method and ICEM method can meet the requirements for CE prediction.

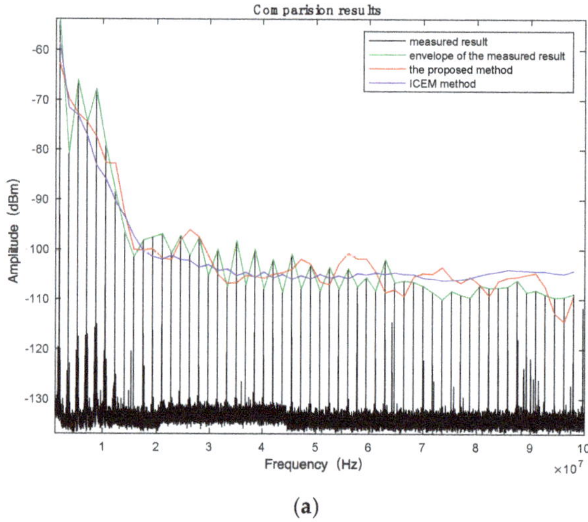

(**a**)

Figure 22. *Cont.*

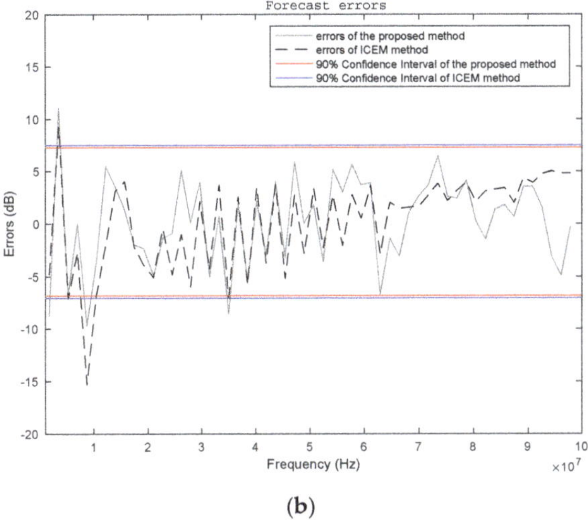

(b)

Figure 22. Comparison between the proposed method and ICEM method for Board 1. (**a**) Predicted results; (**b**) forecast errors.

- **Board 2** and **Board 3:** The usage parameters are different with the ICEM test board.

To make it different from the ICEM test board parameter settings, the author changed the main components of the actual board as Board 2 which is shown in Table 4. CE of the actual board at its voltage input port was predicted by the two methods respectively. The comparison results are shown in Figure 23.

What the comparison results show are as following. Although the accuracy of the ICEM method is similar with the proposed method approximately to the middle of the considered frequency band, it decreases significantly with increasing frequency. As a statistical result, the 90% confidence interval of forecast errors is (-6.54 dB, 9.56 dB) by the proposed method, while (-8.28 dB, 15.29 dB) though the ICEM method.

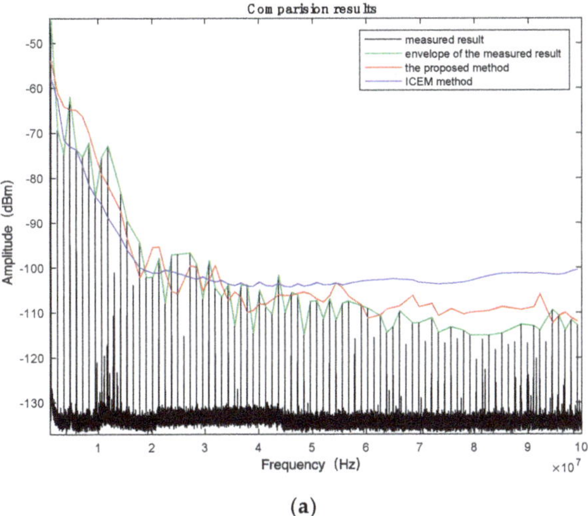

(a)

Figure 23. *Cont.*

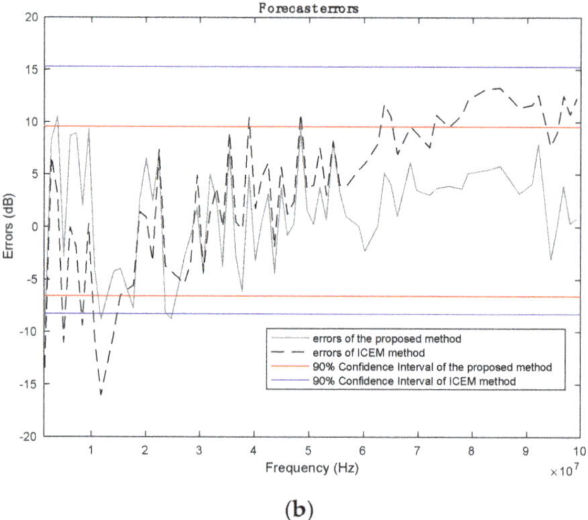

(b)

Figure 23. Comparison between the proposed method and ICEM method for Board 2. (**a**) Predicted results; (**b**) forecast errors.

Another comparative test board parameter setting are as Board 3 shown in Table 4, and the comparison results are shown in Figure 24.

It can be seen from the comparison results that the accuracy of the proposed method is still acceptable after changing the board's design parameter settings. Except for a minimum point near 8 MHz, the forecast error is less than 10 dB in almost the entire considered frequency band. The 90% confidence interval of forecast errors is (−7.53 dB, 6.46 dB). Unfortunately, the ICEM method does not perform well in the CE prediction of Board 3. The forecast error is much larger than 10 dB in the frequency bands below 10 MHz or above 60 MHz. In particular, when the CE has obvious periodic variation characteristics, the ICEM method can only describe the trend roughly but cannot effectively describe its envelope. The 90% confidence interval of forecast errors is (−8.29 dB, 17.6 dB).

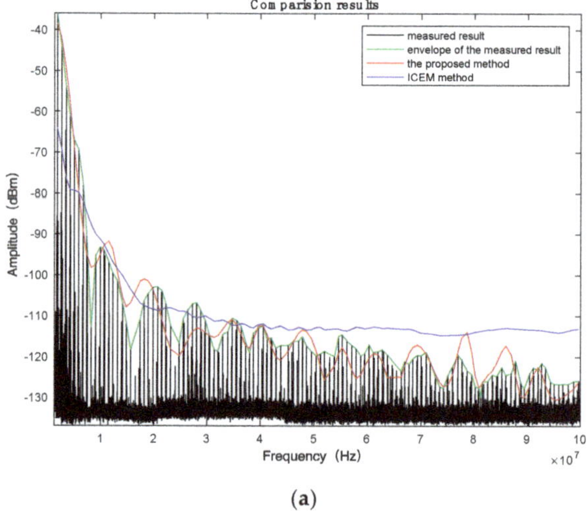

(a)

Figure 24. *Cont.*

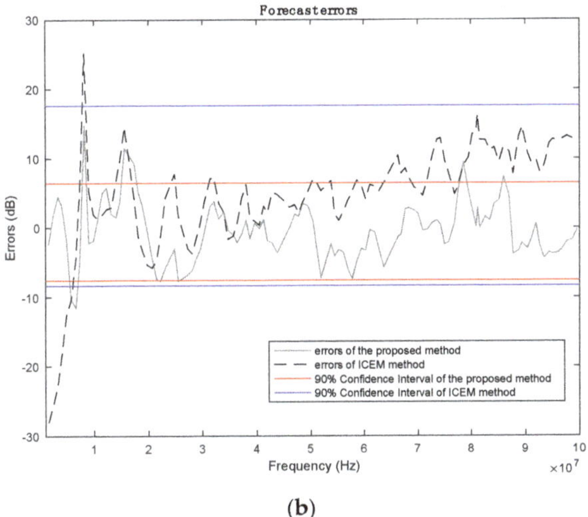

(b)

Figure 24. Comparison between the proposed method and ICEM method for Board 4. (**a**) Predicted results; (**b**) forecast errors.

From the above mentioned comparison results we can find that when the parameters of the actual circuit are different from the test board, the modeling accuracy of ICEM modeling method is greatly reduced. This is because the interference signal generated by the SMPS chip is related to the parameter setting of the external circuit. In the ICEM method, the IA modeling is reliant entirely on the measured results and without considering its association with peripheral parameters. Therefore, when the actual circuit used is inconsistent with the test board parameters, the ICEM model will no longer be applicable. In practical applications, since the circuit parameters have not been determined during the product design stage, a huge number of test boards with different parameter settings are needed when we use the ICEM method to predict the conduction emission. The workload it brings are enormous.

Otherwise, the method proposed in this paper fully considers the mechanism of interference generation. It found the relationship between emission characteristics and peripheral circuit parameters, and established a parametric CE model of the SMPS chip. Therefore, it is more convenient in actual use, since only the model parameters need to be adjusted as required without further measuring.

In summary, compared with the traditional ICEM modeling method, the proposed modeling method has better applicability in the product design stage under the premise of ensuring that the modeling accuracy is not reduced.

5. Conclusions

This paper proposes a modeling method to establish a parametric-conducted emission model of a switching model power supply (SMPS) chip through a developed vector-fitting algorithm. Reference the ICEM standard, the parametric conducted emission model is also divided into two parts: ICIA and ICPDN. The parameters of ICIA are identified by measured data and correlated with key components; an improved vector fitting algorithm is proposed to solve the fitting problem of ICPDN without phase information. The experimental results show that the proposed method could achieve the international standard requirements and has sufficient accuracy and effectiveness.

Author Contributions: X.H. proposed, designed and wrote the paper; S.X. supervised the experiments and write-up of the paper; Z.C. helped in carrying out experiments.

Funding: This research was funded by National Natural Science Foundation of China (No.61271045) and Defense Industrial Technology Development Program (No.JCKY2016601B005).

Conflicts of Interest: The authors declare no conflict of interest.

References

1. Ding, F.; Zhang, J. Analysis and design for VMS EMC based on MPC555. *World Electr. Veh. J.* **2010**, *4*, 754–759. [CrossRef]
2. Krug, F.; Lewke, B. Electromagnetic interference on large wind turbines. *Energies* **2009**, *2*, 1118–1129. [CrossRef]
3. Podbersic, M.; Matko, V.; Segula, M. An EMI filter selection method based on spectrum of digital periodic signal. *Sensors* **2006**, *6*, 90–99. [CrossRef]
4. Gulez, K.; Mutoh, N. Application of Neural Networks to Estimate Common Mode (CM) Model Impedance Parameters in AC Drives. *Math. Comput. Appl.* **2003**, *8*, 225–233. [CrossRef]
5. Ran, L.; Gokani, S.; Clare, J.; Bradley, K.J.; Christopoulos, C. Conducted electromagnetic emissions in induction motor drive systems. I. time domain analysis and identification of dominant modes. *IEEE Trans. Power Electron.* **1998**, *13*, 757–767. [CrossRef]
6. Rondon-Pinilla, E.; Morel, F.; Vollaire, C.; Schanen, J.L. Modeling of a Buck Converter with a SiC JFET to Predict EMC Conducted Emissions. *IEEE Trans. Power Electron.* **2014**, *29*, 2246–2260. [CrossRef]
7. Zhai, L.; Cao, Y.; Lin, L.; Zhang, T.; Kavuma, S. Mitigation Conducted Emission Strategy Based on Transfer Function from a DC-Fed Wireless Charging System for Electric Vehicles. *Energies* **2018**, *11*, 477. [CrossRef]
8. Zhai, L.; Zhang, X.; Bondarenko, N.; Loken, D.; Van Doren, T.; Beetner, D. Mitigation emission strategy based on resonances from a power inverter system in electric vehicles. *Energies* **2016**, *9*, 419. [CrossRef]
9. Bishnoi, H.; Baisden, A.C.; Mattavelli, P.; Boroyevich, D. Analysis of EMI Terminal Modeling of Switched Power Converters. *IEEE Trans. Power Electron.* **2012**, *27*, 3924–3933. [CrossRef]
10. Frantz, G.; Frey, D.; Schanen, J.L.; Revol, B. EMC models of Power Electronics converters for network analysis. In Proceedings of the European Conference on Power Electronics & Applications, Lille, France, 2–6 September 2013.
11. Labrousse, D.; Revol, B.; Gautier, C.; Costa, F. Fast Reconstitution Method (FRM) to Compute the Broadband Spectrum of Common Mode Conducted Disturbances. *IEEE Trans. Electromagn. Compat.* **2013**, *55*, 248–256. [CrossRef]
12. Ghfiri, C.; Durier, A.; Boyer, A.; Dhia, S.B.; Marot, C. Construction of an Integrated Circuit Emission Model of a FPGA. In Proceedings of the Asia-Pacific International Symposium on Electromagnetic Compatibility, Shenzhen, China, 17–21 May 2016; pp. 402–405.
13. Ramanujan, A.; Sicard, E.; Boyer, A.; Levant, J.L.; Marot, C.; Lafon, F. Developing a Universal Exchange Format for Integrated Circuit Emission Model—Conducted Emissions. In Proceedings of the 2015 10th International Workshop on the Electromagnetic Compatibility of Integrated Circuits (EMC Compo), Edinburgh, UK, 10–13 November 2015.
14. Funabiki, N.; Nomura, Y.; Kawashima, J.; Minamisawa, Y.; Wada, O. A LECCS model parameter optimization algorithm for EMC designs of IC/LSI systems. In Proceedings of the 2006 17th International Zurich Symposium on Electromagnetic Compatibility 2006, Singapore, 28 February–3 March 2006.
15. Ramdani, M.; Sicard, E.; Boyer, A.; Dhia, S.B.; Whalen, J.J.; Hubing, T.H.; Coenen, M.; Wada, O. The Electromagnetic Compatibility of Integrated Circuits—Past, Present, and Future. *IEEE Trans. Electromagn. Compat.* **2009**, *51*, 78–100. [CrossRef]
16. I/O Buffer Information Specification (IBIS). 2008. Available online: http://www.eigroup.org/ibis/ibis.htm (accessed on 5 May 2019).
17. IEC 62404. International Electro-Technical Commission: I/O Interface Model for Integrated Circuit (IMIC). IEC Standard, 2003. Available online: www.iec.ch (accessed on 5 May 2019).
18. IEC 62014-3. International Electro-Technical Commission: Models of Integrated Circuits. IEC Standard, 2002. Available online: www.iec.ch (accessed on 5 May 2019).
19. Lochot, C.; Levant, J. ICEM: A new standard for EMC of IC definition and examples. In Proceedings of the 2003 IEEE Symposium on Electromagnetic Compatibility, Boston, MA, USA, 18–22 August 2003.
20. Ghfiri, C.; Boyer, A.; Durier, A.; Dhia, S.B. A new methodology to build the Internal Activity Block of ICEM-CE for complex Integrated Circuits. *IEEE Trans. Electromagn. Compat.* **2018**, *60*, 1500–1509. [CrossRef]

21. Capriglione, D.; Chiariello, A.; Maffucci, A. Accurate Models for Evaluating the Direct Conducted and Radiated Emissions from Integrated Circuits. *Appl. Sci.* **2018**, *8*, 477. [CrossRef]

22. Levant, J.L.; Ramdani, M.; Perdriau, R.; Drissi, M.H. EMC Assessment at Chip and PCB Level: Use of the ICEM Model for Jitter Analysis in an Integrated PLL. *IEEE Trans. Electromagn. Compat.* **2007**, *49*, 182–191. [CrossRef]

23. Park, H.H.; Song, S.H.; Han, S.T.; Jang, T.S.; Jung, J.H.; Park, H.B. Estimation of Power Switching Current by Chip-Package-PCB Cosimulation. *IEEE Trans. Electromagn. Compat.* **2010**, *52*, 311–319. [CrossRef]

24. Labussière-Dorgan, C.; Bendhia, S.; Sicard, E.; Tao, J.; Quaresma, H.J.; Lochot, C.; Vrignon, B. Modeling the Electromagnetic Emission of a Microcontroller Using a Single Model. *IEEE Trans. Electromagn. Compat.* **2008**, *50*, 22–34. [CrossRef]

25. Li, S.; Bishnoi, H.; Whiles, J.; Ng, P.; Weng, H.; Pommerenke, D.; Beetner, D. Development and validation of a microcontroller model for EMC. In Proceedings of the 2008 International Symposium on Electromagnetic Compatibility—EMC Europe, Amsterdam, The Netherlands, 27–30 August 2008.

26. IEC 62433-2. International Electro-Technical Commission: EMC IC modelling: Models of Integrated Circuits for EMI Behavioral Simulation—Conducted Emissions Modelling (ICEM-CE). IEC Standard, 2008. Available online: www.iec.ch (accessed on 5 May 2019).

27. Su, D.; Xie, S.; Chen, A.; Shang, X.; Zhu, K.; Xu, H. Basic Emission Waveform Theory: A Novel Interpretation and Source Identification Method for Electromagnetic Emission of Complex Systems. *IEEE Trans. Electromagn. Compat.* **2018**, *60*, 1330–1339. [CrossRef]

28. LTM8025 Datasheet and Product Info. Available online: https://www.analog.com/en/products/ltm8025.html (accessed on 5 May 2019).

29. DC1379B Product Details. Available online: https://www.analog.com/en/design-center/evaluation-hardware-and-software/evaluation-boards-kits/dc1379b.html (accessed on 5 May 2019).

30. F-33-2 Datasheet. Available online: http://www.fischercc.com/products/f-33-2/ (accessed on 20 May 2019).

31. Ansys SIwave. Available online: https://www.ansys.com/products/electronics/ansys-siwave (accessed on 25 March 2019).

32. Gustavsen, B.; Semlyen, A. Rational approximation of frequency domain responses by vector fitting. *IEEE Trans. Power Deliv.* **2002**, *14*, 1052–1061. [CrossRef]

33. Moré, J.J. The Levenberg-Marquardt algorithm: Implementation and theory. *Lect. Notes Math.* **1978**, *630*, 105–116.

© 2019 by the authors. Licensee MDPI, Basel, Switzerland. This article is an open access article distributed under the terms and conditions of the Creative Commons Attribution (CC BY) license (http://creativecommons.org/licenses/by/4.0/).

Article

A Three-Bridge IPT System for Different Power Levels Conversion under CC/CV Transmission Mode

Bingyang Luo, Yatao Shou, Jianghua Lu, Ming Li, Xiangtian Deng * and Guorong Zhu

School of Automation, Wuhan University of Technology, Wuhan 430000, China
* Correspondence: dengxt@whut.edu.cn; Tel.: +86-189-7140-0950

Received: 25 July 2019; Accepted: 8 August 2019; Published: 9 August 2019

Abstract: This paper proposes an inductive power transfer (IPT) system with three-bridge switching compensation topology. With the proposed IPT topology, the equivalent circuit and the resonant condition are analyzed to achieve the load-independent constant current (CC) and load-independent constant voltage (CV) outputs. On this basis, multiple power levels can be achieved under CC/CV conditions by bridge arm switching, which makes it possible to complete charging tasks for multiple power level electric vehicles (EV) without switching the IPT system. A circuit simulation was built to verify the different power level switching effects of the structure. A 3.3 kW IPT system was designed to verify the proposed structure. At the rated output power, the experimental efficiency was up to 92.04% and 91.21% in CC and CV output modes, respectively.

Keywords: inductive power transfer (IPT); three-bridge switching; constant current (CC); constant voltage (CV); fixed frequency

---

1. Introduction

The IPT (inductive power transfer) technology uses a loosely coupled transformer (LCT) structure with primary and secondary coils to achieve non-contact energy transfer. Due to the safety, stability, and lower maintenance cost of this structure, IPT systems are receiving more and more attention and research in many fields, such as consumer electrics, biomedical implants, and electric vehicles (EVs) [1–3]. As shown in Figure 1, an IPT system mainly consists of five parts: High-frequency inverter, loosely coupled transformer with primary and secondary coils, primary and secondary compensation networks, rectifier, and battery load.

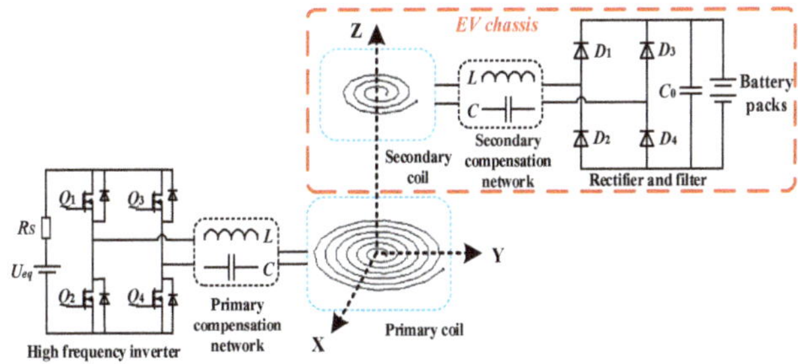

Figure 1. Schematic diagram of an inductive power transfer (IPT) system.

The existence of an air gap between the primary and secondary coils of the IPT system will lead to large leakage inductance. In order to improve the system efficiency and power transfer capability, the compensation network will be used to resonate with the leakage inductance [4]. The four basic topologies (series-series, series-parallel, parallel-series, and parallel- parallel) are the most widely studied, and the SS (series–series) topology is one of the most widely used topologies due to its simple structure and primary capacitance that is independent of the variation of the coupling coefficient [5–7].

As lithium batteries are the main load batteries in the field of electric vehicles, the research on the charging characteristics of lithium batteries is of great significance for the design of IPT systems [8]. The typical charging process of the lithium battery for EVs is shown in Figure 2. From Figure 2, the charging state could be divided into two steps: Constant current (CC) mode and constant voltage (CV) mode [9]. In CC mode, the current is constant and the voltage gradually rises. When the voltage rises to the rated value, it will become stable and the current will gradually decrease until the state of charge ends [10].

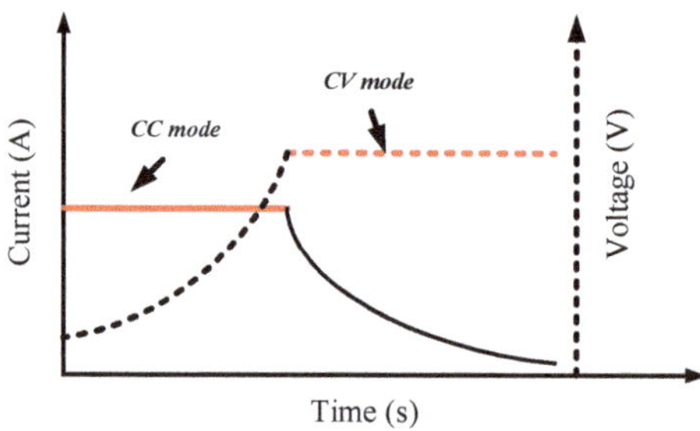

Figure 2. Process of lithium-ion battery cell charging.

In order to obtain load-independent CC/CV transmission characteristics, there are currently three main methods: A back-end DC/DC converter, variable operating frequency control, and special tuning methods for resonant tank network in IPT systems. However, adding a DC/DC converter not only increases the cost, but also introduces additional inevitable losses. The method involving changing switching-frequency increases the complexity of system control and reduces system stability due to frequency bifurcation [11–13]. In J. Lu's paper [14], the resonant conditions for obtaining CC/CV transmission characteristics by T-equivalence and pi-equivalence are introduced in detail. Using this method, the load-independent CC/CV transmission condition of many typical compensation networks can be analyzed easily.

In addition, different charging objects have different charging power level requirements in CC/CV mode [15,16]. For example, the charging power level applied to a bus and a car by the IPT system must be different. The traditional IPT systems only perform charging tasks for a fixed power level, which necessarily has certain limitations [17,18].

To meet multi-power level charging flexibility in CC/CV mode, this paper adopted a three-bridge arm switching concept to perform different modes conversion under constant frequency. Switching from CC to CV mode was achieved by designing two SS compensation networks with the same compensating components on the secondary side, and the primary side compensating components were transformed by switching the working bridge arm. One of the SS compensation networks was designed for the CC transmission mode while the other was designed for the CV transmission mode. Since the voltage gain and transconductance in both CC and CV modes can be specified by parameter design, the specific value of the output power can also be obtained, which means that the structure can achieve two different power-level charging. This paper establishes a mathematical model of system output power in CC/CV mode and illustrates the designability of different power level outputs. The resonance conditions of the CC and CV modes were analyzed by the leakage inductance model. The verification of theoretical analysis was completed by establishing an IPT system under MATLAB/Simulink. Furthermore, an experiment platform was built to verify the simulation results.

2. System Structure and Analysis

2.1. System Structure

As shown in Figure 3, the proposed IPT system includes a three-phase inverter, two SS compensation topologies with a shared secondary compensation capacitor, a three-winding loosely coupled transformer with two primary coils and one secondary coil, and an uncontrolled rectifier with a capacitor output filter. Q_1–Q_6 and D_1–D_4 represent six switching power devices of the primary side inverter and four diodes of the secondary side rectifier, respectively. C_{p1} and C_{p2} are the primary compensation capacitors, while C_s is the secondary compensation capacitor. L_{p1} and L_{p2} represent the self-inductance of the two transmitting coils on the primary side, and L_s represents the self-inductance of the receiving coil. M_{p1p2}, M_{p1s}, and M_{p2s} represent the mutual inductance between L_{p1} and L_{p2}, L_{p1} and L_s, and L_{p2} and L_s, respectively. U_{DC} and U_L are system input and output DC voltages, respectively. R_L represents battery load.

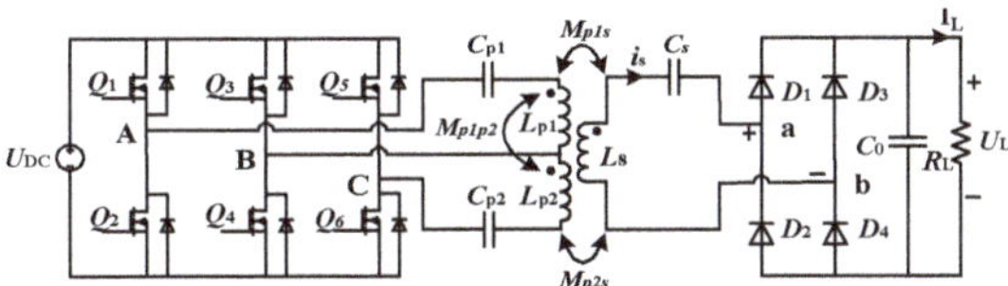

Figure 3. Three-bridge arm-switching IPT system structure.

2.2. CC Mode

The equivalent circuit of Figure 3 for achieving CC output mode is shown in Figures 4 and 5. The system structure shown in Figure 4 was designed to get CC charging mode. This means the power devices Q_5 and Q_6 are OFF all the time in this mode, and Q_1–Q_4 form a single-phase full-bridge inverter. Actually, Figure 4 shows a SS-compensated IPT system. The FHA (fundamental harmonics analysis) method was taken in this paper to simplify the analysis of the output characteristics. In this case, the output voltage of the inverter U_{AB} and the equivalent ac load R_{eq} are derived as

$$\begin{aligned} U_{AB} &= \tfrac{2\sqrt{2}}{\pi} U_{DC} \\ R_{eq} &= \tfrac{8}{\pi^2} R_L \end{aligned} \tag{1}$$

The equivalent model of the SS resonant circuit with reference to the primary side is shown in Figure 5. The symbols " ′ " means that the equivalence was measured from the secondary side to the primary side. L_{kp1} and L'_{ks} means the leakage-inductance of the primary and secondary coil, respectively, and L_{m1} represents the magnetizing inductance. L_{kp1} and C_{p1} are equivalent to capacitor C_{eq1}. C'_s shows the equivalent capacitance of the secondary side compensation capacitance converted to the primary side. R'_{eq} indicates that the equivalent resistance was converted from the secondary side to the primary side. Z_{Rcc}, Z_{m1}, and Z_{incc} represent the impedance.

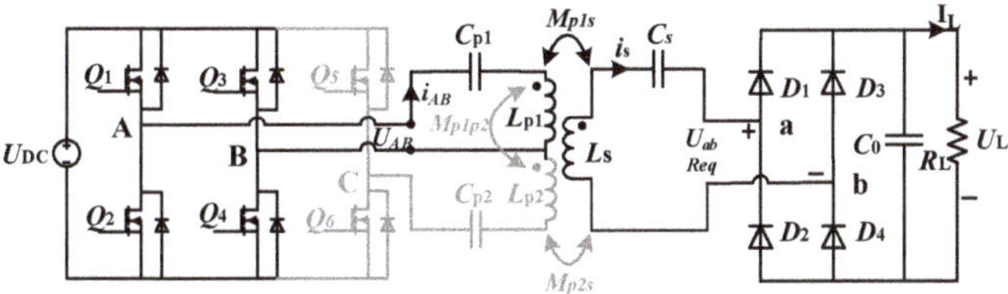

Figure 4. IPT system structure diagram in constant current (CC) mode.

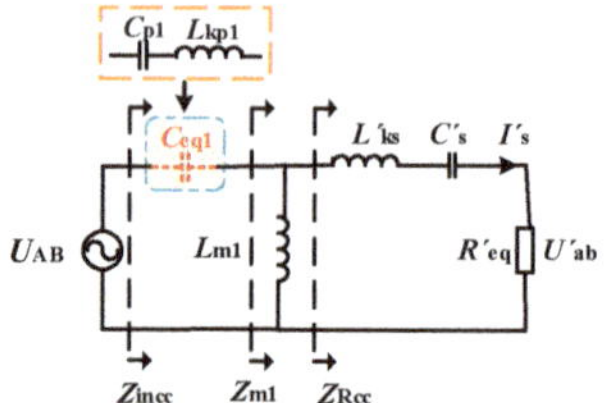

Figure 5. Equivalent circuit of Figure 4 with reference to the primary side.

Some of the parameters are shown in the following formula, such as turns ratio, n_{cc}, and coupling coefficient, k_{cc},

$$n_{cc} = \sqrt{\frac{L_s}{L_{p1}}}, \quad k_{cc} = \frac{M_{p1s}}{\sqrt{L_{p1}L_s}}. \tag{2}$$

The equivalent variables of Figure 5 are expressed as:

$$\begin{aligned}
L_{kp1} &= (1 - k_{cc})L_{p1} \\
L_{m1} &= k_{cc}L_{p1} \\
L'_{ks} &= \frac{(1-k_{cc})L_s}{n_{cc}^2} \\
C'_s &= n_{cc}^2 C_s \\
R'_{eq} &= \frac{R_{eq}}{n_{cc}^2} \\
U'_{ab} &= \frac{U_{ab}}{n_{cc}}
\end{aligned} \tag{3}$$

In order to achieve CC output, the following resonant conditions [19] are designed.

$$\begin{aligned}
\frac{1}{C_{eq1}} &= \frac{1}{C_{p1}} - \omega_{cc}^2 L_{kp1} \\
\omega_{cc}^2 &= \frac{1}{L_{m1}C_{eq1}}
\end{aligned} \tag{4}$$

where ω_{cc} represents the resonant frequency in CC transmission mode.

Then, the voltage gain of the resonant network shown in Figure 5 can be expressed as

$$G_{cc} = \frac{U_{ab}}{U_{AB}} = \frac{n_{cc}U'_{ab}}{U_{AB}} = n_{cc}\frac{|R'_{eq}|}{|Z_{Rcc}|}\frac{|Z_{m1}|}{|Z_{incc}|}. \tag{5}$$

The DC voltage transfer ratio is derived as

$$M_{cc} = \frac{U_L}{U_{DC}} = \frac{\frac{\pi}{2\sqrt{2}}U_{ab}}{\frac{\pi}{2\sqrt{2}}U_{AB}} = G_{cc} = \frac{8R_L}{j\pi^2\omega_{cc}n_{cc}L_{m1}}. \tag{6}$$

Therefore, the system output voltage and output current under the CC mode can be characterized as

$$U_{L(CC)} = \frac{8R_L U_{DC}}{j\pi^2\omega_{cc}n_{cc}L_{m1}}, \tag{7}$$

$$I_{L(CC)} = \frac{8U_{DC}}{j\pi^2\omega_{cc}n_{cc}L_{m1}}. \tag{8}$$

Moreover, the system output power of the CC mode is available.

$$P_{(CC)} = R_L I_{L(CC)}^2. \tag{9}$$

From Equation (8), it can be seen that the output current was constant and load independent.

2.3. CV Mode

The equivalent circuit of CV mode in IPT system is shown in Figures 6 and 7. In this mode, the power devices Q_1 and Q_2 are OFF all the time, and Q_3–Q_6 form a single-phase full-bridge inverter. Furthermore, Figure 6 is modeled as Figure 7 based on the FHA method. It can be seen that Figure 7 is essentially a T-circuit. In this case, the output voltage of the inverter U_{BC} is derived as

$$U_{BC} = \frac{2\sqrt{2}}{\pi} U_{DC}. \tag{10}$$

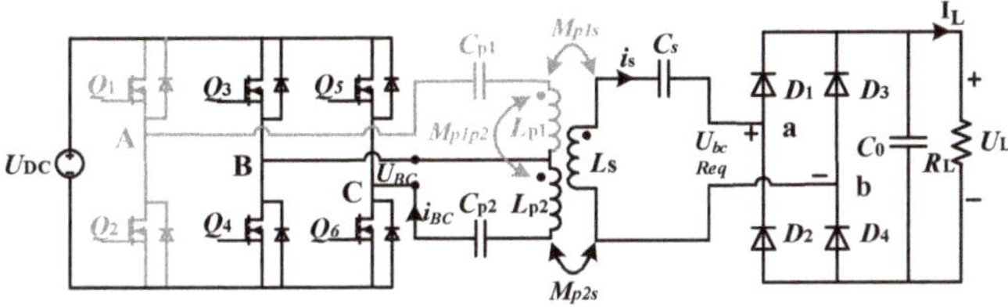

Figure 6. IPT system structure diagram in constant voltage (CV) mode.

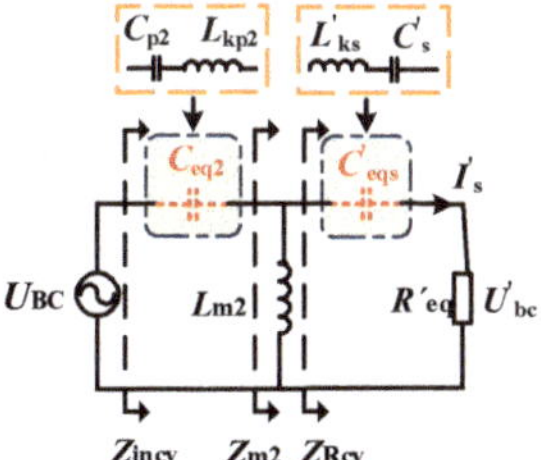

Figure 7. Equivalent circuit of Figure 6 with reference to the primary side.

The turns ratio of secondary to primary side and the coupling coefficient are defined as

$$n_{cv} = \sqrt{\frac{L_s}{L_{p2}}}, \quad k_{cv} = \frac{M_{p2s}}{\sqrt{L_{p2}L_s}}. \tag{11}$$

As shown in Figure 7, the equivalent variables are derived as

$$\begin{aligned}
L_{kp2} &= (1 - k_{cv})L_{p2} \\
L_{m2} &= k_{cv}L_{p2} \\
L'_{ks} &= \frac{(1 - k_{cv})L_s}{n_{cv}^2} \\
C'_s &= n_{cv}^2 C_s \\
R'_{eq} &= \frac{R_{eq}}{n_{cv}^2} \\
U'_{bc} &= \frac{U_{bc}}{n_{cv}}
\end{aligned} \tag{12}$$

According to J. Lu's paper [19], the resonance condition for realizing CV output is given by

$$\begin{aligned}
\frac{1}{C_{eq2}} &= \frac{1}{C_{p2}} - \omega_{cv}^2 L_{kp2} \\
\frac{1}{C'_{eqs}} &= \frac{1}{C'_s} - \omega_{cv}^2 L'_{ks} \\
\omega_{cv}^2 &= \frac{1}{L_{m2}(C_{eq2} + C'_{eqs})}
\end{aligned} \tag{13}$$

where ω_{cv} represents the resonant frequency in CV transmission mode.

Then, the voltage gain of the resonant network shown in Figure 7 can be expressed as

$$G_{cv} = \frac{U_{bc}}{U_{BC}} = \frac{n_{cc}U'_{bc}}{U_{BC}} = n_{cc}\frac{|R'_{eq}|}{|Z_{Rcv}|}\frac{|Z_{m2}|}{|Z_{incv}|}. \tag{14}$$

According to Equation (10), the DC voltage transfer ratio is derived as

$$M_{cv} = \frac{U_L}{U_{DC}} = \frac{\frac{\pi}{2\sqrt{2}}U_{bc}}{\frac{\pi}{2\sqrt{2}}U_{BC}} = G_{cv} = \frac{n_{cv}C_{eq2}}{C'_{eqs}}. \tag{15}$$

Therefore, the system output voltage and output current under the CV mode can be characterized as

$$U_{L(CV)} = \frac{n_{cv}C_{eq2}U_{DC}}{C'_{eqs}}, \tag{16}$$

$$I_{L(CV)} = \frac{n_{cv}C_{eq2}U_{DC}}{R_L C'_{eqs}}. \tag{17}$$

Moreover, the system output power of the CV mode is available.

$$P_{(CV)} = \frac{U^2_{L(CV)}}{R_L}. \tag{18}$$

From Equation (16), it can be seen that the output voltage is constant and load independent.

Combining the theoretical analysis of the above-two transmission modes, it could be known that the three-bridge inverter structure could realize the conversion from CC mode to CV mode by switching the working bridge arm under the fixed-frequency.

Combining the Equations (9) and (18), it can be seen that the output power of both CC mode and CV mode is related to the load resistance R_L, and the specific output powers in CC and CV modes can be obtained by designing the rated load resistance at the switching time. Combined with the parameter design, the CC/CV working mode was switched by changing the bridge arm, and different power level outputs were also completed. Therefore, the structure can achieve different power level switching based on CC/CV conditions.

3. Simulation and Experiment

3.1. Parameter Design

The LCT model parameters based on CC/CV transmission mode selected in this paper were obtained through Maxwell. Figure 8 shows the specific model dimensions of the LCT structure. The transmission coils and the corresponding ferrites were square structures, the side length of the primary side was 250 mm, the side length of the auxiliary section was 330 mm, and the air gap between the primary and secondary coils was 110 mm. To facilitate the parameter design process, both the primary and secondary self-inductance of LCT were designed to be equal, that is, the value of the turns-ratio n was 1. Through the finite element simulation of the LCT structure, the specific parameter values of the LCT were obtained as shown in Table 1.

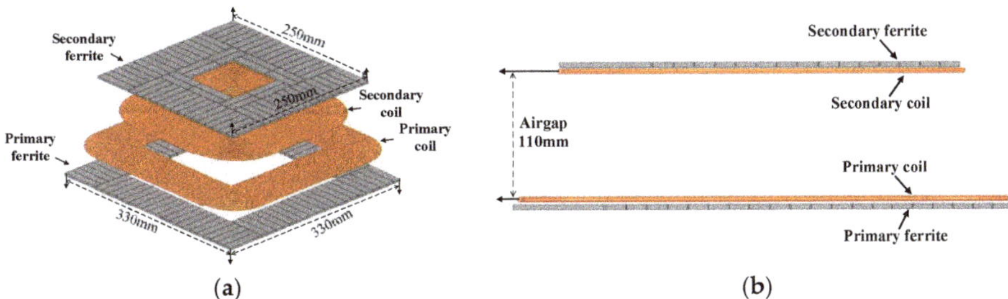

Figure 8. Loosely coupled transformer (LCT) structure in IPT system with (**a**) main view and (**b**) front view.

Table 1. Parameters in IPT system.

Parameters	Value
U_{DC}	220 V
U_{L}	260 V
k	0.2
The LCT coil inductance ($L_{p1} = L_{p2} = L_s$)	120 µH
The resonance frequency f_{cc}/f_{cv}	85 kHz
The system output power (P)	3.3 kW

According to Equation (4), the primary compensation capacitor in the CC mode is calculated as

$$C_{p1} = \frac{1}{\omega_{cc}^2 (L_{m1} + L_{kp1})} = \frac{1}{\omega_{cc}^2 L_{p1}} = 29.216 \text{ nF}. \tag{19}$$

By using Equations (11) and (12), it was found that the SS topology primary compensation capacitor, C_{p2}, in the CV mode and the fixed secondary compensation capacitor, C_s, could be obtained separately.

$$C_{p2} = \frac{1}{\frac{1}{C_{eq2}} + \omega_{cv}^2 L_{kp2}} = 35.138 \text{ nF}. \tag{20}$$

$$C_s' = \frac{1}{\frac{1}{C_{eqs}} + \omega_{cv}^2 L_{ks}'} = 28.308 \text{ nF}. \tag{21}$$

3.2. Simulation Results

Through the parameter design of the previous section, the simulation verification based on transconductance and voltage gain was completed. Figure 9 shows the transconductance versus system operating frequency for various load resistances. It can be seen that the transconductance was independent of the load when the IPT system operated at 85 kHz. Figure 10 shows the AC voltage gain versus system operating frequency for various load resistances. It can be seen that the voltage gain was independent of the load when the IPT system operated at 85 kHz.

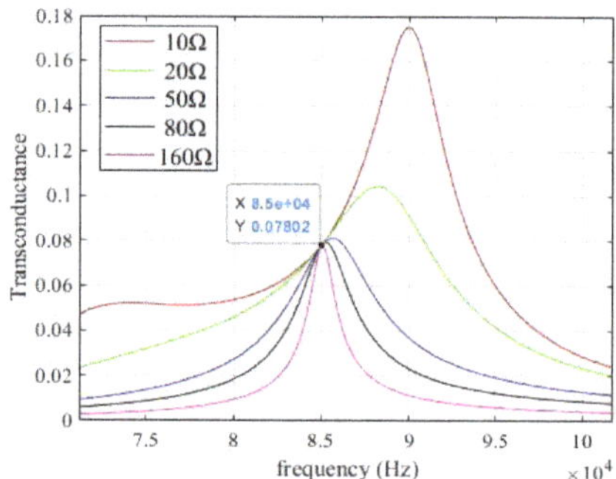

Figure 9. Transconductance with respect to frequency for different load conditions.

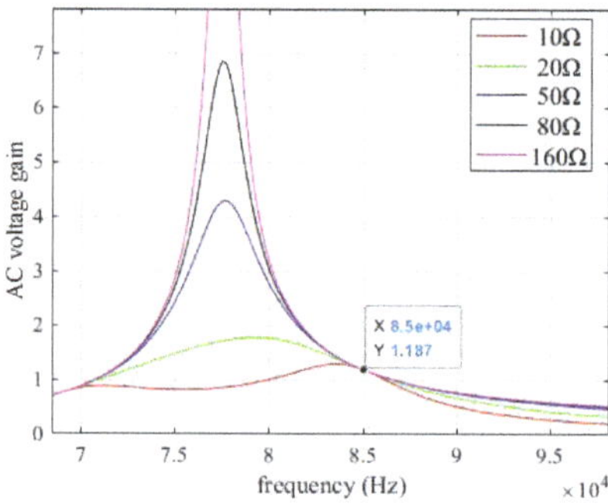

Figure 10. Voltage gain with respect to frequency for different load conditions.

According to the previous analysis, by determining the rated load resistance, R_L, at the time of switching, two different output powers were realized while switching from CC to CV mode. In order to prove the previous analysis, a three-bridge switching IPT system circuit under MATLAB/Simulink was built. The simulation results are shown in Figure 11. From the previous mathematical model of the output power of CC/CV mode, it can be known that the output power of CC/CV mode is different except for the specific load resistance (The specific load resistance value can be calculated when P1 is equal to P2). Figure 11a shows that the bridge arm switching process did not change the system output power under the specific load resistance condition as described above, and the IPT system had a system output power of 3.3 kW in both CC and CV modes. Figure 11b shows that under other rated load resistance condition, the bridge arm switching process changed the system output power from 2.8 to 3.9 kW.

The time (t1) in Figure 11 represents the switching load resistance under the CC mode condition, in which the current changed by 0.5 A. The time (t2) in Figure 11 represents the switching of the working arm so that the system was switched from the CC mode to the CV mode. The time (t3) in Figure 11 represents the switching load resistance under the CV mode condition. It can be seen from the simulation results that the CC-to-CV mode switching was realized by changing the working arm, and two specific system output powers in CC/CV mode were also obtained.

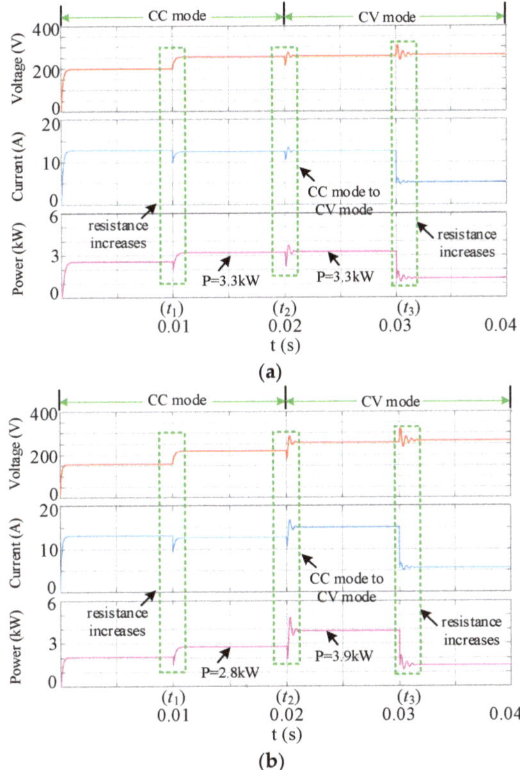

Figure 11. Charging process under t_1: Increasing the resistance. t_2: Switching from CC mode to CV mode. t_3: Increasing the resistance. (**a**) Constant rated output power during switching. (**b**) Changing rated output power during switching.

3.3. Experimental Results

Combined with the previous analysis, this paper established a 3.3 kW experimental platform as shown in Figure 12. The CC and CV transmission modes were realized by two SS topologies based on different combined bridge arms at 85 kHz.

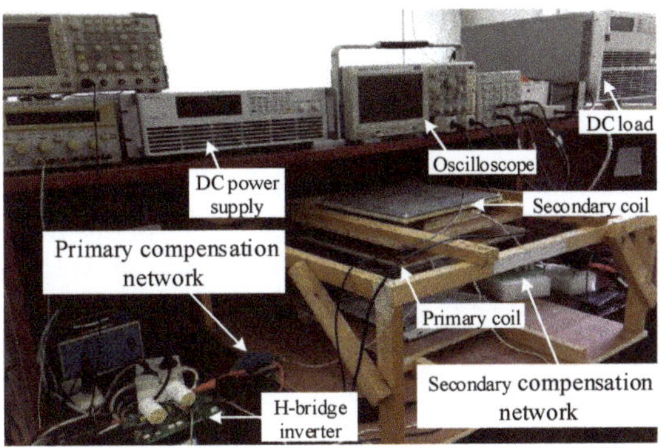

Figure 12. IPT system experiment platform.

The Figure 13a shows the system working in CC mode, u_{AB} represents the inverter output voltage, i_{AB} represents the inverter output current, and u_{gs} represents the drive voltage of the switching tube. The Figure 13b shows the system working in CV mode, u_{BC} represents the inverter output voltage, i_{BC} represents the inverter output current, and u_{gs} represents the drive voltage of the switching tube. It can be seen that Zero Voltage Switch (ZVS) was implemented in the CV mode of the system, and there were ZVS trends in the CC mode.

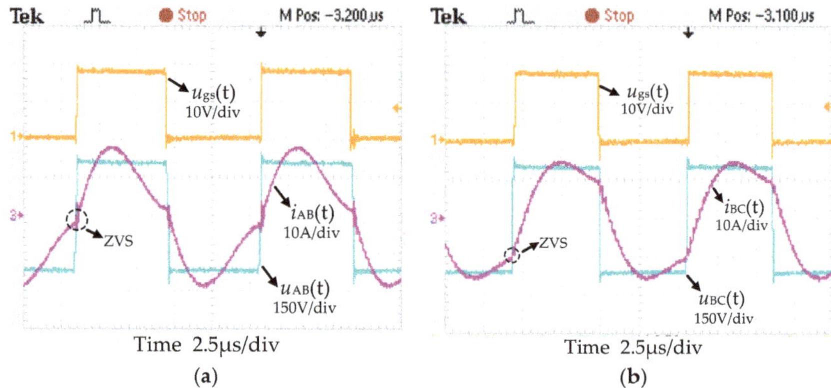

Figure 13. Waveform of inverter output voltage/current and rectifier input voltage under the (a) CC transmission mode and (b) CV transmission mode.

As shown in Figure 14a, the load was switched from half to full at t_{cc1} and then back to half at t_{cc2}. As shown in Figure 14b, the load was switched from half to full at t_{cv1} and then back to half at t_{cv2}. According to the experimental results of the system output voltage and current in Figure 14, the validity of the previous analysis with the CC/CV transmission characteristics can be verified.

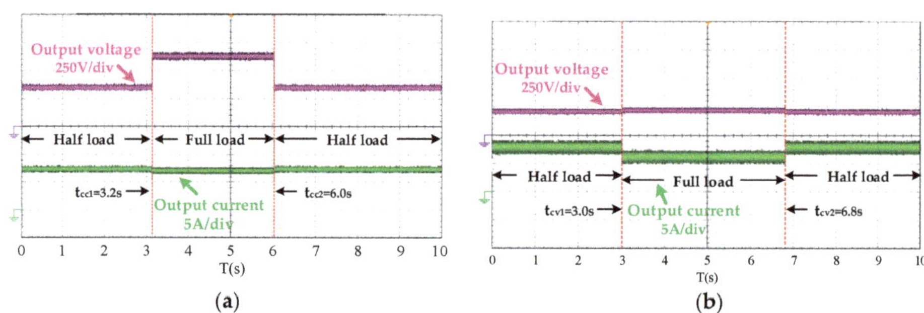

Figure 14. Output voltage and current of IPT system with (a) CC transmission mode and (b) CV transmission mode.

Figure 15 shows the experimental efficiencies of the three-bridge IPT system versus different output powers. It can be seen that in a certain range, the system efficiency was increasing with the output power in both CC and CV mode. In addition, the system efficiencies at rated output power were 92.04% and 91.21% in CC and CV mode, respectively.

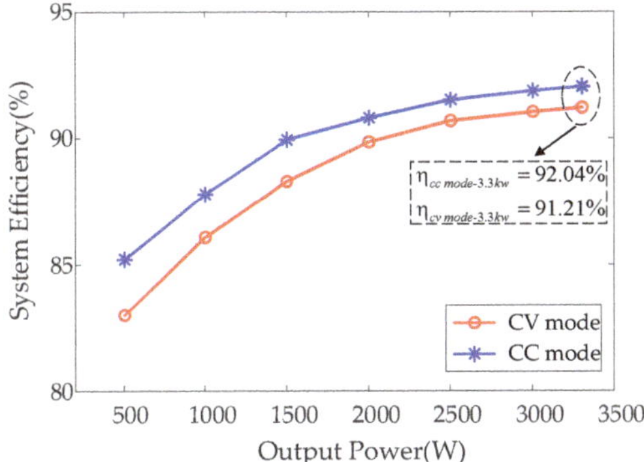

Figure 15. Experimental efficiencies of the IPT system versus output power in both CC and CV output modes.

4. Conclusions

This paper proposed a three-bridge IPT system, which could realize the CC/CV mode transmission under the fixed-frequency by topology parameter design and convert from CC to CV mode by switching the working arm. On this basis, by designing the rated load resistance at the switching time, the specific output powers before and after the switching time were realized. Therefore, the effectiveness of CC/CV mode switching and the output of different powers by switching the bridge arms was verified. By establishing and analyzing the leakage inductance equivalent model based on SS topology, the resonance condition of the CC/CV mode transmission characteristics were obtained. Finally, circuit simulations and the corresponding experiment were built, which verified the theoretical analysis and the effectiveness of this structure.

Author Contributions: Conceptualization, M.L. and J.L.; data curation, Y.S. and M.L.; project administration, X.D. and B.L.; resources, G.Z.; supervision, G.Z.; validation, X.D.; writing—review and editing, Y.S. and J.L.

Funding: This research was funded by National Natural Science Foundation of China grand number 51777146 and the Science and Technology Foundation of Guizhou province in China (No. LH-[2014]7369)

Conflicts of Interest: The authors declare no conflict of interest.

References

1. RamRakhyani, A.K.; Mirabbasi, S.; Chiao, M. Design and Optimization of Resonance-Based Efficient Wireless Power Delivery Systems for Biomedical Implants. *IEEE Trans. Biomed. Circuits Syst.* **2011**, *5*, 48–63. [CrossRef] [PubMed]
2. Li, Q.; Liang, Y.C. An Inductive Power Transfer System with a High-Q Resonant Tank for Mobile Device Charging. *IEEE Trans. Power Electron.* **2015**, *30*, 6203–6212. [CrossRef]
3. Beh, H.Z.Z.; Covic, G.A.; Boys, J.T. Investigation of Magnetic Couplers in Bicycle Kickstands for Wireless Charging of Electric Bicycles. *IEEE J. Emerg. Sel. Top. Power Electron.* **2015**, *3*, 87–100. [CrossRef]
4. Zhang, W.; White, J.C.; Abraham, A.M.; Mi, C.C. Loosely Coupled Transformer Structure and Interoperability Study for EV Wireless Charging Systems. *IEEE Trans. Power Electron.* **2015**, *30*, 6356–6367. [CrossRef]
5. Villa, J.L.; Sallan, J.; Osorio, J.F.S.; Llombart, A. High-Misalignment Tolerant Compensation Topology for ICPT Systems. *IEEE Trans. Ind. Electron.* **2012**, *59*, 945–951. [CrossRef]
6. Khaligh, A.; Dusmez, S. Comprehensive Topological Analysis of Conductive and Inductive Charging Solutions for Plug-In Electric Vehicles. *IEEE Trans. Veh. Technol.* **2012**, *61*, 3475–3489. [CrossRef]
7. Sallan, J.; Villa, J.L.; Llombart, A.; Sanz, J.F. Optimal Design of ICPT Systems Applied to Electric Vehicle Battery Charge. *IEEE Trans. Ind. Electron.* **2009**, *56*, 2140–2149. [CrossRef]
8. Gerssen-Gondelach, S.J.; Faaij, A.P.C. Performance of batteries for electric vehicles on short and longer term. *J. Power Sour.* **2012**, *212*, 111–129. [CrossRef]

9. Qu, X.; Han, H.; Wong, S.; Tse, C.K.; Chen, W. Hybrid IPT Topologies With Constant Current or Constant Voltage Output for Battery Charging Applications. *IEEE Trans. Power Electron.* **2015**, *30*, 6329–6337. [CrossRef]

10. Yilmaz, M.; Krein, P.T. Review of Battery Charger Topologies, Charging Power Levels, and Infrastructure for Plug-In Electric and Hybrid Vehicles. *IEEE Trans. Power Electron.* **2013**, *28*, 2151–2169. [CrossRef]

11. Gati, E.; Kampitsis, G.; Manias, S. Variable Frequency Controller for Inductive Power Transfer in Dynamic Conditions. *IEEE Trans. Power Electron.* **2017**, *32*, 1684–1696. [CrossRef]

12. Buja, G.; Bertoluzzo, M.; Mude, K.N. Design and Experimentation of WPT Charger for Electric City Car. *IEEE Trans. Ind. Electron.* **2015**, *62*, 7436–7447. [CrossRef]

13. Lu, J.; Zhu, G.; Shou, Y.; Liu, F. Coupling- and Load-Independents Output Voltage and ZPA Operation in LCC-Series Compensated IPT System. In Proceedings of the 2018 IEEE Energy Conversion Congress and Exposition (ECCE), Portland, OR, USA, 23–27 September 2018; pp. 1172–1176.

14. Lu, J.; Zhu, G.; Wang, H.; Lu, F.; Jiang, J.; Mi, C.C. Sensitivity Analysis of Inductive Power Transfer Systems With Voltage-Fed Compensation Topologies. *IEEE Trans. Veh. Technol.* **2019**, *68*, 4502–4513. [CrossRef]

15. Rahnamaee, H.R.; Madawala, U.K.; Thrimawithana, D.J. A modified hybrid multi-level converter for high-power high-frequency IPT systems. In Proceedings of the 2014 International Power Electronics and Application Conference and Exposition, Shanghai, China, 5–8 November 2014; pp. 624–629.

16. Li, Y.; Mai, R.; Lu, L.; He, Z.; Liu, S. Harmonic elimination and power regulation based five-level inverter for supplying IPT systems. In Proceedings of the 2015 IEEE PELS Workshop on Emerging Technologies, Wireless Power (2015 WoW), Daejeon, South Korea, 5–6 June 2015; pp. 1–4.

17. Cheng, C.; Zhou, Z.; Li, W.; Zhu, C.; Deng, Z.; Mi, C.C. A Multi-Load Wireless Power Transfer System With Series-Parallel-Series Compensation. *IEEE Trans. Power Electron.* **2019**, *34*, 7126–7130. [CrossRef]

18. Liu, F.; Yang, Y.; Ding, Z.; Chen, X.; Kennel, R.M. A Multifrequency Superposition Methodology to Achieve High Efficiency and Targeted Power Distribution for a Multiload MCR WPT System. *IEEE Trans. Power Electron.* **2018**, *33*, 9005–9016. [CrossRef]

19. Lu, J.; Zhu, G.; Lin, D.; Wong, S.; Jiang, J. Load-Independent Voltage and Current Transfer Characteristics of High-Order Resonant Network in IPT System. *IEEE J. Emerg. Sel. Top. Power Electron.* **2019**, *7*, 422–436. [CrossRef]

© 2019 by the authors. Licensee MDPI, Basel, Switzerland. This article is an open access article distributed under the terms and conditions of the Creative Commons Attribution (CC BY) license (http://creativecommons.org/licenses/by/4.0/).

Article

A Fractional-Order Element (FOE)-Based Approach to Wireless Power Transmission for Frequency Reduction and Output Power Quality Improvement

Guidong Zhang [1], Zuhong Ou [1] and Lili Qu [2],*

[1] School of Automation, Guangdong University of Technology, Guangzhou 510006, China; guidong.zhang@gdut.edu.cn (G.Z.); zuhong.ou@foxmail.com (Z.O.)
[2] School of Automation, Foshan University, Foshan 528225, China
*   Correspondence: qulili@fosu.edu.cn; Tel.: +86-139-2590-6648

Received: 3 September 2019; Accepted: 11 September 2019; Published: 13 September 2019

Abstract: A wireless power transmission (WPT) requires high switching frequency to achieve energy transmission; however, existing switching devices cannot satisfy the requirements of high-frequency switching, and the efficiency of current WPT is too low. Compared with the traditional power inductors and capacitors, fractional-order elements (FOEs) in WPT can realize necessary functions though requiring a lower switching frequency, which leads to a more favorable high-frequency switching performance with a higher efficiency. In this study, a generalized fractional-order WPT (FO-WPT) is established, followed by a comprehensive analysis on its WPT performance and power efficiency. Through extensive simulations of typical FO wireless power domino-resonators (FO-WPDRS), the functionality of the proposed FO-WPT for medium and long-range WPT is demonstrated. The numerical results show that the proposed FOE-based WPT solution has a higher power efficiency and lower switching frequency than conventional methods.

Keywords: fractional order elements; high-frequency switching; wireless power transmission

---

1. Introduction

Wireless power transmission (WPT) is gaining more and more attention in city transportation applications since *Tesla* firstly revealed WPT in the 1880s [1]. It is also revealed that WPT is of great practical significance due to its immunity to fire and electric shock [2]. Recently, the most extensive studies in WPT are inductive coupling WPT (ICWPT), magnetic resonance coupling WPT (MRCWPT) and microwave wireless power transmission (MWPT) [3]. Compared with ICWPT, MRCWPT transfers power over a longer distance and powers several multiple loads via a single transmitter coil simultaneously [4]. Among three techniques mentioned above, MWPT's (also named as radio frequency-based WPT) efficiency is the lowest one but its transfer distance is longest [5,6]. Therefore, when *Soljačić* with the Massachusetts Institute of Technology (MIT) successfully lit up a bulb over 2 m with 40% energy efficiency in 2007 [7], MRCWPT stimulated international research.

In [8], the authors propose a simple method for the analysis of the resonance frequency of a planar spiral coil to evaluate the lumped parameters of the coils. Using the technology of electromagnetic resonant couplings WPT [9], the authors addressed the fact that the output power can reach 100 W. Ref. [10] showed that magnetic resonant coupling WPT normally works in megahertz and its efficiency is improved with increasing frequency. Despite its long-lasting popularity, using coils to achieve WPT through high-frequency actions cannot satisfy today's needs, and power electronics elements are being developed to realize the same function [11]. However, it is well known that current power electronics technologies cannot satisfy such high-frequency switching [12]; therefore, in order to meet the capability of contemporary power switches to realize WPT, it is of great necessity to propose new methods for reducing the resonant frequency. Traditionally, a pair of coil resonators were applied to mid- and long-range wireless power transfer, but this technique has a poor performance in energy transfer efficiency [13]. To solve this problem, a new method of using coil arrays at both the transmitter and receiver sides was introduced employing the technique of mid-range strong magnetic resonant coupling [14]. Although the transmission separation is 50 cm, its efficiency of the system is 50% and it is still too low. Another new method is to determine a model for coupling coefficient to compute optimal frequency for the power transfer [15]. To overcome this problem, relay resonators, which are normally used in meta-materials and waveguide research [16], were implemented into WPT for mid-range or even long-range wireless power transfer [17]. Inspired by this idea, the domino resonators for waveguide applications at 100 MHz have been reported in [18]; however, the switching frequency is too high for existing power electronics elements. Zhong found that cross-coupling effects of non-adjacent resonators can both improve the efficiency and reduce the switching frequency at 520 kHz resulting from the effects of the magnetic coupling of non-adjacent resonators [19]. In contrast to Zhong's strategy, a WPT system which was designed using superconducting coils to boost wireless power transmission efficiency offered higher transmission efficiency at a longer distance [20]. However, this method is not practical because of the high economic cost and difficulty in realizing the superconducting coils [21]. Please note that WPT's resonant frequency is still very high via current power converter technology. Therefore, a methodology is eagerly required to further reduce the frequency and improve the output power and efficiency.

To well solve the aforementioned problems, an optimizable circuit structure was developed in [22], achieving 85% transfer efficiency at a distance of 10 cm and 50% at 20 cm. It is obvious that they solved the problem by tuning parameters with an optimized model, which nonetheless did not resolve the problem at a fundamental level [23]. Therefore, it still lacks a methodology to realize maximum efficiency and obtain other favorable output features.

Recently, Fractional-Order Electrical Elements (FOE) were applied to electrical engineering applications. In general, two elements are required for impedance matching networks of integer circuit, while only a single FOE is required for fractional cases [24,25] since it has the feature of an extra degree-of-freedom to realize functions of both inductors and capacitors. For example, fractional-order capacitor (FOC) with $1 < \alpha < 2$ (α is the order of the FOC) can be used to improve output power features and decrease the resonant frequency [26,27]. In fact, these unique features of FOEs are needed for the applications of WPT. Furthermore, an FOE is mostly constructed into a ladder network consisting of resistors and capacitors (or inductors) [28], which demonstrates that implementation of FOE costs little; however, this still lacks some fundamental studies.

To reduce resonant frequency and improve the output power and efficiency of WPT, this paper proposes a novel WPT implemented with FOEs, which we brand FO-WPT. A typical example, wireless power domino-resonators, or FO-WPDRS, is presented to confirm the flexibility and versatility of the proposed solution. The remainder of this paper is organized as follows. Fundamental analysis and introduction of FOEs are presented in Section 2. It is followed up with modelling of FO-WPT in Section 3. In order to verify this idea, FO-WPDRS is analyzed as an example with detailed derivations of output power and transmission efficiency in Section 4. Compared with WPDRS, the unique features of the proposed FO-WPDRS are demonstrated in Section 5. An FOWPT prototype is built to prove the viability of the proposed method, the experimental results on the characteristics of the FOC and the performance of applying FOC to WPT are presented and discussed in Section 6. Finally, a conclusion is drawn in Section 7.

2. Fundamental Analysis of FOEs

This section presents the fundamental analysis of FOEs. A series-connected fractional-order $RL_\beta C_\alpha$ circuit is shown in Figure 1.

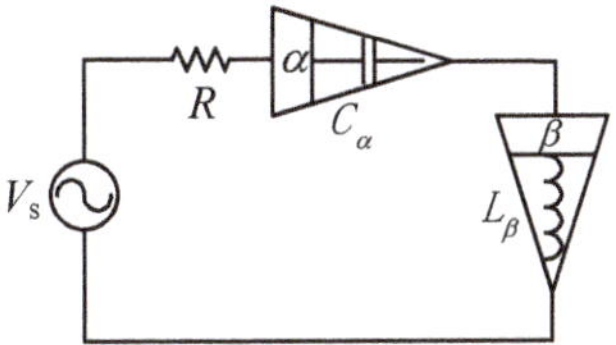

Figure 1. Schematic of fractional-order $RL_\beta C_\alpha$ circuit.

The current-voltage relationship of fractional-order capacitor is defined as [27]

$$i(t) = C_\alpha \frac{d^\alpha v(t)}{dt^\alpha}, \tag{1}$$

where d^α/dt^α is termed as the fractional-order derivative, $v(t)$ is the capacitor voltage, $i(t)$ is the capacitor current, α ($0 < \alpha < 2$) is the order of the capacitor and C_α is the capacitance expressed in $F/s^{1-\alpha}$. The impedance of FOC can be derived from Equation (1) and expressed as $Z_C(s) = 1/(s^\alpha C_\alpha)$, where s is the Laplace operator. With $s = j\omega$, the impedance can be represented by the following equation.

$$Z_C(j\omega) = \frac{1}{(j\omega)^\alpha C_\alpha} = \frac{1}{\omega^\alpha C_\alpha}(\cos(\frac{\alpha\pi}{2}) - j\sin(\frac{\alpha\pi}{2})), \tag{2}$$

where ω is the angular frequency. Similarly, the impedance of fractional-inductor can be expressed as Equation (3).

$$Z_L(j\omega) = (j\omega)^\beta L_\beta = \omega^\beta L_\beta(\cos(\frac{\beta\pi}{2}) + j\sin(\frac{\beta\pi}{2})), \tag{3}$$

where β ($0 < \beta < 2$) is the order of the inductor and L_β is the inductance. In terms of Equations (2) and (3), the impedance of the FOEs includes real part and imaginary part, while the conventional ones only have the imaginary part.

As shown in Figure 2, the order of FOEs varies from -2 to 2, which means that the phase falls in $(-\pi, \pi)$. It can be noted that that FOEs become integer elements if $\alpha(\beta) \to 1$; when $\alpha(\beta) \to 2$ on the other hand, the circuit works as a second-order system.

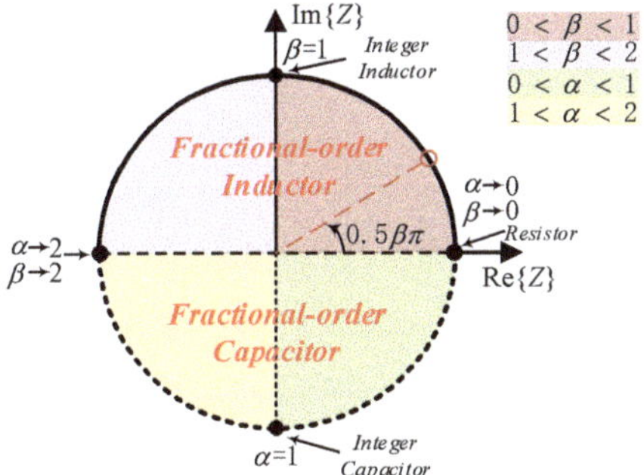

Figure 2. Fractional-order elements.

According to the phasor diagram shown in Figure 3, FOEs act as power-consuming elements with $1 < \alpha < 2$ or $1 < \beta < 2$ where it has the characteristic of a negative resistor and supplies power.

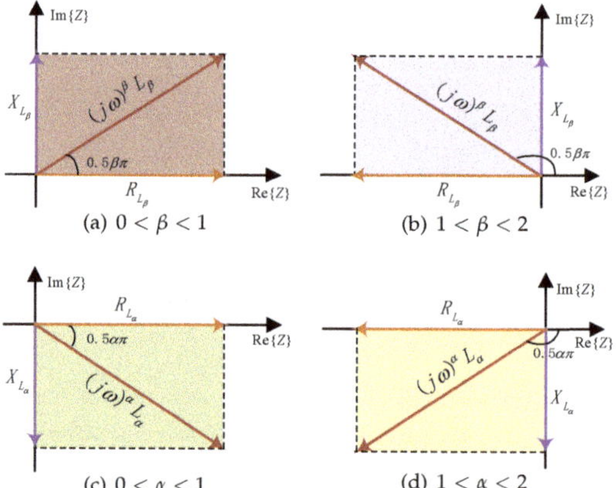

(a) $0 < \beta < 1$ (b) $1 < \beta < 2$

(c) $0 < \alpha < 1$ (d) $1 < \alpha < 2$

Figure 3. Phasor diagram.

3. Modeling the Proposed FO-WPT Strategy

In this section, an FO-WPT model is established and the relationships of fractional resonant frequency, efficiency and output power are derived.

The schematic of FO-WPT is shown in Figure 4a, where the dashed circuit of Figure 4a shows the coil-to-coil resonators, with the detailed model depicted in Figure 4b. In Figure 4a, the primary circuit consists of a high-frequency voltage source V_s, an FOC C_{α_1} whose order is α_1, a fractional-order inductor (FOI) L_{β_1} with order β_1 and a primary resistor R_1. On the secondary side, it is composed of an FOI L_{β_2} with an order β_2, an FOC C_{α_2} with an order α_2 and resistor R_2. Term M_γ is the mutual inductance whose order is γ and R_L represents an equivalent resistance of the load. Then, one can describe the corresponding equation as in Equation (4).

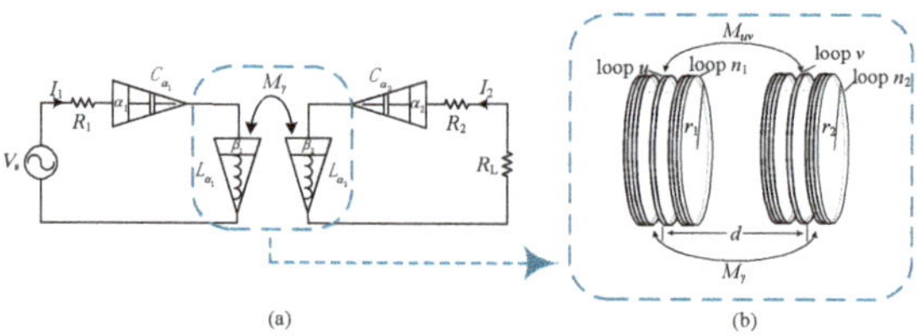

Figure 4. (a) Schematic of FO-WPT and (b) its coil-to-coil model.

$$\begin{bmatrix} Z_{f_1} & (j\omega)^{\gamma_{12}} M_{\gamma_{12}} \\ (j\omega)^{\gamma_{21}} M_{\gamma_{21}} & R_L + Z_{f_2} \end{bmatrix} \cdot \begin{bmatrix} I_1 \\ I_2 \end{bmatrix} = \begin{bmatrix} V_s \\ 0 \end{bmatrix},$$

(4)

where $M_{\gamma_{12}} = M_{\gamma_{21}} = M_\gamma$ is the mutual inductance between winding-1 and winding-2, $\gamma_{12} = \gamma_{21} = \gamma$ is the order of mutual inductance, $Z_{f_n} = R_n + (j\omega)^{\beta_n} L_{\beta_n} + 1/((j\omega)^{\alpha_n} C_{\alpha_n})(n = 1, 2)$.

To simplify the analysis, a system with two coaxial circular filamentary current windings are taken as an example for FO-WPT in this study. *Maxwell* has derived a well-known equation to calculate the mutual inductance. Then, one can obtain the mutual inductance of a pair of single circular loops (loop u and loop v, as shown in Figure 4b) as

$$M_{uv} = \mu_0 \frac{\sqrt{r_1 r_2}}{g}[(2 - g^2)\mathbf{K}(g) - 2\mathbf{E}(g)],$$

(5)

where

$$g = \frac{4r_1 r_2}{d^2 + (r_1 + r_2)^2},\tag{6}$$

where r_1, r_2, and d are the radius of loop u, loop v, and the distance between them, respectively. $\mathbf{K}(g)$ and $\mathbf{E}(g)$ are the first and second kind corresponding complete elliptic integrals. For two coaxial circular thin-wall windings, the mutual inductance can be calculated by

$$M_\gamma = \sum_{u=1}^{n_1} \sum_{v=1}^{n_2} M_{uv}.\tag{7}$$

In terms of Equation (4), currents in primary side and secondary side are

$$\begin{cases} I_1 = \dfrac{(Z_{f_2} + R_L)V_s}{Z_{f_1}(Z_{f_2} + R_L) - (j\omega)^{\gamma_{12}+\gamma_{21}} M_{\gamma_{12}} M_{\gamma_{21}}}, \\[2ex] I_2 = \dfrac{(j\omega)^{\gamma_{21}} M_{\gamma_{21}} V_s}{Z_{f_1}(Z_{f_2} + R_L) - (j\omega)^{\gamma_{12}+\gamma_{21}} M_{\gamma_{12}} M_{\gamma_{21}}}. \end{cases}\tag{8}$$

Then the output power of FO-WPT can be calculated by Equation (9), which is shown at the top of the next page, where $A = Re[Z_{f_1}]$, $B = Im[Z_{f_1}]$, $C = Re[Z_{f_2}] + R_L$, and $D = Im[Z_{f_2}]$.

$$\begin{aligned}
P_o = |I_2|^2 R_L &= \left| \frac{(j\omega)^{\gamma_{21}} M_{\gamma_{21}} V_s}{Z_{f_1}(Z_{f_2} + R_L) - (j\omega)^{\gamma_{12}+\gamma_{21}} M_{\gamma_{12}} M_{\gamma_{21}}} \right|^2 R_L \\
&= \frac{\omega^{2\gamma_{12}} M_{\gamma_{12}}^2 R_L V_s^2}{\left[AC - BD - \omega^{\gamma_{12}+\gamma_{21}} M_{\gamma_{12}} M_{\gamma_{21}} \cos\frac{(\gamma_{12}+\gamma_{21})\pi}{2} \right]^2 + \left[BC + DA - \omega^{\gamma_{12}+\gamma_{21}} M_{\gamma_{12}} M_{\gamma_{21}} \sin\frac{(\gamma_{12}+\gamma_{21})\pi}{2} \right]^2},
\end{aligned}\tag{9}$$

In terms of Equation (4), complex power of winding-1 and winding-2 are expressed as

$$S_{f_1} = V_s I_1^* = Z_{f_1} I_1^2 + (j\omega)^{\gamma_{12}} M_{\gamma_{12}} I_2 I_1^*,\tag{10}$$

and

$$S_{f_2} = 0 = (j\omega)^{\gamma_{21}} M_{\gamma_{21}} I_1 I_2^* + (Z_{f_2} + R_L) I_2^2,\tag{11}$$

where I_1^* and I_2^* are conjugates of I_1 and I_2, respectively.

In terms of Equation (3), the expression of fractional-order mutual inductance Z_{M_γ} can be written as

$$Z_{M_\gamma} = (j\omega)^\gamma M_\gamma = \omega^\gamma M_\gamma \left(\cos\left(\frac{\gamma\pi}{2}\right) + j\sin\left(\frac{\gamma\pi}{2}\right)\right).\tag{12}$$

Substituting Equation (12) into Equations (10) and (11) yields

$$\begin{aligned}
S_{f_1} &= V_s I_1^* \\
&= Z_{f_1} I_1^2 + \omega^{\gamma_{12}} M_{\gamma_{12}} \left(\cos\left(\frac{\gamma_{12}\pi}{2}\right) + j\sin\left(\frac{\gamma_{12}\pi}{2}\right)\right) I_2 I_1^*,
\end{aligned}\tag{13}$$

and

$$\begin{aligned}
S_{f_2} &= \omega^{\gamma_{21}} M_{\gamma_{21}} \left(\cos\left(\frac{\gamma_{21}\pi}{2}\right) \right. \\
&\left. + j\sin\left(\frac{\gamma_{21}\pi}{2}\right)\right) I_1 I_2^* + (Z_{f_2} + R_L) I_2^2 = 0.
\end{aligned}\tag{14}$$

Rewrite Equation (14) as

$$
\begin{aligned}
&- j\omega^{\gamma_{21}} M_{\gamma_{21}} \sin(\frac{\gamma_{21}\pi}{2}))I_1 I_2^* \\
&= \omega^{\gamma_{21}} M_{\gamma_{21}} \cos(\frac{\gamma_{21}\pi}{2})I_1 I_2^* + (Z_{f_2} + R_L)I_2^2 \\
&= (j\omega^{\gamma_{21}} M_{\gamma_{21}} \sin(\frac{\gamma_{21}\pi}{2}))I_1^* I_2)^*,
\end{aligned}
\tag{15}
$$

then, substituting Equation (15) into Equation (13) results in

$$
\begin{aligned}
V_s I_1^* \\
= Z_{f_1} I_1^2 + (Z_{f_2}^* + R_L)I_2^2 + 2 \cdot \omega^{\gamma_{21}} M_{\gamma_{21}} \cos(\frac{\gamma_{12}\pi}{2})I_1^* I_2 \\
= Z_{f_1} I_1^2 + (Z_{f_2}^* + R_L)I_2^2 + 2 \cdot Re[(j\omega)^{\gamma_{12}} M_{\gamma_{12}}] \cdot I_1^* I_2.
\end{aligned}
\tag{16}
$$

Then, one can obtain the FO-WPT efficiency as

$$
\eta = \frac{R_L |I_2|^2}{Re\left[V_s I_1^*\right]}.
\tag{17}
$$

4. An Example: Fractional-Order Wireless Power Domino-Resonators System

In [19], Zhong, who has considered the effect of nonadjacent resonators, established a Wireless Power Domino-Resonator Systems and found that the maximum efficiency operation slightly shifted away from the resonant frequency of the resonators. In this section, an FO-WPDRS, which is equipped with FOEs, is derived from the model proposed by Zhong and its schematic is shown in Figure 5. A general model of FO-WPDRS is depicted in Figure 6.

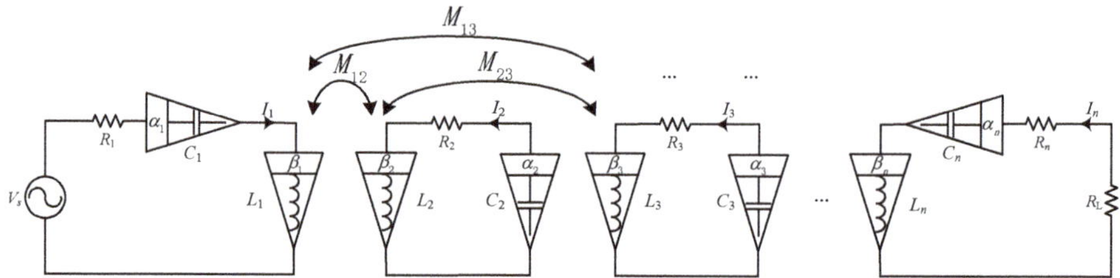

Figure 5. Schematic of FO-WPDRS.

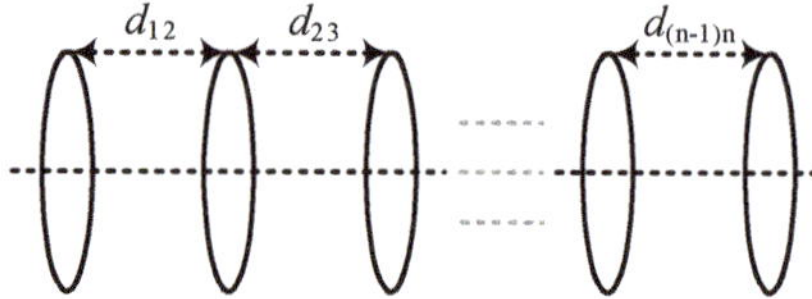

Figure 6. General model of domino-system with n resonators.

$$
\begin{bmatrix}
Z_{f_1} & (j\omega)^{\gamma_{12}} M_{12} & (j\omega)^{\gamma_{13}} M_{13} & \cdots & \cdots & (j\omega)^{\gamma_{1n}} M_{1n} \\
(j\omega)^{\gamma_{21}} M_{21} & Z_{f_2} & (j\omega)^{\gamma_{23}} M_{23} & \cdots & \cdots & (j\omega)^{\gamma_{2n}} M_{2n} \\
\vdots & \vdots & \vdots & \ddots & \vdots & \vdots \\
(j\omega)^{\gamma_{(n-1)1}} M_{(n-1)1} & \cdots & \cdots & & Z_{f_{n-1}} & (j\omega)^{\gamma_{(n-1)n}} M_{(n-1)n} \\
(j\omega)^{\gamma_{n1}} M_{n1} & \cdots & \cdots & & (j\omega)^{\gamma_{n(n-1)}} M_{n(n-1)} & R_L + Z_{f_n}
\end{bmatrix}
\begin{bmatrix} I_1 \\ I_2 \\ \vdots \\ I_{n-1} \\ I_n \end{bmatrix}
$$

$$
= \begin{bmatrix} V_s \\ 0 \\ \vdots \\ 0 \\ 0 \end{bmatrix}
\tag{18}
$$

The circuit equation is expressed in Equation (18), as shown at the top of the next page, where M_{ij} ($i \neq j$) is the mutual inductance between winding-i and winding-j, R_L is the load resistance which is connected to winding-n and $Z_{f_i} = R_i + (j\omega)^{\beta_i} L_i + 1/((j\omega)^{\alpha_i} C_i)(i = 1, 2, \cdots, n)$. Other variables are listed as follows:

I_i is the current in winding-i,
L_i is the fractional-order inductance of winding-i,
C_i is the fractional-order capacitance of winding-i,
R_i is the resistance of winding-i,
α_i is the order of FOC in winding-i,
β_i is the order of FOI in winding-i,
γ_{ij} is the order of mutual inductance M_{ij} and $M_{ij} = M_{ji}$,
ω is angular frequency.

Then, the total complex power of the system reads

$$
S_{\text{total}} = V_s I_1^* = Z_{f_1} |I_1|^2 + \sum_{m=2}^{n} (j\omega)^{\gamma_{1m}} M_{1m} I_m I_1^*.
\tag{19}
$$

The complex power of winding-i ($i \geq 2$) can generally be written as

$$
\begin{aligned}
S_{f_i} &= Z_{f_i} |I_i|^2 + \sum_{m=1}^{i} (j\omega)^{\gamma_{mi}} M_{mi} I_m I_i^* + \sum_{l=i+1}^{n} (j\omega)^{\gamma_{il}} M_{il} I_l I_i^* \\
&= 0,
\end{aligned}
\tag{20}
$$

where I_i^* is the conjugate of current I_i.

Then, Equation (20) can be rewritten as

$$
\begin{aligned}
&- j\omega^{\gamma_{i1}} M_{i1} \sin(\frac{\gamma_{i1}\pi}{2})) I_1 I_i^* \\
&= Z_{f_i} |I_i|^2 + \omega^{\gamma_{i1}} M_{i1} \cos(\frac{\gamma_{i1}\pi}{2}) I_1 I_2^* \\
&+ \sum_{m=2}^{i} (j\omega)^{\gamma_{mi}} M_{mi} I_m I_i^* + \sum_{l=i+1}^{n} (j\omega)^{\gamma_{il}} M_{il} I_l I_i^* \\
&= (j\omega^{\gamma_{i1}} M_{i1} \sin(\frac{\gamma_{i1}\pi}{2})) I_1^* I_i)^*.
\end{aligned}
\tag{21}
$$

Substituting Equation (21) into Equation (22) results in

$$S_{\text{total}} = V_s I_1^*$$
$$= Z_{f_1}|I_1|^2 + R_L|I_n|^2 + \sum_{m=2}^{n} I_m^2 Z_{f_m}^*$$
$$+ 2 \cdot \sum_{m=2}^{n} Re[(j\omega)^{\gamma_{1m}} M_{1m}] I_m I_1^*$$
$$+ 2 \cdot \sum_{l=2}^{n-1} \sum_{m=l+1}^{n} Re[I_l^* I_m] \cdot (j\omega)^{\gamma_{lm}} M_{lm}, \tag{22}$$

where $Z_{f_m}^*$ is the conjugate of Z_{f_m}.

Therefore, the real power P is

$$P = Re[S_{\text{total}}]$$
$$= R_L|I_n|^2 + \sum_{m=1}^{n} I_m^2 \left(R_m + \omega^{\beta_m} L_m \cos(\frac{\beta_m \pi}{2}) + \frac{\cos(\frac{\alpha_m \pi}{2})}{\omega^{\alpha_m} C_m}\right)$$
$$+ 2 \cdot \sum_{m=2}^{n} \omega^{\gamma_{1m}} M_{1m} \cos(\frac{\gamma_{i1}\pi}{2}) \cdot Re[I_m I_1^*] \tag{23}$$
$$+ 2 \cdot \sum_{l=2}^{n-1} \sum_{m=l+1}^{n} Re[I_l^* I_m] \cdot \omega^{\gamma_{lm}} M_{lm} \cos(\frac{\gamma_{lm}\pi}{2}).$$

Furthermore, its efficiency is

$$\eta = \frac{R_L|I_n|^2}{P}. \tag{24}$$

5. Output Characteristics of FO-WPDRS

Based on the aforementioned analysis, the output characteristics of FO-WPDRS are discussed in this section. In order to simplify the analysis, we denote the distance between adjacent resonators as $d_{12} = d_{23} = \cdots = d_{(n-1)n}$, and assign parameters of each practical resonator as follows in the numerical analysis.

- fractional-order inductance: $L_{\beta_i} = L_\beta$,
- fractional-order capacitance: $C_{\alpha_i} = C_\alpha$,
- fractional-order mutual inductance: $M_{ij} = M_{ji} = M \ (i \neq j)$,
- resistance of winding-i: $R_i = R$,
- order of FOI: $\beta_i = \beta$,
- order of FOC: $\alpha_i = \alpha$,
- order of fractional-order mutual inductance: $\gamma_{ij} = \gamma_{ji} = \gamma = \beta$.

Numerical analyses were conducted based on the parameters listed in Table 1, and the corresponding results and analyses are presented below.

Table 1. Simulation Parameters.

Parameter	Value
Voltage source V_s	50 V
Inductance L_β	90.7 µH
Capacitance C_α	1.036 nF
Resistance of winding-i R	0.9 Ω
Mutual inductance M	2.85 µH
Separation $d_{(n-1)n}$	0.3 m
Load resistance R_L	10 Ω

5.1. Fractional Orders of α and β to Improve Resonant Frequency

The resonant frequency of FO-WDPRS is described in Equation (25), while the resonant frequency of the traditional WPT ($\alpha = \beta = \gamma = 1$) is presented in Equation (26).

$$f_{\text{frac}} = \frac{1}{2\pi} \left[\frac{\sin(\alpha\pi/2)}{L_\beta C_\alpha \sin(\beta\pi/2)} \right]^{\frac{1}{\alpha+\beta}} \tag{25}$$

$$f_{\text{trad}} = \frac{1}{2\pi} \frac{1}{\sqrt{L_\beta C_\alpha}} = 520\ kHz \tag{26}$$

According to Figure 7, relevant remarks about the resonant frequency of FO-WDPRS are made as follows:

- α and β are a pair of symmetric parameters which have the same impact on the resonant frequency.
- As depicted in Figure 7a, it is found that $\lg \frac{f_{\text{frac}}}{f_{\text{trad}}} \gg 0$ ($Z \gg 0$), which means that when $\alpha < 1$ and $\beta < 1$, the resonant frequency FO-WDPRS will dramatically increase with an increased $\alpha(\beta)$ compared to the traditional one.
- As shown in Figure 7b,c, it is noted that $\lg \frac{f_{\text{frac}}}{f_{\text{trad}}} < 0$ ($Z < 0$) in some regions of the α-β plane, implying that the resonant frequency of FO-WDPRS can be reduced.
- Similarly, it is found that $\lg \frac{f_{\text{frac}}}{f_{\text{trad}}} < 0$ ($Z < 0$) in Figure 7d, which implies that the resonant frequency of FO-WDPRS can be lowered if we set $1 < \alpha < 2, 1 < \beta < 2$.
- To describe the impact of α and β on resonant frequency, Figure 7a–d are collocated in Figure 7e to find the detailed range of α and β for $f_{\text{frac}} > f_{\text{trad}}$ and $f_{\text{frac}} < f_{\text{trad}}$, respectively. Hence, one can choose the appropriate α and β to realize the desired frequency.

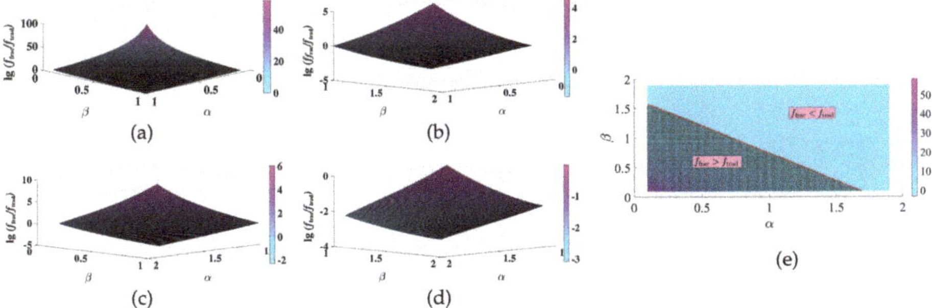

Figure 7. Relationship curves among α, β, and $\lg \frac{f_{\text{frac}}}{f_{\text{trad}}}$: Variations of $0 < \alpha < 2$ and $0 < \beta < 2$. In detail, sub-figures demonstrate the details of 3D plots of (**a**) $0 < \alpha < 1, 0 < \beta < 1$, (**b**) $0 < \alpha < 1, 1 < \beta < 2$, (**c**) $1 < \alpha < 2, 0 < \beta < 1$, (**d**) $1 < \alpha < 2, 1 < \beta < 2$, (**e**) $0 < \alpha < 2, 0 < \beta < 2$ (Top view). [Note: the resonant frequency of fractional-order WPT is normalized by the traditional one, and the logarithm of normalized value based on 10, $\lg \frac{f_{\text{frac}}}{f_{\text{trad}}}$, is taken as Z axis.]

5.2. Fractional Orders of α and β to Improve Output Power and Efficiency

To study the effect of fractional-order inductors, capacitors and mutual inductors on the output power and efficiency of FO-WPDRS, 3-dimension (3D) plots of the relations among α, β, P_o and efficiency η are depicted in Figures 8 and 9, respectively. (Note: all the systems operate at the resonant frequency.)

According to Figure 8a–d, it is obvious that there is a resonance point with the maximum output power in all cases. From the simulation results, it can be concluded that the output power of the above FO-WPDRS is greater than the traditional WPT ($\alpha = \beta = \gamma = 1$), which further validates the effectiveness of the proposed FOE method in increasing the output power.

Compared with Figure 9a–e, it is clear that the efficiency of FO-WPDRS, whose average distance between two adjacent resonators is 0.3 m, is higher than the ones with the optimized loads, frequencies and distances or only loads in [19].

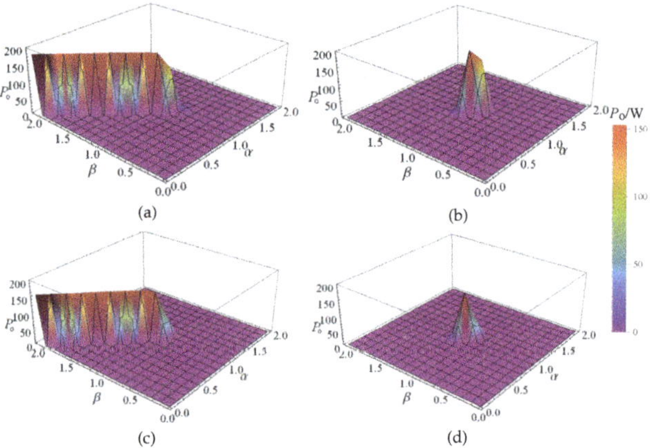

Figure 8. Relationship curves among α, β, and P_o: Variations of $0 < \alpha < 2$ and $0 < \beta < 2$. In detail, sub-figures demonstrate the details of 3D plots of (**a**) FO-WPDRS with 3 resonators, (**b**) FO-WPDRS with 4 resonators, (**c**) FO-WPDRS with 5 resonators, and (**d**) FO-WPDRS with 6 resonators.

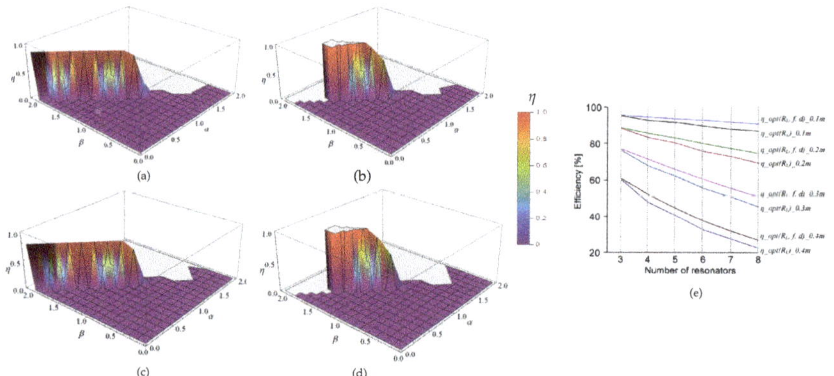

Figure 9. Relationship curves among α, β, and η: Variations of $0 < \alpha < 2$ and $0 < \beta < 2$. In detail, sub-figures demonstrate the details of 3D plots of (**a**) FO-WPDRS with 3 resonators, (**b**) FO-WPDRS with 4 resonators, (**c**) FO-WPDRS with 5 resonators, (**d**) FO-WPDRS with 6 resonators and (**e**) Efficiency comparison between the domino systems with optimized loads, frequencies and distance ($\eta_{opt}(R_L, f, d)$) and the equally spaced systems operating at resonant frequency and with only optimized loads ($\eta_{opt}(R_L)$) with four different average distances: 0.1 m, 0.2 m, 0.3 m, and 0.4 m in [19].

6. Simulations and Experiments

Two experimental setups are established in this study to examine the performance of the proposed FO-WPT mechanism: (1) the FOC is powered by a high-frequency power source for the observation of the order of the FOC, and (2) incorporate the FOC built in (1) into the FO-WPT for performance validation. Detailed measurements and analyses are presented follows.

The theoretical analyses in Sections 4 and S4 studied FOCs with a range of orders. In this section, we build a prototype of an FOC with an order of 0.54 to realize the WPT. Please note that it is of significant technical difficulty to manufacture an FOC with an order higher than one. With the facilities available in our laboratory, we only use the FOC of $0 < \alpha < 1$ to validate our theory. Building FOCs with orders greater than 1 is to be conducted in our future work.

6.1. Implementation of Fractional-Order Capacitor

A practical FOC is implemented as depicted in Figure 10 based on Ref. [28]. To realize fractional order $\alpha = 0.54$, the corresponding parameters are: $R_1 = 10$ kΩ, $R_2 = 3.3$ kΩ, $R_3 = 1.0$ kΩ, $R_4 = 330$ Ω, $R_5 = 100$ Ω, $C_1 = 1.0$ μF, $C_2 = 470$ nF, $C_3 = 330$ nF, $C_4 = 220$ nF, $C_5 = 100$ nF, $R_p = 22$ kΩ, $C_p = 100$ nF.

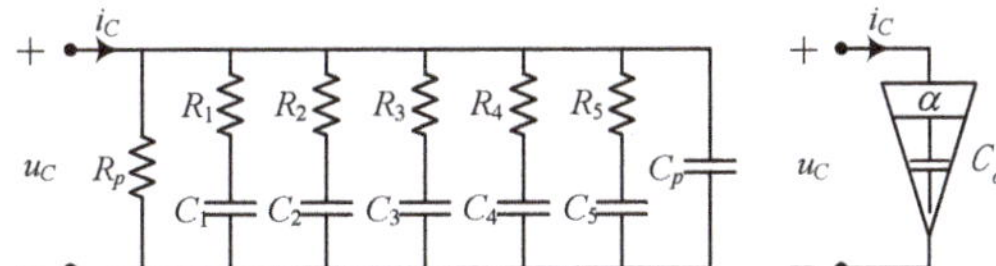

Figure 10. Schematic of the FOC.

To further demonstrate the characteristics of the designed FOC, dynamic behavior of the FOC is tested and the corresponding resultant waveforms are shown in Figure 11. Therein, the actual order α can be obtained by measuring the phase difference between the input voltage and current. As seen in Figure 11, the input current i_C has a leading phase angle of 48.6° from the input voltage u_C, and the actual order α is thus 0.54.

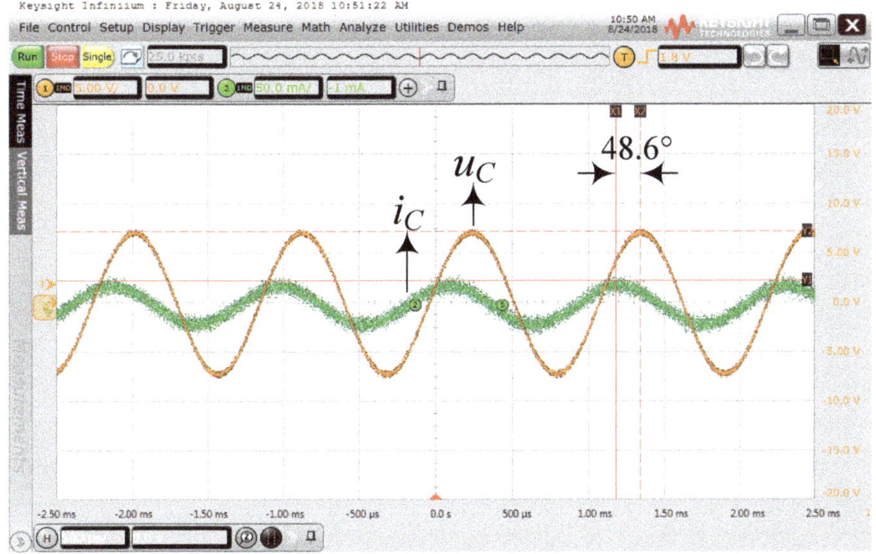

Figure 11. Experimental waveforms of FOC.

6.2. FO-WPT with the Constructed FOC

In this subsection, a prototype of FO-WPT with the constructed FOC is presented in Figure 12, and the corresponding simulation and experimental waveforms are shown in Figures 13 and 14, respectively. The distance between the transmitter and receiver is 0.3 m. It can be seen that FO-WPT is in resonance since the current i_R and voltage u_R have the same phase angle, which indicates a unity power factor, representing the best power use, i.e., a higher energy efficiency. This experiment together with the simulation studies shown before corroborates the advantages of the proposed FO-WPT with fractional order elements, which have more favorable output characteristics and higher energy transfer efficiency.

Figure 12. Experimental prototype of the FO-WPT.

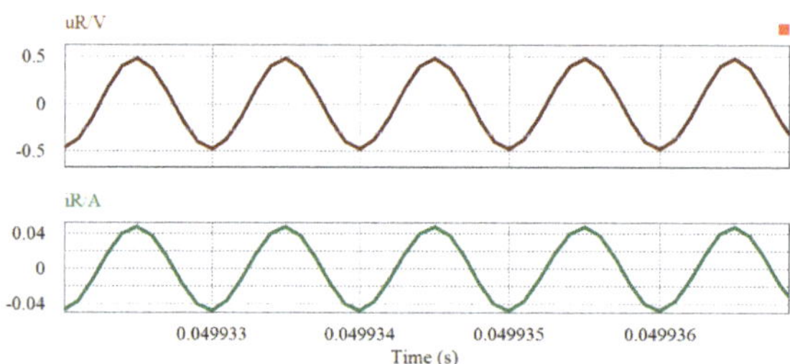

Figure 13. Simulation waveforms of FO-WPT.

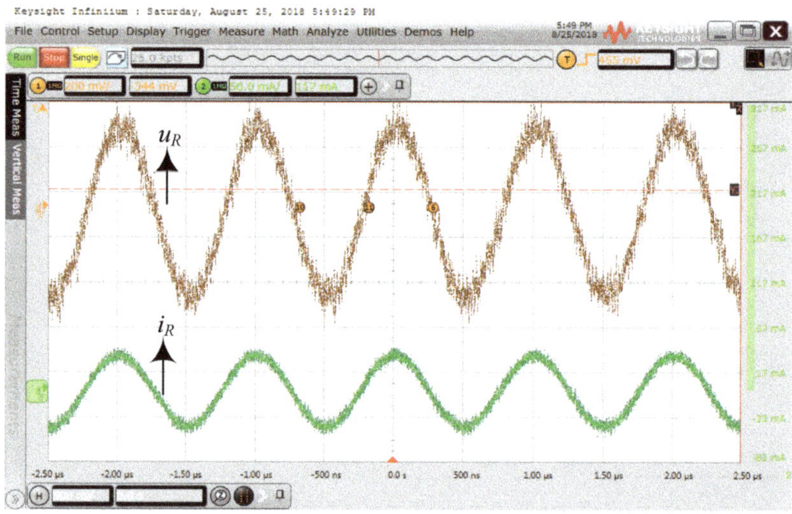

Figure 14. Experimental waveforms of FO-WPT.

It is obvious that the experimental results well agree with the simulations and theoretical analyses according to the figures, it is found that there a little voltage drop in experimental one due to the parasitic parameters of the components.

7. Conclusions

In this study, we proposed a novel fractional-order circuit elements-based wireless power transmission solution, aiming to reduce the resonant frequency and improving the output power quality and efficiency. Detailed mathematical modeling of the proposed FO-WPT has been presented. Simulation studies of a typical example FO-WPDRS validated the theoretical analysis, and shown reduced frequency, improved efficiency and enhanced output power. An experimental prototype has been developed to validate the proposed method, and the results are corresponding with the analyses. This work could lay a solid foundation for further investigations in fractional-order elements-based WPT solutions, which may lead to wide industrial applications in wireless power transfer.

Author Contributions: Conceptualization, G.Z. and Z.O.; methodology, G.Z. and Z.O.; software, Z.O.; validation, G.Z. and Z.O.; formal analysis, L.Q.; investigation, Z.O.; writing—original draft preparation, Z.O.; writing—review and editing, G.Z.; supervision, G.Z.; project administration, L.Q.; funding acquisition, L.Q.

Funding: This research was funded by [National Natural Science Foundation of China] grant numbers [51907032, U1501251, 51277030, 61733015], [Natural Science Foundation of Guangdong Province] grant numbers [2017A030310243, 2018A030313365] and [Science and Technology Planning Project of Guangzhou] grant number [201804010310].

Conflicts of Interest: The authors declare no conflict of interest.

References

1. Jiang, W.; Xu, S.; Li, N.; Lin, Z.; Williams, B.W. Wireless power charger for light electric vehicles. In Proceedings of the IEEE 11th International Conference on Power Electronics and Drive Systems, Sydney, Australia, 9–12 June 2015; pp. 562–566.
2. Yang, Q.; Zhang, P.; Zhu, L.; Xue, M.; Zhang, X.; Li, Y. Key fundamental problems and technical bottlenecks of the wireless power transmission technology. *Trans. China Electrotech. Soc.* **2015**, *30*, 1–8.
3. Musavi, F.; Wilson, E. Overview of wireless power transfer technologies for electric vehicle battery charging. *IET Power Electron.* **2014**, *7*, 60–66. [CrossRef]
4. Liu, X.; Wang, G.; Ding, W. Efficient circuit modelling of wireless power transfer to multiple devices. *IET Power Electron* **2014**, *7*, 3017–3022. [CrossRef]
5. Almohaimeed, A.; Amaya, R.; Lima, J. An adaptive power harvester with active load modulation for highly efficient short/long range RF WPT applications. *Electronics* **2018**, *7*, 125. [CrossRef]
6. Lu, G.; Shi, L.; Ye, Y. Maximum Throughput of TS/PS Scheme in an AF Relaying Network With Non-Linear Energy Harvester. *IEEE Access* **2018**, *6*, 26617–26625. [CrossRef]
7. Kurs, A.; Karalis, A.; Moffatt, R.; Joannopoulos, J.D.; Fisher, P.; Soljačić, M. Wireless power transfer via strongly coupled magnetic resonances. *Science* **2007**, *317*, 83–86. [CrossRef] [PubMed]
8. Nottiani, D.G.; Leccese, F. A simple method for calculating lumped parameters of planar spiral coil for wireless energy transfer. In Proceedings of the 11th International Conference on Environment and Electrical Engineering, Venice, Italy, 18–25 May 2012; pp. 869–872.
9. Houran, M.; Yang, X.; Chen, W. Magnetically Coupled Resonance WPT: Review of Compensation Topologies, Resonator Structures with Misalignment, and EMI Diagnostics. *Electronics* **2018**, *7*, 296. [CrossRef]
10. Liu, X.; Liu, J.; Wang, J.; Wang, C.; Yuan, X. Design Method for the Coil-System and the Soft Switching Technology for High-Frequency and High-Efficiency Wireless Power Transfer Systems. *Energies* **2017**, *11*, 7. [CrossRef]
11. Wang, T.; Liu, X.; Jin, N. Wireless Power Transfer for Battery Powering System. *Electronics* **2018**, *7*, 178. [CrossRef]
12. Zhang, W.; Mi, C.C. Compensation topologies of high-power wireless power transfer systems. *IEEE Trans. Veh. Tech.* **2016**, *65*, 4768–4778. [CrossRef]

13. He, X.; Shu, W.; Yu, B.; Ma, X. Wireless Power Transfer System for Rotary Parts Telemetry of Gas Turbine Engine. *Electronics* **2018**, *7*, 58. [CrossRef]

14. Rong, C.; Tao, X.; Lu, C.; Hu, Z.; Huang, X.; Zeng, Y.; Liu, M. Analysis and Optimized Design of Metamaterials for Mid-Range Wireless Power Transfer Using a Class-E RF Power Amplifier. *Appl. Sci.* **2019**, *9*, 26. [CrossRef]

15. Hernández Robles, I.A.; Lozano García, J.M.; Martínez Juárez, J.J. Simulation for Obtaining Relevant Parameters for Optimal Wireless Power Transfer. *Comput. Y Sist.* **2019**, *23*, 81. [CrossRef]

16. Yin, J.; Lin, D.; Lee, C.K.; Hui, S.R. A systematic approach for load monitoring and power control in wireless power transfer systems without any direct output measurement. *IEEE Trans. Power Electron* **2014**, *30*, 1657–1667. [CrossRef]

17. Tan, L.; Guo, J.; Huang, X.; Wen, F. Output power stabilisation of wireless power transfer system with multiple transmitters. *IET Power Electron* **2016**, *9*, 1374–1380. [CrossRef]

18. Syms, R.R.A.; Solymar, L.; Young, I.R.; Floume, T. Thin-film magneto-inductive cables. *J. Phys. D Appl. Phys.* **2010**, *43*, 55102. [CrossRef]

19. Lee, C.K.; Zhong, W.X.; Hui, S.Y.R. Effects of magnetic coupling of nonadjacent resonators on wireless power domino-resonator systems. *IEEE Trans. Power Electron.* **2012**, *27*, 1905–1916. [CrossRef]

20. Park, N.; Choi, H.S.; Jeong, I.S.; Choi, H.W. Analysis of efficiency characteristics of superconducting coil-applied wireless power transmission systems by transmission distance. In Proceedings of the 2017 16th International Superconductive Electronics Conference (ISEC), Naples, Italy, 12–16 June 2017; pp. 1–3.

21. Jeong, I.S.; Jung, B.I.; You, D.S.; Choi, H.S. Analysis of S-parameters in magnetic resonance WPT using superconducting coils. *IEEE Trans. Appl. Supercond.* **2016**, *26*, 1–4. [CrossRef]

22. Chen, L.; Liu, S.; Zhou, Y.C.; Cui, T.J. An optimizable circuit structure for high-efficiency wireless power transfer. *IEEE Trans. Ind. Electron.* **2013**, *60*, 339–349. [CrossRef]

23. Li, H.; Li, J.; Wang, K.; Chen, W.; Yang, X. A maximum efficiency point tracking control scheme for wireless power transfer systems using magnetic resonant coupling. *IEEE Trans. Power. Electron.* **2015**, *30*, 3998–4008. [CrossRef]

24. Radwan, A.G.; Shamim, A.; Salama, K.N. Theory of fractional order elements based impedance matching networks. *IEEE Microw. Wirel. Compon. Lett.* **2011**, *21*, 120–122. [CrossRef]

25. Radwan, A.G.; Shamim, A.; Salama, K.N. Impedance matching through a single passive fractional elemen. In Proceedings of the 2012 IEEE International Symposium on Antennas and Propagation, Chicago, IL, USA, 8–14 July 2012; pp. 1–2.

26. Chen, M.R.; Zeng, G.Q.; Dai, Y.X.; Lu, K.D.; Bi, D.Q. Fractional-Order Model Predictive Frequency Control of an Islanded Microgrid. *Energies* **2019**, *12*, 84. [CrossRef]

27. Radwan, A.G. Resonance and Quality Factor of the $RL_\alpha C_\alpha$ Fractional Circuit. *IEEE J. Emerg. Sel. Top. Circuits Syst.* **2013**, *3*, 377–385. [CrossRef]

28. Shu, X.; Zhang, B. The Effect of Fractional Orders on the Transmission Power and Efficiency of Fractional-Order Wireless Power Transmission System. *Energies* **2018**, *11*, 1774. [CrossRef]

 © 2019 by the authors. Licensee MDPI, Basel, Switzerland. This article is an open access article distributed under the terms and conditions of the Creative Commons Attribution (CC BY) license (http://creativecommons.org/licenses/by/4.0/).

 *electronics*

Article

A Fast Charging Balancing Circuit for LiFePO$_4$ Battery

Sen-Tung Wu, Yong-Nong Chang *, Chih-Yuan Chang and Yu-Ting Cheng

Department of Electrical Engineering, National Formosa University, No.64, Wunhua Rd., Huwei Township, Yunlin County 632, Taiwan; stwu@nfu.edu.tw (S.-T.W.); 10565104@gm.nfu.edu.tw (C.-Y.C.); 10765128@gm.nfu.edu.tw (Y.-T.C.)
* Correspondence: ynchang@nfu.edu.tw; Tel.: +886-5-631-5637

Received: 5 September 2019; Accepted: 2 October 2019; Published: 10 October 2019

Abstract: In this paper, a fast charging balancing circuit for LiFePO$_4$ battery is proposed to address the voltage imbalanced problem of a lithium battery string. During the lithium battery string charging process, the occurrence of voltage imbalance will activate the fast balancing mechanism. The proposed balancing circuit is composed of a bi-directional converter and the switch network. The purpose of bi-directional is that the energy can be delivered to the lowest voltage cell for charging mode. On the other hand, the energy stored in the magnetizing inductors of the transformer can be charged back to the higher voltage cell in recycling mode. This novel scheme includes the following features: (1) The odd-numbered and even-numbered cells in the string with the maximum differential voltage will be chosen for balancing process directly. In this topology, there is no need to store and deliver the energy through any intermediate or the extra storing components. That is, the energy loss can be saved to improve the efficiency, and the fast balancing technique can be achieved. (2) There is only one converter to complete the energy transfer for voltage balancing process. The concept makes the circuit structure much simpler. (3) The structure has bi-directional power flow and good electrical isolation features. (4) A single chip controller is applied to measure the voltage of each cell to achieve the fast balancing process effectively. At the end of the paper, the practical test of the proposed balancing method on LiFePO$_4$ battery pack (28.8 V/2.5 Ah) is verified and implemented by the experimental results.

Keywords: active balance circuit; bi-directional converter; lithium battery; series-connected battery; fast charging

---

1. Introduction

In recent years, lithium batteries and related techniques have developed and are widely used. The battery industries have experiences and capabilities for mass production of battery packs and modules. Battery packs and modules are composed of several battery cells for high power energy storing applications. However, there is a critical issue about imbalance of electric charge [1–5] within high power battery strings. The issue is caused by the characteristic [6], depth of discharge [7], and aging problem of each cell [8,9]. Based on the reasons above, when the battery string is being charged or discharged, the imbalance of each cell of the battery string becomes more serious. In addition, as the cycle of charging-discharging from the battery string increases, the internal resistance and the capacity of each cell will be varied to shorten the life cycle of battery strings.

In order to increase the efficiency and extend the lifetime of battery strings, a battery management system (BMS) [10–16] is a key feature which is utilized to monitor the parameters of the battery. Also, BMS plays an important role for management and protection of the battery. The main functions of BMS are monitoring, protection, and balancing parts [17,18]. The monitoring is to sense the relative key parameters from the battery packs, like voltage, current, and temperature. The protection is to avoid the situation of over-charging or over-discharging from the battery packs. The last one is about the balancing technique for each cell. In general, the balancing circuits can be divided into passive or active balancing topologies. The most common passive balancing schemes [19–22] utilize series resistors, series diodes, or Zener diodes to be the voltage dividers which are parallel to each cell for balancing purpose. These passive concepts with simple control methods and smaller size of balancing circuits can achieve a voltage equalizer for each cell during the process; however, the power losses and thermal issues will be the key factors to influence the balancing performance, accuracy, and the lifetime of cells. On the other hand, the active balancing topologies [23–27] can deliver the energy from higher voltage cell to lower one by using the storing elements and switches networks. Some of the active balancing schemes adopted multi-windings of the main transformer to fulfill the balancing performance with better electrical isolation feature, but the number of cells is limited by the windings; therefore, the dimension of the transformer will be increased for purpose of more balanced cells. In facts, these active balancing methods help to increase the efficiency and the performance during the balancing process, but they spent more time for delivering the energy from the high voltage cell to the low voltage cell. The reason is that if the highest voltage cell begins to charge to the lowest voltage cell, the energy flow has to be delivered through the other cells one by one to the target cell. Thus, the balancing path has a big problem of time wasting. In this paper, the proposed concept is to shorten the balancing time and to make the balanced cell being charged precisely. To recall the active balancing concept mentioned above, the basic concept of active balancing is to deliver the energy from the higher voltage battery to the lower voltage battery to achieve the balancing function for each cell. Nowadays, BMS is a core of many electric applications powered by batteries.

Based on the descriptions aforementioned, the paper proposed a fast charging balancing circuit for LiFePO4 battery pack. In this study, a digital signal processor (DSP) is adopted to implement the algorithm and to be the main controller.

2. Proposed Structure and Operation Mode Analysis

2.1. The Proposed Balancing Circuit

As shown in Figure 1, the proposed balancing method utilizes a bi-directional converter [28,29] to balance the voltage of each cell in a battery string. This method helps to deliver the energy from higher-voltage battery to lower-voltage battery directly. In other words, there is no more energy loss during the delivering process, and this technique can also shorten the balancing time effectively. In this study, the forward converter has good isolation feature and simple structure for bi-directional function. The strategy is to balance the voltage between the odd-numbered battery and the even-numbered battery which exist maximum differential voltage in the battery string. The switch network shown in blue dotted block is formed by several couples of MOSFETs with back-to-back connection, and each connection also becomes a bi-directional switch set with two MOSFETs which is connected with the cell. Hence, this connection can provide a bi-directional path to deliver the energy. In addition, each of the bi-directional switches set can avoid the other currents flowing through the cell during the balancing process.

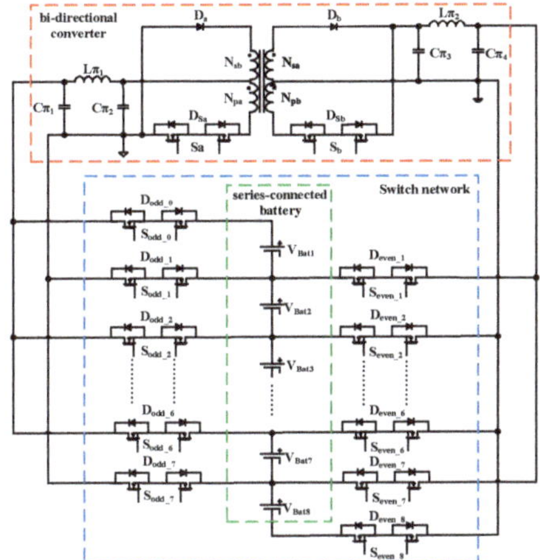

Figure 1. The proposed fast charging balancing circuit for LiFePO$_4$ battery.

2.2. The Operation Mode Analysis

The operation mode can be divided by two different modes. The first one is to balance the voltage from odd-numbered battery to even one. The other mode is to balance the voltage from even-numbered battery to odd one. The operation principle of these two modes will be discusses in detail below. In the following analysis, each battery denoted from V_{Bat1} to V_{Bat8} can be represented as a battery cell in the green dotted block in Figure 1.

2.2.1. The Operation Mode of Balancing Process from the Odd-Numbered Battery to the Even-Numbered Battery

The following analysis is stated for balancing from V_{Bat1} to V_{Bat8}. The theoretical waveforms are shown in Figure 2.

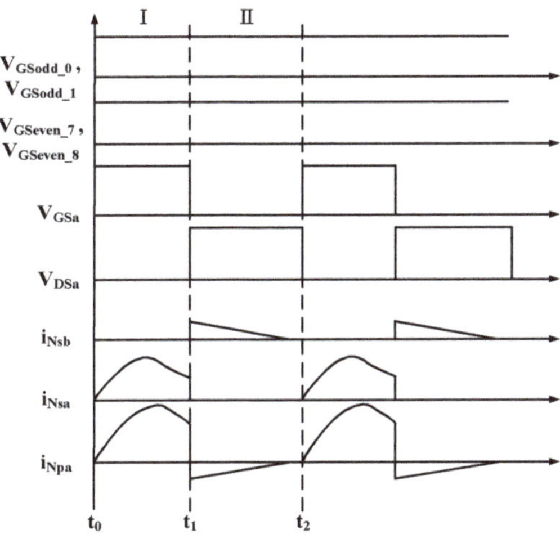

Figure 2. The theoretical waveforms in balancing process from V_{Bat1} to V_{Bat8}.

Mode I—Charging Mode ($t_0 < t < t_1$)

As shown in Figure 3, the bi-directional switch set S_a, S_{odd_0}, S_{odd_1}, S_{even_7}, and S_{even_8} are all turned on. The others are all turned off. In this mode, the current from V_{Bat1} will charge to L_{ma} through π filter I. In the meantime, the energy from the primary winding N_{pa} can be transferred to the secondary winding N_{sa}; thus, D_b is turned on by the forward bias. The current i_{Nsa} begins to charge to V_{Bat8} through π filter II. The red dotted current path shows the charging condition.

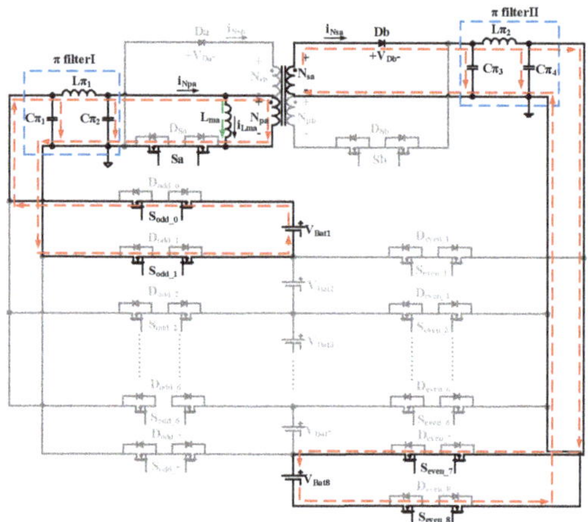

Figure 3. Mode I, the battery balancing process from higher V_{Bat1} to lower V_{Bat8}.

Mode II—Recycling Mode ($t_1 < t < t_2$)

As shown in Figure 4, the bi-directional switch set S_a turns off, but S_{odd_0}, S_{odd_1}, S_{even_7}, and S_{even_8} are all still turned on. The rest of switch sets are turned off. In this interval, the energy stored in L_{ma}, and the current i_{Lma} remains continuously. Thus, i_{Lma} (shown with green dotted current) will flow through the primary winding N_{pa} to induce a current i_{Nsb} from N_{sb}. The induced current i_{Nsb} also flows through π filter I and charge back to V_{Bat1}. During this mode, the energy stored in the magnetizing inductor can be released and also recycled to the battery(V_{Bat1}) effectively. At the right side of the converter, the energy of V_{Bat8} is provided by π filter II.

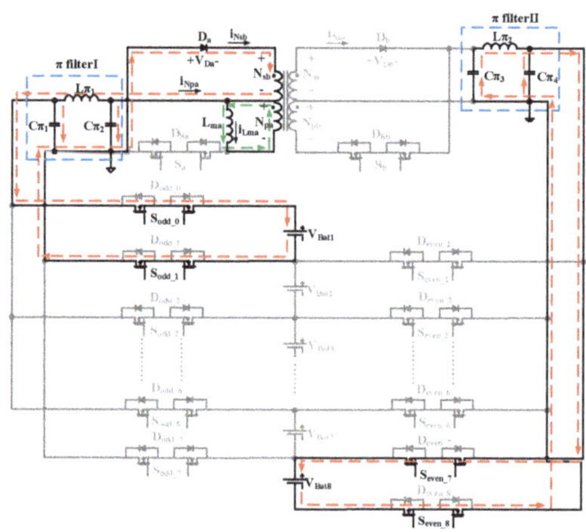

Figure 4. Mode II, the battery balancing process from higher V_{Bat1} to lower V_{Bat8}.

2.2.2. The Operation Mode of Balancing Process from the Even-Numbered Battery to the Odd-Numbered Battery

The following analysis is stated for balancing from V_{Bat8} to V_{Bat1}, and Figure 5 shows the theoretical waveforms during the balancing process.

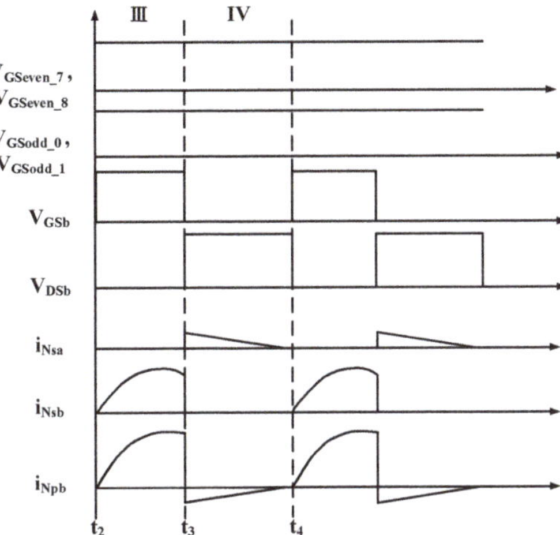

Figure 5. The theoretical waveforms in balancing process from V_{Bat8} to V_{Bat1}.

Mode III—Charging Mode ($t_2 < t < t_3$)

As shown in Figure 6, the bi-directional switch set S_b, S_{odd_0}, S_{odd_1}, S_{even_7}, and S_{even_8} are all turned on. The others are all turned off. The current from V_{Bat8} will charge to L_{mb} through π filter II and also deliver the energy from the primary winding N_{pb} to the secondary winding N_{sb}. In this transferring state, D_a is turned on, and i_{Nsb} starts to charge to V_{Bat1} through π filter I. The charging path is shown in red-dotted current.

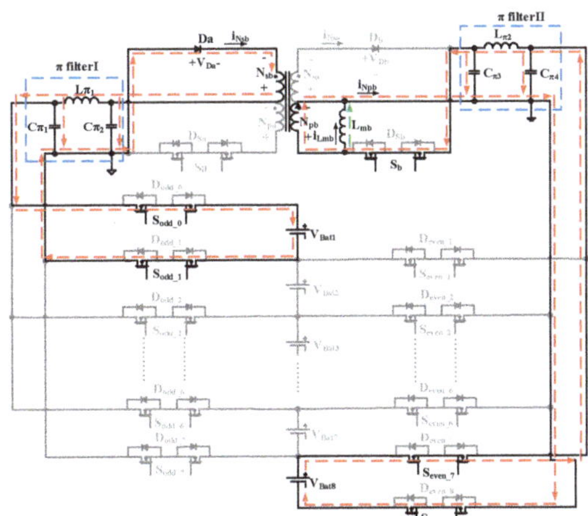

Figure 6. Mode III, the battery balancing process from higher V_{Bat8} to lower V_{Bat1}.

Mode IV-Recycling Mode ($t_3 < t < t_4$)

As shown in Figure 7, the bi-directional switch set S_b turns off, but S_{odd_0}, S_{odd_1}, S_{even_7}, and S_{even_8} are all still turned on. The rest of the switch sets are turned off. In this interval, the energy stored in L_{mb}, and the current i_{Lmb} remains continuously. Thus, i_{Lmb} (shown with green dotted current) will flow thru the primary winding N_{pb} to induce a current i_{Nsa} from N_{sa}. The induced current i_{Nsa} also flows through π filter II and charge back to V_{Bar8}. During this mode, the energy stored in the magnetizing inductor can be released and also recycled to the battery (V_{Bat8}) effectively. At the left side of the converter, the energy of V_{Bat1} is provided by π filter I. In this recycling mode, the energy is saved during this interval.

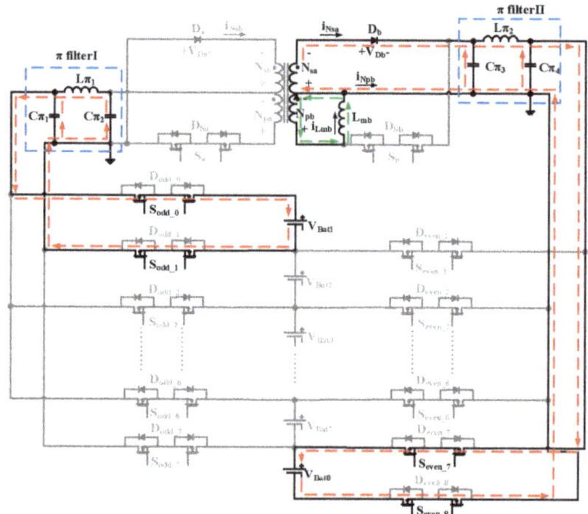

Figure 7. Mode II, the battery balancing process from higher V_{Bat8} to lower V_{Bat1}.

3. Design Consideration and Specification of Cell

In this part, the design consideration is discussed in detail. Table 1 lists the experiment key parameters (switching frequency, duty cycle, turns ratio, capacitance, and inductance) in this study. In addition, Table 2 is the specification of LiFePO$_4$ battery. These parameters of battery help to design the charger and the related components. At first, the turns ratio (N_S/N_P) of the transformer has to be determined by using the nominal voltage of cell. In order to simplify the derivation, all the switches are assumed to be ideal. Besides, this application is operated in low voltage, the power switches and the diodes can be selected for low power rating to decrease the cost of the converter.

Table 1. Experimental design parameters.

Design Parameters	Value
Switching Frequency f_s	20 kHz
Duty Cycle D	45%
Turns Ratio N_{pa}: N_{pb}: N_{sa}: N_{sb}	1: 1: 1.2: 1.2
Filtering Capacitance $C_{\pi1}$, $C_{\pi2}$, $C_{\pi3}$, $C_{\pi4}$	100 μF
Filtering Inductance $L_{\pi1}$, $L_{\pi2}$	33 μH

Table 2. Specification of cell, (Company: A123 System LiFePO$_4$).

Model Number	ANR26650M1B
Charging Voltage	3.6 V
Nominal Voltage	3.3 V
Nominal Capacity	2.5 Ah
Operating Temperature	$-30\,^\circ$C~55 $^\circ$C
Storage Temperature	$-40\,^\circ$C ~60 $^\circ$C

The turns ratio can be derived by the voltage of the charging behavior. Refer to Figure 8, V_{in} is fed in π filter I, and the voltage across the primary side N_{Pa} is also the V_{in} (assuming the filters and the switches are ideal). At the secondary side, the voltage V_{NSa} across N_{Sa} is induced by N_{Pa}. During the charging state from V_{in} to V_o, the condition $(V_{NSa} - V_o) > V_D$ has to be satisfactory to turn on the diode D_b. Therefore, the charging path can be established as red dotted current.

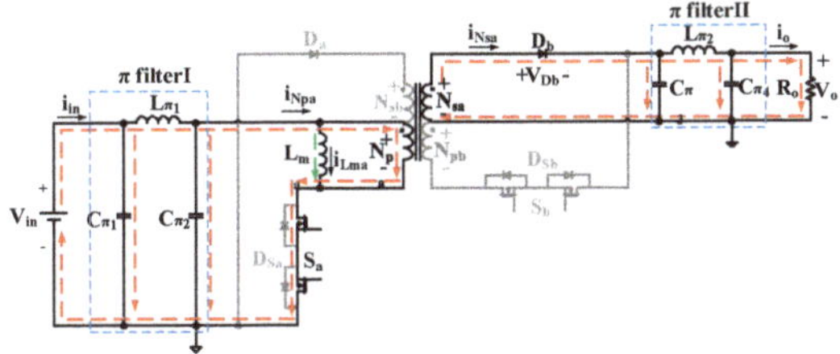

Figure 8. Determine turns ratio by voltage of charging behavior.

Based on (1), the across voltage on the diode has to be higher than the cut-in bias V_D for turning on the diode. That is, the input voltage V_{in} is definitely higher than V_o. After the battery balancing process ends, V_{in} will approach to the charging voltage of the battery and also to be identical to V_o In the steady state, $V_{in} = V_o$. Thus, the equation (1) can be rewritten which is shown in (2).

$$\left(V_{in} \times \frac{N_s}{N_p} - V_o\right) > V_D \tag{1}$$

$$V_{in} \times \left(\frac{N_s}{N_p} - 1\right) > V_D \tag{2}$$

After rearranged (2), the equation (3) can be obtained as below.

$$\frac{N_s}{N_p} > \frac{V_D}{V_{in}} + 1 \tag{3}$$

In order to obtain the turns ratio from (3), the voltage of V_{in} is 3.3V, and the forward bias voltage of V_D is 0.45 V respectively. Based on these parameters which are substituted into (4), the derived turns ratio is 1.14.

$$\frac{N_s}{N_p} > \left[\frac{0.45}{3.3} + 1\right] = 1.14 \tag{4}$$

In this study, the actual turns ratio is chosen for 1.2 in this experiment.

4. Fast Battery Balancing Control Strategy and the Algorithm

The main digital controller utilized in this proposed structure is dsPIC33EP128GM304 from Microchip Technology. The first step of the procedure is to sense the voltage of each battery by the voltage detector circuit and to send the information to the processor with A/D converters. After analyzing from the processor, the battery cell in the string which needs to be activated for balancing will be chosen by the processor. In other words, the related switches (S_1, S_2, S_{odd_0}~S_{odd_7}, and S_{even_1}~S_{even_8}) around the imbalanced battery cells will be turned on or off for balancing process. Figure 9 shows the structure of the proposed fast battery charging balancing circuit.

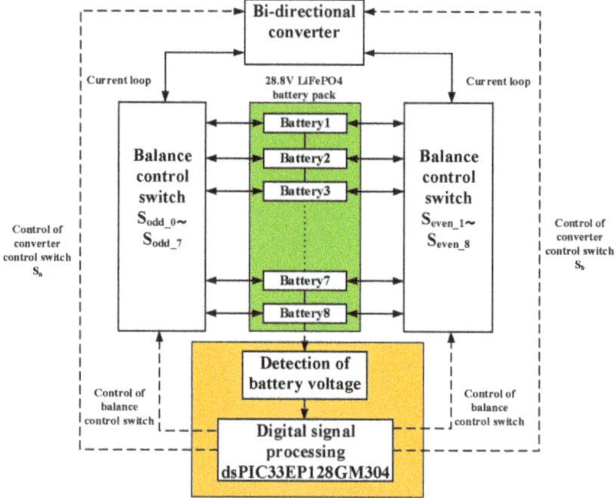

Figure 9. The structure of fast charging balancing circuit.

When the battery string is being charged, the digital processor utilizes A/D converters and the voltage detection circuit to detect the battery voltage from V_{Bat1} to V_{Bat8}. After detecting the actual battery voltage, the average voltage V_{avg} can be calculated by the processor, and the formula is shown in (5).

$$V_{avg} = \frac{V_{bat1} + V_{bat2} + \ldots + V_{bat8}}{8} \tag{5}$$

Besides, the start-up voltage $V_{balance\_start}$ for balancing process is shown in (6), and $\triangle V$ is the threshold voltage which can be determined by users' demand.

$$V_{balance\_start} = V_{avg} + \triangle V \tag{6}$$

If any of the battery voltage is higher than the preset of $V_{balance\_start}$, the proposed balancing mechanism will be activated. Once the battery balancing mechanism enables, these batteries which exist the maximum differential voltage between the odd-numbered and even-numbered will be selected. During the balancing procedure, the digital processor keeps detecting the batteries voltage and refreshing the average voltage. Until the highest battery voltage $V_H \le V_{avg}$ or the lowest battery voltage $V_L \ge V_{avg}$, the balancing process will be finished. Figure 10 is the flow chart of dynamic battery charging balancing strategy.

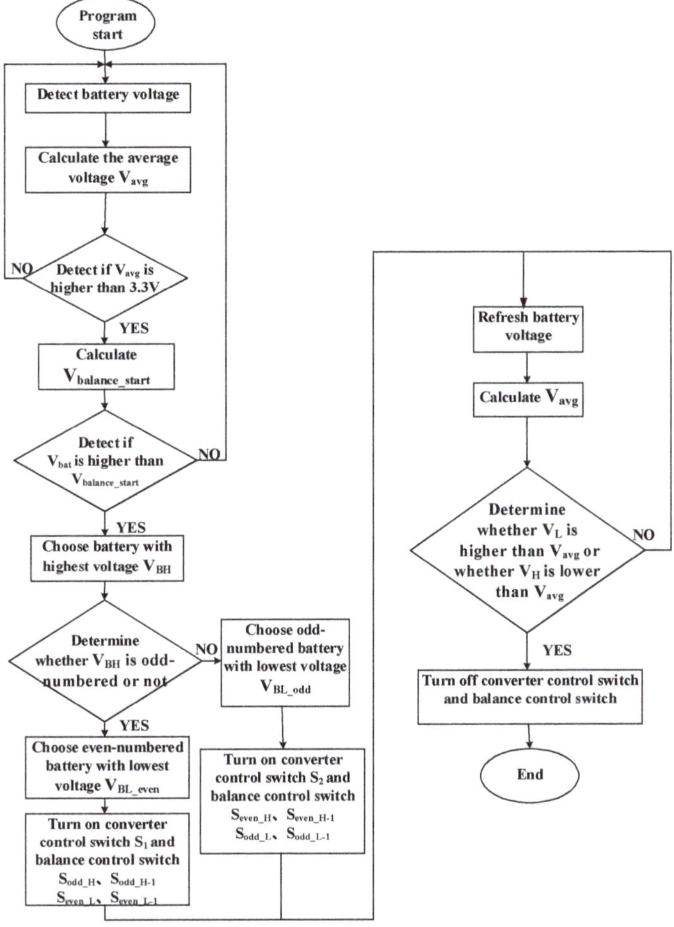

Figure 10. Flow chart of dynamic battery charging balancing strategy.

In the balancing process, the voltage of each cell will be kept detecting and measuring all the time. In order to sense the voltage of each cell precisely, and to avoid the load effect between the cell and the input of the analog to digital converter (ADC), the detection circuit for battery voltage is adopted. From Figure 9, the detection of battery voltage is composed of a differential-voltage operational amplifier (OPA) and a low pass filter (LPF) to achieve the voltage measurement of a cell. The function of LPF is to filter the high frequency noise at the output of OPA. Then, the output of LPF will connect to the ADC's input of the MCU. In facts, the charging voltage of cell, V_{Bat} is 3.6 V, but the maximum input voltage of the ADC is 3 V; therefore, the proportion of the resistors around the OPA needs to be considered ($R_1 = R_3$, $R_2 = R_4$) for full scale voltage as shown in (7). In this circuit, the LPF has no attenuation in low frequency. Thus, $V_1 = V_O$. In the meanwhile, the voltage of cell can also be measured by the voltage recorder. The detection of battery voltage circuit is shown in Figure 11.

$$V_O = \left(\frac{R_2}{R_1}\right) \times V_{bat} \tag{7}$$

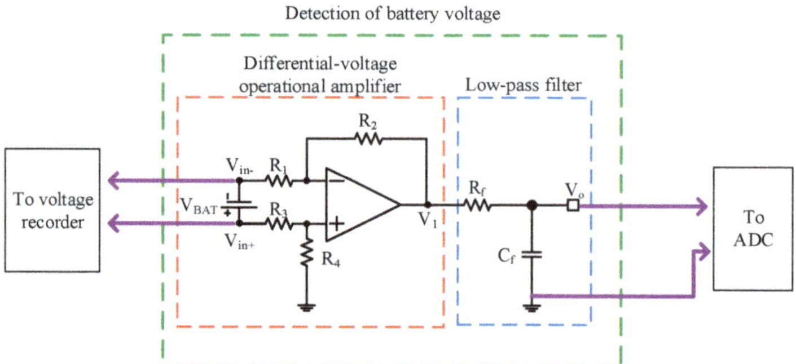

Figure 11. The detection of battery voltage circuit.

5. Experimental and Simulation Results

In order to verify the proposed battery charging balancing circuit with the theoretical derivation, Figures 12–16 show the triggering waveforms for building up the balancing loop and the related waveforms when V_{Bat1} charges to V_{Bat8}. In the opposite, Figures 17–21 provide the waveforms when V_{Bat8} charges to V_{Bat1}. These waveforms are measured and simulated to prove that the balancing process is feasible and implemented.

5.1. Waveforms for V_{Bat1} Charges to V_{Bat8}

In this section, the simulation results are shown to compare with the experiments. Figure 11 shows the gate signals for turning on S_{odd_0}, S_{odd_1}, S_{even_7}, and S_{even_8}.

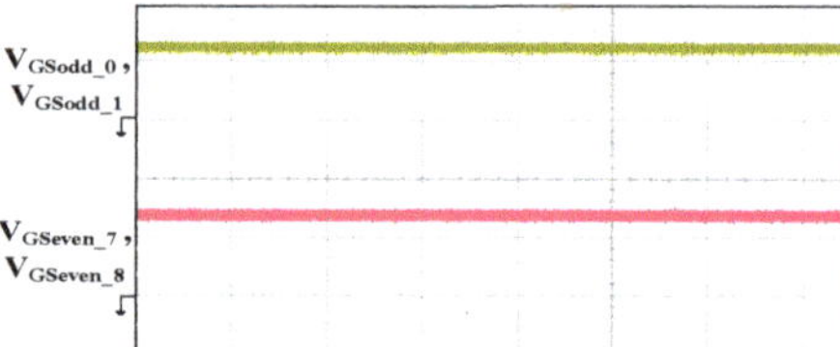

Figure 12. Experiment of V_{GS} triggering waveforms for turning on the switch network (V_{GSodd_0}, V_{GSodd_1}, V_{GSeven_7}, V_{GSeven_8}), V_{GSodd_0}, V_{GSodd_1}, V_{GSeven_7}, V_{GSeven_8}: 10 V/div, Time: 20 μs/div.

When the balancing loop keeps turning on all the time, S_a also turns on (V_{GSa}: high) in the meantime. The converter goes to charging mode when V_{Bat1} charges to V_{Bat8}. If S_a turns off (V_{GSa}: low), the converter goes to recycling mode. In recycling mode, the rest of energy stored in the magnetizing inductor from the previous stage will charge back to V_{Bat1}. The following experiments and simulations are shown from Figures 13–16.

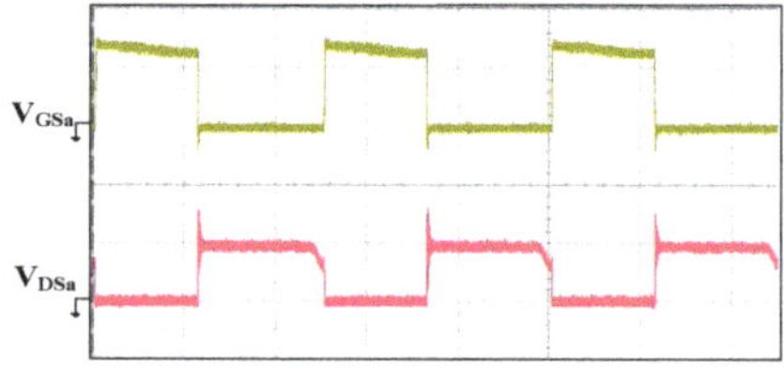

Figure 13. Experiment of voltage waveforms of S_a (V_{GSa}, V_{DSa}), V_{GSa}, V_{DSa}: 10 V/div, Time: 20 μs/div.

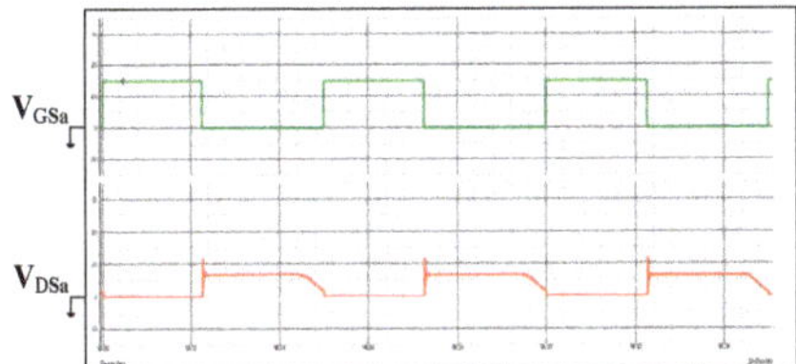

Figure 14. Simulation of voltage waveforms of S_a (V_{GSa}, V_{DSa}), V_{GSa}, V_{DSa}: 10 V/div, Time: 20 μs/div.

Referred to Figures 15 and 16, when the converter is being operated in charging mode, i_{Npa} starts to increases for charging the magnetizing inductor and to deliver the energy to lower voltage cell. Once the converter goes to recycling mode, i_{Nsb} is induced by the magnetizing inductor. Thus, the stored energy is charged back to the higher voltage cell through the filter. From these two figures below, the current direction of i_{Npa} is opposite to i_{Nsb}. This phenomenon proves that the recycling mode is successful.

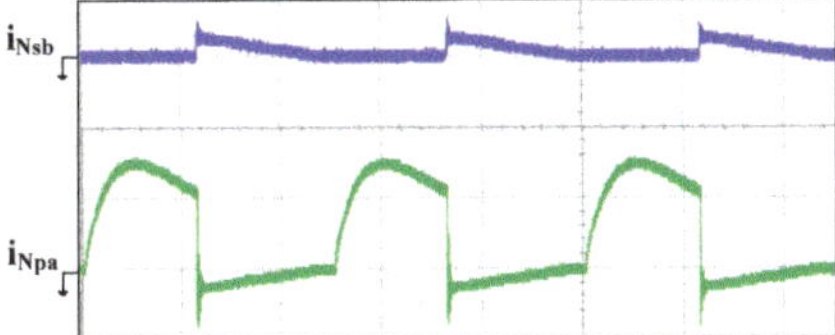

Figure 15. Experiment of current waveforms (i_{Nsb}, i_{Npa}), i_{Nsb}, i_{Npa}: 500 mA/div, Time: 20 μs/div.

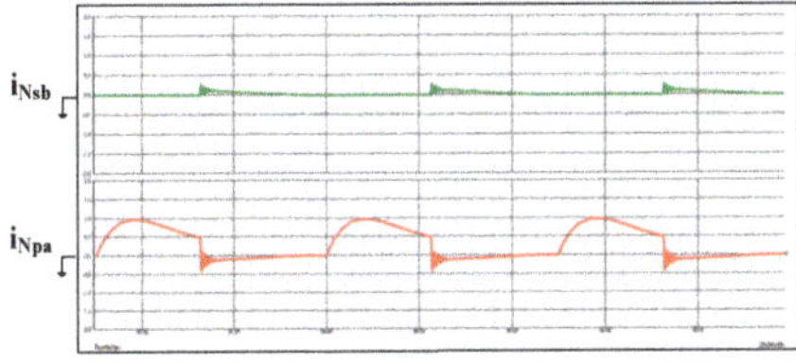

Figure 16. Simulation of current waveforms (i_{Nsb}, i_{Npa}), i_{Nsb}, i_{Npa}: 500 mA/div, Time: 20 μs/div.

5.2. Waveforms for V_{Bat8} Charges to V_{Bat1}

This operation principle is almost the same as the previous section, but the only difference is that the charging direction is inverse (V_{Bat8} charges to V_{Bat1}). The simulation results are shown to compare with the experiments. Figure 17 shows the gate signal for turning on S_{odd_0}, S_{odd_1}, S_{even_7}, and S_{even_8} for building up the charging path. In addition, Figures 18–21 present the related waveforms when V_{Bat8} charges to V_{Bat1}.

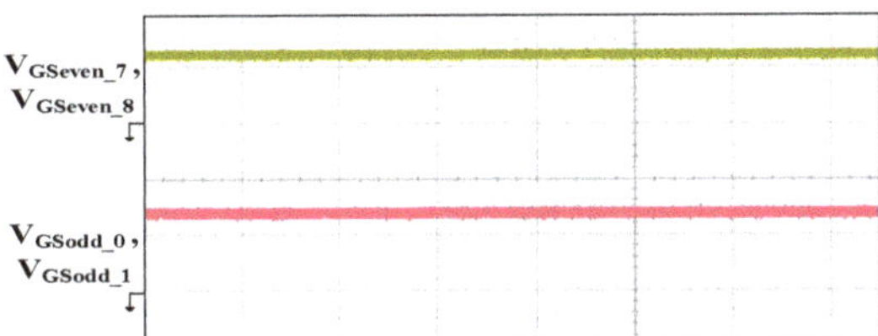

Figure 17. Experiment of V_{GS} triggering waveforms for the switch network (V_{GSodd_0}, V_{GSodd_1}, V_{GSeven_7}, V_{GSeven_8}), V_{GSodd_0}, V_{GSodd_1}, V_{GSeven_7}, V_{GSeven_8}: 10 V/div, Time: 20 μs/div.

After the balancing loop established, S_b turns on in the meantime as well. The converter enters into charging mode when V_{Bat8} charges to V_{Bat1}. However, if S_b turns off, the converter moves to recycling mode and the rest of energy stored in the magnetizing inductor will charge back to V_{Bat8}. The experiments and simulations are shown from Figures 18–21.

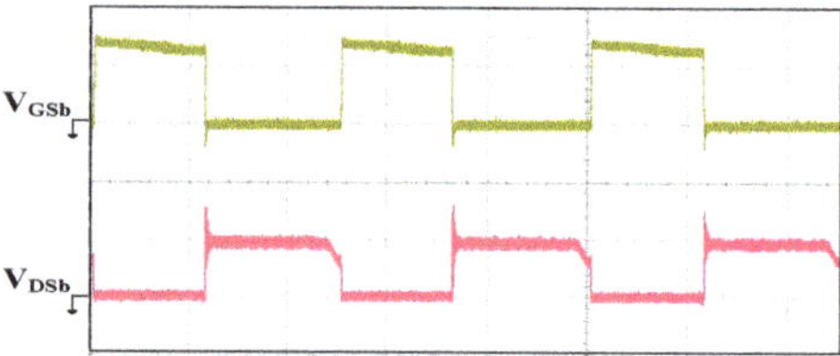

Figure 18. Experiment of voltage waveforms of S_b (V_{GSb}, V_{DSb}), V_{GSb}, V_{DSb}: 10 V/div, Time: 20 μs/div.

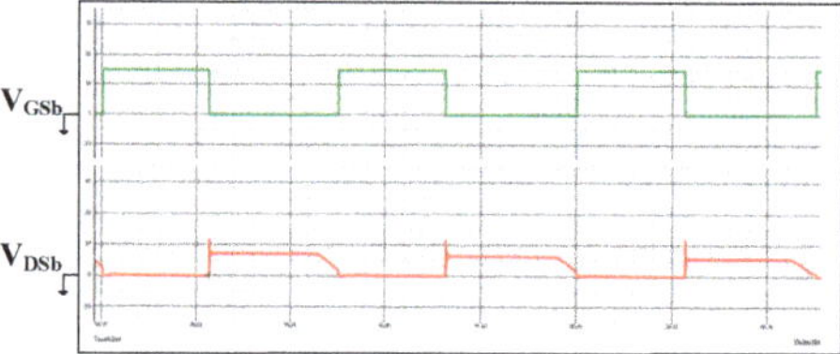

Figure 19. Simulation of voltage waveforms of S_b (V_{GSb}, V_{DSb}), V_{GSb}, V_{DSb}: 10 V/div, Time: 20 μs/div.

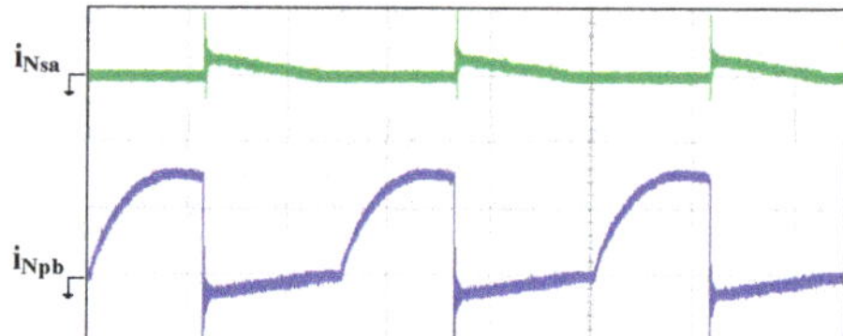

Figure 20. Experiment of current waveforms (i_{Nsa}, i_{Npb}), i_{Nsa}, i_{Npb}: 500 mA/div, Time: 20 μs/div.

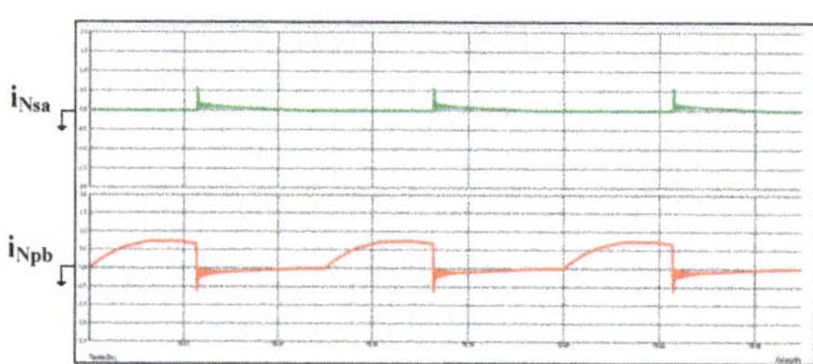

Figure 21. Simulation of current waveforms (i_{Nsa}, i_{Npb}), i_{Nsa}, i_{Npb}: 500 mA/div, Time: 20 μs/div.

To sum up the measurements and simulations above, these results are compared and proved that the proposed fast charging and balancing circuit is feasible.

Before the balancing process starts, each of the battery cells has been discharged for test and experiment. The open loop voltage of each cell is listed in Table 3 from V_{Bat1} to V_{Bat8} individually. When the cell is being charged with 1C current, the ΔV is set for 0.03 V during the balancing interval I to IV. If the voltage reaches to 3.5 V, the ΔV is set from 0.03 V to 0.02 V during the balancing interval V to VII. As the battery voltage rises from 3.5 V to 3.6 V at interval V, the main concept of choosing ΔV is set for lower voltage to make the balancing process much more precise. During the experiments, the voltage and the curve of each cell is measured and drawn by a voltage recorder (model number: midi LOGGER GL800, manufacture: GRAPHTEC).

Table 3. Open loop voltage for discharged cells in the test battery string.

Cell Number	Open Loop Voltage (V)
V_{Bat1},(Cell$_1$)	2.623
V_{Bat2},(Cell$_2$)	2.616
V_{Bat3},(Cell$_3$)	2.592
V_{Bat4},(Cell$_4$)	2.602
V_{Bat5},(Cell$_5$)	2.611
V_{Bat6},(Cell$_6$)	2.625
V_{Bat7},(Cell$_7$)	2.634
V_{Bat8},(Cell$_8$)	2.639

From Figure 22, each of these cells has 7 balancing intervals from interval I to VII. The balancing time and the energy loss of the proposed converter from I to VII are summarized as Table 4. As shown in Table 4, the balancing time becomes shorter when each cell reaches to the charging voltage 3.6 V. Also, the energy loss of the converter gets lower because the balancing process goes to the end.

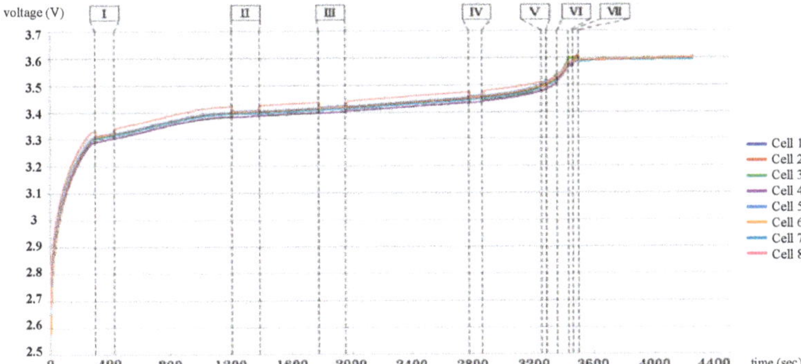

Figure 22. Battery's voltage curve during fast charging balancing process.

Table 4. Balancing time and energy loss in each interval.

Balancing Interval	Balancing Time (sec.)	Energy Losses (J)
I	124	27.42
II	179	42.94
III	178	42.697
IV	90	22.012
V	31	6.989
VI	88	11.391
VII	40	9.395

In Table 5, the measurements and experimental results are listed. The total balancing time is to sum up the time from interval I to VII. To compare with the conventional balancing method, the proposed concept can shorten the balancing time effectively. Besides, after the balancing process, the maximum differential voltage among the balanced cells is 0.018 V.

Table 5. Measurement of battery balancing process.

$\triangle V_1 = 0.03$ V (from I–V), $\triangle V_2 = 0.02$ V(from V–II)	Value
Total balancing time (sec.)	730
Charging time (sec.)	4250
Total energy loss (J)	162.844
Average efficiency of the converter (%)	79.8
Maximum differential voltage (V)	0.018

As mentioned above, $\triangle V$ is determined by user's demand and the balancing performance. If the user requires a lower differential voltage among these cells in the battery string, $\triangle V$ has to be set lower to achieve the better balancing performance; however, it takes more time for the balancing process. Table 6 gives a comparison of different value of $\triangle V$. In the opposite, a higher $\triangle V$ will obtain a worse balancing performance even the balancing time is shorter.

Table 6. Comparison of different single $\triangle V$ for a complete balancing process.

$\triangle V$ (V)	Total Balancing Time (sec.)	Maximum Differential Voltage (V)
0.02	2695	0.014
0.03	411	0.023
0.04	365	0.038
0.05	315	0.061

6. Conclusions

The paper proposes a fast charging balancing circuit for LiFePO$_4$ battery. The main concept is to give a fast voltage balancing strategy for each cell within a single battery string. In this study, a novel bi-directional forward converter is utilized and connected with the balancing bi-directional switches network to form a bi-directional battery balancing circuit.

The feature of the proposed scheme is to balance the maximum differential voltage between the odd-numbered battery and the even-numbered battery in the battery string directly. In order to verify the proposed structure, a fast battery charging circuit for a string with 8 pieces of cell (3.6 V/2.5 Ah for each cell) is implemented. In addition, the merit of this circuit can also avoid over charging situation during the balancing process.

Referred to the experimental and simulation results in previous section, a faster charging balancing circuit for a battery string is achieved. The advantage of this scheme also improves the precision of each cell after balancing procedure. The future research will keep improving the balancing algorithms for random cells in the battery string. Moreover, the proposed study can save more time during the balancing process. At present, industrial applications, electric vehicles batteries, and higher capacity batteries are always composed of many cells connected in series and parallel; therefore, the characteristics and aging phenomenon of each cell have to be concerned with care. In the near future, a battery management system can be added for the proposed balancing circuit to obtain a high precision and faster battery equalizer for each cell.

Author Contributions: S.-T.W. and Y.-N.C. conceptualized and supervised the experiments of the research. S.-T.W. revised and modified the research data into paper format. Y.-N.C. provided the testing bench of the research. C.-Y.C. was in charge of software environment building up. Y.-T.C. was in charge of measurement, and data recording.

Funding: This research received no external funding.

Conflicts of Interest: The authors declare no conflict of interest.

References

1. Huang, W.; Qahouq, J.A.A. Energy Sharing Control Scheme for State-of-Charge Balancing of Distributed Battery Energy Storage System. *IEEE Trans. Ind. Electron.* **2015**, *62*, 2764–2776. [CrossRef]
2. Narayanaswamy, S.; Kauer, M.; Steinhorst, S.; Lukasiewycz, M.; Chakraborty, S. Modular Active Charge Balancing for Scalable Battery Packs. *IEEE Trans. Very Large Scale Integr. Syst.* **2017**, *25*, 974–987. [CrossRef]
3. Chatzinikolaou, E.; Rogers, D.J. Performance Evaluation of Duty Cycle Balancing in Power Electronics Enhanced Battery Packs Compared to Conventional Energy Redistribution Balancing. *IEEE Trans. Power Electron.* **2018**, *33*, 9142–9153. [CrossRef]
4. Baughman, A.C.; Ferdows, M. Double-Tiered Switched-Capacitor Battery Charge Equalization Technique. *IEEE Trans. Ind. Electron.* **2008**, *55*, 2277–2285. [CrossRef]
5. Einhorn, M.; Roessler, W.; Fleig, J. Improved Performance of Serially Connected Li-Ion Batteries with Active Cell Balancing in Electric Vehicles. *IEEE Trans. Veh. Technol.* **2011**, *60*, 2448–2457. [CrossRef]
6. Einhorn, M.; Conte, F.V.; Kral, C.; Fleig, J. Comparison, Selection, and Parameterization of Electrical Battery Models for Automotive Applications. *IEEE Trans. Power Electron.* **2013**, *28*, 1429–1437. [CrossRef]
7. Duggal, I.; Venkatesh, B. Short-Term Scheduling of Thermal Generators and Battery Storage with Depth of Discharge-Based Cost Model. *IEEE Trans. Power Syst.* **2015**, *30*, 2110–2118. [CrossRef]
8. Mejdoubi, A.E.; Chaoui, H.; Gualous, H.; Bossche, P.V.D.; Omar, N.; Mierlo, J.V. Lithium-Ion Batteries Health Prognosis Considering Aging Conditions. *IEEE Trans. Power Electron.* **2019**, *34*, 6834–6844. [CrossRef]
9. Xiong, R.; Zhang, Y.; Wang, J.; He, H.; Peng, S.; Pecht, M. Lithium-Ion Battery Health Prognosis Based on a Real Battery Management System Used in Electric Vehicles. *IEEE Trans. Veh. Technol.* **2019**, *68*, 4110–4121. [CrossRef]
10. Li, L.; Xu, Z.; Zhu, J.; Jing, X.; Shuntao, X. Research on dynamic equalization for lithium battery management system. In Proceedings of the 2017 29th Chinese Control and Decision Conference (CCDC), Chongqing, China, 28–30 May 2017.
11. Steinhorst, S.; Lukasiewycz, M. Formal Approaches to Design of Active Cell Balancing Architectures in Battery Management Systems. In Proceedings of the 2016 IEEE/ACM International Conference on Computer-Aided Design (ICCAD), Austin, TX, USA, 7–10 November 2016.
12. Garche, J.; Jossen, A. Battery Management Systems (BMS) for Increasing Battery Life Time. In Proceedings of the Third International Telecommunications Energy Special Conference, Dresden, Germany, 7–10 May 2000.
13. Lawder, M.T.; Suthar, B.; Northrop, P.W.C.; De, S.; Hoff, C.M.; Leitermann, O.; Crow, M.L.; Santhanagopalan, S.; Subramanian, V.R. Battery Energy Storage System (BESS) and Battery Management System (BMS) for Grid-Scale Applications. *Proc. IEEE* **2014**, *102*, 1014–1030. [CrossRef]
14. Wang, J.B.; Koa, D. Design and Implementation of a Battery Module. In Proceedings of the 2014 International Conference on Intelligent Green Building and Smart Grid (IGBSG), Taipei, Taiwan, 23–25 April 2014.
15. Bonfiglio, C.; Roessler, W. A Cost Optimized Battery Management System with Active Cell Balancing for Lithium Ion Battery Stacks. In Proceedings of the IEEE Vehicle Power Propulsion Conference, Dearborn, MI, USA, 7–10 September 2009.
16. Lan, C.W.; Lin, S.S.; Syue, S.Y.; Hsu, H.Y.; Huang, T.C.; Tan, K.H. Development of an Intelligent Lithium-Ion Battery-Charging Management System for Electric Vehicle. In Proceedings of the 2017 International Conference on Applied System Innovation (ICASI), Sapporo, Japan, 13–17 May 2017.
17. Elsayed, A.T.; Lashway, C.R.; Mohammed, O.A. Advanced Battery Management and Diagnostic System for Smart Grid Infrastructure. *IEEE Trans. Smart Grid* **2016**, *7*, 897–905.
18. Xu, D.; Wang, L.; Yang, J. Research on Li-ion Battery Management System. In Proceedings of the 2010 International Conference on Electrical and Control Engineering, Wuhan, China, 25–27 June 2010.
19. Shibata, H.; Taniguchi, S.; Adachi, K.; Yamasaki, K.; Ariyoshi, G.; Kawata, K.; Nishijima, K.; Harada, K. Management of serially-connected battery system using multiple switches. In Proceedings of the 4th IEEE International Conference on Power Electronics and Drive Systems, Denpasar, Indonesia, 25–25 October 2001.
20. Amin; Ismail, K.; Nugroho, A.; Kaleg, S. Passive balancing battery management system using MOSFET internal resistance as balancing resistor. In Proceedings of the 2017 International Conference on Sustainable Energy Engineering and Application, Jakarta, Indonesia, 23–24 October 2017.

21. Harada, K.; Taniguchi, S.; Adachi, K.; Ariyoshi, G.; Kawata, Y. On the removing of a less quality battery from a series-connected system. In Proceedings of the INTELEC, Twenty-Second International Telecommunications Energy Conference, Phoenix, AZ, USA, 10–14 September 2000.
22. Luo, W.; Lv, J.; Song, W.; Feng, Z. Study on passive balancing characteristics of serially connected lithium-ion battery string. In Proceedings of the 2017 13th IEEE International Conference on Electronic Measurement & Instruments, Yangzhou, China, 20–22 October 2017.
23. Kutkut, N.H.; Wiegman, H.L.N.; Divan, D.M.; Novotny, D.W. Design considerations for charge equalization of an electric vehicle battery system. *IEEE Trans. Ind. Appl.* **1999**, *35*, 28–35. [CrossRef]
24. Wei, X.; Zhu, B. The research of vehicle power Li-ion battery pack balancing method. In Proceedings of the 2009 9th International Conference on Electronic Measurement & Instruments, Beijing, China, 16–19 August 2009.
25. Sooksood, K.; Stieglitz, T.; Ortmanns, M. An Active Approach for Charge Balancing in Functional Electrical Stimulation. *IEEE Trans. Biomed. Circuits Syst.* **2010**, *4*, 162–170. [CrossRef] [PubMed]
26. Li, S.; Mi, C.C.; Zhang, M. A High-Efficiency Active Battery-Balancing Circuit Using Multiwinding Transformer. *IEEE Trans. Ind. Appl.* **2013**, *49*, 198–207. [CrossRef]
27. Wei, X.; Zhao, X.; Dai, H. The application of flyback DC/DC converter in Li-ion batteries active balancing. In Proceedings of the 2009 IEEE Vehicle Power and Propulsion Conference, Dearborn, MI, USA, 7–10 September 2009.
28. Hosseinzadeh, M.; Salmasi, F.R. Robust Optimal Power Management System for a Hybrid AC/DC Micro-Grid. *IEEE Trans. Sustain. Energy* **2015**, *6*, 675–687. [CrossRef]
29. Schonbergerschonberger, J.; Duke, R.; Round, S.D. DC-Bus Signaling: A Distributed Control Strategy for a Hybrid Renewable Nanogrid. *IEEE Trans. Ind. Electron.* **2006**, *53*, 1453–1460. [CrossRef]

© 2019 by the authors. Licensee MDPI, Basel, Switzerland. This article is an open access article distributed under the terms and conditions of the Creative Commons Attribution (CC BY) license (http://creativecommons.org/licenses/by/4.0/).

Article

Add-On Type Pulse Charger for Quick Charging Li-Ion Batteries

Bongwoo Kwak [1,2]**, Myungbok Kim** [1] **and Jonghoon Kim** [2,*]

[1] EV Component & Materials R&D Group, Korea Institute of Industrial Technology, Gwangju 61012, Korea; bwkwak11@kitech.re.kr (B.K.); boks@kitech.re.kr (M.K.)

[2] Department of Electrical Engineering, Chungnam National University, Daejeon 34134, Korea

* Correspondence: wdhgns0422@cnu.ac.kr; Tel.: +82-10-7766-5010

Received: 31 December 2019; Accepted: 28 January 2020; Published: 30 January 2020

Abstract: In this paper, an add-on type pulse charger is proposed to shorten the charging time of a lithium ion battery. To evaluate the performance of the proposed pulse charge method, an add-on type pulse charger prototype is designed and implemented. Pulse charging is applied to 18650 cylindrical lithium ion battery packs with 10 series and 2 parallel structures. The proposed pulse charger is controlled by pulse duty, frequency and magnitude. Various experimental conditions are applied to optimize the charging parameters of the pulse charging technique. Battery charging data are analyzed according to the current magnitude and duty at 500 Hz and 1000 Hz and 2000 Hz frequency conditions. The proposed system is similar to the charging speed of the constant current method under new battery conditions. However, it was confirmed that as the battery performance is degraded, the charging speed due to pulse charging increases. Thus, in applications where battery charging/discharging occurs frequently, the proposed pulse charger has the advantage of fast charging in the long run over conventional constant current (CC) chargers.

Keywords: add-on pulse charger; quick charge; pulse charging; Li-ion battery

---

1. Introduction

Recently, owing to air pollution and global warming issues, interest in electric vehicles (EVs) and plug-in hybrid vehicles (PHEVs) has increased greatly. The batteries used in electric vehicles are mostly lithium-based secondary batteries because of their weight and mileage. Lithium-ion batteries have continued to gain market share in the secondary battery market over the last several years. This is because of their high quality density, low self-discharge and many applications [1,2]. Therefore, the demand for durability of lithium ion batteries is increasing [3,4]. In addition, improving the charging time and efficiency of the battery is an important issue in all battery powered applications.

The demand for fast charging is increasing in the case of batteries for electric vehicles [5]. The use of electric cars is increasing the need for fast chargers. Various charging technologies have been developed to improve the condition, life cycle, charging time and charging efficiency of lithium-ion batteries [2,6–10]. Among various technologies, constant current (CC) and constant voltage (CV) charging methods are widely used [1,11]. CC–CV charging technology, charges with CC below the threshold voltage. When the threshold voltage is reached, the current gradually falls, charging to a constant voltage. The magnitude of the CC and CV depends on the battery specifications. The disadvantage of this charging technique is that it reduces the charge current in the CV section to prevent battery damage owing to overvoltage. Therefore, the charging time is extended in this section [6]. In the case of CC–CV charging, the charging time is long because the charging speed is slow.

Another method is to charge with a current higher than the rated current for fast charging until the battery voltage reaches the set threshold. After the CC period, the charging current should be reduced for battery safety. In this case, the charging speed of the battery increases, but the cost of the charging system increases. This is because a higher charging system is required for the battery capacity [12].

Pulse charging, which charges the battery by controlling the charging current, improves fast charging time and battery charging efficiency without significantly increasing costs [13–15]. However, to achieve the pulse charging effect, the duty and frequency of the pulse current must be appropriately selected [14]. In existing studies [2,15], the battery was charged in the form of a pulse from zero current to the charging current. The pulse frequency, duty and magnitude of the charging current affect the battery charging time. Depending on the frequency and duty of pulse charging, the charging time is shorter than that of the CC–CV charger method. In a previous study [2], adaptive pulse charging reduced the charging time by approximately 12.7%. In the study [15], the charge time was improved by 3.4% by applying the duty-varied voltage pulse charging method. However, these charging technologies require a dedicated charger with additional hardware and software.

In this paper, an add-on pulse type charging system is proposed, which can be connected to a conventional charging system and pulse the charging current. The proposed add-on pulse type charging system can control the amplitude, frequency and duty of the current pulse. By applying the proposed pulse charger, the existing charger can be applied CC–CV charging and the pulse charger method. Based on the results, the faster charging effect of the pulse charging method was verified as the performance of the lithium ion battery decreased.

2. Structure of Conventional Charging System

A conventional charging system consists of a pulse charger as shown in Figure 1a, and a CC–CV charger as shown in Figure 1b. Pulse charging is fast, but it is accompanied by the degradation of the battery. Therefore, selective charging should be available when only a quick charge is needed. Charging requires both pulse chargers and CC–CV chargers for existing systems.

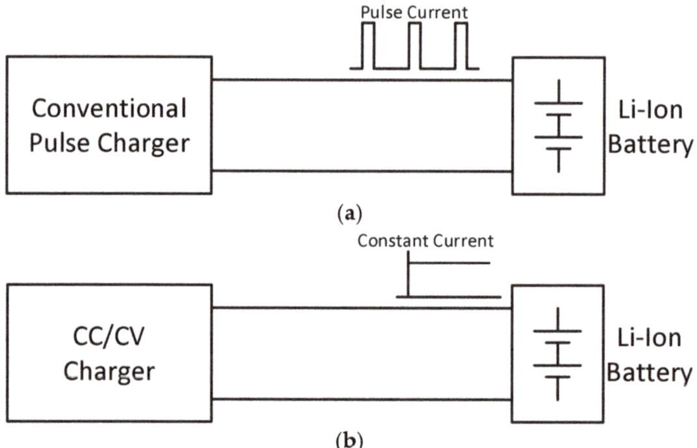

Figure 1. Structure of the conventional charging system. (**a**) Pulse charger (**b**) constant current–constant voltage (CC–CV) charger.

3. Add-On Type Pulse Charger

The proposed add-on pulsed charging system is shown in Figure 2. If the add-on type pulse charger does not operate as shown in Figure 2a, it operates as a CC–CV charger. As shown in Figure 2b, when the pulsed charging circuit is driven, it operates as a pulse charger. Therefore, the pulse charger and CC–CV charging can be selectively performed using the same charger.

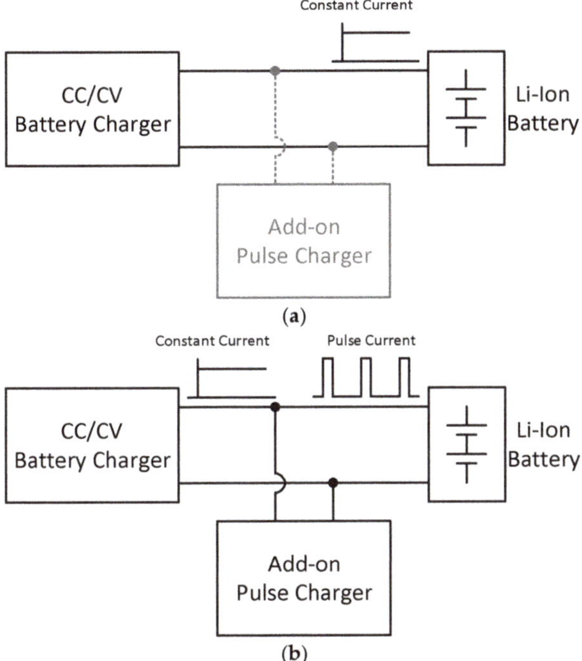

Figure 2. Structure of the conventional charging system. (**a**) Pulse charger (**b**) CC–CV charger.

Pulse charging uses a controlled pulse current to charge the battery. Figure 3a shows the basic structure of the pulse charger pulsed to the battery by controlling the switch (S_1). Figure 3b shows the waveform of the pulse current applied to the battery, which is the control element of the peak current amplitude (ΔI_c), pulse duty ($D_p = t_{on}/T_p$) and frequency ($f_p = 1/T_p$). The pulse frequency, duty cycle and level affect the charging time, battery life cycle and battery parameters. As a result, pulse charging improves the charging speed and energy efficiency [5].

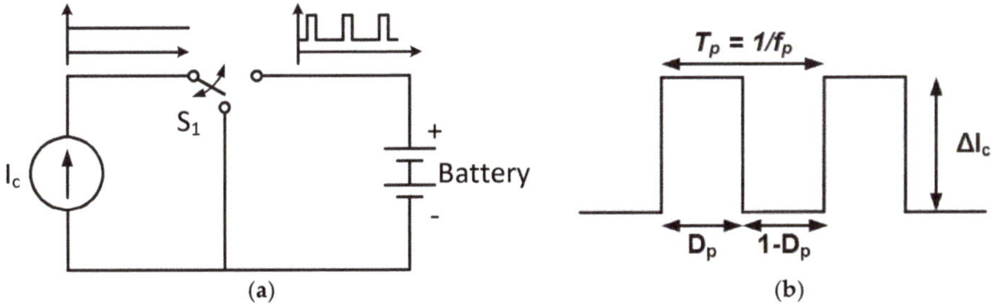

Figure 3. Structure of the pulse charger. (**a**) Pulse charging model and (**b**) pulse waveform.

As shown in Figure 4, a synchronous DC/DC converter circuit is applied to the pulse generator module with pulse parameter control. The basic operation uses a portion of the battery charge current to charge the link capacitor, discharging the charged capacitor energy to supply a pulsed current to the battery.

First, the link capacitor is charged using the energy of the charger. When the lower switch Q_2 is turned on and the upper switch Q_1 is turned off, it works as shown in Figure 5a. When low side switch Q_2 is turned off, current flows through the body diode of high side switch Q_1. As shown in Figure 5b, when Q_1 is turned on, the voltage across Q_1 becomes a zero voltage condition [16]. The battery is charged by the charge current of the CC charger minus the capacitor charge current.

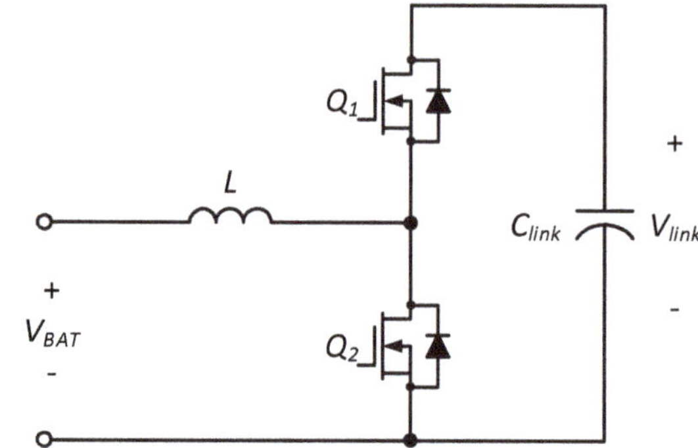

Figure 4. Circuit of the add-on pulse charger.

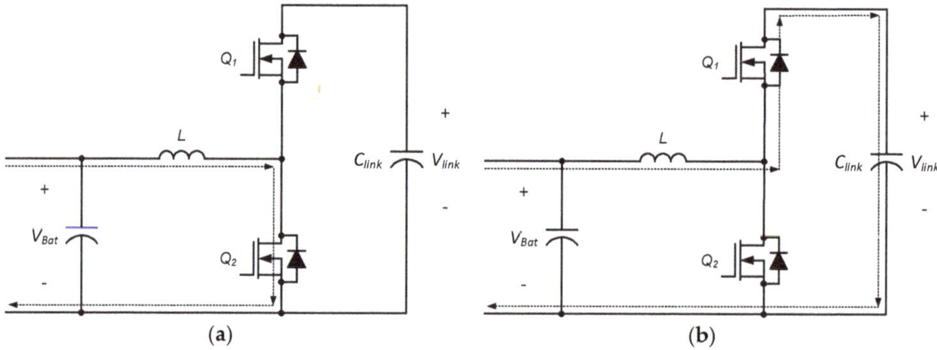

Figure 5. Capacitor charge mode (**a**) Q_1: turn off, Q_2: turn on, (**b**) Q_1: turn on and Q_2: turn off.

Next is the mode for discharging the link capacitor to add charging current to the battery. When the high-side switch Q_1 is turned on and the low-side switch Q_2 is turned off, it works as shown in Figure 6a. When the high-side switch Q_1 is turned off, current flows through the body diode of the low-side switch Q_2. As shown in Figure 6b, when Q_2 is turned on, the voltage across Q_2 becomes a zero voltage condition [16]. The battery is charged by the capacitor discharge current plus the charging current of the CC charger.

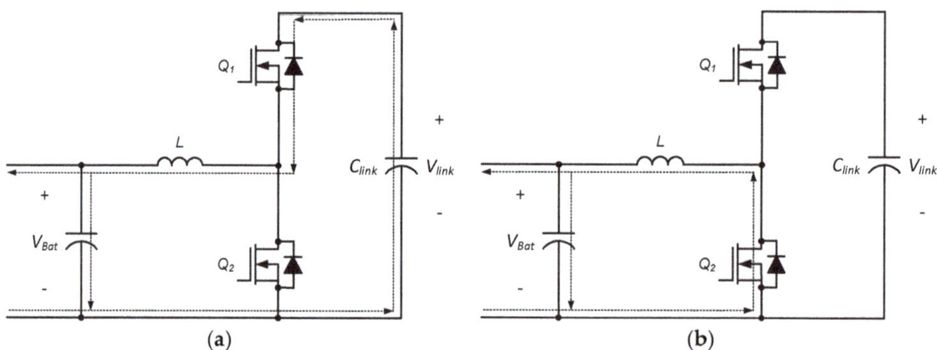

Figure 6. Capacitor discharge mode (**a**) Q_1: turn on, Q_2: turn off, (**b**) Q_1: turn off and Q_2: turn on.

3.1. Design of Circuit Parameters

Target specifications were chosen to design the pulsed charging circuit parameters. The voltage of a typical electric car is 350–400 V. Tesla organizes battery systems in series and in parallel with 18650 cylindrical cells. In this study, the battery voltage was reduced by approximately 10 times for safe experimentation. Thus, 18650 cylindrical cells were used in 10 series and two parallel configurations. The battery used was the INR 18650-35E from Samsung SDI. The battery's nominal voltage was 36 V, and the maximum charging current is 1 C-rate. The maximum voltage of the link capacitor was selected within the operating voltage range of the power semiconductor device (IPT059N15N3). Detailed specifications are listed in Table 1

Table 1. Specifications of the pulse charger.

Parameter	Value	Unit
Battery Voltage (Nominal)	36	V
Charge/Discharge Current	6.8	A
Pulse Frequency	200–1000	Hz
Maximum Link Voltage	120	V
Switching Frequency	100	kHz

The link capacitor design is important in generating the pulse current during battery charging through charging/discharging of the link capacitor. The capacitance can be calculated from the variation of the electric charge of the link capacitor. The change amount of the electric charge can be obtained through Equation (1)

$$\Delta Q = C_{link} \cdot \Delta V_{link} = \int_0^{Tp} I_L \cdot dt \tag{1}$$

where the maximum inductor current is 6.8 A. The charge/discharge pulse time is 1–5 msec. The voltage change of the capacitor through charging/discharging is shown in Equation (2).

$$\Delta V_{link} = V_{link.max} - V_{link.min} \tag{2}$$

The maximum voltage was set to 120 V, considering a 20% margin at 150 V, the rated voltage of the power semiconductor. The minimum voltage matches the battery voltage, so the voltage change on the capacitor is 84 V. Depending on the voltage specification, Equation (2) can be summarized by Equation (3).

$$V_{link.max} = V_{link.min} + \Delta V_{link} = \frac{2 \cdot I_L \cdot t_{pulse}}{C_{link}} + V_{link.min} \tag{3}$$

Equation (3) can be expressed as Equation (4) when summarized by capacitance.

$$C_{link} = \frac{2 \cdot I_L \cdot t_{pulse}}{V_{link.max} - V_{link.min}} \tag{4}$$

Through the above equation, the capacitance of the link capacitor is 160 μF. Therefore, 200 μF is chosen considering a 20% design margin.

In the charge and discharge mode, the inductor current always determines the inductance value to operate in continuous mode. For the inductor design, when the Q_1 turn-on time is $0 < t < DT$ in the steady state, the voltage V_L applied to the inductor is given by Equation (5).

Where D is the duty ratio of pulse width modulation (PWM) supplied to Q_1. T is switching time.

$$V_L = V_{Bat} - V_{link} \tag{5}$$

The relationship between the current I_L of the inductor and the inductor voltage V_L is shown in Equation (6).

$$V_L = L\frac{dI_L}{dt} \tag{6}$$

Through Equations (5) and (6), the inductor current ripple for $0 < t < DT$ can be obtained as shown in Equation (7).

$$\Delta I_L = \frac{V_{link} - V_{Bat}}{L}DT \tag{7}$$

The inductor current ripple can be obtained based on the rated output current. The maximum inductor current is shown in Equation (8), and the minimum inductor current is shown in Equation (9).

$$I_{max} = I_{link} + \frac{\Delta I_L}{2} = \frac{V_{Bat}}{R} + \frac{1}{2}\left(\frac{V_{link} - V_{Bat}}{L}DT\right) \tag{8}$$

$$I_{min} = I_{link} - \frac{\Delta I_L}{2} = \frac{V_{Bat}}{R} - \frac{1}{2}\left(\frac{V_{link} - V_{Bat}}{L}DT\right) \tag{9}$$

The current ripple of the inductor can be calculated as shown in Equation (10).

$$\Delta I_L = \frac{V_{Bat}(V_{link} - V_{Bat})}{V_{link} \cdot L \cdot f_{sw}} \tag{10}$$

Equation (11) is expressed as inductance and can be obtained as follows.

$$L = \frac{V_{Bat}(V_{link} - V_{Bat})}{V_{link} \cdot \Delta I_L \cdot f_{sw}} \tag{11}$$

The inductor current ripple is typically chosen to be 10–20% of the rated current. In this study, we chose 10% to reduce the battery current ripple. Therefore, the calculated value of inductance was 370 μH. It was designed as 400 μH considering a 10% margin.

3.2. Control Algorithm for Add-On Pulse Charger

A control algorithm is established for charging/discharging the link capacitor for pulse current supply. If the command value of the inductor current is greater than zero, the step down operation is performed. Conversely, if the inductor current command value is less than zero, the step up operation is performed. Pulse duty and pulse amplitude commands should be set for voltage balancing of link capacitors during charge/discharge. Pulse duty is limited up to 50% and inductor current command is calculated according to pulse duty and pulse amplitude command. The current reference for the link capacitor charge mode and discharge mode is calculated by the following equation.

$$I_{L\_ref\_discharging} = I_d \times \left(1 - D_p\right) \tag{12}$$

$$I_{L\_ref\_charging} = -I_d \times D_p \tag{13}$$

where I_d is the set pulse magnitude.

The inductor current, battery charge current, and link voltage waveforms for the inductor's set current are shown in Figure 7.

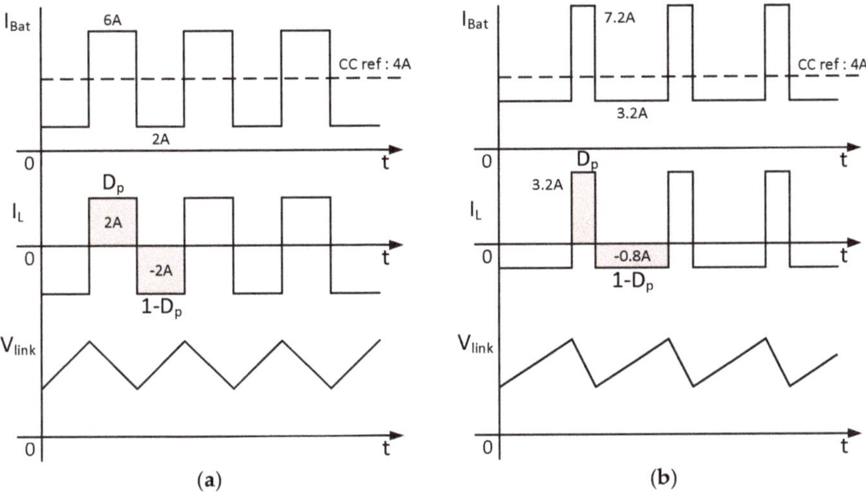

Figure 7. Characteristic waveforms according to the inductor current control: (**a**) D_p: 50% and (**b**) D_p: 20%.

In the $1\text{-}D_p$ section, the link capacitor was charged using the charging current of a conventional charger. The battery was charged by subtracting the link capacitor charge current from the charger current. In the D_p section, the link capacitor was discharged. The capacitor discharge current was added to the charger's charging current to charge the battery. The Proportional-Integral (PI) current controller was applied to equally control the charge and discharge current of the inductor according to the pulse duty. The control block of the pulse charger is shown in Figure 8. The inductor current reference of the pulse charger was set according to the set pulse current specification. A negative setting inductor current command charges the link capacitor. On the other hand, positive numbers discharge the link capacitor. The PI controller was used to control the inductor current for charging and discharging the link capacitor. The add-on type pulse charging system could control the pulse of the battery charging current through the inductor current control.

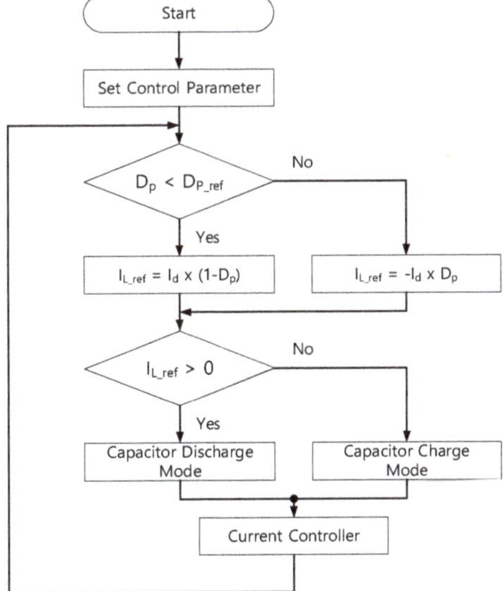

Figure 8. Control block diagram of the pulse charger.

4. Simulation and Experimental Results

This section presents the simulation and experimental results. Additional pulse charging circuits were designed using the circuit parameters selected above. The simulation used PSIM. The pulse charger specifications are listed in Table 2. The battery charge current was selected as 4 A in consideration of the INR 18650-35E maximum charge current. The simulation circuit was implemented as shown in Figure 9.

Table 2. Specifications and circuit parameters of the pulse charger.

Parameter	Value	Unit
Inductance	400	μH
Link Capacitance	200	μF
CC Charge Current	4	A
Pulse Frequency	200–1000	Hz
Pulse Duty	50	%
Depth Pulse Current	0–4	A

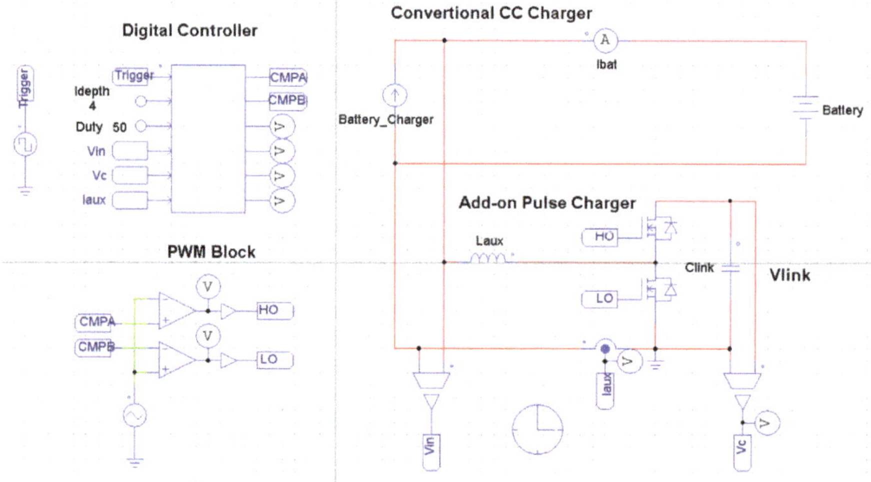

Figure 9. Circuit of the simulation.

The simulation results for the add-on type pulse chargers are shown in Figures 10 and 11. Figure 10 shows simulation results for a pulse frequency of 500 Hz, 50% pulse duty and an inductor current reference of 2 A. Depending on the pulse duty, the inductor current was controlled to be 1 A on charge and −1 A on discharge. Figure 10 shows the simulation results for a pulse frequency of 1000 Hz, 20% pulse duty and inductor current reference of 4 A. Depending on the pulse duty, the inductor current was controlled to be 3.2 A on charge and −0.8 A on discharge. The inverter current was controlled according to the duty and current reference of the pulse.

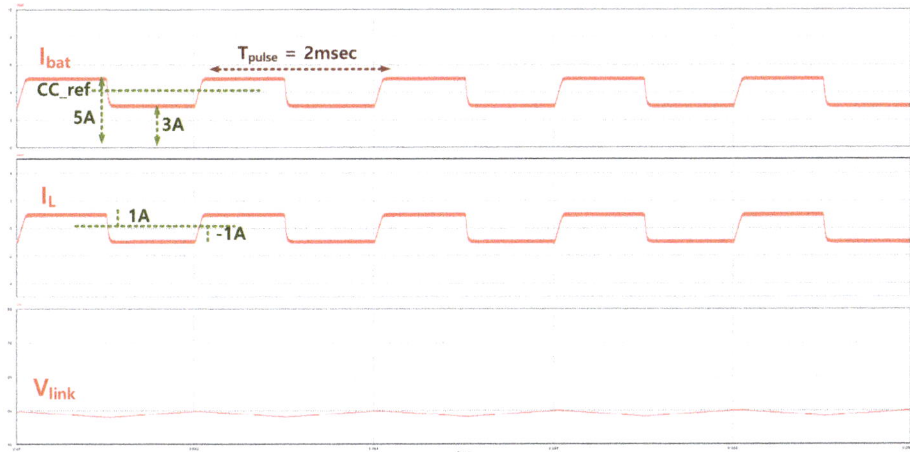

Figure 10. Results of the simulation at a 50% pulse duty ratio and 500 Hz pulse frequency.

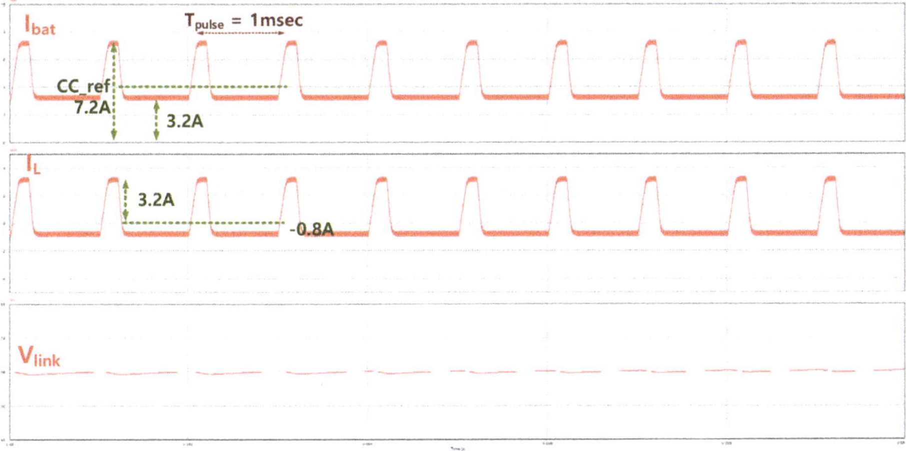

Figure 11. Results of the simulation at a 20% pulse duty ratio and 1000 Hz pulse frequency.

Based on the simulation, the battery was tested for pulse charging. To charge the battery with a CC, a diagram of the test equipment was constructed as shown in Figure 12a. A prototype of an add-on pulse charger is shown in Figure 12b. Figure 12c shows the configured experimental set. DC power supplies and electronic loads were used to construct a constant current charging system. The battery consisted of 10 series and two parallel structures. Oscilloscopes and data loggers were used to collect the characteristic battery data by pulse charging.

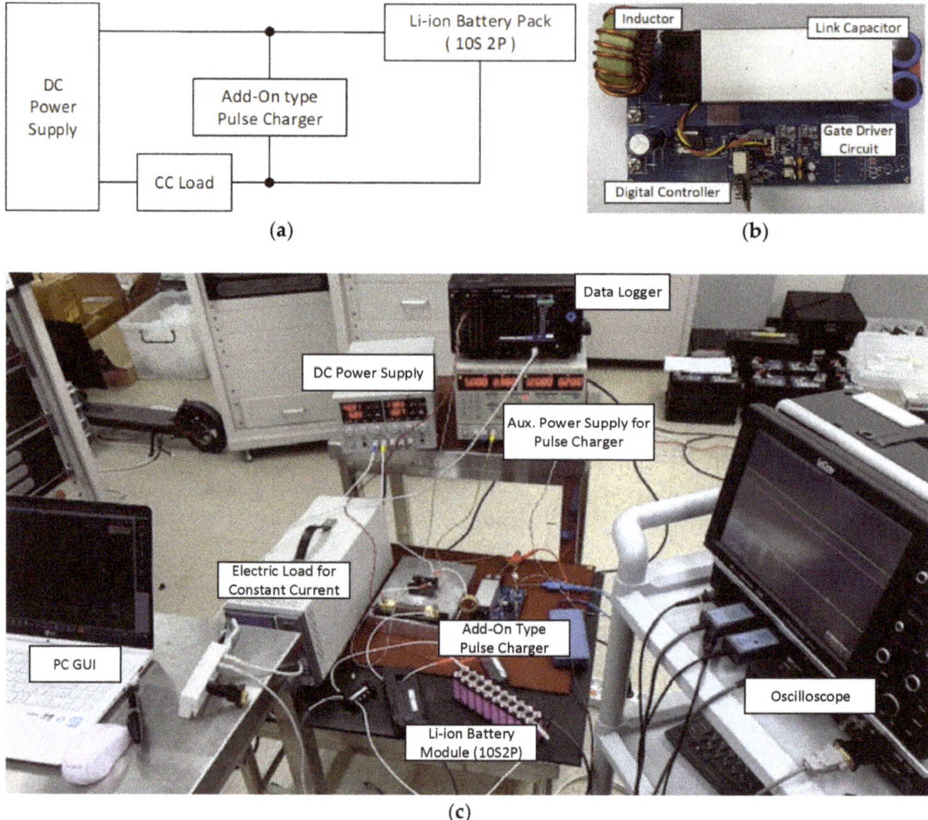

Figure 12. Experiment set configuration. (**a**) Block diagram of the experiment set, (**b**) prototype of the add-on type pulse charger and (**c**) experiment environment.

Pulse charge experiments were performed in the constant current charge range of the battery. CC charging and pulse charging tests were performed for a comparative analysis of conventional charging methods and pulse charging methods. The charging of the pulse was proceeded by changing the frequency, duty and magnitude of the pulse.

Figure 13 shows the steady-state characteristics by a set frequency at 50% pulse duty and 4 A pulse magnitude. Figure 14 shows the steady-state characteristics by a set duty at 1000 Hz pulse frequency and 4 A pulse magnitude. It was confirmed that the link capacitor was charged and discharged according to the set battery charge pulse. The proposed add-on pulse charger verified that the pulse frequency, magnitude and duty of the battery charge current could be selectively changed by charging/discharging the link capacitor. In addition, it could be confirmed that the mean value of the charging current of the battery is the same.

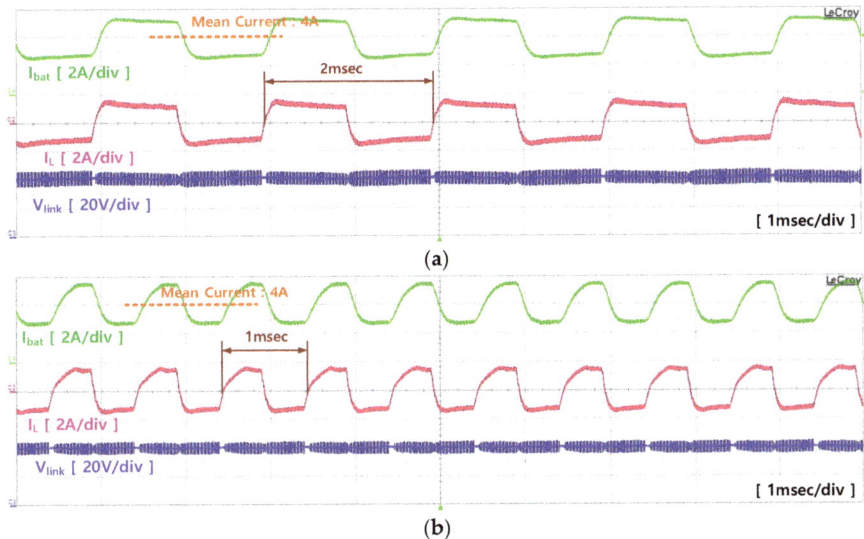

(**a**)

(**b**)

Figure 13. Characteristics of the steady state: (**a**) pulse frequency: 500 Hz and (**b**) pulse frequency: 1000 Hz.

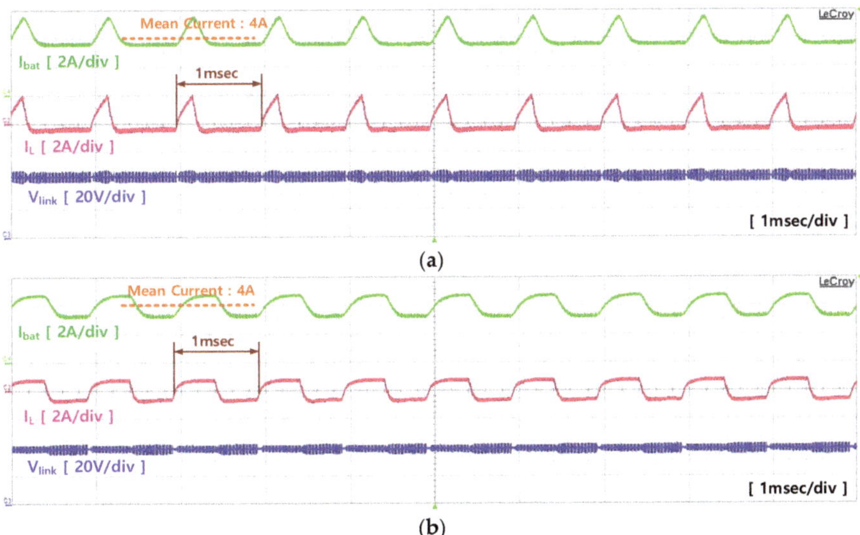

(**a**)

(**b**)

Figure 14. Characteristics of the steady state: (**a**) pulse duty: 20% and (**b**) pulse duty: 50%.

Table 3 shows the experimental results by the frequency change. The initial temperature is slightly different at experimental conditions. However, considering aging according to the experimental sequence, 1000 Hz was identified as the best point.

Table 3. Experimental results by pulse frequency.

Pulse Frequency	Start Voltage	End Voltage	End Temperature	End Temperature	Charge Time
500 Hz	33.1 V	41.5 V	33.1 °C	41.5 °C	76.9 min
1000 Hz	33.1 V	41.5 V	30.0 °C	40.1 °C	77.0 min
2000 Hz	33.1 V	41.5 V	33.1 °C	41.5 °C	77.0 min

Referring to the existing paper, it was confirmed that the optimum pulse frequency changes according to the battery characteristics. Experiments were performed to select the optimal frequency for Li-ion battery cells used in this paper. Experiments were performed in the order of 1000 Hz, 2000 Hz and 500 Hz. At this time, the pulse duty was 50% and the pulse current magnitude was 6 A. Considering the temperature conditions, all experiments were conducted in a chamber.

Next, experiments were conducted to evaluate the characteristics of the proposed pulse charging and CC charging. Pulse charging began first to reduce the effects of battery aging. The experiment proceeds in the order of pulse charging–discharging-CC charging–discharging. Pulse charge was performed at 1000 Hz pulse frequency, 50% pulse duty and 6A pulse current. Table 4 shows the pulse charging results. Table 5 shows the result of CC charging. In the above experiment, CC charging appeared faster. The reason is that due to the characteristics of the pulse controller, the average current of pulse charging was charged as low as about 0.02 A.

Table 4. Experimental results by pulse charging.

No.	Start Voltage	End Voltage	Start Temperature	End Temperature	Charge Time
1	33.1 V	41.5 V	29.6 °C	39.7 °C	4674 sec
2	33.1 V	41.5 V	30.1 °C	39.9 °C	4650 sec
3	33.1 V	41.5 V	30.0 °C	40.1 °C	4620 sec

Table 5. Experimental results by CC charging

No.	Start Voltage	End Voltage	Start Temperature	End Temperature	Charge Time
1	33.1 V	41.5 V	30.0 °C	39.0 °C	4614 sec
2	33.1 V	41.5 V	30.0 °C	39.1 °C	4584 sec
3	33.1 V	41.5 V	30.0 °C	40.2 °C	4578 sec

Experiments show that the CC charge was quickly charged in the new battery state. However, due to the control characteristics of the pulse charge during the experiment, the average battery charge current was 0.02 A less than the CC charge. Therefore, in the fresh battery, it was determined that the speed difference between pulse charging and CC charging was very small.

However, analysis of the experimental data confirmed that the charge rate of the pulse charge was increased when the battery was repeatedly charged/discharged. In the case of pulse charging, the second charging speed was about 24 s faster than the first charging speed. The third charge was about 30 s faster than the second charge. As the test results show above, as battery was aging, the charge speed gradually increased. In contrast, for CC charging, the second charging rate was 30 s faster than the first charging rate. The third charge was 6 s faster than the second charge. As the test results show above, as the battery was aging, the charge rate gradually slowed down. The results are shown in Table 6.

Table 6. Result of the reduction of the charging time by the charging method.

Reduce the Charging Time-Pulse Charging	Reduce the Charging Time-CC Charging
24 sec	30 sec
30 sec	6 sec

Pulse charging was more effective when battery performance deteriorated. In the case of a real lithium-ion battery, the fresh period is very short given the total life. In particular, in applications where charging/discharging occurs frequently, such as in electric vehicles, the proposed pulse charging method was proven to benefit from long-term fast charging rather than CC charging.

5. Conclusions

This paper proposed a pulse charging technique using additional charging circuits for the quick charging of lithium-ion batteries. Pulse charging shortens the charging time, but may degrade the performance of the battery. Therefore, pulse charging should be able to be optionally required. However, conventional charging systems require both pulse chargers and CC–CV chargers. The proposed add-on pulse charging circuit was connected to the existing CC–CV charger, allowing pulse charging. This charging circuit had the advantage of being applicable not only to electric vehicles but also all applications that use batteries.

This paper presented the pulse charging circuit, experimental method and data analysis. Pulse charging technology was applied to 18650 cylindrical lithium ion batteries to analyze the experimental results. Pulse charge and CC charge experiments were performed in the constant current charge section of the battery. Through repeated experiments, the charging rate by pulse charging in fresh battery condition was similar to that of conventional CC charging.

However, as battery performance deteriorated, the charge speed increase of CC charging gradually decreased. On the other hand, the increase of the charging speed by pulse charge gradually increased. To neglect the effects of aging, pulse filling was first performed. Therefore, pulse charging is more effective when battery performance degrades. In a real battery, its life is much longer than a fresh battery. Thus, in applications where charge/discharge frequently occurs, such as in electric vehicles, the proposed pulse charging rather than CC charging has a long-term advantage. In particular, the existing charging infrastructure was constructed using CC–CV chargers. Instead of installing a separate pulse charger for fast charging, using the proposed charger is expected to reduce overall costs.

In the future, we will compare pulse charging and CC charging characteristics while continuing to degrade battery performance.

Author Contributions: Conceptualization, B.K., and M.K.; methodology, validation, analysis, and analyzed the data, B.K., and J.K.; writing—original draft, B.K.; Writing—review and editing, supervision, and project administration, M.K., and J.K. All authors have read and agreed to the published version of the manuscript.

Funding: This study has been conducted with the support of the Ministry of Trade, Industry and Energy as "Future Growth Engine Business project (20003558)".

Conflicts of Interest: The authors declare no conflict of interest.

References

1. Shen, W.; Vo, T.T.; Kapoor, A. Charging algorithms of lithium-ion batteries: An overview. In Proceedings of the 2012 7th IEEE Conference on Industrial Electronics and Applications (ICIEA), Singapore, Singapore, 18–20 July 2012; pp. 1567–1572.
2. Niroshana, S.M.I.; Sirisukprasert, S. Adaptive pulse charger for Li-ion batteries. In Proceedings of the 2017 8th International Conference of Information and Communication Technology for Embedded Systems (IC-ICTES), Chonburi, Thailand, 7–9 May 2017; pp. 1–6.
3. Liu, L.; Park, J.H.; Lin, X.; Sastry, A.M.; Lu, W. A thermal-electrochemical model that gives spatial-dependent growth of solid electrolyte interphase in al Li-ion battery. *J. Power Sources* **2014**, *268*, 482–490. [CrossRef]
4. O'Malley, R.; Liu, L.; Depcik, C. Comparative study of various cathodes for lithium ion batteries using an enhanced Peukert capacity model. *J. Power Sources* **2018**, *396*, 621–631. [CrossRef]
5. Amanor-Boadu, J.M.; Guiseppi-Elie, A.; Sánchez-Sinencio, E. The Impact of Pulse Charging Parameters on the Life Cycle of Lithium-Ion Polymer Batteries. *Energies* **2018**, *11*, 2162. [CrossRef]
6. Chen, L.-R. A Design of an Optimal Battery Pulse Charge System by Frequency-Varied Technique. *IEEE Trans. Ind. Electron.* **2007**, *54*, 398–405. [CrossRef]

7. Surmann, H. Genetic optimization of a fuzzy system for charging batteries. *IEEE Trans. Ind. Electron.* **1996**, *43*, 541–548. [CrossRef]

8. Ullah, Z.; Burford, B.; Dillip, S. Fast intelligent battery charging: neural-fuzzy approach. *IEEE Aerosp. Electron. Syst. Mag.* **1996**, *11*, 26–34. [CrossRef]

9. Wang, S.-C.; Liu, Y.-H. A PSO-Based Fuzzy-Controlled Searching for the Optimal Charge Pattern of Li-Ion Batteries. *IEEE Trans. Ind. Electron.* **2015**, *62*, 2983–2993. [CrossRef]

10. Chen, L.R.; Chen, J.J.; Ho, C.M.; Wu, S.L.; Shieh, D.T. Improvement of Li-ion battery Discharging Performance by pulse and Sinusoidal current strategies. *IEEE Trans. Ind. Electron.* **2013**, *60*, 5620–5628. [CrossRef]

11. Cope, R.C.; Podrazhansky, Y. The art of battery charging. In Proceedings of the Fourteenth Annual Battery Conference on Applications and Advances (Cat. No.99TH8371), Long Beach, CA, USA, 12–15 January 1999.

12. Li, G.; Li, Z.; Zhang, P.; Zhang, H.; Wu, Y. Research on a gel polymer electrolyte for Li-ion batteries. *Pure Appl. Chem.* **2008**, *80*, 2553–2563. [CrossRef]

13. Yin, M.; Cho, J.; Park, D. Pulse-Based Fast Battery IoT Charger Using Dynamic Frequency and Duty Control Techniques Based on Multi-Sensing of Polarization Curve. *Energies* **2016**, *9*, 209. [CrossRef]

14. Amanor-Boadu, J.M.; Guiseppi-Elie, A.; Sanchez-Sinencio, E. Search for Optimal Pulse Charging Parameters for Li-Ion Polymer Batteries Using Taguchi Orthogonal Arrays. *IEEE Trans. Ind. Electron.* **2018**, *65*, 8982–8992. [CrossRef]

15. Chen, L.R. Design of duty-varied voltage pulse charger for improving Li-ion battery charging response. *IEEE Trans. Ind. Electron.* **2009**, *56*, 480–487.

16. Zhang, J.; Lai, J.-S.; Kim, R.-Y.; Yu, W. High-Power Density Design of a Soft-Switching High-Power Bidirectional dc–dc Converter. *IEEE Trans. Power Electron.* **2007**, *22*, 1145–1153. [CrossRef]

© 2020 by the authors. Licensee MDPI, Basel, Switzerland. This article is an open access article distributed under the terms and conditions of the Creative Commons Attribution (CC BY) license (http://creativecommons.org/licenses/by/4.0/).

Article

Individual Phase Full-Power Testing Method for High-Power STATCOM

Qingjun Huang *, Bo Li, Yanjun Tan, Xinguo Mao, Siguo Zhu and Yuan Zhu

State Key Laboratory of Disaster Prevention and Reduction for Power Grid Transmission and Distribution Equipment, State Grid Hunan Electric Company Limited Disaster Prevention and Reduction Center, Changsha 410129, China
* Correspondence: huangqj@hust.edu.cn; Tel.: +86-0731-86332088

Received: 28 May 2019; Accepted: 29 June 2019; Published: 4 July 2019

Abstract: For a high-power static synchronous compensator (STATCOM), a full-power pre-operation test in the factory is necessary to ensure the product quality of a newly manufactured one. But owing to the hardware limitation and cost of test platform, such test is currently too difficult to conduct in the factory, thus it poses great risk to the on-site operation and commissioning. To address this issue, this paper proposes an individual phase full-power testing method for STATCOM. By changing the port connection, three-phase STATCOM was reconstructed into a structure that two phases are in parallel and then in series with the third-phase, and then connected to two phases of the rated voltage grid. Then by rationally matching the voltage and current of three phases, the parallel phases can get a reactive current hedging under both the rated voltage and rated current, meanwhile three phases maintain their active power balance. As a result, STATCOM gets a phase full-power tested phase by phase. The simulation results in Matlab/Simulink show that, under the proposed test system, both the voltage and current of the parallel two phases get their rated values while the grid current is only about 3% of the rated current, meanwhile the DC-link voltage of each phase converter is stabilized. Compared with other testing methods for STATCOM, this method requires neither extra hardware nor high-capacity power supply to construct the test platform, but it can simultaneously examine both the entire main circuit and a large part of the control system in STATCOM. Therefore, it provides a cost-effective engineering method for the factory test of high-power STATCOM.

Keywords: full-power testing; high-power; individual phase; operation test; static synchronous compensator (STATCOM)

---

1. Introduction

Static synchronous compensator (STATCOM) is an effective solution for fast voltage/reactive power support during normal and faulty conditions [1,2]. During the last decades, much research has been carried out on its structure, control, modulation, and application [3–7]. For medium-voltage application, STATCOM usually adopts the multi-level converter topology, including the flying-capacitor multi-level converter [8], diode-clamped multilevel converter [9], and cascaded multi-level converter [10–12]. Since the modular multilevel converter (MMC) topology was proposed and applied [13], MMC-based STATCOM has gained growing attention because of its outstanding performance, such as a low harmonic, low losses, scalability, and redundant design [14–16]. Especially with the development of new energy power generation, it has obtained wide applications and has become the standard configuration in wind farms and photovoltaic power plants to improve low voltage ride through capability [17,18]. In most applications, its capacity is generally 10–100 Mvar. In 2018, South Korea built three ±400 Mvar STATCOMs, which are the world largest [19].

To ensure the safe, reliable, and stable operation of a newly manufactured STATCOM, a complete set of pre-operation tests are necessary. The standard IEEE P1052/08 [20] contains a special chapter to guide STATCOM testing, and specifies the factory test items of STACOM components, such as switching devices and controls. IEC 62927 defined the detailed requirements of electrical testing for STATCOM converter [21]. It lists many test items and objectives of value or value section, but rarely gives the corresponding test circuit or method. For the testing of STATCOM controls, a real-time digital simulator (RTDS) is usually employed to form a hardware- in-the-loop test system where the main circuit part is digitally modeled [22,23]. It can comprehensively test the control and protection systems, but cannot evaluate any main power component. For the testing of STATCOM value, the common method is to employ two submodules to form a back-to-back test platform [24,25]. It can check the converter adequacy with regard to current, voltage, and temperature stresses in various operation conditions. But it requires an extra DC voltage source connected to the DC-bus of one submodule, so that the active power loss of the submodules during operation can be supplemented. It applies to submodules rather than value section, which consist of several submodules, because the latter does not have a common DC-bus. In fact, the most critical difficulty for value section tests is the requirement for energy supplements. A typical test method for value section is that two value sections are connected in parallel to a single-phase source. One section generates inductive reactive current, while the other generates capacitive reactive current; meanwhile the voltage source provides a certain current to supplement the active losses of value sections. Because the medium voltage distribution network is usually three-phase three-wire system, this method requires a single-phase high-power voltage source instead of the grid, resulting in higher test platform cost. In [26], an equivalent testing method was presented for full-power testing of STATCOM converter. It employs an additional high-voltage DC source and two low-voltage DC sources besides two tested STATCOM converters. These DC sources are used to charge the tested converters and compensate their active loss during operation. This also leads to high platform cost and a large footprint.

All these testing methods are for individual component of STATCOM, but not for a complete machine. But the performance of STATCOM depends on, not only individual components, but also component assembly, components wiring, and mutual coordination among devices. Therefore, it is necessary to conduct a whole machine test for a newly manufactured STATCOM. For the testing of whole STATCOM, an intuitive idea is to directly carry out a full-power test in the factory. Since the STATCOM rating is usually far beyond the power supply capacity of a factory, such an idea is not feasible. An alternative option is to employ two STATCOMs to form a back-to-back test platform [27]. One is operated as test object to output the rated reactive current, while the other is used as an accompanying device to output the opposite reactive current to neutralize the former, thereby maintaining the tiny total grid current. It can test the actual performance of the whole STATCOM, but requires an extra STATCOM besides the tested one, resulting in very high cost and poor utilization. Moreover, it has high requirement for synchronous coordinated control of these two STATCOMs. Therefore, such a test is almost impossible in the practical applications. In [28], a current closed-loop test method for ±100 Mvar STATCOM was proposed, based on the principle of equal potential. Firstly, three-phase STATCOM was connected to the grid to charge, and then disconnected from the grid and short-circuit at the three-phase ports to conduct zero-voltage full-current closed-loop operation. It can conveniently verify the capability of current control as well as voltage insulation level of a complete STATCOM. However, under this method, STATCOM operates either, at high voltage or high current and, therefore, cannot test the actual operating characteristics at both the rated voltage and rated current. Moreover, it can only run for a short time so it cannot be used for long-term power assessment of the STATCOM.

To overcome these problems, this paper proposes an individual phase full-power testing method for STATCOM. In this method, three-phase STATCOM was reconstructed into a structure that two phases are in parallel and then series with the third-phase. Then, by rationally matching the voltage and current of each phase, the parallel two phases can steadily operate under both the rated voltage and current, thus realize a phase full-power test. The corresponding math relationship was analyzed and a double-loop control system was designed. A simulation was carried out in Matlab, and the simulation results show that STATCOM can stably operate under the proposed system.

2. Circuit Configuration and Operation Principle

The circuit configuration of the proposed testing method is shown in Figure 1. It can seem that one phase port of the normal STATCOM is disconnected from the grid, and then shorted to one of the other two phases. Its equivalent circuit model is shown in Figure 2, where the A-phase and B-phase arms are connected in parallel, and then in series, with the C-phase to a single-phase voltage source, whose amplitude equals the rated line voltage rather than the rated phase voltage.

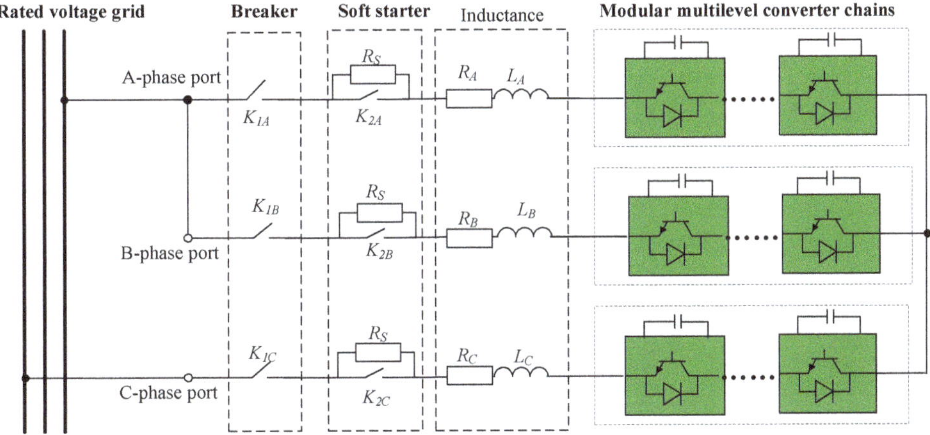

Figure 1. Circuit configuration of the proposed STATCOM testing method.

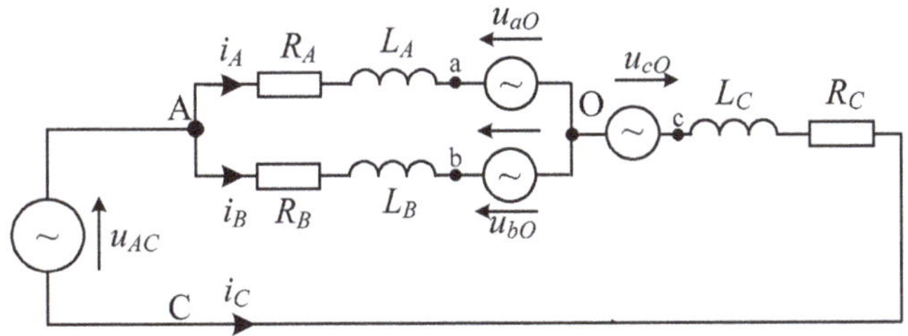

Figure 2. Equivalent simplified circuit diagram of the proposed testing system.

Under the aforementioned structure, three phase converters are individually controlled as follows:

(1) The output voltage of the C-phase converter was properly adjusted so that the amplitude of the A-phase voltage is just equal to its rated phase voltage, namely, the C-phase voltage is equal to the vector difference between the access line voltage and the expected A-phase voltage.

(2) The currents in the parallel two phases were controlled to perform a phase-to-phase reactive current hedging with the rated current amplitude, namely, one phase follows the rated capacitive reactive current, while the other with the rated inductive reactive current.

In this way, for the A-phase or B-phase arm (including the converter value and filter reactance), both its voltage and current reached the rated values, thus achieving a phase full-power operation. Moreover, since the reactive currents form a circulation between A-phase and B-phase arms, the grid current is tiny and close to zero, thus the power supply capacity of such test platform is relatively small and easy to implement.

Similarly, by changing the shorted phase, the full power test for B-phase or C-phase converter can also be realized. Then all of three phases are individually tested at phase full-power condition, so this method can be named as individual phase full-power testing method.

3. Mathematical Model of the Individual Phase Full-Power Testing System

3.1. Basic Relationship and Constraints

In the equivalent circuit model shown as Figure 2, the voltage and current of three phase arms meet the basic relationship as following:

$$\begin{cases} u_{AO} = R_A i_A + L_A \frac{d}{dt} i_A + u_{aO} \\ u_{AO} = R_B i_B + L_B \frac{d}{dt} i_B + u_{bO} \\ u_{CO} = R_C i_C + L_C \frac{d}{dt} i_C + u_{cO} \end{cases} \tag{1}$$

$$\begin{cases} i_C = -(i_A + i_B) \\ u_{CO} = u_{AO} - u_{AC} \end{cases} \tag{2}$$

where u_{AO}, u_{CO} represent the arm voltages of A-phase and C-phase (i.e., u_{AO} is the voltage difference between the A-phase port **A** and the internal neutral point **O** in Figure 2, similar is u_{CO}). i_A, i_B, and i_C are the arm currents. u_{aO}, u_{bO}, and u_{cO} are the output voltage of three converters (i.e., u_{aO} is the voltage difference between the port *a* in Figure 2 and the neutral point **O**, similar are u_{bO} and u_{cO}). R_A, R_B, and R_C are the equivalent resistance of each arm, L_A, L_B, and L_C are the inductance of each arm. While, u_{AC} represents the grid access voltage between the A-phase and C-phase ports, and depends mainly on the grid rather than the STATCOM.

Considering the defined operating conditions of this test system, the above voltages satisfy the following constraints: (1) The amplitude of A-phase voltage u_{AO} (not the A-phase converter output voltage u_{aO}) equals to the rated phase voltage of STATCOM to reach the rated voltage condition; (2) the amplitude of line voltage u_{AC} equals to the rated line voltage; (3) the amplitude of C-phase converter output voltage u_{cO} cannot exceed its maximum allowable range. Since the C-phase current in this test system is far less than its rated value, its voltage drop across resistance and inductance is very tiny. Thus, the amplitude of C-phase arm voltage u_{CO} is very close to that of C-phase converter voltage u_{cO}. In short, the amplitudes of the above voltages satisfy the following constraints:

$$\begin{cases} \|u_{AO}\| = U_{pN} \\ \|u_{AC}\| = U_{lN} = \sqrt{3} U_{pN} \\ \|u_{CO}\| \le U_{cmax} \end{cases} \tag{3}$$

where U_{pN}, and U_{lN}, respectively represent the amplitudes of the rated phase and line voltages of STATCOM. While, U_{cmax} represents the maximum output voltage of every phase converter, generally, it is larger than the rated phase voltage of STATCOM.

Ideally, there is only the rated amplitude of inductive/capacitive reactive current in A-phase and B-phase arms, while C-phase current is zero (because A-phase and B-phase arms constitute a circulating current). In fact, due to the loop resistance and switching device loss in every phase, each phase requires the absorption of a certain active power from the grid to compensate the resistance losses and convertor losses, and then maintain its active power balance, so that the DC-link voltage of each converter can remain stable. Under typical hardware parameters, the active current is about 1% of the rated current when each phase operates at its rated voltage and current. Although the active current has little effect on the amplitude of phase current, it is essential to maintain the active power balance. Therefore, in the steady state operation, both A-phase and B-phase currents contain reactive and active components, while C-phase current contains the negative sum of the active components in the A-phase and B-phase currents, and it would also contain the remaining reactive components that are not completely neutralized.

In terms of amplitude, the A-phase and B-phase currents are approximately equal to their rated value, while C-phase current is much smaller. Since the active loss of each phase is closely related to its current amplitude, in the steady state operation, the absorbed active power of three phases (not the active power of static converters) can be expressed as:

$$\begin{cases} P_A = u_{AO} \cdot i_A = U_{PN} I_{aP} > 0 \\ P_B = u_{BO} \cdot i_B = U_{PN} I_{bP} > 0 \\ P_C = u_{CO} \cdot i_C \approx 0 \end{cases} \tag{4}$$

where I_{aP}, and I_{bP}, respectively represent the active components in A-phase and B-phase currents.

Combining the constraints (2)–(4), we can plot the voltage and current vector diagram of the STATCOM as Figure 3. Here the A-phase voltage vector u_{AO} is selected as the orientation reference, and then according to the voltage constraints (3), the C-phase voltage vector u_{CO} can only be on the circle with point C as the center and the rated line voltage U_{lN} as the radius, moreover it has to be within the circle with point O as the center and the maximum converter voltage U_{cmax} as the radius. Combining the two requirements, the C-phase voltage vector has to be on the arcs CC′ in Figure 3.

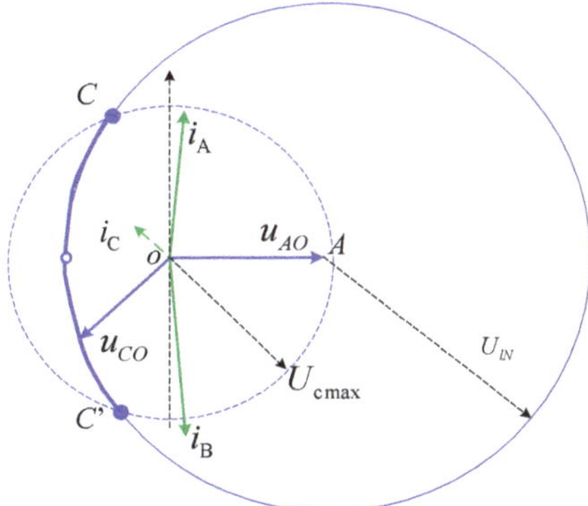

Figure 3. The basic voltage and current vector diagram of the testing system.

According to equation (4), the angle between the A-phase voltage u_{AO} and current i_A must be less than 90°, namely, the A-phase current must contains some positive active component, and the same is true for the B-phase voltage and current. At the same time, because the C-phase voltage and current are not at zero, in order to satisfy the third formula in equation (4), the C-phase voltage vector u_{CO} and current i_C must be basically vertical to each other. Based on the above constraints, we can deduce two conclusions:

(1) The C-phase voltage vector u_{CO} cannot be parallel to A-phase voltage u_{AO}. This is because both A-phase and B-phase currents contain some positive active component to compensate for the resistance loss and convertor loss, if C-phase voltage u_{CO} is parallel to A-phase voltage u_{AO}, the C-phase current cannot be vertical to the C-phase voltage vector u_{CO}, thus C-phase cannot get its active power balance. This shows that C-phase voltage and A-phase voltage cannot be simplified to a simple algebraic superposition relationship, and they have to be a vector superposition relationship.

(2) The reactive components in the A-phase and B-phase currents cannot be exactly offset, in other words, the size of their reactive components cannot be the same. This is because when the reactive components are exactly offset, the sum of A-phase and B-phase currents is in the same direction as the A-phase voltage u_{AO}, in that case the C-phase voltage u_{CO} cannot be vertical to the current i_C unless u_{CO} itself is perpendicular to u_{AO}, thus C-phase cannot get its active power balance.

3.2. Recommended Voltage-Current Combination

In the range satisfying the foregoing constraints, there are many alternative combinations of voltage and current. As an example, here we choose a representative combination as Figure 4, and their math expressions are as follows:

$$\begin{cases} u_C = U_{pN}e^{j(-120°)} \\ i_A = I_{aP} + jI_{pN} \\ i_B = I_{bP} - jI_{pN} + jI_{add} \end{cases} \tag{5}$$

where I_{add} represents an additional reactive component in the B-phase current. It is usually negative in steady state but can also be positive or zero during some dynamic process.

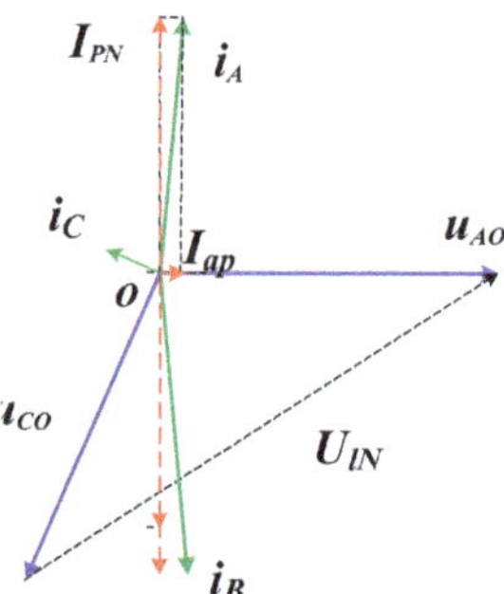

Figure 4. The voltage and current vector diagram of the recommended voltage-current combination.

Substituting Equation (5) into (2), the C-phase current can be calculated as:

$$i_C = -(i_A + i_B) = -(I_{aP} + I_{bP}) - jI_{add} \tag{6}$$

while the C-phase absorbed active power from the outside can be calculated as:

$$\begin{aligned} P_C &= u_{CO} \cdot i_C = \left[U_{pN}\cos(-120°) + jU_{pN}\sin(-120°) \right] \cdot \left[-(I_{aP} + I_{bP}) - jI_{add} \right] \\ &= \tfrac{1}{2}U_{pN}(I_{aP} + I_{bP}) + \tfrac{\sqrt{3}}{2}U_{pN}I_{add} \end{aligned} \tag{7}$$

As in (7), by adjusting the additional component I_{add} in the B-phase current, the active power absorbed by the C-phase arm can be regulated in order to achieve the stability of the DC-bus voltage of C-phase converter. On the other hand, for B-phase arm, this additional component I_{add} is the reactive current component, so it has no effect on the active power balance of B-phase arm. Of course, it has yet no effect on the active power balance of A-phase arm. In other words, we can independently regulate the C-phase convertor DC voltage without affecting the other two phases.

4. Control System Design

4.1. Control System Structure

Based on foregoing analysis, the control structure of the proposed testing system is constructed as Figure 5. Firstly, a phase locked loop (PLL) was employed to extract the phase information of the access port voltage, and then generated a plurality of phase references. Secondly, the individual phase instantaneous control was used to calculate the required voltage of every-phase converter [29,30]. Finally, the carrier phase shifting pulse width modulation (CPS-PWM) [31,32] was used to generate the required drive signals for each MMC submodule. The specific structure is as follows.

(1) In the phase-locked unit, the phase of the access port voltage u_{AC} was directly detected, and then minus 30° to be as the phase reference of A-phase voltage, and further minus 120° to be as the phase reference of C-phase voltage. This is because in the selected voltage-current relationship shown in Figure 4, the A-phase voltage u_{AO} lags the access port voltage u_{AC} by 30°, and it is ahead of C-phase voltage u_{CO} by 120°.

(2) In the individual phase instantaneous control, both A-phase and B-phase adopt a double loop control structure. The A-phase current reference contains two components. One is to generate the required reactive current for power evaluation while the other is to regulate the A-phase total DC voltage. The B-phase current reference consists of three components, the first part is to offset the A-phase reactive current, the second is to regulate the B-phase total DC voltage, and the third is to regulate the C-phase total DC voltage. The C-phase uses an open-loop voltage control to directly calculate the required C-phase output voltage according to the given amplitude and phase.

(3) The CPS-PWM unit distributes the drive signals of every submodule according to the each phase total voltage demand calculated by the individual phase controller, and it also includes a voltage balancing control among different sub-modules in each phase chain, and such balancing control has been studied in many literatures [33–37]. Because this unit is basically the same as the conventional STACOM, here it will not be described again.

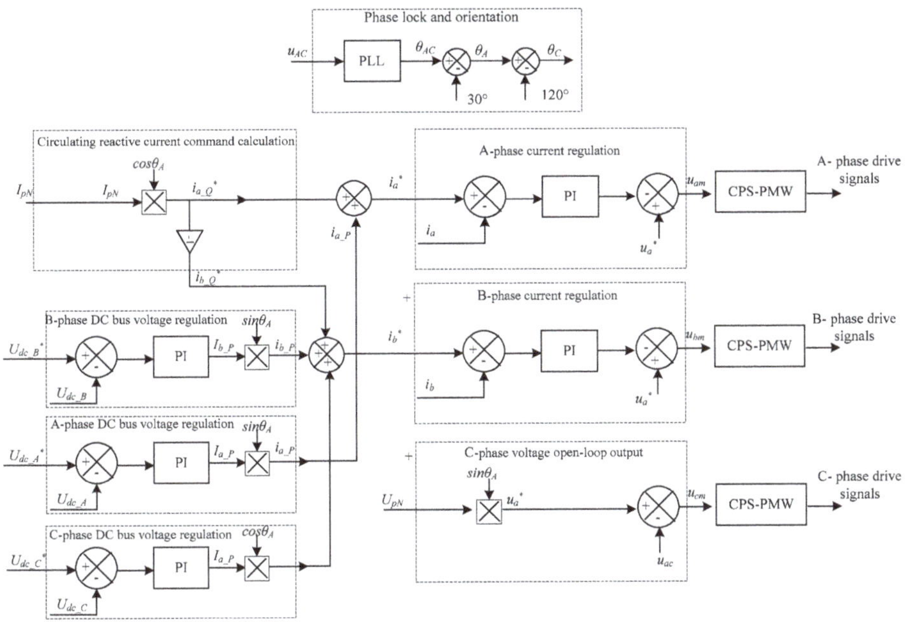

Figure 5. The control system block of the proposed testing system.

4.2. Soft Power-on Process for the Individual Phase Full-Load Testing System

Since the circuit configuration of the individual phase full-power testing system is different from conventional three-phase STATCOM, the conventional soft power-on process is no longer applicable for the individual phase testing system. To solve this problem, the paper proposes a soft power-on method applicable for this configuration, and shown as Figure 6.

(1) Before powering up the main part of the tested STATCOM, firstly block the three-phase converters and put in the three-phase soft power-up resistor.

(2) Then turn on the grid breaker to start the power-on process, thus the grid line voltage charges the DC capacitors of A/B/C three-phase converters by uncontrolled rectification.

(3) When DC-bus voltage of serial phase (here C-phase) converter exceeded a set threshold U_f, bypass C-phase converter while keep both A-phase and B-phase converters blocked. Thus, the grid-line voltage charges the DC-link capacitors of A-phase and B-phase converters by uncontrolled rectification, while the DC voltage of C phase converter keeps its current value.

(4) When the DC-bus voltage of parallel phases converters has exceeded their set threshold Udc, bypass the A/B phase converters while keep C-phase converters blocked. Thus, the grid line voltage charges the DC capacitors of C phase converter by uncontrolled rectification. While, the DC voltage of A/B phase converters keeps its current value.

(5) When the DC-bus voltage of C-phase converter has exceeded its set threshold U_{dcN}, there is a bypass of the soft start resistor by turning on the switches, and then end the soft power-on process.

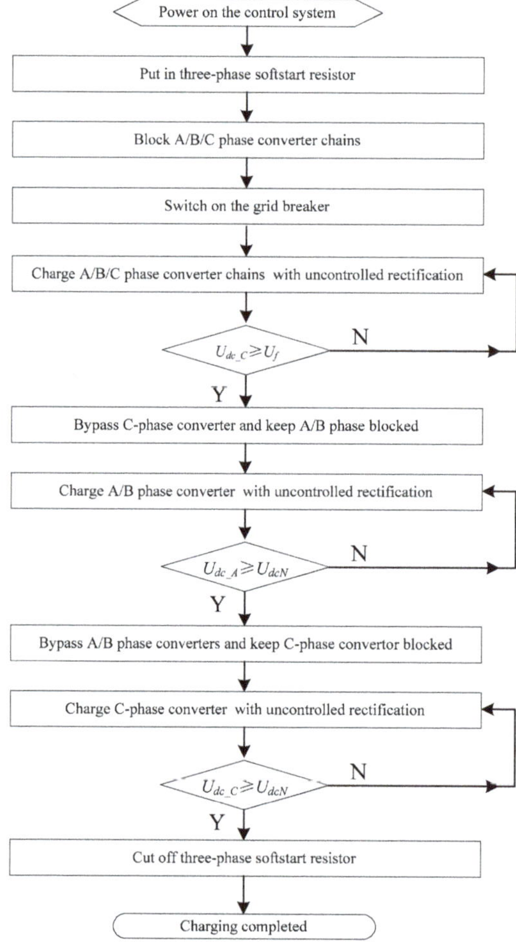

Figure 6. The flow diagram of soft power-on process for the individual phase full-load testing system.

5. Simulation Verification

In order to verify the proposed STATCOM testing system, a simulation model was built in Matlab/Simulink. Its circuit configuration and control structure are shown as Figures 1 and 5, and the main simulation parameters are listed in Table 1.

In the simulation process, three different operating conditions are set:

(1) During $t = 0$–1.6 s, the STATOM performs the proposed soft power-on process as Figure 6.
(2) At $t = 1.6$ s, the proposed control system shown as Figure 5 is put into operation. And the DC-bus voltage reference of each phase is stepped from 10 kV to 10.5 kV at $t = 1.6$ s to verify the proposed DC voltage regulation, while the reactive current of A-phase is still kept zero.
(3) From the time, $t = 1.8$ s, the A-phase reactive current reference steps to its rated value.

The simulation results are shown in Figures 7–10.

Table 1. Parameters in the simulation model.

Parameter	Symbol	Value
Rated rating	S_N	±10 Mvar
Rated line voltage	U_{IN}	10 kV
Rated phase voltage	U_{PN}	5.7 kV
Rated current	I_{PN}	0.57 kA
Arm resistance	$R_A, R_B, R_C,$	0.1 Ω (0.01 pu)
Arm reactance	$L_A, L_B, L_C,$	3.2 mH (0.1 pu)
Submodule number in each phase	N	4
DC-link capacitance of each submodule	C_{dc}	40 mF
Rated DC-bus voltage of each phase converter	U_{dcN}	10.5 kV
Switching frequency of submodule	f_s	500 Hz
Proportional coefficient of the current-loop controller	K_{p_c}	200
Proportional coefficient of the voltage-loop controller	K_{p_v}	4
Integral coefficient of the voltage-loop controller	K_{i_v}	50

As shown in Figure 7, in the soft power-on process of 0–1.6 s, the DC-bus voltage of each phase converter rises smoothly to its expected value, and each current is far below the rated value. During this process, there are four stages: Firstly, both three phase converters were charging; secondly, A-phase converter charging; thirdly, C-phase converter charging; fourthly, no-load standby. This proves the feasibility of the proposed soft power-on method. It should be noted, in order to speed up the simulation so as to show the overall process in a limited time, the starting resistor in the simulation model is set to be small as 25 Ω. Generally, the actual soft-starting resistance is larger, thus the corresponding inrush current is smaller and the power-on process would be smoother.

As shown in Figure 8, during the transient process when the DC voltage reference steps at $t = 1.6$ s, three phase voltages are significantly different from the soft start process, but both their amplitude and phases are in line with their expectations, as shown Figure 4. The DC-bus voltages of three phase converters gradually rise to their set values and then remain stable, and each phase current is also gradually reduced from the initial sinusoidal waveform to near zero. During this period, both the amplitude and phase of A-phase current are different from that of B-phase current. This is because the B-phase current contains an additional component to regulate the C-phase DC-bus voltage.

As shown in Figure 9, during the third stage, three-phase currents quickly reach their steady state and then remain stable, while three-phase DC voltages remain stable. Among them, both A-phase and B-phase current quickly reach their rated value, their amplitudes are substantially equal, and their phases are approximately opposite. At the same time, the C-phase current is far less than the A-phase current. In Figure 9, the A-phase and C-phase voltage deals with a low-pass filter to show the low frequency components. In order to show the harmonic distortion in the proposed system, the original voltage and current without filtering are given in Figure 10. It can be seen that the fundamental component of the C-phase current is about 3% of the rated value (23 A/800 A = 2.9%).

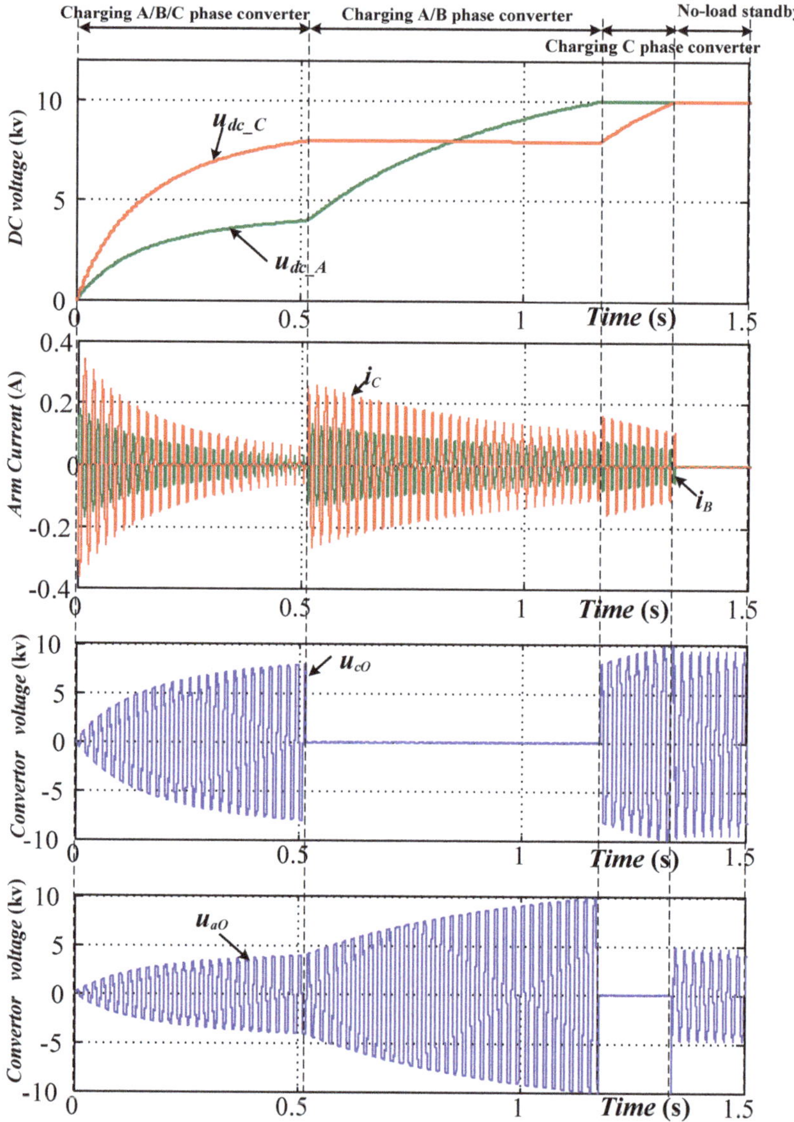

Figure 7. Simulation results of the proposed soft power-on process.

As in Figure 10 shown, the original output voltages of three-phase converters are typical multi-level waveform. Because the number of submodules in this simulation are only four in each phase, there are various high-frequency ripples. It can be expected that when the submodules number increases, the output voltage would be smoother, thus the current ripple will be smaller. Although, there are obvious high-frequency components in A-phase and C-phase voltage, their fundamental components are still consistent with the expected values, indicating that these high-frequency harmonic components do not affect the feasibility of the testing system.

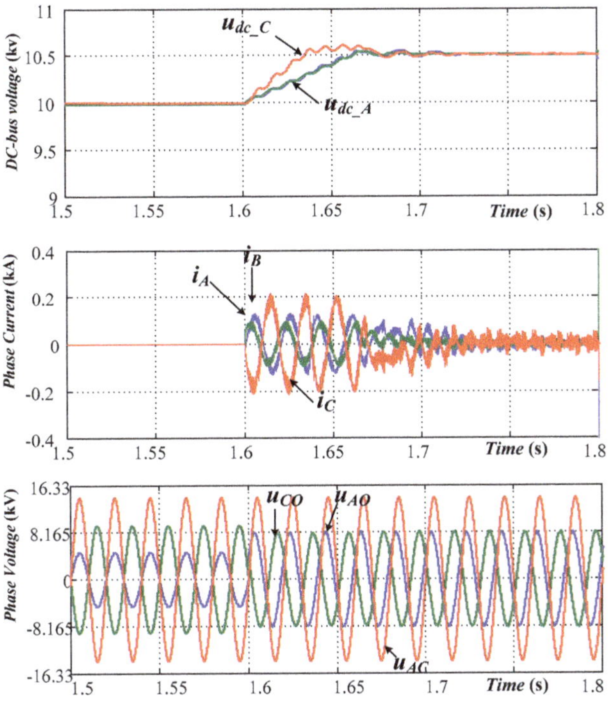

Figure 8. Simulation results when the DC-bus voltage reference is step.

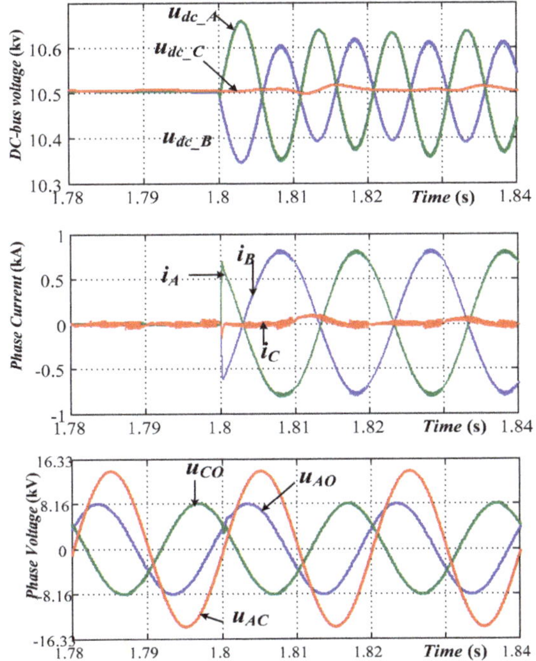

Figure 9. Simulation results at the individual phase full-power operation state.

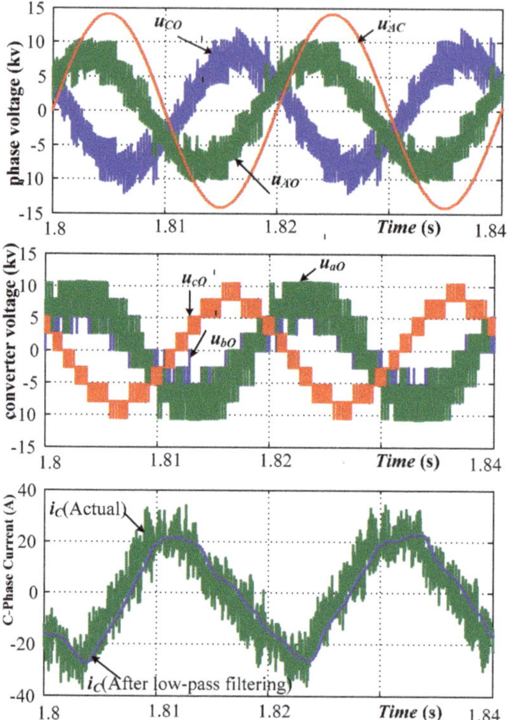

Figure 10. Original voltage and current without filtering at the full-power operation state.

6. Conclusions

To address the factory test of high power STATCOM, a phase full-power testing method is present in this paper. By changing the port connection, three-phase STATCOM was reconstructed into a structure that two phases are in parallel, and then in series with the third-phase, and then connected to two phases of the rated voltage grid. Then, by rationally matching the three-phase voltage and current, the parallel two phases can run a reactive current hedging under both their rated voltage and rated current. The difficulty with this method is in maintaining the active balance of three-phase converters under this special structure. By mathematical modeling and theoretical analysis, it can be concluded that, the output voltage of the series phase converter cannot be simply set to the algebraic difference of the rated line voltage and phase voltage. Moreover, the reactive current components of the two parallel phases cannot be exactly offset, otherwise three phase convertors cannot maintain their active power balance. A specific combination of three-phase voltage and current is designed to maintain the steady operation of the system, and a corresponding control system is designed. Since the circuit configuration of the proposed testing system is different from the conventional STATCOM, this paper proposes a novel soft power-on method for this structure. The testing system was verified by a simulation in Matlab/Simulink. The simulation results show that, the proposed system can run stably, and two phases can operates at both, the rated phase voltage, and rated current, while the grid current is about 3% of the rated current. As a result, the three-phase STATCOM can experience a phase full-power test, phase-by-phase.

The proposed method has a certain similarity with the common method, where the two singe-phase convertors are in parallel and then connected to the single-phase grid, since it also employs two convertors to perform reactive power hedging. But the novelty of this method is that it uses the third-phase convertor of STATCOM to compensate for the voltage difference between the power supply voltage and the required phase voltage, so that the power voltage does not need to be directly equal to the required voltage for STATCOM testing. As a result, the existing power grid can be used to directly provide the test power, thereby avoiding the additional platform hardware and its cost. The main contribution of this paper is to clearly explain the constraint among three-phase voltage, current, and active power under such special structure, and solve the problem of charging and active power balance of three phase converters, and design a control system to realize it. This makes the aforementioned idea achievable.

Compared with other methods for the testing of STATCOM, the most prominent feature of this method is that its test platform only needs one tested STATCOM and a small capacity rated voltage grid, meaning no extra hardware is required and a lower test cost. Therefore, it is suitable for the whole-machine test of high-power STATCOM before its leaving the factory.

Author Contributions: Conceptualization, Q.H.; formal analysis, Q.H.; data curation, X.M.; writing-review and editing, Q.H. and Y.T.; software Y.Z. and S.Z.; project administration, B.L.

Funding: This research was partly funded by the science and technology project of State Grid Electric Corporation grant number 5216A016000P, and partly funded by the science and technology project of State Grid Hunan Electric Company Limited grant number 5216AF17000A.

Conflicts of Interest: The authors declare no conflict of interest.

References

1. Liang, Y.; Nwankpa, C.O. A New type of STATCOM based on cascading voltage source inverters with phase-shifted unipolar SPWM. *IEEE Trans. Ind. Appl.* **1999**, *35*, 1447–1453.
2. Schauder, C.; Gernhardt, M.; Stacey, E.; Lemak, T.; Gyugyi, L.; Cease, T.; Edris, A. Development of a ±100 MVAr static condenser for voltage control of transmission systems. *IEEE Trans. Power Deliv.* **1995**, *10*, 1486–1496. [CrossRef]
3. Jain, A.; Joshi, K.; Behal, A.; Mohan, N. Voltage regulation with STATCOMs: Modeling, control and results. *IEEE Trans. Power Deliv.* **2006**, *21*, 726–735. [CrossRef]
4. Rao, P.; Crow, M.; Yang, Z. STATCOM control for power system voltage control applications. *IEEE Trans. Power Deliv.* **2000**, *15*, 1311–1317. [CrossRef]
5. Barrena, J.A.; Marroyo, L.; Vidal, M.Á.R.; Apraiz, J.R.T. Individual voltage balancing strategy for PWM cascaded H-bridge converter-based STATCOM. *IEEE Trans. Ind. Electron.* **2008**, *55*, 21–29. [CrossRef]
6. Townsend, C.D.; Summers, T.J.; Betz, R.E. Phase-shifted carrier modulation techniques for cascaded H-bridge multilevel converters. *IEEE Trans. Ind. Electron.* **2015**, *62*, 6684–6696. [CrossRef]
7. Singh, B.; Al-Haddad, K.; Saha, R.; Chandra, A. Static synchronous compensators (STATCOM): A review. *IET Power Electron.* **2009**, *2*, 297–324. [CrossRef]
8. Fujii, K.; De Doncker, R.W. Optimization of soft-switched flying capacitor multi-level converters applied to STATCOMs. In Proceedings of the 2007 European Conference on Power Electronics and Applications, Aalborg, Denmark, 2–5 September 2007; pp. 1–10.
9. Marchesoni, M.; Tenca, P. Diode-clamped multilevel converters: A practicable way to balance DC-link voltages. *IEEE Trans. Ind. Electron.* **2002**, *49*, 752–765. [CrossRef]
10. Lv, J.; Gao, C.; Chen, S.; Liu, X.; Chen, Z. A novel STATCOM based on diode-clamped modular multilevel converters. *IEEE Trans. Power Electron.* **2017**, *32*, 5964–5977.
11. Peng, F.; Lai, J. Dynamic performance and control of a static VAr generator using cascade multilevel inverters. *IEEE Trans. Ind. Appl.* **1997**, *33*, 748–755. [CrossRef]
12. Cheng, Y.; Qian, C.; Crow, M.L.; Pekarek, S.; Atcitty, S. A comparison of diode-clamped and cascaded multilevel converters for a STATCOM with energy storage. *IEEE Trans. Ind. Electron.* **2006**, *53*, 1512–1521. [CrossRef]
13. Lesnicar, A.; Marquardt, R. An innovative modular multilevel converter topology suitable for a wide power range. In Proceedings of the 2003 IEEE Bologna Power Tech Conference Proceedings, Bologna, Italy, 23–26 June 2003.

14. Pereira, M.; Retzmann, D.; Lottes, J.; Wiesinger, M.; Wong, G. An MMC STATCOM for network and grid access applications. In Proceedings of the 2011 IEEE Trondheim Power Tech Conference, Trondheim, Norway, 13–23 June 2011; pp. 1–5.

15. Inoue, S.; Yoshii, T.; Akagi, H. Control and performance of a transformerless cascade PWM STATCOM with star configuration. *IEEE Trans. Ind. Appl.* **2007**, *43*, 1041–1049.

16. Bina, M.T. A Transformerless Medium-voltage STATCOM topology based on extended modular multilevel converters. *IEEE Trans. Ind. Electron.* **2011**, *26*, 1534–1545.

17. Popavath, L.; Kaliannan, P. Photovoltaic-STATCOM with low voltage ride through strategy and power quality enhancement in a grid integrated wind-PV system. *Electronics* **2018**, *7*, 51. [CrossRef]

18. Hossain, M.J.; Pota, H.R.; Ugrinovskii, V.A.; Ramos, R.A. Simultaneous STATCOM and pitch angle control for improved LVRT capability of fixed-speed wind turbines. *IEEE Trans. Sustain. Energy* **2010**, *1*, 142–151. [CrossRef]

19. Park, J.; Yeo, S.; Choi, J. Development of ±400Mvar World Largest MMC STATCOM. In Proceedings of the 21st International Conference on Electrical Machines and Systems (ICEMS), Jeju, Korea, 7–10 October 2018.

20. IEEE. *Approved Draft Guide for the Functional Specifications for Transmission Static Synchronous Compensator (STATCOM) Systems*; Institute of Electrical and Electronics Engineers (IEEE): New York, NY, USA; Available online: https://ieeexplore.ieee.org/document/STDAPE23348 (accessed on 3 July 2019).

21. IEC. *Voltage Sourced Converter (VSC) Valves for Static Synchronous Compensator (STATCOM): Electrical Testing*; International Electrotechnical Commission (IEC): Geneva, Switzerland; Available online: https://webstore. iec.ch/publication/33207 (accessed on 3 July 2019).

22. Guan, R.; Xue, Y.; Zhang, X.P. Advanced RTDS-based studies of the impact of STATCOM on feeder distance protection. *J. Eng.* **2018**, *2018*, 1038–1042. [CrossRef]

23. Langston, J.; Qi, L.; Steurer, M.; Sloderbeck, M.; Liu, Y.; Xi, Z.; Mundkur, S.; Liang, Z.; Huang, A.Q.; Bhattacharya, S.; et al. In Testing of a controller for an ETO-based STATCOM through controller hardware-in-the-loop simulation. In Proceedings of the 2009 IEEE Power & Energy Society General Meeting, Calgary, AB, Canada, 1–8 July 2009.

24. Liu, W.H.; Song, Q.; Zhang, D.J.; Chen, T.J.; Teng, L.T.; Zheng, D.R. Equivalent Tests of Links of 50 MVA STATCOM. *Trans. China Electrotech. Soc.* **2006**, *21*, 73–78.

25. Woodhouse, M.L.; Donoghue, M.W.; Osborne, M.M. *Type Testing of the GTO Valves for a Novel STATCOM Converter*; IEE: London, UK, 2001; pp. 84–90.

26. Zhang, Y.; Li, Q.; Zhang, S. Converter chain equivalent experiment for STATCOM of large capacity based on dc power supply. *J. Power Supply* **2017**, *5*, 94–99.

27. Xie, Y.; Zhuo, F. Factory-used energy-saving testing method for STATCOM. *Shanxi Power* **2013**, *11*, 1–4.

28. Liu, D.; Zhang, Y.; Ma, J.; Liu, X.; Gao, J. Current closed-loop test method for ±100 mvar high-voltage STATCOM in weak grid based on principle of equal potential. *J. Power Supply* **2018**, *1*, 119–124.

29. Shi, Y.; Liu, B.; Shi, Y.; Duan, S. Individual phase current control based on optimal zero sequence current separation for a star-connected cascade STATCOM under unbalanced conditions. *IEEE Trans. Power Electron.* **2016**, *31*, 2009–2100. [CrossRef]

30. Hatano, N.; Ise, T. Control scheme of cascaded H-bridge STATCOM using zero-sequence voltage and negative-sequence current. *IEEE Trans. Power Deliv.* **2010**, *25*, 543–550. [CrossRef]

31. Li, B.; Yang, R.; Xu, D. Analysis of the phase-shifted carrier modulation for modular multilevel converters. *IEEE Trans. Power Electron.* **2015**, *30*, 297–310. [CrossRef]

32. Jing, N.; He, Y. Phase-shifted suboptimal pulse-width modulation strategy for multilevel inverter. In Proceedings of the 2006 1st IEEE Conference on Industrial Electronics and Applications, Singapore, 24–26 May 2006.

33. Deng, F.; Chen, Z. Voltage-balancing method for modular multilevel converters under phase-shifted carrier-based pulsewidth modulation. *IEEE Trans. Power Electron.* **2015**, *62*, 4158–4169. [CrossRef]

34. Hagiwara, M.; Akagi, H. Control and experiment of pulse width modulated modular multilevel converters. *IEEE Trans. Power Electron.* **2009**, *24*, 1737–1746. [CrossRef]

35. Qin, J.; Saeedifard, M. Reduced switching-frequency voltage balancing strategies for modular multilevel HVDC converters. *IEEE Trans. Power Deliv.* **2013**, *4*, 2403–2410. [CrossRef]

36. Deng, F.; Chen, Z. A control method for voltage balancing in modular multilevel converters. *IEEE Trans. Power Electron.* **2014**, *1*, 66–76. [CrossRef]
37. Han, C.; Huang, A.Q.; Liu, Y.; Chen, B. A generalized control strategy of per-phase DC voltage balancing for cascaded multilevel converter-based STATCOM. In Proceedings of the 2007 Power Electronics Specialists Conference, Orlando, FL, USA, 17–21 June 2007.

© 2019 by the authors. Licensee MDPI, Basel, Switzerland. This article is an open access article distributed under the terms and conditions of the Creative Commons Attribution (CC BY) license (http://creativecommons.org/licenses/by/4.0/).

Article

Shaping SiC MOSFET Voltage and Current Transitions by Intelligent Control for Reduced EMI Generation

Congwen Xu, Qishuang Ma *, Ping Xu and Tongkai Cui

School of Automation Science and Electrical Engineering, Beihang University, Beijing 100191, China;
xucongwenbuaa@163.com (C.X.); xu_ping@buaa.edu.cn (P.X.); cuitongkai@buaa.edu.cn (T.C.)
* Correspondence: qsma304@126.com

Received: 11 April 2019; Accepted: 2 May 2019; Published: 8 May 2019

Abstract: In power converters, the fast switching of the power conversion components results in rapid changes in voltage and current, which results in oscillations and high-level electromagnetic interference (EMI), so the power components become a source of internal electromagnetic interference. Taking SiC Metal-Oxide-Semiconductor Field-Effect Transistor (MOSFET) as an example, an intelligent control method to suppressing interference sources is proposed in this paper. The combination of open-loop and closed-loop methods can simultaneously reduce the electromagnetic interference generated by voltage and current. Firstly, this paper analyzes how to select a reference signal. The relationship between the time domain and the frequency domain of the noise signal is analyzed. The convolution of the trapezoidal signal and the Gaussian signal is selected as the reference signal, which is named S-shaped signal in this paper. The S-shaped signal has continuous infinitely conductive characteristics, so its spectrum has a large attenuation in the high frequency region. Secondly, a new topology is proposed. Based on the closed-loop gate control, a current control signal is added, which can simultaneously shape the output voltage and control the output current slope. Both the simulation results and the experimental results show that the output voltage can follow the reference signal, S-shaped signal, and the slope and overshoot of output current can be changed. Compared with classical gate driver method, the spectrum of output voltage and output current obtained by the method proposed in this paper has a large attenuation, in other words, the electromagnetic interference is significantly reduced.

Keywords: power converters; EMI; intelligent control; classical gate driver; interference sources

---

1. Induction

Power converters composed of SiC MOSFETs are widely used in home, aerospace, and other fields due to their high operating frequency, high voltage, and high power density. However, due to the high dv/dt and di/dt of SiC MOSFET for high-speed switching, electromagnetic interference (EMI) is generated so that the SiC MOSFET becomes an electromagnetic interference source [1]. High dv/dt may produce common mode interference signals in power electronic circuits, while higher di/dt can cause higher voltage and current overshoot and oscillation due to parasitic factors in the circuit. The EMI signal can easily affect external sensitive components by radiation or conduction. It is common to reduce conducted interference by adding filters in the conduction path and radiated interference by shielding sensitive components. However, these two methods will take up more PCB space and increase the cost greatly. It is a challenge for engineers to balance the suppression effect of electromagnetic interference and costs [2]. Therefore, it is an interseting research direction that is suppressing EMI by controlling the fast switching transient [3]. Changing gate driving current [4], and using a built-in snubber [5] all have limited effect for suppressing EMI. Active voltage control (AVC) method is applied to control the slope dv/dt of the output voltage, which is effective to control the switching transients of IGBT to reduce the EMI [6]. References [7–9] introduced how to predict the EMI for the defined edges of signal and what the influence of the defined reference signal is for the transients of IGBT, which provided the possibility to shape the Gaussian switching transients to suppress EMI. The active gate drive (AGD) method is an effective method for suppressing electromagnetic interference from the viewpoint of suppressing electromagnetic interference sources. The AGD method mainly includes three methods: controlling the gate current, controlling the gate voltage, and controlling the gate resistance. References [10,11] applied a method of controlling the gate current to control the switching transient. During the turn-off period of SiC MOSFET, a large current is injected into the gate to increase dv/dt to reduce the turn-off time and the turn-off loss is reduced thereby. When the output current drops, the rapidly decreasing gate current reduces the di/dt of the output current and the oscillation of the voltage generated by the inductance and parasitic capacitance is reduced. However, the dv/dt of the output voltage is not adjustable.

Reference [12] proposes a method of controlling the gate current by closed-loop and adding a delay compensation device to control the gate source turn-off voltage slope dv/dt and current slope di/dt, thereby achieving the trade-off between switching losses and electromagnetic interference. However, this method also has its limitations, because it can only control dv/dt and di/dt when MOSFET is closed. A new method of active gate drive method is proposed Reference [13], the principle of which is to increase the gate resistance value during the Miller plateau of the gate voltage during the turn-on and turn-off of the SiC MOSFET. This control method can control di/dt during turn-on and dv/dt during turn-off, respectively, and its switching loss is low. This control method can control di/dt during turn-on and dv/dt during turn-off, respectively, and its switching loss is low but this method cannot control dv/dt during turn-on and di/dt during turn-off. The precondition for this method to work properly is to select the correct switching time of gate resistance and the gate voltage cannot have large fluctuations, however, which is difficult to implement in engineering. In [14], the closed-loop control method is applied and di/dt and dv/dt are measured and then compared with reference signals to get the error. The error is input to the buffer pole through a differential op amp to dynamically adjust the gate voltage. The power semiconductor is turned on by the waveform following the reference signal. However, it is impossible to adjust dv/dt without di/dt restrictions because the same reference and the same gain are used [4]. All of the above methods of controlling a single dv/dt or di/dt have limited suppression of high frequency noise, and fail to achieve simultaneous control of voltage and current turn-on and turn-off. The results of reference [1] indicate that any switching waveform can be expressed in the form of convolution that it is get by an ideal square wave signal and a normalized instantaneous function derivative. From a frequency domain perspective, the switching waveform is the product of the square wave signal spectrum and the instantaneous function derivative spectrum. Reference [15] shows that if the transient function can be derived n times, the high-frequency spectral envelope of the switching waveform will be attenuated by a slope of −20(n + 1) dB/dec. The Gaussian signal has continuous infinitely conductive characteristics, so that the high frequency spectrum of the Gaussian waveform is attenuated faster than the high frequency spectrum of the single slope theoretically. Therefore, in [2], the convolution waveform of the square wave signal and the Gaussian signal as the reference signal is used, and the switching voltage following the reference voltage by the method of closed-loop active voltage control is realized.

This paper proposed an intelligent control method based on the characteristic that voltage and current are not synchronized during SiC MOSFET switching. The intelligent control method that include a closed-loop circuit to shape the output voltage and an open-loop circuit to control the slope of the output current control the switching voltage and current simultaneously.

The article is organized as follows: Theoretical analysis of voltage reference signals and current control signals is discussed in the Section 2. An intelligent control method combining closed-loop and open-loop control methods is proposed in the Section 3. Simulation and experimental verification were performed based on the proposed method in Sections 4 and 5. The conclusion of the experiment is presented in Section 6.

2. Theoretical Analysis

2.1. Selection of Voltage Reference Signal

In order to make the output voltage have a smooth transient during turn-on and turn-off period, it is a feasible method to select an appropriate reference signal in closed-loop control circuit. Therefore, the paper discussed how to select the reference signals first. Reference [1] discussed the effects of time domain parameters on the frequency domain envelope of the signal. The relationship between the number of derivatives of the transient signal in the time domain and the envelope of the high frequency component of the signal in the frequency domain is discussed in [15] that plays an important role to guide this paper. According to [16], a brief review of how to select the reference signal is shown as follows.

It is assumed that the switching transient waveform is represented by sw(t), and its image is shown in Figure 1a. Its expression is assumed as follow Equation (1)

$$\begin{aligned} t < 0:\ &sw(t) = 0 \\ 0 < t < \tau:\ &sw(t) = s(t) \\ \tau < t < t_0:\ &sw(t) = A \\ t_0 < t < t_0 + \tau:\ &sw(t) = A - s(t - t_0) \\ t_0 + \tau < t < T:\ &sw(t) = 0 \end{aligned} \tag{1}$$

where s(t) is the switching transient function, A is the amplitude of the switching signal that is a constant, τ is the rise time and fall time, T is the switching period of the signal, t_0 is the signal width of square wave.

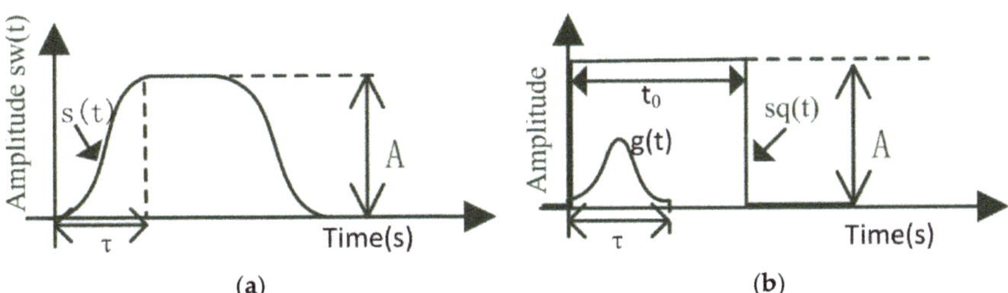

Figure 1. Switching signal (**a**) and structure (**b**).

Figure 1b explains how to get sw(t). The switching signal, sw(t), can be expressed as the convolution of the square wave signal sq(t) with 1/A*ds(t)/dt that is the normalized derivative of the transient function. The equation is

$$sw(t) = sq(t)*(1/A.ds(t)/dt) = sq(t)*g(t) \tag{2}$$

Perform the Fourier transform on Equation (2) and get the following Equation (3)

$$sw(f) = sq(f).g(f) \tag{3}$$

Obviously, it can be seen from Equation (3) that the switching signal sw(f) can be regarded as the product of the Fourier transform of a square wave signal and a shaped smoothing function.

According to [1], in the frequency domain, the slope of the envelope of sw(f) is closely related to the derivative times k of the function in the time domain. It is supposed that $sw^{(k)}(t)$ shows discontinuity and P Dirac pulses will appear in $sw^{(k+1)}(t)$ after the (k + 1) times derivative.

$$sw^{(k+1)}(t) = \sum_{i=1}^{P} A_i \delta(t - \tau_i) \tag{4}$$

τ_i and A_i are the times and amplitudes of the Dirac pulse. The expression of Fourier transform on (4) is

$$F_s^{(k+1)}(n) = \sum_{i=1}^{P} A_i e^{-jnw_0\tau_i} \tag{5}$$

where $w_0 = 2\pi/T$, $F_s(n)$ is the coefficient of the Fourier transform of sw(t). The expression of $F_s(n)$ is as Equation (6)

$$F_s(n) = \frac{\sum_{i=1}^{P} A_i e^{-jnw_0\tau_i}}{T(jnw_0)^{k+1}}, \; n \neq 0 \tag{6}$$

From Equation (6), high-frequency harmonics coefficients is get as Equation (7) when $n \to \infty$.

$$\lim_{n\to\infty} F_s(n) = \frac{\sum_{i=1}^{P} |A_i|}{T(nw_0)^{k+1}} \tag{7}$$

It can be seen from (7) that for an arbitrary waveform, the high-frequency boundary amplitude is determined by the sum of its derivative times k, P Dirac pulses amplitude and its fundamental frequency $f_0 = w_0/2\pi$ in the frequency domain. The attenuation envelope slope is $-20(k + 1)$ in the high-frequency domain. In order to verify the higher frequency, the derivatives times of the function, the better the high-frequency EMI suppression effect [17], the following examples are used.

We define the zero-order signal to be the square signal $q_0(t)$ and two-order signal to be the symmetrical trapezoidal signal. Infinite-order signal, $sw_\infty(t)$, can be expressed as the convolution of a square wave signal and a Gaussian signal. Their expressions are

$$sw_1(t) = q_0(t)^*g_0(t) \tag{8}$$

$$sw_\infty(t) = q_0(t)^*gs(t) \tag{9}$$

where $g_0(t)$ is a square signal, $gs(t) = \frac{1}{\sigma_t \sqrt{2\pi}} e^{-\frac{t^2}{2\sigma_t^2}}$ is a Gaussian signal. The mentioned zero-order signal and two-order signal that we selected both are symmetric signal. The duty cycle is D = 0.5, the switching frequency is 10 kHz, rise time is $t_{off} = 1$ μs, fall time is $t_{fall} = 10$ μs, $\sigma_t = \frac{1}{4}t_s = 0.25 \times 10^{-6}$.

The spectrum of $q_0(t)$, $sw_1(t)$ and $sw_\infty(t)$ are shown in Figure 2. It is worth mentioning that only time t is $|t| \to \infty$, the Gaussian function is $gs(t) \to 0$. In other words, the value of Gaussian function is not strictly equal to 0 in a finite period so there is a theoretical limitation. However, when $|t| \geq 2\sigma_t$, the Gaussian value is close to 0. Therefore, if $\sigma_t = (1/4)t_s$ is taken, the value of Gaussian function can be regarded as 0 at the two endpoints of the finite time.

From Figure 1, it can be seen that the high frequency component of one-order signal (trapezoidal signal) is lower than zero-order signal (square signal) and the high frequency component of infinite-order signal (Gaussian signal) is lower than one-order signal, which is consistent with the theory that the higher the derivative times is, the better the suppression effect on high frequency noise is.

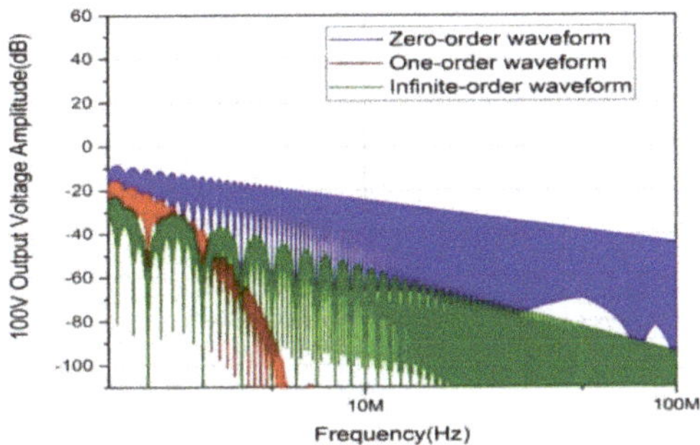

Figure 2. Spectra comparison of zero-order waveform (square signal), one-order waveform (trapezoidal signal) and infinite-order waveform (Gaussian signal).

2.2. Control Principle for Current

According to the literature [18], the derivative of the drain current to time, $\frac{dI_d}{dt}$ is expressed as Equation (10) when the MOSFET is turned on.

$$\frac{dI_d}{dt} = \frac{V_{gg} - \left(\frac{I_d}{g_m} + V_{th} + \frac{L_g C_{iss}}{g_m}\frac{d^2 I_d}{dt^2}\right)}{R_{g,on}\frac{C_{iss}}{g_m} + L_s} \tag{10}$$

where V_{gg} is the gate voltage, V_{th} is the threshold voltage, L_g is the gate parasitic inductance, $C_{iss} = C_{gs} + C_{gd}$ (C_{gs} gate source parasitic capacitance, C_{gd} gate drain parasitic capacitance), g_m is transconductance, $R_{g,on}$ gate resistance at turn-on and L_s is the source parasitic inductance. The drain reverse recovery current is given by (11) [19].

$$I_{rr} = \sqrt{\frac{2Q_{rr}\frac{dI_d}{dt}\Big|I_d = I_l}{S+1}} \tag{11}$$

where Q_{rr} is the reverse recovery charge and S represents the fast factor. It can be seen from (10) (11) that the reverse recovery current and current overshoot can be controlled by changing the V_{gg}.

During turn-on period for MOSFET, the derivative of the drain current to time is given by (12)

$$\frac{dI_d}{dt} = -\frac{V_{ee} + \frac{I_d}{g_m} + V_{th} + \frac{L_g C_{iss}}{g_m}\frac{d^2 I_d}{dt^2}}{R_{g,off}\frac{C_{iss}}{g_m} + L_s} \tag{12}$$

where V_{ee} is the gate voltage at turn-off for MOSFET. From (12), the slope of the drain voltage can be controlled by changing the gate voltage V_{ee} during turn-off period.

3. Proposed Intelligent Control Method and Operation Principle

The total circuit diagram used in the proposed intelligent control method for voltage and current is shown in Figure 3. The circuit consists of an open-loop circuit, a closed-loop circuit and a conventional totem pole drive. The drain-source voltage V_{ds} of the SiC MOSFET is shaped by the closed-loop circuit in the upper side. When Logic1 outputs 1, switches S1 and S2 are turned on at the same time, the closed-loop circuit starts to work and the drain voltage is fed back by the drain voltage feedback circuit to the positive terminal of the operational amplifier according to the ratio β. The feedback signal is compared with a reference signal generated by the reference signal generator 1, and the gate of the SiC MOSFET is controlled by a totem pole drive circuit. When Logic2 outputs 1, the switches S3 and S4 are turned on, and the current control signal V_i control the gate voltage by the operational amplifier

2. It is necessary to add a resistor R_g to the gate to connect the gate to prevent the SiC MOSFET from operating outside the safe zone. Another function for the R_g is to reduce the oscillation of voltage and current when turned on and off. The value of R_g can be selected according to the values recommended in the datasheet. In this paper, its value is 5 ohms.

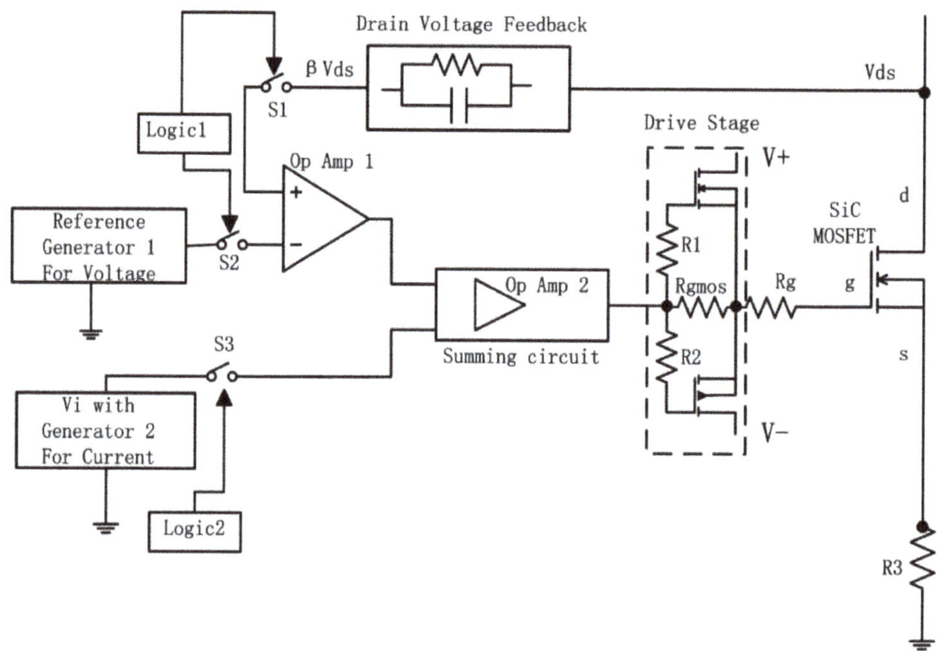

Figure 3. The general scheme of control method for the proposed time division shaped switching transitions.

The control signal selected is the Gaussian signal in this paper, which is generated by arbitrary waveform generator, M3202A, according to Equation (11). The proposed intelligent control method for voltage and current is based on fact that the current and voltage are not synchronized during MOSFET switching period. Figures 4 and 5 show the time domain signal of the SiC MOSFET during turn-on and turn-off period under hard switching conditions respectively. During the period from t2 to t3, the current I_d rises firstly, and the voltage V_d has a small decrease (in order to simplify the analysis, V_d is regarded as unchanged in this paper). At the time t3, the current rises to the maximum, the voltage begins to drop rapidly, and it drops to the lowest moment. During the period from t7 to t8, the voltage V_d rises rapidly, and the current I_d has a small drop (to simplify the analysis, the I_d does not change during this time period). At t8, the voltage V_d (Drain voltage) gets the maximum value. The current I_d drops rapidly after t8 and the current gets the minimum value t9. Therefore, the drain voltage and current of SiC MOSFET are not synchronized during switching period, which provides a possibility for intelligent control.

Time domain diagram of Gaussian control signals for voltage and current is shown in Figure 6. S1(t) and S2(t) are rising edges for drain voltage and current respectively and the falling edge signal is symmetric with the rising edge signal. They can be obtained by Equation (11).

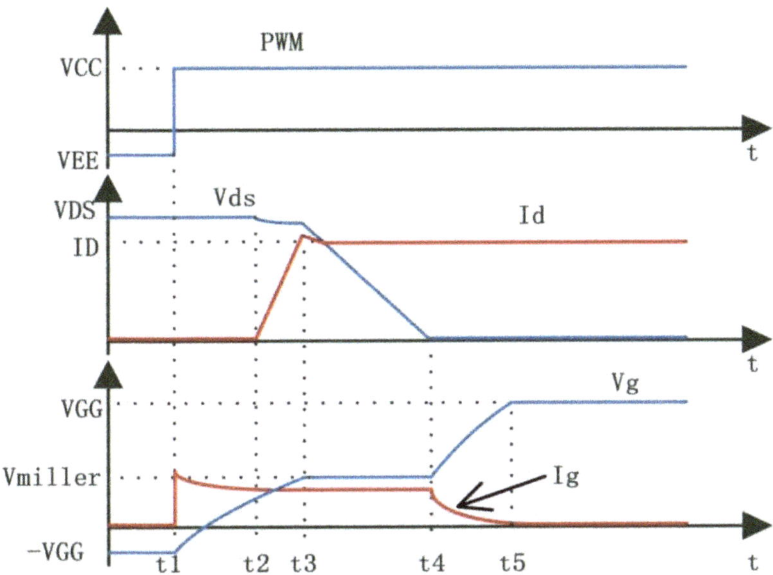

Figure 4. Turn-on voltage and current of SiC MOSFET under hard switching conditions.

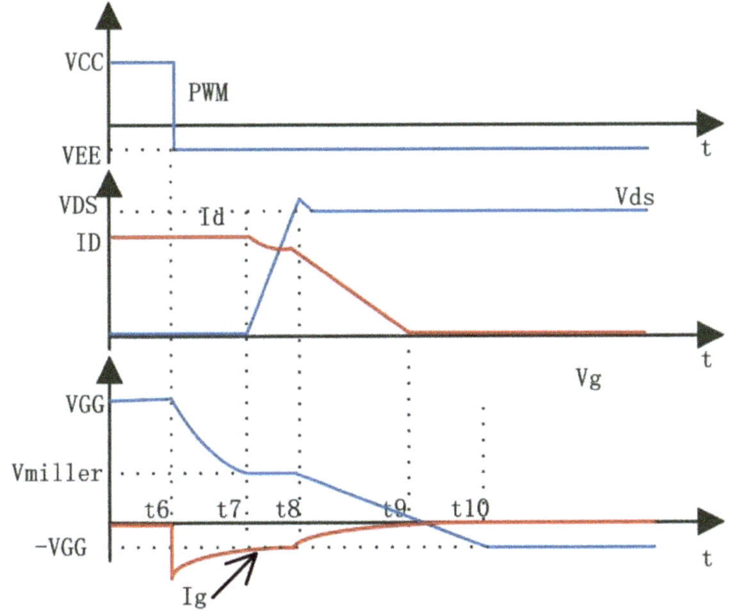

Figure 5. Turn-off voltage and current of SiC MOSFET under hard switching conditions.

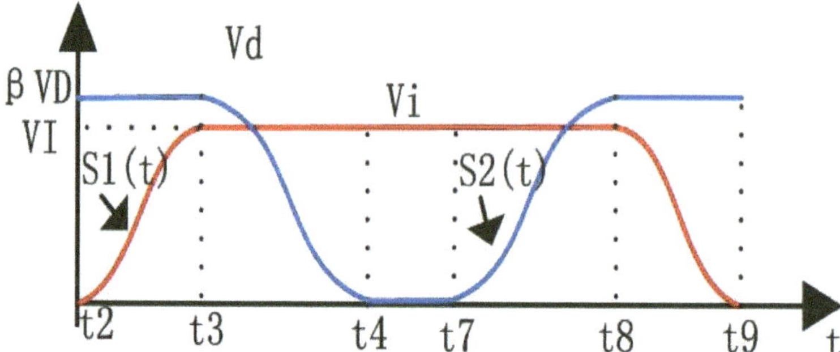

Figure 6. Time domain diagram of Gaussian control signals for voltage and current.

Figures 7 and 8 are the theoretical waveforms of the voltage and current of the proposed intelligent control for turn-on and turn-off, respectively. The switch S3 is turned on and the current control signal is connected to the circuit to control the output current during the 0–t3 period. At time t3, switch S3 is opened and the open-loop circuit stop working. It should be noted that S1 and S2 are closed at time t21 that is before time t3. The delay time for switches S1 and S2 is td1 = t3 − t21. The delay time is caused by MOSFET, operational amplifier, and coupling circuit, which must be taken into account to make the upper part of the circuit achieve the best turn-on transient behavior, when the open-loop stops working. In contrast to the turn-on transient, S1 and S2 are turned on during the t21–t8 period and the voltage closed-loop operates normally. At time t8, S1 and S2 are turned off. However, S3 is turned on at time t_{71} that is before time t8. This is because a delay time td2 = t8 − t71 is required to achieve current control after the switch S3 turned-on. The delay time that we concerned to cause by MOSFET, operational amplifier, and coupling circuit guarantees that MOSFET has an optimal turn-off behavior transient. The values of td1 and td2 can be dynamically adjusted to determine the optimum value, according to the switching voltage and current transients of the MOSFET.

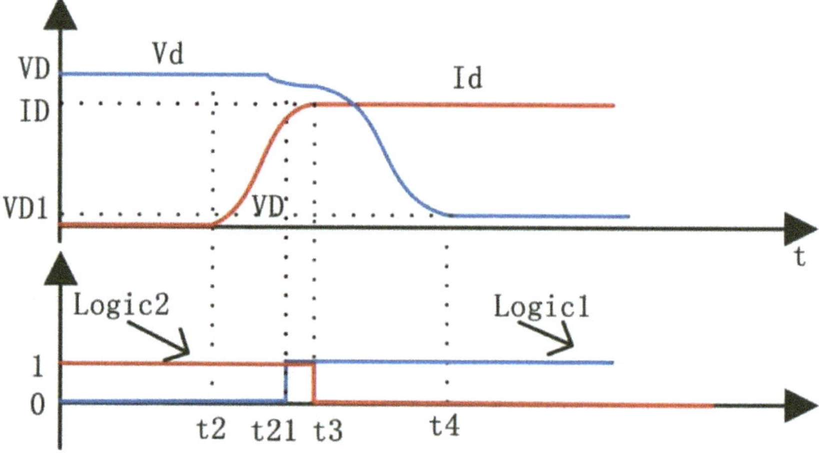

Figure 7. Turn-on voltage and current waveform of SiC MOSFET the in the proposed circuit.

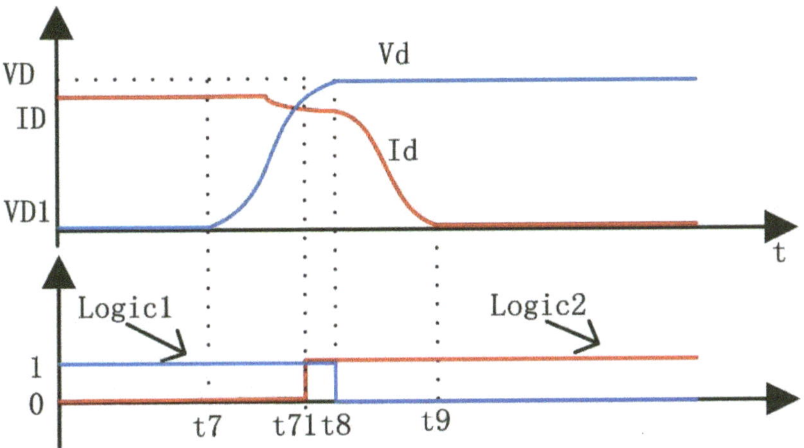

Figure 8. Turn-off voltage and current waveform of SiC MOSFET in the proposed circuit.

4. Circuit Simulation and Discussion

Before the experiment, it is necessary to simulate the proposed circuit, which can improve the experimental guidance and improve the efficiency for the experiment. Pspice is a professional circuit simulation software so it was chosen as the simulation software in this paper. Japan's ROHM semiconductor's SCT2160KE SiC MOSFET was selected and the body manufacturer offered the user a 1200-V/22-A Pspice model of SiC MOSFET. In order to simplify the circuit, the standard chopper unit was selected as the control object with inductive load in this paper. Both the voltage reference signal and the current control signal are Gaussian signals and the rising edge time, 1 µs, is equal to the falling edge time. The amplitudes of the voltage reference signal and the current control signal are 5, 4, respectively. According to Equation (11), the values of the Gaussian signal was calculated by the MATLAB and then imports into Pspice. The period is set to 6 µs and the duty cycle is 2/3. It is worth noting that in the simulation, the Gaussian reference signal is compared with the output signal that is fed back according to a certain ratio, rather than compared with the real output signal. Figures 9 and 10 are simulation diagram for turn-on and turn-off in time domain respectively. The rising edge time and falling edge time of the voltage reference signal are both 1 µs and the rising edge time and falling edge time of the current control signal are also 1 µs in Figures 9 and 10. The rising edge time and the falling edge time of the reference voltage are both 1 us, and the rising edge time rise time and the falling edge time of the current control voltage are both 0.5 µs. Figure 11 has the same working conditions with Figure 12 including the voltage reference signal and the current control signal. Comparing Figures 9–12, it can be seen that the proposed method can not only achieve the output voltage following the reference voltage, but also control the rising and falling edge of the drain current. It is worth mentioning that due to the presence of the on-resistance and the sampling resistor, the minimum value of the output voltage V_{ds} is not zero, and the minimum voltage of V_{ds} is defined as VD1 in this paper.

$$V_{D1} = I_D(R_{in} + R_3),\qquad(13)$$

where I_D is the on-state current, R_{in} is on-state resistance and R_3 is the sampling resistor.

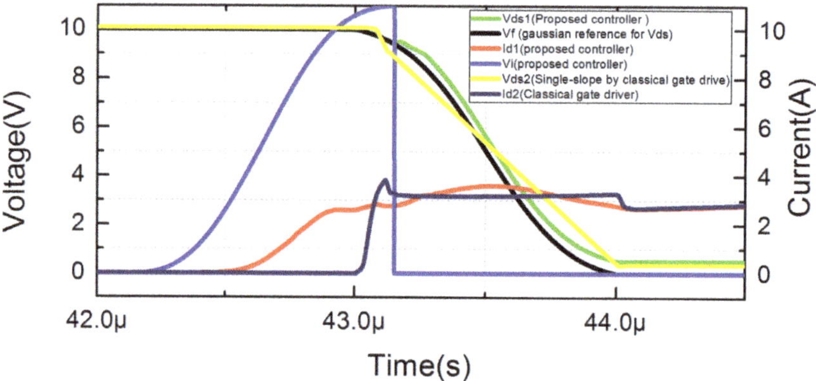

Figure 9. Pspice simulation for SiC MOSFET during turn-on with 1 µs falling edge of V_f (reference voltage) and 1 µs rising edge of V_i (the voltage to control current).

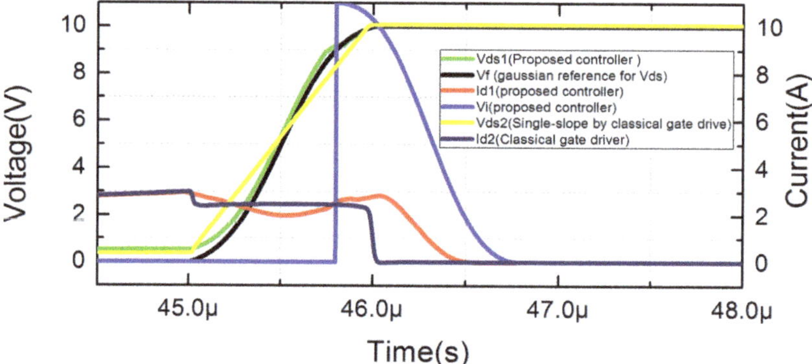

Figure 10. Pspice simulation for SiC MOSFET during turn-off with 1 µs rising edge time of V_f (reference voltage) and 1 µs falling edge time of V_i (the voltage to control current).

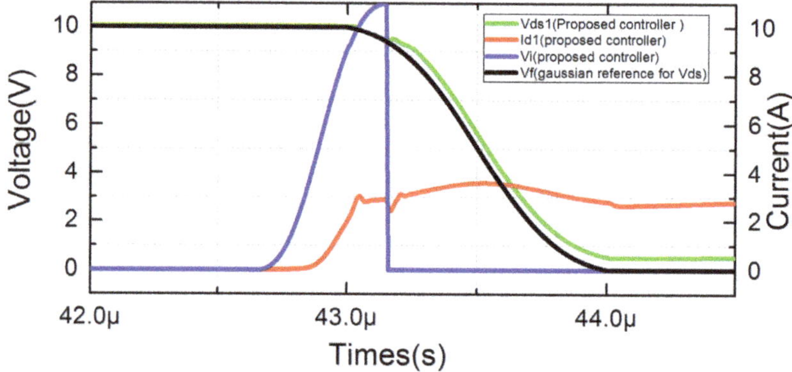

Figure 11. Pspice simulation SiC MOSFET during turn-on with 1 µs falling edge time of V_f (reference voltage) and 0.5 µs rising edge time of V_i (the voltage to control current).

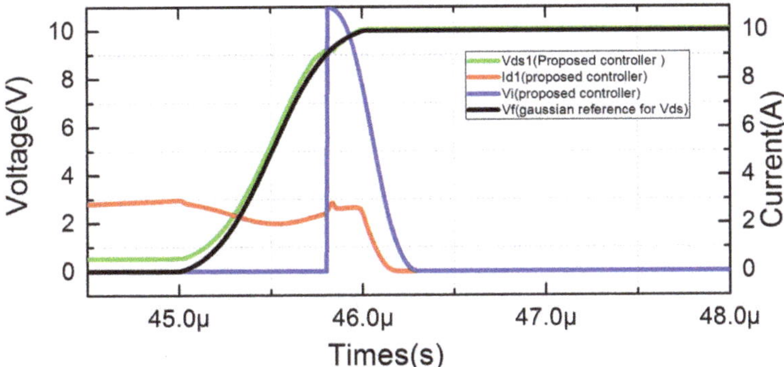

Figure 12. Pspice simulation SiC MOSFET during turn-off with 1 μs decline time of V_f and 0.5 μs rise time of V_i.

It is worth mentioning that the time domain diagrams of the gate-source voltage and gate current of the conventional gate drive method are also shown in Figures 9 and 10. This paper compares the spectrum of the obtained drain voltage and drain current obtained by the conventional gate driving method and the proposed method shown in Figures 13 and 14. Figure 13 shows the spectrum of the drain voltage obtained by the conventional method and the proposed method. It can be seen from Figure 13 that the envelope of the spectrum obtained by the proposed method is lower than the conventional gate driving method in the frequency range of 100 kHz to 70 MHz. However, the proposed method has obvious bumps in the spectrum around 80 M, which may be caused by fluctuations of the drain voltage caused by voltage switching. Figure 14 compared the spectrum of the drain current obtained by the proposed method with the drain current obtained by the conventional method. It can be seen from Figure 14 that when the frequency is less than 200 kHz, their spectral envelopes are basically coincident. When the frequency is higher than 200 kHz, the envelope obtained by the proposed method has a significant attenuation compared with the classical gate driving method and the attenuation is about 30 dB. Therefore, the effect of the proposed control method for suppressing EMI is very obvious.

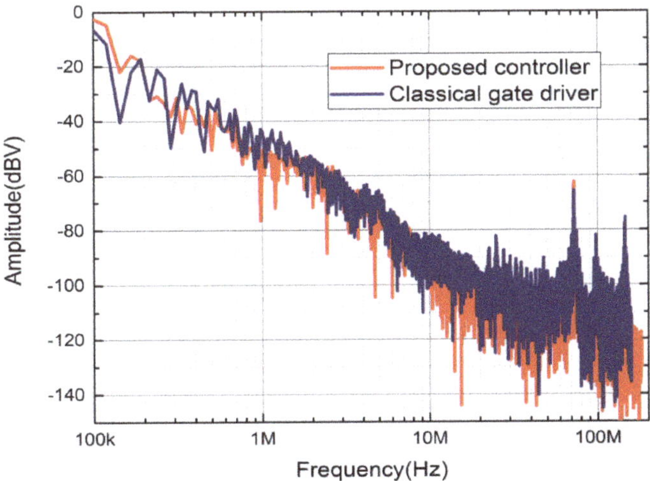

Figure 13. Pspice simulation V_{ds} spectra comparison ($t_{off} = t_{fall} = 1.0$ μs, purple: classic gate driver, red: proposed controller).

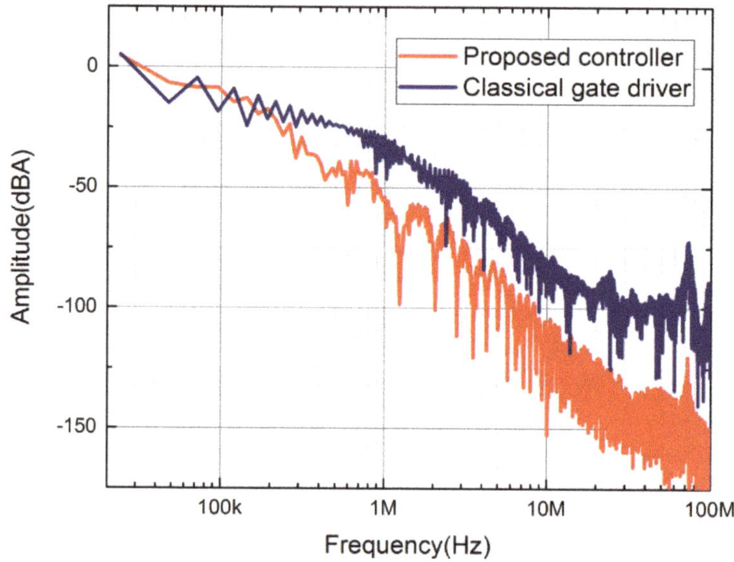

Figure 14. Pspice simulation I_d spectra comparison ($t_{off} = t_{fall} = 1.0$ µs, purple: classic gate driver, red: proposed controller).

5. Experiment Results of Proposed Method and Classic Method and Discussion

Similar to the simulation, the chopper unit with inductive load was also selected as the experimental circuit for simplicity. The op amp TH3091 that has wide Supply Range ±5 v to ±15 v, low voltage noise 2 nV $\sqrt{\text{Hz}}$ and high band 200 MHz was used in this paper. The specific parameters of the proposed circuit and the classical gate drive circuit are listed in Table 1. The time domain data of the Gaussian signal is obtained by MATLAB and imported into the M3202A that is an arbitrary signal generator. The hardware configuration of the proposed intelligent control method is shown in Table 1. In the first experiment, proposed controller 1 was used, assuming that the rising edge time $t_{off} = 1$ µs, the falling edge time $t_{fall} = 1$ µs, cycle T = 6 µs, duty cycle D = 1/3 for the reference signal V_f and the rising edge time $t_{r2} = 1$ µs, the falling edge time $t_{f2} = 1$ µs for the control voltage V_i. The time domain relationship between voltage reference signal and current control signal is shown in Figure 6. In the second experiment, the proposed controller 2 was used, in which the setting of the voltage reference signal V_f was the same as the first experiment, but V_i, $t_{off} = 0.5$ µs, $t_{fall} = 0.5$ µs, is different with the first experiment.

Table 1. Experiment circuit parameters.

	Rg(Ω)	Op Amp	A1 × A2		Driver
Classical gate driver	5	NULL	NULL		MOS
Proposed Intelligent controller	5	THS3091	10 × 1		MOS

In Figure 15, the drain-source voltage V_{ds} for the blue track of the controller 1 and the red track of controller 2 is delayed by 0.1 µs compared with the reference voltage at turn-on. At turn-off the drain-source voltage V_{ds} for the blue track of the controller 1 and the red track of controller 2 follows the reference voltage well. However, for the blue trajectory and the red trajectory, there is a significant fluctuation in the top corner region at turn-on and turn-off, which may be caused by fluctuations in current for switching, parasitic inductance, and sampling resistor parasitic capacitance.

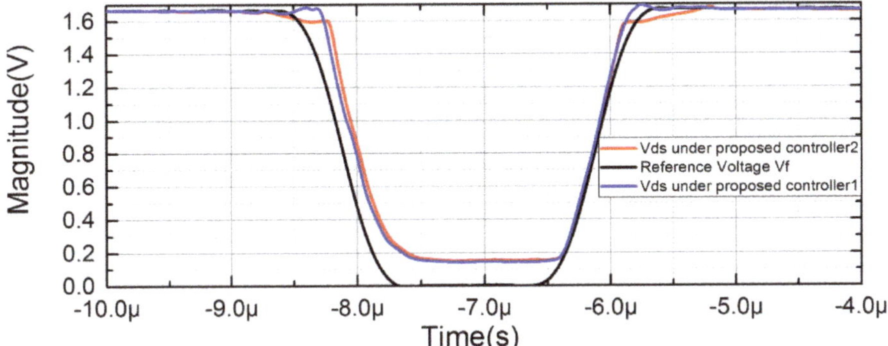

Figure 15. Experiment result that voltage comparison V_{ds} of proposed controller 1 (blue, $t_{off} = t_{fall} = 1$ μs of V_f), proposed controller 2 (red, $t_{off} = t_{fall} = 1$ μs of V_f), and reference voltage V_f (black, $t_{off} = t_{fall} = 1$ μs).

The classic gate driver is also used in the chopper for comparison, in which the hardware configuration is shown in Table 1 and the drive signal is the ideal PWM, the period 6 μs, duty cycle 1/3. The spectrum comparison of the drain-source voltage is shown in Figure 16. As can be seen from Figure 16, the envelopes of the three spectra in the low frequency region are almost coincident. The envelope of the classic gate drive in the 10^6–10^7 Hz range is 20 dB higher than the envelope of the proposed control 1 and control 2. The envelope of the classic gate drive in the range of 10^7–10^8 Hz is 40 dB higher than the proposed envelope of control 1 and control 2. In the analyzed frequency range of 10^5–10^8 Hz, the spectral envelopes of the proposed control 1 and control 2 are almost coincident, which shows that the current control signal V_i has little effect on the drain-source noise when the reference voltage V_f is the same in the proposed circuit.

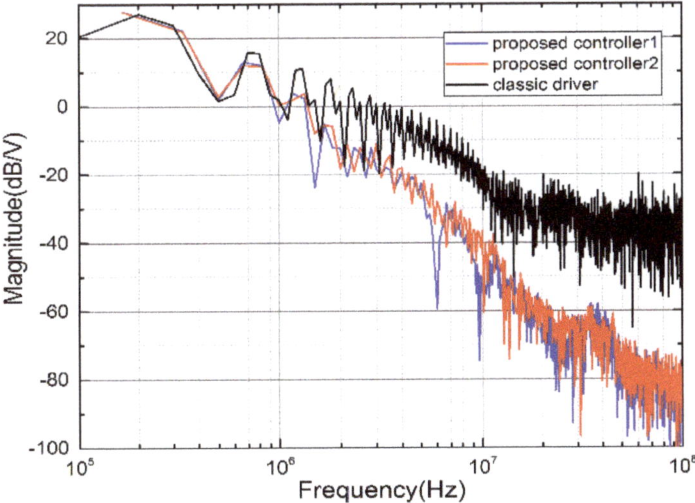

Figure 16. Spectra comparison of V_{ds} between classic gate driver (black), proposed controller 1 (blue) and proposed controller 2 (red). Same reference voltage V_f was used in controller 1 and controller 2.

In Figure 17, the drain current I_d of proposed controller 1 has lower slope di/dt of 2.94 A/us than controller 2 of 6.46 A/us so that the controller 1 reduces the current overshoot 40% at turn-off.

At turn-on, the switching speed di/dt of controller 1 is 4.58 A/us, and the di/dt of controller 2 is 6.87 A/us. Accordingly, the controller 1 reduces the current overshoot about 50%.

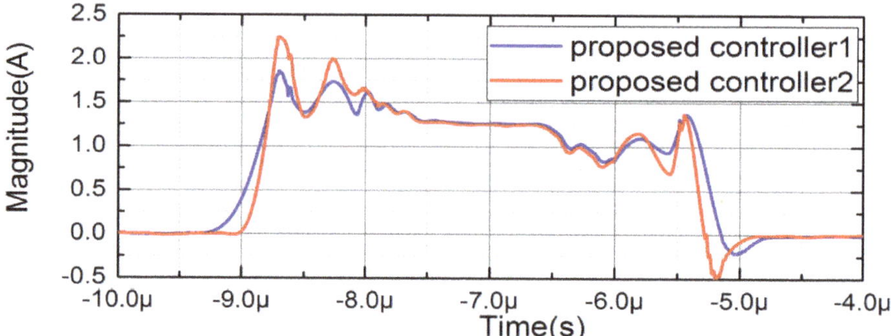

Figure 17. Experiment results of I_d for turn-on and off transition between controller 1 ($t_{off} = t_{fall} = 1$ μs of V_i) and controller 2 ($t_{off} = t_{fall} = 0.5$ μs of V_i). Same reference voltage V_f was used in controller 1 and controller 2.

The spectrum comparison of the drain-source current I_d is shown in Figure 18. It can be seen from the Figure 18 that before 3×10^6 Hz, the spectrum of the classical gate driver, the proposed control 1 and control 2 are almost coincident. When the frequency is greater than 3×10^6 Hz, the spectrum envelope of the classical gate driver is 10–40 dB higher than the proposed controller 1 and controller 2. In the frequency range of 10^7–5×10^7 Hz, control 1 is reduced by 5–10 dB compared with controller 2, since the current slope di/dt of controller 1 is smaller than controller 2.

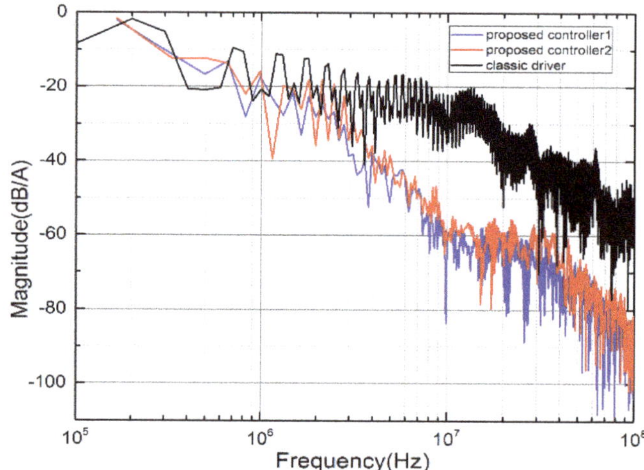

Figure 18. Spectra comparison for I_d between classic gate driver (black), proposed controller 1 (blue) and proposed controller 2 (red).

Switching losses also has been studied in this paper. The switching power of the three experiments are compared in Figure 19. As can be seen from Figure 19, there are mainly two peaks in one cycle. The first power peak is generated when the SiC MOSFET is turned on, and the second power peak is generated when the SiC MOSFET is turned off. The area enclosed by the power curve and the time axis represents the energy of the loss, and the loss at turn-on is higher than it at turn-off significantly.

The typical gate driver switching losses are smaller than the proposed controller 1 and controller 2. Figure 20 shows the curve of switching loss over time in one cycle. It can be seen in Figure 20 that the proposed controller's switching losses are approximately 120% higher than the classic gate drivers at the end of a cycle, since the switching time of the classic gate drive is much shorter than the proposed controller 1 and controller 2. The switching loss of the controller 2 is 6.4% lower than the controller 1, because the current rise and fall time of the controller 2 is shorter than the controller 1. It should be noted that in actual application, we can adjust the parameters to adjust the loss. According to [20], we know that there is compromise between EMI and switching losses. Therefore, we can reduce the switching loss by reducing the rising and falling edge times of the reference voltage within the acceptable EMI range.

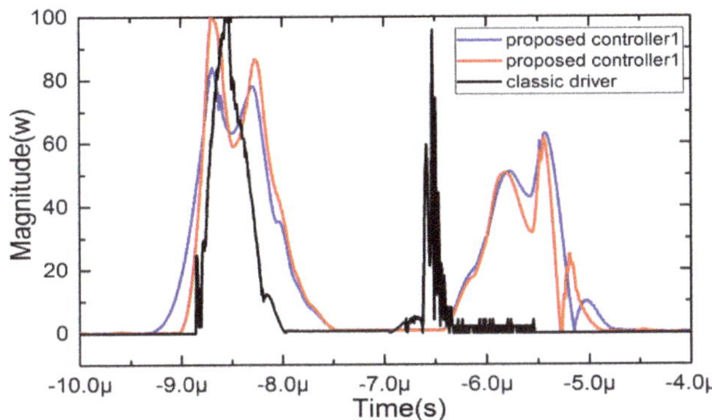

Figure 19. Power comparison between the proposed controller 1, the proposed controller 2, and classical gate driver.

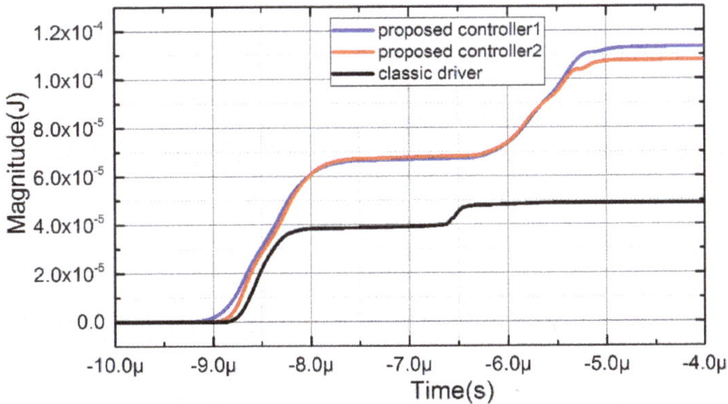

Figure 20. Switching losses comparison between the proposed controller 1, the proposed controller 2, and classical gate driver.

6. Conclusions

The proposed intelligent control method has good EMI suppression for voltage and current compared with the classical gate driving method, the hard switching method. It has been proved that the intelligent control method can shape the SiC MOSFET switching transition effectively. Compared with the common method, it achieves control of drain-source voltage and drain current simultaneously. This method requires a precise fit of the voltage control signal and the current control signal so that a high frequency signal generator is needed. However, the proposed method needs longer switching time than classic driver so that the proposed method causes more power loss. Therefore, it is needed to tradeoff the EMI and switching loss. According to [20], it is an effective way to adjust the rise and fall switching time to reduce switching loss. Control signals can be obtained by convolution calculations and generated by high frequency multi-channel arbitrary signal generators. However, there are certain limitations, for example, voltage and current oscillations during circuit switching. The future work is to study the finite time stability of the circuit when two circuits work at the same time.

Author Contributions: Conceptualization, C.X. and Q.M.; Methodology, C.X.; Validation, C.X., Q.M., P.X. and C.T.; Investigation, C.X.; Writing—original draft preparation, C.X.; Writing—review and editing, C.X., Q.M., P.X. and C.T.

Funding: This research received no external funding.

Acknowledgments: The authors would like to thank Beihang University, Beijing, China for the financial support in executing this work successfully. This work has been carried out as a part of research work in the School of Automation Science and Electrical Engineering, Beihang University, Beijing, China.

Conflicts of Interest: The authors declare no conflict of interest.

References

1. Costa, F.; Magnon, D. Graphical Analysis of the Spectra of EMI Sources in Power Electronics. *IEEE Trans. Power Electron.* **2005**, *20*, 1491–1498. [CrossRef]
2. Yang, X.; Yuan, Y.; Zhang, X.; Palmer, P.R. Shaping High-power IGBT Switching Transitions by Active Voltage Control for Reduced EMI Generation. *IEEE Trans. Ind. Appl.* **2015**, *51*, 1669–1677. [CrossRef]
3. Meng, J.; Ma, W.; Pan, Q.; Zhang, L.; Zhao, Z. Multiple slope switching waveform approximation to improve conducted EMI spectral analysis of power converters. *IEEE Trans. Electromagn. Compat.* **2006**, *48*, 742–751. [CrossRef]
4. Idir, N.; Bausière, R.; Franchaud, J.J. Active gate voltage control of turn-on di/dt and turn-off dv/dt in insulated gate transistors. *IEEE Trans. Power Electron.* **2006**, *21*, 849–855. [CrossRef]
5. Chen, J. Design optimal built-in snubber in trench field plate power MOSFET for superior EMI and efficiency performance. In Proceedings of the 2015 International Conference on SISPAD, Washington, DC, USA, 9–11 September 2015; pp. 459–462.
6. Palmer, P.R.; Rajamani, H.S. Active voltage control of IGBTs for high power applications. *IEEE Trans. Power Electron.* **2004**, *19*, 894–901. [CrossRef]
7. Yang, X.; Zhang, X.; Palmer, P.R. IGBT converters conducted EMI analysis by controlled multiple-slope switching waveform approximation. In Proceedings of the IEEE ISIE, Taipei, Taiwan, 28–31 May 2013; pp. 1–6.
8. Yang, X.; Zhang, X.; Zhang, J.; Palmer, P.R. Predictive optimisation of high-power IGBT switching under active voltage control. In Proceedings of the IECON 2013—39th Annual Conference of the IEEE Industrial Electronics Society, Vienna, Austria, 10–13 November 2013; pp. 1169–1174.
9. Yang, X.; Palmer, P.R. Optimised IGBT turn-off under active voltage control in PEBB applications. In Proceedings of the 5th European Conference on Power Electronics and Applications (EPE), Lille, France, 2–6 September 2013; pp. 1–10.
10. Wang, Z.; Shi, X.; Tolbert, L.M.; Blalock, B.J. Switching performance improvement of IGBT modules using an active gate driver. In Proceedings of the 28th Annual IEEE Applied Power Electronics Conference and Exposition (APEC), Long Beach, CA, USA, 17–21 March 2013; pp. 1266–1273.
11. Witting, B.; Fuchs, F.W. Analysis and comparison of turnoff active gate control methods for low-voltage power MOSFETs with high current ratings. *IEEE Trans. Power Electron.* **2012**, *27*, 1632–1640. [CrossRef]
12. Riazmontazer, H.; Rahnamaee, A.; Mojab, A.; Mehrnami, S.; Mazumder, S.K.; Zefran, M. Closed-loop control of switching transition of SiC MOSFETs. In Proceedings of the 2015 IEEE Applied Power Electronics Conf. and Exposition (APEC), Charlotte, NC, USA, 15–19 March 2015; pp. 782–788.
13. Camacho, A.P.; Sala, V.; Ghorbani, H.; Martinez, J.L.R. A Novel Active Gate Driver for Improving SiC MOSFET Switching Trajectory. *IEEE Trans. Ind. Electron.* **2017**, *64*, 9032–9042. [CrossRef]
14. Logsiger, Y.; Kolar, J.W. Closed-Loop di/dt and dv/dt IGBT Gate Driver. *IEEE Trans. Power Electron.* **2015**, *30*, 3402–3417.

15. Oswald, N.; Stark, B.H.; Holliday, D.; Hargis, C.; Drury, B. Analysis of Shaped Pulse Transitions in Power Electronic Switching Waveforms for Reduced EMI Generation. *IEEE Trans. Ind. Appl.* **2011**, *47*, 2154–2165. [CrossRef]

16. Cui, T.K.; Ma, Q.S.; Xu, P.; Wang, Y.C. Analysis and Optimization of Power MOSFETs Shaped Switching Transients for Reduced EMI Generation. *IEEE Access* **2017**, *5*, 20440–20448. [CrossRef]

17. Patin, N.; Vinals, M.L. Toward an optimal Heisenberg's closed-loop gate drive for power MOSFETs. In Proceedings of the IECON 2012—38th Annual Conference on IEEE Industrial Electronics Society, Montreal, QC, Canada, 25–28 October 2012; pp. 828–833.

18. Xu, C.W.; Ma, Q.S.; Xu, P.; Cui, T.K.; Zhang, P.M. An Improved Active Gate Drive Method for SiC MOSFET Better Switching Performance. In Proceedings of the 2018 IEEE 3rd Advanced Information Technology, Electronic and Automation Control Conference (IAEAC), Chongqing, China, 12–14 October 2018.

19. Wang, J.J.; Chung, H.S.H.; Li, R.T.H. Characterization and Experimental Assessment of the Effects of Parasitic Elements on the MOSFET Switching Performance. *IEEE Trans. Power Electron.* **2013**, *28*, 573–590. [CrossRef]

20. Oswald, N.; Anthony, P.; McNeill, N.; Stark, B.H. An Experimental Investigation of the Tradeoff between Switching Losses and EMI Generation With Hard-Switched All-Si, Si-SiC, and All-SiC Device Combinations. *IEEE Trans. Power Electron.* **2014**, *29*, 2393–2407. [CrossRef]

© 2019 by the authors. Licensee MDPI, Basel, Switzerland. This article is an open access article distributed under the terms and conditions of the Creative Commons Attribution (CC BY) license (http://creativecommons.org/licenses/by/4.0/).